Plant Biochemistry

This textbook is affectionately
dedicated by the editors to
the memory of
PROFESSOR TOM AP REES
1930–1996
A British plant biochemist
internationally recognized for
his many contributions to our
knowledge of carbohydrate
regulation and metabolism in
higher plants

Plant Biochemistry

edited by

P.M. Dey

Division of Biochemistry, School of Biological Sciences,
Royal Holloway, University of London,
Egham Hill, Egham, Surrey TW20 0EX, UK.

and

J.B. Harborne

Department of Botany,
Plant Sciences Laboratories,
University of Reading,
Whiteknights,
Reading RG6 2AS, UK.

ACADEMIC PRESS

San Diego   London   Boston
New York   Sydney   Tokyo   Toronto

Academic Press, Inc.
525 B Street, Suite 1900, San Diego, California 92101–4495, USA
http://www.apnet.com

Academic Press Limited
24–28 Oval Road, London NW1 7DX, UK
http://www.hbuk.co.uk/ap/

ISBN 0-12-214674-3

A catalogue record for this book is available from the British Library

Typeset by Wyvern Typesetting Ltd, Bristol
Printed in Great Britain by Bath Press, Bath.

97 98 99 00 01 02 EB 9 8 7 6 5 4 3 2 1

Contents

Contributors

G. Avigad
Department of Biochemistry, University of Medicine and Dentistry of New Jersey,
Robert Wood Johnson Medical School, 675 Hoes Lane, Pistacataway, NJ 08854-5635, USA

J.R. Bowyer
Division of Biochemistry, School of Biological Sciences, Royal Holloway, University of London, Egham,
Surrey TW20 0EX, UK

Peter M. Bramley
Division of Biochemistry, School of Biological Sciences, Royal Holloway, University of London, Egham,
Surrey TW20 0EX, UK

M.D. Brownleader
Biochemistry Section, School of Applied Science, South Bank University, 103 Borough Road, London
SE1 0AA, UK

P.M. Dey
Division of Biochemistry, School of Biological Sciences, Royal Holloway, University of London, Egham
Hill, Egham, Surrey TW20 0EX, UK

J.A. Gatehouse
Department of Biological Sciences, University of Durham, South Road, Durham DH1 3LE, UK

J.B. Harborne
Plant Science Laboratories, Department of Botany, University of Reading, Whiteknights, Reading
RG6 6AS, UK

John L. Harwood
Department of Biochemistry, University of Wales, Cardiff CF1 1ST, UK

Eric Lam
AgBiotech Center, Waksman Institute, Rutgers University, Pistcataway, NJ 08855-0759, USA

Peter J. Lea
Division of Biological Sciences, University of Lancaster, Lancaster LA1 4YQ, UK

R.C. Leegood
Robert Hill Institute and Department of Animal and Plant Sciences, University of Sheffield, Sheffield
S10 2UQ, UK

J.S. Grant Reid
Department of Biological and Molecular Sciences, University of Stirling, Stirling FK9 4LA, UK

D. Strack
Institut fur Pflanzenbiochemie, Weinberg 306120, Halle(Saale), Germany

Jonathan D. Walton
Department of Energy Plant Research Laboratory, Michigan State University, East Lansing, MI 48824, USA

Michael Wink
Institut fur Pharmazeutische Biologie, Im Neuenheimer Feld 364, D-6900, Heidelberg, Germany

Preface

In recent years, there has been a rebirth of interest in plant biochemistry and global investment in research in this important sector of science is on an upward trend. Plants are a renewable source of energy and their products are essential for the survival of humanity on this planet. The current expansion in plant molecular biology has shown that our present knowledge of basic plant metabolism needs to be refined to keep up with the advances that are possible for developing new and improved crop plants via genetic engineering. This aspect is becoming more important in the current growth rate of world population and an increasing demand for food, pharmaceuticals and other commercially important plant products.

The aim of this book is to provide students and researchers in plant sciences with a concise, general account of plant biochemistry. The edited format of the book has allowed us to invite recognized experts in plant biochemistry to contribute chapters on their special topics. Conn and Stumpf's comprehensive multi-volume series on the 'Biochemistry of Plants' produced by Academic Press provides an excellent, detailed survey of advances in the field. The present volume is an attempt to distil the essence of this series but with updated information. This should suit the more general reader and thus, prove more attractive and of interest to a wide audience. This book is also intended to replace the earlier text, 'Plant Biochemistry', 3rd edition, edited by Bonner and Varner and published by Academic Press in 1976. The recently completed multi-volume series, 'Methods in Plant Biochemistry', also published by Academic Press, should serve as a useful reference for details on techniques in each subject area.

The emphasis in this book is on plant metabolism, but enzymological, functional, regulatory and molecular biological aspects have also been covered as far as possible. Pathways and structural features have been used widely as illustrations. Further reading lists and bibliographies have been included in all chapters. However, the authors have had the freedom of choosing the extent of coverage in their respective chapters. Thus, the reader may find some diversity in style and presentation of articles in this book. As editors, we have endeavoured to integrate and augment, in a meaningful manner, the diverse topics of plant biochemistry and yet maintain the characteristics of each author's contribution.

The topics covered in this book can be divided into four sections: (1) The Cell, (2) Primary Metabolism, (3) Special Metabolism and (4) The Plant and the Environment. The plant cell, its molecular components and function are discussed in chapter 1. The various pathways of primary metabolism are described in chapters 2–9 which include carbohydrates, lipids, nitrogen, nucleic acids, proteins and gene regulation. The special metabolism section has three chapters on: phenolics, isoprenoids and secondary nitrogen compounds. The last three chapters on, pathology, ecology and biotechnology relate to the plants and environment section. However, it has not been possible to include all conceivable topics of plant biochemistry, and we have not reviewed the biochemistry of plant growth substances in detail since there are several excellent texts available in this area.

Prakash Dey
Jeffrey Harborne

1 The Plant, the Cell and its Molecular Components

P.M. Dey, M.D. Brownleader and J.B. Harborne

1.1 INTRODUCTION

This chapter fills in the necessary background on the biology and molecular biology of plants for what follows in later chapters. The first section considers the classification of plants in relation to that of other living organisms. This is a subject which has been revolutionized in recent years as a result of the application of DNA and RNA sequence analysis to phylogenetic ideas derived from morphology and fossil exploration. The second section discusses some of the variations that are known in the biochemistry of plants, knowledge that has been applied successfully to assist in the classification of species, genera and families within the plant kingdom.

The third section of this chapter is concerned with the physical make up of the plant cell, the many different organelles that have been recognized within the cell and the nature of the surrounding cell wall. The fourth section reviews the techniques available for cell fractionation. The application of these techniques has advanced immensely our knowledge of the localization of metabolic pathways within the plant cell. The fifth and final section provides the basic information on molecular aspects of the plant cell, the processes of mitosis and meiosis and the nature of plant DNAs.

1.1.1 Plant classification

Plants are eukaryotic organisms, which dominate the terrestrial world. They are universally green in color, due to the presence of chlorophylls *a* and *b*, the essential catalysts of photosynthesis. They make up one of the five kingdoms into which all living organisms are currently classified (Table 1.1). Their closest relatives are the algae, which provide the vegetation of the aquatic world. Algae were once classified with the green plants, in spite of their variation in color from brown and red to yellow and green, but are now placed in the Protoctista. At one time fungi were also classified as plants, but they differ in so many ways (e.g. they lack chlorophyll) that their separation into a single kingdom (Table 1.1) is a logical one.

Plants as we understand them today fall into two major groups: the Bryophyta, which lack a lignified cell wall and hence are unable to grow upwards more than a few centimetres; and the Tracheophyta, which have a lignified cell wall and may become tall trees (e.g. *Sequoia sempervirens*) over 100 m in height (Table 1.2). The Bryophyta are further subdivided into liverworts and mosses, which are chemically distinctive. Liverworts contain many terpenoid secondary constituents, which protect them from herbivores, whereas mosses produce a variety of phenolic constituents at the cell wall for this purpose.

The Tracheophyta are divided into six classes (Table 1.2), but most have relatively few members. The psilophytes, lycopods and horsetails are relict taxa, more representative of the natural vegetation of the Devonian period some 370 million years ago. A larger group are the ferns with some 8000 members, whereas the gymnosperms (or conifers) have about 1000 species. It is, however, the angiosperms, the flowering plants, to which most green plants belong.

Table 1.1 The five kingdoms of the living universe

Cell type	Kingdom	Examples
Prokaryotes	Monera	Eubacteria, Cyanobacteria, Archeobacteria
Eukaryotes	Protoctista	Algae, Protozoa, Ciliata
	Fungi	Ascomycetes, Zygomycetes, Basiodiomycetes
	Plantae	Bryophyta, Tracheophyta
	Animalia	Insecta, Arthropoda, Mammalia

The angiosperms evolved at the same time in evolutionary history (135 my) as the insects and they characteristically produce brightly colored flowers, which attract insects and other animals for purposes of pollination. Likewise, these plants produce fruits which are nutritionally rewarding to the animals which eat them and then disperse the seeds contained in them. It is the angiosperms which provide the overwhelming proportion of plants useful to humans: the cereals, wheat, barley, oats, rice and maize; the legume peas and beans; and the many other crop plants of world agriculture.

The angiosperms are classified as monocotyledons (monocots), plants which develop as a single cotyledonous structure on seed germination, and as dicotyledons (dicots), where the more familiar double-bladed first 'leaf' is formed as the shoot develops from the seed. Monocot families include the economically important grasses (or Gramineae, 8000 spp.) and palms (or Palmae, 2700 spp.) and the ornamental groups such as the lilies (Liliaceae), irises (Iridaceae) and narcissi (Amaryllidaceae). The largest dicot family is the Compositae (21 000 spp.), plants of which store carbohydrate in the form of fructose polymer (or fructan) rather than as starch (see Chapter 4). Economically, the most important dicot family is the Leguminosae, members of which have the distinctive capacity to fix nitrogen in the roots, in the presence of symbiotic (*Rhizobium*) bacteria. Dicots may be trees, shrubs or herbs, whereas monocots lack regular secondary thickening and are usually herbaceous in habit. The monocots are generally assumed to have arisen from dicot stock, but there is still argument about which dicots are the most likely progenitors. One favored group are water plants in the family Nympheaceae.

1.1.2 Evolutionary origin

Various plausible theories have been proposed to indicate how higher plants have evolved from simpler forms of plant life (Fig. 1.1). The fact that the plant cell contains three different kinds of DNA, mitochondrial, chloroplastic and nuclear (see section 1.5), can be most easily explained through the endosymbiotic theory of cell evolution. The mitochondrial genome is like that of bacteria and hence is thought to be of bacterial origin. The extant ancient bacterial organism *Paracoccus denitrificans* has been proposed as a possible candidate for engulfment by an early 'plant' cell to give rise to the mitochondria that exist today within the cytoplasm. Likewise, a cyanobacterial origin (e.g. *Anabaena flos-aquae*) is the generally accepted route by which the plant cell was furnished with its chloroplasts. This endosymbiotic theory is well supported by biochemical comparisons that have been made between present day mitochondrion and chloroplast and the cellular organization of the relevant prokaryotic ancestors.

It is also possible to trace the origin of individual features of plant metabolism by comparing these features as they occur in more primitive (e.g. bacterial) organisms. For example,

Table 1.2 Classification within the higher plants

Division (number of spp.)	Class	Age of fossils (my)	Example
Bryophyta*	Liverworts	380	*Marchantia polymorpha*
(23 000)	Mosses	280	*Sphagnum rubellum*
Tracheophyta†	Psilophytes	400	*Psilotum nudum*
(300 000)	Lycopods	380	*Lycopodium cernuum*
	Horsetails	380	*Equisetum palustre*
	Ferns	375	*Pteridium aquilinum* (bracken)
	Gymnosperms	370	*Ginkgo biloba* (Maidenhair tree)
	Angiosperms	135	*Pisum sativum* (pea)

*With unlignified cell wall and limited secondary metabolism (e.g. alkaloids absent).
†All with lignified cell wall and with increasing complexity in secondary metabolism.

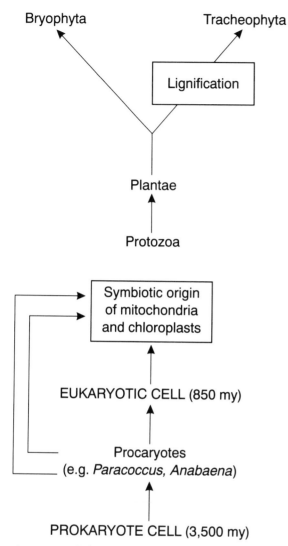

Figure 1.1 Phylogenetic tree for green plants.

with time. In the bread mold fungus, *Neurospora crassa*, there are five enzymes of the pathway which are linked together in a single polypeptide chain, whereas in plants, all the enzymes are clearly coded for separately from each other. In addition, several of the enzymes (e.g. chorismate mutase) exist in more than one form. These isoenzymes vary in their response to end-product inhibition and hence are probably involved in the regulation of the pathway. Again, the last two steps in the synthesis of tyrosine can be catalyzed in a different order in different organisms. Some plants such as *Vigna radiata*, are able to operate both routes, whereas most bacteria and fungi that have been examined can only operate one or other route (Fig. 1.2).

Another interesting comparison that has been made is of the chemistry of fossil and present day plants. Such comparisons generally indicate that the end products of biochemical pathways change little over many millions of years. For example, the hydrocarbon (alkane) pattern of a fossil *Equisetum brougniarti*, 200 million years old, was found to be essentially identical to that of the present day *Equisetum sylvaticum*. Likewise, flavonoids present in fossilized leaves of *Celtis* (Ulmaceae) dating back 20 million years were nearly identical to those of a present day *Celtis* species. Other chemical features of fossil plants that can be recovered and compared with those of living plants include fatty acids, porphyrins and many isoprenoids. Some alterations may occur in the oxidation level of these molecules, but the carbon skeletons usually remain intact. In addition, sporopollenin, part of the outer coating of plant pollens and polymeric material derived from carotenoid, shows identical properties whether it is isolated from fossil or from fresh pollen material.

1.2 COMPARATIVE BIOCHEMISTRY

Plants are traditionally classified into species, genera, tribes, families and orders on the basis of morphological and anatomical characters. There is such a wealth of visual 'eyeball' characters that it is doubtful whether the current systems of plant classification will ever be superseded. Nevertheless, increasingly biochemical evidence, derived from the chromatography or gel separation of plant extracts, is being considered as an adjunct to classical plant taxonomy. Such data are particularly useful in cases where morphology or anatomy fail to provide a satisfactory system of classification.

the three-dimensional structure of the photosynthetic enzyme, rubisco, appears to have been established early in evolution, since its form in the plant chloroplast is very close to that found in photosynthetic bacteria such as *Chromatium vinosum*. Thus the quaternary structure of both these rubiscos has the same standard eight small and eight large subunits (L_8S_8) (see Chapter 2). That rubisco has evolved from simpler forms of the enzyme is apparent from the fact that there are in bacteria forms with fewer components (L_6S_6 in *Pseudomonas oxalaticus*) and one which completely lacks the small subunits (L_2 in *Rhodospirillum rubrum*).

The shikimic acid pathway, which produces the three aromatic amino acids, phenylalanine, tyrosine and tryptophan, also appears to have evolved

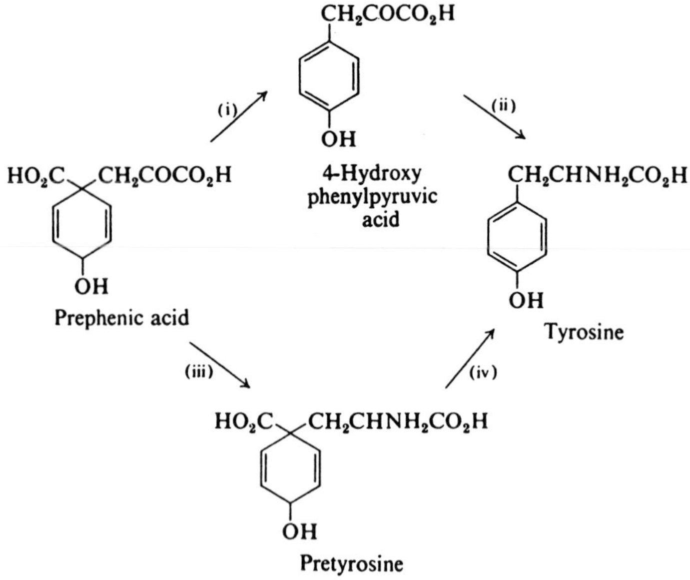

CH_2COCO_2H

4-Hydroxy
phenylpyruvic
acid

(i)

(ii)

HO_2C CH_2COCO_2H

OH
Prephenic acid

$CH_2CHNH_2CO_2H$

OH
Tyrosine

(iii)

(iv)

HO_2C $CH_2CHNH_2CO_2H$

OH
Pretyrosine

Figure 1.2 Alternative pathways to tyrosine biosynthesis in plants. Enzymes: i, prephenate dehydrogenase; NAD^+; ii, 4-hydroxyphenylpyruvate transaminase, pyridoxal 5′-phosphate; iii, prephenate transaminase, pyridoxal 5′-phosphate; iv, pretyrosine dehydrogenase, NAD^+.

It has long been known that plants can vary in their metabolism, even in some of the basic features. Lignin synthesis, for example, in angiosperms differs from that in gymnosperms, since an extra precursor molecule, sinapyl alcohol, is required for the final polymerization (see Chapter 10). The application of biochemistry to plant classification, developing as a new interdisciplinary subject, only became possible on a significant scale after rapid chromatographic methods of surveying plant tissues for their chemical constituents were applied. Plant screening was first developed for the products of secondary metabolism. Such surveys indicated, for example, that increasingly complex organic structures were often associated with other advanced characters derived from biology. In terpenoid synthesis, more complex classes of terpenoids were generally present in the more highly evolved, herbaceous plant families (Fig. 1.3).

At the same time as studies were carried out on secondary metabolites, amino acid sequences were determined for several small plant proteins that were reasonably easy to separate and purify. Three proteins studied in a comparative way were cytochrome *c*, ferredoxin and plastocyanin. The phylogenetic tree derived from the cytochrome *c* sequence variation in 25 plants is not very different (Fig. 1.4) from such trees derived from morphological characters. It is noteworthy (Fig. 1.4) that *Ginkgo biloba*, the oldest living flowering plant, separates expectedly from all the other species analyzed.

More recent experiments in the field of plant biochemical systematics have been concerned with DNA data derived from restriction enzyme analysis of the chloroplast DNA. The results (Fig. 1.5) generally support existing classificatory arrangements, although they occasionally indicate unexpected discontinuities. The chloroplast DNA data suggest that the subtribe Barnadesiinae is differentiated from other members of the same family, the Compositae, although the family is from other viewpoints relatively homogeneous. Similar studies have also been carried out on mitochondrial DNA and ribosomal RNA of plants. Such macromolecular data are making an increasing impact on traditional plant taxonomy.

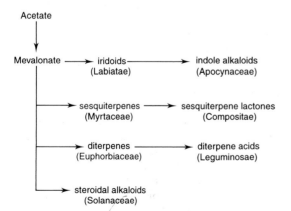

Figure 1.3 Evolution of mevalonate-derived secondary metabolites by gain (illustrative families are shown in parentheses).

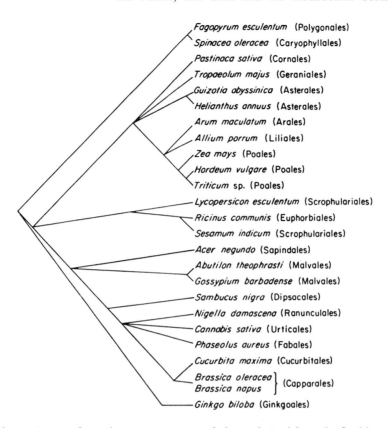

Figure 1.4 A phylogenetic tree of cytochrome *c* sequences of plants, derived from the flexible numerical method.

1.3 THE EUKARYOTIC PLANT CELL

Plant cells vary considerably in their size, shape and structure. It is, therefore, difficult to present a general picture of plant cells. Some are biochemically inactive, such as those in the xylem. Cells of the phloem lack the nucleus and are not regarded as fully active. The tracheids and the vessel elements of the xylem have extensive secondary wall thickening made up of lignin which confers strength to the plant. The parenchymatous cells, which constitute up to 80% of the entire cell complement in many plants, are metabolically active. These cells can be subdivided according to their diverse functions. In these cells two main areas can be distinguished: the thin cell wall and the protoplasm (Fig. 1.6). The cell wall surrounds the protoplast, and confers strength and integrity to the cell. The cell wall comprises the outer primary wall and the inner secondary wall. The actively dividing meristematic cells have predominantly primary walls. Two primary walls of adjoining cells do not interact directly but are separated by a layer called the middle lamella which consists mainly of pectic polysaccharides

rich in galacturonic acid. The protoplast is delineated by the plasma membrane, otherwise referred to as the plasmalemma. This membrane is semi-permeable which means that water can pass across while at the same time there is considerable solute selectivity. The plasma membrane is therefore an important control barrier for the passage of substances between the cell and the external milieu.

The connection between adjacent cells through plasmodesmata is an important feature of plant cells. These are protoplasmic strands or pores lined by plasmalemma and extending through the wall of the cells. They maintain protoplasmic continuity (symplasm), provide a major pathway for the movement of cell materials and are possibly involved in cell communication. The pores are, however, small enough to prevent organelle movements between cells. The protoplast is the metabolically active part of the cell and includes the cytoplasm and vacuole which is surrounded by the tonoplast. Each cell can contain one large vacuole or several smaller vacuoles. The cytoplasm is an aqueous medium (cytosol) of different viscosity and composition which contains

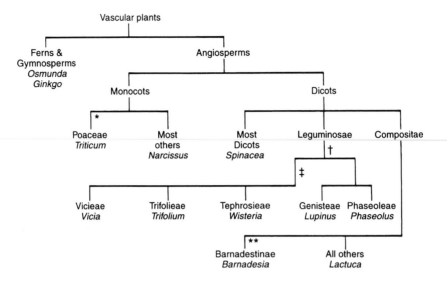

Figure 1.5 A vascular plant phylogeny based on chloroplast DNA rearrangements. Key: *three inversions, †50 kilobase inversion, ‡inverted repeat delete, **22 kilobase inversion.

membrane-bound bodies known as organelles as well as structures without encapsulating membranes (particulate structures) and also certain enzymes that support many metabolic processes. Also embedded in the cytosol are chloroplasts (classified under the general term, plastid), mitochondria, ribosomes, Golgi apparatus, peroxisomes, glyoxysomes, spherosomes, microtubules and microfilaments.

1.3.1 Cell wall

Rapidly growing cells (e.g. meristematic tissue) are surrounded by thin primary cell walls. Once growth has ceased, the secondary wall is laid down between the primary wall and the cell membrane. The primary cell wall is 1–3 μm thick and begins to form between dividing cells. A cell plate formed between the two daughter cells originates from microtubules that are different from those involved in the separation of chromatids and which act as the base for the construction of the new cell wall. The microtubules may direct the cell wall-forming materials to the plate which grows from the centre towards the periphery of the cell and soon becomes the pectin-rich middle lamella.

The composition of the dicotyledonous cell wall is typically, 25–40% cellulose, 15–25% hemicellulose, 15–40% pectic materials, 5–10% proteins (hydroxyproline-rich glycoproteins and enzymes) and a very small proportion of phenolic compounds (Fry, 1988; Brett & Waldron, 1990). The cell wall comprises a crystalline microfibrillar array of cellulose embedded in an amorphous mass of pectic and hemicellulosic materials. An extensin filamentous network is believed to act as a 'cement' within the composite structure and strengthen the cell wall (Fig. 1.7).

The cell wall also contains enzymes and it can be regarded as an 'organelle'. The cell wall accommodates increases in cell size. Cellulose, the main component of the cell wall, is insoluble in water, stable at low pH (dilute acids) and has a flattened ribbon-like shape. Each ribbon can be 1–7 μm long ($M_r > 1000$ kDa) and corresponds to 2000–14 000 D-glucose units joined by $\beta(1 \rightarrow 4)$ linkages. About 40–70 of these glucan chains are held together by intermolecular hydrogen bonds to form crystalline microfibrils that are unbranched, several micrometres long, 5–8 nm wide and of high tensile strength. They are laid down on the outer surface of the plasma membrane in elongating cells, at right angles to the long axis, and supercoiled around the cell in order to permit cell elongation. Microfibrils are not tightly packed and they are separated by a 50–100 nm mass of hydrated amorphous wall material which confers mechanical strength to the cell wall. The newer and older layers of microfibrils overlap each other like layers of cotton thread around a reel.

The amorphous wall matrix consists mainly of hemicelluloses and pectic polysaccharides (Fig. 1.7). Pectins are the most water-soluble component of the cell wall and are readily extracted with either hot water, ammonium oxalate, EDTA or other chelating agents which can complex with calcium ions present in the cell wall. Pectins are a loosely defined group of polysaccharides containing a significant amount of galacturonic acid and some L-arabinose, D-galactose and L-rhamnose. Galacturonic acid residues are linked together by α-$(1 \rightarrow 4)$ linkages to form a linear chain. Calcium ions form crosslinks between carboxyl groups of adjacent pectin chains.

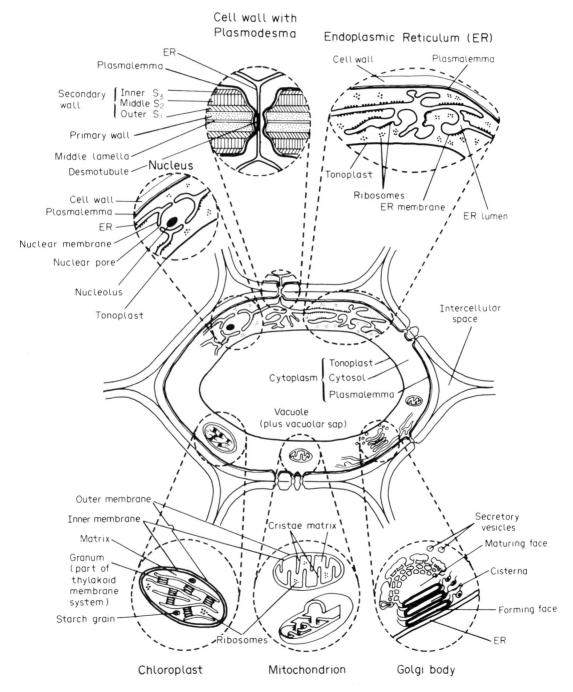

Figure 1.6 Diagrammatic representation of a plant cell. (From Goodwin & Mercer, 1983; with kind permission from Pergamon Press, Oxford.)

Hemicelluloses represent a heterogeneous group of polysaccharides with respect to the sugar residues and nature of the linkages. After the pectin has been selectively removed, these hemicelluloses are extracted from the cell wall with NaOH or KOH. However, this treatment degrades neutral polysaccharides. Therefore, subsequent analysis of the cell wall does not reflect its true state. Three main structural types of hemicelluloses have been characterized: xylans, glucomannans and xyloglucans. Xylans have a β-(1 → 4)-linked xylopyranose main chain attached to L-arabinofuranose, D-glucuronic acid, 4-O-methyl-D-glucuronic acid and D-galactose as single units or in the form of complex branches. Glucomannans contain a random assortment of β-(1 → 4)-linked glucose and mannose residues along the polysaccharide chain. Xyloglucans have a backbone of β-(1 → 4)-linked glucose units with side chains of α-1,6-xylosyl residues and other

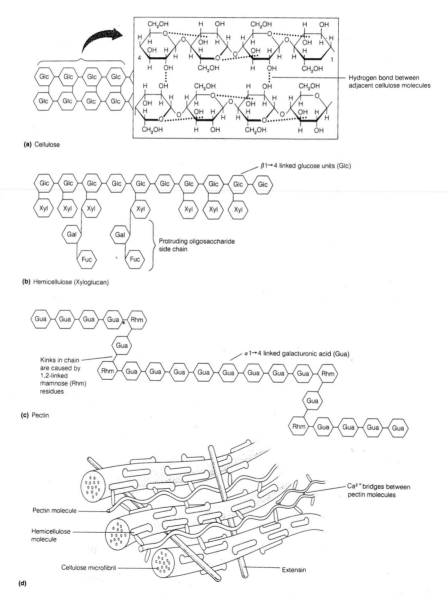

Figure 1.7 Diagrammatic representation of primary cell wall and the structure of polysaccharide constituents: (a) cellulose, (b) hemicellulose, (c) pectic material and (d) proposed assembly of components. (From Taiz & Zeiger, 1991; with permission from the Benjamin/Cummings Publishing Company, California.)

such branches of complex structures containing xylose, galactose, fucose and arabinofuranose units. Additional sugars, β-D-galactose and α-L-arabinose, are attached to O-2 of some xylose residues. The galactose can sometimes be attached to L-fucose. Only part of the surface of the linear xyloglucan is able to hydrogen bond to cell wall cellulose microfibrils.

Extensin is the major structural cell wall protein. It is a hydroxyproline-rich glycoprotein ($\sim$30–50% carbohydrate) with a protein backbone containing significant proportion of Ser-(Hyp)$_{4-6}$ sequence. Arabinose is the major carbohydrate component ($\sim$90–95%) which is linked to hydroxyproline units as $(1 \rightarrow 2)$ or $(1 \rightarrow 3)$-linked stubs of tri- and tetrasaccharide chains. In addition, sporadic serine-linked galactosyl single units are also present. The extensin molecules are crosslinked covalently with each other in the cell wall. Thus, the soluble precursor becomes insoluble at a certain stage of cellular development. A variety of enzymic proteins (hydrolases, transferases, peroxidases etc.) are also present in the cell wall. There is some evidence of the attachment of phenolic components (ferulic and coumaric acids) to galactose and arabinose residues of cell wall polysaccharides resulting in crosslinking.

Further details of the metabolism of structural polysaccharides can be found in Chapter 5. Polysaccharides are presumably not directly transcribed from 'codes'. Their structure critically depends on the specificity and location of glycosyl transferases. The physiological role of cell walls or its specific components is becoming increasingly apparent, for example, in plant growth regulation, development and differentiation, in stress and ethylene production, pathogenesis and elicitor action etc. (Bolwell, 1993).

Exo- and endoglycosidases can hydrolyse particular glycosidic bonds within cell wall polysaccharides. These enzymes can originate from organisms that invade the plants or from the host plant and are involved in the dissolution of the cell wall during the formation of xylem vessels and phloem sieve tubes, leaf abscission, fruit ripening, seed germination and pollen tube formation.

The secondary cell walls are much thicker than the primary walls and consist of 40–45% cellulose, 15–35% hemicellulose, 15–30% lignin and negligible amounts of pectic polysaccharides. The protoplasts of the cells of the primary wall secrete the secondary wall materials when the cells have stopped enlarging. The protoplast then totally diminishes and only the wall remains. The rings, spirals or network in a mature stem cross-section are due to secondary wall deposition. The stretched cellulose microfibrils embedded in lignin gives strength to the wood. Lignins are complex polymeric compounds that are chemically resistant and are not readily soluble in most solvents. Three aromatic alcohols, coniferyl-, sinapyl- and p-coumaryl alcohols, are covalently linked to each other and are the major constituents of lignin. These constituents arise from the shikimic acid pathway and are crosslinked by peroxidase *in vivo*. By altering the levels of specific enzymes of this pathway or those of relevant peroxidases via genetic engineering (e.g. antisense gene technology) it should be possible to either decrease or increase the lignin content of the cell walls. Increased lignin level will add strength to the cell wall, e.g. in wood, and decreased levels of lignin in forage grasses will increase digestibility by cattle.

1.3.2 Cytoplasm

(a) Plasmalemma

Plasmalemma is the membrane that encapsulates the cytoplasm. It is the thickest (9–10 nm) membrane compared to those of Golgi ($\sim$8 nm) and tonoplast (5–6 nm). The outer surface of the membrane has embedded spherical proteinaceous bodies consisting of enzyme complexes responsible for the synthesis of cellulose microfibrils. The fluid mosaic model of the membrane structure (Fig. 1.8) depicts the basic components of the plasmalemma. However, different intracellular membranes have different compositions. The proteins embedded (intrinsic proteins) in the phospholipid bilayer can migrate laterally within its plane. The non-polar domains of the protein associate with the hydrophobic tails of the phospholipid bilayer whereas the polar protein domains associate with the hydrophilic heads. The extrinsic proteins are bound ionically to the hydrophilic heads of the phospholipid bilayer. Two types of proteins can therefore be differentially extracted either by salt solutions or detergents. The membrane proteins may have different functions: some are enzymes and exhibit catalytic activity and others can transport metabolites. Some intrinsic proteins may be attached to microtubules and therefore connected to other cellular structures. Polysaccharides contained in Golgi cisternae are directed to the plasma membrane by microtubules. These secretory vesicles fuse with the plasmalemma and transfer their contents to the apoplast by exocytosis. The

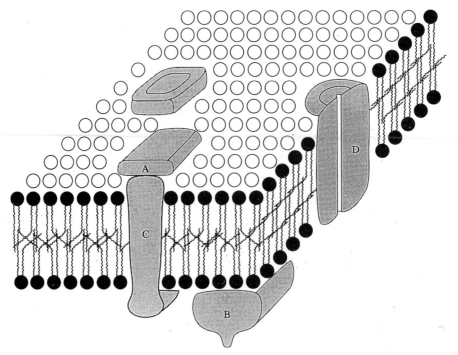

Figure 1.8 Fluid mosaic model of membrane structure. Circular heads (in black) denote polar regions of individual phospholipid molecule, the tails denote lipophilic regions and the grey structures denote proteins. Proteins A and B are extrinsic proteins (non-amphipathic) and C and D are intrinsic proteins (amphipathic). A is attached to the polar region of C, and B is attached to the polar heads of the phospholipid molecule in the bilayer. (From Anderson & Beardall, 1991; with permission from Blackwell Scientific Publications, California.)

plasma membrane also regulates intracellular solute composition by predominantly passive and active transport mechanisms. Other mechanisms referred to as endocytosis and pinocytosis involve ingestion of liquid via invagination of the membrane and subsequent detachment of the vesicles within the cell. Microvesicles of salt gland cells of mangrove species can fuse with the plasma membrane and extrude salt solutions outside the leaf.

(b) *Tonoplast*

The membrane surrounding the vacuole (vacuolar membrane or tonoplast) is thinner and has a different function than the plasma membrane. The permeability of the membrane controls the water potential of the cell, for example, in the guard cells of stomata. Changes in intravacuolar ion concentration cause osmotic flow of water and results in swelling and shrinking of the cell. This causes stoma to close or open. The origin of vacuolar membrane is believed to be either endoplasmic reticulum or as in the case of plasma membrane, the Golgi.

(c) *Vacuoles*

Plant vacuoles are membrane-bound bodies of variable size and number within the protoplast. Within many cells, up to 90% of the cell volume may be occupied by a single vacuole or multiple vacuoles. In meristematic tissue, several smaller vacuoles (provacuoles) fuse together as the cells mature to produce a large vacuole. The surrounding single membrane, the tonoplast, pushes the cytoplasm against the plasma membrane as a result of high turgor within the vacuole. The solute potential in the vacuole promotes water uptake, causing cell rigidity and promoting cell enlargement. The smaller vacuoles contain some oxido-reductases which are typical of the membranes of the endoplasmic reticulum (ER). This led to the suggestion (Matile, 1975, 1978) that these particles originate from the ER and further development and differentiation give rise to fully developed vacuoles.

Solutes in vacuoles may be present in high concentrations (0.4–0.6 M) and include inorganic ions such as sodium, potassium, calcium, magnesium, chloride, sulphate and phosphate. In some

CAM (crassulacean acid metabolism) plants, for example in *Bryophyllum*, all malic acid is present in the vacuole. Citrus fruits store approximately 0.3 M citric acid in vacuoles which can be squeezed out as juice. This compartmentation, therefore, prevents many important enzymes from being denatured. The movement of substances from the cytoplasm into vacuoles takes place by active permease-mediated transport and by fusion of membrane-bound vesicles with the tonoplast. Also stored in the vacuoles are: TCA (tricarboxylic acid) cycle acids; phenolic compounds such as flavonoids and tannins; amides; amino acids, peptides and proteins; betalains and alkaloids; gums and carbohydrates such as sucrose. These substances may occur as aqueous solution, amorphous deposits or crystals (e.g. calcium oxalate).

In addition to the turgor function of the vacuole it also serves as a storage organelle. Some storage substances are waste products of the cell which have been excreted into the vacuole from the surrounding cytoplasm. Other stored substances serve as food (energy) reserves, for example, proteins in the form of globulins, carbohydrates and phosphate in the form of phytin which comprises insoluble salts of phytic acid (inositol hexaphosphoric acid). Flavonoid deposits in vacuoles, for example in flower petals, assist in pollination by insects and poisonous alkaloids, on the other hand, repel predators. In some plants, the endospermic vacuoles (oleosomes) contain large reserve lipid deposits which are mobilized during seed germination.

Vacuoles are also rich in hydrolytic enzymes and therefore function in a similar manner to animal lysosomes. The enzymic activities are capable of hydrolyzing proteins, carbohydrates, lipids reserve phosphates and nucleic acids. They also mobilize the cytoplasmic organic matter incorporated by phagocytosis. Disruption of the vacuoles liberates the enclosed enzymes into the cytoplasm and consequently causes degeneration of intracellular components. The aleurone grains or protein bodies of seeds also display lysosomal properties. The levels of the hydrolytic activities rise during seed germination.

(d) Endoplasmic reticulum

The endoplasmic reticulum (ER) is a network of interconnecting tubular and flattened sac-like structures of membrane extending throughout the cytoplasm. The membrane is continuous with the nuclear membrane and comprises scattered integral and peripheral proteins. Ribosomes are attached to some parts of the cytoplasmic face of the membrane and these regions are known as 'rough endoplasmic reticulum' which are distinct from the ribosome-free 'smooth endoplasmic reticulum'. Protein translation from mRNA occurs at the rough ER and the final protein product enters the lumenal space of the ER. This mature protein has an N-terminal sequence of hydrophobic amino acids (18–30 amino acids) known as the targeting or leader sequence which directs the entire peptide through the hydrophobic ER membrane and into the lumen of the ER. It also helps to anchor the ribosome on the membrane surface. The smooth ER, on the other hand, functions as a site for lipid biosynthesis and membrane assembly.

The lumen of the ER can be regarded as a compartment which is joined to the ER of the neighboring cell via plasmodesmata. ER is also a dynamic entity capable of growth and turnover. It divides during cell division and a broad range of enzyme systems are integrated in the membrane which have an important role in cell metabolism.

(e) Ribosomes

Ribosomes, which are the sites of protein synthesis, are small, spheroidal (oblate or prolate) particles found in large numbers within plant cells. They are not surrounded by membrane. Three types of ribosomes have been characterized: the cytoplasmic ribosomes (20 nm in diameter) have a sedimentation coefficient of about 80S; the chloroplast and mitochondrial ribosomes (15 nm in diameter) have sedimentation coefficients of 70–75S. They are all composed of rRNA and protein. Approximately 90% of the RNA of plant cells is found in the ribosomes of the cytoplasm. In the cytoplasm they occur either in free form or attached to the outer surface of the membranes of the ER, and referred to as rough ER. In many cells they occur as clumps or strings of ribosomes joined by a thin thread of RNA. The aggregates are known as polyribosomes or polysomes and the assembly can be disrupted by treatment with ribonucleases.

The cytoplasmic ribosomes contain 45–50% RNA and 50–60% protein and are composed of two subunits, 60S and 40S, which require Mg^{2+} for structural cohesion. The large subunits have three rRNA molecules (25S, 1300 kDa; 5.8S, 50 kDa; 5S, 40 kDa) and the small subunits have only one type of rRNA (18S, 700 kDa). The larger and smaller subunits contain 45–50% and about 30% protein, respectively. The 70S ribosomes of chloroplasts and mitochondria are also made up of two subunits, 50S and 30S and contain twice as much RNA as protein. The large subunits of the chloroplast and mitochondrial ribosomes have two, 5S and 23S rRNA molecules and the small subunits have one 16S rRNA molecule. The rRNAs are encoded by the mitochondrial and chloroplast genomes. Most of the proteins are, however, coded for by the nuclear DNA and then imported into these organelles.

Proteins of ribosomes can be separated from RNA with ribonucleases. However, they are relatively insoluble in aqueous media. Proteins can be extracted by using high concentrations of salts or urea and further resolved by gel electrophoresis under denaturing conditions. The 45–50 proteins associated with the large subunit of the ribosome and 30 proteins associated with the small subunit have a molecular weight range of 10–30 kDa and are extremely basic in nature. They can be selectively extracted from the subunits by increasing concentrations of salt or urea. This reflects the differences in their binding strengths and probably their intra-subunit location. The protein-synthesizing function of the ribosomes is destroyed upon removal of the proteins. However, mixing the RNA and corresponding proteins under optimal conditions results in self-assembled functional subunits and restores the ability of the ribosomes to synthesize proteins. The ribosomal proteins are synthesized in the cytoplasm and then transferred to the nucleolus for assembly into subunits. The RNA molecules are then incorporated and the complex is transported to the cytoplasm where they are assembled to form intact ribosomes. After the synthesis of proteins of both 80S and 70S ribosomes in the cytoplasm, they must be partitioned into two types, but very little is known of the mechanism.

A three-dimensional image of the intact ribosome depicts a channel between the small subunit and the large subunit. The small subunit has a transverse cleft dividing it superficially into two unequal halves. The mRNA, aminoacyl-tRNA and the factors required for protein synthesis fit into the channel. The process of protein synthesis is shown diagrammatically in Fig. 1.9. The first event is transcription involving synthesis of the RNA molecule complementary to a specific gene in the nucleus. Processing of this RNA yields mRNA which passes through the nuclear pore and into the cytoplasm where it attaches itself to the channel of the ribosome. The process of translation begins and proteins with specific amino acid sequence are synthesized.

The formation of ribosomes begins with the synthesis of a large precursor 45S rRNA molecule in the nucleolus. It is processed to 25S, 5.8S (included in the 60S ribosomal subunit along with a 5S rRNA which has a separate origin) and 18S (constitutes 40S subunit) molecules. The proteins constituting the ribosomes are synthesized in the cytoplasm and selectively taken up by the nucleus. This is achieved with a nuclear targeting signal peptide of basic amino acids. The rRNA molecules and proteins are then assembled within the nucleolus to form the 60S and 40S ribosomal subunits. These are then transported to the cytoplasm where a few other proteins are added and subsequently form the intact ribosome.

(f) *Golgi bodies*

The Golgi body, otherwise referred to as the dictyosome, is about 1–3 μm in diameter and 0.5 μm high and a number of these structures, collectively known as Golgi apparatus, are found throughout the cytoplasm. Usually each body is composed of 4–6 flattened sac-like discs called cisternae within a stack. These are enclosed by typical membranes which appear similar to those of the smooth ER. The cisternae are firmly held together and remain intact after isolating Golgi by sucrose density gradient centrifugation. Bundles of fibres have been observed extending across the spaces between cisternae which presumably strengthen the material in the space and also provide regularity of spacing ($\sim$10 nm apart) within the body. It is rare to find only one cisterna or as many as 20–30 (in some algae).

There are vesicles close to the peripheral regions of cisternae: these bud off from the rim of the outermost *trans* cisternae and can be seen as independent vesicles (Fig. 1.10). These cisternae are closest to the plasma membrane. The cisternae close to the centre of the cell in the stack are known as *cis* cisternae. The proteins enclosed in membrane-bounded vesicles are transferred from the lumen of the ER to the Golgi body. Vesicles from the ER migrate through the cytosol and fuse with the *cis* face (forming face) of the Golgi. The glycoproteins are enzymically modified within the lumen of the Golgi and then leave the cisternae in vesicles that bud off from the *trans* face of the stack. The vesicles migrate through the cytoplasm (directed by the microtubules) and fuse with either the plasmalemma or the tonoplast. Therefore, non-cellulosic cell wall polysaccharides, (hemicellulose and pectin) and protein (extensin) are transferred to the plasma membrane

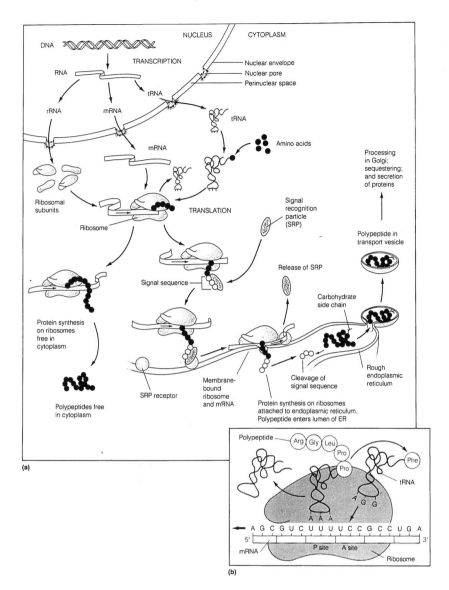

Figure 1.9   (a) Diagrammatic illustration of transcription and translation. SRP, Signal recognition particle. (b) Polymerization of amino acids to form polypeptide chain. (From Taiz & Zeiger, 1991; with permission from the Benjamin/Cummings Publishing Company, California.)

to be incorporated into the growing cell wall architecture.

Two types of vesicles known as smooth and coated vesicles have been identified. The latter have a protein skeleton called clathrin which consists of subunits in the form of a three-pronged structure. These structures are arranged in such a way as to form a cage around the vesicle. The coated vesicles are believed to transport storage proteins to protein-storing vacuoles.

From the previous description it is apparent that the intracellular membranes play important roles with respect to subcellular compartmentation, determining integrity and metabolic functions of the compartments. There exists an intracellular endomembrane system where the membrane undergoes differentiation during its transition from the ER to the plasma membrane and tonoplast.

(g)   Mitochondria

Mitochondria are the cellular sites of respiration and are contained within the cytoplasm. Two of

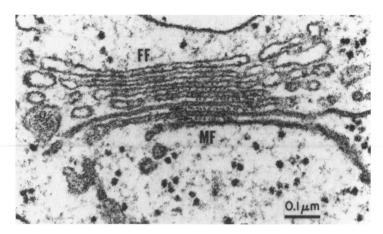

Figure 1.10 Electron micrograph of a Golgi body from maize root tip. The cisternae are separated from one another and develop sequentially across the stack from the forming face (FF or *cis* face) to the maturing face (MF or *trans* face). Inter-cisternal elements (arrows) are present between some of the cisternae and generally increasing in number towards the mature face. (From Mollenhauer & Morré, 1980; with permission from Academic Press, Inc.)

the principal biochemical processes in these organelles include the TCA cycle and oxidative phosphorylation.

As mitochondria are dynamic organelles, they can change their shape rapidly, move about and divide, and it is difficult precisely to gauge their size and number in a cell. However, for a defined type of cell, their number per unit volume of cytoplasm remains approximately constant. They may vary from 20 to several thousand per cell (100–3000 in maize root cap cell depending upon the stage of maturity). The *Chlorella* cell has only one large mitchondrion. Microscopic examination shows a typical structure of mitochondria as a short rod-shaped body with hemispherical ends (1–4 μm in length and 0.5–1.0 μm in diameter). They are bound by a double-membrane envelope encompassing an intermembrane space. The innermost non-membranous compartment is known as the matrix. The outer membrane is smooth and has a larger average thickness (5.5 nm) than the inner membrane (4.5 nm). It has a larger proportion of lipids and is permeable to a wide range of compounds (up to M_r 5 kDa). The inner membrane is highly folded into tubular invaginations known as cristae (Fig. 1.11). The surface area of the inner membrane is therefore substantially increased; the shape and number of the cristae in mitochondria may vary depending upon the respiratory activity. The space within the cristae is continuous with the intermembrane space. The inner membrane acts as a barrier to many ions, small compounds and proteins and has specific transmembrane transport systems.

The matrix consists of lipids and proteins and contains particles of varying electron density. Aggregates of 10 nm diameter and 30–40 nm diameter particles can be observed together with ribosomes (15–25 nm diameter) and DNA filaments. A consequence of Ca^{2+} being imported by a process coupled to electron transport is that some granules contain calcium phosphate. The sedimentation coefficient of mitochondrial ribosomes is 70S and they consist of 50S (containing 23S and 5S rRNA) and 30S (containing 30S rRNA) subunits. Some mitochondrial proteins are synthesized within the mitochondria. The mechanism of protein import is complex and is believed to take place at the sites of close contact between the outer and inner mitochondrial membranes. The cytoplasmic precursor proteins contain targeting information that is recognized by receptors of the outer membrane surface and ultimately translocated into the matrix. Here they are processed and sorted to their submitochondrial destinations (Pfanner *et al.*, 1991).

The enzymes of the TCA cycle are contained in the matrix and probably exist as a loose complex. Only one enzyme of the TCA cycle, succinate dehydrogenase, is bound to the inner membrane. This membrane contains an F_0/F_1 ATP-synthase which catalyzes the synthesis of ATP from ADP and inorganic phosphate. The membrane is impermeable to the passage of H^+ and the formation of electrochemical gradients across the inner mitochondrial membrane is due to the controlled movement of H^+ through the transmembrane protein (H^+-ATP synthase) and

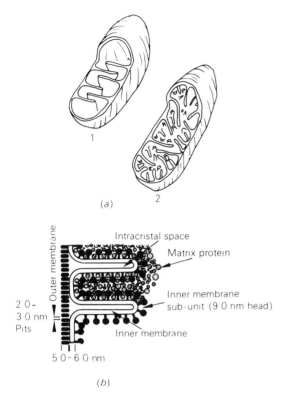

Figure 1.11   (a) Diagrammatic representation of mitochondria from (1) an animal cell showing the regular plate-like cristae and from (2) a plant cell with less regular and tube-like cristae. (b) Mitochondrial membranes and subunits. (From Hall *et al.*, 1982; with permission from Longman Group Ltd., London.)

the membrane is the mobile redox carrier from complexes I and II to complex III. In complex II succinate dehydrogenase lies across the membrane. This enzyme has two subunits, Fp ($\sim$67 kDa) and Ip ($\sim$30 kDa). The former is a flavo-iron–sulfur protein complex with two centers, S1 and S2, and the latter is an iron–sulfur protein complex called S3. Complex II is responsible for electron transfer from $FADH_2$ to S3 via S1 and S2. Complex III consists of cytochrome b (contains b_{560} and b_{565}), cytochrome c, and iron–sulfur protein. It transfers electrons from reduced ubiquinone to cytochrome c, via cytochrome b. Complex IV has five proteins including cytochrome oxidase complex (with copper atoms, Cu_A and Cu_B), cytochrome a and a_3. It is involved in electron transfer from cytochrome c to O_2. In addition to the above complexes, the membrane also has external NADH dehydrogenase on the external face which can oxidize cytoplasmic NADH. This is coupled to ubiquinone reduction within the membrane. The reduced ubiquinone can also be oxidized by an oxidase (alternative oxidase) presumed to be situated on the membrane facing the matrix.

(h)   *Plastids*

Plastids develop from immature plastids, called proplastids. Plastids are double membrane-bound organelles and three types have been characterized: (1) leukoplasts or colorless plastids, including amyloplasts (containing starch); proteinoplasts (containing protein); elaioplasts (containing fat) and other storage plastids; (2) chromoplasts (containing red, yellow, orange and other pigments as in many fruits, flowers and autumn leaves), and (3) chloroplasts (containing chlorophyll and sometimes, starch) which undertake photosynthesis. Photosynthesis requires energy which is supplied by solar radiation and is converted into a chemical form that can be utilized by the cell. Light absorption by suitable pigments and subsequent production of ATP and NADPH effectively reduces CO_2 to carbohydrate.

Chloroplasts arise from proplastids which are colorless with no internal membranes. As the embryo develops, proplastids divide (the sperm cell does not contribute) and turn into chloroplasts when leaves and stems are formed. Young chloroplasts can also divide, especially in the presence of light and each mature cell contains a certain number of chloroplast (up to a few hundred). DNA replication in chloroplast is independent of DNA replication in the nucleus,

dissipation of the gradient is coupled with ATP formation. The enzymes and other components of the electron transport chain constitute a part of the inner membrane structure occupying approximately 10% of the area. Oxidative phosphorylation and energy transduction therefore, takes place at this site. The inner mitochondrial membrane also transports fatty acids via the carrier carnitine into the matrix to undergo β-oxidation. The intermembrane space is known to contain nucleoside phosphokinases which allow the nucleotides to be transferred into the mitochondria from the cytoplasm.

The location of the components of the electron transport chain in the inner mitochondrial membrane is illustrated in Fig. 1.12. They are grouped in distinct complexes (I–IV). Complex I comprises 26 different proteins (combined M_r, 850 kDa) including NADH dehydrogenase, FMN (flavin mononucleotide)-containing protein and three iron–sulfur proteins. The NADH dehydrogenase faces the matrix. The lipid-soluble ubiquinone in

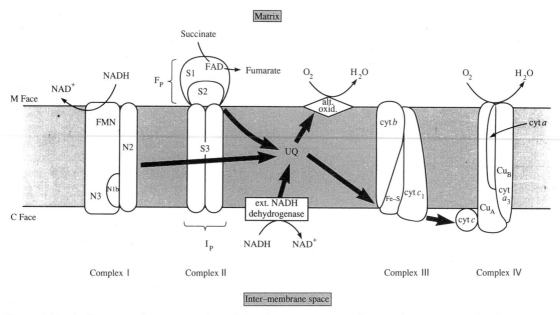

Figure 1.12   A diagrammatic representation of the electron transport chain in the inner mitochondrial membrane. The bold arrows depict pathway of electron flow from complex I or II via the ubiquinone (UQ) pool to complex III and IV; alt, alternative oxidase; ext., external; N1, N2, N3 and S1, S2, S3, iron–sulfur protein centers of complex I and II; Cu_A, Cu_B, copper atoms of cytochrome oxidase; Fe-S, iron–sulfur protein of complex III. (From Anderson & Beardall, 1991; with permission from Blackwell Scientific Publications, Oxford.)

however, the replication in the organelle must remain under cellular regulation. Upon illumination of proplastids, enzymes are formed (or imported from cytosol), light-absorbing pigments appear, membranes proliferate and other structures (stroma, lamellae and grana stacks) are generated. These transformations occur within minutes after exposure to light.

Higher plant chloroplasts are generally concavo-convex, plano-convex or biconvex with a diameter of $5–10\,\mu m$ and a thickness of $2–3\,\mu m$. They exist around the perimeter of photosynthetic cells with their broad faces parallel to the cell wall for maximum absorption of light. However, they may alter their position in relation to the intensity of the incident light and are also influenced by cytoplasmic streaming. The chloroplast is bound by a double membrane or envelope which also controls molecular import or export. An amorphous gel-like, and enzyme-rich material called the stroma is contained within this organelle (Fig. 1.13). This stroma contains various structures and soluble materials. Plastoglobuli (e.g. plastoquinone) are lipid-rich particles that are not encapsulated by membrane and lie close to thylakoid membranes. Starch grains are also found around this site which cease to exist in dark

and reappear in light. Ribosomes (70S) are found either free or as thylakoid membrane-bound forms; they contain 50S and 30S subunits. Chloroplast DNA exists as 20–50 histone-free, double-stranded circles in specific regions of stroma called nucleoids. Plastid genes code for all the tRNAs ($\sim$30) and rRNAs (4). Even though proteins involved in transcription, translation, and photosynthesis are encoded by the plastid genome, most proteins found in the chloroplasts are encoded by nuclear genes and ultimately imported. The transport and routing of proteins into chloroplasts has been reviewed (Keegstra, 1989). The formation of chlorophyll, the haem of chloroplast cytochrome, galactolipids, carotenoids and fatty acids are undertaken within the chloroplast.

Embedded in the stroma is an internal membrane system within mature chloroplasts which is not connected to the envelope in contrast to mitochondria. The membranes are known as stromal thylakoids or stromal lamellae and consist of two membranes ($\sim$7 nm thick) separated by an inter-membrane space (4–70 nm in thickness) called the lumen. This space is filled with protein and plays a special role in photosynthesis. In places these thylakoids are stacked in a multilayered

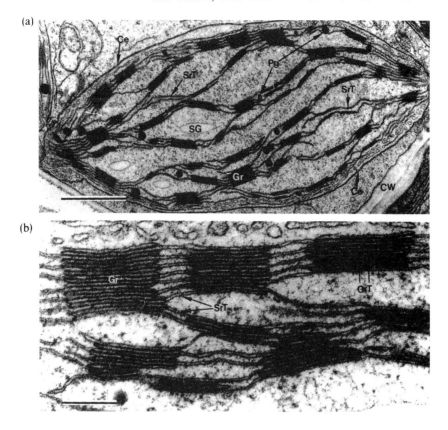

Figure 1.13 Electron micrograph of chloroplasts from mesophyll cell of *Digitaria sanguinales*. (a) Entire chloroplast (bar denotes 1 μm). (b) Details of the internal membrane complex associated with several grana (bar denotes 0.25 μm). Ce, chloroplast envelope; CW, cell wall; Gr, granum; GrT, granal thylakoid; Pg, plastoglobuli; SG, starch grain; SrT, stromal thylakoid. (From Anderson & Beardall, 1991; with permission from Blackwell Scientific Publications, Oxford.)

manner (as a pile of coins) and are known as grana (0.3–2 μm in diameter with 10–100 thylakoids per granum). There could be 40–60 grana within a chloroplast. The stroma thylakoids are longer, extend through the stroma and interconnect grana. The component thylakoids of the grana stack are interconnected with tubular frets. The entire thylakoid system therefore, appears to comprise a single complex cavity separated from stroma (Fig. 1.14). The surfaces joining the adjacent grana sacs are termed appressed regions and those surfaces exposed to the stroma are called non-appressed regions.

When thylakoids are separated from isolated chloroplasts, four major protein complexes can be identified: the light harvesting complex protein (LHCP), photosystem II (PSII) (pigment–protein complex, occurs in the appressed region), cytochrome b_6–cytochrome f complex (at the junction between appressed and non-appressed regions), PSI pigment–protein complex (in the non-

appressed region) and ATP synthase or coupling factor (CF$_0$ and CF$_1$, occurs in the non-appressed region). The arrangement of the complexes and other components of the membrane with regard to the process of energy transduction, especially in the light of separate sites of PSII and PSI, are illustrated in Fig. 1.15.

The fluidity of the membrane (with high content of unsaturated acyl lipid) plays an important regulatory role in the transport of electrons from H_2O to NADP. Plastoquinone with a long half-life for reoxidation of about 20 ms is a good candidate as a lateral shuttle electron carrier between PSII and PSI. The content of plastoquinone in the membrane, its size and lipophilic nature is consistent with the above supposition. The electrons from the cytochrome b_6/f complex can be transferred to PSI via the mobile lateral carrier plastocyanin in the intra-thylakoid space. The absorption of light photons (8; 4 each by PSII and PSI) results in the transfer of electrons from two

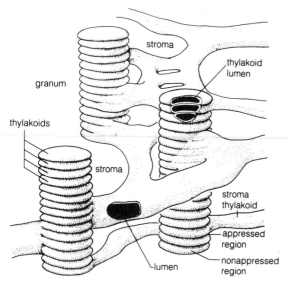

Figure 1.14 Illustration of the internal membranes of chloroplast. (From Salisbury & Ross, 1992; with permission from Wadsworth Publishing Company, California.)

(i) *Nucleus*

The nucleus contains the blue print for determining the structure and function of cells. It is the largest organelle present in the cytoplasm, being about $10 \, \mu m$ in diameter in non-dividing cells. A double membrane encompassing the perinuclear space encapsulates this organelle. The membrane disappears and reappears during mitotic cell division. The membrane has numerous (about 3000) pores of varying size (diameter range 30–100 nm). The two membranes are fused together at the edges of the pores to form annules. Since the membranes are continuous with the membranes of the ER, the perinuclear space is connected with the lumen of the ER. The pores are the communication lines between the nuclear sap (nucleoplasm) and the cytosol (cytoplasm) and assist in the entry of nucleotide precursors of DNA and RNA, histones and proteins into the nucleus while mRNA, tRNA and ribosomal subunits are exported to the cytoplasm. There is some degree of control over the passage of substances because the pore architecture does not allow free communication between the two compartments.

water molecules to form two NADPH. Thus, eight protons are transferred from stroma to the intra-thylakoid space and the pH gradient created drives the formation of ATP.

The nucleus is the major center for the control of gene expression and therefore DNA and RNA

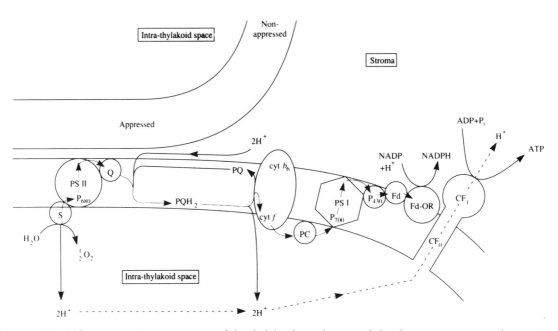

Figure 1.15 A diagrammatic representation of the thylakoid membrane and the electron transport pathway. PSI and PSII, photosystem I and II; S, water oxidizing Mn-protein; PQ/PQH$_2$, oxidized and reduced plastoquinone; Q, quinone; cyt, cytochrome; P$_{680}$, P$_{700}$, chlorophyll-protein dimers; P$_{430}$, a component associated with Fe-S protein; Fd, ferredoxin; Fd-OR, ferredoxin-NADP oxidoreductase; CF, coupling protein factor (ATP synthase). (From Anderson & Beardall, 1991; with permission from Blackwell Scientific Publications, Oxford.)

comprise 25–40% of the total mass. There is usually more DNA present than RNA. The protein component can be subdivided into basic proteins (up to ~30%) and acidic proteins (up to ~70%). The basic proteins are histones which contain a high proportion of basic amino acids such as arginine and lysine, and have isoelectric points of pH 10–12. These belong to five main classes, H1 (M_r, ~21 kDa; Lys $\gg$ Arg), H2A (M_r, ~15 kDa; Lys < Arg), H2B (M_r, ~14 kDa; Lys + Arg), H3 (M_r, ~15.5 kDa; Lys < Arg) and H4 (M_r, ~11 kDa; Lys < Arg). These proteins complex with DNA and the resulting nucleoproteins are responsible for controlling gene activity in the cell. The acidic proteins constitute the ribosomal proteins and to some extent form part of the chromosomal protein.

The DNA–protein complex, chromatin, occupies certain regions of the nuclear sap (nucleoplasm) and during nuclear division in dividing cells, it coils and condenses to form chromosomes. The chromatin is initially in the form of a very long DNA molecule attached to several million histone molecules. The chromatin fibers can be 1–10 m long which are condensed in a small space within the nucleus. At the initial stages of chromatin transformation to chromosomes, the DNA takes up a long filamentous shape with attached histone beads. The condensed coiling relaxes between cell divisions and the chromosomes become indistinguishable.

The nucleoplasm also contains spherical bodies known as nucleoli (up to about four) each about 3–5 μm in diameter bound by membranes. Each nucleolus is a mass of fibers and granules. The nucleoli cease to exist during mitosis (metaphase and anaphase) but reappear during telophase as small bodies. They may combine to form a single nucleolus during interphase. The nucleolar content comprises proteins and RNAs. The granules in the nucleolus resemble the cytoplasmic ribosomes whereas the fibrillar part resembles RNA. The granules are the ribosomal subunits. The RNAs are transcribed at special sites of chromosomes known as nuclear organizing regions. Proteins are synthesized in the cytoplasm and imported into the nucleus where they combine with the RNAs to form subunits that pass through the nuclear pores and assemble as ribosomes in the cytoplasm.

The nuclear sap contains a variety of enzymes and is probably well structured (as in cytoplasm) containing elements which can organize chromatin and nucleoli. Enzymes of glycolysis and the TCA cycle have been detected, but these could be due to cytoplasmic contamination of the

preparations. In animal cells, the nucleus is an important site of NAD production due to the presence of the enzyme NAD pyrophosphorylase which can be correlated with the control of DNA synthesis.

(j) *Microbodies*

Microbodies are ubiquitous organelles found in the majority of eukaryotic plant cells. They are mostly spherical and have a diameter ranging from 0.2 μm to 1.5 μm. The organelles enclose a granular matrix which contains either fibrils or an amorphous nucleoid. These organelles are cytochemically defined by the presence of both catalase, which destroys toxic hydrogen peroxide to oxygen and water, and flavin-linked oxidases that produce hydrogen peroxide from molecular oxygen. Microbodies tend to be associated with the endoplasmic reticulum. In particular, peroxisomes may be closely associated with chloroplasts and glyoxysomes may be associated with lipid bodies. Isolation of microbodies is facilitated by the fact that these protein-rich organelles are denser than other cell organelles of similar size such as mitochondria. However, the thin and delicate 7 nm outer membrane makes isolation of microbodies by homogenization and subsequent centrifugation difficult. The tissue is homogenized gently for a few seconds and then centrifuged at about 10 000–15 000 g. The resuspended pellet is loaded onto a sucrose density gradient and the microbodies separate at a higher density than mitochondria and proplastids.

Two types of microbodies, peroxisomes and glyoxysomes, have been characterized. These organelles differ in their distribution and enzyme composition, although both have the capacity to transform non-carbohydrate material into carbohydrate. Peroxisomes are found in both animal cells and in leaves of higher plants whereas glyoxysomes are located only in plants and particularly in germinating seeds. Peroxisomes act in parallel with chloroplasts in higher plants and are believed to undertake photorespiration whereas glyoxysomes are involved in the formation of sugars by the breakdown of fatty acids in germinating seeds.

(i) Peroxisomes

Peroxisomes are found in the photosynthetic cells of green plants, particularly in the palisade cells of C3 leaves and bundle sheath cells of C4 leaves. They are found close to mitochondria and chloroplasts which is consistent with their putative role in photorespiration. The primary electron donor is glycolate in plants and the consumption of 1 mole of oxygen results in the production of 1 mole of hydrogen peroxide (Fig. 1.16). The catalase then catalyzes the destruction of hydrogen peroxide to oxygen and water. Peroxisomes in close association with chloroplasts and mitochondria contain a large selection of enzymes that participate in the glycolate pathway and are involved in the formation of the amino acids, serine and glycine from specific intermediates of the photosynthetic carbon reduction cycle (Fig. 1.16). Glycolate is formed in chloroplasts from phosphoglycolate that is generated by the oxygenase activity of ribulose bisphosphate carboxylase. On a hot day, the leaf stoma remain closed in order to reduce water loss from the plant. Consequently, as photosynthesis continues in the absence of significant gaseous exchange, the ratio of oxygen to carbon dioxide rises and ribulose bisphosphate carboxylase operates in an oxygenase mode. The infiltration of excised leaves with glycolate demonstrated that subsequent oxidation to glyoxylate and formation of serine and glycine could both occur in the dark and in the presence of an inhibitor of photosynthetic oxygen production, DCMU [3-(3,4-dichlorophenyl 1)-1,1 dimethyl urea]. This suggests that initial metabolism of glycolate occurs in organelles distinct from the chloroplasts, such as peroxisomes. A flavin oxidase referred to as glycolate oxidase converts the glycolate into glyoxylate, with the production of hydrogen peroxide. The glyoxylate, in turn, may return to the chloroplast and be reduced to glycolate by glyoxylate reductase via a glycolate/glyoxylate shuttle. However, only a low glyoxylate reductase activity has been characterized in the chloroplast. Alternatively, glyoxylate may be oxidized by glycolate oxidase to oxalate which remains unmetabolized in the peroxisomes or glyoxylate is transaminated with L-glutamate to yield glycine which can subsequently be converted into serine inside the mitochondria. Transamination of serine with glyoxylate generates hydroxypyruvate by a serine:glyoxylate aminotransferase. This product can be reduced to glycerate which leaves the peroxisomes and enters the chloroplasts and ultimately forms hexose sugars.

The glycolate pathway is not restricted to one organelle. The first and last reactions of the cycle occur in the chloroplasts (glycolate biosynthesis and glycerate kinase, respectively). As a plausible consequence of the reactivity of glyoxylate, its metabolism is confined to the peroxisomes. During the glycolate cycle, there is a net uptake of oxygen and evolution of carbon dioxide. This is referred to as photorespiration because this pathway is stimulated in the light.

The function of peroxisomes remains obscure. In order to address the functional significance of peroxisomes, it is necessary to define the role of the glycolate pathway. Although serine and glycine can be synthesized by the glycolate pathway which utilizes the peroxisomal space, these intermediates can be synthesized independently of 3-phosphoglycerate without intervention of the glycolate pathway. Possibly, the glycolate

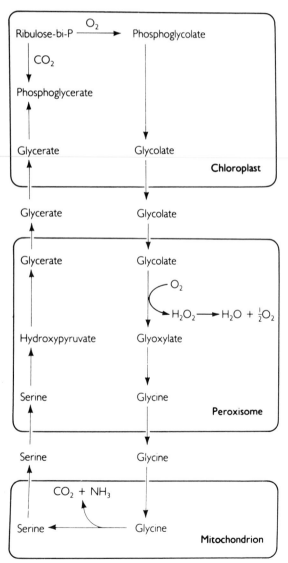

Figure 1.16 Role of peroxisomes in the metabolism of glycolate. (From Sadava, 1993; with permission from Jones & Bartlett Publishers International, Boston.)

pathway generates ATP during the conversion of glycine into serine in the mitochondria. Alternatively, glycolate formation operates as a protective mechanism against toxic and highly diffusible hydrogen peroxide. Photorespiration may have evolved in order to consume excess ATP and NADPH produced in photosynthesis. The glycolate/glyoxylate shuttle between the peroxisome and the chloroplast may destroy excess reducing power formed by the reduction of glyoxylate to glycolate by glyoxylate reductase in the chloroplast.

(ii) Glyoxysomes

Glyoxysomes are temporary in that they occur during transient periods in the life cycle of a plant such as in certain beans and nuts which store fats in their seeds as energy reserves. Glyoxysomes appear in the first few days after seed germination in endosperm cells and associate closely with lipid bodies. They disappear after the storage fats are broken down and converted into carbohydrate. The appearance of these

organelles therefore, appears to coincide with the conversion of fats into carbohydrate during seed germination.

The glyoxysomes contain the enzymes of fatty acid oxidation and the glyoxylate pathway. The metabolic processes that occur within glyoxysomes are characterized in Fig. 1.17. Storage lipids are initially broken down to glycerol and fatty acids in the lipid bodies. The long-chain fatty acids enter the glyoxysomes and are broken down to acetyl-CoA via the β-oxidation pathway. The acetyl-CoA is converted into citric acid from oxaloacetate and acetyl-CoA by citrate synthetase and then converted into isocitrate by aconitase in the glyoxylate pathway, which involves some of the reactions of the TCA cycle. Therefore, the glyoxysomes of higher plants are the site of β-oxidation of fatty acids and the enzymes of the glyoxylate cycle. Succinate is the end product of glyoxysomal metabolism of fatty acid and is not metabolized further in this organelle. The succinate is believed to diffuse out and it is converted into oxaloacetate in the mitochondria. In the

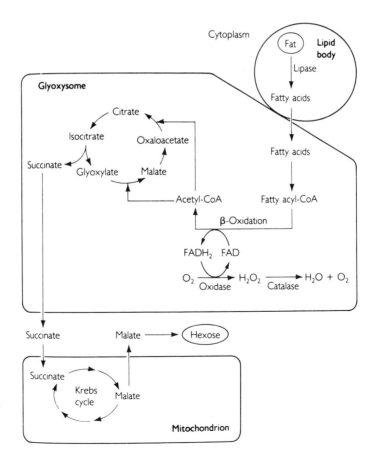

Figure 1.17 Metabolism in glyoxosomes. (From Sadava, 1993; with permission from Jones & Bartlett Publishers International, Boston.)

cytoplasm, glucose is produced by gluconeogenesis. The conversion of fat into carbohydrate is unique to glyoxysomes and therefore, does not occur in mammals.

Other enzymes such as urate oxidase and allantoinase are found in glyoxysomes. These enzymes do not have an obvious connection with the conversion of fat into carbohydrate and catalyse the conversion of uric acid into allantoin and subsequent hydration of allantoin to allantoic acid during the degradation of purine bases. No other enzymes of urate oxidation have been found. Other enzymes such as glycolate oxidase and several aminotransferases exist in glyoxysomes. Their functions remain unclear and may have existed in an ancestral and physiologically more active microbody.

(k) Microtubules

Microtubules form part of the cell cytoskeleton, a three-dimensional network of proteins in the cytoplasm. The cytoskeleton plays an important role in distributing cell organelles, cell shape maintenance, cell differentiation, ordered deposition of cell wall components, mitosis, meiosis and cytokinesis. Microtubules have a cylindrical shape with an outer diameter of 25 nm and a 5 nm thick wall with a hollow core of about 15 nm. Their length varies from a few nanometres up to 1 millimetre. They comprise globular proteins called tubulin (M_r 120 kDa) which consist of two subunits (α and β) of equal size (each 55 kDa) but different structure (Fig. 1.18a). A cross-section of the microtubule shows that the wall of the cylindrical structure consists of an assembly of β circular subunits (5 nm in diameter). Thirteen strands of tubulin form the cylindrical microtubule. The tubulin strand is composed of α and β subunits (5 nm in diameter) arranged lengthwise in an alternate but slightly overlapping manner with an 8 nm periodicity. Adjacent microtubules are sometimes interconnected and this contributes significantly to the cohesion of the cytoskeletal structure and rigidity of the cell, for example, in the cell wall-devoid generative cell of the pollen grain. In vitro, assembly and disassembly of microtubules (Fig. 1.18b) are dependent on various factors such as tubulin concentration and the presence of Ca^{2+} and Mg^{2+}. Disassembly is accelerated at high Ca^{2+}, low temperature, high hydrostatic pressure and the presence of certain external factors such as colchicine. High tubulin concentration promotes growth of microtubule at both ends, lower tubulin concentration causes preferential growth at one end and very low concentration results in net disassembly of the microtubule. Glycerol binds to tubulin and stabilizes the microtubule. GTP is a cofactor for subunit assembly and binds to two sites on a tubulin dimer.

Tubulin undergoes a number of post-translational modification. These include (1) tyrosylation at the C-terminus by an ATP requiring tubulin-tyrosine ligase, (2) phosphorylation of serine and threonine residues of α subunits, (3) acetylation of lysine residues which presumably confers stability to the microtubule, and (4) although purified tubulin is not a glycoprotein, glycosylation by glucosaminyl transferase has been demonstrated in vitro.

Microtubules can be grouped into three functional types: (1) those associated with chromosomes in cell division which constitute the mitotic spindle; (2) those found in the cytoplasm and controlling the direction of cellulose microfibrils and other cellular components, and (3) those constituents of flagella and cilia in motile cells. Microtubules of the first two categories are less stable and are able to disassemble and reassemble according to cell requirements. As colchicine inhibits mitosis by preventing the formation of the mitotic spindle and blocking cell division at metaphase, it may have significant long-term effects on the morphology and physiology of the cell. Immunological techniques can be used as a powerful tool to study tubulin in vivo because antibodies raised against animal tubulin cross-react strongly with plant tubulin.

(l) Microfilaments

Microfilaments are thinner (5–7 nm in diameter) than microtubules, they occur in bundles at different cellular locations and can act alone or in co-ordination with microtubules in intracellular movement and cytoplasmic streaming. They are complex assemblies of complex protein monomers. The protein component is a special form of muscle protein known as β-actin. γ-Actin is found in smaller proportions within non-muscle cells. Both have a very similar molecular size ($M_r \sim 43$ kDa) and amino acid compositions. These proteins are believed to be highly conserved during evolution because antibodies raised against muscle actin cross-react strongly with plant microfilaments. Microfilament can be assembled in vitro into a typical

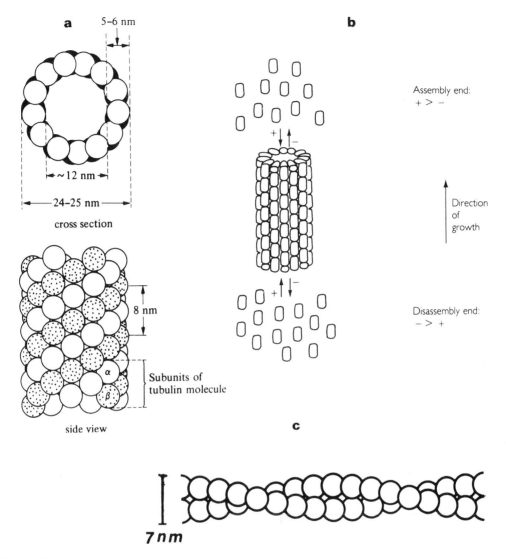

Figure 1.18 (a) Diagrammatic representation of the microtubule. Each microtubule consists of 13 protofilaments; the α and β subunits of each are shown. (From Goodwin & Mercer, 1983; with permission from Pergamon Press, Oxford.) (b) A model showing growth of a microtubule. If the rate of assembly is greater than disassembly there is a net growth. At steady state, the two rates are equal. (From Sadava, 1993; with permission from Jones & Bartlett Publishers, Boston.) (c) Diagrammatic representation of a microfilament, showing two strands of actin subunits. (From Hall *et al.*, 1982; with permission from Longman Group Ltd, London.)

structure of a twisted helical strand (Fig. 1.18c). An actin binding protein called filamin forms cross-links between the strands and this confers a gel-like constituency to the assembly. When myosin is added to such a complex, the assembly can contract reversibly in the presence of ATP. Other proteins are also associated with microfilaments.

The reversible microfilament assembly inhibitor, cytochalasin B is able to disrupt micro-filament assembly and inhibit cellular contract-ile processes such as cytoplasmic streaming and vectorial transport of vesicles to the cell surface. A network of microfilaments is present throughout the cell during mitosis. At the end of cell division, a plate comprising closely associated microfilaments is formed in the center of the cell and therefore microfilament contraction is believed to assist cytokinesis.

1.4 TECHNIQUES OF CELL FRACTIONATION

Cell fractionation proceeds essentially in two consecutive stages: (1) homogenization which disrupts the tissue and releases the cellular components into the resultant homogenate and (2) centrifugation which separates the individual components within the homogenate according to density, size and shape. First, the plant cell has several distinct features that make tissue disruption more difficult. The plant cell is encapsulated within a rigid cell wall that may be strengthened by lignin. The secondary plant cell wall is so tough that shearing forces required to disrupt this cell wall may rupture organelles. Second, the mature plant cells contain a large central vacuole which is formed from the coalescence of small vacuoles in immature and dividing plant cells. The central vacuole contains a large number of organic molecules such as tannins, flavones and alkaloids and lytic enzymes such as proteases, nucleases and phosphatases. Disruption of the vacuole which can occupy 90% of the total cellular volume can release a substantial amount of toxic molecules and hydrolytic enzymes that can destroy the required organelle or the protein. Various precautions are therefore, necessary to ensure that the vacuole is not disrupted by the cell fractionation procedure.

The ultimate objective of the cell fractionation procedure and the nature of the starting material are important parameters in defining homogenization as a semi-empirical method of cell disruption. In order to understand the complex nature of biological membranes and organelles within a cell, the plasma membrane must be disrupted. The ultrastructure and function of each individual separated component can be assessed with minimal physical and biological damage. The challenge in much *in vitro* work is to choose methods of cell disruption and incubation of subcellular fractions, cells, tissues or organs that are appropriate to the particular investigation and which minimizes artefacts.

The ideal homogenate comprises all the desired cellular components in abundance and in an unaltered morphological and metabolic state. In practice, fractionation of selected material often produces components with at least one altered characteristic. The ultrastructure may be preserved to the detriment of metabolic integrity. If the ultimate objective of the cell fractionation is to isolate a particular compound, the retention of morphological and metabolic integrity throughout the preliminary stages of cell disruption may be unnecessary. However, a detailed analysis of the function of individual components within a cell would demand preservation of metabolic and ultrastructural integrity.

1.4.1 Homogenization medium

Cell disruption is usually performed in either a slightly hypo-osmotic or an iso-osmotic medium to preserve morphological integrity. Sucrose is commonly used as an osmoticum to prevent subcellular organelles or vesicles from adverse swelling or shrinkage. Mannitol and sorbitol have also been used when sucrose has been found to interfere with the biochemical analysis of the subcellular component. Sucrose is preferred to salt solutions because the latter tend to aggregate subcellular organelles. Hypo-osmotic media are often used in the isolation of plasma membrane fractions because under these conditions, disrupted cells yield plasma membranes in the form of large ghosts or sheets. This significantly reduces the high degree of variation in size of these fragments prior to centrifugation. Although the homogenization medium is usually aqueous in nature, non-aqueous media such as ether/chloroform have been used to isolate subcellular organelles. However, non-aqueous media may inactivate some enzymes and reduce morphological integrity of some tissues.

Chelating agents (EDTA or EGTA) may be added to the homogenization medium to remove divalent cations, such as Mg^{2+} or Ca^{2+} which are required by membrane proteases. It is important to note that protease inhibitors should still be included in the medium because EDTA and thiol reagents may activate some proteolytic enzymes. However, many membrane marker enzymes require these cations for activity and may be inhibited after the cations have been removed from the extract. The addition of Mg^{2+} or Ca^{2+} to the homogenization medium maintains nuclear integrity. However, these ions can cause membrane aggregation and decrease respiratory control in mitochondria.

Cell disruption of plant tissue often releases phenols which can have several deleterious effects on plant enzymes. Plant phenolic compounds comprise essentially two groups: phenylpropanoid compounds (e.g. hydrolyzable tannins) and the flavonoids (e.g. condensed tannins). The phenolics can hydrogen bond with peptide bonds of proteins, or undergo oxidation by phenol oxidase to quinones. Quinones are powerful oxidizing agents that also tend to polymerize and condense with reactive groups of proteins to form a dark pigment called melanin which forms the basis of the browning of plant tissue. Attachment of melanin to proteins can inactivate many enzymes. Phenolic hydroxyl groups may form ionic interactions with basic amino acids of proteins or interact hydrophobically with hydrophobic regions of proteins. Therefore, it is important to remove the phenolic compounds as quickly as possible and this is best achieved by using adsorbents that bind to the phenols or quinones or by using protective agents such as borate, germanate, sulfites and mercaptobenzothiazole that inhibit the phenol oxidase activity. The phenol adsorbent, polyvinyl-pyrrolidone in either a water-soluble form or as the highly cross-linked insoluble product 'Polyclar' can be added to the isolation medium. Polyvinyl-pyrrolidone binds those phenolic compounds that form strong hydrogen bonds with proteins. Bovine serum albumin has also been reported to react with plant phenolics and is also an effective quinone scavenger.

Cell or tissue fractionation is invariably performed at $+4°C$ to reduce the activity of membrane proteases. Proteases can be subdivided into endopeptidases and exopeptidases. Endopeptidases, which are enzymes that cleave internal peptide bonds, include: acid proteases that have a pH optima of pH 2.5–3.8 and can be inhibited by the transition state inhibitor pepstatin; serine proteases that may be inhibited by diisopropyl-fluorophosphate and phenylmethylsulfonylfluor-ide; and sulfhydryl proteases which are inhibited by N-ethylmaleimide or E-64 (L-trans-epoxysucci-nyl-leucylamide-[4-guanidino]-butane). Examples of exopeptidases include carboxypeptidases that are present in most plant tissues and readily hydrolyze most C-terminal amino acids. All plant carboxypeptidases studied to date are inhibited by diisopropylfluorophosphate. Aminopeptidases, dipeptidases and tripeptidases have also been identified in plants. Other compounds used in homogenization media include disulfide-reducing agents such as 2-mercaptoethanol, dithiothreitol, dithioerythritol, reduced glutathione or cysteine. This is because many enzymes contain an essential active site sulfhydryl group that must remain reduced in order to maintain enzyme activity. Many of the problems of organelle isolation are associated with the rupturing of the fragile membranes, for example the tonoplast surrounding the vacuole. Apart from the physical destruction of these membranes, plant tissues contain active lipolytic enzymes which can attack the lipid components of membranes directly and disrupt organelles. Free fatty acids released by the hydrolytic action of acyl hydrolases can inhibit ATPase activities of mitochondria, chloroplast and plasma membrane. Other deleterious effects of fatty acids are also well established. There are certain conditions that can reduce lipid breakdown by lipases and lipolytic acyl hydrolases in plants. These include using (1) an isolation medium with a pH of 7.5–8 at which most lipolytic enzymes and lipooxygenases exhibit little activity, (2) chelating agents such as EDTA because many phospholipases are Ca^{2+} dependent; many lipolytic enzymes, however, are unaffected by chelating agents and (3) other additives such as anti-oxidants to prevent oxidative breakdown of fatty acid hydroperoxides, disulfide reducing agents, specific enzyme inhibitors and the use of fatty acid-free bovine serum albumin which binds free fatty acids. Each individual homogenization medium is different and may contain specific inhibitors for enzymes such as phospholipases or phos-phatases. Glycerol for example, is often added to the isolation medium in order to inhibit phosphatidic acid phosphatase. Ethanolamine and choline chloride are commonly used to inhibit phospholipase D.

The buffering capacity of the homogenization medium for plant cell fractionation must be high in order to offset the release of organic acids by the disrupted vacuole which constitutes a substantial volume of the cell. The homogenization medium can either be formulated by empirical evaluation or by rational analysis. For example, to ensure that the relevant purified protein remains intact all major inhibitors of proteases may be added to the homogenization medium or alternatively physiological solutions may be copied. Since the cytoplasmic pH of the plant cell is around 7.5–8.0, the homogenization medium should reflect the physiological state of the cell and is therefore weakly buffered to the cytoplasmic pH.

1.4.2 Procedures of cell and tissue disruption

In order to minimize membrane protease activity and the denaturation of proteins by excessive heat, all stages of cell fractionation must be performed at $+4°C$. All media and apparatus should be precooled and maintained at this low temperature throughout the procedure. Unwanted organelle damage may be caused by the considerable shearing forces required to break the rigid cell wall. Enzyme activities may be lost if there is foaming of the homogenate in the high-speed blender. The following physical and non-physical homogenizing procedures impose different degrees of physical stress upon the plant tissue.

(a) *Physical methods of cell disruption*

(i)   Solid shear methods of cell disruption

Disruption of the cells or tissue results from the shearing forces generated between the cells and a solid abrasive. These techniques are severe and tend to damage large subcellular organelles such as mitochondria, nuclei and chloroplasts. The tissue is placed in a ceramic mortar and a heavy round-ended pestle is used to grind the material in the presence of a small amount of abrasive in the form of coarse sand or fine silica sand and alumina for more delicate tissue. The pestle and mortar are usually precooled before use so that the material is maintained at a low temperature during disruption. Plant tissue is sometimes frozen quickly in liquid nitrogen and then ground in a pestle and mortar. An appropriate homogenization medium is then added to the ground tissue.

(ii)   Liquid shear methods of cell disruption

These methods of cell and tissue disruption rely on the shearing forces generated between the tissue and liquid medium. The material is placed in a precooled capped mixing container with the homogenization medium and subjected to efficient blending for a short and defined period of time. The cutting blades in a blender are orientated at different angles to each other to enhance the mixing of the homogenate and rotate at high speed but for only a short period of time because considerable shearing forces can be generated.

In tissue homogenizers (Fig. 1.19) the tissue is ground by the relatively mild shearing forces generated by an upward and downward rotation of a pestle within a glass cylinder. The Potter–Elvehjem homogenizer comprises a power-driven teflon, pyrex, glass or lucite pestle and a glass cylinder. The speed of rotation of the pestle within the glass cylinder can be closely monitored and the chopped tissue is forced between the walls of the glass cylinder and the rotating plunger as it passes up and down the vessel. The homogenization conditions may vary considerably between different tissues and critically depends upon the presence of hard vascular tissue. The Dounce homogenizer operates in an identical manner to that of the Potter–Elvehjem homogenizer. The difference is that the pestle is rotated and passed up and down the glass cylinder by hand and is not driven by a motor. This homogenizer generates a considerably more gentle and controllable disruption of tissue because the operator can adjust the

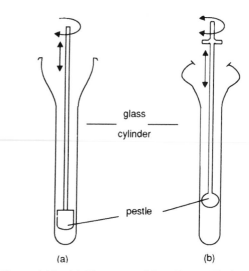

Figure 1.19   (a) The motor-driven Potter–Elvehjem homogenizer. (b) The Dounce homogenizer. (From Brownleader & Simpkin, 1992; with permission from Butterworth-Heinemann Ltd, Oxford.)

ferocity of the shearing forces by hand to suit the tissue in question. In contrast to the Potter–Elvehjem homogenizer, high shearing forces that may generate excessive heat production and possible damage to the tissue are avoided. In addition, the smaller shearing forces generate large sheets of plasma membrane that can sediment in a low centrifugal field.

Microhomogenizers are designed to disrupt a small amount of cells or tissues. They essentially consist of a loop of wire containing the material to be disrupted, connected to a motor, and immersed in the medium within a capillary tube. Razor blade fragments may be added to the medium to enhance the mixing of the material. The mixture can then be vortexed and the razor blade fragments subsequently removed with a magnet after microhomogenization. Motorized homogenizers such as the Polytron and Ultraturrax can be used as an effective replacement for the Potter–Elvehjem or Dounce homogenizer. The cutting blades are rotated at high speed to generate low shearing forces.

Pressure homogenization is based on the principle that cells equilibrated with an inert gas such as argon at high pressure imbibe large amounts of the gas. When the pressure inside the vessel is suddenly returned to atmospheric pressure, the newly formed gas bubbles inside the cytoplasm, rupture the membranes of the cell and form vesicles. This procedure has been successfully used to disrupt cultured cells. There is no thermal damage to the tissue and the speed of disruption and the ease with which the cells are disrupted by only one decompression has proved advantageous. The problems associated with pressure homogenization are that there is a variable degree of subcellular organelle damage and difficulties may arise in separating endoplasmic reticular and plasma membrane vesicles. As a result, this technique does not lend itself well to the isolation of an enriched plasma membrane preparation. In a French press the cell suspension is poured into a chamber which is closed at one end by a needle valve and at the other end by a piston. Extremely high pressures are applied by a hydraulic press against a closed needle valve. When the desired pressure is attained, the needle valve is fractionally opened to marginally relieve the pressure. The cells subsequently expand and rupture, thereby releasing the cellular components through the fractionally open valve. In the Hughes press the cells are disrupted with a solid abrasive. Moist cells are either placed in a chamber with an abrasive and then forced through a narrow annular space by pressure applied to a piston at $-5°C$ or the cells are ruptured by shearing forces generated by ice crystals at $-25°C$ and the frozen material is forced through a narrow orifice. In the osmotic shock method, plant protoplasts are ruptured in a hypo-osmotic medium. The sudden drop in osmotic pressure of the suspending solution causes cell lysis. In the freezing/thawing technique, the protoplasts can be lysed by successive periods of freezing and thawing. Ice crystals are generated during the freezing stage which disrupt the cells during thawing. However, this technique is slow and releases a limited amount of intracellular components.

(b) *Non-physical methods of cell disruption*

Organic solvents, such as chloroform/methanol mixtures are commonly used to dissolve membrane lipids and release the integral proteins and subcellular components. Organic solvent or mineral acid extraction does not, however, preserve morphological or metabolic integrity.

Chaotropic anions such as potassium thiocyanate, potassium bromide and lithium diiodosalicylate are believed to enhance the solubility of hydrophobic groups in an aqueous environment by increasing the disorder of the water molecules around the group. This facilitates the transfer of the apolar groups to the aqueous phase, thereby weakening the hydrophobic interactions which enhance the stability of most membranes, multimeric proteins and the native conformation of

biological macromolecules. This in turn, leads to a destabilization of the lipid membrane releasing the subcellular components and biological macromolecules for further analysis. Chaotropic agents have many applications such as depolymerizing protein aggregates and multimeric enzymes and investigating perturbations of the native structure of proteins and nucleic acids.

Detergents solubilize the integral membrane proteins by interacting with the phospholipid bilayer. They form mixed micelles of the various components of the membrane and the detergent. They can be categorized into three classes. Anionic detergents such as sodium dodecyl sulfate disrupt the membrane. Non-denaturing detergents, such as deoxycholate which effectively disrupts the membrane, and unlike denaturing anionic detergents permit observation of the integral membrane proteins in their native state. The third class of detergents are the non-ionic detergents, such as Triton X-100 that do not destroy enzyme activity of integral membrane proteins.

Enzymatic digestion of the cell wall of the plant cell results in protoplasts (cell wall deficient-cells) which may then be disrupted by mild-shearing forces generated by the Potter–Elvehjem or the Dounce homogenizer. A complex mixture of broad-specificity enzymes may be used for hydrolysing the cell wall. These include the chitinases, pectinases, lipases, proteases and cellulases. Several commercial cellulolytic enzyme preparations are isolated from wood-degrading fungi. Cellulysin (Calbiochem-Boehring Corp.), a common cellulolytic enzyme preparation is widely used in the isolation of plant protoplasts. Studies have suggested that impure enzyme preparations often prove superior in protoplast isolation to the pure cellulase because of the diversity of constituent enzymes. However, batch quality is variable and this may adversely affect protoplast viability and purity.

Isolation of fragile organelles requires tailor-made conditions and the formulation of the homogenization medium tends to arise by intuitive empirical design. The general protocol is fairly well established but the precise components and their concentration must be established for individual plant species. Clearly, less harsh conditions of cell disruption are required to isolate organelle membranes from suspension-cultured cells than the hypocotyl of glasshouse-grown red beet. Gentle razor blade chopping or grinding with a pestle and mortar is suitable for soft plant tissue such as root tips. However, short bursts of blending within a highly protective homogenization medium may be acceptable for isolating organelle membrane from hard plant tissue such as red beet hypocotyl. Once the tissue has been disrupted, the cell wall fragments must be removed by filtration through cheesecloth.

1.4.3 Centrifugation

The separation of subcellular fractions, e.g. mitochondria, plasma membrane, tonoplast and intact cells by centrifugation is one of the most commonly used tools in research. Particles of different density (buoyancy), size and shape sediment at different rates in a centrifugal field. The sedimentation rate of the particle is closely associated with the centrifugal field, the size and molecular shape of the particle and viscosity of the suspension medium and that the particle will remain stationary when the density of the particle and the suspending medium are equal. Non-spherical particles with identical molecular weight but different shape will have different sedimentation characteristics.

(a) *Preparative centrifugation*

This process is concerned with the isolation of a relatively large amount of biological material for further analysis. Analytical centrifugation is primarily concerned with obtaining the molecular weight of molecules and in conjunction with diffusion data, some indication of the shape of the molecule. There are principally three types of centrifugation.

(i) Differential centrifugation

A heterogeneous suspension of particles can be separated into particular fractions by subjecting it to an increasing relative centrifugal field over an extended period of time. This technique separates particles as a function of size and density and is only suitable for materials with markedly different sedimentation rates. A particular centrifugal field is chosen over a period of time in order to pellet predominantly particles of larger mass and retain smaller particles within the supernatant. In order to enhance the purity of the pellet, it is subjected to repeated steps of resuspension in the suspension medium and centrifugation. A typical program to isolate highly enriched organelle fractions from a plant tissue is illustrated in Fig. 1.20.

The main problem associated with this technique is that an impure preparation or enriched fraction is almost invariably obtained. The initial

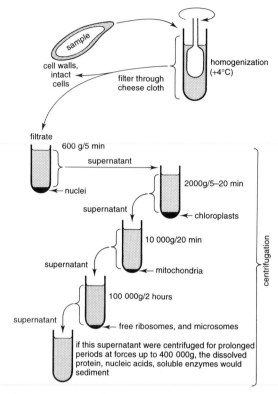

Figure 1.20 Procedure to isolate enriched organelles from plant tissue. (Modified from Simpkin & Brownleader, 1992; with permission from Butterworth-Heinemann Ltd, Oxford.)

suspension comprises a homogenous population of small (e.g. salts, microsomes; 0.05–0.2 μm) and large particles (e.g. large nuclei; 3–12 μm). The applied centrifugal field may pellet small particles suspended near the bottom of the centrifuge tube and large particles suspended near the top of the tube. The resultant pellet will comprise of a heterogeneous mixture of small and large particles. Other difficulties with this technique are that there are significant differences in size of the same organelle within a single cell type. In addition, different organelles may adhere to each other within the pellet and prove exceedingly difficult to separate with repeated washing steps.

(ii) Rate-zonal centrifugation

Rate-zonal centrifugation operates on the same principle as differential centrifugation but applied in a different way. Unlike differential centrifugation where the sample is distributed throughout the medium, in rate-zonal centrifugation, the sample is initially present only on top of the

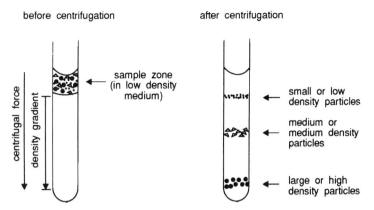

Figure 1.21 Schematic diagram of rate-zonal and isopycnic centrifugation. (From Simpkin & Brownleader, 1992; with permission from Butterworth-Heinemann Ltd, Oxford.)

gradient as a narrow band. The sample is carefully layered on to a pre-formed density gradient. A positive increment in density of the suspending medium prevents premature diffusion of the sample. Centrifugation subsequently proceeds for a fixed period of time so that the particles separate into a series of bands in accordance with parameters such as the centrifugal field, size and shape of the particle and difference in density between the particle and the suspending medium (Fig. 1.21). However, the particles will pellet over a prolonged period of time. Rate-zonal centrifugation is commonly used for the isolation of subcellular fragments such as plasma membrane, tonoplast and Golgi apparatus from plant microsomes. Differential centrifugation of homogenized tissue is routinely conducted in order to remove contaminating cell components such as heavy nuclei, cell debris, cell wall, chloroplasts and mitochondria from microsomes prior to density gradient centrifugation.

(iii)   Isopycnic centrifugation

This technique utilizes the principle that the density of the particle never exceeds the maximum density of the gradient. In contrast to rate-zonal centrifugation, isopycnic centrifugation is based solely on the buoyant density of the particle and is unaffected by the size or shape of the particle. Separation of the particles proceeds until its buoyant density is equal to the density of the gradient. It is important to appreciate that buoyant density of the particle is critically dependent upon the extent of hydration of the anhydrous particle by the density gradient medium. The choice of media is therefore extremely important because high viscosity of the medium can lead to poor resolution of bands. No further sedimentation occurs towards the bottom of the centrifuge tube and as a result, this technique is often referred to as equilibrium isodensity centrifugation and is independent of time. There are principally two types of isopycnic centrifugation. The continuous density gradient is either pre-formed prior to centrifugation or formed during centrifugation.

To prepare self-forming gradients, the sample (DNA or protein) is mixed thoroughly with a concentrated solution of cesium chloride or cesium sulfate to form a uniform and homogenous suspension. Centrifugation generates a density gradient of cesium chloride and subsequent separation of the nucleic acids or proteins to their individual buoyant densities. Although these solutes have low viscosity, they exert a very high osmotic pressure making them unsuitable for separating osmotically sensitive particles. Cesium chloride solutions are not sufficiently dense to band RNA. Cesium sulfate solutions can, however, band DNA, RNA and protein. High-molecular-weight RNA may, however, aggregate and precipitate in the presence of sulfate ions. It is, however, most commonly used as a tool in molecular biology in order to separate nucleic acids.

(b)   *Nature of density gradient material*

The ideal gradient material should obey all of the following criteria:

1. It should be inert towards the biological material, the centrifuge tubes and rotor;
2. It should not interfere with the monitoring of the sample material;

3. It should be easily separated from the fraction after centrifugation;
4. It should be relatively easy to monitor the exact concentration of the gradient medium throughout the density gradient by analysis of the refractive index of the solution for example;
5. It should be stable in solution and be readily available in a pure and analytical form; and
6. It should exert minimal osmotic pressure.

Sucrose is the most suitable gradient medium for rate-zonal centrifugation of particles such as separation of ribosomal subunits. It is extremely stable in solution and inert towards biological material. It is also readily available and inexpensive. However, it exerts a high osmotic pressure at comparatively low concentrations and its high viscosity at concentrations exceeding 40% precludes its use in isopycnic centrifugation where certain particles would simply not be able to reach their buoyant densities. Glycerol has frequently been used as a gradient medium for rate-zonal centrifugation because it can preserve many enzyme activities. The main problem is that a glycerol solution of equal density to that of sucrose is considerably more viscous.

In order to eliminate the problems of high osmotic pressure associated with sucrose, a variety of polysaccharides in the form of glycogen, dextrans and Ficoll have been used. Ficoll (copolymer of sucrose and epichlorohydrin) is the most commonly used polysaccharide gradient material with a molecular weight of 400 kDa. Although Ficoll exerts a comparatively low osmotic pressure at low concentration, it does have acute viscosity problems and therefore, requires longer centrifugation times to achieve equilibrium densities. However, these high-molecular-weight polymers are useful if the organelles are damaged by hypertonic solutions. These polysaccharides have been used for rate-zonal and isopycnic banding of cells and organelles.

Most iodinated compounds used as gradient material are based on the structure of tri-iodobenzoic acid. Examples of this class of gradient media include metrizamide, Nycodenz, Triosil and Urografin. All iodinated compounds except metrizamide and Nycodenz have a free carboxyl group. These two compounds are therefore soluble in many organic solvents and are unaffected by either the ionic environment or the pH of the solution. The non-ionic, iodinated media, metrizamide and Nycodenz have been found to be extremely inert as a gradient medium and can be used to separate many particles such as nucleoproteins, nucleic acids and subcellular organelles. Nycodenz and metrizamide are, in general, very similar. Nycodenz has some advantages over metrizamide in that (1) it can be autoclaved and (2) the absence of the sugar ribose allows many standard chemical assays to be performed. Metrizamide is commonly used to isolate intact vacuoles from plant protoplasts. Iodinated compounds exert a much lower osmotic pressure and viscosity than sucrose and polysaccharides and although cesium salts operate at high density and low viscosity in isopycnic centrifugation, their high osmotic pressure at low concentration precludes their use in the separation of osmotically sensitive samples.

The most widely used colloidal silica solution is Percoll. These solutions are colloidal suspensions of silica particles and are not true solutions. This material has several advantages over other colloidal particles in that it is coated with polyvinylpyrrolidone to prevent adherence to biological material and to enhance the stability of the suspension. Colloidal suspensions exert very little osmotic pressure and can separate large particles such as large subcellular organelles. Unfortunately, these particles absorb strongly within the ultraviolet region of the electromagnetic spectrum and therefore, interfere with optical measurements of proteins and nucleic acids. Silica particles tend to pellet during high speed centrifugation and therefore, only particles significantly larger than colloidal silica can be separated because silica will sediment before the smaller biological particles separate and form bands in density gradient centrifugation.

(c) *Enzyme markers*

There are a number of difficulties associated with cell fractionation. First, one must be aware of artefacts, components that are not present in the intact cell. This can include the loss or elevation of enzyme activity, non-specific complexes of protein and nucleic acid, unnatural adherence of organelles and the degradation of large molecules such as proteins or lipids by proteases and lipases. Several procedures mentioned above should minimize the generation of these laboratory artefacts. Second, contamination of organelles can present a problem with the final purification procedures. Discontinuous sucrose gradient centrifugation is used to separate plasma membrane and tonoplast from red beet. Potential cross-contamination between the two fractions can be caused by the direct removal of the bands with a Pasteur pipet. However, this cross-contamination can be assessed very simply by enzyme marker analysis. The principle of this biochemical analysis is that each cellular fraction comprises a unique or a combination of unique enzyme activities which can be assessed and used to follow purification of a particular organelle or membrane fraction.

The plasma membrane H^+-ATPase and Ca^{2+}-ATPase form a phosphorylated intermediate during ATP hydrolysis. The tonoplast H^+-ATPase, tonoplast H^+-PPase and mitochondrial H^+-ATPase do not form phosphorylated intermediates and are therefore not inhibited by the transition state analog vanadate. Vanadate-inhibited ATPase activity has, therefore, been used as a marker for the plasma membrane. However, soluble acid phosphatases are also inhibited by vanadate and contamination of plasma membrane with these phosphatases can occur if they attach to or are enclosed within plasma membrane vesicles. Acid phosphatases are inhibited preferentially by molybdate. Since this activity can be associated with tonoplast and plasma membrane fractions, acid phosphatase-inhibitory assays are routinely performed in order to ensure that this activity does not obscure any H^+-ATPase activity. Chitin synthetase acts as an additional marker for the plasma membrane.

Nitrate-inhibited Mg^{2+}-ATPase activity is commonly used as a marker for the tonoplast. The tonoplast H^+-ATPase and the mitochondrial H^+-ATPase are inhibited by similar concentrations of nitrate. In order to ensure that inhibition by nitrate can be accounted for by the tonoplast ATPase, the same activity must not be inhibited by azide and/or oligomycin (mitochondrial H^+-ATP synthase inhibitors). Cytochrome oxidase and succinate dehydrogenase can also be used as markers for the mitochondria.

Other enzyme markers can be used to monitor purification of individual cellular fractions. α-Mannosidase is another marker for vacuoles, glucose-6-phosphate dehydrogenase is a marker for the cytosol and NADH-cytochrome c reductase is used as a marker for microsomes.

1.5 MOLECULAR ASPECTS

Some of the diploid vegetative cells that comprise the flowering or sporophyte part of the plant are induced to differentiate into megasporophyte (female) and microsporophyte (male) cells that then undergo meiosis. The haploid spores divide by mitosis to generate the gametophyte. In corn, one of the megaspore nuclei divides by three mitotic divisions to generate an embryo sac containing eight haploid nuclei, one of which

becomes the egg nucleus. The microspore contains three haploid nuclei, two of which become the sperm nuclei in the pollen grain. During fertilization, one of the sperm nuclei fuses with the egg nucleus to form the diploid zygote which develops into the embryo. The remaining sperm nucleus fuses with two remaining egg nuclei in the embryo sac to form the triploid endosperm that nourishes the developing embryo. The combined endosperm and embryo are called the kernel (seed).

1.5.1 Mitosis

(a)   Interphase

In higher plants where most growth occurs by cell elongation, cell division takes place in localized regions called meristems which are found at root and shoot tips. Both primary and secondary growth in plants are associated with zones in which cells are rapidly dividing. The embryonic tissues comprise cells that have characteristically thin walls, prominent nuclei and small vacuoles. There are two types of meristems: apical meristems that occur at the tips of roots and shoots and give rise to primary growth, and lateral meristems that occur in woody stems and roots and give rise to secondary growth.

During interphase nuclear chromatin is dispersed throughout the nucleus and there are no chromosomal features. The nuclear envelope is apparent and various cytological features associated with mitosis are absent. Four phases of interphase have been observed and these include G_0, G_1, S and G_2. Within the G_0 phase, a cell is regarded as being quiescent, in a non-dividing state. The cell does not undergo any DNA replication and there is very little RNA synthesis. A cell may be induced to enter the subsequent G_1 phase when the quiescent cell is challenged with growth factors and hormones. The G_1 phase follows the G_0 phase and is variable in length. The cell commits itself to initiating DNA synthesis and prepares for the following S interval. Nuclei and cells grow throughout the G_1 period. At this point, the cell cycle can continue and the cell will divide, the cell may permanently stop division or alternatively, the cell can become reversibly quiescent. These non-dividing cells where G_1 DNA is constant are referred to as G_0 cells. There are several lines of evidence to suggest that the cell cycle G_0 phase is distinct from an extended G_1 phase. Biochemically, G_1 and G_0 cells differ in non-histone protein and RNA synthesis profiles. The DNA synthesis inhibitor, hydroxyurea is more potent in cells that were in G_1 than in G_0 cells stimulated to divide. The G_1 phase is, thus, an essential phase during which time the cell prepares for DNA replication that occurs in the S phase of the cell cycle. Within the S phase, DNA replication is initiated and the DNA content doubles prior to cell division. The cell requires sufficient 'division potential' for the initiation of DNA replication. This 'division potential' is believed to be a diffusible cytoplasmic factor that promotes the G_1-S transition. The cell cycle cannot continue until DNA replication has been completed. The final stage of interphase is referred to as the G_2 phase and this is a period of preparation for chromosome condensation and mitosis. No DNA synthesis and very little RNA synthesis occur and the cell within the G_2 phase contains twice the number of chromosomes as a normal diploid cell. At this point, chromatin condensation takes place and the nuclear membrane disappears.

In most plant cells entering mitosis, a band of microtubules encircles the nucleus and this is called the preprophase band which lie in the cell cortex near the plasma membrane. As mitosis proceeds, microtubules begin to appear at the surface of the nucleus. The nuclear envelope breaks down and the microtubules invade the nuclear area and form the mitotic spindle. As the mitotic spindle forms, the preprophase band disappears. The function of the preprophase band is unclear. It has been suggested that the preprophase microtubules are precursors of the spindle microtubules and guide the deposition of the extra wall material to mark the division site and positioning or orientation of the nucleus prior to mitosis. The chromosomes condense and become aligned along the metaphase plate at the center of the spindle.

The cell then enters mitosis and this culminates in the formation of two nuclei and subsequently two daughter cells. Mitosis incorporates five stages referred to as prophase, prometaphase, metaphase, anaphase and telophase (Fig. 1.22). Mitosis usually takes about 1–3 h with prophase being the longest (1–2 h) and anaphase the shortest (2–10 min) stage.

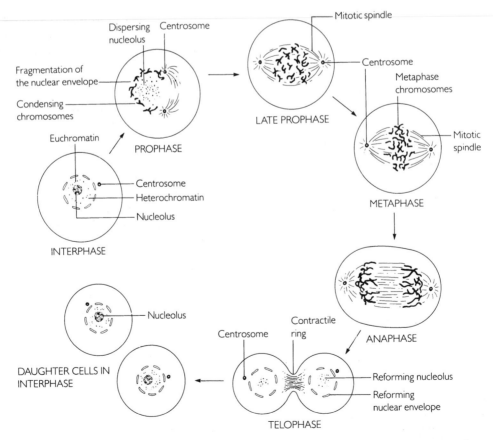

Figure 1.22   Schematic diagram of mitosis. (From Sadava, 1993; with permission from Jones & Bartlett Publishers International, Boston.)

(b) *Prophase*

During prophase, the chromosome undergoes a period of coiling, and the nucleolar membrane and the nuclear membrane disintegrate. Condensation of the chromosome is particularly visible in plants. First, the combined helical DNA and protein structure fold back on themselves to form a secondary coil, in a process that is regulated by the concentration of free Ca^{2+}. The shortened and thickened chromosome divides longitudinally into two chromatids which are joined together by a region with low DNA content and high protein content; this is the centromere. Each chromosome duplicates to form two identical chromatids which remain joined at the centromere. There are three structural and functional domains: the kinetochore domain which attaches to the spindle fibers, a central region that incorporates most of the centromere and a pairing domain where the two chromatids are attached along the inner surface. The orientation of the chromosomes is not random. The telomeres (ends of the chromosome) are orientated in an outer direction and are near the nuclear envelope whereas the centromeres are orientated towards the nucleus, and therefore begin to align the chromosome during metaphase. It must be noted that nuclear breakdown is not essential for mitosis because the nuclear envelope is retained in many organisms of protista and fungi, e.g. yeast.

(c) *Prometaphase*

At prometaphase (late prophase) the chromosomes condense inside the nuclear envelope and asters of fibers appear on the outside of the chromosomes. When the nuclear envelope has disappeared, a spindle forms in prometaphase. The spindle fibers comprise bundles of microtubules radiating from the opposite ends and referred to as poles of the cell. The chromosomes then migrate to the equatorial plane where they attach to one of the spindle fibers. In animal cells, spindle formation occurs by centrosomes, which are composed of twin centrioles at right angles surrounded by amorphous material. The centrosome is the major microtubule-organizing center during interphase. The centrosome replicates in late G_1 and S phases and the pair can be observed just outside the nuclear envelope. However, higher plants have no characterized centrosomes although the nucleation and dynamics of their microtubules suggest that plants possess cell cycle-dependent microtubule organizing center (MTOC) activities. Initiation of microtubule polymerization within a cell usually occurs at specific nucleating sites referred to as MTOCs. In most higher plants, initiation of mitosis is characterized by two successive events involving different microtubule populations. These are the production of the preprophase band and the development of the bipolar spindle. Cytoplasmic microtubules of higher plants radiate from the surface of the nucleus towards the cell cortex. This is one of the particular aspects of the plant cytoskeleton in comparison with other cell types. The plant nuclear surface may comprise an MTOC activity and this may represent one of the important factors in the control of the initiation of mitosis. There is experimental evidence to support the role of the plant nuclear surface as a microtubule nucleation site. When mitosis begins and chromosomes undergo condensation, tubulin incorporation has been shown to increase on the nuclear surface of *Haemanthus* endosperm cells in prophase, potentially a cell-cycle-dependent control of nucleation. The formation of the bipolar spindle (Fig. 1.22) around the nucleus is achieved through a transient convergence of microtubules forming aster-like centers, which subsequently produce spindle poles. This increased microtubule interaction is believed to be mediated by specific microtubule-associated proteins (MAPs). Calmodulin is found at higher concentration at the centriolar polar regions of animal cells. It is also found in these prophase microtubule centers, suggesting that there is a calmodulin-regulated mechanism in higher plants.

The important chromosomal event of prometaphase is the attachment of the chromosomes to the spindle and their movement towards the center of the spindle. Attachment of the chromosome to the spindle occurs at the kinetochore, which contains proteins for chromatid attachment. The breakdown of the nuclear envelope permits the kinetochores to attach to the spindle microtubules.

(d) *Metaphase*

The chromatids are attached to the spindle at the equatorial metaphase plate via the kinetochores. The spindle fiber from one pole attaches to the kinetochore of one chromatid of the chromosome pair and the fiber from the other pole attaches to the other chromatid.

(e) *Anaphase*

The process of chromosome segregation involves the bipolar orientation of sister-kinetophores and appears to be regulated by Ca^{2+}. There is an apparent physical break between the two chromatids and each chromatid of a pair migrates to opposite poles of the spindle. Each chromatid contains a complete copy of the chromosomal DNA, therefore, accurate anaphase migration of the chromatid is essential. If both chromatids of a pair migrate to the same pole, the phenomenon is referred to as non-disjunction. Here, one of the daughter cells contains an extra copy of a chromosome and the other daughter cell is devoid of this chromosome.

(f) Telophase

At the end of anaphase each pole of the spindle has a complete set of chromosomes. In telophase, two new nuclei are formed. The nuclear membrane reforms around the daughter nuclei, between which an array of microtubules referred to as the phragmoplast appears. Within the phragmoplast the new cell wall begins to be assembled and this is known as a cell plate. Cell division is complete when the cell plate fuses with the parental walls. Spindle dissolution proceeds so that MTOCs become surrounded by progressively shorter spindle fibers and the chromosomes decondense to form thinner threads. In higher plants, cytokinesis, the division of cell contents into two separate daughter cells begins in early telophase with an aggregation of Golgi-derived vesicles at the region of the equatorial plate. In addition, spindle fibers and ER cisternae congregate in the region and may be responsible for the assembly of the plasma membrane and cell wall. As the cell plate is formed and the new cell walls and membrane are completed the cells are then separated. Plasmodesmata apparently form during cytokinesis when ER cisternae, traversing the region of the cell plate prevent the fusion of vesicles containing membrane and cell wall material. After cytokinesis, the nuclei return to the interphase stage. Both daughter nuclei contain an identical chromosome complement, the result of the equal separation of identical sister chromatids at anaphase.

1.5.2 Meiosis

Mitosis takes place when transmission of chromosomes from the parent cell to its daughter cells is an asexual process. This vegetative process of cell proliferation produces genetically identical daughter cells. Meiosis (Fig. 1.23) is concerned with sexual transmission of chromosomes where the genes from two different parent cells eventually reside in a single daughter cell. In meiosis, haploid gametes with n amount of DNA such as pollen are formed from diploid cells (2n) in the germinal tissue of higher plants.

There are some common aspects of meiosis and mitosis. There is a G_1 phase followed by the S phase within interphase prior to both phenomena, thereby ensuring that a diploid cell enters both meiosis and mitosis with a 4n amount of DNA. Chromosomes condense during prophase, reside at an equatorial plate during metaphase, migrate to the poles during anaphase, and decondense during telophase. Asters and a spindle form during prometaphase. In addition, migration of the chromosomes occur along the spindle fibers attached to kinetochores. However, there are important differences between the two processes. Meiotic chromosomal division is followed by two nuclear divisions in sequence rather than the one nuclear division observed in mitosis. This halves the chromosomal number of the daughter cells. Consequently, this process is a reductive division. Mitosis generates two cells with two chromosomes each. However, meiosis produces four cells with one chromosome each. If a diploid cell (2n) replicates its DNA (4n) and then undergoes meiosis, the first division will produce two cells with 2n amount of DNA and the second meiotic division will result in four progeny cells with a haploid number of chromosomes. Another difference is that mitosis occurs in many types of cells and tissues whereas plants have defined regions where meiosis occurs; for example, in a bud at the shoot apex or within meristematic cells. In addition, there is only one period of DNA replication followed by a single cell division in mitosis whereas meiosis comprises one DNA replication followed by two cell divisions. In diploid cells there are two sets of chromosomes in each somatic cell. These chromosomes are referred to as homologous because their genetic material is near-identical. In heterozygous chromosomes, one chromosome carries a different allele for a genetic trait than the other chromosome. In mitosis, the two homologous chromosomes are randomly arranged on the metaphase plate. However, in meiosis, homologous chromosomes are paired during the first division and line up on the metaphase plate non-randomly. Another difference is that recombination of homologous chromosomes occurs in meiosis, producing genetic variability within the progeny. Finally, mitosis is completed within hours or days and meiosis can take many months to complete in the higher plant.

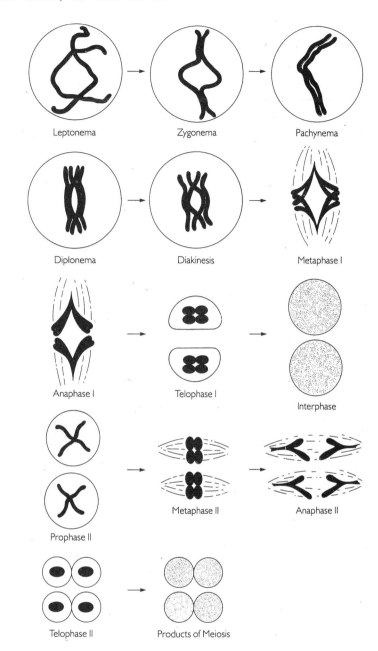

Figure 1.23 Schematic diagram of meiosis. (From Sadava, 1993; with permission from Jones & Bartlett Publishers International, Boston.)

Meiosis begins with the G_1 and S phases and it is the relatively short G_2 period when meiosis becomes irreversible. The cell then enters prophase I. Prophase I is a relatively complex phase and can be broken down into five distinct subphases known as leptonema, zygonema, pachynema, diplonema and diakinesis. Prophase I is followed by metaphase I, anaphase I and telophase I and these stages are followed by prophase II, metaphase II, anaphase II and telophase II. A short interphase period usually separates the first meiotic step from the second meiotic stage in plants.

(a) *Prophase I*

During the prophase I period, homologous chromosomes pair and there is recombination between the chromosomes.

(i) Liptonema

Once the cells enter the liptonema phase, they are committed to meiosis. It represents the end of interphase, where chromosome replication has occurred. The chromosomes appear as long threads and in fact consist of a pair of sister chromatids which are identical and held together by a centromere. In addition, a complex of RNA and protein associated with DNA called a lateral component or lateral element is deposited along the chromatids during leptonema. Lateral element synthesis is conducted by both sister chromatids. In addition to chromosome condensation and lateral element synthesis, the telomeres of the chromosomes begin to associate with a region of the nuclear envelope called the attachment site, which is devoid of nuclear pores and slightly thickened. This site may align homologous chromosomes.

(ii) Zygonema

In this phase, the homologous chromosomes become paired longitudinally to form the synapsis. A synaptonemal complex (Fig. 1.24) exists between the chromosomes and arises from the movement of aligned lateral elements. This complex comprises two lateral elements and a central region which is approximately 100 nm wide and is bisected by a narrow band called the central component which, like the lateral element is composed primarily of RNA and protein. The formation of the synaptonemal complex should not be overlooked because the complex is inextricably involved in recombination or crossing over between homologous chromosomes. A possible hypothesis to account for the function of the synaptonemal complex in meiotic genetic exchange is that the lateral components may serve as scaffold complexes through which important DNA sequences may be carefully threaded during leptonema in preparation for the zygonema phase where the two homologous chromatids must be aligned precisely. Interestingly, in mutant strains of maize that are devoid of the synaptonemal complex, meiosis stops at pachynema.

(iii) Pachynema

The synapsed homologous chromosomes become shortened and thickened during this stage. Synaptonemal complex formation and subsequent pairing of the chromosomes is complete and that recombination of genetic material is also conducted during this stage in meiosis. The synaptonemal complex gradually disappears towards the end of the pachynema phase.

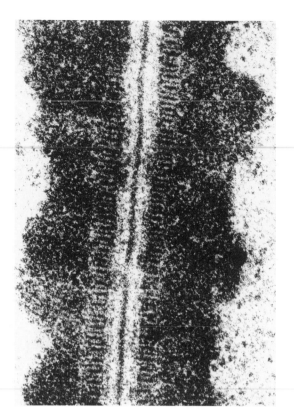

Figure 1.24   The synaptonemal complex of *Neotiella*. (From Sadava, 1993; with permission from Jones & Bartlett Publishers International, London.)

(iv) Diplonema

There appears to be separation between the homologous chromatids during diplonema. There are two types of attachment that prevent the chromosomes from separating completely. First, the sister chromatids are joined by the centromeres and second, the non-sister chromatids are usually held together at one of more sites that are referred to as chiasmata, a cross-like conformation.

(v) Diakinesis

In diakinesis the chromosomes are totally condensed and connected to each other through the chiasmata. At this stage, the nucleolus and the nuclear envelope disappear.

(b) *Prometaphase I and metaphase I*

The spindle is formed during this phase and the chromosomes subsequently become attached. The homologous chromosomes migrate towards the poles and remain attached to each other through the chiasmata and not as in mitosis through centromeres. In mitosis, the two centromeres of each chromosome separate so that each chromatid migrates to a pole, whereas in the first division of meiosis, and apparently due to the different attachment of the kinetochore with the spindle fibers, the centromeres between the homologous chromosomes do not separate so that both chromatids migrate to the pole.

(c)   Anaphase I and telophase I

The chromatids continue to be joined through chiasmata until the anaphase I stage. In this phase, separation of the chromosomal pairs is complete, the chiasmata is destroyed and each pair of chromosomes migrate to each pole. There is no centromere division in anaphase I, in contrast to mitosis where there is an apparent separation of centromeres. At telophase I, the chromosomes unfold and this occurs concomitantly with the reappearance of the nuclear membrane and the nucleoli.

(d)   Interphase

This phase separates the two meiotic divisions in higher plants. There is no replication of chromosomal DNA in contrast to interphase preceding mitosis.

(e)   Prophase II, metaphase II, anaphase II and telophase II

The events in the second meiotic division parallel those observed in mitosis. One chromatid of each pair migrates to opposing poles to form four haploid cells which return to an interphase state.

Meiosis is therefore responsible for the generation of four haploid cells containing genetically different chromosomes. The genetic variability is attributed to chromosome migration and recombination of genetic material during prophase I of meiosis.

1.5.3   The nuclear genome

It is during interphase that the chromosomes are dispersed within the nucleus, cannot be distinguished by staining with basic dyes and are transcriptionally active. In metaphase when the chromosomes are highly condensed, the RNA polymerase is unable to reach the DNA and RNA synthesis ceases.

The size of the eukaryotic genome can only be used as an extremely rough guide of species complexity. The perceived minimum DNA content for some taxa does increase to a certain extent with increasing complexity of the species and one may suggest that this minimum value indicates the amount of DNA essential for genes and associated controlling elements. However, there is a significant difference in DNA content between organisms of similar perceived complexity. For example, the human cell contains 700 times more DNA than *Escherichia coli* and yet the salamander cell contains 40 000 times more DNA than *E. coli*. However, the large and variable size of eukaryotic nuclear genomes, especially in plants suggest that there is a lot of genetic material that is not needed for development.

(a)   Gene frequency

Analysis of reassociation data of plant DNA demonstrated that a lot of the genome reanneals more rapidly than that expected if the genome was composed of single copy or unique sequences. There appears to be three frequency classes of DNA: unique DNA or single copy DNA, moderately repetitive and highly repeated DNA.

Most of the structural genes are composed of single copy sequences in all examined organisms, and therefore this sequence is considered to be of greatest importance for coding proteins. Any correlation between minimum DNA content and complexity is unclear because there can be no obvious explanation to account for the reason why broad beans have nearly twice as much structural DNA as peas and four times as much structural DNA as mung beans, especially because they all belong to the same Leguminosae family. Therefore, it is believed that much of the single copy DNA is not required for mRNA synthesis. Only a small fraction of mRNA is hybridized to unique DNA under saturating DNA conditions. Indeed, typically 1–10% of the unique DNA is transcribed into mRNA. In addition, a significant portion of the single copy sequences in all higher plant genomes contain less than 2000 base pairs. In mung bean and cotton, sequences in excess of 1200–1800 base pairs constitute most of the single copy DNA and about 40–60% of the nuclear genome.

Low and moderately repetitive DNA can be repeated from several to thousands of times. Some of the DNA in this category encode histones and rRNA and most repetitive sequences are interspersed within single copy sequence elements. Highly repeated sequences of DNA consist of short sequences of 6–8 base pairs which are repeated from 100 000–10 000 000 times in each genome. These highly repeated sequences are often associated with the centromere during cell division. Only a few repetitive sequences have defined functions. These include the following.

1. *Multigene families.* Even though most of the structural genes contain single copy sequences, a number of multigene families have been identified where the gene sequences are present in enough copies to be included with moderately repetitive DNA. Examples include the rRNA, tRNA genes and histone genes.
2. *Regulatory genes.* The interspersion of short repeated sequences suggests that these sequences are control elements of transcription. These interdispersed sequences may function as receptors to control transcription of the adjacent structural genes because different repetitive DNA sequence families are transcribed differently during different developmental stages. However, this does not account for the 20-fold variation in DNA content between peas and mung beans. One would equally expect that these regulatory control elements would be conserved in evolution. However, there is very little conserved homology in repetitive sequences within the same genus. Therefore, most of the repetitive sequences do not behave in a way expected for regulatory elements. This does not preclude the capacity of some repetitive DNA sequences to control gene expression. However, the large amount of interspersed repetitive DNA must have alternative functions in the nuclear genome.

Other possible roles for these repetitive sequences include influencing chromosomal behavior during mitosis and meiosis. For example, these sequences may function in chromosomal condensation or homologous chromosome recognition and pairing in addition to providing specific attachment sites for spindle fibers. Repetitive DNA may function as transposable insertion sequence elements. Finally, these repetitive sequences may simply add a large amount of redundant DNA to the genome and therefore control the rate and duration of various developmental processes. Preliminary studies on the *lac* repressor in *E. coli* show that it has a definite affinity for DNA in general, as well as a high affinity for the operator sequence. Regulation of the *lac* operon is dependent on the concentration of repressor and the delicate balance between the amount of operator and non-operator DNA. Assuming that the relative affinity for the operator and non-operator is constant, then increasing the overall DNA concentration will decrease the operator/non-operator DNA and decrease the probability that the repressor will bind to the operator region. Consequently, the *lac* genes are expressed to a much higher extent.

(b) Introns

Eukaryotes unlike prokaryotic cells contain introns. Classical introns are found only in the nuclear genome and not in the mitochondrial or plastid genome. They are non-coding stretches of DNA interspersed between the coding regions (exons) of the gene. They begin with the dinucleotide GT and end with the dinucleotide AG. Group I introns are highly conserved and occur in the mitochondrial, chloroplast and nuclear genomes of lower eukaryotes. These introns are apparently not essential for gene expression because they are often present in one strain and not in a closely related strain. Group II introns are also found in mitochondrial and chloroplast genomes and have a conserved but different secondary structure from those of group I introns. Group III introns have been found in chloroplasts of *Euglena* and are extremely rich in AT dinucleotides.

However, their function is unclear. They may have no function and simply represent evolutionary remnants in eukaryotes that most prokaryotes have been able to discard fairly quickly. The nucleotide sequence of the introns with the exception of the splicing regions have poor overall homology in comparison with exons. However, different recognition of the sequence elements within the intron may result in differential splicing of a single heterogenous nuclear mRNA species to generate more than one type of mRNA molecule.

(c) Chromatin structure

The total length of DNA in pea is 3 m. This must be packed into a nucleus less than 10μm in diameter. Chromatin is a complex of DNA, RNA and protein and plays a critical role in the packaging of nuclear DNA. It must play an important role in regulating gene expression because local variations in chromatin structure are believed to alter the accessibility of the DNA for transcription. Chromatin is a less effective template than purified DNA for transcription *in vitro*.

Chromatin actually consists of double-stranded DNA wrapped around a core of proteins called histones to form a nucleosome. The length of the DNA is reduced by 5–6-fold when the DNA is condensed within a nucleosome. A nucleosome consists of about 200 base pairs of DNA wrapped in a supercoil around eight histone proteins. The histones are, in brief, a class of basic proteins characterized by a high content of lysine and arginine residues. They have a molecular weight of approximately 11 300–25 000 daltons and there are five types: H1, H2A, H2B, H3 and H4. Each nucleosome is composed of two molecules of H2A, H2B, H3 and H4. H1 is peripherally associated with the nucleosome and is not considered to be an integral part of the nucleosome. After extensive micrococcal nuclease treatment, the nucleosome is digested to a wedge-shaped disk 'core' particle containing 140 base pairs of DNA. This core DNA content is constant and therefore any variation in total nucleosomal DNA is attributed to the length of the linker or spacer DNA between the core particles. *In vivo* the chromatin fiber consists of a linear chain of closely packed nucleosomes. The negatively charged DNA supercoils around the positively charged surfaces of the histones and proteins. The chain of nucleosomes is arranged into successively higher ordered structures reducing the total length of the complex even further. Histone H1 plays an important role in the association of adjacent nucleosomes. In its absence, nucleosome and polynucleosomes are soluble over a wide range of ionic strengths. The nucleosomes associate with each other to form a solenoid structure and therefore provide a further level of coiling of DNA to the 20–30 nm chromatin fibers observed during interphase. There are about six nucleosomes per turn of the solenoid. This form of chromatin is thought to be transcriptionally inactive. However, the difference in length of simple and extended chromatin and highly condensed chromatin is 5000–10 000-fold. A further factor of 100–200 must still be accounted for. There is still uncertainty as to how the solenoid structure is further condensed to generate the size of interphase or metaphase chromatin. Chromosomal scaffolding proteins may participate in the folding of the chromatin to the highly condensed forms observed in the above phases. Long loops of DNA (30 000–90 000 bp) connected to a central 'scaffold' have been identified in histone-depleted metaphase chromosomes and this enhanced level of folding contains about 150–450 nucleosomes. Also, some non-histone proteins have been implicated in the maintenance of the shape of the metaphase chromosome. The metaphase chromosome scaffold has been isolated and found to contain only about 0.1% of total DNA, no detectable histone and about 25–30 high-molecular-weight non-histone proteins. The scaffolds to which loops of chromatin are probably bound consists of two main proteins of 135 and 170 kDa. The larger protein is topoisomerase II suggesting that the formation of superhelical coils is important in chromosomal condensation.

(d)   *Differential gene expression*

There is a substantial amount of evidence to suggest that expression of particular developmentally regulated genes is closely associated with a particular stage of plant development. However, this must be interpreted in view of the fact that only a small fraction of structural genes expressed in plant cells have been examined.

The mature sporophyte has three vegetative organs, the leaf, stem and root and three reproductive organs, the petal, pistil and stamen. Studies on the gene expression of tobacco stem, root, anther, ovary and petal mRNA with leaf DNA have demonstrated that each organ has a substantial amount of common mRNAs. Many genes activated within these distinct organs are the same. However, each sporophyte organ also expresses a unique set of structural genes. There are therefore developmental-specific mRNAs in some organs that are not expressed in other sporophyte organs. The plant cell is also perceived to express genes constitutively that are encoded by all cells. These genes are referred to as 'housekeeping' genes and they probably encode proteins that are involved in respiration, cell growth, replication and gene expression, for example.

There are at least 50 different cell and tissue types in a flowering plant. A particular class of cells has a particular function which requires a unique set of structural genes. In the leaf, for example, mesophyll cells harvest light energy and guard cells regulate gas exchange between the plant and the external milieu. Hybridization *in situ* has demonstrated that expression of the S2-incompatibility glycoprotein gene in *Nicotiana alata* is restricted to the transmitting tissue of the style. Most plant organ-specific genes are believed to exhibit tissue or cell-type specific patterns of gene regulation. Gene activity is complex and highly regulated. It is important to recognize that 25–40% of structural genes expressed in the plant organ are organ-specific and about 30% of the gene expressed in each organ are 'housekeeping' genes.

1.5.4 Extranuclear DNA

The cytoplasm of higher plants contain two organelles with distinct and autonomous extranuclear genomes. These are the mitochondrial and chloroplast genomes and they are distinct from the nuclear DNA. Even though these genomes are a fundamental feature of these organelles, they cannot code for all the proteins required for their complete development. Effective interaction between the nucleus, mitochondria and the chloroplast is essential for the correct functioning of the cell.

The mitochondrion and chloroplast are believed to have descended from free-living prokaryotes which entered into an endosymbiotic relationship with a host having a nuclear genome. It can be assumed that there may have been considerable gene transfer between the nucleus, chloroplast and mitochondria if these organelles had existed in the same cell. This would surely have been undertaken if the endosymbiotic theory is correct because a lot of the original mitochondria and chloroplast gene must have transferred to the nucleus in order to account for the size and coding capacity of the current mitochondrial and chloroplast genomes.

(a) *The genome of the mitochondria*

The plant mitochondrial genome is mostly composed of double-stranded and circular DNA. The size of the mitochondrial genome is variable, ranging from 200 to 2500 kb, and in some species, contains low-molecular-weight circular and linear DNA and single- and double-stranded RNA. No relationship has been established between the size of the mitochondrial genome and its complexity.

(i) Structure of mitochondrial DNA

The DNA content of the mitochondrial genome is fairly constant in higher plants. Higher plants appear to have the largest DNA molecules; *Pisum* mitochondrial DNA is 30 μm in circumference. No unusual bases have been detected in this genome and based on the use of restriction endonuclease isoschizomers that can distinguish between methylated and non-methylated restriction sites, there does not appear to be a significant level of methylation. The mitochondrial genome contains primarily high-molecular-weight DNA along with a smaller amount of circular and linear DNA or RNA.

The mitochondria contain their own transcription and translation apparatus. However, most mitochondrial proteins are encoded by nuclear chromosomal DNA, synthesized by cytosolic ribosomes and then imported into the mitochondria. The mitochondrial genome encodes the large 26S and small 18S ribosomal RNA, possibly the complete set of tRNAs and proteins involved in the respiratory chain and ATP synthesis. There are however, a number of open reading frames that have not been assigned to a protein, gene chimaeras, non-functional (pseudogenes) genes and non-transcribed sequences within the genome. The genes encoding various proteins such as subunit I and II of the cytochrome oxidase complex, the cytochrome $bc1$ complex, apocytochrome B and the F_1 subunit of the F_0/F_1-ATP synthase have been identified in higher plant mitochondria. There are also mitochondrial restriction fragments with homology to chloroplast DNA sequences. In maize (*Zea mays*), an uninterrupted sequence of 12 kb from the inverted repeat of the chloroplast genome was found to be part of the maize mitochondrial genome and this sequence includes the 16S ribosomal RNA gene. The gene (*rbc* L) that encodes the large subunit of ribulose bisphosphate carboxylase and the entire sequence of the chloroplast tRNA^{his} is known to reside within the mitochondrial genome. Considerable homology between the two mitochondrial and chloroplast genome sequences has been observed in other plant species such as cauliflower (*Brassica oleracea*) where a 550 bp mitochondrial sequence was almost 100% identical to a region of the chloroplast genome and contains a tRNA^{leu} gene. The genetic code used by mitochondria is almost identical to that used by nuclear DNA. However, UGA, a stop codon for nuclear DNA is read as tryptophan in mitochondria. In addition, mitochondrial AUA is read as methionine, whereas AUA is read as isoleucine in nuclear DNA. These differences are due to the relatively small number of different tRNAs encoded by the mitochondrial DNA. Only 22 different tRNAs are encoded by the mitochondrial genome whereas more than 40 tRNAs are synthesized by nuclear chromosomal DNA. Interestingly the α-subunit of the mitochondrial ATPase is encoded by a nuclear gene in animals and fungi; however, the mitochondrial genome encodes this subunit in higher plants.

(b) *Plastid DNA*

DNA has been found in various plastid types such as proplastids, elaioplasts, chromoplasts, leucoplasts and chloroplasts. The DNA fiber is located in more than one region within these organelles. Moreover, electron microscope autoradiography suggests that there is a preferred association of higher plant chloroplast DNA with the photosynthetic membranes. The chloroplast genome is the best-studied plastid genome.

(i) Genome of the chloroplast

The higher plant chloroplast genome does not apparently contain any genetic sequences that are related to the mitochondria or nucleus, apart from those sequences which have functional sequence homology such as ribosomal RNAs and tRNAs.

(ii) Structure of the chloroplast genome

The DNA of the chloroplast is double stranded, circular and ranges in size from 120 to 169 kb (Fig. 1.25). Chloroplast DNA from *Pisum sativum* is approximately 40 μm in circumference. Covalent, closed and superhelical chloroplast DNA of around 40–45 μm have been found in all higher plants examined to date.

Pea chloroplast DNA contains approximately 18 ribonucleotides. This evidence is primarily based on the fact that (1) there are alkali-labile sites in the closed circular DNA. Alkali-sensitivity is a characteristic feature of ribonucleotide linkages (2) the closed circular chloroplast genome was converted to open circular DNA upon treatment with pancreatic RNase and RNase T1. The significance of these ribonucleotides is unknown, although they may function as specific recognition sites. Chloroplasts and mitochondrial DNA can exist in multiple circular forms of dimers and catenated monomers. These forms represent about 3 and 1.5%, respectively, of the genome in higher plants. These dimers appear to have been formed from the fusion of two circular monomers.

One of the noticeable features of the chloroplast genome is the presence of the large inverted repeats (IR). The inverted repeats are separated by a large and small single copy region (LSC and SSC, respectively). In the liverwort, *Marchantia*

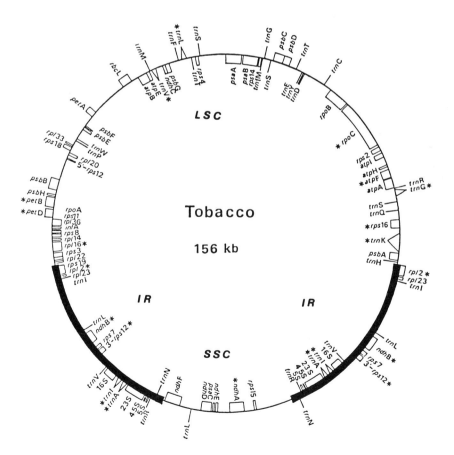

Figure 1.25 Gene map of tobacco (*Nicotiana tabacum*) chloroplast genome. Genes named inside the circle are transcribed clockwise. Split genes are highlighted with an asterisk. IR, inverted repeat sequences; SSC, short single copy region; LSC, large single copy region. (From Sugiura, 1989; with permission from Academic Press, Inc., London.)

polymorpha, the small single copy region contains 19 813 bp and the large single copy region contains 81 095 bp. The pea and broad bean chloroplast genomes are devoid of these inverted repeats. The chloroplast genome contains all the chloroplast rRNA genes (3–5 genes), tRNA genes (about 30 genes) and all the genes for the 'housekeeping' proteins synthesized in the chloroplast (100–150 genes). About one fifth of the tobacco chloroplast genome is transcribed and the majority of RNA formed is believed to represent mRNA and not tRNA or rRNA. Major proteins synthesized in chloroplasts include a soluble large subunit-ribulose bisphosphate carboxylase and a 32 kDa membrane protein associated with photosystem II. A number of unassigned open reading frames have been identified in the plastid genome. Most of the genome encodes proteins involved in plastid transcription and translation or in photosynthesis. Therefore, the great majority of the plastid genome tends to be expressed in photosynthetically active cells. Interestingly, all plastid genomes contain the gene that encodes the respiratory chain NADH dehydrogenase which is similar to that found in human mitochondria. This gene is expressed highly in the plastid and may point to the existence of a respiratory chain in chloroplasts.

There are no histones associated with the chloroplast DNA and there is also no 5-methylcytosine which is a characteristic feature of nuclear DNA. The DNA is believed to be attached to the chloroplast membranes and also extends into the stroma. The complete chloroplast genome from liverwort and tobacco has been sequenced (Fig. 1.25). Each chloroplast contains a single DNA molecule present in multiple copies. The number of copies varies between species; however, the pea chloroplasts from mature leaves normally contain about 14 copies of the genome. There can be in excess of 200 copies of the genome per chloroplast in very young leaves.

The chloroplasts of photosynthetic plant cells are believed to have arisen from a symbiotic association between a photosynthetic bacterium and a non-photosynthetic eukaryotic organism. The eukaryotic host cell presumably engulfed the endosymbiotic organism by endocytosis, and this would have accounted for the double membrane of the present day chloroplast. At one point the endosymbiotic organism must have had a genome that encoded all the proteins required for independent existence. Perhaps many of the genes were already present within the nuclear genome of the host cell. The gene encoding the small subunit of ribulose bisphosphate carboxylase may have been present in the endosymbiont.

1.5.5 Recombinant DNA technology

Plant breeding has been developed in order to introduce and maintain desirable phenotypic traits. However, the technique of crossing plants is slow, unpredictable, demands that the plants can reproduce sexually and will transfer the entire gene complement. Recombinant DNA technology overcomes these limitations and allows plant geneticists to identify and clone specific genes for desirable traits such as pesticide resistance, and to introduce these genes into agriculturally important plant varieties. Sexual compatibility becomes irrelevant. Plants however, have certain drawbacks. First, they have very large genomes and sometimes exhibit genome polyploidy, e.g. two-thirds of grasses are polyploid. Polyploidy may contribute to somaclonal variation in tissue culture. Plants regenerated from single cells are not genetically homogeneous, because plant cells growing in tissue culture are genetically unstable. Second, agriculturally important monocotyledonous plants are comparatively more difficult to transform with DNA vector systems than dicotyledonous systems.

With the advent of recombinant DNA technology or genetic engineering, it is possible genetically to manipulate plant cells in culture and to regenerate complete plants with desirable phenotypic characteristics such as pathogen resistance. It is possible to study the expression and inheritance of the manipulated trait in the whole plant and increase our understanding of the molecular, biochemical and physiological basis of the plant genotype and their regulation within a highly differentiated and complex whole plant. Genetic manipulation of the whole plants may help to increase the yield of crop plants or to produce flowers with different colors by the transgenic expression of pigment genes for example.

Simply, genetic manipulation of somatic plant cells is divided into three principal areas: first, the isolation of the mutants by cell selection, second, the synthesis of full or partial somatic hybrids by protoplast fusion and third, the production of transgenic plants by transformation.

(a) *Antisense RNA and DNA*

A potential way to prevent expression of cellular gene proteins is to target the RNA. This technology involves introducing RNA or single-stranded DNA, complementary to the mRNA of the target gene, into a cell. The so-called antisense molecule base pairs with the mRNA from the cellular gene, thereby preventing translation of the mRNA species. The mechanism of inhibition of gene expression by the antisense molecule is unknown, but it is thought that the hybridization of the sense and antisense molecule may block passage of the mRNA to the cytoplasm, enhance the turnover of the RNA or prevent the translation of the RNA species. There are three principal ways of introducing the antisense RNA or DNA into cells. First, the antisense RNA can be synthesized *in vitro* using bacteriophage RNA polymerase and then micro-injected into cells. Second, expression vectors can be constructed in order to produce high levels of antisense RNA in transfected cells (Fig. 1.26). In this instance, the cells are transformed with a plasmid that contains the cloned gene downstream of a strong promoter. The gene in the plasmid is orientated backwards so that the RNA transcribed from the plasmid is complementary to the mRNA transcribed from the corresponding cellular gene. If the antisense RNA is present in large excess, it will base pair to virtually all this mRNA to form double-stranded DNA that cannot be translated into protein. Third, synthetic single-stranded DNA oligonucleotides can be used. The short oligonucleotide complementary to the AUG codon (translation initiation site) is introduced into the cell and base pairs with the mRNA, thereby preventing initiation of translation. The oligonucleotide can be introduced into the suspension cells by simply adding high concentrations of the oligonucleotides to the cell cultures. Alternatively, the oligonucleotide must be chemically modified in order to enhance the uptake and stability of the oligonucleotide.

An example of the use of antisense RNA exploits the effect of ethylene on fruit ripening.

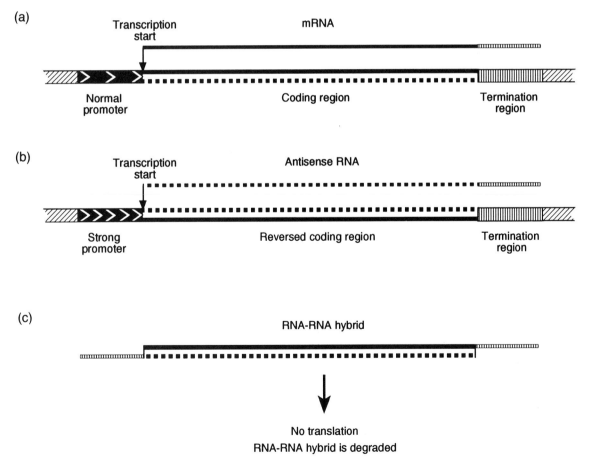

(a)

Transcription start

mRNA

Normal promoter   Coding region   Termination region

(b)

Transcription start

Antisense RNA

Strong promoter   Reversed coding region   Termination region

(c)

RNA-RNA hybrid

No translation
RNA-RNA hybrid is degraded

Figure 1.26 Antisense RNA in preventing cellular gene expression. (a) Transcription of the target gene under control of its own promoter. (b) Transcription of the reversed gene under control of a strong artificial promoter to generate many copies of antisense RNA. (c) mRNA-antisense RNA hybrid. (By courtesy of Dr Peter Sterk, Royal Holloway, University of London.)

Expression of an antisense RNA for an enzyme in the metabolic pathway of ethylene has successfully inhibited fruit ripening. Tomatoes can now have a longer shelf life.

(b)   *Oligonucleotide-directed mutagenesis*

In order to understand gene function, it is important to change the DNA sequence and assess this effect on structure and function. Tedious screening of randomly generated mutations can only be applied effectively to relatively simple organisms such as fruit fly and bacteria. Random methods will put a mutation at any location in a plasmid for example, and is often used when very little is known about the function of a particular DNA fragment. It can only identify broad functional domains and cannot really define structure/function relationships at a molecular level. Specific and site-directed mutagenesis allows the gene to be isolated, modified and then returned to the organism where the function can be examined. Site-directed mutagenesis can be used to precisely insert a mutation into a specific region of DNA. As a consequence, this powerful procedure can be used to analyze the effect of changing particular amino acids and assessing protein function. However, site-directed mutagenesis requires that the sequence of DNA is known before it can be changed. Oligonucleotide-directed mutagenesis begins with the synthesis of an oligonucleotide of about 7–20 nucleotides long which contains a base mismatch with the complementary wild-type sequence. The procedure requires that the DNA is available in single-stranded form and this can be conveniently achieved by cloning the gene in a M13 phage or in a phagemid clone. The synthetic oligonucleotide primes DNA synthesis by a DNA polymerase and it will base pair with the complementary sequence template to form a heteroduplex (Fig. 1.27). Once the primer has been extended around the whole of the template the ends of the newly synthesized strand are ligated to form a double-stranded circular DNA molecule. The heteroduplex is then transformed into *E. coli* where either strand can be replicated to form homoduplexes. The plasmid that grows up in the colony usually contains only one type of plasmid. In order to pick out the mutant colonies from the wild-type colonies the clones can be screened by nucleic acid hybridization with ^{32}P-labeled oligonucleotide used originally as a probe. The mutagenic oligonucleotide is radioactively labeled and hybridized to DNA from bacterial colonies on nitrocellulose paper disks. Under specific temperatures, the labeled oligonucleotide will hybridize to the mutant DNA and not to the wild-type DNA. The DNA of the mutant clone must be sequenced in order to verify that the desired mutant has been constructed.

Another way of introducing mutations at specific points is by synthesizing genes from longer synthetic oligonucleotides (40–80 nucleotides in length). The oligonucleotides can be annealed and ligated *in vitro* to assemble an entire double-stranded DNA molecule with mutations at discrete points.

(c)   *Heterologous expression*

Cell selection and protoplast fusion are discussed before evaluating the potential benefits of transgenic plant species. Cell selection involves actively selecting cell lines with an altered phenotype. The new phenotype could arise from either genetic or non-genetic events and so the selected cell lines are referred to as variants. The term mutant is retained for those cell lines where a genetic event has taken place. In brief, the variant must be screened and then selected. This involves imposing a selection pressure on the cell population so that only the individuals with the altered phenotype can survive or grow. Genetic variations in regenerated plants have a chromosomal bias and may involve complete chromosomal sets (polyploidy), loss or gain of individual chromosomal sets (aneuploidy), or structural rearrangements within individual chromosomes (deletions, duplications, inversions and translocations). The modified cell line with the stable phenotype can then be isolated from the cell culture. The altered phenotype may include increased resistance to stress such as drought, freezing and heat, resistance to herbicides, fungicides, heavy metals, aluminum, salinity, pathogens and viruses. Unfortunately, the molecular basis for most agriculturally important phenotypes are so poorly understood that it is not really possible to design effective selection protocols and define potential selection parameters.

Protoplast fusion permits the production of the somatic hybrids. One cell can be induced to fuse with another cell to form the heterokaryon which consists of cytoplasms containing chloroplasts, mitochondria and nuclei from both parent cells. Generation of the protoplast from the plant cell can be performed simply by incubating the tissue in a cocktail of enzymes that include cellulases, hemicellulases and pectinases. After the cell wall has been removed from the cell, a hypo-osmotic external milieu will lyse the protoplast unless a suitable osmoticum such as mannitol and sorbitol is present in the incubation medium. Protoplast fusion may be induced by an electric field or chemically, by addition of 20–40% polyethylene glycol which results in protoplasts aggregation and dilution of polyethylene glycol which causes protoplast fusion. The somatic hybrid must be selected after fusion of the protoplasts. The selection procedure can for example, be based on biochemical marker analysis or a differential effect of a metabolic inhibitor. One protoplast could be irreversibly inhibited by one inhibitor and another protoplast could be irreversibly inhibited by another inhibitor. However, the somatic hybrid

will survive and grow in the presence of the inhibitors because both parent cells will contribute enzymes that remain unaffected by each inhibitor. At present, it has proved difficult to regenerate crop plants from protoplasts.

Transformation of plants with foreign DNA has proved difficult in the past. One of the most common ways of introducing genes into plants is based on the natural gene transfer ability of the Gram-negative soil bacterium *Agrobacterium tumefaciens*. This bacterium infects wounded plant tissue of many dicotyledonous and some

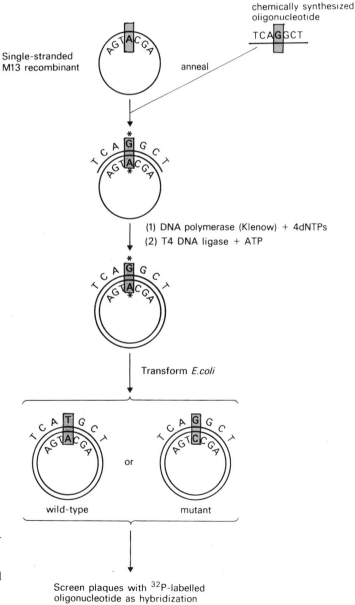

Figure 1.27 Oligonucleotide-directed mutagenesis. The asterisk indicates the mismatched bases. (From Old & Primrose, 1986; with permission from Blackwell Scientific Publications, Oxford.)

monocotyledonous plants and causes crown gall disease. The crown gall is a plant tumor, a lump of undifferentiated tissue that represents true oncogenic transformation. This bacterium contains large plasmids and the virulence trait is known to be plasmid-borne and they are therefore referred to as Ti (tumour-inducing) plasmids. Complete Ti-plasmid DNA is not found in plant tumor cells, but a small segment of DNA (approximately 20 kb) is found integrated in the plant genome. This DNA segment is called T-DNA. The plant cells express the T-DNA genes that encode enzymes responsible for phytohormone biosynthesis (and consequently tumor growth) and opine production.

The Ti-plasmid is, therefore, an excellent natural vector for genetically engineering plant cells because it can transfer its T-DNA from the bacterium into the plant genome. It is possible to insert foreign DNA into T-DNA which would then incorporate into the plant genome and the foreign gene(s) would then be expressed by the plant cell. It is important to note that the wild type Ti-plasmids are too large to act as vectors. Smaller plasmid vectors containing the T-DNA have been constructed that permit considerably simpler transfer of foreign genes to plant cells. This small plasmid known as mini-Ti is transferred conjugatively to *A. tumefaciens* and can induce opine tumors if the bacterium contains a plasmid that carries the virulence genes but not the T-DNA. Thus the *vir* functions can be donated by a separate plasmid and confer complete tumorigenicity. The plant cell can be regenerated from the transformed cells and the foreign DNA can then be transmitted normally by sexual reproduction to subsequent generations. One of the problems with the transfer of foreign DNA by the T-DNA is that transfer is often closely associated with transformation of the cells to tumors. It is sometimes difficult to induce regeneration into normal plantlets or into normal tissue that can be grafted on to healthy plants. Some researchers have overcome this problem by deleting one or more of the tumor-inducing genes.

In addition, the Ti-plasmid approach cannot often be applied to cereal plants because these plants are insensitive to crown gall disease. One way to overcome the restrictions imposed by the host range of *Agrobacterium tumefaciens* is to introduce DNA directly into cells using a physical, rather than a biological, process such as electroporation. A high concentration of plasmid DNA containing the cloned gene is added to a suspension of protoplasts and the mixture is applied to an electric field of 200–600 V cm^{-1}.

After electroporation, the cells are grown in tissue culture for 1–2 weeks and then the cells that have taken up the DNA can be selected. Maize and rice protoplasts have been successfully transformed with 0.1–1% efficiency. However, electroporation still requires the use of protoplasts with the comcomitant difficulties associated with regenerating whole plants from protoplasts and problems of somaclonal variation resulting from prolonged periods of tissue culture.

Genes have been transferred into plant cells with intact cell walls by shooting extremely small metal beads of 1 μm in diameter coated with DNA at velocities of about 430 m s^{-1}. This procedure has been used with suspension cultures of embryonic cells plated on filters and intact leaves. Embryonic maize cells have been transformed by this procedure with reporter genes and the *bar* gene that encodes the enzyme phosphinothricin acetyltransferase which inactivates phosphinothricin, a component of herbicides. Whole plants regenerated from these transformed cells were found to be resistant to a herbicide applied directly to leaves. The *E. coli* gene for the enzyme β-glucuronidase (GUS) is particularly important as a reporter gene in plants. Plants have virtually undetectable levels of GUS. When plant cells expressing β-glucuronidase are incubated with a chromogenic glucuronide substrate, a blue color is produced which can be observed histochemically or if a different substrate is used, GUS can be determined quantitatively with a fluorimeter. Alternatively, the living cells expressing a cloned gene can be detected by using the luciferase gene as a potential reporter gene. A particular problem with the use of reporter genes is that they give no indication of either mRNA stability or post-translational modification of the co-transformed protein.

These powerful molecular biological tools have and are anticipated to make a substantial impact on biotechnology. Recombinant DNA technology can enhance our understanding of the molecular basis of physiological processes. Traditional plant physiology has for too long been poorly defined at the molecular level. Today, there are many techniques that can be employed to investigate the role of proteins involved in stress, photosynthesis and defense.

Plants activate a large array of inducible defense responses during microbial infection. These include accumulation of phytoalexins, production of hydrolytic enzymes (e.g. invertases, chitinases), enhanced activity of peroxidases, deposition of callose and enhanced insolubilization of hydroxy-proline-rich glycoproteins in the cell wall. An important enzyme in flavanoid (phytoalexin) biosynthesis is chalcone synthase. The gene encoding this protein has been isolated. A recombinant gene consisting of the chalcone synthase promoter from *Antirrhinum majus*, the neomycin phosphotransferase III reporter gene and the termination region of the chalcone synthase gene from parsley (*Petroselinum hortense*) have been inserted into tobacco. The promoter functioned normally in tobacco and was induced by UV-B irradiation. Two regulatory regions were identified by deletion analysis of the promoter sequence. One region participated in UV-B light responsivity and the other region was essential for optimal expression of the gene.

Genetically manipulating plants can have significant importance in agriculture. Plants can be genetically altered so that they express a bacterial toxin which is toxic to larvae of a number of insects. This could have desirable effects on reducing the use of chemical pesticides. Plants can be used to express heterologous proteins because they are cheap and a large quantity of protein can be obtained from a single field. Some heterologous proteins such as enkephalin, a human neuropeptide and human serum albumin have been expressed in plants. Perhaps potato tubers could be engineered to synthesize large amounts of therapeutically important proteins.

REFERENCES

Anderson, J.W. & Beardall, J. (1991) *Molecular Activities of Plant Cells: An Introduction to Plant Biochemistry*. Blackwell Scientific Publications, Oxford.

Bolwell, G.P. (1993) *Int. Rev. Cytol.* **146**, 261–324.

Brett, C. & Waldron, K. (1990) *Physiology and Biochemistry of Plant Cell Walls*. Unwin Hyman, London.

Brownleader, M.D. & Simpkin, I. (1992) In *Biotechnology by Open Learning: Techniques used in Bioproduct Analysis*, pp. 48–57. Butterworth-Heinemann, Oxford.

Fry, S.C. (1988) *The Growing Plant Cell Wall: Chemical and Metabolic Analysis*. Longman, London.

Goodwin, T.W. & Mercer, E.I. (1983) *Introduction to Plant Biochemistry*. Pergamon Press, London.

Hall, J.L., Flowers, T.J. & Roberts, R.M. (1982) *Plant Cell Structure and Metabolism*. Longman, UK, London.

Keegstra, K. (1989) *Cell* **56**, 247–253.

Matile, Ph. (1975) In *Plant Biochemistry*, 2nd edn. (J. Bonner & J.E. Varner, eds), pp. 189–224. Academic Press, New York.

Matile, Ph. (1978) *Annu. Rev. Plant Physiol.* **29**, 193.

Mollenhauer, H.H. & Morré, D.J. (1980) In *The Biochemistry of Plants*, Vol. 1 (N.E. Tolbert, ed.), pp. 437–488. Academic Press, New York.

Old, R.W. & Primrose, S.B. (1986) *Principles of Gene Manipulation*, 3rd edn. Blackwell Scientific Publications, Oxford.

Pfanner, N., Sollner, T. & Neupert, W. (1991) *Trends Biochem. Sci.* **16**, 63–67.

Sadava, D.E. (1993) *Cell Biology*. Jones and Bartlett Publishers, Boston, London.

Salisbury, F.B. & Ross, C.W. (1992) *Plant Physiology*, 4th edn. Wadsworth Publishing Company, California.

Simpkin, I. & Brownleader, M.D. (1992) In *Biotechnology by Open Learning: Techniques used in Bioproduct Analysis*, pp. 59–77. Butterworth-Heinemann, Oxford.

Sugiura, M. (1989) In *The Biochemistry of Plants*, Vol. 15 (A. Marcus, ed.), pp. 133–150. Academic Press, New York.

Taiz, L. & Zeiger, E. (1991) *Plant Physiology*. The Benjamin/Cummings Publishing, California.

FURTHER READING

Caprita, N.C. & Gibeaut, D.M. (1993) *Plant Journal* **3**, 1–30.

Harborne, J.B. & Turner, B.L. (1984) *Plant Chemosystematics*. Academic Press, London.

Margulis, L. (1981) *Symbiosis in Cell Evolution*. W.H. Freeman, San Francisco.

Nitecki, M.H. (ed.) (1982) *Biochemical Aspects of Evolutionary Biology*. Chicago University Press, Chicago.

Overall, R.L. & Blackman, L.M. (1996) *Trends in Plant Science* **1**, 307–311.

Palmer, J.D. (1987) *Am. Nat.* **130**, 56–529.

Stumpf, P.K. & Conn, E.E. (1980–1990) *The Biochemistry of Plants: A Comprehensive Treatise*, Vols 1–16. Academic Press, New York.

Swain, T. (1974) Biochemical Evolution of Plants. In *Comparative Biochemistry, Molecular Evolution*, Vol. 29, A comprehensive biochemistry (M. Florkin & E.H. Stotz, eds), pp. 125–299. Elsevier, Amsterdam.

Watson, J.D., Gilman, M., Witkowski, J. & Zoller, M. (1993) *Recombinant DNA*, 2nd edn. W.H. Freeman, New York.

2 Photosynthesis

J.R. Bowyer and R.C. Leegood

2.1 INTRODUCTION

Photosynthesis is the process by which organisms convert light energy into chemical energy in the form of reducing power (as NADPH or NADH) and ATP, and use these chemicals to drive carbon dioxide fixation and reduction to produce sugars. In oxygenic photosynthetic organisms, including higher plants, the source of reducing equivalents is H$_2$O, releasing O$_2$ as a by-product. The overall reaction of oxygenic photosynthesis can be represented as:

$$CO_2 + 2H_2O \rightarrow (CH_2O) + H_2O + O_2$$

This process is responsible for producing virtually all the O$_2$ in the atmosphere and for fixing about 10^{11} tons of carbon from CO$_2$ into organic compounds annually.

The sugars produced by the photosynthetic fixation of CO$_2$ provide raw material for the biosynthesis of all the organic molecules found in plants. They are also the source of the chemical fuel which is oxidized by oxygen in the mitochondria in order to generate ATP for use in a wide variety of energy-consuming processes in the plant such as biosynthesis, active transport of ions and metabolites across membranes, and intracellular movement of organelles.

In plants, photosynthesis occurs primarily in leaf cells in organelles called chloroplasts, which are about 5 μm long and bound by two membrane envelopes. The number of chloroplasts per leaf cell varies from 1 to over 100, depending on cell type, species and growth conditions.

The utilization of light energy to drive the synthesis of NADPH and ATP takes place in a complex system of membrane-enclosed sacs within the chloroplast referred to as thylakoid membranes, and the reactions involved in CO$_2$ fixation and reduction to sugar are catalyzed by soluble enzymes in the chloroplast matrix referred to as the stroma.

The first part of this chapter will address the current state of knowledge on the structure, function and regulation of the thylakoidal light energy-conversion apparatus and the second part will deal with the stromal reactions involved in CO$_2$ fixation and reduction. For excellent introductions to these topics, the reader is referred to a number of standard undergraduate biochemistry teaching texts (Stryer, 1995; Zubay, 1993), and to Nicholls & Ferguson (1992) and Ho (1995) for more advanced reading.

2.2 LIGHT ENERGY UTILIZATION TO PRODUCE ATP AND NADPH

2.2.1 Introduction and background information

The principal light-absorbing pigments in the thylakoid membranes are the cyclic tetrapyrrole derivatives chlorophylls (Chl) *a* and *b*. Their structures are shown in Fig. 2.1 and absorption spectra in Fig. 2.2. Carotenoids (linear polyenes)

PLANT BIOCHEMISTRY
ISBN 0-12-214674-3

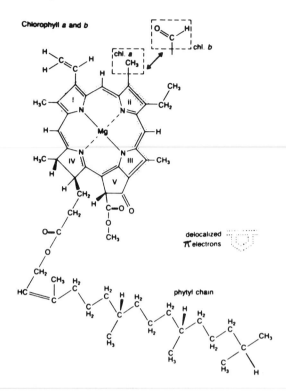

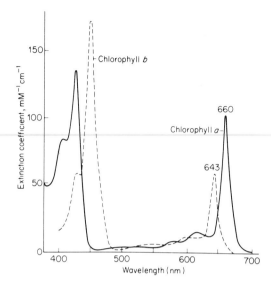

Figure 2.2 Absorption spectra of Chl$a$ and Chl$b$ in ether. The blue absorption bands are associated with excitation to the second excited singlet state (S_2) and the red absorption bands are associated with excitation to the first excited singlet state (S_1). Reproduced from Hall & Rao, (1994).

Figure 2.1 Structure of Chl$a$. The dotted lines indicate the extent of the delocalized π orbital system which is responsible for the absorption of visible light. Chl$b$ differs from Chl$a$ in having a formyl (-CHO) group instead of methyl (-CH$_3$) on ring II. Modified from Lawlor (1993).

are also important light-absorbing pigments found in the thylakoid membranes. Structures of the major carotenoids found in plants are shown in Chapter 11. Chl$a$ performs two functions, in light harvesting and in photochemistry. These processes take place in integral membrane protein complexes referred to as light-harvesting (or antenna) complexes and reaction centers respectively.

Electron transfer from water (E_mO_2/H_2O at pH 7 is 815 mV) to NADP$^+$ (E_m NADP$^+$/NADPH at pH 7 is -340 mV) is driven by two successive photochemical reactions in two different types of reaction center called photosystem II (PSII) and photosystem I (PSI). (E_m is the midpoint oxidation-reduction potential relative to the standard hydrogen electrode.) The two photosystems are connected via an electron transport chain which includes a further integral membrane complex, the cytochrome b_6f complex (Fig. 2.3). This is similar in structure and function to the mitochondrial cytochrome bc_1 complex.

Electron transfer from water to NADP$^+$ is coupled to proton translocation across the thylakoid membranes from the stromal phase into the

lumenal space enclosed by the thylakoid membranes. These membranes are relatively impermeable to protons so that proton translocation establishes an electrochemical potential difference of protons ($\Delta\mu_{H^+}$) between the two aqueous phases on each side of the membrane which is used to drive ATP synthesis (Fig. 2.4).

There have been enormous advances in our understanding of photosynthesis in recent years through the application of molecular biology techniques, X-ray and electron crystallography, and spectroscopic techniques of increasing sensitivity and resolution. Molecular biology techniques have not only enabled the primary sequences of most of the proteins involved in photosynthesis to be elucidated, but have also provided the means to alter single amino acids so that the function of specific residues can be probed. Some of the fruits of these developments are outlined in this chapter.

The evolution of oxygenic photosynthesis and biogenesis of the photosynthetic apparatus, including the import of nuclear gene-encoded, cytoplasmically-synthesized proteins into the thylakoid membrane system are outside the scope of this chapter. For further information on biogenesis see Barkan *et al.* (1995), Mullet (1988), Theg & Scott (1993) and Keegstra *et al.* (1995), and on evolution see Cavalier-Smith (1992), Morden *et al.* (1992) and Wolfe *et al.* (1994). For a discussion on possible evolutionary relationships between

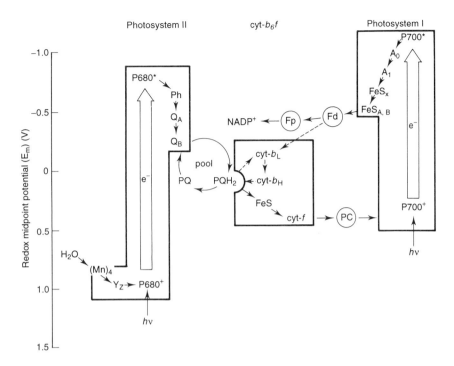

Figure 2.3 The 'Z-scheme' for linear electron flow through PSII and PSI in chloroplast thylakoids, showing the redox midpoint potential (E_m) at pH 7 of the components. The two vertical arrows indicate formation of the excited singlet state of the primary electron donor in each reaction center. The boxes enclose the three integral membrane protein complexes involved in photosynthetic electron transport. All the electron transport components are defined in the text. Reproduced with minor modifications from Ho (1995).

different types of reaction centers, see Nitschke & Rutherford (1991).

(a) Light absorption, energy transfer and photochemistry

(i) Photochemical properties of chlorophyll

Chl a and b can be excited by visible light to the first (S_1, absorption of red light in the wavelength range 650–700 nm) or second (S_2, absorption of blue light in the wavelength range 420–460 nm) excited electronic singlet states (Fig. 2.5). This occurs within < 10 fs $(1\,\text{fs} = 10^{-15}\,\text{s})$. The electronic energy levels are split into a number of vibrational levels which broaden the two major absorption bands. Further broadening is achieved in Chl–protein complexes because the molecular environment of the Chl affects its electronic energy levels.

Relaxation to the lowest vibrational state of either S_2 or S_1 occurs within ~1 ps $(1\,\text{ps} = 10^{-12}\,\text{s})$ by thermal emission. Chl in the S_2 state decays by internal conversion into the S_1 state on a similar time scale. Chl$a$ is a much more powerful reductant in the S_1 state than in the ground state;

excitation by a 700 nm photon generating the S_1 state lowers the E_m by 1.8 V in comparison to the ground state. Electron transfer from Chl$a$ to an acceptor occurs in the reaction centers, and it is this process which drives photosynthetic electron transport in the PSI and PSII complexes. The photochemical electron transfer occurs in <20 ps, considerably faster than the other pathways for dissipation of the S_1 state, namely fluorescence, thermal emission and intersystem crossing to the triplet state ^{3}Chl* (see also Fig. 2.25), which occur in a few ns $(1\,\text{ns} = 10^{-9}\,\text{s})$.

(ii) Light-harvesting and energy transfer

The solar energy flux of photons is too low to provide reaction centers with photons at a sufficient rate to justify the energy costs involved in synthesizing the reaction centers in the first place. Additional pigment–protein complexes are synthesized which function to absorb light over a broad range of wavelengths and transfer the energy to the reaction center (reviewed in Van Grondelle et al., 1994). The light-harvesting antenna pigment–protein complexes typically

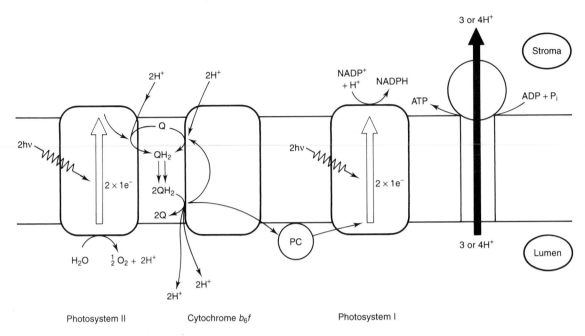

Figure 2.4 Proton pumping stoichiometry of the photosynthetic electron transfer complexes and the proton-translocating ATP synthase in the chloroplast thylakoid membrane. For the electron transport chain, the proton stoichiometry is shown per two electrons passing down the chain. For the ATP synthase, the proton stoichiometry is shown per ATP synthesized. Open arrows indicate photochemical charge separation reactions.

contain about 300 Chl per reaction center with a Chla/b ratio of 2–3. When a Chl molecule in the antenna is excited following photon absorption, the excitation energy is transferred between many Chl molecules before being trapped at the reaction center. For energy transfer to the reaction center to be efficient, the individual transfer steps must be highly optimized, requiring Chl–Chl distances of $< 2\,\text{nm}$ and optimal relative orientation of the interacting molecules.

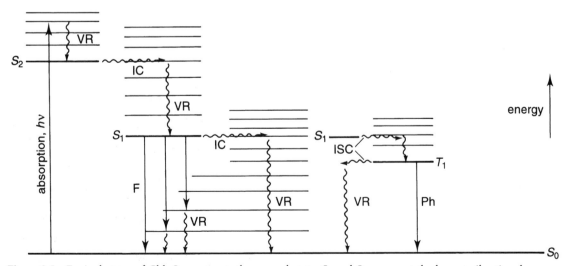

Figure 2.5 Excited states of Chl. S_0 represents the ground state, S_1 and S_2 represent the lowest vibrational energy level of the first and second excited singlet states, respectively, and T_1 represents the lowest vibrational energy level of the triplet state. The other horizontal lines represent vibrational energy levels associated with each electronic state. The arrows represent various non-photochemical processes for energy dissipation. VR, vibrational relaxation; IC, internal conversion; ISC, intersystem crossing; F, fluorescence; Ph, phosphorescence. The multiple decay routes of S_1 are shown next to each other. See text for further explanation. Reproduced with minor modifications from Ho (1995).

Plate 1 Structure of PSII. (a) Three-dimensional representation of the PSII complex, based on electron microscopy, viewed from a direction slightly above the thylakoid membrane plane. The lumenal (L) and stromal (S) sides of the membrane are shown with the lipid head groups represented by white circles and the fatty acid chains are drawn as black lines in the hydrophobic region of the membrane, shown in white. The complex is clearly asymmetric with respect to the membrane, with much more mass on the lumenal side of the complex. This side contains an intramolecular cavity, indicated by an asterisk, that is surrounded by the 33, 23 and 17 kDa extrinsic proteins of PSII. The positions of these proteins are marked by I, II and III (Ford *et al.*,1995), where the 33 kDa protein has been firmly identified as protein I, but the 23 and 17 kDa proteins have not yet been distinguished (R.C Ford, personal communication).Protein density is represented by netting, with blue, yellow/ green and orange representing increasing density. The scale bar is 2nm. Based on Holzenburg *et al.*, (1993), with additional information added by Dr R. Ford, who kindly provided this figure. (b) Model of the PSII reaction center proteins D1 and D2 showing the predicted locations of some of the prosthetic groups. The lumen-facing side of the complex is at the bottom. D1is shown in green and D2 in blue.The prosthetic groups are shown in red. The Chl*a* phytyl side chains and quinone isoprenoid side chains have been omitted for clarity. The scale bar is 2nm. This diagram was kindly provided by Dr J. Nugent.

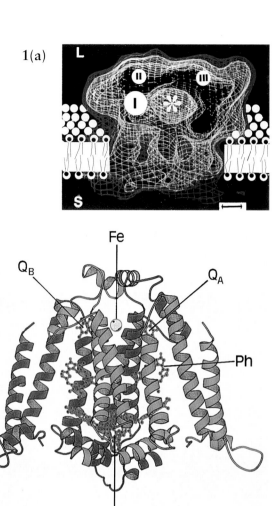

1(a)

1(b)

Plate 2 Structure of LHCIIb based on electron crystallography. (a) The LHCIIb monomer in the membrane, viewed from the side. Chl*a* molecules are shown in dark green, and Chl*b* in light green. Phytyl chains have been omitted for clarity. Two lutein molecules are shown in yellow. Membrane-spanning helices (A-C) are in purple. The assumed position of the membrane surfaces is indicated by blue bands. The maximal thickness of the complex, from the stromal to lumenal side, is 4.8nm. Reproduced from Kühlbrandt (1994).

2(a)

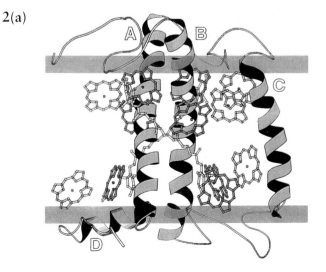

2(b)

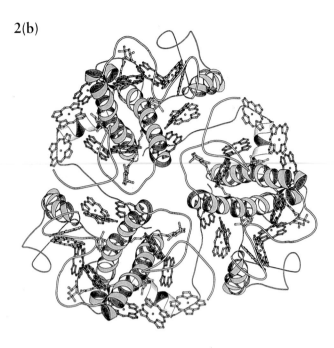

Plate 4 (Below) Arrangement of the structural elements of the PSI core complex of *Synechococcus* sp., based on the X-ray crystal structure. Side view of the complex showing transmembrane helices as blue cylinders, horizontal helices as pale blue cylinders, antenna Chl*a* head groups as green disks and groups involved in electron transfer as yellow discs. The two lowermost disks are thought to be P700, the two disks above these are accessory Chl*a* molecules, and the two uppermost disks represent approximate locations for A_0 and A_1. Only about half the total number of Chl*a* molecules in the complex have been located to date. Iron-sulfur clusters are represented by red spheres. The lowermost cluster, partially concealed behind a helix in this picture, is FeS_X. The upper two clusters are FeS_A and FeS_B, but they cannot be individually assigned. Figure kindly provided by Dr P. Fromme, and based on Krauss *et al.*, 1993.

3

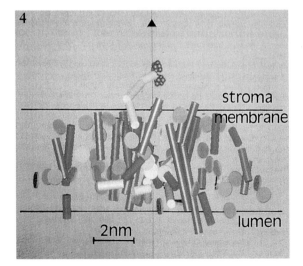

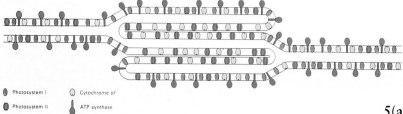

Photosystem I   Cytochrome *bf*

Photosystem II   ATP synthase

5(a)

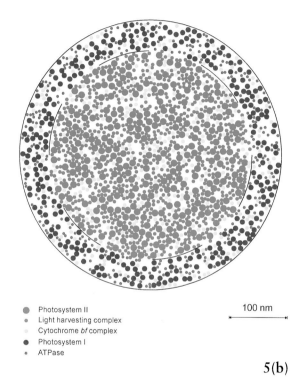

● Photosystem II
● Light harvesting complex
○ Cytochrome *bf* complex
● Photosystem I
• ATPase

100 nm

5(b)

Plate 6 (Below) Diagram illustrating the electronic spin states of Chl, carotenoid (Car) and oxygen involved in Chl excitation, intersystem crossing to produce $^3Chl^*$ (a), and its reaction with either O_2 to produce $^1O_2^*$, which is highly reactive (b), or with Car to produce $^3Car^*$ (c), which decays rapidly to the ground state by thermal emission. Reproduced from Zubay (1993).

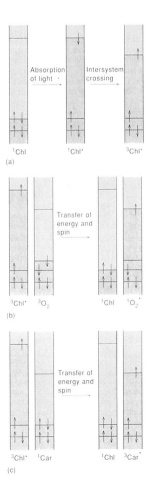

Plate 5 (Above) (a) Schematic representation of the distribution of integral membrane protein complexes in appressed and non-appressed membranes. Reproduced from Stryer (1995), based on a drawing by J.M Anderson and B. Anderson. (b) Scale model of the size and packing density of protein complexes in the thylakoid membrane, based on freeze fracture electron microscopy and biochemical studies. The areas inside and outside the broken circle represent the appressed grana and non-appressed membrane regions respectively. PSII complexes are shown located only in the appressed region (i.e. PSIIβ centers (see section 2.2.5.1) are not shown) and PSI and ATP synthase complexes are restricted to the non-appressed region. Cytochrome b_6f complex is assumed to be uniformly distributed. PSII is shown as a 16 nm diameter particle, through to be attributed to the PSII core complex surrounded by eight LHCII monomers (seealso section 2.2.2.1). The model also assumes three LHCIIb trimers per PSII, each approximately 8 nm in diameter. These assumptions and others result in 50% of the appressed granal membranes being occupied by integral proteins. Figure kindly provied by Prof. W. Haehnel.

6

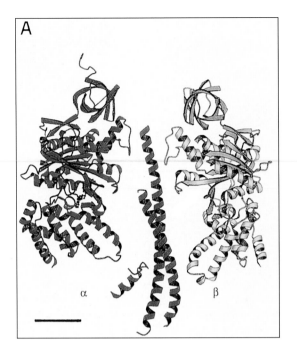
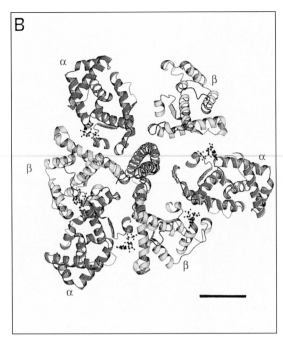

Plate 7 Structure of bovine F_1 ATPase from X-ray crystallography. α subunits are shown in red and β subunits in yellow. The γ subunit is shown in purple. Nucleotides are in black, in a 'ball and stick' representation. Scale bar = 2nm. (a) Vertical section through the complex, showing diametrically opposed α and β subunits. (b) Lateral view of the whole complex from the membrane-facing side, showing the asymmetrical structure, and nucleotides bound at interfaces between the subunits. Reproduced from Abrahams *et al.*, 1994.

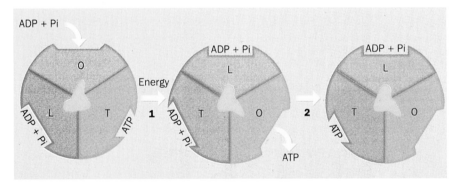

Plate 8 Binding change mechanism for ATP sysnthesis. The central asymmetric mass (yellow), representing $\gamma\delta\varepsilon$, rotates in step 1 relative to the three $\alpha\beta$ pairs (blue/green) which, for the purposes of illustration, remain stationary. Rotation is driven by the transfer of protons chanelled through CF_0 from low pK_a sites facing the positive (lumenal) side of the thylakoid membrane to high pK_a stites facing the negative (stromal) side. Rotation forces the three catalytic sites to undergo conformational changes associated with substrate binding and release. T,L and O stand for tight, loose and open conformations of the nucleotide bindings. See text for further details. Reproduced from Cross (1994).

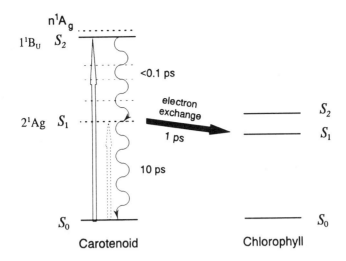

Figure 2.6 Schematic diagram of the excited singlet states of Chl and carotenoids and the proposed mechanism of singlet energy transfer from carotenoid to Chl$a$. ($\Rightarrow$) Absorption of light by carotenoid; ($\rightsquigarrow$) internal conversion (thermal emission); ($\rightarrow$) singlet energy transfer via an electron exchange mechanism. Transition from the carotenoid ground state (S_0) to S_1 (2^1Ag) ($\Rightarrow$) is dipole forbidden; absorption of a photon cannot promote this transition. Assumed time constants for each process are indicated (0.1 ps for excitation, 10 ps for internal conversion and 1 ps for energy transfer). Reproduced with minor modification from Owens (1994).

At intermolecular separations greater than ~1.5 nm, energy transfer from molecule to molecule proceeds via a mechanism called Förster resonance energy transfer. This involves dipole–dipole interaction and depends on an overlap between the electronic excited state energy levels of the donor and acceptor molecules, with relaxation of the donor coupled to excitation of the acceptor. The efficiency of transfer is heavily dependent on the separation of the molecules (decreasing with the sixth power of the separation) and their relative orientation, requiring the dipole moments of the interacting molecules to be parallel for optimal efficiency. Over a mean distance of 2 nm, the transfer time is about 1 ps. At intermolecular separations of less than ~ 1.5 nm, direct interactions between molecular orbitals can occur, and the excitation energy is effectively shared between the two molecules in a process known as delocalized exciton coupling. This process is more rapid than resonance energy transfer, such that the overall average single step energy transfer time is 0.2 ps or less.

The other light-absorbing pigments in higher plant chloroplasts are carotenoids. These are linear polyenes with absorption maxima in the wavelength range 420–500 nm. The overall stoichiometry of total carotenoid to total Chl (a plus b) is approximately 0.3–0.4. Light absorption by carotenoids generates the 1^1B$_U$ excited singlet state which decays within a few hundred fs to the 2^1Ag excited singlet state. The latter state cannot be reached by photon absorption, only by internal conversion from the 1^1B$_U$ state. It has a lifetime of ~10 ps and it is from this state that energy transfer to Chl, generating the S_1 singlet excited state of Chl, occurs (Fig. 2.6). Because the 2^1Ag state has a low dipole strength and because its lifetime is very short, singlet–singlet energy transfer from carotenoid to Chl is proposed to be *via* an electron exchange mechanism rather than Förster transfer (Frank & Cogdell, 1993).

2.2.2 Photosystem II

The PSII complex utilizes light energy to oxidize water and reduce plastoquinone. These two reactions occur on opposite sides of the thylakoid membrane, involving proton binding on the stromal side as plastoquinone is reduced, and proton release on the lumenal side as water is oxidized. Through these reactions, PSII contributes to the generation of the transthylakoidal electrochemical gradient of protons which drives ATP synthesis.

(a)   Structure of PSII

The PSII complex consists of over 20 polypeptides, 13 of which are encoded by the chloroplast genome (reviewed in Erickson & Rochaix, 1992;

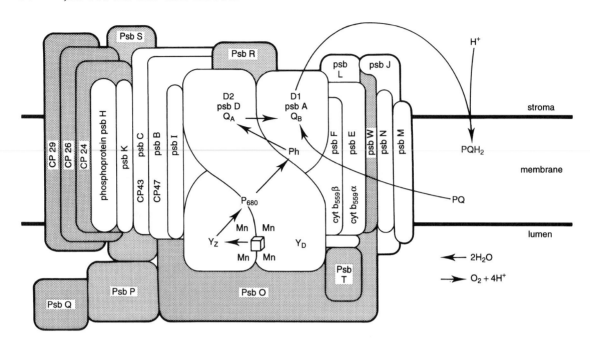

Figure 2.7   Subunit composition of PSII. Subunits are identified by their trivial name and/or encoding gene; LHCIIb subunits are not included. The subunits are also listed in Table 2.1. Subunits encoded by the chloroplast genome are shown in white and nuclear-encoded subunits are shown in grey. The lumen-facing side of the complex is at the bottom of the diagram and the position of the membrane is indicated by the double line. The precise location of the Mn cluster with respect to the D1 and D2 subunits is not yet established, although it is likely to be more closely associated with D1 (see section 2.2.2b(ii)). Y_Z and Y_D refer to Tyr$_Z$ and Tyr$_D$, respectively. The arrows indicate electron flow. Electron transfer between the Mn cluster and Y_D (Tyr$_D$) is slow. One molecule of phaeophytin is bound to both D1 and D2; only the phaeophytin bound to D1 is involved in electron transport. P680 is a Chl*a* dimer ligated between D1 and D2. There is at least one other Chl molecule bound to each of D1 and D2 (see text, Fig. 2.8 and Plate 1b for further details). Based on a diagram provided by R. Herrmann, H. Pakrasi and R. Oelmüller.

Fig. 2.7 and Table 2.1). All the prosthetic groups involved in electron transport from water to plastoquinone are bound to a heterodimer of two homologous integral membrane proteins, D1 and D2 (Fig. 2.8). These two polypeptides show limited homology to the L and M subunits of the purple photosynthetic bacterial reaction center, the high resolution structure of which has been determined (for reviews see Michel & Deisenhofer, 1988; Schiffer & Norris, 1993). The structure of PSII at high resolution has not yet been elucidated because crystals suitable for X-ray diffraction to high resolution are not yet available. However, a structure at 2 nm resolution has been obtained (Holzenburg *et al.*, 1993; Ford *et al.*, 1995, see below and Plate 1(a)) and the structure of the D1/D2 heterodimer has been modeled based on the bacterial reaction center structure (Ruffle *et al.*, 1992; Plate 1(b)). It is predicted that the PSII reaction center shows the same pseudo-twofold symmetry as the bacterial reaction center.

Closely associated with the D1/D2 heterodimer are three other low-molecular-weight subunits with membrane-spanning domains (reviewed in Seibert, 1993). One of these, the *psb* I gene product, which encodes a polypeptide of 4.5 kDa, has no prosthetic groups associated with it and its function is unknown. The other two polypeptides are the α (10 kDa) and β (4.3 kDa) subunits of cytochrome *b*-559. Both have membrane-spanning helical domains and probably associate to form a heterodimer binding a single haem group via a histidine ligand from each subunit (Cramer *et al.*, 1986). The cytochrome complex can exist in the membrane independently of PSII (Malnoe *et al.*, 1988). The number of cytochrome *b*-559 hemes per PSII is not yet firmly established and may be one or two (MacDonald *et al.*, 1994).

There are two Chl*a*-binding subunits, termed CP47 (or CPa-1) and CP43 (or CPa-2), based on their apparent molecular weights, in the PSII complex. The number of Chl*a* molecules bound

Table 2.1 Polypeptide components of the PSII core complex

Protein	Gene	Location of gene[a]	Molecular mass (kDa)	Function
Hydrophobic polypeptides				
D1	*psb*A	C	34	Reaction center apoprotein
D2	*psb*D	C	51	Reaction center apoprotein
CP47 (CPa-1)	*psb*B	C	47	Binds 10–15 antenna Chl*a*
CP43 (CPa-2)	*psb*C	C	34	Binds 10–15 antenna Chl*a*
Cytochrome *b*-559 α-subunit	*psb*E	C	9	Heme-binding
Cytochrome *b*-559 β-subunit	*psb*F	C	4	Heme-binding
PsbH	*psb*H	C	10	Unknown
PsbI	*psb*I	C	5	Unknown
PsbJ	*psb*J	C	< 5	Unknown
PsbK	*psb*K	C	4.3	Unknown
PsbL	*psb*L	C	4.3	Unknown
PsbM	*psb*M	C	3.8	Unknown
PsbN	*psb*N	C	4.7	Unknown
Hydrophilic polypeptides				
33 kDa protein	*psb*O	N	33	Stabilization of water oxidation
23 kDa protein	*psb*P	N	23	Stabilization of water oxidation
17 kDa protein	*psb*Q	N	17	Stabilization of water oxidation
10 kDa protein	*psb*R	N	10	Unknown
22 kDa protein	*psb*S	N	22	Chl*a*/*b* binding

[a]C chloroplast-encoded; N nuclear-encoded.

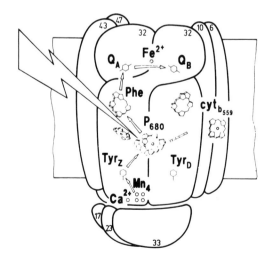

Figure 2.8 Schematic representation of the prosthetic groups and electron transfer pathways involved in the water–plastoquinone oxidoreductase activity of PSII. Each subunit shown is marked with its apparent molecular mass deduced from gel electrophoresis. The locations of the monomeric Chl*a* molecules adjacent to P680 and the inactive phaeophytin molecule on D2 are based on structural comparisons with the purple bacterial reaction center. From Rutherford *et al.* (1992).

per subunit is uncertain, but the consensus of recent estimations is 10–15. The primary function is assumed to be in light harvesting and in energy transfer from the more distal antenna to the reaction center (Bricker, 1990). Both polypeptides also bind β-carotene and CP43 may also bind the carotenoid lutein. Both polypeptides are predicted to contain six membrane-spanning helical domains and to have a long hydrophilic loop projecting into the thylakoid lumen connecting membrane-spanning helices V and VI. CP47 appears to be more closely associated with the reaction center core than CP43 (Fotinou and Ghanotakis, 1990).

Several other low-molecular-weight membrane-spanning polypeptides of unknown function are found in the PSII complex. A 22 kDa polypeptide component (PsbS protein) predicted to have four membrane-spanning helices, is a Chl*a*/*b*-binding protein related to light-harvesting chlorophyll–protein complex II (LCHII) (Funk *et al.*, 1994; see also section 2.2.2d).

Three extrinsic polypeptides with apparent molecular masses of 33, 23 and 17 kDa are associated through electrostatic interactions and hydrogen-bonding with the lumenal surface of the

PSII complex, the 33 kDa polypeptide being bound more tightly than the other two (reviewed in Vermaas et al., 1993). The binding site for the 33 kDa protein appears to include D1, D2, cytochrome b-559, the 22 kDa PsbS polypeptide, the 10 kDa PsbR polypeptide and CP47. The PsbR polypeptide is a hydrophobic protein on the lumenal side of the PSII complex. The 23 kDa is thought to be anchored to the membrane primarily by the PsbR polypeptide, and the 17 kDa polypeptide is attached to the membrane predominantly via the 23 kDa protein (Anderson & Åkerlund, 1987).

Digital image analysis of electron micrographs of negatively stained ordered two-dimensional arrays of PSII complexes have provided a low resolution (2 nm) structure of the PSII complex (Holzenburg et al., 1993; Plate 1(a)). The complex is highly asymmetric with respect to the membrane, with much more mass on the lumenal side of the complex. This side contains an intramolecular cavity that is surrounded by the 33 kDa, 23 kDa and 17 kDa polypeptides (Holzenburg et al., 1994). The involvement of these polypeptides in stabilizing the water-splitting active site at the lumenal side of the PSII complex (section 2.2.2b(ii)) makes it tempting to speculate that the cavity seen in the PSII structure contains the water-splitting active site. There are eight protrusions around the periphery of the complex which are thought to be monomeric LHCII polypeptides (section 2.2.2d).

(b) Functional properties of PSII

(i) Electron transfer from P680 to Q_B

The functional properties of PSII have been reviewed by Ghanotakis & Yocum (1990), Hansson & Wdrzynski (1990), Rutherford (1992) and Evans & Nugent (1993). The primary electron donor in PSII, P680, is probably a dimer of Chla molecules which are more weakly coupled than the equivalent dimeric donor Chla in PSI and the bacteriochlorophyll dimer in bacterial reaction centers (Van Gorkom & Schelvis, 1993). In order to be able to oxidize water, P680$^+$ must generate an oxidizing potential greater than +0.8 V. Its midpoint potential has been estimated to be ∼1.2 V, which is far higher than the primary donor in any other type of reaction center and must arise from the particular binding environment of P680 in the PSII complex. In the bacterial reaction center, the special pair of bacteriochlorophyll molecules constituting the primary donor is ligated by a pair of histidines, one from each of the L and M subunits. The equivalent histidines are

also present in the D1/D2 heterodimer (His198 in each subunit) and it is proposed that they have an equivalent function in ligating the Chla dimer of P680 (Michel & Deisenhofer, 1988).

In the excited singlet state, ^{1}P680* transfers an electron to a molecule of pheophytin a (Ph) (E_m −610 mV, Rutherford et al., 1981) generating the radical pair P680$^+$ Ph$^-$. The time constant of this reaction is reported to be 3 ps (Wasielewski et al., 1989) and 21 ps (Durrant et al., 1993); a consensus has not yet been reached. The Ph acceptor appears to be bound to the D1 subunit (Nixon, 1994), as expected from the structural similarities between PSII and the purple bacterial reaction center. It is very likely that the second molecule of Ph in PSII is bound in a symmetrical but not identical location in the D2 subunit. Differences in the environment of the two Ph molecules and their relationship to P680 mean that electron transfer to the Ph bound to D1 is overwhelmingly favored. The P680$^+$Ph$^-$ radical pair is in equilibrium with the excited state ^{1}P680*Ph (Roelofs et al., 1991) and normally decays by electron transfer from Ph$^-$ to a molecule of plastoquinone (PQ) called Q_A with a half time of ∼300 ps (Eckert et al., 1988). The Q_A binding site is on the D2 subunit. P680 is located towards the lumenal side of the membrane and Q_A is located towards the stromal side so that electron transfer from P680 to Q_A generates a membrane potential which is converted into a transmembrane ∆pH following proton binding/release and counter-ion flow (see below and section 2.2.7c). The Q_A semiquinone anion (Q_A^-, E_m for Q_A/Q_A^- −80 mV, Krieger et al., 1995) formed by electron transfer from Ph$^-$ then reduces a second molecule of plastoquinone bound to an equivalent site on the D1 protein, called Q_B. Reduction of Q_B to the semiquinone anion has a half time of ∼150 µs (Bowes & Crofts, 1980). It is accompanied by slow binding of one proton (half-time 2.7 ms, Polle & Junge, 1986) to one or more amino acid residues in the vicinity of Q_B^-, in response to pK increases resulting from an electrostatic effect of the negative charge on the semiquinone (Crofts et al., 1984).

Following re-reduction of P680$^+$ (see section 2.2.2b(ii)), the reaction center is able to undergo another light-induced turnover. The acceptor side reaction on this next turnover is the reduction of Q_B^-(H$^+$) by Q_A^-, with a half time of ∼600 µs (Bowes & Crofts, 1980), considerably slower than the reduction rate of Q_B. On the second turnover, a second proton is bound in a few ms (Polle & Junge, 1986) but proton binding from a membrane-adsorbed dye (Neutral Red) is considerably faster (half time of 310 µs, Haumann & Junge,

1994a), i.e. a second proton can be bound in response to formation of $Q_A^- Q_B^-(H^+)$. Following the second electron transfer step, the doubly-reduced quinol becomes directly protonated to form $Q_B H_2$, with concomitant deprotonation of the amino acid residues which bound protons in response to formation of $Q_A^- Q_B^-$. The plastoquinol species so formed has a relatively low affinity for the Q_B site and diffuses into the membrane lipid where it becomes part of a plastoquinone/plastoquinol (PQ/PQH_2) pool (E_m at pH 7 $\sim$ 120 mV) of approximately eight molecules per PSII (McCauley & Melis, 1986). In this way the PSII reaction center, which can only produce one electron per turnover, efficiently reduces PQ to PQH_2 with the required protons derived from the stromal side of the membrane (reviewed in Crofts & Wraight, 1983).

The sequence of reactions on the acceptor side may be represented as:

$$Q_A\,PQ \rightleftharpoons Q_A Q_B \xrightarrow{h\nu} Q_A^- Q_B \underset{H^+}{\overset{H^+}{\rightleftharpoons}} Q_A Q_B^-(H^+)$$

$$\xrightarrow{h\nu} Q_A^- Q_B^-(H^+) \underset{H^+}{\overset{H^+}{\rightleftharpoons}} Q_A Q_B H_2 \rightleftharpoons Q_A\,PQH_2$$

where PQ and PQH_2 represent plastoquinone and plastoquinol molecules in the pool.

The Q_A and Q_B sites, although showing partial symmetry, have quite different properties. Thus at the Q_A site, plastoquinone cannot exchange with the pool of plastoquinone in the membrane, whereas plastoquinone at the Q_B site binds relatively weakly from the membrane pool. Protonation reactions associated with Q_A^- formation are relatively slow and do not occur on the timescale of normal electron transport (Vass et al., 1992), whereas protonation on formation of Q_B^- occurs rapidly. Q_A^- is not readily reduced to Q_A^{2-} (Rutherford & Zimmermann, 1984) whereas Q_B^- is rapidly reduced and protonated to $Q_B H_2$, which also binds weakly. In contrast, Q_B^- binds tightly, stabilizing the semiquinone state. This brings the E_m values for the Q_B/Q_B^- (H^+) and Q_B^- (H^+)/$Q_B H_2$ redox couples much closer together than for PQ in the lipid bilayer (E_m at pH 7 for PQ/PQ^- would be lower than -200 mV), facilitating two-step reduction by Q_A^- (reviewed in Crofts & Wraight, 1983). Putative hydrogen-bonding ligands to the quinone

oxygens at the Q_A and Q_B sites have been proposed from comparative studies with the bacterial reaction center (Ruffle et al., 1992). A large number of chemicals, many of which are herbicides, bind to the Q_B site and block Q_A^- oxidation by preventing PQ from binding to this site. They include the phenylureas and triazines, notable examples of which are diuron (3-(3,4-dichlorophenyl)-1,1-dimethylurea) and atrazine (2-chloro-4(ethylamino)-6-(isopropyl-amino)-s-triazine (Bowyer et al., 1991; Oettmeier, 1992).

Located between the Q_A and Q_B sites is an iron atom which has an E_m of 400 mV and is normally in the Fe^{2+} state (Diner & Petrouleas, 1987, 1988). By analogy with the bacterial reaction center, it is thought to be ligated by four histidine residues, two from each of the D1 and D2 polypeptides (Michel & Deisenhofer, 1988). Although this iron atom can undergo redox reactions with Q_A and with exogenous quinones in the Q_B site (Zimmermann & Rutherford, 1986), it is not thought to participate normally in electron transfer between the two plastoquinones and may have a structural role in the PSII complex.

The quinone reactions on the acceptor side of PSII are similar to the equivalent reactions of ubiquinone or menaquinone in bacterial reaction centers, but unlike the latter, bicarbonate binding is required for this reaction to proceed in PSII (Govindjee & van Rensen, 1993). Bicarbonate depletion results in a 2–3-fold decrease in the rate of the first electron transfer from Q_A^- to Q_B, a more marked inhibition of the second electron transfer from Q_A^- to Q_B^-(H^+), and a significant inhibition of the release of $Q_B H_2$ from the reaction center. The bicarbonate appears to provide two further oxygen ligands to the iron atom (Diner & Petrouleas, 1990). There may be a second binding site for bicarbonate to an Arg residue on D1 or D2 which facilitates protonation of a histidine proposed as one of the residues which undergoes a pK shift on Q_B^- formation (Govindjee & van Rensen, 1993). The physiological significance of this dependence on bicarbonate is unknown.

(ii) Electron donation to P680⁺ and water oxidation

The normal reductant for $P680^+$ is a tyrosine residue on the D1 protein, identified by site-directed mutagenesis as Tyr 161 (Debus et al., 1988b), and referred to as Y_Z (E_m estimated to be 950–990 mV (Vass & Styring, 1991)). Y_Z reduces $P680^+$ with half times varying from 20 ns to 200 ns, depending on the redox state of the water-splitting manganese cluster which re-reduces Y_Z^+, increasing as the manganese cluster becomes more oxidized, possibly due to electrostatic effects (Brettel *et al.*, 1984).

The water-oxidizing active site is a cluster of four manganese atoms bound to the lumenal side of the PSII complex (reviewed by Rutherford *et al.*, 1992). The oxidation of two molecules of water to produce one molecule of oxygen requires the removal of four electrons. Electrons are removed one at a time by Y_Z^+ from the manganese cluster following successive turnovers of PSII. This generates a sequence of redox states of the manganese cluster referred to as the S states (S_0–S_4). The kinetics of Y_Z^+ reduction vary from 30 to 1300 μs, depending on the S state, becoming slower as the cluster becomes more oxidized (Dekker *et al.*, 1984). When four electrons have been removed, generating the S_4 state, two molecules of water are oxidized in a concerted reaction, reducing S_4 back to S_0. The S_1 state is the most stable state and predominates in dark-adapted centers.

Neither the structure of the manganese cluster, its ligands, its redox properties in each S state or the stage(s) of the S-state cycle in which the water molecules bind, have been unequivocally established (reviewed in Rutherford *et al.*, 1992). The cluster consists of four Mn atoms which are probably linked in dimers by di-μ-oxo bridges. Several models have been proposed for the higher order arrangement of the dimers. A structure based on the spectroscopic data available is illustrated in Fig. 2.9; it shows the two non-equivalent Mn dimers linked by a mono-μ-oxo bridge (Klein *et al.*, 1993; Yachandra *et al.*, 1993).

The inclusion of Ca and Cl atoms (in the +2 and −1 states, respectively) close to the Mn cluster is based on spectroscopic studies on PSII in which these ions are depleted and substituted with Sr^{2+} and F^-, respectively. The other Mn ligands are thought to be provided by water and by proteinaceous amine and carboxyl groups, the most likely candidates being Asp, Glu, His and Tyr. Spectroscopic studies have identified His as an Mn ligand (Tang *et al.*, 1994). Site-directed mutagenesis has been used in an attempt to identify specific amino acid residues acting as Mn ligands. The studies to date indicate that Asp 170, His 332, Glu 333 and Asp 342 of the D1 protein, and the α-carboxyl group of the carboxyl-terminal residue of D1 (Ala 344) are ligands (Nixon & Diner, 1992; Nixon *et al.*, 1992a,b; Prakrasi & Vermaas, 1992). A folding model for D1 based on comparison with the bacterial reaction center L subunit places these groups on the lumenal surface of the PSII complex (Fig. 2.10), and more detailed molecular models have been developed in which these residues are located in sufficiently close proximity to ligate the Mn cluster (Ruffle *et al.*, 1992). Glu-69 on the D2 subunit may also be a Mn ligand (Vermaas *et al.*, 1990).

Possible redox states of the water-oxidizing Mn complex in each S state, based on spectroscopic data, are shown in Fig. 2.11. There is general agreement that the S_0 to S_1 and S_1 to S_2 transitions involve Mn oxidation but the S_2 to S_3 transition is more controversial. Some spectroscopic evidence indicates that a histidine ligand to Mn is oxidized in the S_2 to S_3 transition (Boussac *et al.*, 1990; Berthomieu & Boussac, 1995), but there is also evidence which suggests that this transition

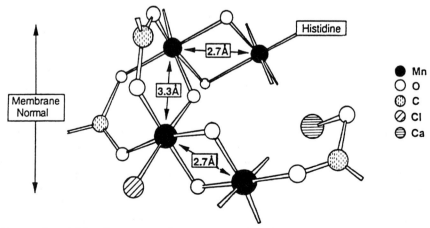

Figure 2.9 Proposed model for the water-oxidizing Mn complex of PSII based on X-ray absorption spectroscopy and electron paramagnetic resonance spectroscopy. From Yachandra *et al.* (1993).

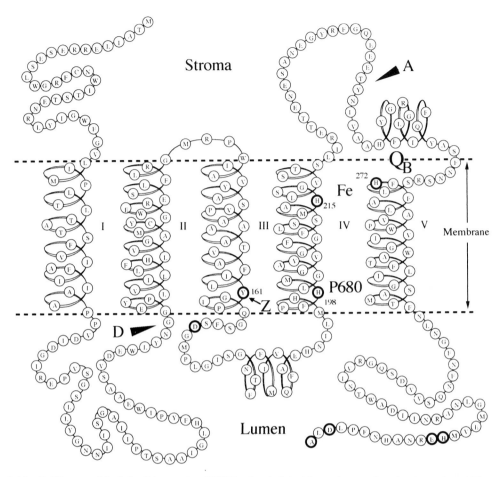

Figure 2.10 Folding model of the D1 protein of PSII in the thylakoid membrane, showing its amino acid sequence. Residues implicated in Mn ligation (Asp 170, His 332, Glu 333, Asp 342, Ala 344), Tyr 161 (Z) donor to P680, His 198 ligand to P680 and His 215 and His 272 ligands to the non-heme iron atom are indicated with bold circles. Approximate positions of primary D1 cleavage sites associated with donor side (D) and acceptor side (A) photoinactivation are also indicated. Modified from a figure provided by J. Barber.

involves Mn oxidation (MacLachlan *et al.*, 1994). No spectroscopic data are available on S_4.

The two substrate water molecules show marked heterogeneity in their exchange rates with the Mn cluster. One molecule of water exchanges slowly with the S_3 state ($t_{1/2} \sim 500$ ms) and the other exchanges much faster ($t_{1/2} \sim 25$ ms) with the S_3 state or enters in the S_4 state. Thus the two water molecules bind in different ways to the Mn cluster (Messinger *et al.*, 1995).

Molecular models of the PSII reaction center indicate the location and environment of Y_Z and an equivalent tyrosine residue in a symmetrical position on the D2 subunit, called Y_D (Debus *et al.*, 1988a). The latter can be oxidized by P680$^+$ and react with the Mn cluster, but far more slowly than Y_Z (Babcock & Sauer, 1973; Styring & Rutherford, 1987). Modeling studies predict that

Y_D is in a more hydrophobic environment than Y_Z, which insulates Y_D from the lumenal aqueous phase. Distance measurements based on effects of the Mn cluster on the electron paramagnetic resonance properties of Y_D and Y_Z indicate that the Mn cluster is 2.4–2.7 nm from Y_D (Kodera *et al.*, 1992) and approximately 1 nm from Y_Z. Y_D and Y_Z are separated by ~ 3 nm (Astashkin *et al.*, 1994) and are symmetrically placed with respect to the proposed pseudo twofold symmetry axis of the reaction center (Koulougliotis *et al.*, 1995). Molecular models predict that this axis places Y_D and Y_Z symmetrically on each side of the P680 Chl*a* dimer, both ~ 1.4 nm from their nearest Chl Mg atom (Svensson *et al.*, 1990) with a closest approach to Chl of ~ 0.8 nm (Ruffle *et al.*, 1992). Taken as a whole, the distance and mutagenesis data suggest that the binding site for the Mn

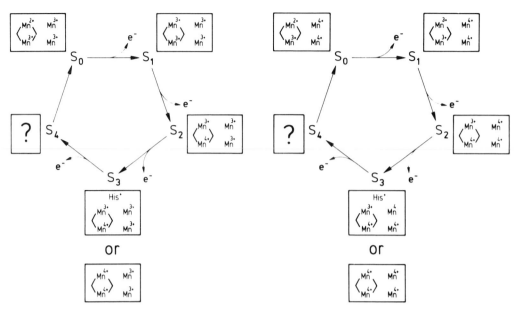

Figure 2.11   Two possible schemes for the redox states (S_0–S_4) of the water-oxidizing Mn complex formed on successive turnovers of PSII. Electrons are withdrawn one at a time by P680$^+$, via Tyr$_Z^+$. The S_4 state converts spontaneously into S_0 with concomitant release of oxygen. No spectroscopic data are available for S_4. The dark-stable state is S_1. From Rutherford *et al.* (1992).

cluster is located primarily on the lumenal surface of the D1 protein.

Both Ca^{2+} and Cl^- are required for oxygen evolution (reviewed in Debus, 1992). Sequential removal of the three extrinsic polypeptides (17, 23 and 33 kDa) results in an increasing requirement for Cl^- and the absence of the 23 kDa polypeptide leads to a specific requirement for Ca^{2+} (reviewed in Ghanotakis & Yokum, 1990; Andersson & Styring, 1991). The extrinsic polypeptides are considered to stabilize the water-oxidizing manganese cluster but are not directly involved in the catalytic cycle. The 33 kDa protein appears to be the most important. Mutants of the cyanobacterium *Synechocystis* PCC 6803 lacking the protein are able to grow photoautotrophically but are more sensitive to photoinhibition (see section 2.2.6) (Mayes *et al.*, 1991) and there appear to be some changes in the kinetics and thermodynamics of at least some of the S-states of the water-splitting system (Burnap *et al.*, 1992).

There are thought to be two Ca^{2+} bound per PSII complex (Han & Katoh, 1992). One of the two Ca^{2+} atoms appears to be associated with the peripheral LHCII complex (see section 2.2.2d) where it probably has a structural role. The other Ca^{2+} atom is assumed to be the Ca^{2+} located close to the Mn cluster. A proteinaceous carboxylate group serves as a bridging ligand between an Mn and the Ca^{2+} ion in the S_1 state, which is selectively broken on formation of the S_2 state (Noguchi *et al.*, 1995). Residues Asp-59, Asp-61 and Glu-65 of D1 may have a role in ligating this Ca^{2+} ion (Nixon & Diner, 1994). Chemical treatments to remove the Ca^{2+} block S-state turnover after formation of the S_3 state. The block on further electron transfer appears to be between Y_Z and P680$^+$ (Boussac *et al.*, 1992). Proton release which normally occurs on S_3 formation does not take place in Ca^{2+}-depleted material, so that the block on electron transfer may be due to an electrostatic effect of the net positive charge in the water-splitting cluster (Boussac *et al.*, 1990). There are two proposed roles for Ca^{2+}:

1. in regulating protonation/deprotonation events associated with water oxidation and/or
2. regulating access of water to the active site.

Cl^- depletion has a similar effect to Ca^{2+} depletion, i.e. proton release on S_3 formation is blocked and electron transfer from Y_Z to P680$^+$ is prevented once the S_3 state is reached (Lubbers *et al.*, 1993). It has been proposed that Cl^- may also play a role in regulating access of water to the active site.

(iii) Proton-release associated with water oxidation

A total of four protons must be released as a direct consequence of water oxidation:

$$2H_2O \rightarrow O_2 + 4H^+ + 4e^-$$

Although it is well established that a molecule of oxygen is released after every four turnovers of the PSII complex, the pattern and kinetics of proton release on successive turnovers are complex, depending on pH and the experimental material used. In most situations, however, one or more protons are released per turnover, accumulating to four protons per four turnovers (reviewed in Lavergne & Junge, 1993).

It is generally considered that the protons released when water is oxidized re-protonate groups which are de-protonated during the removal by Y_Z^+ of four electrons from the water oxidation system. Proton release on each turnover results from a downward pK shift of one or more ionizable groups in response to the removal of an electron on each turnover. The ionizable groups involved have not been identified but are likely to be amino acid side chains and Mn ligands such as μ-oxo bridges thought to link Mn atoms in the water-splitting cluster. A fast component of proton release ($<20\,\mu s$) is associated with rapid deprotonation triggered by the oxidation of Y_Z by P680$^+$ (Haumann & Junge, 1994b). This deprotonation may be reversed by a proton originating from a group closer to the Mn cluster as Y_Z^+ is reduced by the cluster. There is some evidence that the protons released during successive turnovers of the Mn cluster are not directly liberated into the thylakoid lumen but into membrane-sequestered domains (Jahns & Junge, 1992; Lavergne & Junge, 1993).

(iv) Other aspects of the function of PSII

The function of cytochrome b-559 is not fully understood but it appears to play a crucial role in the assembly and/or stability of PSII (Cramer et al., 1993). It does not appear to have a direct role in the water–plastoquinone oxidoreductase activity of PSII but there is some evidence to suggest that it may be involved in electron transfer reactions protecting PSII from photoinactivation (see section 2.2.6a(vii)).

A minor population of PSII centers ($\sim30\%$) are unable to reduce Q_B (Guenther & Melis, 1990). This may be attributed to centers which are unable to gain access to PQ due to tight packing of complexes in the membrane, generating local PQ-free domains (Joliot et al., 1992) or to centers in which the water-splitting Mn complex has not been ligated. In the latter centers, the E_m of Q_A/Q_A^- is raised (see section 2.2.6a(vii)) to a value close to that of Q_B, thus retarding forward electron transfer (Johnson et al., 1995).

Under optimal conditions, light-driven electron transfer from H_2O to the PQ pool is unidirectional, but under various conditions, it can be demonstrated that the intermediate charge separated states spontaneously recombine. The recombination kinetics become slower as the physical distance between the reacting partners becomes greater and/or the redox span becomes smaller, varying from $150\,\mu s$ for P680$^+$ Q_A^- recombination (Boussac et al., 1992) to 20–$30\,s$ for $S_2Q_B^-$ recombination (Robinson & Crofts, 1983). Under certain states of PSII which may be generated in vivo, recombination is thought to proceed via P680$^+$ Ph$^-$, generating ^{3}P680* in high yield (van Gorkom, 1985), with potentially damaging consequences (see section 2.2.6a(i)).

(c) *PSII reaction center complex*

A PSII reaction center complex consisting of D1, D2, PsbI protein and cytochrome b-559 has been isolated (Nanba & Satoh, 1987) which undergoes the primary photochemistry of PSII but does not bind Q_A or Q_B or the Mn cluster (reviewed in Seibert, 1993; van Grondelle et al., 1994). The consensus view on the pigment content of the reaction center is six Chl$a$, two Pha, two β-carotene and one cytochrome b-559 (Gounaris et al., 1990), but recent data suggest that two of the Chl$a$ are more peripherally bound to the surface of the complex, and that only four Chl$a$ are bound within the reaction center complex (Aured et al., 1994). Such a Chl$a$ content is identical to the bacteriochlorophyll content of purple bacterial reaction centers which are thought to be evolutionarily related to PSII (Michel & Deisenhofer, 1988). Two of the Chl$a$ would then form the P680 dimer and the other two would act as accessory Chl molecules, channeling excitation energy to P680.

(d) *Light-harvesting chlorophyll–protein complexes of PSII*

Much of the antenna Chl$a$ and all the Chl$b$ serving PSII and PSI is bound by a group of related polypeptides which form integral membrane protein complexes referred to as LHCII (Light-Harvesting Complex II) serving PSII, and LHCI

Table 2.2 Light-harvesting Chla/b protein complexes associated with PSII

Complex	Chla/b ratio[a]	Polypeptide components	Gene[b]	Molecular mass (kDa)
LHCIIa (CP29)	4.0	Lhcb4	*lhcb4*	29
LHCIIb	1.35	Lhcb1	*lhcb1*	27–28
		Lhcb2	*lhcb2*	25–27
		Lhcb3	*lhcb3*	25
LHCIIc } (CP26)	2.9	Lhcb5	*lhcb5*	26.5
LHCIIc' }				29
LHCIId (CP24)	1.51	Lhcb6	*lhcb6*	24
LHCIIe				13

[a]Data for spinach from Ruban *et al*. (1994).
[b]All the genes are nuclear.

serving PSI. The polypeptides are encoded in the nuclear genome by a large family of *lhc* (*cab*) genes (*lhcb* for PSII and *lhca* for PSI) (Green *et al*., 1991) and synthesized in the cytosol, and must therefore be imported into the chloroplast.

At least four distinct complexes serve PSII, designated LHCIIa, LHCIIb, LHCIIc and LHCIId (Table 2.2). The antenna size of PSII in higher plants varies with growth conditions (section 2.2.8) but is normally about 250 Chl per P680 (Ghiradi *et al*., 1986), of which at least 80% is bound in the LHCII complexes, the remainder being bound by CP47, CP43, D1 and D2. The major complex, which binds about 65% of the Chl serving PSII and 40% of the total Chl in the thylakoid membrane, is LHCIIb (Peter & Thornber, 1991). LHCIIa, c and d, formerly known as CP29, CP26 and CP24, respectively, each bind about 5% of the PSII chlorophyll (Peter & Thornber, 1991). Each LHCII complex has a characteristic Chla/b ratio (Table 2.2) and carotenoid content. Lutein is the main carotenoid in all the complexes, varying from over 60% in LHCIIb to less than 40% in LHCIIa. β-carotene is completely absent in LHCIIb. Neoxanthin accounts for 30% of the carotenoid in LHCIIb but only 10–15% in LHCIId. The other carotenoids found in LHCII are violaxanthin, antheraxanthin and zeaxanthin. They are participants in the so-called xanthophyll cycle which is involved in regulating non-photochemical energy dissipation (section 2.2.5f). One or two copies of each of the LHCIIa, c and d subunits are estimated to be present per PSII complex (Dainese & Bassi, 1991). LHCIIa and LHCIIc are more closely associated with the PSII core than LHCIIb and LHCIId (Bassi & Dainese, 1992).

The Lhcb polypeptide components of LHCIIb, which account for one third of the total thylakoid protein (Peter & Thornber, 1991), are known to

associate in the thylakoid membranes as mixed trimers (Simpson, 1979; Cashmore, 1984; Ide *et al*., 1987; Butler & Kühlbrandt, 1988). Distinct subcomplexes of LHCIIb have been proposed to constitute a mobile peripheral antenna and an inner pool tightly bound to PSII (Larsson *et al*., 1987a). The mobile pool is involved in a phosphorylation-dependent mechanism to regulate energy distribution between PSII and PSI (see section 2.2.5c). The mobile pool is thought to be enriched in the Lhcb2 polypeptide and the inner pool is thought to be enriched in the Lhcb1 polypeptide (Larsson *et al*., 1987a). The Lhcb3 polypeptide is present at a similar stoichiometry to LHCIIa, c and d and is not phosphorylated (Peter & Thornber, 1991). Approximately twelve LHCIIb subunits (i.e. four trimers) are associated with each PSII complex in the appressed membranes (see section 2.2.5a; Dainese & Bassi, 1991).

The structure of the LHCIIb trimer has been solved by using electron crystallography to 0.34 nm resolution (Kühlbrandt & Wang, 1991; Kühlbrandt, 1994; Kühlbrandt *et al*., 1994; Plate 2). Each polypeptide of molecular mass 25 kDa binds non-covalently a minimum of 12 molecules of Chl (probably 7 Chl$a$ and 5 Chl$b$, giving a Chla/b of 1.4) and two lutein molecules. Each polypeptide has three transmembrane helices and an amphipathic helix at the lumenal membrane surface. Two of the helices, designated A and B, form a symmetrical central core binding six symmetrical Chl$a$ molecules and the two xanthophylls.

The Chl chlorin rings in each monomer of LCHIIb are arranged on two levels, corresponding roughly to the upper and lower leaflets of the lipid bilayer. Their ring planes are oriented approximately perpendicular to the membrane plane, at an average angle of 10° from the membrane normal. The Chl molecules are attached to the polypeptide by co-ordination of the central magnesium atom to polar amino acid side chains (His, Gln, Asn, Glu charge-compensated by ion pairing to Arg) and by main-chain peptide carbonyls. The Chl chlorin ring center–center separation within a monomer on each level falls in the range 0.9–1.4 nm, and 1.3–1.4 nm between the two planes. Thus energy transfer between Chl molecules within a monomer proceeds *via* delocalized exciton coupling within about 0.1 ps. The separation of Chl molecules between monomers in the same trimer is greater than the separation for effective energy transfer by exciton coupling so that energy transfer probably proceeds via Förster transfer. The average Chl concentration in a trimer is 0.3 M. The minimum separation of trimers in the membrane is ~1.5 nm but may be considerably greater, so that energy transfer between them proceeds via the Förster mechanism. The minimum distance between chlorophylls in trimers in adjacent appressed membranes is ~3 nm, which is still close enough for Förster transfer.

The lowest vibrational energy level of S_1 is higher for Chl*b* than Chl*a* and energy absorbed by Chl*b* in the LHCIIb complex is transferred to Chl*a* within less than 1 ps. If the energy is not transferred on to another molecule, the Chl*a* remains excited for 1–3 ns, sufficient time for conversion to ^{3}Chl* which can then react with O_2 to produce the highly reactive species $^1O_2^*$ (see section 2.2.6b(i)). The xanthophyll molecules in the complex function both to absorb light and to quench ^{3}Chl*, and it is therefore expected that the Chl*a* molecules are located closest to the two xanthophylls. The two xanthophylls are situated on either side of helices A and B with their polyene chains at an angle of 50° to the membrane normal. They form an internal cross-brace in the center of the complex, providing a strong link between the peptide loops at the stromal and lumenal surfaces of the complex. The other carotenoids found in LHCIIb are present in substoichiometric amounts and may be located at the periphery of the complex. Helix C and the membrane surface loops show greater sequence variability in the *lhc* (*cab*) gene family than the helix A/B core. These loops appear to bind Chl*b* and might represent later additions to an ancestral Chl*a* protein, and account for the variable Chl*b* content of related complexes (Tables 2.2 and 2.4).

2.2.3 Cytochrome b_6f complex

The cytochrome b_6f complex is an integral membrane protein complex which catalyzes electron transfer from plastoquinol to plastocyanin and pumps protons across the thylakoid membrane. Its turnover rate in chloroplasts can be up to 300 s^{-1} (reviewed in Hauska *et al.*, 1983; Hope, 1993; Cramer *et al.*, 1994). Under saturating light, the oxidation of plastoquinol represents the rate-limiting step in photosynthetic electron transport (half time 2–3 ms (Velthuys, 1979)). The rate slows with decreasing lumenal pH due to the increased midpoint potential of the PQ/PQH$_2$ couple (which increases by 60 mV per pH unit decrease), hence reducing the thermodynamic driving force for PQH$_2$ oxidation.

The cytochrome b_6f complex consists of four major polypeptides: cytochrome f (31 kDa, *pet*A gene product), cytochrome b_6 (23 kDa, *pet*B gene product), the 'Rieske' iron–sulfur protein (20 kDa, *pet*C gene product) and subunit IV (17 kDa, *pet*D gene product) (Cramer *et al.*, 1991). All except the Rieske iron–sulfur protein are chloroplast encoded. A 4.5 kDa protein (*pet*G gene product) is also an integral component (Haley & Bogorad, 1989).

(a) *Structure of the cytochrome* b_6f *complex*

Cytochrome f is the redox protein from which electrons leave the cytochrome b_6f complex to reduce the soluble copper-containing protein plastocyanin in the thylakoid lumen (reviewed in Gray, 1992). It is a *c*-type cytochrome in which the heme prosthetic group is covalently linked via two of its pyrrole rings to cysteinyl residues. Most of the protein is located on the lumenal side of the thylakoid membrane, into which it is anchored by a single transmembrane segment at the carboxyl terminus. The structure of the soluble lumen-exposed polypeptide has been determined at 0.23 nm resolution (Martinez *et al.*, 1994; Plate 3, Fig. 2.12). The heme is in a predominantly hydrophobic environment, and has a midpoint redox potential of 350–380 mV. One of the two axial heme Fe ligands is a histidine (His25). The other heme iron axial ligand is the α-amino group of the amino-terminal residue of the polypeptide, Tyr 1. A signal peptide is cleaved from the amino terminus of cytochrome f during its translocation across the thylakoid membrane; this processing event must precede complete heme ligation.

A patch of lysine residues at one end of the elongated lumenal portion of cytochrome f may represent a binding surface for plastocyanin. Plastocyanin has a conserved patch of negatively charged residues about 2 nm from its electron-accepting copper atom. The other end of the cytochrome f protein may interact with the Rieske iron–sulfur protein. The tertiary structure of the

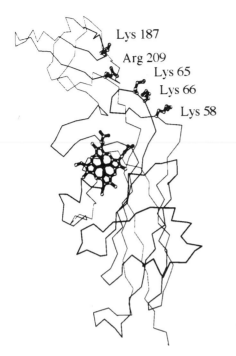

Figure 2.12 Polypeptide backbone of cytochrome *f* showing solvent-exposed basic residues that are thought to interact with acidic residues on the surface of plastocyanin. Reproduced from Martinez *et al.* (1994).

Rieske iron–sulfur protein is still the subject of debate. It has a lumen-located portion which contains a 2Fe–2S cluster and up to two possible membrane-spanning sections on each side of a hairpin bend (Widger & Cramer, 1991), or it may be entirely extrinsic (Gonzalez-Halphen *et al.*, 1988; Breyton *et al.*, 1994). Site-directed mutagenesis studies indicate that one of the Fe atoms has two His ligands and the other has two Cys ligands. The E_m of the cluster is 310–370 mV.

The tertiary structure of the cytochrome *b* subunit has been modeled by identifying long hydrophobic sequences which could form transmembrane helical domains. On the basis of this, the cytochrome *b* subunit is proposed to have four transmembrane helices and two amphiphilic helices located on the lumenal side of the subunit (Crofts *et al.*, 1987). The subunit binds two heme groups non-covalently, with the heme planes perpendicular to the membrane plane and located within the complex such that electron transfer between the two partially spans the membrane (Kramer & Crofts, 1994). The axial ligands for both heme groups are histidine residues, and candidate pairs of conserved histidines have been identified on helices II and IV (Degli Eposti *et al.*,

1993). The heme towards the lumenal side is called cytochrome b_L (L for low potential, E_m at pH 7 is −150 mV; Kramer & Crofts, 1994) and the other heme is called cytochrome b_H (H for high potential, E_m at pH 7 is −45 mV).

An additional subunit of the cytochrome b_6f complex, called subunit IV, has three transmembrane helices. In the related mitochondrial cytochrome bc_1 complex, the cytochrome *b* subunit appears to be equivalent to a fusion of the cytochrome b_6 subunit and subunit IV. The amphiphilic helix in the lumenal loop linking transmembrane helices II and III of subunit IV is thought to be involved in quinol binding, with acidic residues Asp-155 and Glu-166 facilitating H^+ translocation to the membrane surface associated with quinol oxidation (the Q_p site, p for positive, associated with proton release, see next section) (Beattie, 1993). There is a high level of sequence identity between species in the stromal side peripheral segments of the cytochrome b_6 subunit and subunit IV, and these are thought to be required to establish the binding site for plastoquinone at the quinone reduction site (Q_n site; n for negative, associated with proton binding, see next section) (Cramer *et al.*, 1994).

The cytochrome b_6f complex when isolated by detergent solubilization exists in a dimeric state (80–90% of the total complexes), and the isolated monomeric complexes are virtually inactive (Huang *et al.*, 1994). The functional significance of a dimer–monomer equilibrium in the intact membrane system is not yet understood. One Chl*a* molecule per cytochrome *f* is bound in a specific orientation in the complex, but its functional significance is unknown (Huang *et al.*, 1994).

(b)   *Functional properties of the cytochrome* b_6f *complex*

The cytochrome b_6f complex catalyzes electron transfer from plastoquinol to plastocyanin and couples this exergonic reaction to the translocation of protons across the thylakoid membrane (Fig. 2.13). The overall stoichiometry of the reaction is:

$$PQH_2 + 2PC_{ox} + 2H_{out}^+ \rightarrow PQ + 2PC_{red} + 4H_{in}^+$$

The complex has two sites at which plastoquinone/plastoquinol bind and undergo redox reactions. At the Q_p site, towards the lumenal side of the complex, plastoquinol is oxidized in a concerted reaction in which the first electron reduces the high potential Rieske iron–sulfur cluster and the other electron reduces the low

(a)

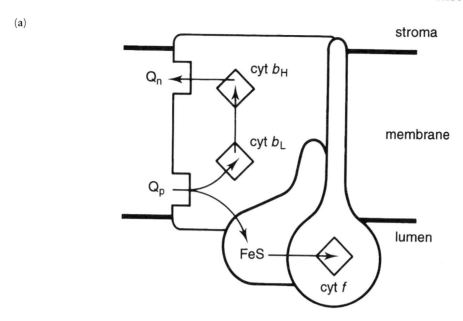

(b)

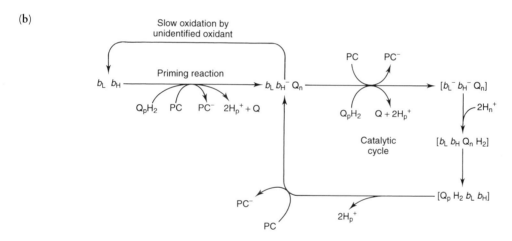

Figure 2.13 Electron transfer reactions in the cytochrome b_6f complex. (a) Approximate location of prosthetic groups and PQ-reactive sites (Q_n and Q_p) in the complex. The stroma-facing side of the complex is at the top of the diagram. Arrows show the direction of electron flow within the complex. (b) Working model for turnover of the cytochrome b_6f complex from the dark-adapted state. The enzyme is initially in a state in which there are no reducing equivalents in the ($b_Lb_HQ_n$) segment. A first turnover of the quinol oxidation site results in a primed state with a single electron in the $b_Lb_HQ_n$ segment, residing primarily on b_H. In the absence of further turnovers, this state oxidizes in a slow process which is not part of the catalytic cycle in the steady state. Further turnovers from the primed state regenerate it and involve: turnover of the quinol oxidation site (Q_p) to produce a transient unstable species $(b_Lb_HQ_n)^{2-}$; uptake of stromal phase protons and formation of a quinol at the Q_n site; and quinol migration and oxidation at the Q_p site, leading to regeneration of the primed intermediate. In this catalytic cycle, net oxidation of one quinol results in transfer of two electrons to two plastocyanins (PC) and two charge translocations across the membrane. Reproduced with minor modification from Rich *et al.* (1992).

potential cytochrome b heme (cytochrome b_L). The two protons produced on plastoquinol oxidation are released into the thylakoid lumen. The reduced iron–sulfur cluster transfers the electron to cytochrome f which is the exit point for electrons delivered to plastocyanin. The electron on cytochrome b_L is transferred rapidly ($>10^4 \, \text{s}^{-1}$) to the second heme group of higher potential, b_H, which is located towards the stromal side of the complex. Cytochrome b_H is in contact with the second quinone binding site, Q_n. Plastoquinone bound at this site is partially reduced by cytochrome b_H, generating a tightly bound semiquinone with the extent of reduction reflecting the redox equilibrium between the two. Following oxidation of a second molecule of plastoquinol at the Q_p site, the electron transferred to cytochrome b_L, together with the electron already on cytochrome b_H, complete the two-electron reduction of plastoquinone to plastoquinol at the Q_n site, with concomitant uptake of two protons from the stromal side of the membrane (Rich *et al.*, 1992; Hope, 1993; Kramer & Crofts, 1994; Fig. 2.13).

One complete turnover of the complex therefore involves the oxidation of two molecules of plastoquinol at the Q_p site and reduction of one molecule of plastoquinone at the Q_n site. This sequence of quinone reactions is comparable with that occurring in the mitochondrial cytochrome bc_1 complex and is referred to as a Q-cycle. In the steady state, it has been suggested that in the cytochrome b_6f complex the molecule of plastoquinol generated at the Q_n site moves rapidly to the Q_p site, where it is re-oxidized, without exchanging with the free quinone pool (Rich *et al.*, 1992).

Plastocyanin (PC) is a small ($\sim$10 kDa) soluble protein in the thylakoid lumen, encoded by the nuclear *pet*E gene. It functions as a mobile electron carrier for transferring electrons from the cytochrome b_6f complex to PSI. Its sites of oxidation and reduction may be separated by relatively large distances of up to 200 nm (see section 2.2.5d). Its prosthetic group is a Cu atom which undergoes one electron reduction from Cu(II) to Cu(I) with an E_m at pH 7 of 370 mV. The half time of oxidation of cytochrome f by plastocyanin is 200–300 μs (Bouges-Bouquet, 1977). The structure of plastocyanin has been determined at atomic resolution, revealing that the copper atom has four ligands (Cys, Met and two His) in a distorted tetrahedral geometry (Collyer *et al.*, 1990).

2.2.4   Photosystem I

The PSI complex functions to transfer electrons from plastocyanin, in the thylakoid lumen, to ferredoxin (Fd), on the stromal side of the membrane, which in turn reduces $NADP^+$, thus completing the electron transfer chain from H_2O to $NADP^+$. For reviews on the structure and function of PSI, see Almog *et al.* (1992), Bryant (1992), Golbeck (1992, 1993) and Sétif (1992).

(a)   *Structure of PSI*

Most of the prosthetic groups involved in electron transfer through PSI are bound to a heterodimer of two chloroplast-encoded intrinsic membrane polypeptides, each of molecular mass $\sim$83 kDa, designated PsaA and PsaB. The heterodimer also binds 12–15 molecules of β-carotene and each subunit binds approximately 45 molecules of Chl$a$ which act as a proximal antenna serving P700. In addition to the heterodimeric core, there are six small intrinsic membrane polypeptides (PsaG, I, J, K, L, M, of 4–18 kDa, see Fig. 2.14 and Table 2.3). Extrinsic subunits PsaC, D and E (molecular masses 8–16 kDa) are located on the stromal side of the complex, and subunit PsaF (15 kDa) is located on the lumenal side. The PsaN subunit (10 kDa) is also thought to be located on the lumenal side. The total molecular mass of PSI including 90 Chl$a$ molecules, but excluding the peripheral LHCI complexes, is 340 kDa.

The PsaA and PsaB proteins are highly hydrophobic and may have as many as 11 transmembrane helices per subunit. They are 95% homologous in higher plants. The PsaB polypeptide has a highly conserved sequence of 100 residues including a leucine-zipper motif which may be involved in binding the PsaA and PsaB subunits together (Kössel *et al.*, 1990). There are two molecules of vitamin K_1 (phylloquinone) per heterodimer, assumed to be bound one per subunit. The PsaA and PsaB subunits also ligate a 4Fe–4S iron–sulfur cluster, called FeS_X, at an interface region, via four cysteinyl residues. The role of the numerous low-molecular-weight hydrophobic polypeptides associated with PSI is unknown, but they may be involved in stabilizing the complex or promoting protein–lipid interactions. The 9 kDa PsaC polypeptide binds two 4Fe–4S clusters called FeS_A and FeS_B. The PsaC protein appears to be an essential component for assembly of the PSI complex, since in its absence, the PsaA and PsaB proteins rapidly turn over. PsaD facilitates the binding of the PsaC protein in a preferential orientation, and may also serve to stabilize the PsaE protein on the PSI core. The PsaD polypeptide is also thought to be involved in docking ferredoxin (Xu *et al.*, 1994). Ferredoxin is a soluble polypeptide encoded by the nuclear *pet*F gene. It binds a 2Fe–2S cluster (E_m −420 mV). The PsaD protein carries a positive charge, whereas ferredoxin is highly negatively charged, and electrostatic interactions between the two proteins are thought to facilitate correct binding of ferredoxin to the PSI complex. PsaE

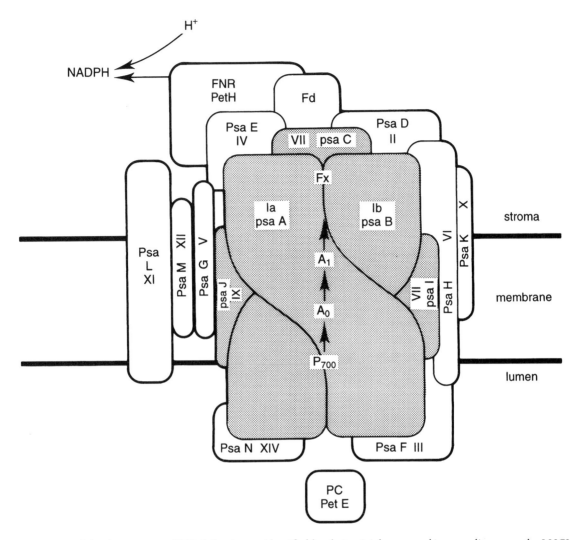

Figure 2.14 Subunit structure of PSI. Subunits are identified by their trivial name and/or encoding gene; the LHCI subunits are not included. The subunits are also listed in Table 2.3. Chloroplast-encoded subunits are shown in grey and nuclear-encoded subunits in white. The stroma-facing side of the complex is at the top of the diagram and the position of the membrane is indicated by the double line. The pathway of electron transfer through PSI is indicated by solid arrows. Fd (at the top of the complex) is able to dissociate to react with ferredoxin–thioredoxin oxidoreductase. From a diagram provided by R. Herrmann, H. Pakrasi and R. Oelmüller.

may be involved in ferredoxin reduction (Xu *et al.*, 1994), in the interaction of ferredoxin with ferredoxin-NADP$^+$ oxidoreductase (FNR) (Andersen *et al.*, 1992) and in cyclic electron flow around PSI (Yu *et al.*, 1993; see section 2.2.5e).

The lumenal PsaF protein has six lysine-rich positively charged regions on its surface which may participate in orienting the negatively charged plastocyanin molecule so that the copper atom is in the optimal position for fast electron transfer to P700$^+$ (Golbeck, 1992, 1993).

(b) *Functional properties of PSI*

The primary donor in PSI, P700, is a dimer of Chl*a* molecules oriented with their ring planes perpendicular to the membrane plane. The E_m of P700$^+$/ P700 is 490 mV (Sétif & Mathis, 1980), considerably lower than that of Chl*a*$^+$/Chl*a* *in vitro* (740–930 mV). The initial charge separation involves electron transfer from P700 to a specific monomeric Chl*a*, A$_0$ (Fig. 2.15) with a half time of 14–22 ps (Wasielewski *et al.*, 1987; Hecks *et al.*, 1994), generating the radical pair P700$^+$ A$_0^-$. Subsequently, the electron is transferred on to a

Table 2.3 Polypeptide components of the PSI core complex

Protein	Gene	Location of gene[a]	Molecular mass (kDa)	Function
Hydrophobic polypeptides				
PsaA	*psa*A	C	83 ⎫	Bind ~90 Chl*a*, 12–15 β-carotene,
PsaB	*psa*B	C	82 ⎭	P700, A_0, A_1, FeS_X
PsaG	*psa*G	N	10.8	Unknown
PsaI	*psa*I	C	4	Unknown
PsaJ	*psa*J	C	5	Unknown
PsaK	*psa*K	N	8.4	Unknown
PsaL	*psa*L	N	18	Unknown
PsaM	*psa*M	C	3.5	Unknown
Hydrophilic polypeptides *Stromal*				
PsaC	*psa*C	C	8.9	$FeS_{A/B}$ apoprotein
PsaD	*psa*D	N	17.9	Ferredoxin-docking protein, stabilizes PsaC
PsaE	*psa*E	N	9.7	Ferredoxin-docking? Ferredoxin-NADP⁺ reductase docking?
PsaH	*psa*H	N	10.2	LHCI linker protein
Lumenal				
PsaF	*psa*F	N	17.3	Plastocyanin-docking protein
PsaN	*psa*N	N	9.8	Unknown

[a]C, chloroplast-encoded; N, nuclear-encoded.

secondary acceptor A_1 with a half time of 40–200 ps (Sétif, 1992; Hecks *et al.*, 1994). Optical and electron paramagnetic resonance (EPR) spectroscopy and quinone extraction and reconstitution experiments indicate that A_1 is a molecule of vitamin K_1 (phylloquinone), and its one-electron reduction therefore generates a semiquinone intermediate. One of the two molecules of vitamin K_1 per reaction center is more readily extracted than the other. Whether both are involved in electron transfer is not yet established (discussed in Golbeck, 1992; Evans & Nugent, 1993). The next electron acceptor is the FeS_X iron–sulfur cluster ligated by the PsaA and PsaB subunits. It has a midpoint potential of −705 mV. The transfer time for electrons from A_1 to FeS_X is about 200 ns (Brettel, 1988). From FeS_X, electrons are transferred to FeS_A and FeS_B and then on to ferredoxin. FeS_A has a midpoint potential of −530 mV$_X$ and FeS_B has a midpoint potential of −590 mV. At present, it is not known in what sequence electrons flow at room temperature from FeS_X to FeS_A and FeS_B and then on to ferredoxin. It is likely that the center closest to FeS_X will be its immediate electron acceptor, and it seems likely that this is the lower potential component FeS_B. Electron transfer from P700 to ferredoxin spans the thylakoid membrane and generates a membrane potential which is converted more slowly into a transthylakoidal $\Delta\mu_{H^+}$ following secondary ion movements. Kinetic components in membrane potential formation with time constants of <1 μs and 30 μs have been tentatively assigned to P700 FeS_B transfer (via A_0, A_1 and FeS_X) and FeS_B to FeS_A transfer, respectively (Sigfridsson *et al.*, 1995).

The final reaction in the electron transport chain from water to NADP⁺ is the reduction of NADP⁺ by ferredoxin which is catalyzed by the FAD-containing enzyme ferredoxin-NADP⁺ oxidoreductase. This has a molecular mass of ~40 kDa and an E_m of −380 mV (Zanetti *et al.*, 1984, 1988). It is encoded by the nuclear *pet*H gene.

P700⁺ is re-reduced by plastocyanin. There are 2–3 molecules of plastocyanin per P700 and in the dark, one of these is bound to PSI. If P700⁺ is

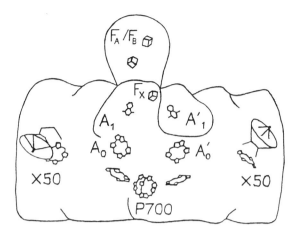

Figure 2.15 Schematic representation of the prosthetic groups involved in electron transfer in PSI. The stroma-facing side of the complex is at the top of the diagram. P700, F_X(FeS_X) and the components between them are bound to the PsaA/PsaB heterodimer and $F_{A/B}$ (FeS_A/FeS_B) are bound to PsaC. PsaA and PsaB each bind 45–50 antenna Chl$a$ (illustrated by the 'radio dish' symbols). The two Chl$a$ molecules closest to P700 are assumed to be accessory Chl (see below). It is not yet known whether electron transfer proceeds uniquely via one set of components in the symmetry-related axis or via both sides. Reproduced with minor modification from Nitschke & Rutherford (1991).

generated by flash excitation, the bound plasto-cyanin reduces P700$^+$ with a half time of 4–11 μs. P700$^+$ without a bound plastocyanin is reduced with a half time of 200–400 μs following diffusion of reduced plastocyanin to the binding site (Haehnel *et al.*, 1980).

(c)   High resolution structure of PSI

Crystals of PSI from the cyanobacterium *Synecho-coccus* sp. have been obtained, and these have enabled the structure of the PSI core complex to be solved at 0.6 nm resolution (Krauss *et al.*, 1993; Plate 4). The three 4Fe–4S clusters in PSI have been located enabling accurate measurement of their separation. The two FeS centers in the PsaC subunit are 1.4 nm and 2.1 nm, respectively, from FeS_X, but it is not yet possible to distinguish which is FeS_A and which is Fes_B. Of the 28 α helical regions that have been identified, most are transmembrane with an angle of between 3° and 30° to the membrane normal, and seven are nearly parallel to the membrane plane. Eight of the transmembrane helices are symmetrically related to eight other helices and these two sets of helices are tentatively assigned to the PsaA and PsaB subunits. Forty-five of the ~90 Chl$a$ molecules in

PSI that have been identified have their porphyrin planes roughly perpendicular to the membrane. Most of the Chl$a$ molecules are separated by center-to-center distances of 0.8–1.5 nm. The location of the 12–16 β-carotene molecules in the PsaA and PsaB subunits have not yet been identified.

The structure at 0.45 nm resolution clearly shows the P700 Chl$a$ dimer with the two Chl molecules parallel to each other and perpendicular to the membrane plane. Between P700 and FeS_X, two Chl molecules in symmetrical positions about the twofold axis and close to P700 are assumed to be accessory Chl molecules equivalent to those present in the bacterial reaction center. Further into the complex are two more symmetrically located Chl molecules, one of which is assumed to be A_0 (P. Fromme, personal communication). There is evidence, therefore, that the striking symmetry seen in the bacterial reaction center is also apparent in PSI. Whether electron transfer from P700 to FeS_X uniquely follows only one of the two possible pathways is not yet established.

(d)   *Light-harvesting chlorophyll–protein complexes of PSI*

Like PSII, the PSI complex also has LHC complexes associated with it, which have a similar structure to LHCII and belong to the same superfamily of proteins (Dreyfuss & Thornber, 1994 and references therein). Thus the LHCI polypeptides, like LHCII, contain three transmembrane helices, but show relatively low (~34%) sequence homology to LHCII. Like LHCII, they bind Chl$a$, Chl$b$ and xanthophylls. There are four different polypeptides which associate to form two distinct LHCI complexes designated LHCI-680 (LHCIa) and LHCI-730 (LHCIb) based on their absorption maxima (Table 2.4). It has been estimated that there are about 100 Chl molecules in the LHCI antenna system of each PSI complex. If, like LHCIIb, each polypeptide binds 12 Chl, this suggests that there are 8 LHCI polypeptides per PSI, possibly in a stoichiometry of two dimeric LHCI-680 and two dimeric LHCI-730 complexes per PSI, which may form a shell around the reaction center (Boekema *et al.*, 1990). It has, however, been suggested that since the LHCI polypeptides show sequence homology to the LHCIIb polypeptides, they are more likely to associate to form trimers. In this case, it is proposed that there are two trimeric LHCIa complexes and two trimeric LHCIb complexes per PSI, each binding 8–10 Chl per polypeptide

Table 2.4 Light-harvesting Chla/b protein complexes associated with PSI

Complex	Chla/b ratio	Polypeptide components	Gene[a]	Molecular mass (kDa)
LHCIa (LHCI-680)	2.0–3.1	Lhca3	lhca3	20.5
		Lhca2	lhca2	18
LHCIb (LHCI-730)	2.2-4.4	Lhca1	lhca1	20
		Lhca4	lhca4	20

a All the genes are nuclear.

(Dreyfuss & Thornber, 1994). The lower Chl content compared with LHCIIb is primarily due to a lower Chl$b$ complement and would be consistent with the lower molecular weight of the polypeptides in comparison to LHCIIb.

The LHCI-680 may allow phosphorylated LHCIIb which has migrated into the nonappressed membrane regions under conditions of over-excitation of PSII (see section 2.2.5c) to couple energetically with PSI (Bassi & Simpson, 1987).

2.2.5 Regulation of light energy utilization

(a) Organization of the thylakoid membranes in higher plant chloroplasts

The thylakoid membranes of higher plant chloroplasts have a unique organization into stacks of flattened disk-shaped sacs or lamellae in which the membranes of adjacent lamellae are closely appressed (granal lamellae) and lamellae in which the membranes are not appressed and thus their outer surfaces are in direct contact with the stroma (stromal lamellae) (Fig. 2.16). The two membrane systems and the lumenal spaces they enclose are continuous.

The thylakoid membrane lipids comprise about 80% glycolipids, with neutral hydrophobic head groups, consisting of monogalactosyldiacyglycerol and digalactosyldiacyglycerol in a 2:1 ratio. The remainder of the lipid is mainly phospholipid (10%) and sulfolipid (5%), both of which carry a negative charge at neutral pH. Linolenic acid (C18:3) is the predominant fatty acid (~70%) and its high degree of unsaturation ensures that the membrane is very fluid. There is a partially asymmetric distribution of the various lipids between appressed and non-appressed membranes (Gounaris et al., 1983) but the functional significance of this is unknown.

The complex thylakoid membrane organization is reflected in a marked lateral heterogeneity in the distribution of the membrane–protein complexes (Andersson & Anderson, 1980; Plate 5a). The ATP synthase and PSI with its associated LHCI are located only in the non-appressed membrane regions, whereas the majority of the PSII and LHCII are located in the appressed membranes. This pool of PSII is referred to as PSIIα. A minor pool of PSII centers with a smaller population of LHCII per PSII (PSIIβ) is located in the stromal lamellae (Melis & Anderson, 1983). The cytochrome b_6f complex is distributed throughout the thylakoid membrane system (Anderson, 1982). The stoichiometry of complexes in the membrane is discussed in section 2.2.8.

Membrane appression is thought to result from Van der Waal's interactions between LHCII complexes in opposing membranes and between LHCII and PSII in the same membrane (reviewed in Allen, 1992a,b). In shade-grown plants, the amount of LHCII per PSII increases, leading to an

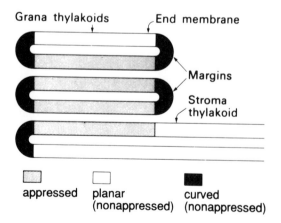

Figure 2.16 Distribution of integral membrane protein complexes in thylakoid membranes. Schematic representation of a cross-section through thylakoids showing the different membrane regions. Reproduced from Anderson (1992).

increase in the extent of the appressed granal membrane system (see also section 2.2.8). Although the membrane lipid itself is fluid, the high concentration of integral membrane–protein complexes, particularly in the appressed membranes (where they occupy ~50% of the membrane area; Plate 5b), severely restricts diffusion of these complexes within the plane of the membrane. The diffusion coefficient of phosphorylated LHCII (see section 2.2.5c) in the appressed membranes has been estimated to be in the range $(2-4) \times 10^{-12}\,cm^2\,s^{-1}$ (Drepper et al., 1993) which is two orders of magnitude lower than that reported for membrane proteins of similar size in artificial lipid bilayers.

(b) *Functional significance of the heterogenous distribution of the photosystems*

There has been considerable debate about why the highly heterogeneous distribution of PSII and PSI has evolved. The formation of grana stacks in which LHCII and PSII are concentrated facilitates light-harvesting and energy transfer to PSII, but the principal reason for the heterogeneous distribution of PSI and PSII appears to lie in the different photochemical properties of the two types of reaction center (discussed in Trissl & Wilhelm, 1993).

The energy of the first excited singlet state of much of the antenna chlorophyll serving PSII (i.e. LHCII, CP47, CP43) is close to that of P680. Because of the large excess of light-harvesting chlorophyll over P680, the distribution of singlet states favors the antenna chlorophylls (Chl_n). The free energy drop between the excited state $(Chl_nP680)^*Ph$ and the radical pair state $Chl_nP680^+Ph^-$ is small, so that the latter is in quasi-equilibrium with the excited state $(Chl_nP680)^*Ph$ (Van Grondelle et al., 1994). If Q_A is already reduced to Q_A^-, the equilibrium disfavors radical pair formation, possibly due to an electrostatic effect of the negative charge on Q_A^- (Van Mieghem et al., 1989). The excited state normally depopulates biphasically with time constants of 300–500 ps reflecting forward electron transfer from Ph^- to Q_A, i.e.

$$Chl_n^* P680\,Ph\,Q_A \rightleftharpoons Chl_n\,P680^*\,Ph\,Q_A$$

$$\downarrow$$

Energy losses (fluorescence and heat)

$$\rightleftharpoons Chl_n\,P680^+\,Ph^-\,Q_A \rightarrow Chl_n\,P680^+\,Ph\,Q_A^-$$

The relatively slow energy trapping time by PSII results in a quantum yield of 85% owing to competing processes of energy dissipation (fluorescence and heat emission).

In PSI, the energy of the first excited singlet state of at least some of the antenna chlorophyll is close to that of P700. However, the free energy difference between $(Chl_nP700)^*A_0$ and $Chl_nP700^+A_0^-$ is much greater than the corresponding value for PSII. For this reason, energy trapping leading to radical pair formation is much more rapid than in PSII, with the excited state Chl_nP700^* depopulating essentially irreversibly with a time constant of 60–90 ps i.e.

$$Chl_n^* P700\,A_0 \rightleftharpoons Chl_n\,P700^*\,A_0 \rightarrow Chl_n\,P700^+\,A_0^-$$

$$\downarrow$$

Energy losses (fluorescence and heat)

The much faster energy trapping in PSI leads to a higher quantum yield of >95%.

If PSI and PSII and their associated antenna chlorophylls were in excitonic contact, i.e. if energy transfer between the two was possible, PSI would drain off excitation energy from PSII because it has longer wavelength-absorbing chlorophylls which would favor energy transfer from PSII antenna chlorophylls to the PSI antenna, and it has faster energy-trapping kinetics than PSII. Physical separation of PSII and PSI in different thylakoid membrane domains therefore prevents PSI draining away excitation energy from PSII.

The physical separation of the photosystems provides the basis of mechanisms for regulating energy distribution between them (see below) and for controlling the relative rates of linear and cyclic electron flow (see section 2.2.5e). The physical separation of the photosystems also has consequences for electron transfer between them. PSII and PSI may be separated by distances of 500–2000 nm, and for linear electron transfer from H_2O to $NADP^+$ to take place, there must be a mobile electron carrier linking the two photosystems (see section 2.2.5d).

(c) *Regulation of energy distribution between the photosystems*

In order that maximal rates of electron transport through PSII and PSI are achieved at a particular light intensity and spectral composition, the delivery of light energy to the two types of reaction center must be balanced so that they are excited at equal rates. The relative light-harvesting capacity of the two photosystems must therefore be controlled (reviewed in Anderson & Andersson, 1988; Allen, 1992a, b).

A pool of LHCIIb located towards the outer edges of the appressed granal membranes is phosphorylated under conditions in which there is an imbalance in excitation of the photosystems in favor of PSII (Bennett, 1991). The phosphorylation site is a threonine residue close to the amino terminus, located on the stromal side of the membrane (Michel *et al.*, 1991). Phosphorylation of this pool of LHCIIb results in its lateral migration in the membrane away from the appressed membrane regions where it is associated with PSII centers and with other LHCIIb complexes, into the non-appressed membranes where it is able to interact with and deliver excitation energy to PSI. This lateral migration is thought to occur either as a result of non-specific electrostatic repulsion arising from the increased negative surface charge associated with added phosphate groups (Barber, 1982), or as a result of a decreased binding affinity of phospho-LHCII at specific sites of interaction laterally with the PSII core complexes and transversely with LHCII in the adjacent appressed membrane, with a corresponding increase in affinity for PSI (Allen, 1992a,b).

Although it appears that both the inner pool of LHCIIb tightly bound to PSII and the mobile peripheral pool (see section 2.2.2d) can be phosphorylated, it is the outer mobile pool which migrates from the appressed to the non-appressed membranes on phosphorylation (Larsson *et al.*, 1987a). The LHCII kinase is activated when plastoquinone is reduced and inactivated when it is oxidized (Allen *et al.*, 1981; Fig. 2.17). Plastoquinone reduction occurs when PSII is turning over faster than PSI. Phospho-LHCII then transfers energy to PSI at the expense of PSII, balancing the energy distribution. If PSI is over-excited relative to PSII, the intersystem electron carriers are oxidized and the kinase is inactivated. A LHCII phosphatase, which appears not to be under redox control, dephosphorylates LHCII causing it to migrate back into the appressed membrane regions (Bennett, 1979). The LHCII kinase is thought to be a 64 kDa protein localized in the grana membranes, particularly at the edges of the grana stacks (Gal *et al.*, 1990).

The kinase must associate with the cytochrome $b_6 f$ complex in order to display activation by plastoquinol (Frid *et al.*, 1992). Redox titrations of kinase activity give an E_m value of 40 mV at pH 7.6 with an n value (number of electrons transferred) of 1 (Silverstein *et al.*, 1993). Both the E_m and n value are incompatible with redox control being exerted by the bulk plastoquinone pool. An attractive candidate for the site of redox control is the Q_n site of the cytochrome $b_6 f$ complex, but

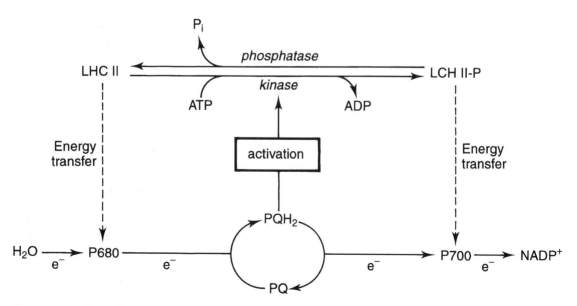

Figure 2.17 Scheme for redox control of the LHCII kinase by the redox state of plastoquinone. An imbalance of excitation between the two photosystems in favor of PSII results in increased reduction of PQ which leads to activation of LHCII kinase. Phospho-LHCII decouples from PSII and transfers energy to PSI. Increased energy transfer to PSI will tend to oxidize PQH_2, inactivate the kinase and allow the LHCII phosphatase reaction to predominate. Reproduced from Allen (1992) with minor modification.

the molecular mechanism linking kinase activity to the redox state of this site is unknown.

LHCIIa (CP29) is also subject to phosphorylation by a redox-controlled kinase, but phosphorylated forms of LHCIIc and LHCIId have not been found (Peter & Thornber, 1991). The D1, D2, PsbH and CP43 (PsbC) polypeptides of PSII are also phosphorylated reversibly under conditions when plastoquinone is reduced. The kinase and phosphatase responsible for phosphorylation/dephosphorylation of PSII polypeptides are probably different from the LHCIIb kinase/phosphatase (Allen, 1992a; Ebbert & Godde, 1994). The physiological significance of phosphorylation of PSII polypeptides is not fully understood. However, photoinhibitory light induces maximal phosphorylation of D1 and it is thought that this process is involved in regulation of the repair cycle of damaged PSII centers (see section 2.2.6a(vi); Rintamäki et al., 1995).

(d) Consequences of physical separation of the photosystems on electron transfer between them

There are two mobile electron carriers which could be involved in transferring electrons over long distances from PSII to PSI, plastoquinol transferring electrons from PSII to the cytochrome $b_6 f$ complex, or plastocyanin transferring electrons from the cytochrome $b_6 f$ complex to PSI. Cytochrome $b_6 f$, being a much larger complex, would be unable to diffuse in the membrane at the required rates.

The plastoquinone pool contains approximately eight molecules per PSII complex (McCauley & Melis, 1986). Under conditions in which its reoxidation is blocked, 50–70% of the pool is reduced under saturating light with a half time of 25–60 ms, the remainder with a half time of 0.8–1.0 s. The fast pool is interpreted to be the PQ molecules located in the granal membranes, and the slow pool is thought to represent PQ in the stromal membranes which is reduced more slowly by PSIIβ (Joliot et al., 1992). The half time for redistribution of PQ from the fast pool to the slow pool is ~6 s reflecting slow diffusion from the granal to stromal membranes due to membrane crowding by the integral membrane protein complexes.

The very slow transfer time of plastoquinol from the granal to stromal lamellae means that linear electron transfer from PSII to PSI must involve cytochrome $b_6 f$ complexes in the appressed membranes rather than those in the non-appressed membranes. This in turn means

that plastocyanin must be the mobile electron carrier responsible for long distance electron transfer from cytochrome $b_6 f$ complexes in the appressed membranes to PSI in the non-appressed membranes. Consistent with this conclusion is that millisecond oxidation of all cytochrome f by PSI through plastocyanin has been reported (Whitmarsh, 1986). Thus plastocyanin must be able to move efficiently throughout the entire lumenal space, in spite of its complex geometry, the narrowness of the lumen and the proteins protruding into it (Lavergne & Joliot, 1991). Under conditions in which the plastoquinone pool is oxidized, the half time for electron transfer from PSII to PSI is 10–20 ms. From the above account and section 2.2.3, it is evident that this rate is limited primarily by the diffusion time of plastoquinol from PSII to the cytochrome $b_6 f$ complex within the appressed membrane regions, rather than by the rate of electron transfer through the cytochrome $b_6 f$ complex or the rate of plastocyanin diffusion from cytochrome $b_6 f$ to PSI.

(e) Cyclic electron transport and regulation of the distribution of cytochrome $b_6 f$ complexes

The operation of a proton pumping Q-cycle in the cytochrome $b_6 f$ complex means that the overall $\Delta\mu_{H^+}$-generating capacity of the linear electron transport chain should meet both the ATP and NADPH requirements of CO_2 fixation, i.e. 3ATP per 2NADPH (Osborne & Geider, 1988). It can be demonstrated in vitro, however, that reducing equivalents from the acceptor side of PSI can be cycled back to P700$^+$ via the cytochrome $b_6 f$ complex and plastocyanin. This cycle contributes to generation of the transmembrane $\Delta\mu_{H^+}$, and hence to ATP synthesis, without concomitant NADP$^+$ reduction.

The mechanism by which electrons are delivered from the acceptor side of PSI to the cytochrome $b_6 f$ complex is not fully understood but appears to proceed via ferredoxin and PQ and may involve ferredoxin-NADP$^+$ oxidoreductase (Cleland & Bendall, 1992).

Evidence for the operation of cyclic electron transport in C_3 plants in vivo is limited but it has been demonstrated under physiological conditions in vivo when there is an additional ATP requirement, as in the carbon fixation pathway of C_4 plants (Osmond, 1994). It may also have an important role to play in the synthesis of ATP required for protein synthesis during PSII repair following photoinhibition (Canaani et al., 1989; see section 2.2.6a(vi)). Variability in demand for ATP and NADPH is met by flexibility in the ratio of linear to cyclic electron flow (Foyer *et al.*, 1990; Anderson, 1992).

Cytochrome b_6f complexes are distributed throughout the thylakoid membrane system (appressed granal membranes, non-appressed stroma thylakoids, end grana membranes and grana margins) (Anderson, 1982, 1992; Fig. 2.16; Plate 5a, b). Only the complexes in the non-appressed membranes containing PSI participate in cyclic electron transport, since the long diffusion distances for the mobile electron carriers linking cytochrome b_6f complexes in the appressed membranes with PSI in the non-appressed membranes make the involvement of cytochrome b_6f complexes in the appressed membranes kinetically unfavorable (Anderson, 1989).

Under conditions in which the LHCII kinase is active, there is a marked redistribution of cytochrome b_6f complexes from appressed granal membranes to non-appressed stromal membranes (Vallon *et al.*, 1991). This may be driven by phosphorylation of the cytochrome b_6 polypeptide, which has been demonstrated *in vitro* (Gal *et al.*, 1992). It is proposed that the lateral migration of both cytochrome b_6f complex and LHCII into membrane regions containing PSI is designed to increase the rate of cyclic electron flow, relative to linear electron flow, in order to increase the rate of ATP synthesis relative to NADPH formation.

Increased demand for ATP would lead to a decrease in $\Delta\mu_{H^+}$ since this is the energy source for ATP synthesis. It is not clear whether the decrease in $\Delta\mu_{H^+}$ directly activates LHCII kinase (Fernyhough *et al.*, 1984) in addition to redox control, or whether the effect is indirect, for example by altering the conformation of LHCII to make it a better substrate for the kinase, or by an increased $NADPH/NADP^+$ ratio leading to reduction of PQ.

(f)   *Regulation of energy dissipation*

When the light intensity exceeds that needed to saturate the rate of photosynthetic electron transport, excess absorbed energy (Fig. 2.18) can lead to damage to the PSII reaction center, antenna chlorophyll triplet formation and pigment photo-oxidation. To minimize this, a mechanism has evolved to dissipate the excess absorbed energy harmlessly as heat (Demmig-Adams & Adams, 1992). Although not yet fully understood, it appears to involve two related phenomena, a de-epoxidation of a minor xanthophyll, violaxanthin, in the LHCII complexes under high light, converting it into zeaxanthin (Demmig-Adams, 1990; Demmig-Adams & Adams, 1993; Fig. 2.19), and protonation of amino acid residues in LHCII as a

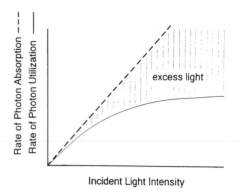

Figure 2.18   Relationship between rate of photon absorption and rate of photon utilization in photosynthesis. Photon absorption (– – –) is linear with light intensity over the physiological range, whereas photon utilization for CO_2 fixation (——) exhibits saturation kinetics. The shaded area between the curves represents the region of light absorption which is in excess of the capacity of photosynthesis. Reproduced from Owens (1994).

result of increased acidification of the thylakoid lumen (Horton & Ruban, 1992). The latter situation arises when light-driven electron transport obligatorily coupled to proton pumping into the lumen exceeds the rate of ATP synthesis, which dissipates the $\Delta\mu_{H^+}$ across the membrane. Two hypotheses have been proposed to explain how zeaxanthin de-epoxidation and protonation of LHCII may lead to energy dissipation, as outlined below.

The energy of the lowest excited singlet state of xanthophylls is normally above that of Chl$a$, so that in the photosynthetic pigment–protein complexes in which these molecules are bound in close proximity, xanthophylls transfer energy to Chl$a$ within a few ps in a non-Förster mechanism (see section 2.2.2d). The predicted energy levels for the lowest excited singlet states of violaxanthin and zeaxanthin suggest that the former lies above that of Chl$a$ whereas that of zeaxanthin is close to or below Chl$a$. Thus if Chl$a$ and zeaxanthin are in close proximity, rapid reversible energy transfer between Chl$a$ and zeaxanthin would occur. Singlet-excited carotenoids decay to the ground state more than two orders of magnitude more rapidly than Chl$a$ (typically 10–40 ps for carotenoids and 5 ns for Chl$a$) (Owens *et al.*, 1993). Conversion of violaxanthin into zeaxanthin could therefore lead to reversible energy transfer from Chl$a$ to zeaxanthin which would accelerate the rate of energy dissipation from singlet-excited Chl$a$. In this mechanism, it is proposed that increased protonation of amino acid residues of

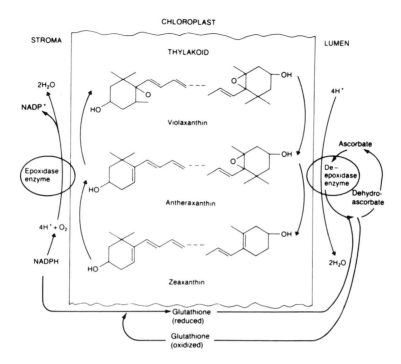

Figure 2.19 Scheme showing the intermediates, enzymes and co-substrates of the xanthophyll cycle involved in regulating energy dissipation in LHCII. The epoxidase has a pH optimum of 7.5 and its activity is favored under limiting light. The de-epoxidase has a pH optimum of 5.1 and its activity is favored under excessive light. Reproduced from Lawlor (1993) with minor modification.

LHCII in the vicinity of the zeaxanthin and Chl*a* could accelerate the rate of energy transfer through electric-field induced shifts in the Chl*a* spectrum, leading to increased spectral overlap, or enhancing the rate of Förster resonance transfer by increasing the dipole character of the zeaxanthin lowest excited singlet state to ground state transition (Owens, 1994).

Non-photochemical energy dissipation at high light intensities has been observed in the *absence* of zeaxanthin (Noctor *et al.*, 1991) and an alternative hypothesis proposes that protonation of amino acid residues leads to structural distortions in the LHCII which promote specific Chl–Chl *and* zeaxanthin–Chl*a* interactions leading to non-radiative energy dissipation (Horton & Ruban, 1994). Zeaxanthin aggregates more effectively than violaxanthin (Ruban *et al.*, 1993) and it is proposed that the conversion of violaxanthin into zeaxanthin plays an important role in facilitating the protonation-triggered changes in pigment–pigment interactions.

The conversion of violaxanthin (two epoxides) via antheraxanthin (one epoxide) into zeaxanthin (no epoxides) is catalyzed by a de-epoxidase which has a pH optimum of 5 and is probably located on the lumenal side of the thylakoid membrane

(Hager & Holocher, 1994; Fig. 2.19). The reverse reaction is catalyzed by an epoxidase with a neutral pH optimum and is probably located on the stromal side of the membrane (Hager, 1980). Thus under conditions of excessive illumination, the resulting low lumenal pH would activate the de-epoxidase. The reversible interconversion of violaxanthin and zeaxanthin is known as the xanthophyll cycle. The mechanism by which the xanthophyll cycle enzymes gain access to their substrates is unknown. In particular, it is not known whether the enzymes act on protein-bound xanthophylls or whether their dissociation into the membrane lipid is required.

The proportion of xanthophyll-cycle carotenoids varies between the different LHCII complexes, with LHCIIb having the lowest (~9%) and LHCIIa (CP29) the highest (~40%) (Ruban *et al.*, 1994). The extent of light-dependent de-epoxidation is highly variable between the LHCII complexes. LHCIIc (CP26) shows the highest conversion of violaxanthin into zeaxanthin (70%), and probably has a particularly important role in regulating energy dissipation. This is commensurate with its probable role in transferring energy absorbed by the peripheral LHCIIb towards the PSII core.

2.2.6 Photoinhibition and related phenomena

(a) *Photoinactivation and repair of photosystem II*

Although a number of mechanisms have evolved to dissipate excess absorbed light energy, the energy-transducing photosynthetic apparatus is susceptible to damage by light. When the rate of photodamage exceeds the rate of repair, there is a drop in photosynthetic efficiency, a phenomenon known as photoinhibition (Osmond, 1994).

The main site of photodamage is the PSII reaction center. PSI is relatively immune from photodamage, and the particular sensitivity of PSII is thought to be due to its unusual chemical properties, as explained in more detail below. Based on *in vitro* studies, two different sequences of chemical reactions leading to photodamage of PSII have been identified, referred to as acceptor side and donor side photoinactivation (Barber & Andersson, 1992). Photoinactivation of PSII is thought to be responsible for the high turnover rate of the D1 protein of PSII which, at high light intensities, can be 50–80-fold higher than any other thylakoid protein (Prasil *et al.*, 1992; Andersson *et al.*, 1994).

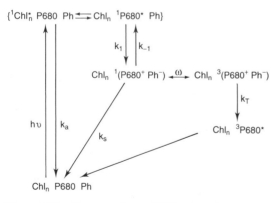

Figure 2.20 Photoreactions of PSII when electron transfer from Ph^- to Q_A is blocked due to prereduction or extraction of Q_A. Chl_n represents antenna Chl, k_a is the intrinsic decay rate of the quasi-equilibrium between excited antenna pigments and P680, excluding primary charge separation (rate k_1). The singlet state of $P680^+Ph^-$ can reversibly evolve to a triplet spin state (indicated by ω). Charge recombination produces either the singlet ground state of P680 (rate k_S) or triplet state (rate k_T) depending on the spin state of the primary pair when they recombine. Based on Van Mieghem *et al.* (1995).

(i) Acceptor side photoinactivation

During normal turnover of PSII, the primary plastoquinone acceptor Q_A is reduced to the semiquinone Q_A^- and then reoxidized by Q_B. It has been demonstrated *in vitro* that if the plastoquinone pool is over-reduced, so that electron transfer from Q_A^- to Q_B is blocked, Q_A^- is slowly doubly reduced and protonated to form Q_AH_2 (Van Mieghem *et al.*, 1989; Vass *et al.*, 1992; Styring & Jegerschöld, 1994). It appears that formation of the doubly reduced and protonated form of Q_A is reversible but Q_AH_2 can be irreversibly lost from its binding site on PSII (Koivuniemi *et al.*, 1993). In centers containing Q_AH_2 or an unoccupied Q_A site, photogenerated $P680^+$ Ph^- decays on a nanosecond time scale, partly by recombination to produce P680 Ph. However, a major competing decay pathway involves spin dephasing of the two unpaired electrons in $P680^+$ Ph^- to a triplet spin configuration $^3(P680^+Ph^-)$ (Van Mieghem *et al.*, 1995; Fig. 2.20). This recombines to form $^3P680^*$. Studies with PSII reaction center preparations show that $^3P680^*$ reacts with O_2 to produce $^1O_2^*$ (see also section 2.2.6b(i)), which is then thought to destroy the chlorophyll of P680, and possibly also interacts with histidine, methionine and tryptophan residues of the protein matrix (Telfer *et al.*, 1994a). The β-carotene in the reaction center is unable to quench $^3P680^*$, probably because if it was located close enough to P680 to achieve this, it would be oxidized by $P680^+$. It does, however, quench a considerable amount of $^1O_2^*$ before the latter escapes from the reaction center (Telfer *et al.*, 1994b).

(ii) Donor side photoinactivation

Donor side photoinactivation is observed under conditions in which the donor side is unable to deliver electrons quickly to $P680^+$. $P680^+$, with a redox potential of more than $1\,V$, oxidizes components in its environment including β-carotene, accessory chlorophyll and perhaps other amino acids (De Las Rivas *et al.*, 1993; Telfer & Barber, 1994).

(iii) Polypeptide cleavage

Both acceptor side and donor side photoinactivation trigger conformational changes and degradation of the D1 protein, with initial cleavage at a site in the loop connecting transmembrane helices IV and V on the stromal side of the complex following acceptor side photoinactivation and in a loop connecting helices I and II on the lumenal side of the complex following donor side photoinactivation (De Las Rivas et al., 1992) (Fig. 2.10). Further proteolysis of the primary breakdown products is at least as rapid as the initial proteolytic events. At high light intensity, degradation of the D2 protein is also observed (Virgin et al., 1990). The molecular mechanism(s) of D1 and D2 degradation and the mechanistic basis of the link between photodamage and D1 degradation are not yet understood. D1 degradation appears to require occupancy of the Q_B site by plastoquinone and is blocked by binding of certain classes of herbicide to this site (Gong & Ohad, 1992).

(iv)   Mechanism of PSII photoinactivation in vivo

The principal mechanism of PSII photoinactivation at high light intensity in vivo is not established, although the D1 degradation fragment resulting from acceptor side (^{3}P680-linked) photoinactivation in vitro has been detected after in vivo light treatment (Greenberg et al., 1987). Complete reduction of the PQ pool leaving the Q_B sites inoperational is unlikely to occur in vivo (e.g. see Krause & Weis, 1991) and it is necessary to argue that at high light intensity, the probability of formation of doubly reduced Q_A species in centers where the Q_B site is transiently unoccupied by plastoquinone is sufficiently high to account for the observed turnover rate of D1. The donor side mechanism can be observed in vivo under conditions in which P680$^+$ reduction is slowed (Gong & Ohad, 1992).

(v)   Photodamage to PSII in low light

Although net damage to PSII occurs at high light intensity, turnover of D1 in PSII also occurs at light intensities as low as 10–30% of saturation. At low light intensity, the probability of charge recombination between the Q_B^- and the S_2 or S_3 states of the water-splitting system is greater than the probability of consecutive reduction of Q_B to Q_B^{2-}, protonation and exchange with the PQ pool. D1 degradation can be correlated with the formation of Q_B^- and its decay by recombination. It is proposed that $Q_B^- S_{2/3}$ recombination proceeds via P680$^+$ Ph$^-$, which can recombine to generate ^{3}P680* leading to $^1O_2^*$ formation (Keren et al., 1995).

(vi)   Repair of damaged PSII

Following cleavage and breakdown of D1, the three extrinsic polypeptides (33, 23 and 17 kDa) and the four Mn atoms of the water-splitting cluster are released into the thylakoid lumen (Hundal et al., 1990). Damage to PSII occurs mainly in the appressed membranes whereas D1 is synthesized by ribosomes attached to the stromal surface of non-appressed membranes. It appears that partially disassembled PSII complexes migrate laterally in the thylakoid membranes to the non-appressed regions where they may act as acceptors for newly synthesized D1 proteins (reviewed in Aro et al., 1993; Fig. 2.21). Assembly of newly synthesized D1 into PSII must also involve religation of some of the cofactors (Van Wijke et al., 1994), and there is some evidence that this occurs concomitantly with D1 translation (Kim et al., 1991). The D1 protein is proteolytically processed by the removal of a short peptide from its carboxyl terminus following its assembly into PSII. This step is an essential pre-requisite for religation of the complete water-splitting manganese cluster (Diner et al., 1988; Taylor et al., 1988), apparently because the carboxyl group of the carboxyl terminal residue of the mature protein acts as one of the Mn ligands. PSIIβ centers in the stromal lamellae may represent recently repaired PSII centers in transit into the appressed membrane regions.

(vii)   Mechanisms for protecting PSII from photodamage

A potential consequence of the generation of a very low pH in the thylakoid lumen is the reversible release of Ca^{2+} from PSII. This has been demonstrated in vitro at pH values below 5.5 (Krieger & Weis, 1992). It is not certain that the pH in the lumen in vivo reaches a sufficiently low value to cause Ca^{2+} release, but in the Ca^{2+}-depleted state, the S-state cycle does not proceed beyond S_3 and electron transfer from Y_Z to P680$^+$ is blocked (see also section 2.2.2b(ii)). The E_m of Q_A/Q_A^- is raised from around -80 mV to $+60$ mV in Ca^{2+}-depleted PSII, bringing it close to that of $Q_B/Q_B^-(H^+)$ (Krieger & Weis, 1992). In such centers, P680$^+$ Q_A^- recombination predominates, thus dissipating absorbed energy and preventing P680$^+$ from oxidizing other components nearby which would lead to donor side inactivation. The shift to higher potential of Q_A^- may reduce the probability of charge recombination proceeding via P680$^+$ Ph$^-$, thus reducing the probability of generating ^{3}P680* (Johnson et al., 1995). This

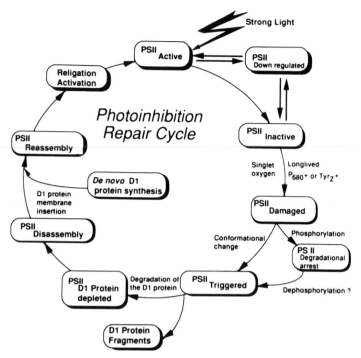

Figure 2.21 Hypothetical scheme showing the different processes involved in photoinactivation and repair of PSII. Reproduced from Aro *et al.* (1993).

may also be important for protecting PSII centers which have not yet ligated an Mn cluster (see section 2.2.2b(iv)).

The presence of Q_A^- reduces both the yield and lifetime of P680$^+$ Ph$^-$ in comparison to PSII centers with Q_A, probably because of an electrostatic effect of the negative charge on the quasi-equilibrium between the excited state and radical pair (Van Mieghem *et al.*, 1989, 1995). This in itself would reduce the probability of ^{3}P680* formation. It has, however, been demonstrated at cryogenic temperature that the presence of Q_A^- reduces the lifetime of ^{3}P680* by two orders of magnitude in comparison to samples containing Q_AH_2 (Van Mieghem *et al.*, 1995). If this occurs at room temperature, it would ensure that in centers containing Q_A^- any ^{3}P680* formed following ^{3}(P680$^+$ Ph$^-$) recombination would decay before it had the chance to react with oxygen to form $^1O_2^*$. This may, therefore, represent a physiologically relevant protection mechanism. $^1O_2^*$ would then only be generated from ^{3}P680* in centers containing Q_A (low light photodamage, see section 2.2.6a(v)), Q_AH_2 or an unoccupied Q_A site (Vass *et al.*, 1992).

Cytochrome *b*-559 can exist in high potential, (HP, E_m 400 mV) and low potential (LP, E_m 60 mV) forms. The HP form is reduced at ambient redox potential and it has been argued that if at high light intensity, a second charge separation has

occurred before Y_Z^+ has been reduced by the Mn cluster, cytochrome *b*-559$_{HP}$ may act as an alternative electron donor to P680$^+$ via a redox-active Chl molecule, preventing it from oxidizing nearby amino acids (Thompson & Brudvig, 1988). It has also been proposed that cytochrome *b*-559$_{LP}$, which would be oxidized at ambient redox potential, can accept electrons from Ph$^-$ in centers in which onward electron transfer is slowed due to prior reduction of Q_A, thus preventing ^{3}P680* formation (Whitmarsh *et al.*, 1994). Cytochrome *b*-559 switches from the HP to LP forms under particular illumination conditions (Styring *et al.*, 1990) but the molecular basis of this switch is unknown.

Finally, P680$^+$ may itself convert chlorophyll excited states to heat, in a manner analogous to that exhibited by P700$^+$ (Weis & Berry, 1987). Whether any of these reactions are involved in protecting PSII against photodamage *in vivo* remains to be proven.

(b)   *Scavenging reactive molecules in chloroplasts*

As discussed above, a number of mechanisms have evolved to enable the photosynthetic apparatus to respond to wide fluctuations in light intensity, both to maximize light utilization at low light and to dissipate excess absorbed light energy as heat at high light intensities. In spite of these mechanisms, the production of reactive by-products, capable of damaging the photosynthetic apparatus, is unavoidable. These fall into two categories, singlet-excited oxygen and reduced oxygen species (superoxide, hydrogen peroxide and hydroxyl radicals).

(i) Formation and scavenging of singlet oxygen

When the absorbed light energy is in excess of the capacity of photosynthesis to utilize the energy to drive photosynthesis, the lifetime of $^1Chl^*$ in the antenna is increased and this increases the probability of $^3Chl^*$ formation through intersystem crossing.

In the singlet excited state, the excited electron and the unpaired electron in the orbital from which it was excited have opposite spin (Plate 6). Conversion to the triplet state involves spin dephasing so that the two electrons have parallel spins. $^3Chla^*$ has a lifetime of ~ 1 ms but it reacts rapidly with oxygen, reducing its lifetime to $\sim 30\,\mu s$ (Durrant et al., 1990) to produce the very reactive species singlet oxygen, $^1O_2^*$. This stimulates peroxidation of unsaturated membrane lipids and oxidizes methionine, tryptophan, histidine or cysteine residues (Halliwell, 1991). It is essential, therefore, to scavenge $^1O_2^*$ at its site of production.

As well as being involved in light harvesting and energy transfer to Chla and thence to the reaction centers, carotenoids play a vital role in scavenging $^3Chl^*$ and $^1O_2^*$. Rapid energy transfer to carotenoid from either species generates $^3Car^*$ which decays rapidly and harmlessly to the ground state by thermal emission on a μs time scale (reviewed in Frank & Cogdell, 1993).

(ii) Formation and scavenging of superoxide

The normal electron acceptor from PSI is $NADP^+$ but PSI is also capable of reducing oxygen to superoxide (E_m at pH 7 for $O_{2(aq)}/O_2^-$ couple is -160 mV; Wood, 1988), and the probability of this occurring increases when the rate of reduction of ferredoxin by PSI exceeds the rate of NADPH utilization by the Calvin cycle (see section 2.3). The maintenance of electron transport under these conditions is particularly important, since the resulting over-acidification of the thylakoid lumen down-regulates excitation of PSII (see section 2.2.5). Typically the production rate of superoxide corresponds to 10–20% of the total flow of electrons through PSI (Asada,

1994). Superoxide dismutase in the chloroplasts disproportionates O_2^- to produce hydrogen peroxide (H_2O_2). O_2^- and H_2O_2 react with each other in a reaction catalyzed by transition metal ions (Haber–Weiss reaction) to produce hydroxyl radicals (OH^*), a highly reactive oxygen species capable of causing oxidative damage to lipids, proteins, pigments and nucleic acids (Halliwell, 1991). To prevent this from occurring, the O_2^-, H_2O_2 and radicals derived from their reaction with other chloroplast components must be effectively scavenged near their site of production.

Although O_2^- disproportionates spontaneously (5×10^5 M^{-1} s^{-1} at pH 7 (Bielski, 1978)), the rate constant decreases markedly with increasing pH owing to involvement of H^+ in the reaction:

$$2O_2^- + 2H^+ \rightarrow H_2O_2.$$

Within the thylakoid membrane where O_2^- is produced by PSI, probably by reaction of FeS_X and $FeS_{A/B}$ with O_2, the rate of disproportionation is slowed by the deficiency of protons in this hydrophobic environment. Superoxide dismutases in the chloroplast accelerate the rate of disproportionation, and it has been estimated that they lower the steady-state concentration of O_2^- from 2.4×10^{-5} M to 3×10^{-9} M. There are three types of superoxide dismutase which differ in their prosthetic metal ligands, a CuZn form found in the chloroplast stroma of most higher plants, an Fe form found in the stroma of several plants, and an Mn form which is thylakoid-bound, in close proximity to PSI, where the O_2^- is generated (Asada et al., 1980).

(iii) Scavenging of hydrogen peroxide

Removal of hydrogen peroxide is essential both to prevent it from reacting with superoxide to form hydroxyl radicals, and because hydrogen peroxide directly inactivates several Calvin cycle enzymes involved in CO_2 fixation (Kaiser, 1976). Hydrogen peroxide is scavenged in chloroplasts by reduction to water in a process ultimately driven by PSI.

The immediate reductant for hydrogen peroxide is ascorbate, with monodehydroascorbate radical (MDA^*) as the oxidation product:

$$2\text{ ascorbate} + H_2O_2 \rightarrow 2\text{ MDA}^* + 2H_2O$$

This reaction is catalyzed by the heme-containing enzyme ascorbate peroxidase. The enzyme exists in two isoforms in roughly equal proportions – a soluble stromal form and a thylakoid-bound form localized in the stromal lamellae, where it is able to scavenge almost all the hydrogen peroxide generated by PSI (Miyake & Asada, 1992).

Although the ascorbate content of chloroplasts is over $10\,mM$, it would all be oxidized by hydrogen peroxide within $100\,s$ at the estimated rate of hydrogen peroxide formation. There are three chloroplastic pathways for regeneration of ascorbate from $MDA^{\bullet}$.

1. A stromal FAD-containing enzyme, monodehydroascorbate reductase, catalyzes the reduction of $MDA^{\bullet}$ by NAD(P)H (Hossain & Asada, 1985):

$$2MDA^{\bullet} + NADPH \rightarrow 2\ ascorbate + NADP^{+}$$

The enzyme has a much lower K_m for NADH than for NADPH and the former is produced from the latter by transhydrogenation.

2. Ferredoxin reduced by PSI can reduce $MDA^{\bullet}$ directly (Grace et al., 1995):

$$MDA^{\bullet} + reduced\ ferredoxin$$
$$\rightarrow ascorbate + oxidized\ ferredoxin$$

3. $MDA^{\bullet}$ radicals spontaneously disproportionate:

$$2MDA^{\bullet} \rightarrow ascorbate + dehydroascorbate$$

The dehydroascorbate generated by $MDA^{\bullet}$ disproportionation is reduced to ascorbate by reduced glutathione in a two-electron reduction catalyzed by the stromal thiol enzyme dehydroascorbate reductase (Hossain & Asada, 1984). The oxidized glutathione is reduced by NADPH generated by PSI in a reaction catalyzed by glutathione reductase (Halliwell & Foyer, 1978).

The thylakoid-associated scavenging systems for superoxide and hydrogen peroxide (thylakoid-bound superoxide dismutase, ascorbate peroxidase and ferredoxin-mediated $MDA^{\bullet}$ reduction) are assumed to represent the primary scavenging processes, suppressing the diffusion of reactive molecules into the stroma. The stromal system (stromal superoxide dismutase, ascorbate peroxidase, $MDA^{\bullet}$ reductase and dehydroascorbate reductase) then acts as a back-up mechanism to scavenge any reactive molecules which escape thylakoidal scavenging.

2.2.7 ATP synthesis

The ATP synthase couples exergonic proton flux across the thylakoid membrane from the lumen side of the membrane to the stroma, to the endergonic phosphorylation of ADP. The proton flux is driven by a difference in proton electrochemical potential between the aqueous phases on each side of the membrane ($\Delta\mu_{H^+}$). This proton electrochemical potential difference is created by proton pumping from the stromal side of the thylakoid membrane to the lumenal side coupled to light-driven electron transport.

(a)   Structure of ATP synthase

In common with similar enzymes from mitochondria and bacteria, the enzyme has a molecular weight of about $550\,kDa$ and consists of a globular hydrophilic unit, called CF_1 (for Chloroplast F_1), which carries the active sites for ATP synthesis, attached to a hydrophobic integral membrane protein unit, called CF_0, which functions as a transmembrane proton channel (reviewed in Boyer, 1993). The enzyme is located in the non-appressed thylakoid membranes, with the CF_1 unit projecting into the aqueous phase on the stromal side of the membrane. The CF_1 unit is an approximate sphere, $9–10\,nm$ in diameter, and is linked to CF_0 by a slender stalk about $4.5\,nm$ long. The CF_1 unit consists of five different polypeptides with the stoichiometry $\alpha_3\beta_3\gamma\delta\epsilon$ and has a molecular mass of $\sim370\,kDa$ (McCarty & Hammes, 1987) (Fig. 2.22). The CF_1 unit is very similar to mitochondrial F_1, the structure of which has recently been solved at $0.28\,nm$ resolution, dramatically extending earlier structural studies on this complex enzyme (Abrahams et al., 1994; Plate 7). The α and β subunits are weakly related in sequence and form a ring, with α and β subunits alternating, surrounding a central cavity. There are six nucleotide binding sites in the $\alpha_3\beta_3$ structure located at the interfaces between the subunits (Bruist & Hammes, 1981). Three of the sites are involved in ADP phosphorylation; each of these is located primarily on a β subunit, but with some contribution from the adjoining α subunit. At the other three sites, the bound nucleotides do not exchange during catalysis, and their function is unknown. Each of these non-catalytic sites is located primarily on an α subunit, with some contribution from the adjoining β subunit. The γ subunit has two extended α helical segments which form a coiled coil passing through the internal cavity in the centre of the $\alpha_3\beta_3$ ring, and forming part of the stalk protruding from the bottom of the ring. The ϵ and δ subunits are thought to be located in the stalk region, in association with the γ subunit

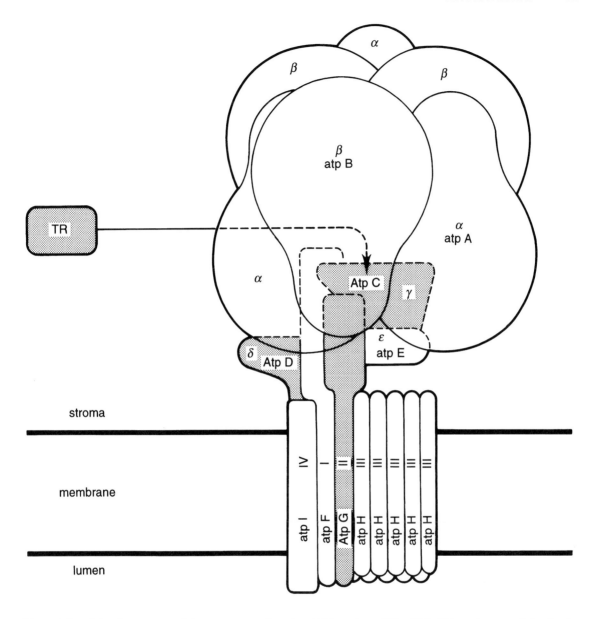

Figure 2.22   Subunit structure of the proton-translocating ATP synthase (CF_1–CF_0). Chloroplast-encoded subunits are shown in white and nuclear-encoded subunits in grey. Subunits are identified by their trivial name and/or encoding gene. The stroma-facing side of the complex is at the top of the diagram and the position of the membrane is indicated by the double line. TR is thioredoxin, which reduces the γ subunit as part of the activation mechanism of the enzyme. From a diagram provided by R. Herrmann, H. Pakrasi and R. Oelmüller.

and an $\alpha\beta$ pair, with ϵ closest to β (Dunn *et al.*, 1990).

The CF_0 unit contains four different polypeptides referred to as subunits I–IV (Grotjohann & Gräber, 1990). The generally accepted subunit stoichiometry is I II III$_{12}$ IV. Subunits III and IV are highly hydrophobic proteins. Subunits I and II have large hydrophilic domains that protrude

beyond the stromal surface of the membrane (Otto & Berzborn, 1989) and help to form the stalk region interacting with components of CF_1.

(b)   *Mechanism of ATP synthesis*

It is proposed that the three catalytic nucleotide binding sites are non-identical and switch between three different states in a cycle driven by proton flux through the CF_1–CF_0 complex (Boyer, 1993). The three states are referred to as open (O), loose (L) and tight (T) (Plate 8). Protons derived from the thylakoid lumen bind initially to low affinity sites accessible to the lumenal side of the membrane. The protons transfer to high affinity sites facing the stromal side of the membrane, causing conformational changes which are proposed to convert the O-site with a very low affinity for nucleotides into an L-site capable of loose binding of ADP and P_i. Concomitantly, the T-site with a bound ATP is converted into an O-site, releasing the bound nucleotide. Simultaneously the third, L, site with loosely bound ADP and P_i is converted into a T-site in which these molecules are tightly bound. At this site, the model predicts that formation of ATP occurs with an equilibrium constant near to unity, and the major energy requirement is involved in the conversion of the T-site into an O-site. Proton release to the stromal side of the membrane must occur to complete one cycle of the enzyme but it is not known what events this is associated with.

This three-site model has obvious attractions given the $\alpha_3\beta_3$ composition of the catalytic assembly, but only two catalytic sites have been identified in kinetic studies and a two-site mechanism has also been proposed (Berden et al., 1991). It has been suggested that the transition between the different catalytic states, implied by the cycle of binding changes, could be achieved by rotation of the $\alpha_3\beta_3$ complex relative to the $\gamma\delta\epsilon$ assembly. The asymmetry in the conformations of the $\alpha\beta$ pairs which both the two- and three-site models require is apparent in the high resolution structure of the bovine mitochondrial F_1-ATPase, as is an asymmetrical association of the $\gamma\delta\epsilon$ complex with the $\alpha_3\beta_3$ subassembly (Abrahams et al., 1994). The γ subunit extends into a central cavity within the $\alpha_3\beta_3$ subassembly and probably plays a key rôle in transmission of conformational changes associated with proton binding to the catalytic subunits.

The function of the CF_0 component and its coupling to CF_1 are not yet understood. When the CF_1 portion is removed, the CF_0 component acts as a highly specific proton channel. It is proposed that proton flux through CF_0 drives the rotation of the γ subunit and that the $\alpha_3\beta_3$ complex is fixed to the CF_0 unit by components in the stalk. Thus CF_0 acts as the motor rotating the γ subunit within the $\alpha_3\beta_3$ complex (Sabbert et al., 1996). The intact CF_1–CF_0 complex is leaky to protons in the absence of nucleotides.

Binding of an ADP and P_i at a potentially catalytic high affinity site blocks this 'proton slip' by blocking proton transfer through the complex and proton release but not proton intake (Groth & Junge, 1993).

(c) Energetics of ATP synthesis and regulation of ATP synthase activity

The electrochemical potential difference of protons between the stromal and lumenal sides of the thylakoid membrane generated by light-driven electron transport is quantitatively expressed in J mol^{-1} as $\Delta\mu_{H^+} = -2.303\,RT\,\Delta pH + F\Delta\psi$ where ΔpH is the pH difference across the membrane (acidic in the lumen) and $\Delta\psi$ is the membrane potential (positive side in the lumen) generated as a result of the transfer of positively charged protons. In thylakoids, counter-ion flow (Cl^- into the lumen and Mg^{2+} efflux into the stroma (Hind et al., 1974; Krause, 1977)) means that $\Delta\psi$ is negligible in the steady state, so that $\Delta\mu_{H^+}$ is largely made up of the ΔpH component. R is the gas constant (8.315 JK^{-1} mol^{-1}), T the absolute temperature and F is the Faraday constant (96 480 JV^{-1} mol^{-1}).

If the proton-translocating ATP synthase couples the transport of n protons across the thylakoid membrane to the synthesis of one molecule of ATP, net synthesis of ATP will occur providing that:

$$\Delta G_{\text{ATP synthesis}} < n\Delta\mu_{H^+}$$

where

$$\Delta G_{\text{ATP synthesis}} = \Delta G^0_{\text{ATP synthesis}}$$
$$+ RT \ln [ATP]/[ADP][P_i].$$

The equilibrium constant at neutral pH for the reaction $ADP + P_i \rightleftharpoons ATP + H_2O$ is $\sim10^{-6}$ M, which ensures that there will exist an energetic threshold value of $\Delta\mu_{H^+}$ for ATP synthesis in all biological environments at least as large as the $\Delta G^0_{\text{ATP synthesis}}$. $\Delta G^0_{\text{ATP synthesis}}$ is ~36 kJ mol^{-1} and the H$^+$/ATP ratio of the ATP synthase appears to be 3 (see Strotmann & Lohse, 1988; a value of 4 has been reported more recently), so that the minimum energetic threshold value of $\Delta\mu_{H^+}$ for net ATP formation is 12 kJ mol^{-1}, equivalent to a ΔpH across the thylakoid membrane of 2.1 units.

Evidence suggests that $\Delta G_{\text{ATP synthesis}}$ in the chloroplast stroma varies from $\sim$42 kJ mol^{-1} in the dark to $\sim$46 kJ mol^{-1} in steady illumination (Giersch *et al.*, 1980). To maintain $\Delta G_{\text{ATP synthesis}}$ at its dark level would require a threshold ΔpH of at least 2.5 units. In fact, because the ΔpH generating system is not active in darkness, ΔpH drops to zero after the cessation of illumination in a few seconds, so that a much higher $\Delta G_{\text{ATP synthesis}}$ is maintained in darkness than would be predicted from thermodynamic considerations. If the ATP synthase was active in darkness, because it catalyzes the *reversible* phosphorylation of ADP to ATP, ATP hydrolysis would be expected to proceed to equilibrium, at which $\Delta G_{\text{ATP synthesis}}$ would be zero. To prevent this wasteful hydrolysis of ATP from occurring, the ATP synthase must be regulated so that it is deactivated in darkness, and activated under illumination (reviewed in Ort & Oxborough, 1992).

The driving force for reversible conversion of ATP synthase from an inactive to a catalytically competent form is the light-generated $\Delta\mu_{\text{H}^+}$. The threshold value for this $\Delta\mu_{\text{H}^+}$ depends on the redox state of a disulfide bond between two cysteine residues in the γ subunit (Ketcham *et al.*, 1984; Mills & Mitchell, 1984; Fig. 2.23). In the dark-deactivated enzyme, the cysteine bridge is in the oxidized state and the threshold $\Delta\mu_{\text{H}^+}$ is $\sim$51 kJ mol^{-1} (corresponding to a ΔpH of $\sim$2.9 units). Since $\Delta G_{\text{ATP synthesis}}$ in the stroma is maintained at 42 kJ mol^{-1} in darkness, the threshold $\Delta\mu_{\text{H}^+}$ for activation of the ATP synthase ensures that it only becomes active on illumination when the $\Delta\mu_{\text{H}^+}$ is large enough to drive net synthesis of ATP. When chloroplasts are illuminated, a low molecular-weight cysteine-containing protein, thioredoxin, is reduced by electrons from PSI, via ferredoxin and ferredoxin–thioredoxin oxidoreductase (see Fig. 2.22). The thioredoxin reduces the disulfide bridge of the γ subunit of the ATP synthase. The threshold $\Delta\mu_{\text{H}^+}$ for activation of reduced ATP synthase is $\sim$45 kJ mol^{-1} (corresponding to a ΔpH of $\sim$2.6 units), i.e. lower than the value for the oxidized enzyme (Hangarter *et al.*, 1987). However, the kinetics of activation *in vivo* suggest that activation precedes reduction (Ort & Oxborough, 1992).

After a light to dark transition, loss of ATP synthase catalytic activity takes several minutes, whereas, at least in higher plants, reoxidation of the γ subunit (probably ultimately by molecular oxygen) can take well over an hour (Ortiz-Lopez *et al.*, 1991). It is assumed, therefore, that after the cessation of illumination, the ATP synthase

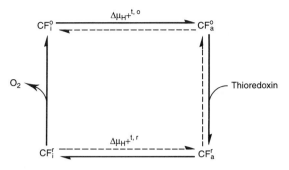

Figure 2.23   Model showing the proposed activation states of chloroplast coupling factor (CF). The activation state of CF is indicated by the subscripts: i, inactive; a, active. The superscripts indicate the redox state of CF: r, reduced; o, oxidized. $\Delta\mu_{\text{H}^+}^{\text{t,o}} > \Delta\mu_{\text{H}^+}^{\text{t,r}}$. The bold arrows indicate the transitions that are proposed to occur normally *in vivo*, and the dashed arrows represent transitions that can be induced *in vitro* or under special conditions *in vivo*. Based on a figure in Ort & Oxborough

catalyzes ATP hydrolysis until the $\Delta G_{\text{ATP synthesis}}$ falls to the threshold value of $\Delta\mu_{\text{H}^+}$ below which the reduced enzyme is inactivated. This value is $\sim$45 kJ mol^{-1}, close to the actual $\Delta G_{\text{ATP synthesis}}$ found in dark-adapted chloroplasts. The inactive enzyme then oxidizes slowly during extended periods of darkness.

Activation and subsequent reduction of the ATP synthase saturates at very low light intensity and these processes are therefore unlikely to be involved in the kinetic regulation of photophosphorylation during steady-state photosynthesis. Whereas reduction of the γ subunit of ATP synthase increases the efficiency of the enzyme in terms of ATP produced per electron transported (Hangarter *et al.*, 1987), it is not yet clear what are the benefits, if any, of the effect of oxidation on the enzyme's properties.

Activation of ATP synthase by $\Delta\mu_{\text{H}^+}$ is associated with the release of, on average, one molecule of tightly bound ADP per enzyme molecule (Strotmann *et al.*, 1979). It now seems certain that this molecule is bound at one of the catalytic sites on CF$_1$ and the conformational changes associated with activation convert the site from one of high binding affinity to a low affinity site. This change in affinity is reversed on deactivation, leading to the rebinding of ADP. The ϵ subunit is also thought to reorient itself relative to the other subunits of CF$_1$ during the $\Delta\mu_{\text{H}^+}$-driven conformational rearrangements associated with enzyme activation.

2.2.8 Overall stoichiometry of components of the thylakoidal photosynthetic apparatus and long-term responses to changes in light intensity

In addition to the regulatory phenomena described above, which enable the plant to respond to short-term changes in light intensity and spectral distribution, plants are also able to make long-term adjustments to their photosynthetic apparatus through synthesis and degradation of specific components (Anderson, 1986).

A typical stoichiometry of components in relatively high light grown spinach plants (Melis & Anderson, 1983) is shown in Table 2.5. Plants grown under low light have more photosynthetic pigments to maximize light harvesting. The ratio of electron transport components and ATP synthase to chlorophyll is much lower so that shade-grown plants have much lower rates of light-saturated photosynthesis than high-light grown plants. Specific responses to growth in low light are an increase in LHCII/PSII (at least twofold) and decreases in the PSII/PSI reaction center ratio (from ~1.6 in high light to ~1.2 in low light) and cytochrome b_6f/PSI (from ~1.0 to 0.7) (Leong & Anderson, 1983; Anderson et al., 1988; Anderson 1992). These changes are designed to ensure maximum rates of ATP and NADPH synthesis under the ambient light conditions. An increase in the PSII antenna size in low-light acclimated plants appears to be associated primarily with an increase in LHCIIb enriched in the Lhcb2 polypeptide (Larsson et al., 1987b, see Table 2.2). During acclimation of plants from low light to high light, this population of LHCIIb is selectively proteolytically degraded (Andersson et al., 1994).

Changes in the stoichiometry of components of the photosynthetic apparatus are regulated at the level of transcription and translation of the genes involved. It has been proposed that a key

mechanism linking changes in light intensity and quality to changes in gene expression is via a redox sensing mechanism like that involved in short-term regulation of energy distribution between the two photosystems (Allen, 1992a,b; 1993). In support of this hypothesis, it has recently been demonstrated that transcription of a nuclear-encoded lhc gene is regulated by the redox state of the PQ pool, initiated by activation of a chloroplast protein kinase, and requiring a signal transduction pathway from the chloroplast to the nucleus (Escoubas et al., 1995).

A mechanism for light regulation of translation of chloroplast mRNAs involving thioredoxin has recently been discovered in the green alga Chlamydomonas reinhardtii. Reduced thioredoxin activates translation of psbA mRNA encoding the D1 protein by reduction of a disulfide bridge on a translational activator protein (Danon & Mayfield, 1994). Thioredoxin is photoreduced by PSI and it is now apparent that as well as mediating light activation of the ATP synthase (see Fig. 2.22) and certain Calvin cycle enzymes (see section 2.3.1b(i)), it is also involved in light regulation of chloroplast translation. It can be anticipated that many more examples of regulation of photosynthetic genes through redox-sensing mechanisms will be reported in the next few years.

2.3 THE BENSON–CALVIN CYCLE

The Benson–Calvin cycle (or Reductive Pentose Phosphate Pathway) is the only pathway in plants which can catalyze the net fixation of carbon dioxide. The overall reaction can be described as the fixation of three molecules of CO_2 into a three-carbon sugar phosphate, triose-P, with the incorporation of one molecule of P_i. The reactions of the cycle occur in the chloroplast stroma and the cycle itself comprises 13 reactions, catalyzed by eleven enzymes (Fig. 2.24). It has three phases:

Table 2.5 Stoichiometry of thylakoidal components in chloroplasts from relatively high light grown spinach plants

	PSIIα	PSIIβ	PSI
Relative concentration of reaction centers	1.43	0.48	1.0
Chl/reaction center	234	100	209
Relative proportion of total Chl serving each photosystem	57	8	35
Chl$a$/b	1.73	5.6	6.0

Data from Melis and Anderson (1983).

1. *carboxylation*: the addition of CO_2 to ribulose-1,5-P_2 (RuBP), with the formation of two molecules of glycerate-3-P, catalyzed by ribulose 1,5-bisphosphate carboxylase-oxygenase (Rubisco).

$$RuBP + CO_2 + H_2O \rightarrow 2(\text{glycerate-3-P}) + 2H^+$$

2. *reduction* of the two molecules of glycerate-3-P deriving from RuBP to triose-P at the expense of 2ATP and 2NADPH. This reaction is catalyzed by glycerate-3-P kinase and glyceraldehyde-P dehydrogenase. Five of the six carbon atoms in glycerate-3-P must then be recycled to regenerate the acceptor molecule, RuBP.

3. *regeneration* of the primary acceptor, RuBP, from triose-P, by which five C_3 molecules are rearranged to three C_5 molecules in the 'sugar phosphate shuffle'. Each molecule of ribose-5-P must be converted into ribulose-5-P which is, in turn, converted into RuBP at the expense of a molecule of ATP. Each molecule of CO_2 fixed

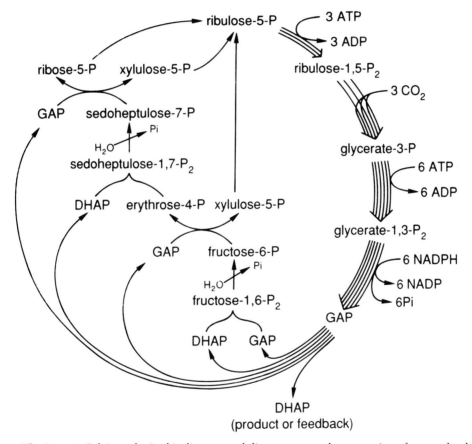

Figure 2.24 The Benson–Calvin cycle. In this diagram each line represents the conversion of one molecule of each metabolite. The cycle can be divided into three phases. The first phase is carboxylation, catalysed by ribulose-1,5-bisphosphate oxygenase (Rubisco). The second is the reductive phase, in which glycerate-3-P is reduced to triose-P (glyceraldehyde-3-P (GAP) and dihydroxyacetone-P (DHAP)) by the actions of glycerate-3-P kinase and NADP-dependent glyceraldehyde-P dehydrogenase. Triose-P isomerase interconverts GAP and DHAP. The third phase is regeneration of the acceptor, ribulose-1,5-P_2 in the 'sugar phosphate shuffle', in which five C_3 units are converted into three C_5 units. One of the six molecules of triose-P, deriving from three CO_2 molecules incorporated by Rubisco, is available for product synthesis or for regeneration. Aldolase converts both DHAP and GAP into fructose-1,6-P_2, and DHAP and erythrose-4-P to the seven carbon sugar, sedoheptulose-1,7-P_2. Fructose and sedoheptulose bisphosphatases convert the bisphosphates into fructose-6-P and sedoheptulose-7-P, and transketolase catalyzes the reactions of these with GAP to generate pentose-P and erythrose-4-P. Interconversion of pentose-P is then catalyzed by ribose-P isomerase (ribose-5-P to ribulose-5-P) and ribulose-5-P 3-epimerase (xylulose-P to ribulose-5-P). Finally, ribulose-5-P kinase catalyzes the ATP-dependent conversion of ribulose-5-P into ribulose-1,5-P_2. Glyceraldehyde-P dehydrogenase, the bisphosphatases and product synthesis recycle the P_i required for continued ATP synthesis.

in the Calvin cycle therefore requires 3ATP and 2NADPH to be provided by photosynthetic electron transport.

Regeneration and autocatalysis are vital properties of the Calvin cycle. Autocatalysis means that the product, triose-P, can be recycled to generate more substrate. If the cycle turns over five times, the amount of the primary acceptor, RuBP, can be doubled. Alternatively triose-P can be utilized in the synthesis of starch within the chloroplast or can be exported via the phosphate translocator, in exchange for P_i from the cytosol. In the cytosol it can be used for sucrose synthesis (Chapter 4) or can be converted into glycerate-3-P, thereby generating ATP and reductant in the cytosol. During steady-state photosynthesis, one-sixth of the triose-P generated from CO_2 is available for product synthesis.

2.3.1 Regulation of the Benson–Calvin cycle

Three factors determine the rate at which the Benson–Calvin cycle turns over. These are (a) the rate at which carbon is withdrawn for the synthesis of carbohydrates and other compounds, (b) light, mediated by electron transport and its interactions with events in the stroma, and (c) in the case of Rubisco, CO_2.

(a) Regulation by transport

The chloroplast envelope forms a barrier to the passage of the majority of proteins and metabolites. The wide range of metabolic activities within the chloroplast, coupled with the need for import of most of its proteins, the majority of which are encoded by the nucleus, means that it possesses a range of translocators which mediate the passage of a diverse range of compounds, including phosphorylated intermediates, sugars, adenylates, dicarboxylic acids and amino acids (Heber & Heldt, 1981; Flügge & Heldt, 1991), proteins (Keegstra et al., 1989) and lipids.

As far as CO_2 fixation is concerned, the chloroplast is a P_i-importing, triose-P exporting organelle. These fluxes of carbon and P_i are mediated by the phosphate translocator, which catalyzes the obligatory counter-exchange of triose-P for P_i across the chloroplast envelope (Flügge & Heldt, 1991). Like other metabolite transporters, the phosphate translocator is situated on the inner membrane of the double chloroplast envelope, where it constitutes some 15% of the total protein. In C_3 plants it is a dimer of two identical subunits (M_r 36 kDa) which transports three carbon compounds in which phosphate is attached to the end of the chain (e.g. glycerate-3-P, glyceraldehyde-3-P or dihydroxyacetone-P). Compounds such as phosphoenolpyruvate (PEP) and glycerate-2-P are not transported (although the situation is different in plastids from C_4 plants or from non-photosynthetic cells). The phosphate translocator does not transport other intermediates of the Benson–Calvin cycle, such as hexose-P, pentose-P or bisphosphates. Compounds are transported in the form carrying a double negative charge, which means that glycerate-3-P is retained in the stroma during photosynthesis because, at the pH of the illuminated stroma, it carries a triple negative charge. Export of triose-P and conversion into glycerate-3-P in the cytosol can also serve to export ATP and reducing equivalents from the chloroplast. The phosphate translocator may also serve in the export of the products of starch degradation, although there are also specific carriers for glucose and maltose (see Heldt & Flügge, 1992). There is also evidence for the transport of pyrophosphate into chloroplasts via a specific translocator. However, this is comparatively slow (e.g. relative to the rate of PP_i generation by sucrose synthesis) and may simply serve to replenish the stromal P_i pool (Lunn & Douce, 1993).

The rate at which carbon is withdrawn from the Benson–Calvin cycle is controlled by the rate at which starch synthesis and sucrose synthesis operate (Chapter 4). The link between events in the chloroplast and sucrose synthesis in the cytosol is provided by the phosphate translocator. In its simplest terms, if sucrose synthesis in the cytosol runs too slowly (for example, in the short term the rate of sucrose synthesis is particularly sensitive to sudden reductions in temperature), this will slow the release of P_i in the cytosol and this will restrict export of triose-P from the chloroplast and lead to a decline of P_i in the stroma, which will, in turn, limit photophosphorylation and hence reduce the rate of photosynthesis and the production of triose-P. The ensuing decline in ATP will lead to the accumulation of glycerate-3-P, and the rise in glycerate-3-P and fall in P_i will activate ADPglucose pyrophosphorylase (Preiss, 1982) and hence the synthesis of starch. There will also be feedback on electron transport (Pammenter et al., 1993).

(b) Regulation by light

Although, in principle, turnover of the Benson–Calvin cycle could be governed entirely by the availability of ATP and NADPH generated by electron transport, in practice there is an extremely complex regulatory network connecting electron transport and CO_2 fixation. Enzymes of carbon assimilation are linked to electron transport at three levels: (1) redox regulation by the thioredoxin system; (2) modulation of their activities by ATP and NADPH; and (3) electron-transport-driven changes in stromal pH and Mg^{2+}. For example, the darkened stroma has a pH of around 7 and a Mg^{2+} concentration of 1–3 mM, whereas the illuminated stroma has a pH around 8 and a Mg^{2+} concentration of 3–6 mM (Leegood et al., 1985). These light-induced changes provide an environment which is closer to the optimum for the operation of the enzymes of the Benson–Calvin cycle. Light-induced changes in Ca^{2+} concentration may also play a role in regulation of chloroplastic and cytosolic processes associated with photosynthesis (Miller & Sanders, 1987; Kreimer et al., 1988).

In vivo studies of the Benson–Calvin cycle have revealed that the reactions catalyzed by Rubisco, the bisphosphatases and ribulose-5-P kinase are displaced from equilibrium in Chlorella (Bassham & Krause, 1969) and many studies of the behavior of the substrates of Rubisco and of the fructose- and sedoheptulose-bisphosphatases suggest that these enzymes are regulatory. In vitro studies have shown that the above enzymes, as well as glyceraldehyde-P dehydrogenase, are regulated by pH, Mg^{2+} and metabolites (see Leegood, 1990 for review) and are also regulated either by the thioredoxin system or, in the case of Rubisco, by carbamylation by CO_2 and by a naturally occurring inhibitor.

(i)　The thioredoxin system

The thioredoxin system provides a crucial link between the activities of electron transport and carbon assimilation. The activities of a number of enzymes in the chloroplast are linked to the availability of the photosynthetically generated reductant, thioredoxin (Buchanan, 1980). Thioredoxin ($E_m - 300$ mV) reduction by ferredoxin ($E_m - 420$ mV) is catalyzed by ferredoxin–thioredoxin reductase, an iron–sulfur protein with a reducible disulfide bridge. Reduced thioredoxin is able to reduce disulfide bridges on the target enzymes, modulating their activities, a process which can be mimicked in vitro by incubation with the reducing agent, dithiothreitol. There are two types of thioredoxin. Thioredoxin f preferentially reduces fructose bisphosphatase and thioredoxin m preferentially reduces NADP–malate dehydro-

genase. Both have a molecular mass of 12 kDa and possess a single reducible disulfide bridge per monomer. Four enzymes of the Benson–Calvin cycle are regulated by thioredoxins: fructose bisphosphatase and sedoheptulose bisphosphatase (both $E_m - 310$ mV), ribulose-5-P kinase ($E_m - 270$ mV) and glyceraldehyde-P dehydrogenase. Other chloroplastic enzymes regulated by thioredoxin include NADP–malate dehydrogenase (in both C_3 and C_4 plants ($E_m - 330$ mV)), the chloroplast ATP synthase ($E_m - 270$ mV) (Ort & Oxborough, 1992) (see p. 83), phenylalanine ammonia lyase, and glucose-6-P dehydrogenase ($E_m - 300$ mV), which is inactivated by reduction in the light. Oxidation of these enzymes in darkness can be achieved by molecular O_2 in vivo, but the precise route of such oxidation has not been determined. Although this and other mechanisms of light modulation can be viewed simply as 'on-off' switches, it is much more likely that they operate in the fine adjustment of enzyme activity to variations in light intensity or temperature. They are also integrated with a range of regulatory properties of these enzymes because light activation can change affinities for substrates as well as V_{max}. The fact that enzymes have different midpoint redox potentials means that each enzyme will have a different sensitivity to light, since a difference in E_m of 30 mV will represent an equilibrium constant of 10 for a two-electron transfer. Thus the ATP synthase could remain substantially active while the activities of the bisphosphatases change markedly (Kramer et al., 1990). It also explains why enzymes such as ribulose-5-P kinase are fully activated at very low light intensities. Such differences may be particularly important in chilling sensitive plants, in which the operation of the thioredoxin system may be impaired at low temperatures (Sassenrath et al., 1990).

Regulation of the reduction of glycerate-3-P

$$\text{glycerate-3-P} + \text{ATP} \longrightarrow \text{glycerate-1,3-P}_2 + \text{ADP}$$
glycerate-3-P kinase

$$\text{glycerate-1,3-P}_2 + \text{NADPH} \xrightarrow{\hspace{3cm}}$$
glyceraldehyde-P dehydrogenase

$$\text{glyceraldehyde-3-P} + \text{NADP}^+ + \text{H}^+$$

Evidence that the reactions catalyzed by glycerate-3-P kinase and glyceraldehyde-P dehydrogenase are close to equilibrium in vivo remains controversial (Heber et al., 1986; Heineke et al., 1991) as estimations of the mass-action ratio in vivo may be complicated by binding of pyridine nucleotides.

Glyceraldehyde-P dehydrogenase is activated by thioredoxin and also undergoes reversible inter-conversion between different oligomeric forms which differ in their pryridine nucleotide specificity. The enzyme shows an increase in NADP-, but not NAD-linked, activity upon light activation. Reductive activation of a 600 kDa form by the thioredoxin system is an essential prerequisite for subsequent dissociation to the smallest active form, which is a tetramer of two different subunits (A_2B_2; 150 kDa). This dissociation process has a low K_a for glycerate 1,3-bisphosphate (1 μM) and the activated enzyme has a higher affinity for glycerate-1,3-bisphosphate (see Baalman et al., 1994). The concentration of glycerate 1,3-bisphosphate, in turn, depends on the amounts of ATP and glycerate-3-P in the chloroplast stroma. The 600 kDa form predominates in darkened leaves but it has a low affinity for NADPH and a high K_a for dissociation (to an oxidized 150 kDa form) by glycerate 1,3-bisphosphate (K_a 20 μM). An intermediate 300 kDa form also exists, particularly during darkening (Baalman et al., 1994). P_i inhibits the catalytic activity of the activated enzyme (Baalman et al., 1994). Baalman et al. (1994) suggest that the 600 kDa enzyme will be completely inactive in the darkened stroma.

Regulation of the bisphosphatases

$$\text{fructose-1,6-}P_2 + H_2O \longrightarrow \text{fructose-6-P} + P_i$$
$$\text{\textit{fructose bisphosphatase}}$$

$$\text{sedoheptulose-1,7-}P_2 + H_2O \longrightarrow$$
$$\text{\textit{sedoheptulose bisphosphatase}}$$

$$\text{sedoheptulose-7-P} + P_i$$

Fructose and sedoheptulose bisphosphatase show appreciable similarity of primary structure, particularly in the residues comprising the active site (Raines et al., 1992). Both enzymes are regulated by the thioredoxin system.

The chloroplastic fructose bisphosphatase contains a unique sequence of 12 amino acids that contains three cysteine residues (Raines et al., 1988). It is also unlike the cytosolic enzyme in that it is insensitive to inhibition by AMP. Two disulfide bridges of the tetrameric enzyme are cleaved upon activation. Reduction of the enzyme leads to no change in V_{max}, but changes in K_m for the substrate, the $[FBP^{4-} \cdot Mg^{2+}]^{2-}$ complex, from 130 to 6 μM. In the case of sedoheptulose bisphosphatase the K_m for the $[SBP^{4-} \cdot Mg^{2+}]^{2-}$ complex falls from 180 to 50 μM upon activation. The nature of the substrate means that activity of both enzymes is strongly dependent on both the pH and Mg^{2+} concentration (see Leegood, 1990). Both enzymes are sensitive to inhibition by phosphate and glycerate, and fructose bisphosphatase is inhibited by fructose-6-P. Reduction of both enzymes by thioredoxin is promoted by the presence of substrate.

Regulation of ribulose-5-P kinase

$$\text{ribulose-5-P} + ATP \longrightarrow \text{ribulose-1,5-}P_2 + ADP$$
$$\text{\textit{ribulose-5-P kinase}}$$

Thioredoxin reduces two disulfide bridges per dimer of ribulose-5-P kinase. Reduction leads to a large increase in V_{max} but no changes in the affinity of the enzyme for ribulose-5-P or ATP. Light activation is extremely effective: the oxidized enzyme at pH 6.8 has an activity only 2% of that of the reduced enzyme at pH 7.8 (the pH values of the darkened and illuminated stroma, respectively). The activity of the enzyme is inhibited by gluconate-6-P, RuBP and P_i. Inhibition by glycerate-3-P is pH-dependent because only glycerate-3-P^{2-} inhibits the enzyme. At the pH of the illuminated stroma (pH 7.8), most glycerate-3-P is in the non-inhibitory glycerate^{3-} form. ADP inhibits the oxidized form of the enzyme.

(ii)   Regulation of Rubisco

In the 1970s it was discovered that, in vitro, Rubisco was activated by preincubation with CO_2 and Mg^{2+} to form an active carbamylated enzyme. This carbamylation occurs on Lys-201, and the CO_2 molecule concerned is different from the CO_2 molecule involved in catalysis.

$$\text{E-lys} + CO_2 + Mg^{2+} \rightarrow E \cdot CO_2 \cdot Mg^{2+}$$

The carbamylation state of Rubisco was also shown to change in vivo, in isolated chloroplasts or in leaves. However, carbamylation in vitro occurred only in the presence of millimolar concentrations of CO_2, whereas in vivo the concentration of CO_2 in the leaf cell would only be about 10 μM. How, then, could Rubisco possibly be carbamylated in vivo? The answer came from a mutant of Arabidopsis thaliana isolated in the early 1980s (Portis, 1992). This mutant was able to survive when grown in high CO_2, but not in air, because it was unable to carbamylate its Rubisco. This mutation was shown to be due to the absence of an enzyme, Rubisco activase, which enhances the carbamylation of Rubisco in the presence of physiological concentrations of CO_2 ($K_a(CO_2) = 4\,\mu M$). Binding of RuBP to the active site prevents carbamylation. The activase appears to be involved in removing bound RuBP from the active site to allow carbamylation (Wang & Portis, 1992).

A second enigma concerning the regulation of Rubisco emerged in the early 1980s. It was found that in some plants, such as soybean, Rubisco in extracts from darkened leaves could not be activated by CO_2 and Mg^{2+}, whereas Rubisco could readily be activated in leaf extracts made from plants in the middle of the day (Vu et al., 1983). It was subsequently shown that this was due to the presence of a tight-binding inhibitor, 2-carboxyarabinitol-1-P (CA1P) (K_d 32 nM), which is an analog of the transition state intermediate, 3-keto-2-carboxyarabinitol bisphosphate (Fig. 2.25). Rubisco activase may also be involved in removing bound CA1P from the enzyme.

CA1P appears to be present in all plants, but it is probably only important in regulation of Rubisco when present in the large amounts found in legumes (soybean, Phaseolus vulgaris), tomato and sunflower. It is, nevertheless, present in small amounts in leaves of plants such as spinach, wheat or maize (Moore et al., 1991). CA1P is gradually degraded with increasing photon flux density (PFD) and is rapidly degraded on exposure to saturating light (Vu et al., 1983; Kobza & Seemann, 1989). At present, comparatively little is known about the mode of synthesis and degradation of this inhibitor. However, carboxyarabinitol has been found to be present in large amounts in leaves of a range of species (Moore et al., 1992), and carboxyarabinitol fed to leaves of plants highly active in CA1P metabolism is readily converted into CA1P in the dark. CA1P is then degraded to carboxyarabinitol in the light, suggesting the operation of a metabolic cycle

between the two (Moore & Seemann, 1992). A CA1P phosphatase activity has been purified (Holbrook et al., 1989).

(c) Control of the Benson–Calvin cycle

The above are just some mechanisms of control of the Benson–Calvin cycle, but it will be obvious that the properties of the system are so complex that they defy simple analysis. Despite study over some 40 years, even our knowledge of the regulation of parts of the Benson–Calvin cycle is still weak. Regulation of the sequence between fructose-6-P and sedoheptulose-7-P and pentose-P, involving the two reactions catalyzed by transketolase, and by xylulose-5-P epimerase and ribose-5-P isomerase, has been poorly characterized in vivo, partly because the substrates are difficult to measure and partly because both enzymes and substrates are shared with other pathways. However, metabolite effectors have been identified for these enzymes (Leegood et al., 1985).

Even with a knowledge of regulatory properties and how concentrations of regulators change, we cannot tell how important an enzyme is in the control of photosynthesis. Such questions can be answered by flux control analysis, first developed by Kacser & Burns (1973). The essence of such analysis is to ask how much the flux changes when the amount of an enzyme is changed by a known amount. The change in the amount of an enzyme can be achieved by the use of specific inhibitors, heterozygous mutants, or transgenic plants which have undergone 'sense' or 'antisense' transformation. Recent work in this area has utilized plants with decreased Rubisco (Fig 2.26; Quick et al., 1991; Stitt et al., 1991; Hudson et al., 1992; Stitt & Schulze 1994), stromal FBPase

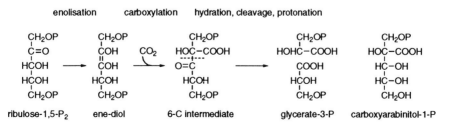

Figure 2.25 Simplified reaction mechanism of Rubisco. Enolization of ribulose-1,5-P_2 yields a 2,3-enediol intermediate which is then carboxylated to yield a six-carbon intermediate, 3-keto-2-carboxyarabinitol bisphosphate. This intermediate undergoes hydration at carbon 3, and then cleavage (---) to give one molecule of glycerate-3-P ($CH_2OP.CHOH.CO^{2-}$) from the lower portion, followed by protonation of the upper portion to yield a second molecule of glycerate-3-P. The similarity between the six-carbon intermediate and the naturally occurring inhibitor, carboxyarabinitol-1-P is also shown.

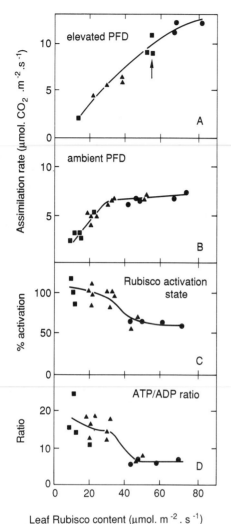

Figure 2.26 Control of photosynthesis by Rubisco in transgenic tobacco plants with altered amounts of Rubisco. (a) Photosynthesis at $1050 \mu mol \, m^{-2} \, s^{-1}$, which is above the PFD at which plants were grown ($340 \mu mol \, m^{-2} \, s^{-1}$). In this situation, Rubisco exerts strong control over photosynthesis, with a large slope (i.e. a large control coefficient), reaching 1 at lower contents of Rubisco. (b,c,d) Photosynthesis, Rubisco activation state and the ATP/ADP ratio measured at the growth PFD. A substantial reduction in Rubisco is accompanied by no reduction in photosynthesis. In this region the ATP/ADP ratio, which may influence the activity of Rubisco via Rubisco activase, increases in parallel with the activation state of Rubisco. Below about 50% of the wild-type activity, Rubisco is fully activated and again exerts strong control over the rate of CO_2 fixation. Redrawn from Quick et al. (1991) and Stitt et al. (1991). The different symbols represent plants from different transgenic lines. The arrow represents the wild-type average.

(Koßmann et al., 1994), ribulose-5-P kinase (Paul et al., 1995), glyceraldehyde-P dehydrogenase (Price et al., 1995), aldolase (Sonnewald et al., 1994), the phosphate translocator (Heineke et al., 1994), carbonic anhydrase (Price et al., 1994) and, in photorespiration, decreased glutamine synthetase, glutamate synthase and P-glycolate phosphatase (see Leegood et al., 1995).

2.4 PHOTORESPIRATION

2.4.1 The oxygenase activity of Rubisco

Rubisco is a bifunctional enzyme. It catalyzes both the carboxylation and the oxygenation of RuBP. Oxygenation of RuBP leads to the production of one molecule of glycerate-3-P and one of glycolate-2-P.

$$RuBP + O_2 \rightarrow glycerate\text{-}3\text{-}P + glycolate\text{-}2\text{-}P + 2H^+$$

Evolution has led to improvements in the specificity factor (Ω) of Rubisco. Ω is a measure of the relative specificity for CO_2 and O_2. $\Omega = V_c K_o / V_o K_c$, where V_c, V_o are V_{max} values for carboxylation and oxygenation and K_c, K_o are K_m values for CO_2 and O_2, and varies from about 10 in the bacteria, to 50 in the cyanobacteria, and about 80 in higher plants (Jordan & Ogren, 1981). However, oxygenation has clearly not been eliminated during the course of evolution. Oxygenation occurs by direct attack of oxygen on the 2,3-enediol intermediate (Fig. 2.25) and is viewed as an inevitable consequence of the reaction mechanism of Rubisco.

The K_m (CO_2) of Rubisco is about $10 \mu M$, which is about the same as the concentration of CO_2 dissolved in water at 20°C, but the CO_2 concentration would be rather less in the mesophyll cells of a leaf in air (ca. $5 \mu M$; von Caemmerer & Evans, 1991). The $K_m(O_2)$ is about $535 \mu M$, which is about double the O_2 concentration in a leaf at 20°C. Under these conditions, Rubisco catalyzes both carboxylation and oxygenation of RuBP. The two substrates are competitive, so that raising the CO_2 concentration will inhibit oxygenation and *vice versa*.

(a) *Influence of temperature on the rates of oxygenation and carboxylation*

The principal factor influencing the rate of oxygenation, apart from the ambient concentrations of O_2 and CO_2, is temperature. High temperatures promote oxygenation, and hence photorespiration, in two ways. First, the solubility of CO_2 in water declines more rapidly than that of O_2 as the temperature is increased. For example, at 10°C, the ratio of the solubilities of O_2 to CO_2 in water is 20, whereas at 40°C it is 28. Second, the specificity factor (Ω) of Rubisco decreases with increasing temperature in the range 7°C to 35°C. This is because the reaction of the 2,3-enediol intermediate (Fig. 2.25) with O_2 has a higher free energy of activation than the reaction with CO_2. This means that oxygenation is more sensitive to temperature and increases faster than carboxylation as the temperature rises (Chen & Spreitzer, 1992).

2.4.2 The photorespiratory pathway

Glycolate-2-P produced by Rubisco when it oxygenates RuBP cannot be utilized within the Calvin cycle. Instead, it is salvaged, albeit inefficiently, in the photorespiratory pathway. This pathway involves three subcellular compartments, the chloroplasts, peroxisomes and mitochondria (Fig. 2.27). The key features of this pathway are the conversion of the two-carbon molecule, glycolate-2-P, to glycine and decarboxylation of two molecules of glycine to serine, CO_2 and NH_3. The three-carbon molecule, serine, is then converted into glycerate-3-P, which re-enters the Benson–Calvin cycle. CO_2 release results in the release of one-quarter of the carbon in glycolate (hence the term photo*respiration*) and decreases the efficiency of photosynthesis. The principal factor influencing the rate of photorespiration is temperature. In air (350 ppm CO_2, 21% O_2) the

ratio of the rate of photorespiration to photosynthesis is around 0.1 at 10°C, rising to about 0.3 at 40°C, but low intercellular concentrations of CO_2, which may occur, for example, under water stress, can result in ratios of 0.6 or higher at high temperatures.

The photorespiratory pathway emphasizes the close integration of carbon and nitrogen metabolism in a leaf, because NH_3 is also released during photorespiration. Since NH_3 is a more valuable resource than CO_2, NH_3 must be efficiently refixed by glutamine synthetase (GS) and glutamine synthase (GOGAT) in the chloroplast. The rate of release of NH_3 by photorespiration is so high that refixation by GS and GOGAT occurs at rates up to ten times the rate of primary NH_3 assimilation (e.g. NH_3 deriving from nitrate reduction).

(a) Enzymes of the photorespiratory pathway

Within the chloroplasts, glycolate-2-P generated by Rubisco is dephosphorylated, a reaction which is catalyzed by glycolate-2-P phosphatase. This recycles P_i within the chloroplast and prevents build up of glycolate-2-P, which is a potent inhibitor of triose-P isomerase.

Glycolate exported from the chloroplast enters the peroxisome and is oxidized by glycolate oxidase to glyoxylate. The H_2O_2 generated is then decomposed by catalase, which is abundant within the peroxisomes. Glyoxylate is transaminated by two enzymes, serine : glyoxylate aminotransferase and glutamate : glyoxylate aminotransferase, to form glycine. Glycine then moves

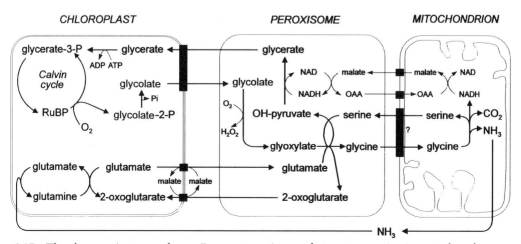

Figure 2.27 The photorespiratory pathway. For most reactions, cofactors, transaminations etc. have been omitted for clarity. OAA, oxaloacetic acid. Translocators are indicated by solid blocks.

to the mitochondria where conversion into serine occurs. Glycine decarboxylase is extremely abundant in mitochondria isolated from leaves, accounting for about one-third of soluble protein (Oliver et al., 1990). It has four different subunits (P, H, T and L) which catalyze glycine decarboxylation, transferring a methylene (C1) group to polyglutamyl tetrahydrofolate, with the generation of NH_3, CO_2 and NADH. Serine hydroxymethyltransferase then transfers this C1 group to glycine to generate serine. Serine returns to the peroxisomes for conversion into hydroxypyruvate by serine : glyoxylate aminotransferase, a reaction whose equilibrium lies in favor of products. Reduction to glycerate by NADH-dependent hydroxypyruvate reductase occurs in the peroxisomes. The role of an NADPH-dependent cytosolic hydroxypyruvate reductase has not been elucidated. Glycerate returns to the chloroplast where it is phosphorylated to glycerate-3-P by glycerate kinase, an enzyme that catalyzes a reaction that lies in favor of glycerate formation ($K_{eq} = 300$) and that appears to be regulated solely by substrate availability (Kleczkowski et al., 1985).

Mutants of barley and *Arabidopsis thaliana* have been isolated which lack particular enzymes of the photorespiratory pathway. The screening for these mutants has been done by growing mutagenized plants in 2% CO_2 and then transferring them to air. Plants lacking enzymes of the photorespiratory pathway show visible symptoms and die rapidly. However, rescue of these plants by return to elevated CO_2 has allowed identification of a range of mutants. These include different plants lacking glycolate-2-P phosphatase, catalase, glycine decarboxylase, serine hydroxymethyltransferase and serine:glyoxylate aminotransferase as well as the enzymes involved in NH_3 assimilation (glutamine synthetase and glutamate synthase) and the chloroplast dicarboxylate transporter. All these mutants indicate that, since growth in high CO_2 in such plants is normal, the photorespiratory pathway is not required for synthesis of intermediates for biosynthesis.

Transport processes in the photorespiratory pathway are complex (Fig. 2.27). On the chloroplast envelope there is a glycolate/glycerate carrier (Howitz & McCarty, 1991) which can also transport glyoxylate and lactate. Glycerate and glycolate can either be counter-exchanged or transported in symport with H^+ or in antiport with OH^-. It should be emphasized that operation of the photorespiratory pathway requires that two glycolate molecules are exported from the chloroplast for each molecule of glycerate which is imported.

During photorespiration, the chloroplast takes up 2-oxoglutarate and releases glutamate (Fig. 2.27). In spinach chloroplasts, dicarboxylate transport involves the exchange of dicarboxylic acids such as malate, succinate, 2-oxoglutarate, aspartate and glutamate. This appears to be catalyzed by two separate processes, involving a 2-oxoglutarate translocator, which exchanges 2-oxoglutarate for succinate, fumarate and malate, but not glutamate, and a general dicarboxylate translocator, which can exchange glutamate for malate (Fig. 2.27; Woo et al., 1987; Yu & Woo, 1992). There is also evidence for a separate glutamine translocator, which also translocates glutamate, but no other dicarboxylic acids.

The single peroxisomal membrane appears to have no specific transport properties and it has been suggested that, like animal peroxisomes, it may contain simple channels. The peroxisomal enzymes appear to be arranged in a matrix which favors metabolite channeling, preventing the release of intermediates, such as glyoxylate and hydroxypyruvate, from the peroxisome, but allowing access of glycolate, malate and serine from the cytosol (Heldt & Flügge, 1992).

There is some evidence for carrier-mediated transport of glycine and serine into the mitochondria, and ammonia may have its own transporter (Ninnemann et al., 1994). Although cytosolic and chloroplastic forms of glutamine synthetase have been identified in leaves, it now seems likely that all the ammonia is reassimilated via chloroplastic glutamine synthetase, because the cytosolic isoform of this enzyme extracted from leaves appears to be confined to the phloem companion cells (e.g. Edwards et al., 1990).

(b) *Interaction between photorespiration and other processes*

Numerous shuttles exist to support transamination and the supply or export of reductant generated or consumed in the photorespiratory pathway. One of the most important questions is the source of NADH for hydroxypyruvate reduction. The fate of the NADH generated during glycine decarboxylation *in vivo* is not yet clear. In isolated mitochondria, glycine decarboxylation is coupled to ATP synthesis (Wiskich et al., 1990; Krömer & Heldt, 1991). There is also evidence that glycine decarboxylation increases cytosolic ATP/ADP ratios in protoplasts (Gardeström and Wigge, 1988). However, it is doubtful whether the respiratory chain would have sufficient capacity to oxidize all the NADH at high photorespiratory rates. An alternative fate *in vivo* is that NADH is oxidized by an oxaloacetate/malate shuttle and malate is then exported and oxidized within the peroxisomes, which contain malate dehydrogenase. NADH generated in this reaction would readily be oxidized, since the equilibrium for the reaction catalyzed by hydroxypyruvate reductase lies far in favor of glycerate formation.

Comparatively few mechanisms have been elucidated for feedback interactions between the photorespiratory pathway and the Benson–Calvin cycle. This is perhaps because, once glycolate has been formed, there is no alternative but to recycle it to glycerate. However, it has been shown that glycerate inhibits the stromal FBPase and SBPase (Schimkat *et al.*, 1990) and that glyoxylate regulates the activation state of Rubisco (Campbell & Ogren, 1990), so that there may be circumstances in which photorespiration feeds back on the capacity of the Benson–Calvin cycle.

2.5 C$_4$ PHOTOSYNTHESIS

The C$_4$ pathway is an adjunct to the Benson–Calvin cycle. The C$_4$ pathway occurs only in conjunction with structural modifications which allow it to operate as a CO$_2$ concentrating mechanism. In all C$_4$ plants, phosphoenolpyruvate (PEP) is carboxylated to C$_4$ acids and these are the first products of photosynthesis, in contrast to glycerate-3-P in C$_3$ plants. These C$_4$ acids are formed in one compartment, the mesophyll cells, and are then transferred, by diffusion, to a relatively gas-tight compartment, the bundle-sheath, where they are decarboxylated (Fig. 2.28). This generates a high concentration of CO$_2$ within the bundle-sheath. Rubisco and the majority of the enzymes of the Benson–Calvin cycle are confined to the bundle-sheath. The CO$_2$ concentration in the bundle-sheath is sufficient to suppress the oxygenase activity of Rubisco. A C$_3$ compound returns to the mesophyll from the bundle-sheath. In a plant such as maize, the net reaction catalyzed by the C$_4$ pathway is to transfer CO$_2$ from the mesophyll to the bundle-sheath at the expense of 2ATP per CO$_2$ transferred. It is, therefore, an ATP-driven CO$_2$ pump. In air it achieves concentrations of inorganic carbon (CO$_2$ + HCO$_3^-$ in the bundle-sheath of the order of 150 μM, equivalent to a CO$_2$ concentration of about 70 μM (Jenkins *et al.*, 1989). This is about 20 times the concentration of CO$_2$ in the mesophyll cells and is sufficient to saturate photosynthesis and to inhibit photorespiration more or less completely. PEP carboxylase has an affinity for inorganic carbon comparable to that of Rubisco (K_m(HCO$_3^-$) 30 μM, equivalent to

6.4 μM CO$_2$ at pH 7) but, unlike Rubisco, it possesses no oxygenase activity.

In the present atmosphere, photorespiration in C$_3$ plants results in substantial losses of fixed carbon, particularly at higher temperatures. However, it was only at the end of the Cretaceous (63 million years ago), during the rise of the angiosperms, that atmospheric CO$_2$ concentrations declined to present levels, with an intervening rise during the Eocene and Oligocene (30–50 million years ago) (Ehleringer *et al.*, 1991; Ehleringer & Monson, 1993). At 200 ppm CO$_2$ (pre-industrial levels), photorespiration could easily reach 50% of net photosynthesis at elevated temperatures in C$_3$ plants. It is during this period of low atmospheric CO$_2$ concentrations that the CO$_2$ concentrating mechanism of C$_4$ photosynthesis is thought to have evolved, probably not more than 30 million years ago, perhaps as recently as 7 million years ago (the date of the oldest known fossilized C$_4$ material). There is evidence, from δ^{13}C measurements of mammalian tooth enamel and paleosol carbonate deposits, for a considerable expansion of C$_4$ ecosystems between 5 and 7 million years ago, perhaps as a response to a decrease in atmospheric CO$_2$ (Cerling *et al.*, 1993).

The C$_4$ pathway represents a series of variations on a biochemical theme that has almost certainly evolved independently many times in response to these low CO$_2$ concentrations. This was possible because none of the enzymes or anatomical structures involved in C$_4$ photosynthesis (or indeed Crassulacean Acid Metabolism (CAM)) is unique to these plants. C$_4$ plants are most abundant in warmer regions of the world and are important components of tropical grasslands and C$_4$ crop plants are among the most important economically (e.g. tropical grasses, such as maize and sugar cane), but in general, C$_4$ plants are not a large group within the plant kingdom (<1% of vascular plants).

2.5.1 Leaf anatomy of C$_4$ plants

C$_4$ plants are unique in possessing two types of photosynthetic cell (Fig. 2.29). A layer of cells surrounding the vascular bundle, the bundle-sheath, is a common structural feature, but only in C$_4$ plants does it contain chloroplasts. The bundle-sheath is thick-walled, sometimes suberized and there is no direct access from the intercellular spaces of the mesophyll. The appearance of a wreath of cells surrounding the vasculature gives rise to the term 'Kranz' (German: wreath) anatomy. In contrast, the

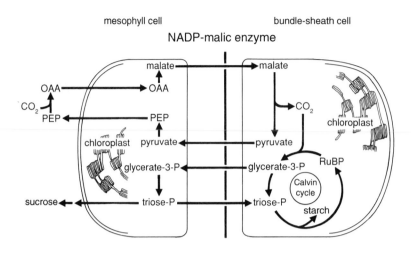

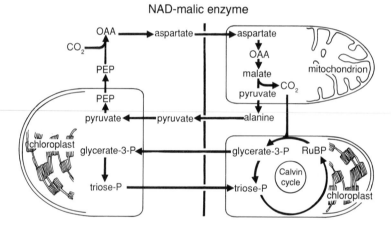

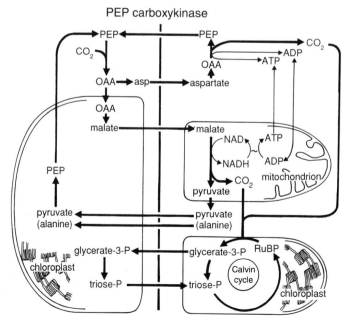

Figure 2.28 Operation of the C₄ pathway and its intracellular compartmentation in each of the three subgroups of C₄ plants, showing the relative complexity of the C₄ cycle in each and the shuttle of glycerate-3-P and triose-P between the bundle-sheath and mesophyll chloroplasts. For most reactions, cofactors, transaminations etc. have been omitted for clarity. Note that in PEP carboxykinase types, NAD-malic enzyme also operates and that alanine is also likely to return from the bundle-sheath to the mesophyll in order to maintain the balance of amino groups between the two compartments. NADH generated by malate decarboxylation is used to generate the ATP required by PEP carboxykinase. Note that NADP-malic enzyme species generally have agranal chloroplasts in the bundle sheath.

classified in accordance with the enzyme which is employed to decarboxylate C_4 acids in the bundle-sheath (Fig. 2.28). These are $NADP^+$-malic enzyme (in the bundle-sheath chloroplast):

$$malate + NADP^+ \rightarrow pyruvate + CO_2 + NADPH$$

NAD^+-malic enzyme (in the mitochondria):

$$malate + NAD^+ \rightarrow pyruvate + CO_2 + NADH$$

and the so-called PEP carboxykinase type plants which actually employ both PEP carboxykinase (in the cytosol):

$$oxaloacetate + ATP \rightarrow PEP + ADP + CO_2$$

and NAD^+-malic enzyme to generate CO_2. In these plants, the NADH generated by NAD-malic enzyme is respired to make ATP to drive the reaction catalyzed by PEP carboxykinase (Carnal *et al.*, 1993). It should be noted that in both NAD-malic and PEP-carboxykinase type plants, the mitochondria play an important part in the photosynthetic process, a feature which is shared by the photorespiratory pathway and by CAM plants. Indeed, photosynthetic fluxes through these organelles greatly exceed fluxes through respiratory pathways. Functionally speaking, the mesophyll cells all perform the same task of converting pyruvate into PEP and carboxylate it to a C_4 acid, although this may be malate or aspartate, or a mixture of the two.

There are also differences in the pathways of electron transport which operate in the bundle-sheath. $NADP^+$-malic enzyme species, such as maize and sugar cane, export malate from the mesophyll to the bundle-sheath. Malate decarboxylation provides reductant in the form of NADPH. In these plants there is a deficiency of PSII in the bundle-sheath and PSI-driven cyclic electron transport predominates, generating only ATP. This means that there is insufficient NADPH to drive the reductive phase of the Benson–Calvin cycle (it is sufficient to reduce half the glycerate-3-P formed in the cycle). The alternative source of reductant is the mesophyll, to which the other half of the glycerate-3-P is exported for reduction to triose-P. In practice, all the C_4 subtypes export glycerate-3-P from the bundle-sheath for reduction in the mesophyll (Fig. 2.28), despite the presence of PSII in the bundle-sheath, and hence the ability to generate sufficient reductant, in NAD^+-malic enzyme and PEP carboxykinase-type plants. The reason for this may be that a reduction in the requirement for NADPH in the bundle-sheath will

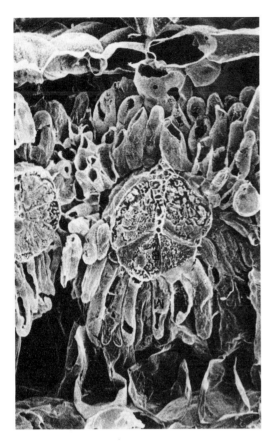

Figure 2.29 Scanning electron micrograph showing the leaf anatomy of the C_4 plant *Atriplex spongiosa*. From Troughton & Donaldson (1972).

mesophyll is typical of the type of photosynthetic tissue found in leaves of most C_4 plants and comprises thin walled cells with abundant intercellular spaces. The distance between bundle-sheath cells is normally only two or three mesophyll cells, so that no mesophyll cell is more than one cell away from a bundle-sheath cell. Mesophyll cells are also connected to bundle-sheath cells by large numbers of plasmodesmata. These features are both necessary for rapid fluxes of metabolites between the two cell types, which is an essential feature of the CO_2 pump. There are also distinct anatomical features in the arrangement of chloroplasts and other organelles at the subcellular level, but the biochemical significance of these differences remains unclear.

2.5.2 Subtypes of the C_4 pathway

There are three distinct biochemical subtypes of C_4 plants in 18 families of higher plants. These are result in a reduction in linear electron transport, and hence O_2 evolution, which may, in turn, be necessary to prevent O_2 building up in a bundle-sheath to concentrations which would favor photorespiration. The remarkable consequence of this arrangement is that the Benson–Calvin cycle is split between two cell types in C_4 plants. Bundle-sheath chloroplasts thus export glycerate-3-P and import triose-P, in contrast to their C_3 or mesophyll counterparts, while regulation of triose-P consumption in the mesophyll is a function which is largely removed from the mesophyll P_i translocator and passes entirely to the enzymes of sucrose synthesis.

2.5.3   Metabolite transport in C_4 plants

C_4 photosynthesis involves the co-operation of two cell types. This involves intercellular transport of metabolites between the mesophyll and bundle-sheath. The most obvious comparison is the close co-operation between vegetative cells and hetero-cysts in the cyanobacteria. In C_4 plants, inter-cellular transport between the mesophyll and bundle-sheath is driven entirely by diffusion via numerous plasmodesmata, which have an exclusion limit of about 900 Da. The permeability of the bundle-sheath cell interface to metabolites is, therefore, about ten times that of C_3 mesophyll cells, whereas the conductance to CO_2 is more than a hundred times lower (Hatch, 1992). In some C_4 species, this is probably due to the presence of a layer of suberin in the bundle-sheath cell wall. Diffusion of metabolites from one cell to another requires the presence of millimolar concentration gradients (Hatch & Osmond, 1976). Such gradients of malate, glycerate-3-P, triose-P and pyruvate have been measured in maize leaves which have been fractionated rapidly into the two cell types (Leegood, 1985; Stitt & Heldt, 1985a). It has been shown subsequently that the ratio of glycerate-3-P to triose-P in the extrachloroplastic compartment of bundle-sheath cells (8.3) is about 20-fold higher than that of the mesophyll cells (0.42), which directly supports the existence of gradients of these metabolites between the cytosols of the two cell types (Weiner & Heldt, 1992).

In the case of pyruvate the gradient between the bundle-sheath and the mesophyll cytosol is assisted by active uptake of pyruvate into the mesophyll chloroplasts. Both bundle-sheath and mesophyll cells contain a pyruvate transporter. In the latter, transport is active. It appears to be driven by a H^+ gradient in $NADP^+$-malic enzyme species and a Na^+ gradient (or perhaps Na^+

activation of the transporter (Murata et al., 1992)) in NAD^+-malic enzyme and PEP carboxykinase species (Heldt & Flügge, 1992; Ohnishi & Kanai, 1990; Ohnishi et al., 1990).

Pyruvate is converted into PEP by pyruvate-P_i dikinase which must then be exported to the cytosol for carboxylation to oxaloacetate (Fig. 2.28). The phosphate translocator in chloroplasts of C_4 plants is distinguished from that found in chloroplasts of C_3 plants by its additional ability to transport PEP and glycerate-2-P. The phosphate translocators of the bundle-sheath and mesophyll of *Panicum milaceum*, an NAD^+-malic enzyme species, have also been shown to differ in their affinities for PEP, triose-P and glycerate-3-P. Oxaloacetate formed by PEP carboxylase is reimported into the chloroplast for reduction to malate. Oxaloacetate (OAA) is carried on a separate, high affinity (K_m(OAA) 50 μM) specific translocator in maize mesophyll chloroplasts, which is also present in chloroplasts of the C_3 plant, spinach (Hatch et al., 1984). This is necessary because other dicarboxylates, such as malate, are present at concentrations many orders of magnitude higher than that of oxaloacetate and would largely inhibit its transport on the dicarboxylate transporter. In both C_3 and C_4 plants, reduction of oxaloacetate and export of the malate produced results in the export of reducing equivalents from the chloroplast.

The occurrence of metabolite gradients also has important consequences for the location and regulation of starch and sucrose synthesis in C_4 plants. In maize, sucrose is normally manufactured in the mesophyll cells and starch in the bundle-sheath. The demands of diffusion-driven intercellular transport mean that the amount of triose-P in the leaves of C_4 plants can be 10–20 times that in the leaves of C_3 plants (it may be as high as 20 or 30 mM in the mesophyll cells of maize). Consequently the concentration of fructose-1,6-bisphosphate in the cytosol will also be high. In C_3 plants in the presence of fructose-2,6-bisphosphate the K_m (FBP) of the cytosolic fructose 1,6-bisphosphatase is rather low at 20 μM. If this enzyme were to function in C_4 plants it would discharge the triose-P pool in the cytosol into hexose-P. Instead, the K_m (FBP) in C_4 plants is between 3 and 5 mM (Stitt & Heldt, 1985b). The properties of the kinase synthesizing fructose-2,6-bisphosphate mean that it is less sensitive to triose-P and glycerate-3-P than in C_3 plants (Soll et al., 1983).

The properties of a key enzyme of starch synthesis, ADPglucose pyrophosphorylase, in the bundle-sheath are also different in C_4 plants. In maize, ratios of glycerate-3-P to P_i of between 7 and 10 are required to activate the enzyme, compared to ratios of 1.5 in a C_3 plant (Spilatro & Preiss, 1987). Again, this reflects high concentrations of glycerate-3-P in the bundle-sheath and the fact that a gradient of P_i must exist between the mesophyll and bundle-sheath, following its release in sucrose synthesis (Fig. 2.28).

2.5.4   Regulation of C_4 photosynthesis

The C_4 pathway requires separate controls from the Benson–Calvin cycle, but the principles are much the same. First, it must be co-ordinated with the rest of the photosynthetic process. CO_2 pumping must occur at rates similar to that at which Rubisco fixes CO_2 and the rate at which electron transport can supply ATP and reductant. The C_4 cycle is also metabolically connected to the C_3 cycle, via the interconversion of glycerate-3-P and PEP, so that withdrawal of carbon for product synthesis must also be regulated.

Mechanisms of control within the C_4 cycle include control by the thioredoxin system ($NADP^+$-malate dehydrogenase), which is a feature shared with the Benson–Calvin cycle, but also complex control of two enzymes, PEP carboxylase and pyruvate-P_i dikinase, by phosphorylation.

(a)   Regulation of events in the mesophyll

PEP carboxylase, which catalyzes:

$$PEP + HCO_3^- \rightarrow oxaloacetate + P_i$$

is probably present in all plants, but its activity is greatly elevated in the cytosol of plants with the C_4 and CAM mechanisms. Its activity is modulated by a wide range of effectors such as phosphorylated intermediates (triose-P and hexose-P), amino acids and organic acids. Activation of the enzyme by hexose-P and triose-P may act as a mechanism that co-ordinates high output from the Calvin cycle with increased turnover of the C_4 cycle (Doncaster & Leegood, 1987). Malate is a strong inhibitor of PEP carboxylases from all plants. In CAM plants the enzyme was observed to be more sensitive to malate when extracted from illuminated leaves than was the enzyme extracted from darkened leaves (Winter, 1982; Nimmo et al., 1984). The converse was true of the enzyme extracted from the leaves of C_4 plants (Huber et al., 1986). This change in malate sensitivity was subsequently shown to be associated with a change in the phosphorylation state of the enzyme, such

that the phosphorylated (active) enzyme is less sensitive to inhibition by malate than the dephosphorylated enzyme (K_i increases from ca. 0.2 mM to 1 mM malate) (Nimmo et al., 1987). Phosphorylation may also increase V_{max}. Activation occurs rather slowly in vivo (30–60 min, or more) in comparison with thioredoxin-activated enzymes (5–10 min) or pyruvate-P_i dikinase. Inhibitors of photosynthesis, such as diuron, gramicidin and DL-glyceraldehyde, prevent phosphorylation of PEP carboxylase in leaves, which may indicate the involvement of both electron transport and carbon metabolism in the signal transduction chain (Bakrim et al., 1992; Jiao & Chollet, 1992).

PEP carboxylase is phosphorylated on a serine residue (Ser-8 in sorghum, Ser-15 in maize (see review by Jiao & Chollet, 1991). Bakrim et al., (1992) have identified two kinases that phosphorylate Sorghum leaf PEP carboxylase, but only one of these resulted in a change in K_i (malate). Extracts from illuminated leaves had a higher capacity to phosphorylate PEP carboxylase, but this was nullified by pretreatment with cycloheximide, suggesting that protein turnover may be important in regulating the activity of the kinase. Pierre et al. (1992) have suggested that changes in cytosolic pH and in Ca^{2+} may also be components of the light-transduction pathway. However, Wang & Chollet (1993) have purified a 30 kDa protein kinase, which is inhibited by malate and is Ca^{2+}-insensitive. PEP carboxylase in CAM plants has been shown to be dephosphorylated by an okadaic acid-sensitive type 2A protein phosphatase.

Pyruvate-P_i dikinase, which catalyzes:

$$pyruvate + Pi + ATP \rightarrow PEP + AMP + PP_i$$

is also regulated by phosphorylation, but this enzyme is unusual in that regulation is part of the catalytic mechanism of the enzyme (Edwards et al., 1985). Regulation is accomplished by a regulatory protein that is a bifunctional enzyme catalyzing the interconversion between an active form phosphorylated on a catalytic site histidine and an inactive form phosphorylated on a regulatory threonine residue. The mechanism allows response to changing photon flux density (PFD) (Hatch, 1981). Removal of the products of the reaction by the actions of adenylate kinase and pyrophosphatase favors the forward reaction.

NAD P-malate dehydrogenase, which catalyzes:

$$malate + NADP^+$$

$$\rightarrow oxaloacetate + NADPH + H^+$$

is regulated by the thioredoxin system in a manner that allows it to respond to a range of PFDs. $NADP^+$ also inhibits the activation of NADP-malate dehydrogenase (Edwards *et al.*, 1985).

(b)   Regulation of decarboxylation in the bundle-sheath

All three decarboxylases generate CO_2. Jenkins *et al.* (1987) argue that this is vital to the operation of the C_4 pathway not only because CO_2 is the substrate for Rubisco but because carbonic anhydrase is absent from the bundle-sheath. There is evidence for the regulation of all three decarboxylases.

Hatch & Slack (1966) showed that transfer of ^{14}C from the C-4 carboxyl group of C_4 acids ceased when leaves of sugar cane were transferred to darkness, indicating efficient regulation of decarboxylation by $NADP^+$-malic enzyme. The regulatory properties of $NADP^+$-malic enzyme from maize leaves are pH and Mg^{2+}-dependent (Asami *et al.*, 1979). Thus $NADP^+$-malic enzyme is likely to be activated by the change in stromal conditions occurring upon illumination, reminiscent of regulation of certain enzymes of the Benson–Calvin cycle, such as FBPase and ribulose-5-P kinase.

The activity of NAD^+-malic enzyme shows a sigmoidal response to malate concentration and regulation by effectors, including CoA, acetyl CoA and fructose-1,6-bisphosphate (Hatch *et al.*, 1974). It is also inhibited by NADH so that the enzyme could be regulated by the $NADH/NAD^+$ ratio. NAD^+-malic enzyme from *Atriplex spongiosa* and *Panicum milaceum* (NAD^+-malic enzyme type) is inhibited by physiological concentrations of ATP, ADP and AMP, with ATP the most inhibitory species at subsaturating concentrations of the activator, malate and Mn^{2+}. Hence if the mitochondrial malate concentration were to fall in the dark or in low light, malic enzyme activity would be reduced substantially in the presence of ATP, but much less in the absence of adenylates. The very high carbon fluxes proceeding via NAD^+-malic enzyme would appear to be freed of respiratory control by engagement of the alternative, cyanide-insensitive pathway of respiration (Gardeström & Edwards, 1985). NAD^+-malic enzyme from *Urochloa panicoides* (PEPCK-type) is inhibited by organic acids (Burnell, 1987) and is strongly activated by ATP and inhibited by ADP and AMP (Furbank *et al.* 1991). Furbank *et al.* (1991) suggest that, if the flux of C_4 acids from the mesophyll exceeds the capacity of PEPCK, ATP and oxaloacetate will accumulate. This could lead

to an increase in oxaloacetate in the mitochondria, and thus increase malate and NAD which, together with the activating effect of ATP, would increase the activity of NAD-malic enzyme, providing an additional route for decarboxylation of C_4 acids.

Although the activity of PEPCK is inhibited by glycerate-3-P, fructose-6-P, fructose-1,6-bisphosphate and dihydroxyacetone phosphate (Burnell, 1986), this enzyme in particular would be expected to be switched off in the dark by some mechanism (Carnel *et al.*, 1993). It has recently been shown that the enzyme from all plants, including *U. panicoides*, is subject to rapid proteolytic cleavage (Walker *et al.*, 1995) which may result in a loss of regulatory properties. The enzyme from a number of plants has recently been shown to be phosphorylated both *in vitro* and *in vivo* (Walker & Leegood, 1996).

2.5.5   Gas exchange in C_3 and C_4 plants and the advantages of C_3 and C_4 photosynthesis

In C_3 plants at low intercellular CO_2 (low p_i), the response of the assimilation rate to CO_2 (Fig. 2.30) is determined by the kinetic properties of Rubisco (Farquhar *et al.*, 1980). This initial slope is often known as the carboxylation efficiency. Lowering the O_2 concentration decreases oxygenation by Rubisco, increases the slope and decreases the CO_2 compensation point. In C_4 plants the initial slope is determined by the response of PEP carboxylase to CO_2 (Collatz *et al.*, 1991). At higher concentrations of CO_2, the maximum rate of photosynthesis is determined by the rate of RuBP regeneration, that is, the supply of ATP and NADPH by electron transport. This limitation can take two forms. Either the maximum capacity of the electron transport system limits synthesis of ATP and reductant or else the rate at which triose-P is utilized in product synthesis limits the rate of P_i-recycling to the chloroplast and that, in turn, limits photophosphorylation (see p. 80; also Harley & Sharkey, 1991).

The CO_2-concentrating mechanism in C_4 plants results in a low CO_2 compensation point (less than 5 ppm, compared to 45 ppm in C_3 plants in air) because C_4 plants are able to saturate Rubisco with CO_2 even at low intercellular concentrations of CO_2 (ca. 100 ppm) (Fig. 2.30). This gives C_4 plants three advantages over C_3 plants. First, it improves carbon gain, with photosynthetic rates and growth rates that are often higher than those of C_3 plants, particularly at the higher tempera-

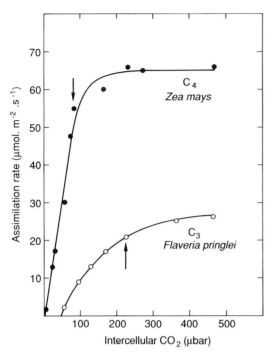

Figure 2.30 Characteristics of gas-exchange in leaves of C_3 and C_4 plants. The arrows show the intercellular CO_2 partial pressure under ambient conditions, when the external CO_2 partial pressure is 350 μbar.

tures which favor photorespiration (see Osmond *et al.*, 1980). Second, it means that the intercellular concentration of CO_2 (and hence stomatal conductance) can be lower in C_4 plants than in C_3 plants for a given rate of photosynthesis (or, alternatively, photosynthesis can be higher in C_4 plants for a given intercellular CO_2 concentration and stomatal conductance). The consequence of a lower stomatal conductance is lower transpiration and an increase in instantaneous water-use efficiency. C_4 plants also have about double the long-term water-use efficiency of C_3 plants (2–5 g CO_2 fixed per kg H_2O transpired in C_4 plants compared with values of between 1 and 3 in C_3 plants). Third, C_3 plants do not fully utilize their Rubisco. Rubisco is a very inefficient enzyme, with a maximum specific activity of about 3.6 μmol min^{-1} mg^{-1} protein. C_3 plants therefore possess large amounts of Rubisco, which can account for as much as 50% of soluble protein in a leaf and which represents a major investment of nitrogen. However, the capacity of Rubisco in a C_3 plant is not fully utilized *in vivo* because: (1) its activity is regulated downwards (p. 88 and Fig. 2.26); (2) it is not saturated with CO_2 and (3) part of it is used to oxygenate, rather than carboxylate, RuBP. Accordingly, the nitrogen-use efficiency of

C_3 plants can be lower than that of C_4 plants, particularly at higher temperatures, which favor high rates of photorespiration and hence inefficient use of the nitrogen invested in Rubisco (see Edwards *et al.*, 1985). Similar considerations apply when the implications of global environmental change for photosynthesis are considered. A projected doubling of atmospheric CO_2 by the year 2050 would result in a substantial decrease in the rate of photorespiration, and give C_3 plants considerable increases in carbon gain, and in water-use and nitrogen-use efficiencies. This is because less Rubisco would be required as the oxygenase activity diminishes and the enzyme approaches saturation with CO_2. By contrast, the benefits accruing to C_4 plants will be fewer.

The CO_2-concentrating mechanism is an energy-consuming process. The C_4 cycle can be considered as a fixed cost appendage to the Benson–Calvin cycle, which, in the absence of photorespiration, also has a fixed cost. This results in a quantum yield (mol CO_2 per absorbed photon) which is constant (ca. 0.054) at different temperatures, light intensities and CO_2 concentrations (Björkman, 1981). In C_3 plants the rate of photorespiration, and hence the cost of fixing CO_2, varies considerably, particularly with changes in temperature. In air at a temperature of 15°C the quantum yield is about 0.07, whereas at 40°C, the quantum yield falls to about 0.04. Clearly, C_3 plants have the advantage at lower temperatures, in which the rate of photorespiration, and hence wastage of carbon and energy, are low and C_4 plants have the advantage at higher temperatures, in which photorespiratory rates in C_3 plants are high. However, the reasons underlying the distribution of C_4 plants to warmer climates are certainly more complex than this simple view (Osmond *et al.*, 1980). The increased cost of fixing CO_2 in C_4 plants could also be disadvantageous under strongly light-limited conditions, and the majority of C_4 species are found in open, high light environments, such as grasslands.

2.5.6 C_3–C_4 intermediates

A number of plants have characteristics that are intermediate between C_3 and C_4 plants. These are called C_3–C_4 intermediates (Rawsthorne *et al.*, 1992). They have intermediate CO_2 compensation points (8–30 ppm) and have a chlorenchymatous bundle-sheath, but lack the strict compartmentation of Rubisco within the bundle-sheath seen in C_4 plants. The P subunit of glycine decarboxylase is specifically absent from the mitochondria of mesophyll cells in C_3–C_4 intermediate species (Rawsthorne, 1992). Hence these plants lack the ability to decarboxylate glycine in the mesophyll cells, but not in the bundle-sheath, so that the only route for glycine generated during photorespiration is transfer to the bundle-sheath for decarboxylation. The CO_2 released can then be refixed by Rubisco. Modeling of this system has shown that such a mechanism could account for the observed reductions in CO_2 compensation point (von Caemmerer, 1989). Interestingly, glycine decarboxylase is also confined to the bundle-sheath of true C_4 plants (Ohnishi & Kanai, 1983). Some C_3–C_4 intermediates, particularly in the genus *Flaveria*, show appreciable labeling of C_4 acids when exposed to $^{14}CO_2$ for short periods, implying fixation by PEP carboxylase in addition to glycine decarboxylation in the mesophyll. It has been suggested that these plants represent genuine evolutionary intermediates between C_3 and C_4 modes of photosynthesis (Monson & Moore, 1989).

2.6   CRASSULACEAN ACID METABOLISM

Crassulacean acid metabolism (CAM) is a photosynthetic adaptation to periodic drought. It allows gas-exchange to occur at night, when air temperatures are cooler and water vapor pressure deficits are lower. Water loss through open stomata at night is lower, by as much as an order of magnitude, than it would be during the day. Long-term water-use efficiencies in CAM plants (10–40 g CO_2 fixed per kg H_2O transpired) are correspondingly higher than in C_3 or C_4 plants (with values between 1 and 3, and 2 and 5, respectively). About 10% of vascular plants have developed CAM photosynthesis. Although it is characteristically found in plants in many arid regions of the world, such as the cacti and euphorbias, it is also a feature of many tropical epiphytes that also experience erratic supplies of water, such as bromeliads (e.g. *Tillandsia* spp.) and the orchids. Indeed, the orchids are the largest plant family, of which a high proportion are epiphytes, and most of these are probably CAM plants. Few CAM plants are of economic importance, but include pineapple (a bromeliad), vanilla (an orchid) and *Agave* (manufacture of tequila etc). The principal metabolic feature of CAM plants is assimilation of CO_2 at night into malic acid which is stored in the vacuole. Malate is generated in the reaction catalyzed by PEP carboxylase and PEP is, in turn, generated by degradation of starch

or soluble sugars. During the day, malate is released from the vacuole and is decarboxylated to provide CO_2 for fixation in the Benson–Calvin cycle behind closed stomata. Starch and sugars are then resynthesized (Fig. 2.31). The cycle of carboxylation and decarboxylation is thus spread out over time, rather than spatially, as in C_4 plants.

As in C_4 plants, CAM represents a variation on a C_3 biochemical theme and has certainly evolved independently many times. It is also associated with distinctive anatomical characteristics, such as succulence at the cellular level (large, thin-walled cells with large vacuoles for storage of organic acids), as well as with succulence in leaves, stems or pseudobulbs. Tissue succulence is defined by low surface area to volume ratio and high water storage capacity, resulting in high tissue water potentials even under water-stressed conditions. A low stomatal frequency and the presence of a thick cuticle also restrict gas-exchange.

2.6.1   Diurnal changes in gas-exchange and metabolism during CAM

Diurnal patterns of gas-exchange and metabolism in well-watered plants can be divided into four phases (Osmond, 1978) (Fig. 2.31).

Phase 1 is typified by the nocturnal degradation of carbohydrate to provide the substrate for PEP carboxylase. Malic acid accumulates in the vacuoles and titratable acidity in extracts increases. Titratable acidity in plants such as the strangling fig (*Clusia* spp.) has been recorded as high as 1.5 M H^+ at dawn, with a sap pH of less than 3. Some CAM plants are also known to accumulate large amounts of citrate, in addition to malate.

Phase 2 is a transitional period that occurs at the onset of illumination. Both PEP carboxylase and Rubisco can be active and lead to a transient increase in CO_2 fixation. Stomata remain partially open during this period.

Phase 3 begins with complete stomatal closure and malate is released from the vacuole and decarboxylated in the cytosol. Elevated CO_2 concentrations within the tissue (as high as 4% or 40 000 ppm) allow CO_2 fixation without photorespiration.

Phase 4 is another transitional period that sees cessation of malate decarboxylation, a decline in intercellular CO_2 concentration and stomatal opening. During this period both Rubisco and PEP carboxylase may be active in CO_2 fixation, and photorespiration may occur.

Phase 2 and, in particular, phase 4 are suppressed under dry conditions. Terrestrial plants

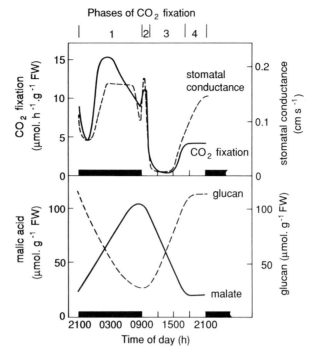

Figure 2.31 Phases of CO_2 fixation in a well-watered CAM plant (*Kalanchoe daigremontiana*), showing stomatal conductance and inverse changes in malate and glucan content. The blocked areas indicate the dark period. Phases 1–4 of gas exchange (Osmond, 1978) are indicated and are discussed in section 2.6.1. Redrawn from Osmond & Holtum (1981).

may seal themselves off completely by stomatal sealing and shedding of roots if the water potential of the soil is lower than that of the plant tissues. CAM idling then occurs, with refixation of respiratory CO_2. Rainfall can lead to rapid development of shallow roots and resumption of gas exchange. In many CAM plants, such as *Mesembryanthemum crystallinum*, CAM is inducible. *M. crystallinum* germinates in early spring in conditions of adequate water supply and completes the seedling stage as a C_3 plant. Water (or salt) stress induces CAM with rapid changes in gene expression and synthesis of the necessary enzymes (Holtum & Winter, 1982; Cushman et al., 1989; Cushman, 1993). On the other hand, many CAM plants will perform typical C_3 patterns of gas-exchange and metabolism under well-watered conditions. There is, therefore, considerable metabolic flexibility in CAM plants. This contrasts with the C_4 plants that have a fixed mode of photosynthetic metabolism. The only known exception to this is repression of the C_4 mode of photosynthesis in *Eleocharis vivipara* under submersed aquatic conditions (Ueno et al., 1988).

2.6.2 Regulation of CAM

CAM requires complex regulation, during the diurnal cycle, over carbohydrate metabolism and

export, PEP carboxylase and activity of the Benson–Calvin cycle, and storage and release of malic acid.

(a) Malate efflux from the vacuole, its decarboxylation and gluconeogenesis

Efflux of malate from the vacuole during the day is probably a passive process, although it must clearly be regulated during the accumulation of malate in the vacuole at night. Once malate has been released from the vacuole, three pathways of malate decarboxylation are found in CAM plants, as in C_4 plants (see p. 95). One group of plants decarboxylate malate via NAD^+ or $NADP^+$-malic enzyme. These include the cacti. In most species both enzymes probably contribute to malate decarboxylation. $NADP^+$-malic enzyme is cytosolic in CAM plants, whereas NAD^+-malic enzyme is mitochondrial. The pyruvate formed must then be transported into the chloroplasts for conversion into PEP, which is exported via the phosphate translocator. This phosphate translocator is, therefore, similar to that in C_4 plants in that it exchanges glycerate-3-P, P_i and PEP (Neuhaus et al., 1988). These plants form starch as their carbohydrate reserve, which is synthesized by import of glycerate-3-P into the chloroplasts, formed from PEP in the cytosol. A distinguishing feature of starch synthesis in CAM plants is to be found in the properties of ADPglucose pyrophosphorylase. It is very sensitive to activation by glycerate-3-P, which also lowers the K_m for glucose-1-P from around 1 mM to 50 μM, in *Hoya carnosa* and *Xerosicyos danguyi* (Singh *et al.*, 1984). The enzyme is virtually insensitive to inhibition by P_i in the presence of glycerate-3-P. High ratios of glycerate-3-P to P_i probably occur during deacidification (Osmond & Holtum, 1981) and will be particularly effective in promoting starch synthesis in these plants.

Another group of CAM plants utilize PEP carboxykinase as the decarboxylase. These plants include the bromeliads and euphorbias. They do not require pyruvate-P_i dikinase, since PEP is formed directly in the cytosol. Another distinguishing feature of these plants is that they store soluble carbohydrates, such as glucose, in the vacuole. However, regulation and transport processes in these plants are poorly understood.

(b) *Glycolysis, PEP carboxylation and malate uptake into the vacuole*

Malic enzyme-type CAM plants store starch in the chloroplasts, whereas PEP carboxykinase-type CAM plants store much of their carbohydrate as hexose, probably in the vacuole. These two pathways have different requirements for transport and for energy (Smith & Bryce, 1992). In starch-storers, carbohydrate degradation results in formation of glycerate-3-P that can be exported to the cytosol via the phosphate translocator. The phosphate translocator thus operates in both directions in these plants, depending on the time of day. In hexose-storers, PP_i dependent phosphofructokinase (PFP) is present (Fahrendorf *et al.*, 1987; Carnal & Black, 1989), but its role in glycolysis in these plants has yet to be unequivocally established.

In CAM plants, PEP carboxylase was observed to be more sensitive to malate when extracted from illuminated leaves than was the enzyme extracted from darkened leaves (Winter, 1982). As in C_4 plants, this change in malate sensitivity has been shown to be caused by nocturnal phosphorylation of an N-terminal regulatory serine residue, resulting in a diminished sensitivity to inhibition by malate compared to the dephosphorylated enzyme (K_i(malate) = 3.0 mM in the dark and 0.3 mM in the light) (Nimmo *et al.*, 1984, see review by Jiao & Chollet, 1991). PEP carboxylase kinase in CAM plants appears to be under the control of a circadian rhythm and involves protein turnover (Carter *et al.*, 1991). Rubisco also appears to be strongly regulated by light in CAM plants (Vu *et al.*, 1984).

Malate generated by PEP carboxylation cannot be allowed to accumulate in the cytosol for two reasons. First, it would result in acidification of the cytosol and second, it would inhibit PEP carboxylase even in the phosphorylated, low K_i(malate) state (although such inhibition may well occur towards the end of phase 1 and in phase 2, as malate is released from the vacuole). Malate uptake into the vacuole occurs against concentration gradient. Malate accumulates in the vacuole as the free acid ($2H^+$ per malate). This may be powered either by an H^+-ATPase (which transports $2H^+$ per ATP hydrolyzed) (Lüttge *et al.*, 1982) or by an H^+-pyrophosphatase at the tonoplast. The contributions of these alternative mechanisms have been clarified during CAM induction in *M. crystallinum*, during which the activity of the tonoplast H^+-ATPase increases, whereas that of the H^+-pyrophosphatase declines (Bremberger *et al.*, 1988; Bremberger & Lüttge, 1992). The energy for malate uptake into the vacuole probably only partly derives from glycolysis, and would need to be supplemented by oxidative phosphorylation (Smith & Bryce, 1992).

ACKNOWLEDGEMENTS

JRB would like to thank Robert Ford, Petra Fromme, Wolfgang Haehnel, Reinhold Herrmann, Werner Kühlbrandt, Jim Barber and Jonathan Nugent for providing photographs and drawings used in this chapter.

BACKGROUND READING

Ho, M-W. (1995) *Living Processes, Book 2, Bioenergetics*. Open University, Milton Keynes.

Nicholls, D.G. & Ferguson, S.J. (1992) *Bioenergetics 2*, pp. 157–187. Academic Press, London.

Stryer, L. (1995) *Biochemistry*, 4th edn, pp. 653–662. W.H. Freeman, New York.

Zubay, G. (1993) *Biochemistry*, 3rd edn, pp. 414–443. W.C. Brown, Dubuque, USA.

FURTHER READING

Abrahams, J.P., Leslie, A.G.W., Lutter, R. & Walker, J.E. (1994) Structure at 2.8 Å resolution of F_1-ATPase from bovine heart mitochondria. *Nature* 370, 621–628.

Allen, J.F. (1992a) Protein phosphorylation in the regulation of photosynthesis. *Biochim. Biophys. Acta* 1098, 275–335.

Almog, O., Shoham, G. & Nechushtai, R. (1992) Photosystem I: composition, organisation and structure. In *The Photosystems: Structure, Function and Molecular Biology*, Topics in Photosynthesis, Vol. 11 (J. Barber, ed.), pp. 443–469. Elsevier, Amsterdam.

Anderson, J.M. (1992) Cytochrome b_6f complex: Dynamic molecular organization, function and acclimation. *Photosynth. Res.* **34**, 341–357.

Anderson, J.M. & Andersson, B. (1988) The dynamic photosynthetic membrane and regulation of solar energy conversion. *Trends Biochem. Sci.* **13**, 351–355.

Aro, E-M., Virgin, I. & Andersson, B. (1993) Photoinhibition of Photosystem II. Inactivation, protein damage and turnover. *Biochim. Biophys. Acta* **1143**, 113–134.

Asada, K. (1994) Mechanisms for scavenging reactive molecules generated in chloroplasts under light stress. In *Photoinhibition of Photosynthesis: from Molecular Mechanisms to the Field* (N.R. Baker & J.R. Bowyer, eds), pp. 129–142. Bios, Oxford.

Barber, J. & Andersson, B. (1992) Too much of a good thing: light can be bad for photosynthesis. *Trends Biochem. Sci.* **17**, 61–66.

Bowyer, J.R., Camilleri, P. & Vermaas, W.F.J. (1991) Photosystem II and its interaction with herbicides. In *Herbicides, Topics in Photosynthesis*, Vol. 10 (N.R. Baker & M.P. Percival, eds), pp. 27–85. Elsevier, Amsterdam.

Boyer, P.D. (1993) The binding change mechanism for ATP synthase – some probabilities and possibilities. *Biochim. Biophys. Acta* **1140**, 215–250.

Bryant, D. (1992) Molecular biology of photosystem I. In *The Photosystems: Structure, Function and Molecular Biology, Topics in Photosynthesis*, Vol. 11 (J. Barber, ed.), pp. 501–549. Elsevier, Amsterdam.

Buchanan, B.B. (1980) Role of light in the regulation of chloroplast enzymes. *Annu. Rev. Plant Physiol.* **31**, 341–374.

Collatz, G.J., Ribas-Carbo, M. & Berry, J.A. (1992) Coupled photosynthesis-stomatal conductance model for leaves of C_4 plants. *Aust. J. Plant Physiol.* **19**, 519–538.

Cramer, W.A., Martinez, S.E., Huang, D., Tae, G-S., Everly, R.M., Heymann, J.B., Cheng, R.H., Baker, T.S. & Smith, J.L. (1994) Structural Aspects of the Cytochrome b_6f Complex; Structure of the Lumen-Side Domain of Cytochrome *f*. *J. Bioenerg. Biomembr.* **26**, 31–47.

Edwards, G.E., Ku, M.S.B. & Monson, R.K. (1985) C_4 photosynthesis and its regulation. In *Photosynthetic Mechanisms and the Environment* (J. Barber & N.R. Baker, eds), pp. 287–327. Elsevier, Amsterdam.

Edwards, G.E. and Walker, D.A. (1983) C_3,C_4: *Mechanisms, and environmental regulation of, photosynthesis.* Blackwell, Oxford.

Ehleringer, J.R. & Monson, R.K. (1993) Evolutionary and ecological aspects of photosynthetic pathway variation. *Annu. Rev. Ecol. Syst.* **24**, 411–439.

Ehleringer, J.R., Sage, R.F., Flanagan, L.B. & Pearcy, R.W. (1991) Climate change and the evolution of C_4 photosynthesis. *Trends Ecol. Evol.* **6**, 95–99.

Farquhar, G.D., von Caemmerer, S. & Berry, J.A. (1980) A biochemical model of photosynthetic CO_2 assimilation in leaves of C_3 species. *Planta* **149**, 78–90.

Flügge, U.I. & Heldt, H.W. (1991) Metabolite translocators of the chloroplast envelope. *Annu. Rev. Plant Physiol. Plant Mol. Biol.* **42**, 129–144.

Golbeck, J.H. (1992) Structure and function of Photosystem I. *Annu. Rev. Plant Physiol. Plant Mol. Biol.* **43**, 293–324.

Gray, J.C. (1992) Cytochrome *f*: Structure, function and biosynthesis. *Photosynth. Res.* **34**, 359–374.

Green, B.R., Pichersky, E. & Kloppstech, K. (1991) Chlorophyll *a/b*-binding proteins: an extended family. *Trends Biochem. Sci.* **16**, 181–186.

Halliwell, B. (1991) Oxygen radicals: their formation in plant tissues and their role in herbicide damage. In *Herbicides, Topics in Photosynthesis*, Vol. 10 (N.R. Baker & M.P. Percival, eds), pp. 87–129. Elsevier, Amsterdam.

Hatch, M.D. (1992) C_4 photosynthesis: An unlikely process full of surprises. *Plant Cell Physiol.* **33**, 333–342.

Hatch, M.D. (1987) C_4 photosynthesis: a unique blend of modified biochemistry, anatomy and ultrastructure. *Biochim. Biophys. Acta* **895**, 81–106.

Hatch, M.D. & Osmond, C.B. (1976) Compartmentation and transport in C_4 photosynthesis. *Encyclopedia of Plant Physiol.* **3**, 144–184. Springer Verlag, Berlin.

Heber, U. & Heldt, H.W. (1981) The chloroplast envelope: structure, function, and role in leaf metabolism. *Annu. Rev. Plant Physiol.* **32**, 139–168.

Heldt., H.W. & Flügge, U.I. (1992) Metabolite transport in plant cells. In *Plant Organelles* (A.K. Tobin, ed.), pp. 21–47. Cambridge University Press., Cambridge.

Hope, A.B. (1993) The chloropast cytochrome *bf* complex: a critical focus on function. *Biochim. Biophys. Acta* **1143**, 1–22.

Horton, P. & Ruban, A. (1994) The role of light-harvesting complex II in energy quenching. In *Photoinhibition of Photosynthesis: from Molecular Mechanisms to the Field* (N.R. Baker & J.R. Bowyer, eds), pp. 111–128. Bios, Oxford.

Jiao, J.A. & Chollet, R. (1991) Posttranslational regulation of phosphoenolpyruvate carboxylase in C_4 and CAM plants. *Plant Physiol.* **95**, 981–985.

Kacser, H. & Burns, J.A. (1973) The control of flux. *SEB Symposium* **27**, 65–104.

Klein, M.P., Sauer, K. & Yachandra, V.K. (1993) Perspectives on the structure of the photosynthetic oxygen evolving manganese complex and its relation to the Kok cycle. *Photosyn. Res.* **38**, 265–277.

Kühlbrandt, W. (1994) Structure and function of the plant light-harvesting complex, LHCII. *Curr. Opin. Struct. Biol.* **4**, 519–528.

Lavergne, J. & Junge, W. (1993) Proton release during the redox cycle of the water oxidase. *Photosynth. Res.* **38**, 279–296.

Leegood, R.C. (1990) Enzymes of the Calvin cycle. In *Methods in Plant Biochemistry* (P.J. Lea, ed.), pp. 15–37. Academic Press, London.

Leegood, R.C. & Osmond, C.B. (1990) The flux of metabolites in C$_4$ and CAM plants. In *Plant Physiology, Biochemistry and Molecular Biology* (D.T. Dennis & D.H. Turpin, eds), pp. 274–298. Longman, London.

Leegood, R.C., Walker, D.A. & Foyer, C.H. (1985) Regulation of the Benson-Calvin cycle. In *Photosynthetic Mechanisms and the Environment* (J. Barber & N.R. Baker, eds), pp. 189–258. Elsevier, Amsterdam.

Lüttge, U. (ed.) (1989) *Vascular plants as epiphytes.* Springer Verlag, Berlin.

Nobel, P.S. (1988) The Environmental Biology of Agaves and Cacti. Cambridge University Press, New York.

Oettmeier, W. (1992) Herbicides of photosystem II. In *The Photosystems: Structure, Function and Molecular Biology, Topics in Photosynthesis*, Vol. 11 (J. Barber, ed.), pp. 349–408. Elsevier, Amsterdam.

Ohad, I., Keren, N., Zer, H., Gong, H., Mor, T.S., Gal, A., Tal, S. & Domovich, Y. (1994) Light-induced degradation of the photosystem II reaction centre D1 protein *in vivo*: an integrative approach. In *Photoinhibition of Photosynthesis: from Molecular Mechanisms to the Field* (N.R. Baker & J.R. Bowyer, eds), pp. 161–177. Bios, Oxford.

Ort, D.R. & Oxborough, K. (1992) In situ regulation of chloroplast coupling factor activity. *Annu. Rev. Plant Physiol. Plant Mol. Biol.* 43, 269–291.

Osmond, C.B. (1978) Crassulacean acid metabolism: A curiosity in context. *Annu. Rev. Plant Physiol.* 29, 379–414.

Osmond, C.B. & Holtum, J.A.M. (1981) Crassulacean acid metabolism. In *Biochemistry of Plants* (M.D. Hatch & N.K. Boardman, eds), 8, 283–328. Academic Press, New York.

Osmond, C.B., Winter, K. & Ziegler, H. (1980) Functional significance of different pathways of CO$_2$ fixation of photosynthesis. *Encyclopedia of Plant Physiol.* 12B, 480–547.

Owens, T.G. (1994) Excitation energy transfer between chlorophylls and carotenoids. A proposed molecular mechanism for non-photochemical quenching. In *Photoinhibition of Photosynthesis: from Molecular Mechanisms to the Field* (N.R. Baker & J.R. Bowyer, eds), pp. 95–109. Bios, Oxford.

Pakrasi, H.B. & Vermaas, W.F.J. (1992) Protein engineering of photosystem II. In *The Photosystems: Structure, Function and Molecular Biology, Topics in Photosynthesis*, Vol. 11 (J. Barber, ed.), pp. 231–257. Elsevier, Amsterdam.

Portis, A.R. (1992) Regulation of ribulose 1,5-bisphosphate carboxylase/oxygenase activity. *Annu. Rev. Plant Physiol. Plant Mol. Biol.* 43, 415–437.

Raines, C.A., Lloyd, J.C. and Dyer, T.A. (1991) Molecular biology of the C$_3$ photosynthetic carbon reduction cycle. *Photosynth. Res.* 27, 1–14.

Rawsthorne, S., von Caemmerer, S., Brooks, A. & Leegood, R.C. (1992) Metabolic interactions in leaves of C$_3$–C$_4$ intermediate plants. In *Plant Organelles* (A.K. Tobin, ed.), pp. 113–139. Cambridge University Press, Cambridge.

Rich, P.R., Madgwick, S.A., Brown, S., von Jagow, G. & Brandt, U. (1992) MOA-stilbene: A new tool for investigation of the reactions of the chloroplast cytochrome *bf* complex. *Photosynth. Res.* 34, 465–477.

Rochaix, J-D. and Erickson, J.M. (1992) The molecular biology of photosystem II. In *The Photosystems: Structure, Function and Molecular Biology*. Topics in Photosynthesis, Vol. 11 (ed. J. Barber), pp. 101–175, Elsevier, Amsterdam.

Rutherford, A. W., Zimmerman, J-L. & Boussac, A. (1992) Oxygen evolution. In *The Photosystems: Structure, Function and Molecular Biology, Topics in Photosynthesis*, Vol. 11 (J. Barber, ed.), pp. 179–229. Elsevier, Amsterdam.

Smith, J.A.C. and Bryce, J.H. (1992) Metabolite compartmentation and transport in CAM plants. In *Plant Organelles* (A.K. Tobin, ed.), p. 141–167, Cambridge University Press.

Styring, S. and Jegerschöld, C. (1994) Light-induced reactions impairing electron transfer through photosystem II. In *Photoinhibition of Photosynthesis: from Molecular Mechanisms to the Field* (N.R. Baker and J.R. Bowyer, eds.), pp. 51–73, Bios, Oxford.

Telfer, A. & Barber, J. (1994) Elucidating the molecular mechanisms of photoinhibition by studying isolated photosystem II reaction centres. In *Photoinhibition of Photosynthesis: from Molecular Mechanisms to the Field* (N.R. Baker & J.R. Bowyer, eds), pp. 25–49. Bios, Oxford.

Trissl, H-W. & Wilhelm, C. (1993) Why do thylakoid membranes from higher plants form grana stacks? *Trends Biochem. Sci.* 18, 415–419.

Van Grondelle, R., Dekker, J.P., Gillbro, T. & Sundstrom, V. (1994) Energy transfer and trapping in photosynthesis. *Biochim. Biophys. Acta* 1187, 1–65.

Vermaas, W.F.J., Styring, S., Schröder, W.P. & Andersson, B. (1993) Photosynthetic water oxidation: the protein framework. *Photosynth. Res.* 38, 249–263.

Winter, K. (1985) Crassulacean acid metabolism. In *Photosynthetic Mechanisms and the Environment* (J. Barber & N.R. Baker, eds), Vol. 6, pp. 329–387. Elsevier, Amsterdam.

OTHER REFERENCES

Allen, J.F. (1992b) *Trends Biochem. Sci.* 17, 12–17.

Allen, J.F. (1993) *Photosynth. Res.* 36, 95–102.

Allen, J.F., Bennett, J., Steinback, K.E. & Arntzen, C. J. (1981) *Nature* 291, 25–29.

Andersen, B., Scheller, H.V. & Møller, B.L. (1992) *FEBS Lett.* 311, 169–173.

Anderson, J.M. (1982) *FEBS Lett.* 138, 62–66.

Anderson, J.M. (1986) *Annu. Rev. Plant Physiol.* 37, 93–136.

Anderson, J.M. (1989) *Physiol, Plant.* 76, 243–244.

Anderson, J.M., Chow, W.S. & Goodchild, D.J. (1988) *Aust. J. Plant. Physiol.* 15, 11–26.

Andersson, B. & Anderson, J.M. (1980) *Biochim. Biophys. Acta* 593, 427–440.

Andersson, B. & Åkerlund, H-E. (1987) In *Topics and Photosynthesis*, Vol. 7 *The Light Reactions* (J. Barber, ed.), pp. 379–420. Elsevier, Amsterdam.

Andersson, B. & Styring, S. (1991) In *Current Topics in Bioenergetics*, Vol. 16, (C.P. Lee, ed.), pp. 1–81. Academic Press, San Diego.

Andersson, B., Ponticos, M., Barber, J., Koivuniemi, A., Aro, E-M., Hagman, Å., Salter, A.H., Dan-Hui, Y. & Lindahl, M. (1994) In *Photoinhibition of Photosynthesis: from Molecular Mechanisms to the Field* (N.R. Baker & J.R. Bowyer, eds), pp. 143–159. Bios, Oxford.

Asada, K., Kanematsu, S., Okada, S. & Hayakawa, T. (1980) In *Chemical and Biochemical Aspects of Superoxide Dismutase* (J.V. Bannister & H.A.O. Hill, eds), pp. 128–135. Elsevier, Amsterdam.

Asami, S., Inoue, K. & Akazawa, T. (1979) *Arch. Biochem. Biophys.* 196, 581–587.

Astashkin, A.V., Kodera, Y. & Kawamori, A. (1994) *Biochim. Biophys. Acta* 1187, 89–93.

Aured, M., Moliner, E., Alfonso, M., Yruela, I., Toon, S.P., Seibert, M. & Picorel, R. (1994) *Biochim. Biophys. Acta* 1187, 187–190.

Baalman, E., Backhausen, J.E., Kitzman, C. & Scheibe, R. (1994) *Bot. Acta* 107, 313–320.

Babcock, G.T. & Sauer, K. (1973) *Biochim. Biophys. Acta* 325, 483–503.

Bakrim, N., Echevarria, C., Cretin, C., Arrio-Dupont, M., Pierre, J.N., Vidal, J., Chollet, R. & Gadal, P. (1992) *Eur. J. Biochem.* 204, 821–830.

Barber, J. (1982) *Annu. Rev. Plant Physiol.* 33, 261–295.

Barkan, A., Voelker, R., Mendel-Harting, J., Johnson, D. & Walker, M. (1995) *Physiol. Plant.* 93, 163–170.

Bassham, J.A. & Krause, G.H. (1969) *Biochim. Biophys. Acta* 189, 207–221.

Bassi, R. & Dainese, P. (1992) *Eur. J. Biochem.* 204, 317–326.

Bassi, R. & Simpson, D. (1987) *Eur. J. Biochem.* 163, 221–230.

Beattie, D.S. (1993) *J. Bioenerg. Biomemb.* 25, 233–234.

Bennett, J. (1979) *Eur. J. Biochem.* 104, 133–137.

Bennett, J. (1991) *Annu. Rev. Plant Physiol. Plant Mol. Biol.* 42, 281–311.

Berden, J.A., Hartog, A.F. & Edel, C.M. (1991) *Biochim. Biophys. Acta* 1057, 151–156.

Berthomieu, C. & Boussac, A. (1995) *Biochemistry* 34, 1541–1548.

Bielski, B.H.J. (1978) In *Ascorbic Acid: Chemistry, Metabolism and Uses* (P.A. Sieb & B.M. Tolbert, eds), pp. 81–100. American Chemical Society, Washington.

Björkman, O. (1981) In *Encyclopedia of Plant Physiology*, Vol. 12A (O.L. Lange, P.S. Nobel, C.B. Osmond & H. Ziegler, eds), pp. 57–107. Springer Verlag, Berlin.

Boekema, E.J., Wynn, R.M. & Malkin, R. (1990) *Biochim. Biophys. Acta* 1017, 49–56.

Bouges-Boquet, B. (1977) *Biochim. Biophys. Acta* 462, 371–379.

Boussac, A., Sétif, P. & Rutherford, A.W. (1992) *Biochemistry* 31, 1224–1234.

Boussac, A., Zimmermann, J-L. & Rutherford, A.W. (1990) *FEBS Lett.* 277, 69–74.

Bowes, J.M. & Crofts, A.R. (1980) *Biochim. Biophys. Acta* 590, 373–384.

Bremberger, C. & Lüttge, U. (1992) *Planta* 188, 575–580.

Bremberger, C., Haschke, H.P. & Lüttge, U. (1988) *Planta* 175, 465–470.

Brettel, K. (1988) *FEBS Lett.* 239, 93–98.

Brettel, K., Schlodder, E. & Witt, H.T. (1984) *Biochim. Biophys. Acta* 766, 403–428.

Breyton, C., de Vitry, C. & Popot, J-L. (1994) *J. Biol. Chem.* 269, 7597–7602.

Bricker, T.M. (1990) *Photosynth. Res.* 24, 1–13.

Bruist, M.F. & Hammes, G.G. (1981) *Biochemistry* 20, 6298–6305.

Burnap, R.L., Shen, J.R., Jursinic, P.A., Inoue, Y. & Sherman, L.A. (1992) *Biochemistry* 31, 7404–7410.

Burnell, J.N. (1986) *Aust. J. Plant Physiol.* 13, 577–587.

Burnell, J.N. (1987) *Aust. J. Plant Physiol.* 14, 517–525.

Butler, P.J.G. & Kühlbrandt, W. (1988) *Proc. Natl. Acad. Sci. U.S.A.* 85, 3797–3801.

Campbell, W.J. & Ogren, W.L. (1990) *Photosynth. Res.* 23, 257–268.

Canaani, O., Schuster, G. & Ohad, I. (1989) *Photosynth. Res.* 20, 129–146.

Carnal, N.W., Agostino, A. & Hatch, M.D. (1993) *Arch. Biochem. Biophys.* 306, 360–367.

Carnal, N.W. & Black, C.C. (1989) *Plant Physiol.* 90, 91–100.

Carter, P.J., Nimmo, H.G., Fewson, C.A. & Wilkins, M.B. (1991) *EMBO J.* 10, 2063–2068.

Cashmore, A.R. (1984) *Proc. Natl. Acad. Sci. U.S.A.* 81, 2960–2964.

Cavalier-Smith, T. (1992) *BioSystems* 28, 91–106.

Cerling, T.E., Wang, Y. & Quade, J. (1993) *Nature* 361, 344–345.

Chen, Z. & Spreitzer, R.J. (1992) *Photosynth. Res.* 31, 157–164.

Cleland, R.E. & Bendall, D.S. (1992) *Photosynth. Res.* 34, 409–418.

Collyer, C.A., Guss, J.M., Sugimura, Y., Yoshizaki, F. & Freeman, H.C. (1990) *J. Mol. Biol.* 211, 617–632.

Cramer, W.A., Theg, S.M. & Widger, W.R. (1986) *Photosynth. Res.* 10, 393–403.

Cramer, W.A., Furbacher, P.N., Szczepaniak, A. & Tae, G.S. (1991) *Curr. Top. Bioenerg.* 16, 179–222.

Cramer, W.A., Tae, G-S., Furbacher, P.N. & Böttger, M. (1993) *Physiol. Plant.* 89, 705–711.

Crofts, A.R. & Wraight, C.A. (1983) *Biochim. Biophys. Acta* 726, 149–185.

Crofts, A.R., Robinson, H.H. & Snozzi, M. (1984) In *Advances in Photosynthesis Research* (C. Sybesma, ed.), pp. 461–468. Martinus Nijhoff, Dordrecht.

Crofts, A.R., Robinson, H., Andrews, K., Van Doren, S. & Berry, E. (1987). In *Cytochrome Systems: Molecular Biology and Bioenergetics*, (S. Papa, B. Chance & L. Ernster, eds), pp. 617–624. Plenum Press, New York.

Cushman, J.C. (1993) *Photosynth. Res.* 35, 15–27.

Cushman, J.C., Meyer, G., Michalowski, C.B., Schmitt, J. & Bohnert, H.J. (1989) *Plant Cell* 1, 715–725.

Dainese, P. & Bassi, R. (1991) *J. Biol. Chem.* **266**, 8136–8142.

Danon, A. & Mayfield, S.P. (1994) *Science* **266**, 1717–1719.

Debus, R.J. (1992) *Biochim. Biophys. Acta.* **1102**, 269–352.

Debus, R.J., Barry, B.A., Babcock, G.T. & McIntosh, L. (1988a) *Proc. Natl. Acad. Sci. U.S.A.* **85**, 427–430.

Debus, R.J., Barry, B.A., Sithole, I., Babcock, G.T. & McIntosh, L. (1988b) *Biochemistry* **27**, 9072–9074.

Degli Eposti, M., DeVries, S., Crimi, M., Ghelli, A., Patarnello, T. & Meyer, A. (1993) *Biochim. Biophys. Acta* **1143**, 243–271.

Dekker, J.P., Plijter, J.J., Ouwehand, L. & van Gorkom, H.J. (1984) *Biochim. Biophys. Acta* **766**, 176–179.

De Las Rivas, J., Telfer, A. & Barber, J. (1992) *FEBS Lett.* **301**, 246–252.

De Las Rivas, J., Telfer, A. & Barber, J. (1993) *Biochim. Biophys. Acta* **1142**, 155–164.

Demmig-Adams, B. (1990) *Biochim. Biophys. Acta* **1020**, 1–24.

Demmig-Adams, B. & Adams, W.W. (1992) *Annu. Rev. Plant Physiol. Plant Mol. Biol.* **43**, 599–526.

Demmig-Adams, B. & Adams, W.W. (1993) In *Carotenoids in Photosynthesis* (A. Young & G. Britton, eds), pp. 206–251. Chapman & Hall, London.

Diner, B.A. & Petrouleas, V. (1987) *Biochim. Biophys. Acta* **893**, 138–148.

Diner, B.A. & Petrouleas, V. (1988) *Biochim. Biophys. Acta* **895**, 107–125.

Diner, B.A. & Petrouleas, V. (1990) *Biochim. Biophys. Acta* **1015**, 141–149.

Diner, B.A., Ries, D.F., Cohen, B.N. & Metz, J.G. (1988) *J. Biol. Chem.* **263**, 8972–8980.

Doncaster, H.D. & Leegood, R.C. (1987) *Plant Physiol.* **84**, 82–87.

Drepper, F., Carlberg, I., Andersson, B. & Haehnel, W. (1993) *Biochemistry* **32**, 11915–11922.

Dreyfuss, B.W. & Thornber, J.P. (1994) *Plant Physiol.* **106**, 841–848.

Dunn, S.D., Tozer, R.G. & Zardarozny, V.D. (1990) *Biochemistry* **29**, 4335–4340.

Durrant, J.R., Giorgi, L.B., Barber, J., Klug, D.R. & Porter, G. (1990) *Biochim. Biophys. Acta* **1017**, 167–175.

Durrant, J.R., Hastings, G., Joseph, D.M., Barber, J., Porter, G. & Klug, D.R. (1993) *Biochemistry* **32**, 8259–8267.

Ebbert, V. & Godde, D. (1994) *Biochim. Biophys. Acta* **1187**, 335–346.

Eckert, H.J., Wiese, N., Bernarding, J., Eichler, H.J. & Renger, G. (1988) *FEBS Lett.* **240**, 153–158.

Edwards, J.W., Walker, E.L. & Coruzzi, G.M. (1990) *Proc. Natl. Acad. Sci. U.S.A.* **87**, 3459–3463.

Erickson, J.M. & Rochaix, J-D. (1992) The molecular biology of photosystem II. In *The Photosystems: Structure, Function and Molecular Biology. Topics in Photosynthesis*, Vol. 11 (J. Barber, ed.), pp. 101–175. Elsevier, Amsterdam.

Escoubas, J-M., Lomas, M., LaRoche, J. & Falkowski, P.G. (1995) *Proc. Natl. Acad. Sci. U.S.A.*, **92**, 10237–10241.

Evans, M.C.W. & Nugent, J.H.A. (1993) In *The Photosynthetic Reaction Centre*, Vol I (J. Deisenhofer & J.R. Norris, eds), pp. 391–415. Academic Press, San Diego.

Fahrendorf, T., Holtum, J.A.M., Mukherjee, U. & Latzko, E. (1987) *Plant Physiol.* **84**, 182–187.

Fernyhough, P., Foyer, C.H. & Horton, P. (1984) *FEBS Lett.* **176**, 133–138.

Ford, R.C., Rosenburg, M.F., Shepherd, F.H., McPhie, P. & Holzenburg, A. (1995) *Micron* **26**, 133–140.

Fotinou, C. & Ghanotakis, D.F. (1990) *Photosynth. Res.* **25**, 141–145.

Foyer, C., Furbank, R., Harbinson, J. & Horton, P. (1990) *Photosynth. Res.* **25**, 83–100.

Frank, H.A. & Cogdell, R.J. (1993) In *Carotenoids in Photosynthesis* (A.J. Young & G. Britton, eds), pp. 252–326. Chapman & Hall, London.

Frid, D., Gal, A., Oettmeier, W., Hauska, G., Berger, S. & Ohad, I. (1992) *J. Biol. Chem.* **267**, 25908–25915.

Funk, C., Schröder, W.P., Green, B.R., Renger, G. & Andersson, B. (1994) *FEBS Lett.* **342**, 261–266.

Furbank, R.T., Agostino, A. & Hatch, M.D. (1991) *Arch. Biochem. Biophys.* **289**, 376–381.

Gal, A., Hauska, G., Herrmann, R. & Ohad, I. (1990) *J. Biol. Chem.* **265**, 19742–19749.

Gal, A., Herrmann, R.G., Lottspeich, F. & Ohad, I. (1992) *FEBS Lett.* **298**, 33–35.

Gardeström, P. & Edwards, G.E. (1985) In *Encyclopedia of Plant Physiology*, New Series Vol. 18 (R. Douce & D.A. Day, eds), pp. 314–346. Springer-Verlag, Berlin.

Gardeström, P. & Wigge, B. (1988) *Plant Physiol.* **88**, 69–76.

Ghanotakis, D.F. & Yocum, C.F. (1990) *Annu. Rev. Plant Physiol. Plant Mol. Biol.* **41**, 225–276.

Ghirardi, M.L., McCauley, S.W. & Melis, A. (1986) *Biochim. Biophys. Acta* **851**, 331–339.

Giersch, C., Heber, U., Kobayashi, Y., Inoue, Y., Shibata, K. & Heldt H.W. (1980) *Biochim. Biophys. Acta* **590**, 59–73.

Golbeck, J.H. (1993) *Curr. Opin. Struct. Biol.* **3**, 508–514.

Gong, H. & Ohad, I. (1992) *J. Biol. Chem.* **266**, 21293–21299.

Gonzalez-Halphen, D., Lindarfer, M.A. & Capaldi, R.A. (1988) *Biochemistry* **27**, 7021–7031.

Gounaris, K., Sundby, C., Andersson, B. & Barber, J. (1983) *FEBS Lett.* **156**, 170–174.

Gounaris, K., Chapman, D.J., Booth, P., Crystall, B., Giorgi, L.B., Klug, D.R., Porter, G. & Barber, J. (1990) *FEBS Lett.* **265**, 88–92.

Govindjee & van Rensen, J.J.S. (1993) In *The Photosynthetic Reaction Centre*, Vol. I (J. Deisenhofer & J.R. Norris, eds), pp. 357–389. Academic Press, San Diego.

Grace, S., Pace, R. & Wydrzynski, T. (1995) *Biochim. Biophys. Acta* **1229**, 155–165.

Greenberg, B.M., Gaba, V., Mattoo, A.K. & Edelman, M. (1987) *EMBO J.* **6**, 2865–2869.

Groth, G. & Junge, W. (1993) *Biochemistry* **32**, 8103–8111.

Grotjohann, I. & Gräber, P. (1990) *Biochim. Biophys. Acta* **1017**, 177–180.

Guenther, J.E. & Melis, A. (1990) *Photosynth. Res.* **23**, 105–109.

Haehnel, W. (1976) *Biochim. Biophys. Acta* **440**, 506–521.

Haehnel, W., Pröpper, A. & Krause, H. (1980) *Biochim. Biophys. Acta* **593**, 384–399.

Hager, A. (1980) In *Pigments in Plants* (F-C. Czygan, ed.), pp. 57–79. Fischer, Stuttgart.

Hager, A. & Holocher, K. (1994) *Planta* **192**, 581–589.

Haley, J. & Bogorad, L. (1989) *Proc. Natl. Acad. Sci. U.S.A.* **86**, 1534–1538.

Halliwell, B. & Foyer, C.H. (1978) *Planta* **139**, 9–16.

Han, K-C. & Katoh, S. (1992) In *Research in Photosynthesis*, Vol. II (N. Murata, ed.), pp. 365–368. Kluwer, Amsterdam.

Hangarter, R.P., Grandoni, P. & Ort, D.R. (1987) *J. Biol. Chem.* **262**, 13513–13519.

Hansson, Ö. & Wydrzynski, T. (1990) *Photosynth. Res.* **23**, 131–162.

Harley, P.C. & Sharkey, T.D. (1991) *Photosynth. Res.* **27**, 169–178.

Hatch, M.D. (1981) In *Photosynthesis IV*, pp. 227–236. Balaban International Science Services, Philadelphia.

Hatch, M.D. & Slack, C.R. (1966) *Biochem. J.* **101**, 103–111.

Hatch, M.D., Mau, S-L. & Kagawa, T. (1974) *Arch. Biochem. Biophys.* **165**, 188–200.

Hatch, M.D., Dröscher, L. & Heldt, H.W. (1984) *FEBS Lett.* **178**, 15–19.

Haumann, M. & Junge, W. (1994a) *FEBS Lett.* **347**, 45–50.

Haumann, M. & Junge, W. (1994b) *Biochemistry* **33**, 864–872.

Hauska, G., Hurt, E., Gabellini, N. & Lockau, W. (1983) *Biochim. Biophys. Acta* **726**, 97–133.

Heber, U., Neimanis, S., Dietz, K-J. & Viil, J. (1986) *Biochim. Biophys. Acta* **852**, 144–155.

Hecks, B., Wulf, K., Breton, J., Leibl, W. & Trissl, H-W. (1994) *Biochemistry* **33**, 8619–8624.

Heineke, D., Riens, B., Grosse, H., Hoferichter, P., Ute, P., Flügge, U.I. & Heldt, H.W. (1991) *Plant Physiol.* **95**, 1131–1137.

Heineke, D., Kruse, A., Flügge, U.I., Frommer, W.B., Riesmeier, J.W., Willmitzer, L. & Heldt, H.W. (1994) *Planta* **193**, 174–180.

Hind, G., Nakatani, H.Y. & Izawa, S. (1974) *Proc. Natl. Acad. Sci. U.S.A.* **71**, 1484–1488.

Ho, M-W. (1995) *Living Processes, Book 2, Bioenergetics*. Open University, Milton Keynes.

Holbrook, G.R., Bowes, G. & Salvucci, M.E. (1989) *Plant Physiol.* **90**, 673–678.

Holtum, J.A.M. & Winter, K. (1982) *Planta* **155**, 8–16.

Holzenburg, A., Bewley, M.C., Wilson, F.H., Nicholson, W.V. & Ford, R.C. (1993) *Nature* **363**, 470–472.

Holzenburg, A., Shepherd, F.H. & Ford, R.C. (1994) *Micron* **25**, 447–451.

Horton, P. & Ruban, A. (1992) *Photosynth. Res.* **34**, 375–385.

Hossain, M.A. & Asada, K. (1984) *Plant Cell Physiol.* **25**, 85–92.

Hossain, M.A. & Asada, K. (1985) *J. Biol. Chem.* **260**, 12920–12926.

Howitz, K. & McCarty, R.E. (1991) *Plant Physiol.* **96**, 1060–1069.

Huang, D., Everly, R.M., Cheng, R.H., Heymann, J.B., Schägger, H., Sled, V., Ohnishi, T., Baker, T.S. & Cramer, W.A. (1994) *Biochemistry* **33**, 4401–4409.

Huber, S.C., Sugiyama, T. & Akazawa, T. (1986) *Plant Physiol.* **82**, 550–554.

Hudson, G.S., Evans, J.R., von Caemmerer, S., Arvidsson, Y.B.C. & Andrews, T.J. (1992) *Plant Physiol.* **98**, 294–302.

Hundal, T., Virgin, I., Styring, S. & Andersson, B. (1990) *Biochim. Biophys. Acta* **1017**, 235–241.

Ide, J.P., Klug, D.R., Kühlbrandt, W., Giorgi, L. & Porter, G. (1987) *Biochim. Biophys. Acta* **893**, 349–364.

Jahns, P. & Junge, W. (1992) *Biochemistry* **31**, 7398–7407.

Jenkins, C.L.D., Burnell, J.N. & Hatch, M.D. (1987) *Plant Physiol.* **85**, 952–957.

Jenkins, C.L.D., Furbank, R.T. & Hatch, M.D. (1989) *Plant Physiol.* **91**, 1372–1381.

Jiao, J. & Chollet, R. (1992) *Plant Physiol.* **98**, 152–156.

Johnson, G.N., Rutherford, A.W. & Krieger, A. (1995) *Biochim. Biophys. Acta* **1229**, 202–207.

Jordan, D.B. & Ogren, W.L. (1981) *Nature* **291**, 513–515.

Kaiser, W.M. (1976) *Biochim. Biophys. Acta* **440**, 476–482.

Keegstra, K., Olsen, L.J. & Theg, S.M. (1989) *Annu. Rev. Plant Physiol. Plant Mol. Biol.* **40**, 471–501.

Keegstra, K., Bruce, B., Hurley, M., Li, H-M. & Perry, S. (1995) *Physiol. Plant.* **93**, 157–162.

Keren, N., Gong, H. & Ohad, I. (1995) *J. Biol. Chem.* **270**, 806–814.

Ketcham, S.R., Davenport, J.W., Woneke, K. & McCarty, R.E. (1984) *J. Biol. Chem.* **259**, 7286–7293.

Kim, J., Klein, P.G. & Mullett, J.E. (1991) *J. Biol. Chem.* **266**, 14931–14938.

Kleczkowski, L.A., Randall, D.D. & Zahler, W.L. (1985) *Arch. Biochem. Biophys.* **236**, 185–194.

Kobza, J. & Seemann, J.R. (1989) *Plant Physiol.* **89**, 918–924.

Kodera, Y., Takura, K. & Kawamori, A. (1992) *Biochim. Biophys. Acta* **1101**, 23–32.

Koivuniemi, A., Swiezewska, E., Aro, E-M., Styring, S. & Andersson, B. (1993) *FEBS Lett.* **327**, 343–346.

Koßmann, J., Sonnewald, U. & Willmitzer, L. (1994) *Plant J.* **6**, 637–650.

Kössel, H., Döry, I., Istran, G. & Maier, R. (1990) *Plant Mol. Biol.* **15**, 497–499.

Koulougliotis, D., Tang, X-S., Diner, B.A. & Brudvig, G.W. (1995) *Biochemistry* **34**, 2850–2856.

Kramer, D.M. & Crofts, A.R. (1994) *Biochim. Biophys. Acta* **1184**, 193–201.

Kramer, D.M., Wise, R.R., Frederick, J.R., Alm, D.M., Hesketh, J.D., Ort, D.R. & Crofts, A.R. (1990) *Photosynth. Res.* **26**, 213–222.

Krause, G.H. (1977) *Biochim. Biophys. Acta* **460**, 500–510.

Krause, G.H. & Weis, E. (1991) *Annu. Rev. Plant Physiol. Plant Mol. Biol.* **42**, 313–349.

Krauss, N., Hinrichs, W., Witt, I., Fromme, P., Pritzkow, W., Dauter, Z., Betzel, C., Wilson, K., Witt, H. & Saenger, W. (1993) *Nature* **361**, 326–331.

Kreimer, G., Melkonian, M., Holtum, J.A.M. & Latzko, E. (1988) *Plant Physiol.* **86**, 423–428.

Krieger, A. & Weis, E. (1992) *Photosynthetica* **27**, 89–98.

Krieger, A., Rutherford, A.W. & Johnson, G.N. (1995) *Biochim. Biophys. Acta* **1229**, 193–201.

Krömer, S. & Heldt, H.W. (1991) *Biochim. Biophys. Acta* **1057**, 42–50.

Kühlbrandt, W. & Wang, D.N. (1991) *Nature* **350**, 130–134.

Kühlbrandt, W., Wang, D.N. & Fujiyoshi, Y. (1994) *Nature* **367**, 614–621.

Larsson, U.K., Sundby, C. & Andersson, B. (1987a) *Biochim. Biophys. Acta* **894**, 59–68.

Larsson, U.K., Andersson, J.M. & Andersson, B. (1987b) *Biochim. Biophys. Acta* **894**, 69–75.

Lavergne, J. & Joliot, P. (1991) *Trends Biochem. Sci.* **16**, 129–134.

Leegood, R.C. (1985) *Planta* **164**, 163–171.

Leegood, R.C., Lea, P.J., Adcock, M.D. & Häusler, R.E. (1995) *J. Exp. Bot.* **46**, 1397–1414.

Leong, T-Y. & Anderson, J.M. (1983) *Biochim. Biophys. Acta* **723**, 391–399.

Lubbers, K., Drevenstedt, W. & Junge, W. (1993) *FEBS Lett.* **336**, 304–308.

Lunn, J.E. & Douce, R. (1993) *Biochem. J.* **290**, 375–379.

Lüttge, U., Smith, J.A.C. & Marigo, G. (1982) In *Crassulacean Acid Metabolism* (I.P. Ting & M. Gibbs, eds), pp. 69–91. American Society of Plant Physiologists, Rockville, Maryland.

MacDonald, G.M., Boerner, R.J., Everly, R.M., Cramer, W.A., Debus, R.J. & Barry, B.A. (1994) *Biochemistry* **33**, 4393–4400.

MacLachlan, D.J., Nugent, J.H.A., Bratt, P.J. & Evans, M.C.W. (1994) *Biochim. Biophys. Acta* **1186**, 186–200.

Malnoe, P., Mayfield, S.P. & Rochaix, J.D. (1988) *J. Cell. Biol.* **106**, 609–616.

Martinez, S.E., Huang, D., Szczepaniak, A., Cramer, W.A. & Smith, J.L. (1994) *Structure* **2**, 95–105.

Mayes, S.R., Cook, K.M., Self, S.J., Zhang, Z.H. & Barber, J. (1991) *Biochim. Biophys. Acta* **1060**, 1–12.

McCarty, R.E. & Hammes, G.G. (1987) *Trends Biochem. Sci.* **12**, 234–237.

McCauley, S.W. & Melis, A. (1986) *Photosynth. Res.* **8**, 3–16.

McNaughton, G.A.L., Fewson, C.A., Wilkins, M.B. & Nimmo, H.G. (1989) *Biochem. J.* **261**, 349–355.

Melis, A. & Anderson, J.M. (1983) *Biochim. Biophys. Acta* **724**, 473–484.

Messinger, J., Badger, M. & Wydrzynski, T. (1995) *Proc. Natl. Acad. Sci. U.S.A.* **92**, 3209–3213.

Michel, H. & Deisenhofer, J. (1988) *Biochemistry* **27**, 1–7.

Michel, H.P., Griffin, P.R., Shabanowitz, J., Hunt, D.F. & Bennett, J. (1991) *J. Biol. Chem.* **266**, 17584–17591.

Miller, A.J. & Sanders, D. (1987) *Nature* **326**, 397–400.

Mills, J.D. & Mitchell, P. (1984) *Biochim. Biophys. Acta* **764**, 93–104.

Miyake, C. & Asada, K. (1992) *Plant Cell Physiol.* **33**, 541–553.

Monson, R.K. & Moore, B.D. (1989) *Plant Cell Env.* **12**, 689–699.

Moore, B.D. & Seemann, J.R. (1992) *Plant Physiol.* **99**, 1551–1555.

Moore, B.D., Kobza, J. & Seemann, J.R. (1991) *Plant Physiol.* **96**, 208–213.

Moore, B.D., Sharkey, T.D., Kobza, J. & Seemann, J.R. (1992) *Plant Physiol.* **99**, 1546–1550.

Morden, C.W., Delwiche, C.F., Kuhsel, M. & Palmer, J.D. (1992) *BioSystems* **28**, 75–90.

Mullet, J.E. (1988) *Annu. Rev. Plant Physiol. Plant Mol. Biol.* **39**, 475–502.

Murata, S., Kobayashi, M., Matoh, T. & Sekiya, J. (1992) *Plant Cell Physiol.* **33**, 1247–1250.

Nanba, O. & Satoh, K. (1987) *Proc. Natl. Acad. Sci. USA* **84**, 109–112.

Neuhaus, H.E., Holtum, J.A.M. & Latzko, E. (1988) *Plant Physiol.* **87**, 64–68.

Nimmo, G.A., Nimmo, H.G., Fewson, C.A. & Wilkins, M.B. (1984) *FEBS Lett.* **178**, 199–203.

Nimmo, G.A., McNaughton, G.A.L., Fewson, C.A., Wilkins, M.B. & Nimmo, H.G. (1987) *FEBS Lett.* **213**, 18–22.

Ninnemann, O., Jauniaux J.-C. & Frommer, W.B. (1994) *EMBO J.* **13**, 3464–3471.

Nitschke, W. & Rutherford, A.W. (1991) *Trends Biochem, Sci.* **16**, 241–245.

Nixon, P.J. (1994) Abstracts *BBSRC Second Robert Hill Symposium on Photosynthesis*, p. 8. Imperial College, London.

Nixon, P.J. & Diner, B.A. (1992) *Biochemistry* **31**, 942–948.

Nixon, P.J. & Diner, B.A. (1994) *Biochem. Soc. Trans.* **22**, 338–343.

Nixon, P.J., Chisholm, D.A. & Diner, B.A. (1992a) In *Plant Protein Engineering* (P. Shewry & S. Gutteridge, eds), pp. 93–141. Cambridge University Press, Cambridge.

Nixon, P.J., Trost, J.T. & Diner, B.A. (1992b) *Biochemistry* **31**, 10859–10871.

Noctor, G., Rees, D., Young, A. & Horton, P. (1991) *Biochim. Biophys. Acta* **1057**, 320–330.

Noguchi, T., Ono, T. & Inoue, Y. (1995) *Biochim. Biophys. Acta* **1228**, 189–200.

Ohnishi, J. & Kanai, R. (1983) *Plant Cell Physiol.* **24**, 1411–1420.

Ohnishi, J. & Kanai, R. (1990) *FEBS Lett.* **269**, 122–124.

Ohnishi, J., Flügge, U.I., Heldt, H.W. & Kanai, R. (1990) *Plant Physiol.* **94**, 950–959.

Oliver, D.J., Neuburger, M., Bourguignon, J. & Douce, R. (1990) *Plant Physiol.* **94**, 833–839.

Ortiz-Lopez, A., Ort, D.K. & Boyer, J.S. (1991) *Plant Physiol.* **96**, 1018–1025.

Osborne, B.A. & Geider, R.J. (1988) *Photosynth. Res.* **16**, 291–292.

Osmond, C.B. (1994) In *Photoinhibition of Photosynthesis: from Molecular Mechanisms to the Field* (N.R. Baker & J.R. Bowyer, eds), pp. 1–24. Bios, Oxford.

Otto, J. & Berzborn, R.J. (1989) *FEBS Lett.* **250**, 625–628.

Owens, T.G., Shreve, A.P. & Albrecht, A.C. (1993) In *Research in Photosynthesis*, Vol. I (N. Murata, ed.), pp. 179–186. Kluwer, Dordrecht.

Pammenter, N.W., Loreto, F. & Sharkey, T.D. (1993) *Photosynth. Res.* **35**, 5–14.

Paul, M.J., Knight, J.S., Habash, D., Parry, M.A.J., Lawlor, D.W., Barnes, S.A., Loynes, A. & Gray, J.C. (1995) *Plant J.* **7**, 535–542.

Peter, G.F. & Thornber, J.P. (1991) *J. Biol. Chem.* **266**, 16745–16754.

Pierre, J.N., Pacquit, V., Vidal, J. & Gadal, P. (1992) *Eur. J. Biochem.* **210**, 531–537.

Polle, A. & Junge, W. (1986) *Biochim. Biophys. Acta* **848**, 265–273.

Prasil, O., Adir, N. & Ohad, I. (1992) In *The Photosystems: Structure, Function and Molecular Biology* (J. Barber, ed.), pp. 295–348. Elsevier, Amsterdam.

Preiss, J. (1982) *Annu. Rev. Plant Physiol.* **33**, 431–454.

Price, G.D., von Caemmerer, S., Evans, J.R., Yu, J-W., Lloyd, J., Oja, V., Kell, P., Harrison, K., Gallagher, A. & Badger, M.R. (1994) *Planta* **193**, 331–340.

Price, G.D., Evans, J.R., von Caemmerer, S., Yu, J-W. & Badger, M.R. (1995) *Planta* **195**, 369–378.

Quick, W.P., Schurr, U., Scheibe, R., Schulze, E-D., Rodermel, S.R., Bogorad, L. & Stitt, M. (1991) *Planta* **183**, 542–554.

Raines, C.A., Lloyd, J.C., Willingham, N.M., Potts, S. & Dyer, T.A. (1992) *Eur. J. Biochem.* **205**, 1053–1059.

Rawsthorne, S. (1992) *Plant J.* **2**, 267–274.

Rintamäki, E., Kettunen, R., Tyystjärvi, E. & Aro, E-M. (1995) *Physiol. Plant.* **93**, 191–195.

Robinson, H.H. & Crofts, A.R. (1983) *FEBS Lett.* **153**, 221–226.

Roelofs, T.A., Gilbert, M., Suvadov, V.A. & Holzworth, A.R. (1991) *Biochim. Biophys. Acta* **1060**, 237–244.

Ruban, A.V., Horton, P. & Young, A.J. (1993) *J. Photobiol. Photobiochem.* **21**, 229–234.

Ruban, A.V., Young, A.J., Pascal, A.A. & Horton, P. (1994) *Plant Physiol.* **104**, 227–239.

Ruffle, S.V., Donnelly, D., Blundell, T.L. & Nugent, J.H.A. (1992) *Photosynth. Res.* **34**, 287–300.

Rutherford, A.W. & Zimmermann, J.L. (1984) *Biochim. Biophys. Acta* **767**, 168–175.

Rutherford, A.W., Mullet, J.E. & Crofts, A.R. (1981) *FEBS Lett.* **123**, 235–237.

Sabbert, D., Engelbrecht, S. & Junge, W. (1996) *Nature* **381**, 623–625.

Sassenrath, G.F., Ort, D.R. & Portis, A.R. (1990) *Arch. Biochem. Biophys.* **282**, 302–308.

Schiffer, M. & Norris, J.R. (1993) In *The Photosynthetic Reaction Centre*, Vol. I (J. Deisenhofer & J.R. Norris, eds.), pp. 1–12. Academic Press, San Diego.

Schimkat, D., Heineke, D. & Heldt, H.W. (1990) *Planta* **181**, 97–103.

Sétif, P. (1992) In *The Photosystems: Structure, Function and Molecular Biology* (J. Barber, ed.), pp. 471–499. Elsevier, Amsterdam.

Sétif, P. & Mathis, P. (1980) *Arch. Biochem. Biopys.* **204**, 477–485.

Seibert, M. (1993) In *The Photosynthetic Reaction Centre*, Vol. 1 (J. Deisenhofer & J.R. Norris, eds), pp. 319–356. Academic Press, San Diego.

Sigfridsson, K., Hansson, O. & Brzezinski, P. (1995) *Proc. Natl. Acad. Sci. U.S.A.* **92**, 3458–3462.

Sigrist, M. & Staehelin, L.A. (1994) *Plant Physiol.* **104**, 135–145.

Silverstein, T., Cheng, L. & Allen, J.F. (1993) *Biochim. Biophys. Acta* **1183**, 215–220.

Simpson, D.J. (1979) *Carlsberg Res. Commun.* **44**, 305–336.

Singh, B.K., Greenberg, E. & Preiss, J. (1984) *Plant Physiol.* **74**, 711–716.

Soll, J., Wörzer, C. & Buchanan, B.B. (1983) In *Advances in Photosynthesis Research* (C. Sybesma, ed.), Vol. 3, pp. 485–488. Martinus Nijhoff, The Hague.

Sonnewald, U., Lerchi, J., Zrenner, R. & Frommer, W. (1994) *Plant Cell Environ.* **17**, 649–658.

Spilatro, S.R. & Preiss, J. (1987) *Plant Physiol.* **83**, 621–627.

Stitt, M. & Heldt, H.W. (1985a) *Biochim. Biophys. Acta* **808**, 400–414.

Stitt, M. & Heldt, H.W. (1985b) *Planta* **164**, 179–188.

Stitt, M. & Schulze, D. (1994) *Plant Cell Environ.* **17**, 465–487.

Stitt, M., Quick, W.P., Schurr, U., Schulze, E-D., Rodermel, S.R. & Bogorad, L. (1991) *Planta* **183**, 555–566.

Strotmann, H., Bickel-Sandkötter, S. & Shoshan, V. (1979) *FEBS Lett.* **101**, 316–320.

Strotmann, H. & Lohse, D. (1988) *FEBS Lett.* **229**, 308–312.

Styring, S. & Rutherford, A.W. (1987) *Biochemistry* **26**, 2401–2405.

Styring, S., Virgin, I., Ehrenberg, A. & Andersson, B. (1990) *Biochim. Biophys. Acta* **1015**, 269–278.

Svensson, B., Vass, I., Cedergren, E. & Styring, S. (1990) *EMBO J.* **7**, 2051–2059.

Tang, X-S., Diner, B.A., Larsen, B.S., Lane Gilchrist, M., Lorigan, G.A. & Britt, R.D. (1994) *Proc. Natl. Acad. Sci. U.S.A.* **91**, 704–708.

Taylor, M.A., Packer, J.C.L. & Bowyer, J.R. (1988) *FEBS Lett.* **237**, 229–233.

Telfer, A., Bishop, S.M., Phillips, D. & Barber, J. (1994a) *J. Biol. Chem.* **269**, 13244–13253.

Telfer, A., Dhami, S., Bishop, S.M., Phillips, D. & Barber, J. (1994b) *Biochemistry* **33**, 14469–14474.

Theg, S.M. & Scott, S.V. (1993) *Trends Cell Biol.* **3**, 186–190.

Thompson, L.K. & Brudvig, G.W. (1988) *Biochemistry* **27**, 6653–6658.

Ueno, O., Samejima, M., Muto, S. & Miyachi, S. (1988) *Proc. Natl. Acad. Sci. U.S.A.* **85**, 6733–6737.

Vallon, O., Bulté, L., Dainese, P., Olive, J., Bassi, R. & Wollman, F.A. (1991) *Proc. Natl. Acad. Sci. U.S.A.* **88**, 8262–8266.

Van Gorkom, H.J. (1985) *Photosynth. Res.* 6, 97–112.

Van Gorkom, H.J. & Schelvis, J.P.M. (1993) *Photosynth. Res.* 38, 297–301.

Van Mieghem, F.J.E., Nitschke, W., Mathis, P. & Rutherford, A.W. (1989) *Biochim. Biophys. Acta* 977, 207–214.

Van Mieghem, F.J.E., Brettel, K., Hillmann, B., Kamlowski, A., Rutherford, A.W. & Schlodder, E. (1995) *Biochemistry* 34, 4798–4813.

Van Wijke, K.J., Nilsson, O. & Styring, S. (1994) *Biochemistry* 269, 28382–28392.

Vass, I. & Styring, S. (1991) *Biochemistry* 30, 830–839.

Vass, I., Styring, S., Hundal, T., Koivuniemi, A., Aro, E-M. & Andersson, B. (1992) *Proc. Natl. Acad. Sci. U.S.A.* 89, 1408–1412.

Velthuys, B.R. (1979) *Proc. Natl, Acad. Sci., U.S.A.* 76, 2765–2769.

Vermaas, W.F.J., Charité, J. & Shen, G. (1990) *Biochemistry* 29, 5325–5332.

Virgin, I., Ghanotakis, D.F. & Andersson, B. (1990) *FEBS Lett.* 269, 45–48.

von Caemmerer, S. (1989) *Planta* 178, 376–387.

von Caemmerer, S. & Evans, J. (1991) *Aust. J. Plant Physiol.* 18, 287–000

Vu, C.V., Allen, L.H. & Bowes, G. (1983) *Plant Physiol.* 73, 729–734.

Vu, C.V., Allen, L.H. & Bowes, G. (1984) *Plant Physiol.* 76, 843–845.

Walker, R. P. & Leegood, R. C. (1996) *Biochem. J.* 317, 653–658.

Walker, R.P., Trevanion, S.J. & Leegood, R.C. (1995) *Planta* 196, 58–63.

Wang, Y-H. & Chollet, R. (1993) *Arch Biochem. Biophys.* 304, 496–502.

Wang, Z.Y. & Portis, A.R. Jr (1992) *Plant Physiol.* 99, 1348–1353.

Wasielewski, M.K., Fenton, J.M. & Govindjee (1987) *Photosynth. Res.* 12, 181–190.

Wasielewski, M.K., Johnson, D.G., Seibert, M. & Govindjee (1989) *Proc. Natl. Acad. Sci. U.S.A.* 86, 524–528.

Weiner, H. & Heldt, H.W. (1992) *Planta* 187, 242–246.

Weiss, E. & Berry, J.A. (1987) *Biochim. Biophys. Acta* 894, 198–208.

Whitmarsh, J. (1986) In *Photosynthesis III, Encyclopaedia of Plant Physiology*, Vol. 19 (L.A. Staehelin & C.J. Arntzen, eds), pp. 508–525. Springer-Verlag, Berlin.

Whitmarsh, J., Samson, G. & Poulson, M. (1994) In *Photoinhibition of Photosynthesis: from Molecular Mechanisms to the Field* (N.R. Baker & J.R. Bowyer, eds), pp. 75–93. Bios, Oxford.

Widger, W.R. & Cramer, W.A. (1991) In *Cell Culture and Somatic Cell Genetics of Plants: The Molecular Biology of Plastids* (L. Bogorad & I.K. Vasil, eds), pp. 149–176. Academic Press, San Diego.

Winter, K. (1982) *Planta* 154, 298–308.

Wiskich, J.T., Bryce, J.H., Day, D.A. & Dry, I.B. (1990) *Plant Physiol.* 93, 611–616.

Wolfe, G.R., Cunningham, F.X., Durnford, D., Green, B.R. & Gantt, E. (1994) *Nature* 367, 566–568.

Woo, K.C., Flügge, U.I. & Heldt, H.W. (1987) *Plant Physiol.* 84, 624–632.

Wood, P.M. (1988) *Biochem. J.* 253, 287–289.

Xu, Q., Jung, Y.S., Chitnis, V.P., Guikema, J.A., Golbeck, J.H. & Chitnis, P.R. (1994) *J. Biol. Chem.* 269, 21512–21518.

Yachandra, V.K., De Rose, V.J., Latimer, M.J., Mukerji, I., Sauer, K. & Klein, M.P. (1993) *Science* 260, 675–679.

Yu, J-W. & Woo, K.C. (1992) *Aust. J. Plant Physiol.* 19, 653–658.

Yu, L., Zhao, J., Mühlenhoff, U., Bryant, D.A. & Golbeck, J.H. (1993) *Plant Physiol.* 103, 171–180.

Zanetti, G., Aliverti, A. & Curti, B. (1984) *J. Biol. Chem.* 259, 6153–6157.

Zanetti, G., Morelli, D., Ronchi, S., Negri, A., Aliverti, A. & Curti, B. (1988) *Biochemistry* 27, 3753–3759.

Zimmermann, J.L. & Rutherford, A.W. (1986) *Biochim. Biophys. Acta* 851, 416–423.

3 Carbohydrate Metabolism: Primary Metabolism of Monosaccharides

M.D. Brownleader, J.B. Harborne and P.M. Dey

3.1 INTRODUCTION

This chapter refers to the use of monosaccharides as energy reserves in plants. Although a large plethora of substrates such as proteins and lipids can be oxidized in plants, respiration tends to be dominated by carbohydrate oxidation through the glycolytic pathway and the tricarboxylic acid (TCA) cycle. Carbohydrate is converted into pyruvate and malate by two major pathways, glycolysis and the oxidative pentose phosphate pathway. The primary roles of the pentose phosphate pathway, for example, to generate NADH for use in biosynthetic reactions and to provide ribose-5-phosphate for nucleotide synthesis and erythrose-4-phosphate for the synthesis of shikimic acid derivatives are discussed. We also refer to the regulation of the pentose phosphate pathway and interconversions between the glycolytic pathway. Lipids are established storage reserves in plants. During germination, lipid is converted into sugar by fatty acid β-oxidation, the glyoxylate cycle and gluconeogenesis. Gluconeogenesis and its regulation are discussed in this chapter.

Under aerobic conditions, pyruvate is oxidized still further to CO_2 by the TCA cycle generating $NADH/FADH_2$ equivalents. Its regulation and physiological roles are discussed. Both NADH and $FADH_2$ are oxidized by O_2 in the mitochondrial electron transport chain which is coupled to ADP phosphorylation. The structural configuration of the electron transport chain along with hypothesis accounting for the proposed coupling mechanism between electron transport and ATP synthesis are discussed.

This chapter highlights the major pathways of carbohydrate metabolism that play important roles in plant physiology such as improved crop yield.

3.2 GLYCOLYSIS

Glycolysis or the Embden–Meyerhof–Parnas pathway, first recognized in yeast cells and mammalian muscle tissue, is now one of the best established pathways of carbohydrate metabolism in plants (Fig. 3.1). It is one of two major pathways by which carbohydrate is degraded ultimately to CO_2, the other being the pentose phosphate pathway (section 3.3). These pathways are catabolic and are concerned with converting the carbon in reduced form, which has been captured from the atmosphere by photosynthesis, into an oxidized form with the release of energy. Glycolysis is a key metabolic feature of the respiratory process of the plant cell (section 3.4) and is particularly important in the cells of germinating seedlings and in non-photosynthetic cells of the mature plant.

PLANT BIOCHEMISTRY
ISBN 0-12-214674-3

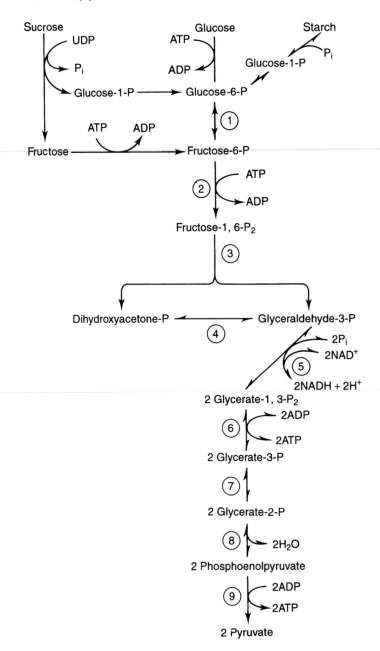

Figure 3.1. Reactions of glycolysis. Enzymes of the glycolytic sequence are: (1) hexose-P isomerase; (2) phosphofructokinase; (3) aldolase; (4) triose-P isomerase; (5) glyceraldehyde-3-P dehydrogenase; (6) glycerate-3-P kinase; (7) glycerate-P mutase; (8) enolase; (9) pyruvate kinase.

3.2.1 The pathway

The starting material for glycolysis is glucose-6-phosphate, which is available from the phosphorylation of glucose, catalyzed by a hexokinase. This reaction requires one molecule of ATP and the divalent metal cation Mg^{2+}, which forms a complex with ATP before it participates in the reaction. The phosphorylation of glucose at the 6-position is essentially irreversible, and the change in standard free energy $\Delta G^{\circ\prime}$ is $-16.7\,kJ\,mol^{-1}$.

Glucose-6-phosphate is also available to the plant from the isomeric glucose-1-phosphate, which is produced either by the hydrolysis of sucrose (Chapter 4) or as one of the products of starch breakdown. The interconversion of glucose-1-phosphate to the 6-phosphate is catalyzed by phosphoglucomutase (or glucose phosphate mutase) in a reversible reaction which shows only a small change in standard free energy ($-7.0 \, \text{kJ mol}^{-1}$). The optimum pH is 7.5 for this reaction, during which a phosphate ion is transiently attached to a serine side chain of the enzyme.

The first committed step of glycolysis is the conversion of glucose-6-phosphate into fructose-6-phosphate in a reversible reaction catalyzed by hexose phosphate isomerase:

Glucose-6-phosphate   Fructose-6-phosphate

$\Delta G^{\circ\prime}$ +1.7 kJ mol^{-1}

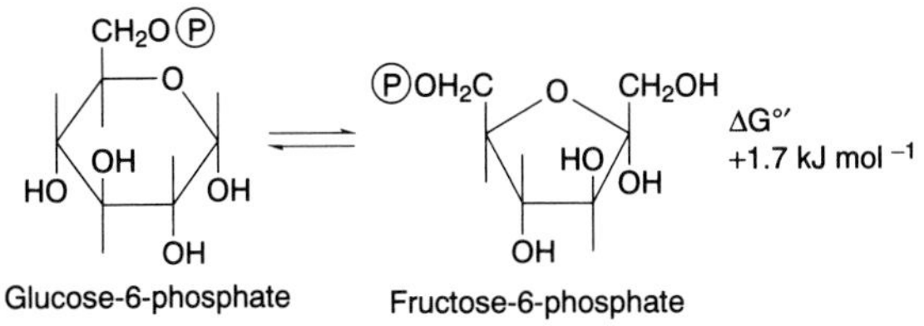

Scheme 1

This is an interesting interconversion by which a hexose, glucose, which takes up the pyranose configuration, is changed to a keto sugar, fructose, which has a furanose configuration.

The second step in glycolysis is the phosphorylation with organic phosphate, of fructose-6-phosphate to produce fructose 1,6-bisphosphate. This is catalyzed by phosphofructokinase in the presence of ATP in an irreversible reaction (Scheme 2).

There is a second enzyme, fructose-6-phosphate: pyrophosphate phosphotransferase, present in the cytosol which is capable of carrying out the same reaction, with pyrophosphate instead of ATP as the other substrate. It operates with only a small change in standard free energy ($\Delta G^{\circ\prime} - 2.9 \, \text{kJ mol}^{-1}$) and hence is readily reversible. Whether it has a role in glycolysis is still debatable. Its importance to carbohydrate metabolism appears to lie more in the reverse reaction, by which fructose-1,6-bisphosphate is hydrolyzed to supply pyrophosphate for sucrose breakdown via UDP-glucose pyrophosphorylase (see Chapter 4).

The third step in glycolysis is a key one in that it involves cleaving a carbon–carbon bond and dividing the molecule of fructose-1,6-bisphosphate into two three-carbon fragments. This reaction is catalyzed by an aldolase and yields two closely related products, dihydroxyacetone phosphate and glyceraldehyde-3-phosphate (Scheme 3).

The products of this reaction are in equilibrium with each other as keto-enol tautomers and the fourth step of glycolysis is the interconversion of one isomer into the other to give two molecules of glyceraldehyde-3-phosphate from the original fructose bisphosphate. This interconversion is catalyzed by triose phosphate isomerase, according to the reaction (Scheme 4).

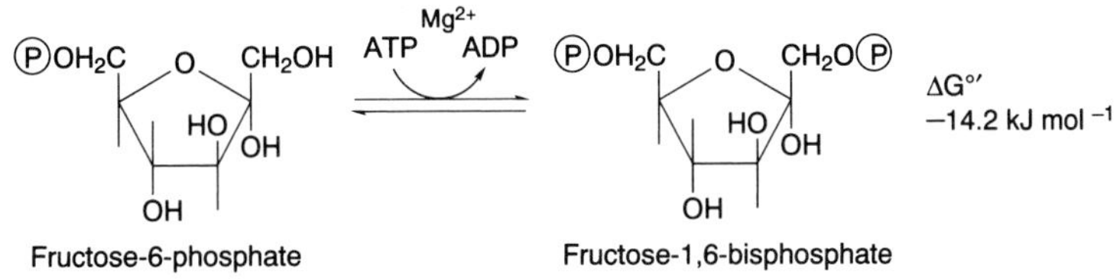

Fructose-6-phosphate   Fructose-1,6-bisphosphate

$\Delta G^{\circ\prime}$ −14.2 kJ mol^{-1}

Scheme 2

Fructose-1,6-bisphosphate   Dihydroxyacetone Glyceraldehyde-
                              phosphate   3-phosphate

$\Delta G^{\circ\prime}$ +29.9 kJ mol^{-1}

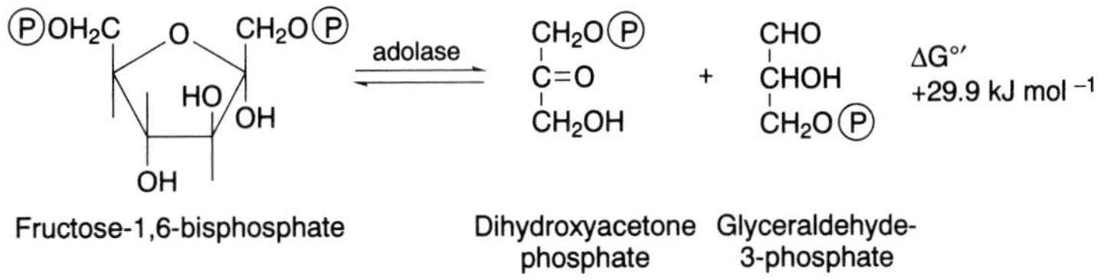

Scheme 3

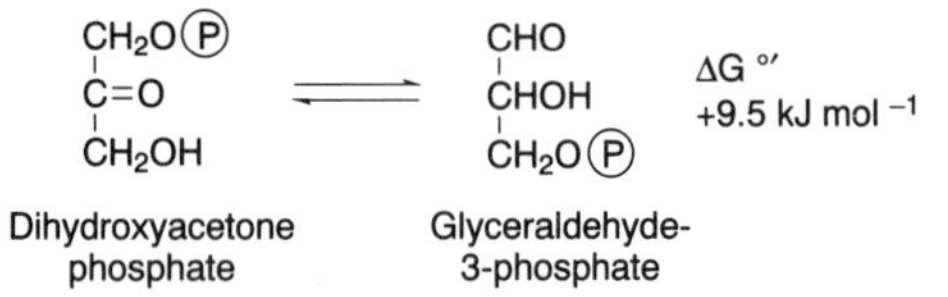

$$
\begin{array}{ccc}
\text{CH}_2\text{O}\,\text{P} & & \text{CHO} \\
\text{C=O} & \rightleftharpoons & \text{CHOH} \\
\text{CH}_2\text{OH} & & \text{CH}_2\text{O}\,\text{P}
\end{array}
\qquad
\begin{array}{c}
\Delta G^{\circ\prime} \\
+9.5 \text{ kJ mol}^{-1}
\end{array}
$$

Dihydroxyacetone   Glyceraldehyde-
phosphate   3-phosphate

Scheme 4

The effectiveness of the isomerase enzyme depends on the continual removal of the product in the next step of the reaction, since there is only a small change in standard free energy.

The fifth step in glycolysis is the conversion of glyceraldehyde-3-phosphate into glycerate-1,3-bisphosphate, a reaction requiring a molecule of NAD^+. A phosphate residue is added to the carboxyl residue of phosphoglyceric acid, according to the scheme:

$$
\begin{array}{ccc}
\text{CHO} & \overset{\text{Pi}}{} & \text{CO}_2\,\text{P} \\
\text{CHOH} & \longrightarrow & \text{CHOH} \\
\text{CH}_2\text{O}\,\text{P} & \underset{\text{NAD}^+\;\text{NADH}}{} & \text{CH}_2\text{O}\,\text{P}
\end{array}
\qquad
\begin{array}{c}
\Delta G^{\circ\prime} \\
+6.3 \text{ kJ mol}^{-1}
\end{array}
$$

Glyceraldehyde-   H$^+$
3-phosphate   Glycerate-
1,3-bisphosphate

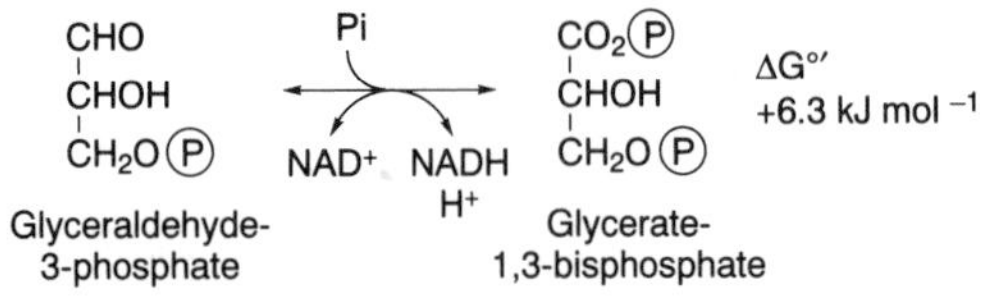

Scheme 5

The enzyme catalyzing this reaction, which occurs in two stages, is called glyceraldehyde-3-phosphate dehydrogenase. The reaction is freely reversible and effectively conserves the energy that is released in the dehydrogenation of aldehyde to give an acid.

The sixth step in glycolysis is the hydrolysis of the glycerate-1,3-bisphosphate, formed as indicated above, to give glycerate-3-phosphate. This reaction provides the plant cell with a molecule of ATP from ADP according to the scheme:

$$
\begin{array}{ccc}
\text{CO}_2\,\text{P} & \overset{\text{ADP}\quad\text{ATP}}{} & \text{CO}_2^- \\
\text{CHOH} & \longrightarrow & \text{CHOH} \\
\text{CH}_2\text{O}\,\text{P} & \underset{\text{Mg}^{2+}}{} & \text{CH}_2\text{O}\,\text{P}
\end{array}
\qquad
\begin{array}{c}
\Delta G^{\circ\prime} \\
-18.9 \text{ kJ mol}^{-1}
\end{array}
$$

Glycerate-   Glycerate-
1,3-bisphosphate   3-phosphate

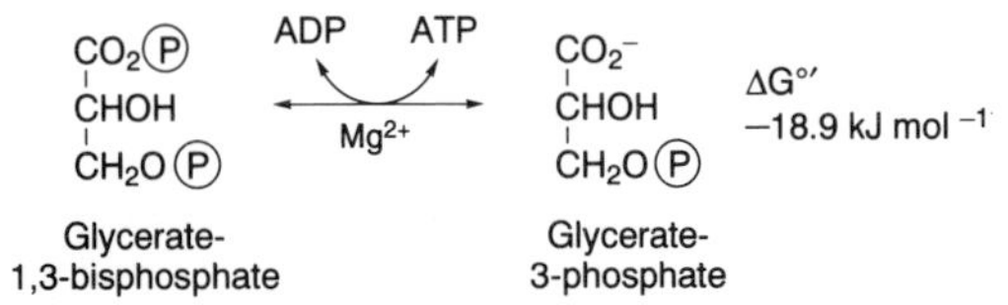

Scheme 6

This oxidation–hydrolysis is catalyzed by glycerate-3-phosphate kinase and there is a requirement for Mg^{2+}.

The seventh step in glycolysis is the equilibration of glycerate-3-phosphate with glycerate-2-phosphate, by which the phosphate group is moved from the 3- to the 2-hydroxyl group. The enzyme catalyzing this step is glycerate phosphate mutase and this is a reversible reaction, requiring little energy change:

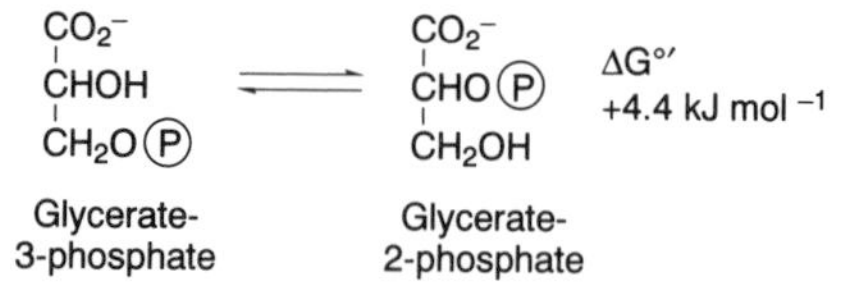

$$
\begin{array}{ccc}
\text{CO}_2^- & & \text{CO}_2^- \\
\text{CHOH} & \rightleftharpoons & \text{CHO}\,\text{P} \\
\text{CH}_2\text{O}\,\text{P} & & \text{CH}_2\text{OH}
\end{array}
\qquad
\begin{array}{c}
\Delta G^{\circ\prime} \\
+4.4 \text{ kJ mol}^{-1}
\end{array}
$$

Glycerate-   Glycerate-
3-phosphate   2-phosphate

Scheme 7

The eighth and penultimate step is the conversion of glycerate-2-phosphate into phosphoenolpyruvate (PEP), with the splitting out of a molecule of water, a reaction catalyzed by an enolase in the presence of Mg^{2+}, according to the scheme:

$$
\begin{array}{ccc}
\text{CO}_2^- & \overset{\text{H}_2\text{O}}{\underset{}{}} & \text{CO}_2^- \\
\text{CHO}\,\text{P} & \underset{\text{Mg}^{2+}}{\rightleftharpoons} & \text{C-O}\,\text{P} \\
\text{CH}_2\text{OH} & \underset{\text{H}_2\text{O}}{} & \text{CH}_2
\end{array}
\qquad
\begin{array}{c}
\Delta G^{\circ\prime} \\
+1.8 \text{ kJ mol}^{-1}
\end{array}
$$

Glycerate-   PEP
2-phosphate   (phosphoenolpyruvate)

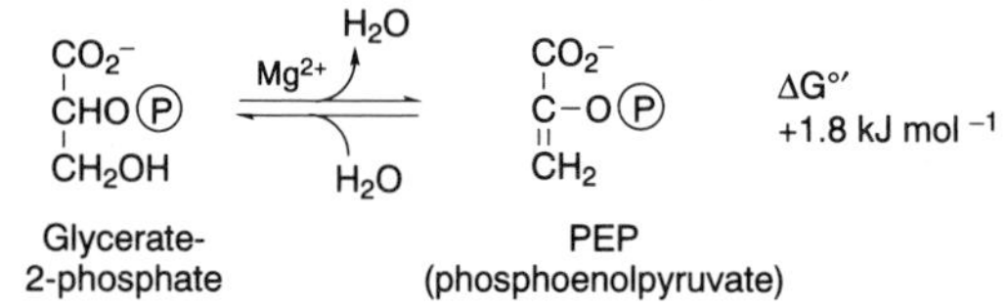

Scheme 8

The ninth and final step is the ketolization of PEP to pyruvate, catalyzed by pyruvate kinase, with the production of another molecule of ATP from ADP according to the reaction:

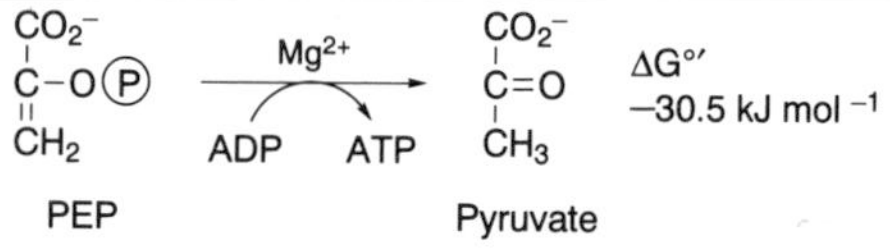

$$
\begin{array}{ccc}
\text{CO}_2^- & \overset{\text{Mg}^{2+}}{} & \text{CO}_2^- \\
\text{C-O}\,\text{P} & \longrightarrow & \text{C=O} \\
\text{CH}_2 & \underset{\text{ADP}\quad\text{ATP}}{} & \text{CH}_3
\end{array}
\qquad
\begin{array}{c}
\Delta G^{\circ\prime} \\
-30.5 \text{ kJ mol}^{-1}
\end{array}
$$

PEP   Pyruvate

Scheme 9

The overall process thus provides a considerable amount of energy to the plant cell. It requires one molecule of ATP at the beginning to produce fructose-1,6-bisphosphate from the monophosphate, but during the pathway, it yields four molecules of ATP and two molecules of NADH for every hexose molecule provided to the system (Fig. 3.1).

Pyruvate, the product of glycolysis, can be transformed in one of two directions, depending on the oxygen supply. Under aerobic conditions, it is converted into acetyl coenzyme A and carbon dioxide, a reaction catalyzed by pyruvate dehydrogenase. The acetyl coenzyme A so formed then enters the Krebs tricarboxylic cycle and the two carbon atoms of the acetyl moiety are finally returned to the atmosphere by the plant as respired CO_2 (section 3.4).

By contrast, under anaerobic conditions, e.g. in water-logged roots of a plant, pyruvate is metabolized by reductive enzymes. It is first decarboxylated to acetaldehyde, CH_3CHO, and this is then reduced to ethanol, CH_3CH_2OH, by alcohol dehydrogenase. Alternatively, pyruvate can be reduced to lactate, $CH_3CHOHCO_2^-$, catalyzed by lactate dehydrogenase. This is, however, likely to be only a minor pathway, since the lowering of the cytosolic pH caused by the product lactate, is likely to switch off lactate formation, allowing ethanol to accumulate instead. For further details of the metabolism that can take place in plant tissues under anaerobic conditions, see Crawford (1989).

Much evidence now indicates that the glycolytic pathway can operate both in the cytoplasm and in the chloroplasts within plant leaves. Most of the enzymes have been detected in chloroplasts, and all of them have been found in the cytoplasm. Furthermore, the enzymes of glycolysis occur in most plants investigated as pairs of isoenzymes, with different properties according to whether they are obtained from the chloroplast or from the cytoplasm. They are immunologically distinct and may be coded for by independent nuclear genes (Copeland & Turner, 1987).

3.2.2 Regulation

Our understanding of the control of glycolysis in plants is still incomplete (Turner & Turner, 1980; Copeland & Turner, 1987). However, there is good evidence that the major points of regulation involve the reactions catalyzed by phosphofructokinase and by pyruvate kinase. Three sorts of observation support this view.

First, of the nine steps in glycolysis outlined above, only the reactions catalyzed by phosphofructokinase and pyruvate kinase have ratios of product:substrate (mass action ratios) that are very much lower than their equilibrium constants. This indicates that these reactions are displaced from equilibrium and hence are irreversible under normal physiological conditions. By contrast, measurements of reactants and products and calculated mass action ratios for all the other steps in the pathway show that these reactions will operate reversibly and close to equilibrium.

Second, the rate of glycolysis can be upset by transferring plant tissues carrying out glycolysis from a gas phase of nitrogen to air or by exposing them to gaseous ethylene. Measurements of changes in the concentrations of intermediates in the pathway are then consistent with regulation of the phosphofructokinase and pyruvate kinase steps. For example, moving carrot root tissues from a nitrogen to an oxygen gas phase induces a threefold increase in the concentration of fructose-1,6-bisphosphate and a 60% increase in pyruvate, with smaller decreases in the concentrations of PEP and glycerate-3-phosphate.

Third, both the putative control enzymes, phosphofructokinase and pyruvate kinase, are sensitive to a range of positive and negative effectors (Copeland & Turner, 1987) and both have strategic positions in the pathway (Fig. 3.1).

In considering the regulation of glycolysis in plants, it is important to realize that under certain physiological conditions, the pathway can operate in the reverse direction with the formation of hexose sugar from PEP. This reversal is known as gluconeogenesis and takes place in germinating oil seeds (e.g. avocado) following the breakdown of storage lipid. It utilizes acetyl coenzyme A which is the product of fatty acid degradation (Chapter 6) and this enters a modified Krebs cycle to yield oxaloacetate which in turn is converted into PEP which is the starting point of gluconeogenesis. This metabolic process of converting lipid into carbohydrate is an essential feature in those plants which store most of the energy of the seed in the form of triglyceride fat rather than as starch. It should now be apparent that the enzymology of gluconeogenesis has an obvious bearing on the possible points of regulation in the glycolytic pathway.

Indeed, the reactions of glycolysis and gluconeogenesis differ distinctively in those two steps catalyzed by phosphofructokinase and pyruvate kinase, emphasizing once again their key roles in regulation. In fact, gluconeogenesis requires two different enzymes at these positions along the reverse pathway. Thus, the first step in gluconeogenesis is the conversion of oxaloacetate into PEP, catalyzed by PEP carboxykinase according to the reaction:

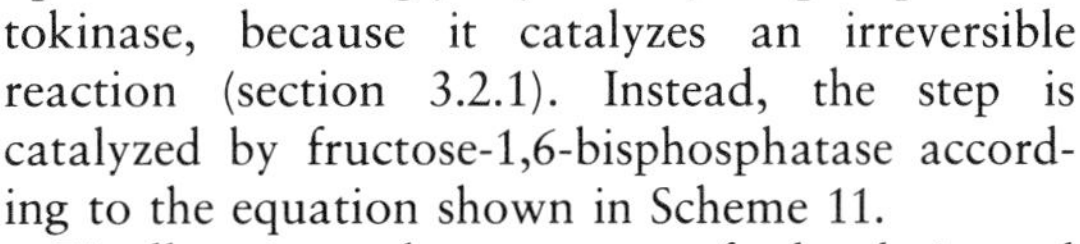

$$
\begin{array}{ccc}
\begin{array}{c} CO_2^- \\ | \\ CH_2 \\ | \\ COCO_2^- \end{array} + ATP & \rightleftharpoons & \begin{array}{c} CO_2^- \\ | \\ C-O(P) \\ | \\ CH_2 \end{array} + ADP + CO_2 \quad \begin{array}{c} \Delta G^{o\prime} \\ +4.2\ kJ\ mol^{-1} \end{array}
\end{array}
$$

Oxaloacetate   PEP

Scheme 10

Again, the eighth step in gluconeogenesis cannot operate with the glycolytic enzyme phosphofructokinase, because it catalyzes an irreversible reaction (section 3.2.1). Instead, the step is catalyzed by fructose-1,6-bisphosphatase according to the equation shown in Scheme 11.

Finally, since the enzymes of glycolysis and gluconeogenesis are located together in cytosol, a mechanism must exist to ensure control over these enzymes so that the flow of carbon operates in one or the other direction, according to the need of the plant at any particular stage of its life cycle.

It is now possible on the basis of what has gone before to suggest a scheme for the regulation of

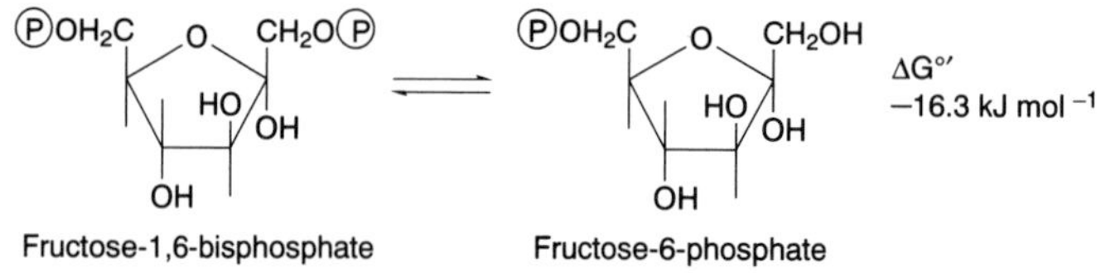

Fructose-1,6-bisphosphate   Fructose-6-phosphate $\Delta G^{\circ\prime}$ -16.3 kJ mol^{-1}

Scheme 11

glycolysis and this has been done by Turner & Turner (1980) (Fig. 3.2). One point of regulation, as argued above, is the conversion of PEP into pyruvate catalyzed by pyruvate kinase. As can be seen in Fig. 3.2, this reaction is inhibited by ATP but stimulated by ADP, so that a change in the ATP:ADP ratios in the cytoplasm could determine its effectiveness. Pyruvate kinase is also inhibited by citrate or Ca^{2+} and stimulated by K^+ or Mg^{2+}. The regulation of this enzyme by ATP and citrate can be observed when plant tissue is moved from air to nitrogen; levels of ATP and citrate decrease so that inhibition of this step is relieved and the path of carbon flows downward. Likewise, a decrease in ADP concentrations lowers pyruvate kinase activity, which causes an accumulation of PEP, which in turn causes an increase in glycerate-2-phosphate and glycerate-3-phosphate (Fig. 3.2). All these intermediates then inhibit the enzyme phosphofructokinase, turning off the flow of carbon from glucose-6-phosphate to pyruvate.

The regulation of glycolysis at the step in the pathway involving phosphofructokinase is rather complex and involves a further sugar phosphate not so far mentioned, fructose-2,6-bisphosphate. As already described, regulation of phosphofructokinase can be affected by a build-up in concentration of PEP, glycerate-2-phosphate or glycerate-3-phosphate. In addition, however, its catalysis is strongly inhibited by this new metabolite, fructose-2,6-bisphosphate, which has also been shown to occur in the cytosol. It is formed from fructose-6-phosphate by the action of the enzyme fructose-6-phosphate-2-kinase. It can likewise be dephosphorylated back to fructose-6-phosphate via the enzyme fructose-2,6-bisphosphatase, which has also been detected in the cytosol:

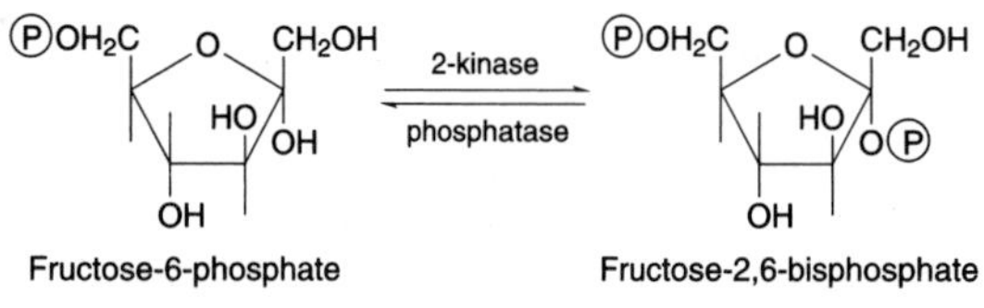

Hence the concentration of fructose-2,6-bisphosphate, which can vary between 1 and $30\,\mu M$, is controlled by the relative activities of the enzymes of its synthesis and degradation.

In this situation, a high concentration of triose phosphate stimulates its dephosphorylation and concomitantly inhibits its synthesis. By contrast, a high concentration of phosphate ion and of fructose-6-phosphate increases its synthesis and inhibits its degradation. Hence, carbon flow of glycolysis can be regulated through the fine control of this new metabolite.

If triose phosphate and glycerate-3-phosphate concentrations are high, the inhibition of fructose-2,6-bisphosphatase will be relieved, phosphofructokinase activity will be inhibited and carbon will now flow in the reverse direction towards gluconeogenesis. The reverse of these conditions will ensure that glycolysis continues in the direction towards oxidation, with the eventual production of acetyl coenzyme A. The metabolic cycle by which this product of glycolysis is converted into carbon dioxide is described in section 3.4.

3.3 PENTOSE PHOSPHATE PATHWAY

The main pathway for the oxidation of carbohydrates, synthesized during photosynthetic 'carbon fixation', is glycolysis followed by the TCA cycle. However, alternative oxidative pathways such as the oxidative pentose phosphate pathway (OPP), formerly called the hexose monophosphate shunt or phosphogluconate pathway, also exist. This pathway is involved in the interconversions and rearrangements of sugar phosphates with the net production of NADPH.

3.3.1 Location of the pathway

Both glycolysis and the pentose phosphate pathway occur in the same cytosolic compartment and can occur simultaneously resulting in precisely controlled complex and dynamic interactions. The cytosol of both non-photosynthetic (e.g. cauliflower buds) and photosynthetic cells (e.g. pea and spinach leaves) are believed to contain all the enzymes involved in both oxidative

Fructose-6-phosphate   Fructose-2,6-bisphosphate

Scheme 12

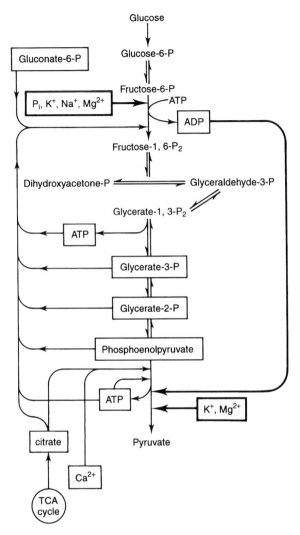

Figure 3.2. A scheme for the regulation of glycolysis. Stimulation of enzyme activity is indicated by bold lines whereas inhibition is shown by grey lines. From Turner & Turner (1980). P = phosphate.

and non-oxidative components of the pentose phosphate pathway. Work by Schnarrenberger *et al.* (1995) with spinach leaves has shown a predominant localization of the enzyme activities of the pentose phosphate pathway in the chloroplast fraction. Activity was also found in the cytosolic fraction and this was attributed to the cytosol-specific isoforms known to exist for these enzymes. However, further analysis appeared to show that chloroplasts of spinach leaf cells possess the complete complement of enzymes of the oxidative pentose phosphate pathway whereas the cytosol contains the enzymes of the first two reactions of the pentose phosphate pathway, contrary to the widely held view that plants generally possess a cytosolic oxidative pentose phosphate pathway capable of cyclic function.

3.3.2 The pathway

There are two primary reaction sequences in this pathway (Fig. 3.3), an oxidative phase (phase 1) from glucose-6-phosphate to ribulose-5-phosphate and a non-oxidative phase (phase 2) from ribulose-5-phosphate to a hexose phosphate and a triose phosphate.

(a) Phase 1

The first step is the oxidation of glucose-6-phosphate to δ-glucono-1,5-lactone-6-phosphate and is catalyzed by glucose-6-phosphate dehydrogenase with the simultaneous reduction of $NADP^+$. This reaction with $\Delta G^{\circ\prime}$: $-0.42\,kJ\,mol^{-1}$ is freely reversible. The enzyme

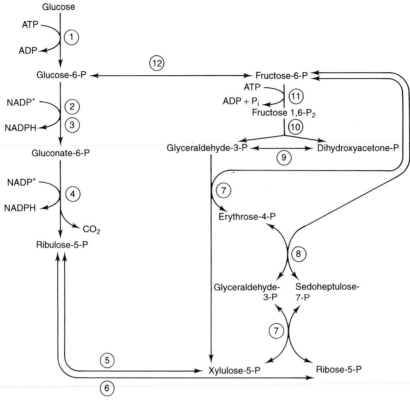

Figure 3.3. The pentose phosphate pathway and its links to glycolysis. (1) Hexokinase, (2) glucose-6-phosphate dehydrogenase, (3) gluconate-6-phosphate lactonase, (4) gluconate-6-phosphate dehydrogenase, (5) ribulose phosphate 3-epimerase, (6) ribose phosphate isomerase, (7) transketolase, (8) transaldolase, (9) triose phosphate isomerase, (10) aldolase, (11) phosphofructokinase, (12) hexose phosphate isomerase.

has high specificity for $NADP^+$ rather than NAD^+. Its minimum active oligomeric form is a dimer, although oligomeric and higher order forms have been observed. Cytosolic and plastidic isoforms of glucose-6-phosphate dehydrogenase exist in plant cells. This enzyme has been purified from pea seedlings and comprises a heterotetramer with a molecular weight of 244 kDa. Interestingly, the enzyme has a K_m for $NADP^+$ of 14 μM and a K_m for glucose-6-phosphate of 120 μM and appears to be absolutely specific for these two substrates. Antibodies raised against the cytosolic isoenzyme of glucose-6-phosphate dehydrogenase and the plastidic isoform do not cross-react with their respective isoforms. Non-enzymic hydrolysis of gluconolactone-6-phosphate can occur spontaneously. However, formation of gluconate-6-phosphate takes place in the next step. This second step is catalyzed by gluconate-6-phosphate lactonase which requires Mg^{2+} and the reaction is irreversible ($\Delta G^{o\prime}$: $-20.9\,kJ\,mol^{-1}$) yielding gluconate-6-phosphate (Scheme 13).

The next step of this pathway is the irreversible oxidative decarboxylation of gluconate-6-phosphate to ribulose-5-phosphate releasing CO_2 with the concomitant reduction of $NADP^+$ and is catalyzed by gluconate-6-phosphate dehydrogenase.

Glucose-6-phosphate dehydrogenase   Gluconate-6-phosphate lactonase

$$NADP^+ + H^+ \quad NADPH \qquad\qquad Mg^{2+}$$

Glucose-6-P   Gluconolactone-6-P   Gluconate-6-P

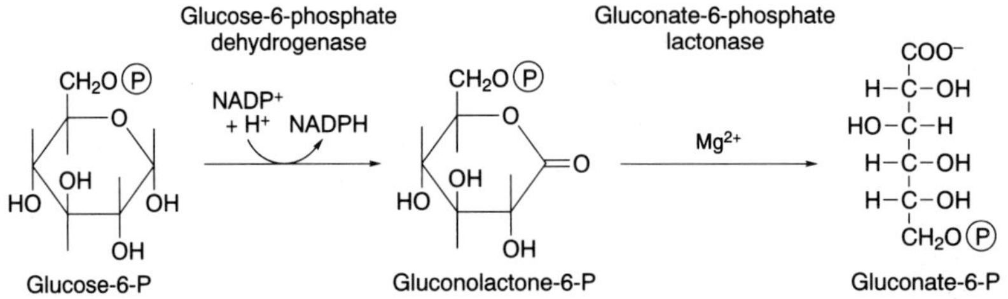

Scheme 13

This enzyme (1) requires divalent ions for maximum catalytic activity, (2) is specific for $NADP^+$ and (3) is believed to form 3-keto-gluconate-6-phosphate as a reaction intermediate. Several isoenzymes have been reported in the cytosolic and plastidic compartments.

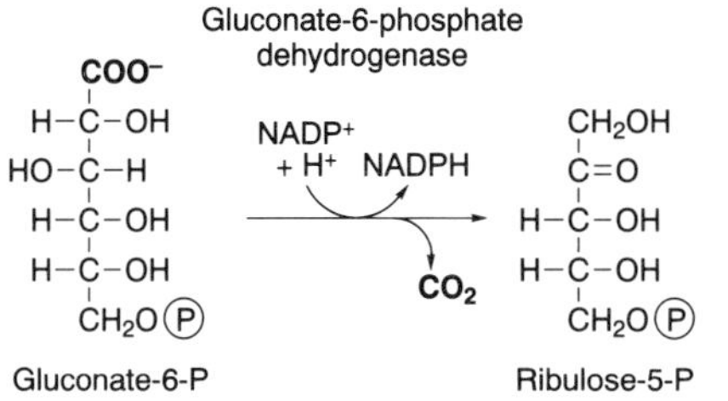

Gluconate-6-phosphate
dehydrogenase

$$\begin{array}{ccc}
COO^- & & CH_2OH \\
H-C-OH & & C=O \\
HO-C-H & NADP^+ + H^+ \quad NADPH & H-C-OH \\
H-C-OH & \longrightarrow & H-C-OH \\
H-C-OH & CO_2 & CH_2O\text{\textcircled{P}} \\
CH_2O\text{\textcircled{P}} & & \\
\end{array}$$

Gluconate-6-P   Ribulose-5-P

Scheme 14

(b) Phase 2

The following reactions are a series of reversible interconversions and rearrangements (little change in $\Delta G^{o\prime}$) of the sugar phosphates.

Ribulose-5-phosphate can be converted either into ribose-5-phosphate in an isomerization reaction catalyzed by ribose-5-phosphate isomerase or into xylulose-5-phosphate (the hydroxyl group of ribulose-5-phosphate at the C-3 position is inverted) by ribulose-5-phosphate-3-epimerase.

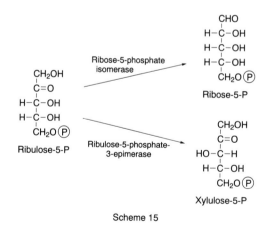

Scheme 15

Ribose phosphate isomerase has been purified from alfalfa shoots and spinach leaf chloroplasts and appears to exist as dimers, trimers or tetramers with molecular weights in the range 40–228 kDa and a K_m value of 0.5–5.0 mM for ribose-5-phosphate. Plant cells typically contain cytosolic and plastidic isoforms. Only limited amount of data is available for ribulose-5-phosphate epimerase.

The next sequence of reactions is catalyzed by a transketolase and a transaldolase. The transketolase transfers a two-carbon unit from a ketose substrate to an aldehyde whereas transaldo-lase transfers a three-carbon glyceraldehyde-3-phosphate unit. Enzymatic reactions that form or break carbon–carbon bonds such as those present in transaldolase and transketolase reactions usually generate a stabilized carbanion which will be added to an electrophilic center such as an aldehyde.

In this pathway, the transketolase transfers a C_2-moiety from xylulose-5-phosphate to ribose-5-phosphate yielding a C_7 keto-sugar-phosphate, sedoheptulose-7-phosphate and a C_3 aldo-sugar-phosphate, glyceraldehyde-3-phosphate. The reaction is reversible. Tightly bound cofactors thiamine pyrophosphate (TPP) and Mg^{2+} are required for catalytic activity. A covalent adduct between xylulose-5-phosphate and TPP is formed in this reaction.

Transketolase has been purified to homogeneity from spinach leaves and there are both cytosolic and plastidic isoenzymes. However, it appears that most of the enzyme activity resides in the plastid. The enzyme has a K_m of about $100\,\mu M$ for xylulose-5-phosphate, erythrose-4-phosphate and ribose-5-phosphate and its native molecular weight is about $150\,kDa$ with a monomeric subunit of 37.6 kDa (Scheme 16).

Transaldolase catalyzes the freely reversible interconversion between the pair, sedoheptulose-7-phosphate and glyceraldehyde-3-phosphate, and the pair, erythrose-4-phosphate and fructose-6-phosphate. A three carbon unit is transferred from sedoheptulose-7-phosphate to glyceraldehyde-3-phosphate with the formation of a C_4 product and hexose phosphate (Scheme 17).

Transaldolase operates in a similar manner to class I aldolase and utilizes a base-catalyzed aldol cleavage reaction in which a Schiff base intermediate is formed between an ϵ-amino group of an essential enzyme-lysine residue and the carbonyl group of sedoheptulose-7-phosphate. The plant enzyme does not require any cofactor and has not yet been well characterized.

A second transketolase reaction transfers a C_2-moiety to erythrose-4-phosphate from xylulose-5-phosphate (formed as shown earlier) to form a second fructose-6-phosphate and glyceraldehyde-3-phosphate which can enter glycolysis. (Scheme 18).

Relatively little attention has been paid to the study of the oxidative pentose phosphate pathway in plant cells. There are clearly two discrete pathways, in the cytosolic and the plastid compartments, which appear to be controlled by a different complement of isoforms.

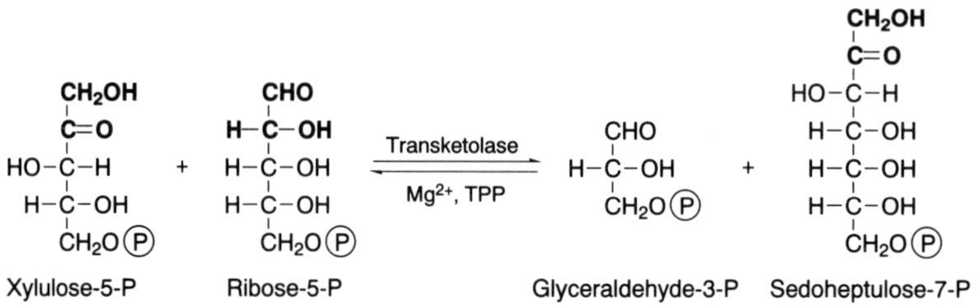

Scheme 16

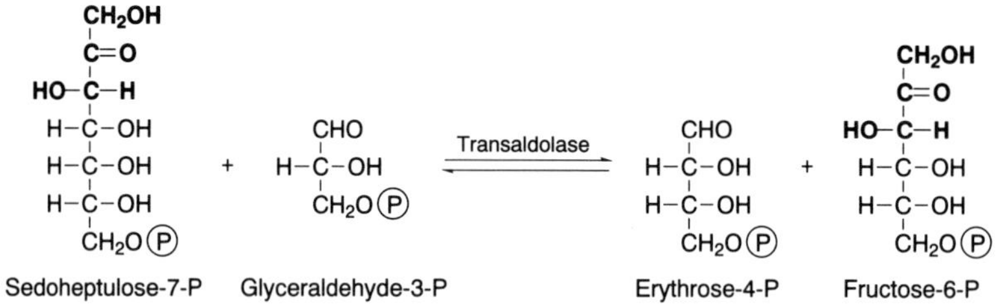

Xylulose-5-P + Ribose-5-P $\xrightleftharpoons[\text{Mg}^{2+}, \text{TPP}]{\text{Transketolase}}$ Glyceraldehyde-3-P + Sedoheptulose-7-P

Scheme 17

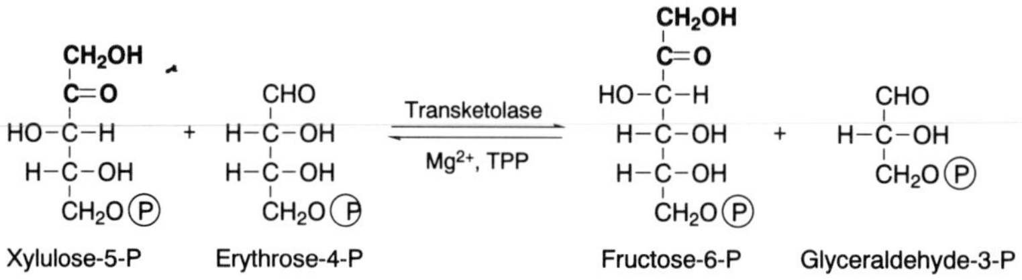

Sedoheptulose-7-P + Glyceraldehyde-3-P $\xrightleftharpoons{\text{Transaldolase}}$ Erythrose-4-P + Fructose-6-P

Scheme 18

Xylulose-5-P + Erythrose-4-P $\xrightleftharpoons[\text{Mg}^{2+}, \text{TPP}]{\text{Transketolase}}$ Fructose-6-P + Glyceraldehyde-3-P

Three molecules of glucose-6-phosphate entering the oxidative pentose phosphate pathway are oxidized to three molecules of the pentose phosphate, ribulose-5-phosphate, and CO_2 with the formation of six NADPH molecules. The three molecules of ribulose-5-phosphate are used in the non-oxidative phase of the pathway to regenerate two molecules of fructose-6-phosphate and one molecule of glyceraldehyde-3-phosphate which are both glycolytic intermediates.

3.3.3 Regulation

Although the pentose phosphate pathway can completely convert glucose-6-phosphate into CO_2 (see Fig. 3.3; recycling the product, fructose-6-phosphate to glucose-6-phosphate), the more usual products are glyceraldehyde-3-phosphate and fructose-6-phosphate that will then enter glycolysis. The pathway can therefore operate as a cycle depending upon cellular requirements.

The oxidative pentose phosphate pathway converts between 15 and 30% of hexose phosphate to glyceraldehyde-3-phosphate and CO_2 in pea and spinach chloroplasts. Glucose-6-phosphate dehydrogenase and a lactonase catalyze the first committed step of the oxidative pentose phosphate pathway which is a strategic control point. It is the major branch point between glycolysis and the oxidative pentose phosphate pathway. The products of the pentose phosphate pathway depend critically on cellular requirements because epimerase, isomerase, transketolase- and transaldolase-catalyzed reactions are freely reversible.

Glucose-6-phosphate dehydrogenase activity is under coarse and fine regulatory control. A marked increase in its activity in sliced potato root during aerobic respiration was also observed. Substantial increases in activity of glucose-6-phosphate dehydrogenase have also been observed after ageing of carrot, swede and potato disks. The chloroplast isoenzyme is affected by the NADPH/NADP$^+$ ratio, pH, Mg^{2+} and levels of glucose-6-phosphate. Ashihara & Komamine (1976) purified glucose-6-phosphate dehydrogenase from the hypocotyls of *Phaseolus mungo* seedlings and showed that inhibition by NADPH was inversely related to pH. Glucose-6-phosphate dehydrogenase is also inhibited by ribulose-1,5-bisphosphate. Control of the chloroplast isoform by the NADPH/NADP$^+$ ratio may therefore be amplified by ribulose-1,5-bisphosphate. Both the cytoplasmic and chloroplastic isoforms of glucose-6-phosphate dehydrogenase from pea leaves are activated in the dark (low NADPH/NADP$^+$); light reactions of photosynthesis generate NADPH. Gluconate-6-phosphate dehydrogenase is also strongly inhibited by NADPH and fructose-2,6-bisphosphate.

In the oxidative pentose phosphate pathway, NADPH is formed through the reactions catalyzed by glucose-6-phosphate dehydrogenase and gluconate-6-phosphate dehydrogenase. The K_i values of NADPH for both enzymes are 11 μM and 20 μM, respectively and the pentose phosphate pathway is therefore regulated by the NADPH/NADP$^+$ ratio. Treatment of plant tissues with methylene blue and nitrate, which accepts electrons from NADPH stimulates the oxidative pentose phosphate pathway. The ratio of NADPH/NADP$^+$ appears to be the principal factor regulating the flux through the pentose phosphate pathway. A reduced NADPH/NADP$^+$ ratio should, in principal, signal an increase in cellular demand for NADPH, activation of glucose-6-phosphate dehydrogenase and an increase in the flux through the oxidative pentose phosphate pathway.

The final reactions of the pentose phosphate pathway, catalyzed by ribose phosphate isomerase, ribulose phosphate 4-epimerase, transketolase and transaldolase are close to equilibrium.

Whether glucose-6-phosphate enters the oxidative pentose phosphate pathway or the glycolytic pathway in plant cells is critical to our understanding of respiratory glucose metabolism. Experiments measuring $^{14}CO_2$ yields and labeling patterns of various intermediates suggest that 5–15% of respiratory glucose metabolism in plant cells proceeds through the oxidative pentose phosphate pathway and will probably not exceed 30% relative to glycolysis. However, the relative amount of glucose metabolized in the oxidative pentose phosphate pathway and glycolysis remains unclear.

3.3.4 Role

The role of the oxidative pentose phosphate pathway appears to be the production of reducing equivalent, NADPH for cytosolic biosynthetic reactions and the sugar phosphates used in the synthesis of nucleotides, lipids and cell wall polymers. The oxidation of NADPH formed in this pathway is not coupled to ADP phosphorylation. In chloroplasts, NADPH is believed to maintain sufficient levels of reduced glutathione which has several important cellular functions such as the H_2O_2 detoxification. Here, it is important to note that NADPH generated in light reactions in chloroplasts is not freely available to the cytosol. Ribose-5-phosphate and erythrose-4-phosphate generated in the pathway are available for the synthesis of nucleotides and shikimic acid (phenolic) derivatives (precursors of aromatic amino acids and lignin in plant cell walls), respectively.

3.4 THE CITRIC ACID CYCLE

The citric acid cycle has a key role in carbohydrate metabolism, since it completes the catabolism of glucose to carbon dioxide. It effectively 'burns up' acetyl coenzyme A the final product of glycolysis which operates in the cytosol, and this occurs specifically in the mitochondria. The cycle is also known as the tricarboxylic acid (TCA) cycle, to emphasize the fact that organic acids are the intermediates of the cycle, or as the Krebs cycle, in honour of Hans Krebs, who first isolated the enzymes of the cycle from animal tissues in 1937.

3.4.1 Outline of the cycle

The cycle (Fig. 3.4) has been described as 'a clever way to cleave a reluctant carbon–carbon bond', alluding to the great stability of acetate, which in the form of the coenzyme A ester is the starting material. It involves as the first step, the linking of acetate (C$_2$) to oxaloacetate (C$_4$) to yield the C$_6$ tricarboxylic acid citrate. Two enzyme-catalyzed rearrangements then take place to produce in turn two α-keto acids, oxalosuccinate (C$_6$) and α-ketoglutarate (C$_5$), both of which are readily decarboxylated in turn with the loss of CO$_2$. The product, a C$_4$ acid, succinate, then undergoes regeneration in three steps (dehydrogenation, hydration and dehydrogenation) to produce oxaloacetate to complete the cycle. As the cycle turns, energy is liberated in the form of NADH, FADH$_2$ and ATP.

Before discussing the enzymology of the cycle in more detail, it is necessary to mention the reaction that comprises the oxidation of pyruvate to acetyl coenzyme A, since this provides a direct link between glycolysis and the citric acid cycle. This is a complex three-step process, involving two carrier molecules, thiamine pyrophosphate (TPP) and lipoic acid, which is also present as the dihydro derivative, dihydrolipoate (DHL).

Thus, the oxidation of pyruvate to acetyl coenzyme A is catalyzed by an enzyme complex, pyruvate dehydrogenase, and occurs as shown in Scheme 19:

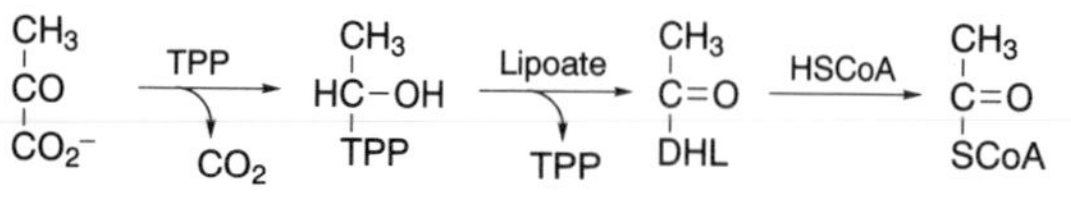

$$
\begin{array}{ccccc}
\underset{\substack{|\\ CO \\ |\\ CO_2^-}}{CH_3} & \xrightarrow[CO_2]{TPP} & \underset{\substack{|\\ HC-OH \\ |\\ TPP}}{CH_3} & \xrightarrow[TPP]{Lipoate} & \underset{\substack{|\\ C=O \\ |\\ DHL}}{CH_3} & \xrightarrow{HSCoA} & \underset{\substack{|\\ C=O \\ |\\ SCoA}}{CH_3}
\end{array}
$$

Scheme 19

In the first step, decarboxylation occurs and the acetyl moiety is transferred to thiamine pyrophosphate with reduction of the carboxyl group, then this intermediate, α-hydroxyethyl TPP, undergoes another transfer to lipoate, during which the two hydrogen atoms of α-hydroxyethyl TPP reduce lipoate to dihydrolipoate (DHL). The final stage in the process involves another transfer reaction with coenzyme A, with the release of DHL and the formation of the coenzyme A ester of acetate. The enzyme complex has a tightly bound FAD coenzyme which is linked as an electron acceptor to NAD^+ and energy is released from the reaction as NADH. The overall change in standard free energy is $-9.4\,kJ\,mol^{-1}$. This then is how acetyl coenzyme A is supplied to the citric acid cycle via glycolysis. However, acetyl coenzyme A is also available to the cell as the product of the β-oxidation spiral, i.e. the oxidative degradation of fatty acids. The source of acetyl coenzyme A for the citric acid cycle will then vary from time to time within the plant, depending on what metabolic processes are in operation at any one time.

3.4.2 The enzymes of the cycle

The eight enzymes of the citric acid cycle, and the reactions they catalyze, are shown in Fig. 3.5. The first enzyme, citrate synthase, controls the condensation of the precursor, acetyl coenzyme A, with oxaloacetate; at the same time the coenzyme A ester group is hydrolyzed and coenzyme A is released into the medium. This reaction has a large negative change in standard free energy (Fig. 3.5) so that the formation of citrate is essentially irreversible.

Citrate synthase is not completely substrate specific and will accept monofluoroacetate $CH_2FCO_2^-$, as the coenzyme A ester, instead of acetyl coenzyme A. Fluoroacetate is a naturally occurring toxin, present notably in the plant

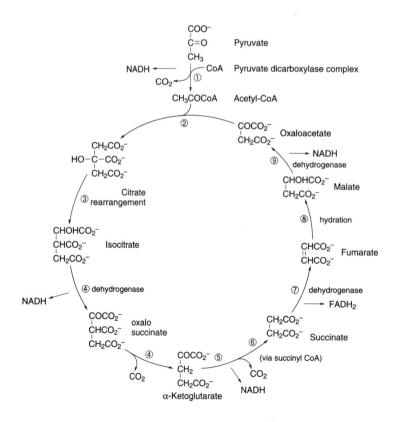

Figure 3.4 The citric acid cycle. The enzymes (numbers in circles) are shown in Fig. 3.5. The sites of NADH and NADH$_2$ release are shown with arrows.

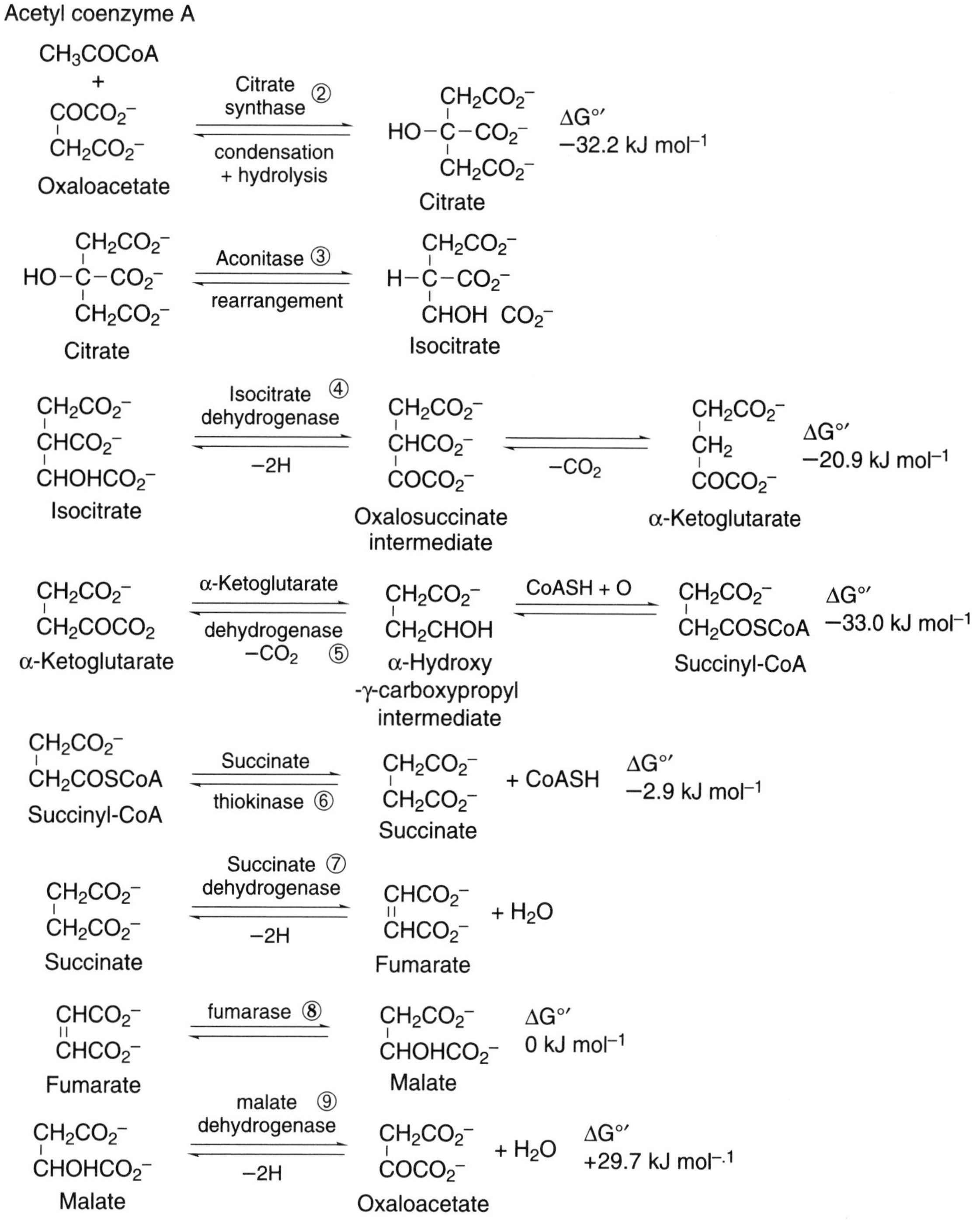

Acetyl coenzyme A

CH_3COCoA
$+$
$\underset{\overset{|}{CH_2CO_2^-}}{COCO_2^-}$
Oxaloacetate

$\xrightarrow[\substack{\text{condensation} \\ \text{+ hydrolysis}}]{\text{Citrate} \quad ② \\ \text{synthase}}$

$HO-\underset{\overset{|}{CH_2CO_2^-}}{\overset{\overset{|}{CH_2CO_2^-}}{C}}-CO_2^-$
Citrate

$\Delta G^{\circ\prime}$
-32.2 kJ mol^{-1}

$HO-\underset{\overset{|}{CH_2CO_2^-}}{\overset{\overset{|}{CH_2CO_2^-}}{C}}-CO_2^-$
Citrate

$\xrightarrow[\text{rearrangement}]{\text{Aconitase} \quad ③}$

$H-\underset{\overset{|}{CHOH \ CO_2^-}}{\overset{\overset{|}{CH_2CO_2^-}}{C}}-CO_2^-$
Isocitrate

$\underset{\overset{|}{CHOHCO_2^-}}{\overset{\overset{|}{CHCO_2^-}}{\overset{\overset{|}{CH_2CO_2^-}}{}}}$
Isocitrate

$\xrightarrow[\text{$-2H$}]{\text{Isocitrate} \quad ④ \\ \text{dehydrogenase}}$

$\underset{\overset{|}{COCO_2^-}}{\overset{\overset{|}{CHCO_2^-}}{\overset{\overset{|}{CH_2CO_2^-}}{}}}$
Oxalosuccinate
intermediate

$\xrightleftharpoons{-CO_2}$

$\underset{\overset{|}{COCO_2^-}}{\overset{\overset{|}{CH_2}}{\overset{\overset{|}{CH_2CO_2^-}}{}}}$
α-Ketoglutarate

$\Delta G^{\circ\prime}$
-20.9 kJ mol^{-1}

$\underset{\overset{|}{CH_2COCO_2}}{\overset{\overset{|}{CH_2CO_2^-}}{}}$
α-Ketoglutarate

$\xrightarrow[\substack{\text{dehydrogenase} \\ -CO_2 \quad ⑤}]{\text{α-Ketoglutarate}}$

$\underset{\overset{|}{CH_2CHOH}}{\overset{\overset{|}{CH_2CO_2^-}}{}}$
α-Hydroxy
-γ-carboxypropyl
intermediate

$\xrightleftharpoons{CoASH + O}$

$\underset{\overset{|}{CH_2COSCoA}}{\overset{\overset{|}{CH_2CO_2^-}}{}}$
Succinyl-CoA

$\Delta G^{\circ\prime}$
-33.0 kJ mol^{-1}

$\underset{\overset{|}{CH_2COSCoA}}{\overset{\overset{|}{CH_2CO_2^-}}{}}$
Succinyl-CoA

$\xrightarrow[\text{thiokinase} \quad ⑥]{\text{Succinate}}$

$\underset{\overset{|}{CH_2CO_2^-}}{\overset{\overset{|}{CH_2CO_2^-}}{}}$ $+$ CoASH
Succinate

$\Delta G^{\circ\prime}$
-2.9 kJ mol^{-1}

$\underset{\overset{|}{CH_2CO_2^-}}{\overset{\overset{|}{CH_2CO_2^-}}{}}$
Succinate

$\xrightarrow[\text{$-2H$}]{\text{Succinate} \quad ⑦ \\ \text{dehydrogenase}}$

$\underset{\overset{||}{CHCO_2^-}}{\overset{\overset{|}{CHCO_2^-}}{}}$ $+$ H_2O
Fumarate

$\underset{\overset{||}{CHCO_2^-}}{\overset{\overset{|}{CHCO_2^-}}{}}$
Fumarate

$\xrightleftharpoons{\text{fumarase} \quad ⑧}$

$\underset{\overset{|}{CHOHCO_2^-}}{\overset{\overset{|}{CH_2CO_2^-}}{}}$
Malate

$\Delta G^{\circ\prime}$
0 kJ mol^{-1}

$\underset{\overset{|}{CHOHCO_2^-}}{\overset{\overset{|}{CH_2CO_2^-}}{}}$
Malate

$\xrightarrow[\text{$-2H$}]{\text{malate} \quad ⑨ \\ \text{dehydrogenase}}$

$\underset{\overset{|}{COCO_2^-}}{\overset{\overset{|}{CH_2CO_2^-}}{}}$ $+$ H_2O
Oxaloacetate

$\Delta G^{\circ\prime}$
$+29.7$ kJ mol$^{-.1}$

Figure 3.5 The enzymes of the citrate cycle. The numbers in circles refer to the participating enzymes shown in Fig. 3.4.

Dichapetalum cymosum and the fatal dose in humans is 2–5 mg per kg body weight. Fluorocitrate produced from the reaction of fluoroacetyl coenzyme A and oxaloacetate is not itself toxic, but it is a potent inhibitor of the next enzyme of the cycle, aconitase. Hence, the fatal effects of fluoroacetate in humans or farm animals ingesting this material are due to the inhibition of the cycle at the aconitase step. The plant *Dichapetalum* must protect its TCA cycle from the fluoroacetate it produces, by compartmentation.

The next enzyme, aconitase, rearranges citrate to isocitrate, via a dehydration to *cis*-aconitate, and a subsequent hydration of this transient intermediate to isocitrate. Aconitase is a stereospecific enzyme, catalyzing the *trans* removal of water to give a *cis*-acid and the *trans* addition of water to yield *threo*-D$_s$-isocitrate.

The next enzyme of the cycle, isocitrate dehydrogenase, catalyzes the oxidative decarboxylation of isocitrate to α-ketoglutarate (or 2-oxoglutarate) using NAD$^+$ as oxidant. The reaction requires the metal ions Mg^{2+} or Mn^{2+} and supplies energy to the mitochondrion in the form of NADH. Energy is given out and the reaction is essentially irreversible.

The fourth enzyme, α-ketoglutarate dehydrogenase, also catalyzes an irreversible reaction. It is a complex, rather similar to that which oxidizes pyruvate to acetyl coenzyme A (section 3.4.1). It carries out the oxidative decarboxylation of α-ketoglutarate, at the same time esterifying the product with coenzyme A. Two carrier molecules, thiamine pyrophosphate and lipoic acid, are involved in the final production of succinyl coenzyme A.

The fifth enzyme, succinate thiokinase, is a simple hydrolyzing enzyme, breaking the thioester bond and at the same time generating a molecule of ATP from ADP and inorganic phosphate.

The last three enzymes of the cycle catalyze the standard reactions of dehydrogenation, hydration and dehydrogenation and require less comment. The conversion of succinate into fumarate, catalyzed by succinate dehydrogenase, generates a molecule of the flavoprotein FADH$_2$ from FAD. This can be reoxidized by the electron transport chain, where it is linked to the generation of approximately 1.5 molecules of ATP by oxidative phosphorylation. Succinate dehydrogenase is also of interest, because it is competitively inhibited by the addition of malonic acid, HO$_2$C CH$_2$CO$_2$H, the lower homolog of succinate. This provides a possible point of control of the cycle, since malonyl coenzyme A is a common cell constituent, required, for example, in the biosynthesis of fatty acids.

For each turn of the cycle, three molecules of NADH are produced as well as one FADH$_2$ and one ATP, thus yielding approximately 10 ATP equivalents (NADH $\simeq$ 2.5 ATP; FADH$_2 \simeq$ 1.5 ATP). Hence, the citric acid cycle is a very important source of energy for the plant cell and it recovers the remaining energy bound up in sugar which has not yet been released by glycolysis. It is possible to calculate that the catabolism of one glucose molecule to six CO$_2$ molecules via glycolysis and the citric acid cycle will provide the plant cell with 32 ATPs. If the standard free energy change for the oxidation of glucose by O$_2$ is -2823 kJ mol^{-1} and the hydrolysis of ATP to ADP is 30.5 kJ mol^{-1}, then the free energy conservation of the biological oxidation of glucose is nearly 34.5%.

Glycolysis: Glucose + 2NAD$^+$ + 2ADP + 2Pi

↓

2NADH + 2 pyruvate + 2ATP + 2H$_2$O + 4H$^+$

Acetyl-CoA formation: 2 pyruvate + 2CoA + 2NAD$^+$

↓

2 acetyl-CoA + 2CO$_2$ + 2NADH

Citric acid cycle:

2 acetyl-CoA + 6NAD$^+$ + 2(FAD) + 2ADP + 2Pi + 4H$_2$O

↓

2CO$_2$ + 2CoASH + 6NADH + 2(FADH$_2$) + 2ATP

Scheme 20

3.4.3   Regulation

It has been suggested that the citric acid cycle in plants is regulated by enzyme turnover, by metabolite transport from within and from outside the mitochondrial membrane, by ADP:ATP ratios or by pH (Wiskitch, 1980). How far these different factors are involved is still not entirely clear. However, it is known that many of the enzymes of the cycle are regulated by various metabolites, as indicated in Fig. 3.6. Obvious points of regulation are the supply of acetyl coenzyme A from the decarboxylation of pyruvate, and the inhibition of the enzyme pyruvate dehydrogenase by its end products, acetyl coenzyme A and NADH. Several enzymes are strongly affected by the levels of AMP, ADP or ATP present, e.g. citrate synthase and α-ketoglutarate dehydrogenase. Therefore,

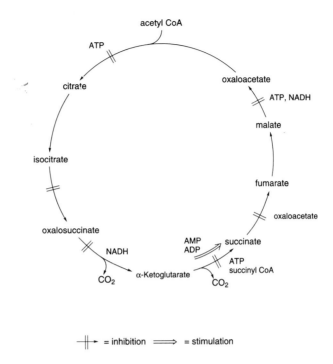

Figure 3.6 Sites of regulation of the citric acid cycle. $\dashv\!\!\mid\!\!\vdash$ = inhibition $\Longrightarrow$ = stimulation

these enzymes (Fig. 3.6) may be particularly important in controlling the flow of carbon through the cycle. It may also be noted that some of the intermediates in the cycle are capable of inhibiting certain enzymes, e.g. oxaloacetate inhibits succinate dehydrogenase. Finally, it is apparent that citrate indirectly exerts a regulatory effect on glycolysis (section 3.2) and hence the amount of pyruvate being supplied to the cycle. Much further work, however, is still needed to determine precisely how this key metabolic feature of carbohydrate degradation is controlled in plants.

3.4.4 Roles

Although the citrate cycle mainly operates as a catabolic pathway, it can be employed synthetically to provide intermediates for example for amino acid synthesis. For example, aspartic acid is produced by transamination of oxaloacetate (Chapter 7). The oxaloacetate required could be taken out of the cycle, although it has to be remembered that some more oxaloacetate would have to be added (e.g. generated from pyruvate) in order to keep the cycle turning.

The most important role of the TCA cycle from the synthetic viewpoint, however, is its place in the conversion of lipid into carbohydrate, which occurs in germinating oil seed (see Chapter 6). Here the citrate cycle is essentially modified by by-passing those steps in the cycle involving the loss of CO_2 (Fig. 3.7). Two new reactions are introduced: the conversion of isocitrate into glyoxylate and succinate and the synthesis of malate from glyoxylate and external acetyl coenzyme A. This ingenious new metabolic pathway, known as the glyoxylate cycle, allows the cycle to keep on turning over and at the same time provides the cell with succinate, which is converted into oxaloacetate in the mitochondria and then into PEP in the cytosol. PEP is the starting material for gluconeogenesis (section 3.2.2) and for every two molecules of succinate taken out of the glyoxylate cycle, one molecule of glucose-6-phosphate will be produced.

The two new enzymes, isocitrate lyase and malate synthase, are specifically induced in the germinating seed as lipid breakdown occurs. They are produced in special organelles, the glyoxysomes, so that the glyoxylate cycle is located away from other metabolic reactions and is free to produce glucose-6-phosphate for the seedling until the plant is able to produce sugar more directly by photosynthesis.

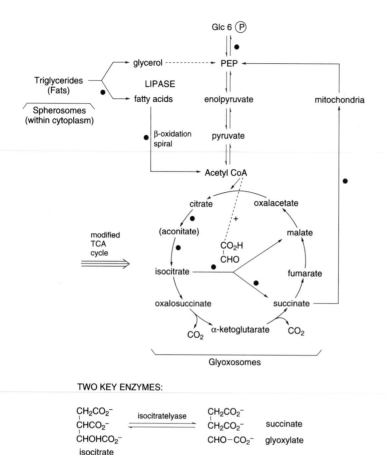

Figure 3.7 The glyoxylate cycle. Solid dots indicate the pathway of carbon in the cycle.

3.5 ELECTRON TRANSPORT AND OXIDATIVE PHOSPHORYLATION

As described in sections 3.2 and 3.4, NADH and FADH$_2$ are formed in the glycolytic and TCA cycles. Each of these energy-rich molecules has a pair of electrons with high transfer potential, and the free energy liberated during transfer can be used for synthesizing ATP. The energy harnessed in ATP is used for driving numerous biochemical processes in the cell. The transfer of electrons is mediated by the electron carriers present in the inner mitochondrial membrane. The electron transport chain, otherwise referred to as the respiratory chain, has a series of four large protein complexes through which electrons pass from lower to higher redox potentials. Only two complexes, however, are common to the oxidation of both NADH and FADH$_2$. The flow of electrons causes pumping of protons from the matrix side to the intermembrane space. This maintains a proton motive force composed of a pH gradient and a membrane potential. The protons flow back into the matrix along various pathways but particularly through the ATP-synthesizing complex embedded in the membrane during ATP synthesis. The central feature of this process of oxidative phosphorylation is the generation of proton gradient and its accompanying membrane potential.

3.5.1 Electron transport

The sequence of events that mediate the flow of electrons from NADH or FADH$_2$ to O$_2$ via cytochrome oxidase (the cyanide-sensitive pathway) appears to be similar in both animal and plant mitochondria. Some important differences between animal and plant mitochondria have been observed and these are discussed at greater depth later in this chapter. It is now sufficient to state simply that these differences are the presence in plants of (1) the respiratory-linked oxidation of external NADH, (2) a cyanide- and antimycin A-insensitive alternative electron transport pathway, (3) a second NADPH dehydrogenase that oxidizes external NADPH produced mainly from the oxidative pentose phosphate pathway and (4) a rotenone-insensitive oxidation of internal NADH (distinct from complex I).

The main components of the respiratory chain are arranged into multi-protein units as shown in Fig. 3.8:

Complex I, otherwise referred to as the NADH dehydrogenase complex or NADH-ubiquinone (UQ) oxidoreductase;
Complex II, otherwise referred to as succinate dehydrogenase complex or succinate-UQ reductase;
Complex III, otherwise referred to as the cytochrome b/c_1 complex or UQH_2-cytochrome c reductase;
Complex IV, otherwise referred to as the cytochrome oxidase complex.

The above-mentioned complexes are hydrophobic in nature. However, the cytochrome c component between complex III and IV is a water-soluble peripheral protein on the outside of the mitochondrial membrane.

(a)   Complex I

The first complex of the respiratory chain (Fig. 3.8) is the internal NADH dehydrogenase which is responsible for the oxidation of NADH and reduction of ubiquinone (also known as Coenzyme Q, CoQ, UQ). NADH may be derived from sources such as TCA cycle, oxidation of pyruvate by pyruvate dehydrogenase or oxidation of glycine in green leaf mitochondria by glycine decarboxylase. The site of NADH oxidation faces the matrix of the mitochondrial membrane; therefore, only the matrix NADH is oxidized.

Ubiquinone is a substituted 1,4-benzoquinone containing 9 or 10 isoprene (C_5) units and therefore are referred to as Q-9 or Q-10, although ubiquinone from different sources contains different numbers of isoprene units. Oxidized ubiquinone can accept electrons and protons to form a

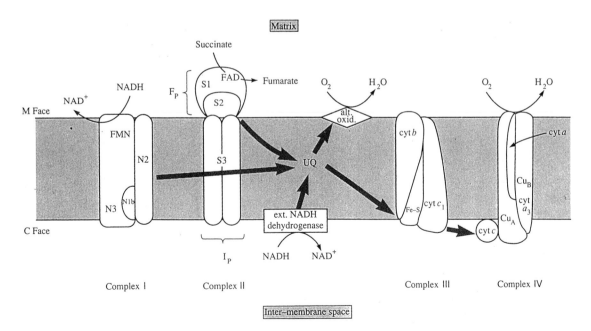

Figure 3.8 A schematic representation of the distribution of redox components within the inner mitochondrial membrane. The bold arrows indicate the general pathway of electron flow from complexes I or II via the mobile pool of ubiquinone (UQ) to complexes III and IV. Abbreviations: alt. oxid. = 'alternative oxidase'; ext. NADH dehydrogenase = external NADH dehydrogenase located on the C face of the inner mitochondrial membrane; N1, N2, N3 and S1, S2, S3 = iron–sulfur protein centers of complex I and complex II respectively; Cu_A, Cu_B = copper atoms associated with cytochrome oxidase; cyt = cytochrome; Fe–S = iron–sulfur protein of complex III (Rieske center). From Anderson & Beardall, (1991), with permission.

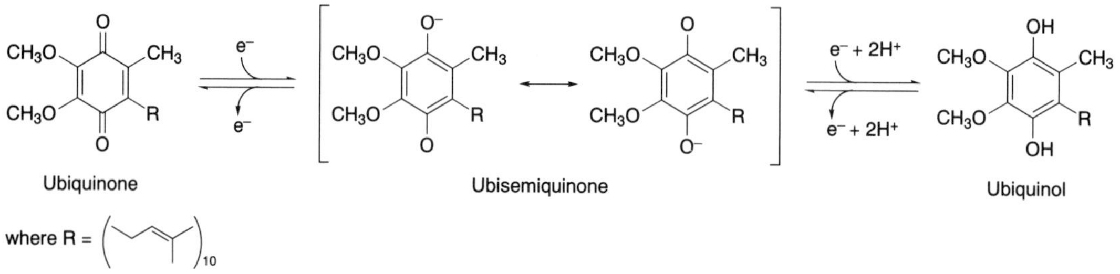

Figure 3.9   Redox states of ubiquinone.

reduced ubiquinone referred to as either ubiquinol or hydroquinone via a stable ubisemiquinone intermediate (Fig. 3.9). Ubiquinone is lipid-soluble and, therefore, is able to act as a mobile redox carrier between the complexes in the inner mitochondrial membrane.

The NADH dehydrogenase–UQ complex is composed of phospholipids and polypeptides; as many as 30 polypeptides may be present with only a few participating in the redox reactions of the respiratory chain. It is as large as the large subunit of a ribosome and is probably the largest protein complex of the inner mitochondrial membrane. Complex I contains a non-covalently bound flavin mononucleotide (FMN), as many as seven iron–sulfur centers and possibly an internal ubiquinone. The iron is acid-labile, unlike the tetrapyrrole-bound Fe in cytochromes. Three of these iron–sulfur centers referred to as N1b, N2 and N3 are well characterized. A predicted pathway of electron transfer in the complex is: $NAD \rightarrow FMN \rightarrow N3 \rightarrow N1b \rightarrow N2 \rightarrow UQ$. The process tends to be inhibited by rotenone. Iron–sulfur centers are able to transfer only electrons through the oxidation/reduction of Fe^{2+}/Fe^{3+}, respectively. As electrons pass through the complex I, protons are translocated from the matrix to the intermembrane space (IM – space). The H^+/e^- ratio is probably 2.

(b) *Complex II*

Succinate is oxidized by the membrane-bound enzyme succinate dehydrogenase (complex II) which comprises flavin adenine dinucleotide (FAD), and iron–sulfur proteins. This complex is not involved in H^+ translocation from the matrix into the inter-membrane space but does reduce ubiquinone (which is mobile) as its end-point. Complex II contains two major subcomplexes. The largest water-soluble subcomplex contains two subunits referred to as Fp and Ip with molecular weights of 65–67 kDa and 26–30 kDa,

respectively. The FP (flavo-iron-sulfur protein) contains FAD (covalently bound) and two iron–sulfur proteins of 2Fe–2S configuration, called centers S1 and S2. The Ip subunit contains an iron–sulfur protein of 4Fe-4S composition and is called S3. The other subcomplex is hydrophobic and consists of two small polypeptides (7 kDa and 13 kDa). A cytochrome *b* is associated with one of the polypeptides; its function is unknown but it is equimolar with FAD. It is suggested that one of these hydrophobic subunits carries the semiquinone-binding site(s). The subunits are required for binding succinate dehydrogenase to the membrane and for the reduction of ubiquinone. They do not, however, participate in succinate oxidation in the presence of artificial electron acceptors. Electrons are transferred from $FADH_2$ through S1, S2 and S3 to ubiquinone.

$$Succinate + UQ \rightarrow Fumarate + UQH_2$$

Complex II is tightly regulated and exists as an inactive form stabilized in a 1:1 stoichiometry by binding with oxaloacetate. Incubation with either ATP or ubiquinol activates the enzyme. The larger subcomplex contains the inhibitor binding site (binds oxaloacetate) and the catalytic active site (for reversible dehydrogenation of succinate).

(c) *Complex III*

The transfer of electrons from ubiquinol to cytochrome *c* occurs through complex III which is also termed cyt bc_1 complex or ubiquinol–cytochrome *c* oxidoreductase. This complex can be compared to the plastoquinol–plastocyanin oxidoreductase, or cyt $b_6 f$ complex of thylakoids (see Chapter 2). Complex III contains three polypeptides with the redox groups: cyt c_1, cyt *b* and an iron–sulfur protein known as the Rieske protein. This protein has a 2Fe-2S cluster attached to the polypeptide via chelation of one Fe to two cysteines and the other Fe to two histidine

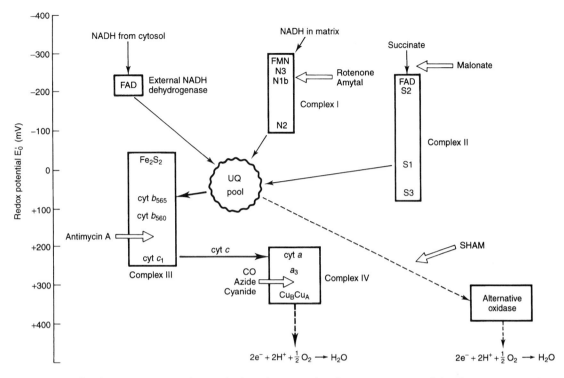

Figure 3.10 The electron transport chain in higher plant mitochondria. Components of the chain are grouped into their presumed complexes. Also indicated are the sites of action of some inhibitors commonly used in investigations of respiratory electron transport (bold arrows) and the approximate standard redox potential (E_0') of the components. Abbreviations: N1b, N2, N3 = iron–sulfur protein complexes of complex I; S1, S2, S3 = iron–sulfur protein complexes of complex II; Cu_A, Cu_B = copper atoms associated with cytochrome oxidase in complex IV; Fe_2S_2 = Rieske iron–sulfur protein center; UQ = ubiquinone. From Anderson & Beardall (1991), with permission.

residues. The globular structure with the 2Fe-2S center extends into the aqueous layer on the P-phase of the mitochondrial membrane but anchors to the membrane by the hydrophobic N-terminus. Cytochrome c_1 is also globular but the hydrophobic C-terminus is the anchoring site. The cyt b subunit binds two hemes. One heme, known as b_L (also called b_{565} because of its α-absorption peak) is located towards the P-phase of the membrane and the second heme, b_H (also called b_{560} for its α-absorption peak) is on the matrix side of the membrane. These will be addressed below (see also Fig. 3.10) as cyt b_{565} and cyt b_{560}, respectively.

The complex also contains core proteins which have no apparent redox function but may have a role in H^+ translocation. Complex III is inhibited by low concentrations of antimycin A (1 mole complex/1 mole inhibitor). This complex isolated from sweet potato mitochondria can be resolved into eight subunits with molecular weights of 51, 49, 33, 32, 27, 17, 10 and 10 kDa. Two polypeptides of intermediate molecular weights have been characterized as cyt b (32 kDa) and cyt c_1 (33 kDa).

Complex III transfers electrons to complex IV (cytochrome c oxidase) via the redox component, cytochrome c. Electron transfer is believed to occur through a 'protonmotive-Q cycle' (see section 3.5.6). The electron carriers are so arranged across the membrane that the electrons are transported in one direction and hydrogen in the other resulting in the pumping of protons. The observed H^+/e^- ratio is 2.

(d) Complex IV

The cytochrome c oxidase complex IV spans the inner mitochondrial membrane. It is the terminal oxidase of the respiratory chain in the transfer of electrons from cyt c to O_2. Cytochrome c is not an integral part of complex IV, but is stoichiometrically associated with it and is believed to be spatially associated with subunit II of cytochrome oxidase. Cytochrome c is a water-soluble electron carrier and exists between the internal and external mitochondrial membranes. It can diffuse freely in this space, thus acting as a mobile shuttle carrying electrons between cytochrome c_1 of complex III and cytochrome *a* of complex IV. The electron flow rate can therefore be regulated by the frequency of collisions of cyt *c* with the complexes. The flow is retarded by artificially lowering its concentration (dilution by adding phospholipids) of endogenous cyt *c*. The structure of this complex has been well studied from a number of sources; the heme is situated in a hydrophobic cleft.

Complex IV is inhibited by azide, cyanide or carbon monoxide. Sweet potato mitochondria have a complex IV containing cytochrome *a*, cytochrome a_3 and two copper atoms (designated Cu_A and Cu_B). Ferricytochrome and the cupric atom Cu_A are paramagnetic and are 'visible' by electroparamagnetic resonance (EPR) spectroscopy and Cu_B is antiferrimagnetically coupled to cytochrome a_3 at the active site of the enzyme and is therefore 'invisible' by EPR spectroscopy. The subunit structure of cyt *c* oxidase is complex: more

than ten subunits may be present, only two of the three largest subunits containing the redox centers are important for an operative electron transfer. Both hemes and Cu_B are in the largest subunit and Cu_A in the smaller subunit, subunit II (Fig. 3.11). As shown in Fig. 3.11, cyt *c* on the cytoplasmic side of the membrane is the source of electrons which are transferred to Cu_A. Cu_B and heme a_3 are the twin centers (close to each other) of subunit I located at the matrix side of the membrane and accept one electron from Cu_A (or heme *a*). The acceptance of the second electron leads to binding of O_2. The oxygen species is not released as it would be toxic. Protons are taken from the matrix side and the center accepts a third electron yielding a Fe^{4+} heme. The fourth electron forms the Fe^{3+} species and further protons are taken from the matrix side to yield water. Thus, protons are not extruded from the matrix, as at coupling sites 1 and 2 (complex I and III, respectively), but are removed as covalently linked to oxygen. The H^+/e^- ratio is considered as 1.

3.5.2 Oxidation of endogenous NADH

The plant inner mitochondrial membrane has an external NADH dehydrogenase which is separate from the dehydrogenase of complex I or internal NADH dehydrogenase. Unlike in mammalian mitochondria, oxidation of endogenous NADH by internal NADH dehydrogenases in plants is only partially inhibited by rotenone. This is because there are two separate NADH dehydrogenases on the inner surface of the inner mitochondrial membrane and only one of these (of complex I) is sensitive to rotenone. The rotenone-insensitive enzyme is distinct from complex I and has a K_m of $80\,\mu M$ for NADH as compared to $8\,\mu M$ for the rotenone-sensitive enzyme. The inhibitor-sensitive enzyme is referred to as the 'first coupling site' because the reversible electron flux from NADH to ubiquinone is coupled to the generation of H^+ electrochemical gradient (and the accompanying membrane potential) across the inner mitochondrial membrane. It has been suggested that the rotenone-insensitive activity simply represents a second UQ reduction site for complex I and not an additional dehydrogenase. This notion has experimental support and a model of complex I was put forward in which there is one NADH-binding site and two sites for the reduction of UQ, only one of which is insensitive to rotenone. The existence of a common intermediate will suggest that its level of reduction can influence the NADH-binding site.

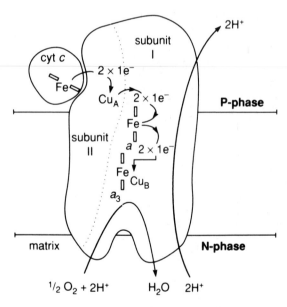

Figure 3.11 A tentative outline structure for cytochrome oxidase. The general tooth-like shape has been deduced from electron microscopy. The positions of the redox groups have been deduced from a range of experiments, including model building of the likely folding patterns of the two largest subunits that are thought to carry the main centers. It is usual to show the site of oxygen reduction at the matrix side of the inner mitochondrial membrane. The protons required in the reduction are taken from the matrix. Thus the charge translocation is achieved by a combination of the inward movement of the electrons, and outward movement of protons to the bimetallic center, in addition to the proton pumping by the enzyme. From Nicholls & Fergusson (1992), with permission.

A high K_m for NADH of the rotenone-insensitive enzyme may be due to the restricted access (or steric hindrance) of UQ to a rotenone-insensitive site. These aspects remain debatable.

The rotenone-insensitive enzyme is not involved in proton translocation from the matrix into the inner IM space, therefore, is not coupled to the change in H^+ electrochemical gradient and associated membrane potential. This enzyme may be involved with the cyanide-resistant electron pathway. It has been suggested that the rotenone-insensitive dehydrogenase assists the turnover of the citric acid cycle intermediates at high cytosolic ATP/ADP ratio. This pathway is regulated by the actual NADH concentration (high K_m) in the mitochondrial matrix and not the ratio of NADH/NAD^+. Unlike the external NADH dehydrogenase, this enzyme is not affected by EGTA or Ca^{2+}. The physiological importance of this enzyme is little understood. However, it is possible that in the event of inhibition of complex I, which will stop proton translocation, the enzyme will ensure continuation of electron transfer.

The possibility of the presence of other NADP-dependent enzymes has been suggested within the mitochondrial matrix, for example, glutathione reductase, isocitrate dehydrogenase and malate dehydrogenase. This highlights a more important role for NADP(H) in plant mitochondrial respiration than earlier anticipated.

3.5.3 Oxidation of exogenous NAD(P)H

The mitochondria of higher plants, in contrast to mammalian mitochondria, catalyze a rapid oxidation of exogenous NAD(P)H in the absence of added cytochrome c. External NADH is oxidized by an external NADH dehydrogenase located on the outer surface of the inner mitochondrial membrane and donates electrons directly to complex III + ubiquinone (by-passing complex I and the first site of H^+ translocation). This pathway is inhibited by antimycin A, is insensitive to rotenone (compare with complex I) and does not seem to be linked with the alternative oxidase. There are no specific inhibitors of this NADH dehydrogenase. However, a number of flavonoids (e.g. platanetin: 3,5,7,8-tetrahydroxy-6-isoprenyl-flavone) inhibit the oxidation of external NADH. Calcium in micromolar concentration (often found bound to mitochondrial membranes) ensures maximal activity of the enzyme. However, no calmodulin is involved. It was, therefore, not surprising to observe inhibition of the enzyme by Ca^{2+} chelators such as EGTA. The inhibition of the enzyme is higher if EGTA is added prior to NADH addition; a low inhibition results if the sequence is reversed. This suggests that oxidation of NADH in some way (conformational change in the enzyme) makes Ca^{2+} less accessible to EGTA.

Exogenous NADPH is also oxidized, apparently by a Ca^{2+}-dependent dehydrogenase. Although there are similarities between the oxidation of external NADH and NADPH (ADP-O ratios below 2 and rotenone-insensitive), pH optimum for NADPH is lower than for NADH and response to the chelators is different. Here, it is important to note that neither phosphatase (converting NADPH into NADH) nor a nicotinamide transdehydrogenase takes part in mitochondrial NADPH oxidation. It is therefore presumed that in plants two separate dehydrogenases are present on the outer surface of the inner mitochondrial membrane.

The subunit structures of the external NADH or NADPH dehydrogenases are not clear. They appear to be flavoproteins and do not seem to contain iron–sulfur centers; thus it would be interesting to study their mechanism of interaction with complex III or ubiquinone.

Relatively little information is available on the physiological role of the external NADH dehydrogenase which is capable of oxidizing cytosolic NADH. The K_m for NADH is low which means that it could efficiently use cytoplasmic NADH (from the glycolytic pathway, β-oxidation in peroxisomes, etc) and function as a redox shuttle to the respiratory chain. Note that peroxisomes are permeable to small solutes. This enzyme could also be implicated in reducing lactate level in plant tissues under conditions of hypoxia. NADPH oxidation by mitochondria on the other hand, could enhance the rate of cytosolic pentose phosphate pathway (cytosolic NADPH is an effective regulator). The NADPH oxidation could also enhance citrate metabolism (citrate converted into α-ketoglutarate in the presence of $NADP^+$-dependent isocitrate dehydrogenase and aconitase). Thus, rapid re-oxidation of NADPH would supply the carbon skeleton for biosynthetic purposes and active cell growth. It is hence implied that the electron flux through the external NADPH dehydrogenase must be well regulated.

3.5.4 Alternative cyanide-resistant oxidase

A unique feature of plant mitochondria is the presence of an alternative mechanism of electron transport from ubiquinol to oxygen which does not involve the cytochrome oxidase complex. This pathway is inhibited by SHAM (salicylhydroxamic acid) but is not inhibited by cyanide or antimycin (see Fig. 3.10). No proton translocation occurs through this pathway and the protons required for the reduction of oxygen are derived from the matrix. Oxidation of ubiquinol releases the protons to the matrix.

Although the cyanide-resistant alternative respiration was identified as early as 1927, the purification of the alternative oxidase has been difficult because of its extremely labile nature in solubilized preparations. Several purification attempts have been reported; in most of them purified samples displayed multiple protein bands on SDS-PAGE. The best example of purification is from skunk cabbage mitochondria which showed four major protein bands of 29, 36, 57–59 and 65 kDa. A native molecular weight of 160 kDa was identified. Redox difference spectra (borohydride reduced minus air oxidized) showed an absorbance increase in the range of 290–335 nm and a decrease below 285 nm with a minimum around 275 nm. There is no firm evidence for the presence or identification of redox centers in any of the purified samples. The association with any alternative oxidase of electron transfer cofactors, such as, flavoproteins, b-type cytochromes and iron–sulfur centers, have been suggested but no conclusive evidence supports the view.

The product of oxygen reduction is believed to be H_2O rather than H_2O_2 or superoxide ion; the latter would be energetically more facile. This supports the possibility of a transition metal center in the alternative oxidase. An 'engaging protein factor' is essential for coupling electron flow between the ubiquinone pool and the alternative pathway. As the cyanide-resistance varies with developmental stage, tissue type and physiological state of the plant, the engaging factor may act as a rate-limiting component in determining the degree of cyanide resistance. The details as to how electrons branch into the alternative pathway and its mechanism are unknown.

The physiological function of the alternative oxidase is not clear. Electron flow from the quinol pool through the pathway does not produce an electrochemical gradient and no ATP is generated. The potential energy is lost as heat. In skunk cabbage the heat production allows growth at low temperatures. In arum lilies the heat generated volatilizes insect attractants to aid pollination. The cyanide-resistant respiration can also be regarded as an 'over-flow' to drain off energy from excess carbohydrates in the system. Thus, the pathway is wasteful with respect to the carbon budget of the plant and energy conservation.

3.5.5 Mitochondrial states

Mitochondria exhibit a low requirement of O_2 in the absence of substrate (state 1). The rate of O_2 consumption increases when substrate, such as, malate is added (state 2). When freshly isolated intact mitochondria are incubated with ADP, there is a rapid increase in oxygen consumption (state 3). Upon conversion of all ADP into ATP, the respiratory activity of the mitochondria decreases to the rate found before addition of ADP when all ADP has been phosphorylated (state 4). The mitochondria will revert to state 3 when incubated with more ADP which will be followed by 'state 4' after ADP is depleted. Respiratory activity is effectively controlled by the availability of ADP. This 'respiratory control' is quantified by the 'respiratory control ratio' or the ratio of state 3:state 4 (the ratio of the rate of O_2 consumption at saturating concentrations of substrate in the presence of ADP to the slow, resting rate after depletion of ADP). The respiratory control ratio is used to characterize the 'tightness' of coupling of substrate oxidation to energy conservation (higher the ratio, tighter the coupling). Depending on the particular substrate, many plant mitochondria preparations give respiratory control ratios between 3 and 7.

3.5.6 Mechanism of electron transport, Q-cycle

The transport of electrons and the associated translocation of protons from the matrix of the mitochondria to the inter-membrane space during passage of electrons from NADH to O_2 is summarized in Fig. 3.12.

The nature and location of the various components of the electron transport chain makes the transport possible. Electrons from the internal NADH (produced in the matrix via oxidation of malate, pyruvate, oxoglutarate and glycine) enter complex I, the first coupling site. When electrons are transferred from an electron carrier to a component which accepts both electrons and protons, it causes proton uptake from the matrix. These are called redox loops. As shown in Fig. 3.12, protons and electrons are transferred to FMN in complex I at the matrix face of the membrane. This follows transfer of electrons to several iron–sulfur proteins and then to ubiquinone. The transfer is coupled to the translocation of protons from the matrix to the inter-membrane space (see Q-cycle). Electrons then enter complex IV via cyt c (see section 3.5.1).

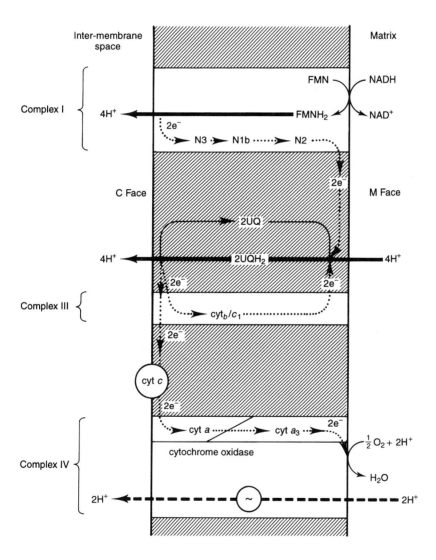

Figure 3.12 Representation of the carriers involved in mitochondria in the movement of electrons from NADH down to the mitochondrial electron transport chain to O_2 and the associated transport of protons (proton translocation) from the mitochondrial matrix to the inter-membrane space. Proton translocation is associated with: (i) the transfer of electrons from endogenous NADH to UQ; (ii) a proton motive Q-cycle; and (iii) a cytochrome oxidase proton pump shown as $\ominus$. Movement of electrons only are shown by dotted lines whereas combined H^+/e^- transport is shown by the bold black arrows. The bold dashed arrow indicates the conformational proton pump associated with cytochrome oxidase. From Anderson & Beardall (1991), with permission.

The mechanism of the participation of ubiquinone in the electron transport process was proposed by Peter Mitchell and termed as 'proton motive Q-cycle' (Fig. 3.13). Ubiquinones exist in excess in the membrane as a pool of UQ and UQH$_2$. They are hydrophobic and uncharged and hence can migrate along the hydrophobic core of the membrane. Diffusion takes place of one

UQH$_2$ to the Q$_p$ binding site adjacent to the Rieske protein (iron–sulfur protein) at the P face of the mitochondrial membrane (Fig. 3.13). One electron is transferred to Rieske protein, the semiquinone anion UQ$^{\cdot-}$ is formed, two protons are released to the P phase and the second electron is transferred to the b_{565} heme (b$_L$; complex III), and UQ is formed. The Rieske protein transfers

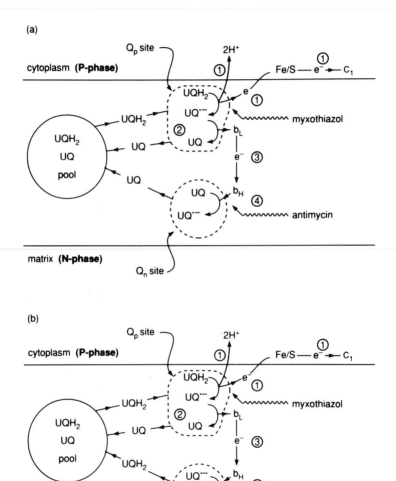

Figure 3.13   The Q-cycle in mitochondria. (a) The electron-transfer events that follow the oxidation of a ubiquinol at the P-side of the inner mitochondrial membrane under conditions in which the quinone binding site at the N-side is initially either vacant or occupied by a ubiquinone molecule. (b) The electron-transfer events that follow the oxidation of a second ubiquinol at the P side of the membrane when the Q$_n$ is occupied by a ubisemiquinone radical. Note that the Q$_n$ site has also been termed the Q$_i$ or Q$_c$ site (c indicating the cytoplasmic side of the membrane in bacterial cytochrome bc_1 complexes) in various systems and the Q$_p$ site is also known as the Q$_o$ or Q$_z$ site. The inhibitory sites of action of myxothiazol and antimycin are also shown by wavy arrows. From Nicholls & Fergusson (1992), with permission. (c) A simplified outline of the Q-cycle. Inhibitory sites of action: 1, antimycin; 2, myxothiazol; 3, 5-n-undecyl-6-hydroxy-4,7-diobenzothiazol (UHDBT).

the electron along the chain to $cyt\,c_1$ and cytochrome oxidase. The electron moves from b_L to b_H (on the $cyt\,b$). UQ then binds to b_H at the Q_n site and electron from the reduced b_H forms $UQ^{\cdot-}$ at this site. Now a second UQH_2 molecule is oxidized at the Q_p site (Fig. 3.13): the process follows as described above and the second electron formed completes the reduction of $UQ^{\cdot-}$ to UQH_2. Two protons are taken from the matrix for this purpose and released to the P phase. The reduced ubiquinone then goes back to the pool and thus the Q-cycle is completed.

Proton translocation can take place by a 'redox loop' mechanism (crossing via electron carriers in one direction and flavin or ubiquinone in the other) and/or redox-linked proton pump. The redox pump is based on protonation/deprotonation of the protein amino acids during the oxidation/reduction processes in a cyclic manner. This repetitive process is believed to be linked to the alternate changes in the orientation of the proton-binding site and also the proton-binding

affinity of the protein. This results in a sequential uptake and release of protons on opposite sites of the membrane, thereby bringing about a pumping action.

The proton pumping mechanism of complex IV is not clear; pumping is not reversible. However, $2H^+/2e^-$ appear in the cytoplasm and $4H^+/2e^-$ disappear from the matrix. No 'redox loop' operates at this site but the oxidase shows the features of proton pump: (a) it spans the membrane and is in contact with the matrix and the inner-mitochondrial space, (b) a conformational change in the protein occurs on reduction of the oxidase and (c) the standard redox potentials between the intermediates are pH dependent. The subunit 3 (see section 3.5.1) was suggested to be involved in the conformational pumping mechanism. This is inhibited by N,N'-dicyclohexyl carbodiimide (DCCD, a known inhibitor of proton translocation in F_1F_0-ATP synthase). However, there are also suggestions that this subunit is not important; it has no Fe or Cu.

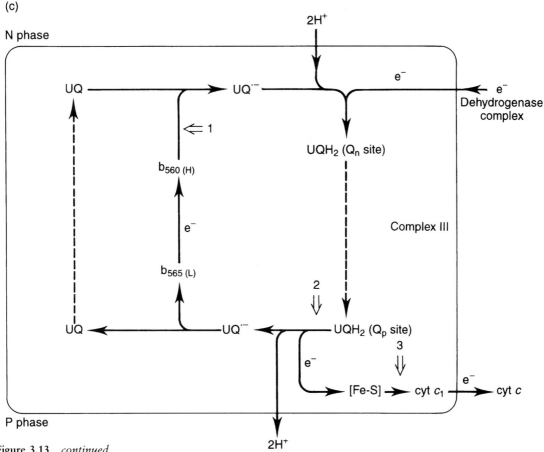

(c)

Figure 3.13   *continued*

The consensus value of the stoichiometry for complexes I and III is $4H^+/2e^-$ and $2H^+/2e^-$ for complex IV. Taken together, they yield 10 protons at the inter-membrane space for every two electrons passing down the electron transport chain.

3.5.7   Proton electrochemical gradient

Oxidation of NADH by O_2 represents the overall reaction of the mitochondrial electron transport chain. The tendency of the reduced compound to lose its reducing electrons as,

$$NADH \rightarrow NAD^+ + H^+ + 2e^-$$

is given by the standard redox potential $E^{\circ\prime}$. This is the force generated by the flow of electrons to a suitable electron acceptor under standard conditions (pH 7.0, 1 M, 25°C). Electrons flow from more electronegative (give up electrons and become oxidized) to electropositive molecules (accept electrons and become reduced). The carriers of the electron transport chain are the redox carriers as shown in Fig. 3.10 with their redox potential values. The carriers are arranged in the inner membrane in a specific order of increasingly positive redox potential.

The oxidation of NADH by oxygen is accompanied by a large negative change in free energy. Some of this energy is conserved by the phosphorylation of ADP to ATP, a transportable, versatile chemical store of the free energy. The mechanism of coupling the electron transport chain to ATP synthesis was proposed by Peter Mitchell in 1961 and is known as the chemiosmotic hypothesis. It proposes a vectorial transport of protons derived from the mitochondrial matrix through the membrane which is otherwise impermeable to protons. The transport takes place as the electrons of NADH are passed to O_2. A proton electrochemical gradient (proton motive force, Δp) is set up as the protons are pumped out creating a lower pH outside the matrix than inside. The relationship is as follows,

$$\Delta p = \Delta \Psi - Z\Delta pH$$

where $\Delta \Psi$ is the membrane potential (outside more positive than inside), ΔpH is the pH gradient and the term Z (60 at 30°C) is used to convert ΔpH into volts, the units of the other two terms. Free energy change ($\Delta G^{\circ\prime}$) can be related to electrochemical gradient:

$$\Delta G^{\circ\prime}/F = n\Delta \mu_{H^+}$$

where F is the Faraday constant, n is the number of protons pumped per phosphorylation site and $\Delta \mu_{H^+}$ is the proton electrochemical gradient. A proton pump, ATP synthase, located at the inner membrane of mitochondria transfers the energy of the proton electrochemical gradient into the formation of ATP (see section 3.5.9).

3.5.8   Inhibitors and uncouplers of electron transport chain and oxidative phosphorylation

Respiratory inhibitors have been important tools in determining the organization of individual components of the respiratory pathway (see Fig. 3.10). Rotenone is the most specific inhibitor of complex I. Other inhibitors include amytal, piericidin A and benzyladenine. However, internal NADH can still be oxidized by a rotenone-insensitive dehydrogenase when complex I is completely inhibited. Succinate oxidation can be completely inhibited by malonate, a competitive inhibitor of succinate dehydrogenase. Antimycin inhibits complex III and cyanide, azide and carbon monoxide inhibit complex IV. However, due to the presence of the alternative CN-resistant pathway, the oxidation of NAD-linked substrates may proceed even when the established cytochrome chain is completely inhibited. Substituted hydroxamic acids, such as salicylhydroxamic acid (SHAM), are frequently used with isolated mitochondria as specific inhibitors of the alternative pathway. However, SHAM is not absolutely specific and has been found to inhibit a variety of enzymes.

Protonophoric uncouplers are useful tools for studying plant respiration. They make the membrane permeable to protons and consequently inhibit oxidative phosphorylation. In the presence of an uncoupling agent such as 2,4-dinitrophenol or FCCP (carbonylcyanide-p-trifluoromethoxy-phenylhydrazone), electron transport and H^+-pumping proceed at a maximal rate with no concomitant generation of a H^+ electrochemical gradient. The uncoupling agents are lipid-soluble weak acids that act as H^+ carriers across the inner mitochondrial membrane preventing H^+ passage through the ATP synthase. The H^+ electrochemical gradient is completely dissipated and no further ATP is synthesized. Incubating an uncoupler with mitochondria causes a large increase in oxygen consumption because of the increased activity of the electron transport chain. This effect takes place because electron transport is lower when the H^+ electrochemical gradient increases across the inner mitochondrial membrane. The respiratory chain is able to 'sense' changes in ATP synthase activity: the energy-yielding oxidation and energy-utilizing phosphory-lation communicate with each other through the common H^+ electrochemical gradient.

Reagents that disrupt membranes can also act as potent uncouplers, for example, detergents, phospholipases (yield phospholipids which act as detergents) and pore-forming proteins (mellitin can create holes in membranes). Transmembrane proteins which allow proton movement will also dissipate the H^+ electrochemical gradient causing uncoupling.

3.5.9 ATP synthesis

Section 3.5.6 describes the formation of the H^+ electrochemical gradient whereby some of the free energy is used for the synthesis of ATP from ADP, mediated by ATP synthase. Most of the detailed studies have been carried out with mammalian ATP synthase. ATP synthesis in relation to photosynthesis has been discussed in Chapter 2.

It is important to emphasize that continuation of ATP synthesis requires that the proton gradient is maintained by the carriers of electron transport and that the synthesized ATP is continually removed from the site of synthesis.

Electron microscopy of the energy-transducing inner mitochondrial membrane negatively stained with phosphotungstate shows 'knob-like' features projecting into the matrix. This knob-like feature is one of the components of the ATP synthase, a trans-membrane complex. The structure of this enzyme is highly conserved and the features are remarkably common in all organisms. This is to be expected because of its key role in energy conservation in the cell. The enzyme consists of a water-soluble F_1 complex (as projecting out knobs) with catalytic activity that phosphorylates ADP to ATP, and a membrane-bound F_0 complex comprising hydrophobic proteins embedded in the inner mitochondrial membrane and acts as a H^+ channel through which protons pass from the P phase to the N phase and induce ATP synthesis by the F_1 complex (Fig. 3.14). F_0 serves as the binding base for F_1. The F_0F_1-ATP synthase can also

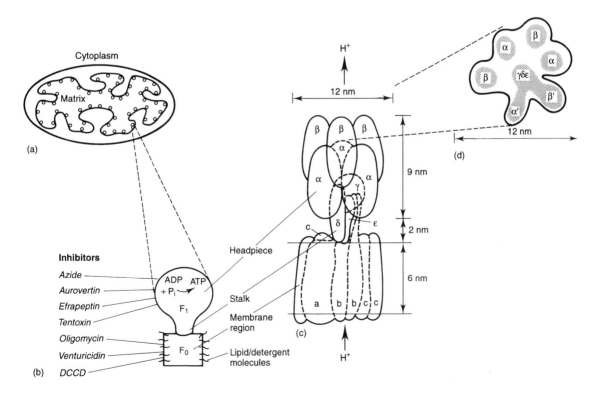

Figure 3.14 ATP synthase in mitochondria. (a) F_1 observed as spheres on inner mitochondrial membrane (in negative stained electron micrographs). (b) Detergent-extracted F_1F_0 ATP synthase. (c) Side view of ATP synthase, showing positions of subunits. (d) Top view of F_1 portion of synthase. One pair of α and β subunits (designated $\alpha'\beta'$) is associated with central mass ($\gamma\delta\epsilon$ subunits), inducing asymmetry. (c and d are derived from electron micrographs using image reconstruction techniques.). From Harris (1995), with permission.

hydrolyze ATP to ADP. Due to this reversibility, this enzyme is referred to as the F_0F_1-ATPase. ATP is synthesized or hydrolyzed on the side of the membrane from which the knobs project out (N phase).

The F_1 moiety has an overall molecular mass of about 350 kDa and can be detached from the membrane by mild treatments, such as salt solution of low ionic strength. It comprises five main subunits. These polypeptides are the α-, β-, γ-, δ- and ϵ-subunits (molecular masses of 56 kDa, 53 kDa, 33 kDa, 16 kDa and 11 kDa, respectively) with a stoichiometry believed to be $\alpha_3\beta_3\gamma\delta\epsilon$ (subscripts represent number of polypeptides in the F_1 complex). The catalytic activity of F_1 and the binding of nucleotides and phosphate reside within the $\alpha\beta$ subunit pair. The subunit pairs are arranged like the segments of a peeled orange such that the α and β subunits alternate each other. There is evidence that the β subunits have one catalytic site each. The α subunit is considered to be involved to some extent with this site. Hydrolysis of ATP requires the presence of α, β and γ subunits whereas synthesis requires all five subunits. The nucleotide-binding site is considered to be at the interface of α/β subunits. The Glu-181 of the catalytic domain of the β subunit when chemically modified with DCCD, results in the loss of catalytic activity. The γ, δ and ϵ subunits constitute the central well/stalk part of the complex and are arranged in an asymmetrical manner. There is considerable sequence homology between the α, β and γ subunits between bacterial, plant and mammalian ATP synthases but comparatively very little homology exists between the δ and ϵ subunits of different species. The minor subunits have regulatory and/or structural functions. The δ and ϵ subunits are believed to be involved in the precise binding of the F_1 to the F_0 sector, the ϵ subunit also displaying regulatory function (inhibits F_1 activity in bacteria and chloroplasts). The γ subunit acts as a pivot in the organization of the α, β subunits around the central axis. Its position is asymmetric with respect to the $\alpha_3\beta_3$ assembly. This subunit also acts as a proton gate maintaining a stoichiometry of the passage of three protons coupled to the synthesis of one ATP. Structural alterations to this protein lead to the passage of protons without ATP synthesis and the membrane becomes uncoupled. Thus, the γ subunit plays a central role in the mechanism of energy transfer from F_0 to F_1 catalytic sites of the complex. Sweet-potato F_1 complex is composed of six subunits, α, β, γ, δ, δ^1, ϵ. The δ^1 subunit may be a proteolytic-degradation product of a larger subunit. ATP synthase from

Escherichia coli and those from other sources have subunit similarities. However, those from mitochondria and thylakoids have extra subunits; their functions are not clear. The δ subunit from *E. coli* can be compared to the mitochondrial oligomycin-sensitivity-conferring protein (OSCP) of F_1. The mitochondrial δ subunit is comparable to the ϵ subunit of *E. coli*. A comparison of subunits is presented in Table 3.1.

The F_0 component of F_1F_0-ATP synthase, embedded in the membrane, has three subunits: a subunit-a without a known function or structure, a subunit-b which is responsible for linking F_0 to F_1, and a subunit-c which is hydrophobic and believed to be involved in proton translocation to F_1 (in *E. coli*) (the nomenclature of the subunits is different for chloroplast ATP synthase, see Table 3.1). The DCCD-binding site resides in subunit-c. DCCD is an inhibitor of ATP synthesis. All three subunits are required to create a proton channel in the lipid bilayer of the membrane; an approximate stoichiometry is ab_2c_{10-12}. DCCD or oligomycin blocks the F_0 proton channel. Soluble F_1-ATPase is insensitive to these inhibitors. Although the isolated F_1 component hydrolyzes ATP rapidly, both F_1 and F_0 are required for ADP phosphorylation. DCCD is generally effective on both bacterial and thylakoid membranes but oligomycin inhibits F_1F_0-ATP synthase (ATPase)

Table 3.1   A comparison of the nomenclature of ATP synthase subunits

Mitochondria	Chloroplast	*E. coli*
F_1		
α	α	α
β	β	β
γ	γ	γ
OSCP	δ	δ
δ	ϵ	ϵ
ϵ	–	–
Inhibitor protein	–	some homology with ϵ
F_0		
a, or subunit 6	IV	a
b	I, II[a]	b
c, or subunit 9[b]	III[b]	c[b]
d	–	–
F_6 (coupling factor 6)	–	–
A6L	–	–
e	–	–
f	–	–
g	–	–

[a] In some photosynthetic bacteria and chloroplasts.
[b] Also known as DCCD-binding protein.

from mitochondria only. The actual mechanism of proton passage through F_0 is unclear. It is possible that protons are carried by amino acid group in subunit-c (relayed by protonated/deprotonated groups of varying pK). Alternatively, protons may pass as H_3O^+ via a hydrophilic channel. The subunit-c has two helical domains traversing the hydrophobic membrane and joined by a hairpin loop at the N phase on the side of F_1 subunit. The two termini (N- and C-) are in the P phase. The Asp-61 residing in the helix 2 reacts with DCCD terminating proton passage. The details of the assembly of c subunits in F_0 are not known. It is considered that subunit-c is in contact with subunit-b which is in contact with γ of F_1.

The precise mechanism as to how ATP synthase functions for both hydrolysis and synthesis of ATP has been debated for some time. It is considered that the proton motive force across the membrane causes conformational change in F_1 and alters its affinities for the substrates and the products. As shown in Fig. 3.14, protons pass through F_0 channel from P phase to N phase; a large proton concentration gradient exists. The affinity of proton binding is co-ordinated with conformational changes which makes ATP synthesis and its release possible. The low-affinity proton binding site is assumed to exist facing to the P phase and the high-affinity site to the N phase of the membrane.

Each of the three catalytic sites of the β subunits can adopt three conformations: a tight conformation (T) where ATP binds tightly, an open conformation (O) from which ATP has been expelled and the site is catalytically inactive, and a loose site (L) which binds ATP and ADP + P_i with equal affinity. Conformational changes due to proton passage causes T site to release ATP and become O site and induce changes to L site (loosely bound ADP + P_i and has no preference for ATP) which becomes T site, a catalytically active site, leading to the synthesis of ATP which remains tightly bound with this site until the next cycle of conformational changes. The binding changes during transition between the catalytic sites is suggested to be due to rotation of the α_3,β_3 subunit assembly in relation to the γ subunit. The association of the assembly with the γ subunit in the axis is asymmetric (one α,β pair making contact at a time, see Fig. 3.14) during rotation. The signals for conformational changes to a particular active site are therefore relayed via γ subunit in association with other stalk components. ATP synthesis in a bound form at T site hardly requires energy, but the energy from proton movement is mainly used for conformational changes (for transformation of T to O site) to expel ATP in order to continue the cycle of ATP synthesis. The conformational changes are probably caused by protonation of protein acidic groups. The mechanism described above is only suggestive and is based on studies by X-ray crystallography, electron micrography and use of specific reagents. These methods provided only static details from which dynamic predictions were made. A schematic representation of ATP synthesis is shown in Fig. 3.15.

In summary, ten protons are translocated for two electrons from endogenous NADH passing through the electron transport chain. Three protons are required for the synthesis of one ATP and one proton is necessary for the transport of ADP + P_i into the mitochondrion and ATP out of it. Two translocating carriers are involved, one for the uptake of ADP and export of ATP, and the other for the uptake of P_i (a proton-P_i symport). The nucleotide carrier does not transport AMP. Considering that four protons are required for the yield of one ATP, 2.5 ATP would result from each NADH oxidized. The oxidation of succinate will result in the synthesis of 1.5 ATP.

With regard to regulation of ATP synthesis, the existence of a regulatory protein in mammalian ATP synthase is interesting. This protein binds to β subunit in the absence of Δp and slows down ATP hydrolysis. The presence of Δp overcomes the inhibition. Thus, under conditions of oxygen deficiency when ATP synthesis slows down, the control mechanism retards ATP hydrolysis. The ϵ subunit of bacteria also has regulatory function in the F_1 complex; it has an inhibitory domain in addition to a domain for interaction with the β subunit.

In oxidative phosphorylation, the pathway of NADH oxidation remains in equilibrium but excess of ATP should reverse this equilibrium. However, the terminal step in electron transport, cytochrome c oxidase, is irreversible. This site is, therefore, a point of control where the availability of reduced cytochrome c is important. This, in turn, is dependent upon a higher [NADH]/[NAD$^+$] ratio and lower [ATP]/[ADP][P_i] ratio. An increased demand of ATP will increase ADP concentration and the above ratio (ATP mass action ratio) will decrease which will then boost oxidative phosphorylation. As ATP is synthesized in mitochondria and utilized in cytosol, it is the mass action ratio in the mitochondria that will directly control ATP synthesis. The inner mitochondrial membrane is impermeable to diffusion of ADP and P_i except via specific transporters.

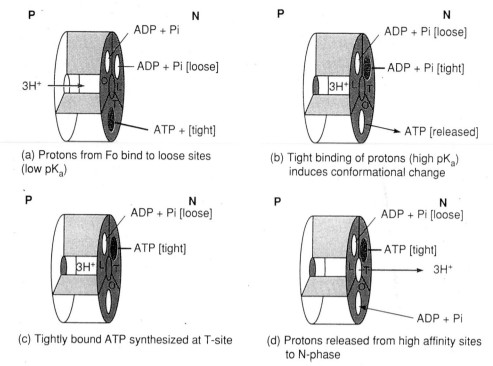

Figure 3.15   A model for the coupling of Δp to the synthesis of ATP by the F_1,F_0-ATP synthase. As described in the text, transfer of protons from low affinity sites (low pK_a) to high affinity sites (high pK_a) is shown as inducing conformational changes in catalytic subunits (here assumed to be three in number and on B chains. O (open), L (loose) and T (tight) indicate three different conformational states of the β subunits. Each subunit experiences each conformation sequentially. (a) Protons from F_0 bind to loose sites (low pK_a); (b) tight binding of protons (high pK_a) induces conformational change; (c) tightly bound ATP synthesized at T-site; (d) protons released from high affinity sites to N-phase. From Nicholls & Fergusson (1992), with permission.

It should be noted in this context that the use of oxygen by mitochondria is related to the availability of ADP and P_i. At mitochondrial state 3 (i.e. when ADP is added; see section 3.5.5), there is high oxygen consumption until ADP is converted into ATP (reaches state 4 with low oxygen consumption). This dependence on ADP concentration is known as respiratory control.

REFERENCES AND FURTHER READING

Abrahams, J.P., Leslie, A.G.W., Lutter, R. & Walker, J.E. (1994) *Nature* **370**, 621–628.

Anderson, J.W. & Beardall, J. (1991) *Molecular Activities of Plant Cells*. Blackwell, Oxford, 384pp.

apRees, T., Entwistle, T.G. & Dancer, J.E. (1991) In *Compartmentation of Plant Metabolism in Non-photosynthetic Tissues* (M.J. Emes, ed.), pp. 95–110. Cambridge University Press, Cambridge.

Ashihara, H. & Komamine, A. (1976) *Physiol. Plantarum* **36**, 52–59.

Brand, M.D. & Murphy, M.P. (1987) *Biol. Rev.* **62**, 141–193.

Bryce, J.H. & Hill, S.A. (1993) In *Plant Biochemistry and Molecular Biology* (P.J. Lea & R.C. Leegood, eds), pp. 1–26. Wiley, Chichester.

Copeland, L. & Turner, J.F. (1987) In *The Biochemistry of Plants*, Vol. 11, *Biochemistry of Metabolism* (D.D. Davies, ed.), pp. 107–127. Academic Press, New York.

Crawford, R.M.M. (1989) *Studies in Plant Survival*. Blackwell, Oxford, 296pp.

Douce, R., Brouquisse, R. & Journet, E-P. (1987) In *Biochemistry of Plants*, Vol. 11, *Biochemistry of Metabolism* (D.D. Davies, ed.), pp. 177–211. Academic Press, New York.

Douce, R. & Neuberger, M. (1989) *Annu. Rev. Plant Physiol. Mol. Biol.* **40**, 371–424.

Harris, D.A. (1995) *Bioenergetics at a Glance*. Blackwell, Oxford, 116pp.

Hinkle, P.C., Kumar, M.A., Resetar, A. & Harris, D.L. (1991) *Biochemistry* **30**, 3576–3582.

Lambers, H. (1990) In *Plant Physiology, Biochemistry and Molecular Biology* (D.T. Dennis & D.H. Turpin, eds), pp. 124–143. Longmans, Harlow.

Miernyk, J.A. (1990) In *Plant Physiology, Biochemistry and Molecular Biology* (D.T. Dennis & D.H. Turpin, eds), pp. 77–100. Longmans, Harlow.

Mitchell, P. (1961) *Nature* **191**, 144–148.

Moore, A.L. & Siedow, J.N. (1991) *Biochim. Biophys. Acta* **1059**, 121–141.

Moser, C.C., Keske, J.M., Warncke, K., Faird, R.S. & Dutton, P.L. (1992) *Nature* **355**, 796–802.

Nicholls, D.G. (1982) *Bioenergetics: An Introduction to the Chemiosmotic Theory.* Academic Press, London, 190pp.

Nicholls, D.G. & Fergusson, S.J. (1992) *Bioenergetics 2.* Academic Press, London, 255pp.

Schnarrenberger, C., Flechner, A. & Martin, W. (1995) *Plant Physiol.* **108**, 609–614.

Storey, B.T. (1980) In *Biochemistry of Plants*, Vol. 2, *Metabolism and Respiration* (D.D. Davies, ed.), pp. 125–195. Academic Press, New York.

Trumpower, B.L. (1990) *J. Biol. Chem.* **265**, 11409–11412.

Trumpower, B.L. & Gennis, R.B. (1994) *Annu. Rev. Biochem.* **63**, 675–716.

Turner, J.F. & Turner, D.H. (1980) In *The Biochemistry of Plants*, Vol. 2, *Metabolism and Respiration* (D.D. Davies, ed.), pp. 279–316. Academic Press, New York.

Walker, J.E. (1992) *Q. Rev. Biophys.* **25**, 253–324.

Wiskitch, T. (1980) In *The Biochemistry of Plants*, Vol. 2, *Metabolism and Respiration* (D.D. Davies, ed.), pp. 244–278. Academic Press, New York.

4 Carbohydrate Metabolism: Storage Carbohydrates

G. Avigad and P.M. Dey

4.1 INTRODUCTION

One of the most abundant groups of organic compounds in the plant kingdom are the carbohydrates which exist as monosaccharides, disaccharides, higher oligosaccharides, polysaccharides and their derivatives. The nature of linkages in the complex carbohydrates and the structure of individual component saccharides have important physiological consequences in the plant. The basic structural features of simple sugars will not be covered in this chapter. The accumulated carbohydrate reserves function as providers of monosaccharides which are metabolically utilized during growth and development of the plant. These carbohydrates also have varied commercial applications. This chapter focuses on biosynthesis, catabolism and biological function of the major storage carbohydrates. The reader is referred to Dey & Dixon (1985) for other storage carbohydrates such as, cyanogenic glycosides, phenolic and flavonoid glycosides, glucosinolates, galactosyl glycerol derivatives (floridoside and isofloridoside), C- and N-glycosides and some algal polysaccharides.

4.2 SUCROSE

The disaccharide sucrose is a principal product of carbon fixation during the photosynthetic reaction. In addition to its central position in the metabolism of photosynthetic plants and the realization that a huge portion of the organic matter in nature had at one time passed through a sucrose molecule, this sugar has an enormous economical importance as a leading agricultural commodity and as a nutrient to most living organisms. It is, therefore, not surprising that there has always been an intensive interest in sucrose chemistry, the technology of its production, use as a chemical raw material and as food and a sweetener. However, historically, advances in understanding the biochemistry and metabolism of sucrose in plants were very slow, much of the difficulties in this endeavor were created by the action of invertase, whose almost ubiquitous presence in plant extracts caused any sucrose in the system studied to be immediately hydrolyzed. The introduction in the late 1940s of radioisotope labeling to study metabolism, the first success in enzymatically synthesizing sucrose (accomplished with bacterial sucrose phosphorylase by Hassid and Doudoroff in 1944) and the key discovery of the sucrose synthesizing enzymes in plants (by Leloir and Cardini in 1955) were probably the most outstanding milestones on the road that brought us to understand sucrose biochemistry as we know it today (see Pontis, 1977 and Avigad, 1982 for a general review).

Sucrose (**1**) is a non-reducing, sweet, highly soluble molecule. It is chemically inert when in

PLANT BIOCHEMISTRY
ISBN 0-12-214674-3

(1)
Sucrose
β-D-Fructofuranosyl α-D-Glucopyranoside

contact with proteins, because it can not form covalent adducts with free amino groups. The sucrose molecule retains a high free energy of hydrolysis ($\Delta G^{o\prime} = -7.0\,\text{kcal mol}^{-1}$) which is the highest known for a glycosidic bond. Consequently, the α-glucosyl residue in sucrose can be viewed as a very efficient mode for energy conservation by the plant.

In plants, sucrose is a product into which a major portion of the CO_2 fixed by photosynthesis is captured. It is the form in which most fixed organic carbon is translocated from the source photosynthetic tissue to non-photosynthetic organs. Sucrose provides substrates for energy and synthesis of cell matter and other storage glycosides such as starch and fructans. Excess sucrose accumulates sometimes in the leaves, but predominantly in storage tissues. Stored carbohydrates including sucrose are mobilized and utilized during germination and growth of the developing plant. Sucrose is mostly stored in the vacuole which comprises close to 70% of the cell volume (Winter et al., 1994). The concentration of sucrose in the leaf vacuole at peak light periods reached 11 mM. In contrast, the cytosol which accounts for only 28% of the cell volume accumulated almost 55 mM sucrose.

With all the enormous knowledge that accumulated about sucrose biochemistry, it is apparent that patterns and fluxes of sucrose metabolism are extremely complex and not so simple to explain or predict. Only several simple enzymic reactions are directly involved with creating or degrading the sucrose molecule, and they are subject to complex genetic and metabolic controls at different levels. These enzymes reside in different cellular compartments and they often appear as isoenzymes, encoded by a number of distinct genes. Expression of these individual isoform enzymes may be controlled in different patterns both at gene transcription and post-translational levels. Many signals, environmental or intracellular, physical or chemical, could exert varying degrees of modulating sucrose metabolism because they can somehow interact with specific responsive elements arrayed along the promoter regions of the sucrose genes

involved. This type of regulation has probably a most enduring effect on sugar metabolism in the life of the whole plant over long periods measured in hours or days. This is in contrast to the relatively rapid response of enzyme activities to allosteric metabolite effectors or to post-translational regulatory mechanisms.

The classically, well recognized view of sucrose as a metabolite serving the net syntheses of organic matter in the plant has to be supplemented by emerging information about a more general direct role this disaccharide has as a signal for gene expression. This effect, discussed later in this chapter, is associated with enzymes not only involved with sucrose itself as the substrate but with the synthesis of many other proteins. In view of this abundant information about an extremely overall complex process that varies dramatically from species to species and from organ to organ under exposure to a variety of physiological conditions, we can only manage to present here some simplified models for sucrose metabolism and its control.

4.2.1. Sucrose synthesis

(a)   The pathway

Triose phosphate as the product of the photosynthetic carbon fixation in the chloroplasts is transported into the cytoplasm and transformed there into hexose phosphates by enzymes of the gluconeogenic pathway. A major portion of these hexoses is used for the synthesis of sucrose which is transported and distributed throughout the plant. Sucrose in target growing tissue is used as the carbon source for energy and the synthesis of cellular matter. Some sucrose is stored in the source photosynthetic tissues, mostly in the vacuole, but the bulk of excess sucrose which reaches sink organs provides the substrate for formation of other storage glycosides, predominantly starch and fructans, stored in the tonoplast. After peak accumulation of sucrose is reached, such as in the ripe fruit, in seed or storage roots and tubers, the disaccharide is rapidly utilized to sustain initiation of tissue growth and development. There is obviously a basic academic interest to define and characterize the enzymes responsible for sucrose biochemistry and to understand the dynamic and regulation of its metabolism, quantitatively one of the major biochemical processes in nature. In addition, as recent developments show, there is a practical value in these studies as they benefit the creation of transgenic plants where sucrose synthesis is intensified to a degree that can dramatically increase the biomass (Foyer & Ferrario, 1994).

A set of several enzymes is responsible for sucrose biosynthesis in plants (Fig. 4.1). The first enzyme, sucrose phosphate synthase (SPS), catalyzes the synthesis of sucrose-6'-phosphate by a transglucosylation from UDP-glucose to fructose-6-phosphate:

$$\text{UDP-glucose} + \text{fructose-6-phosphate}$$
$$\rightleftharpoons \text{sucrose 6'-phosphate} + \text{UDP}$$

A second enzyme, sucrose phosphatase (SPP), hydrolyzes sucrose-6'-phosphate to release free sucrose and P_i:

$$\text{sucrose-6'-phosphate} + H_2O \rightarrow \text{sucrose} + P_i$$

This irreversible reaction by an enzyme that abundantly accompanies SPS in the cell prevents sucrose-6'-phosphate accumulation and provides an energetically favored and efficient production of sucrose even at low UDP-glucose and fructose-6-phosphate or at high sucrose concentrations.

Another enzyme which can produce sucrose is sucrose synthase (SS), which catalyzes transglucosylation from UDP-glucose to fructose in a reaction which is reversible under normal physiological conditions:

$$\text{UDP-glucose} + \text{fructose} \rightleftharpoons \text{sucrose} + \text{UDP}$$

As discussed later, SS under most physiological conditions is usually assigned a role in sucrose cleavage, rather than its synthesis. Numerous studies of these three exclusively cytoplasmic enzymes, particularly of SPS and SS, have been carried out to learn about their structure, genetics, regulation, subcellular distribution and their role in overall carbohydrate metabolism and partitioning. Some of these will be summarized here.

The glucoside donor for sucrose biosynthesis is UDP-glucose which is produced by the UDP-glucose pyrophosphorylase reaction:

$$\text{UTP} + \text{glucose-1-phosphate} \rightleftharpoons \text{UDP-glucose} + PP_i$$

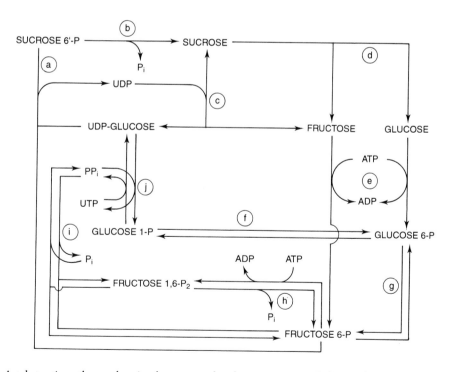

Figure 4.1 A schematic pathway showing key steps related to sucrose metabolism. The reactions shown occur in the cytoplasm, but the vacuole is an additional location for invertase. As described in the text, several of these steps are catalyzed by isoactive genetically distinct enzyme isomers. a, sucrose-P synthase; b, sucrose-P hydrolase; c, sucrose synthase; d, invertase; e, hexokinase and fructokinase; f, phosphoglucoisomerase; g, phosphoglucomutase; h, phosphofructokinase and fructose-1,6-bisphosphatase; i, pyrophosphate : fructose-6-phosphate 1-phosphotransferase; j, UDP-glucose pyrophosphorylase. In the photosynthetic source tissues, fructose-1,6-bisphosphate supplies the bulk of photoassimilated carbon towards the synthesis of sucrose.

The rate of the reaction and its direction is determined by the concentration of the various substrates. This enzyme is located in the cytoplasm of photosynthesizing as well as growing cells. An ample supply of glucose-1-phosphate, the rapid utilization of UDP-glucose, the elimination of pyrophosphate by transphosphorylation and to some extent by pyrophosphatase reactions all serve as forces driving the reaction to the right. At some physiological condition (discussed elsewhere in this book) this reaction may proceed to the left, for example when a significant amount of PP$_i$ is provided by an increased rate converting fructose-1,6-bisphosphate to fructose-6-phosphate in the pyrophosphoryl: fructose-6-phosphate phosphoryltransferase (PFP) reaction. Conditions which promote glucose-1-phosphate formation also exist when UDP-glucose concentration is elevated because its utilization for biosynthetic reactions, e.g. the production of various glycoside and cell wall polymers, has slowed down. A source for UDP-glucose which is of quantitative importance in storage tissues where sucrose concentration is high can be provided by reversal of the SS reaction according to the following scheme:

1. Sucrose + UDP $\rightleftharpoons$ UDP-glucose + Fructose

2. Fructose + ATP
    $\rightarrow$ Fructose-6-phosphate + ADP

3. UDP-glucose + PP$_i$
    $\rightleftharpoons$ Glucose-1-phosphate + UTP

4. Fructose-1,6-P$_2$ + P$_i$
    $\rightleftharpoons$ Fructose-6-phosphate + PP$_i$

catalyzed by: 1, sucrose synthase; 2, fructokinase (UTP may to some extent replace ATP); 3, UDP-glucose pyrophosphorylase and 4, PP$_i$-fructose-6-phosphate 1-phosphotransferase (PFP). The net result of these reactions will result in the conversion of sucrose into hexose phosphates, some of which will enter the plastids and be used for the synthesis of ADP-glucose. It is important to note that the cytoplasmic UDP-glucose pyrophosphorylase is a completely different protein entity from the ADP-glucose pyrophosphorylase which is localized in the plastids (Feingold & Avigad, 1980; Avigad, 1982; apRees, 1988; Tyson & apRees, 1989; Nakano et al., 1989; Wagner & Backer, 1992).

(b)   *The enzymes and their regulation*

(i)   Sucrose phosphate synthase

SPS which has been purified and isolated from several sources such as spinach and corn leaves, is a dimeric or tetrameric molecule assembled from a 138 kDa protein subunit (Kalt-Torres et al., 1987; Bruneau et al., 1991). Some fragmentation of this monomer that may have occurred during purification or because of an *in vivo* post-translational cleavage are probably the reasons for the isolation of a 120 kDa subunit described in many studies (Walker & Huber, 1989a,b; Huber & Huber, 1991, 1992b; Worrel et al., 1991; Klein et al., 1993; Salvucci & Klein, 1993; Reimholz et al., 1994). The enzyme is present predominantly in all green leaf tissue and also in the scutellum of germinating seeds. It is found only at low levels, or sometimes undetected in non-photosynthetic storage tissues. Levels of SPS activity in the leaves vary markedly during phases of growth and development and go through cycles which correspond to the diurnal light/ dark periods. Peak activity is usually reached at noon and declines when light is eliminated.

The level of SPS activity, the enzyme responsible for net synthesis of sucrose in nature, is controlled by a complex array of regulatory mechanisms (Huber & Huber, 1992a) (Fig. 4.2). One type of regulation, sometimes called the 'coarse' control, represents the changes in the level of SPS dictated by gene transcription (a relatively slow response) or by post-translational covalent modifications of the enzyme protein (a relatively faster response). A second class of regulation, or 'fine' control mechanisms (an almost instantaneous response) modulate the activity of the already existing enzyme by non-covalent allosteric inhibition or activation exerted by specific metabolite effectors and by the rate and concentration at which substrates are supplied. The pathway by which these supplies are provided may also be co-ordinately controlled with SPS by the same effectors. One of the more detailed 'coarse' control mechanisms identified for SPS in some plants is the light activation of the enzyme, an event involving a reversible protein phosphorylation/dephosphorylation cycle. Although the exact molecular mechanisms that link the light signal to protein alteration are only partly understood (Harter et al., 1994), it was clearly established that the most active form of the enzyme is a dephosphorylated species. Tissues exposed to the protein phosphatase-2A inhibitors okadaic acid and levamisole prevented SPS activation. In contrast, tissue

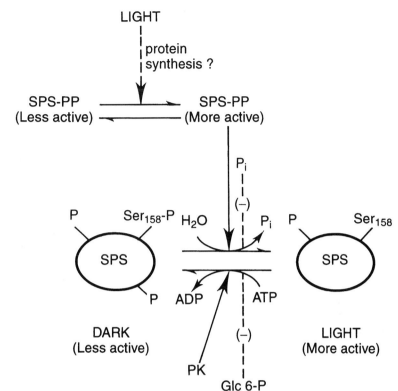

Figure 4.2 Schematic representation of the regulation of spinach SPS by light/dark signals. Light increases activity of a SPS-protein phosphatase (PP), which by removing phosphate from serine-158 on SPS causes the enzyme to be more active. SPS is inactivated when it is phosphorylated by protein kinase (PK). P_i and Glc-6-P are non-covalent allosteric modulators of the enzyme (from Huber *et al.*, 1994*b*).

incubated with high mannose concentrations which caused an intracellular P_i sequestration resulted in the formation, even in the dark, of SPS species of high activity. Lowering the amount of phosphorylated enzyme under such conditions resulted both from a diminished rate of protein kinase activity (ATP depletion) and by activation of a protein phosphatase which is normally inhibited by P_i (Walker & Huber, 1989a,b; Siegl & Stitt, 1990; Quy & Champigny, 1992; Weiner *et al.*, 1993; Huber *et al.*, 1994b). Such effects are also expected to be exerted by the strong protein phosphatase-2A inhibitor calyculin A and by the signals from the powerful elicitor molecules produced by plant pathogens which strongly increase the level of protein phosphorylation (Felix *et al.*, 1994). As discussed later, prolonged deprivation of P_i, such as that which occurs by mannose loading, can also have a marked effect on the rate of gene transcription and increase specific protein synthesis including SPS. It was also noticed that increased osmotic stress elevates the SPS activity in a photosynthetic tissue. Whether this change is the consequence of a change in gene expression or on protein phosphorylation or both, has yet to be determined.

Phosphorylated and non- or less-phosphorylated enzyme monomers of SPS were isolated from plants exposed to dark or light periods before extraction. Protein kinase phosphorylation *in vitro* established that several serine residues on the protein may be phosphorylated, but possibly only two (one is a ser-158 site) are directly associated with induction of changes in catalytic activity (Weiner *et al.*, 1992; Huber & Huber, 1992a). Although studied in detail, it should be emphasized that activation of SPS by light is not a universal process for all photosynthetic plants (Huber *et al.*, 1989, 1990, 1994b). Three groups were broadly defined: group I, light activation results in an increase in V_{max} of enzyme, e.g. in barley and maize; group II, light activation does not change V_{max} but alters other kinetic parameters, such as K_m for substrates, e.g. in spinach, swiss chard, broad bean and sugar beet; group III, SPS activity is not light regulated, e.g. in soybean, pea, *Arabidopsis*, tobacco, cucumbers and melon. It is not clear whether in this latter group phosphorylation of SPS occurs at all. In these species, the enzyme is 'coarse' regulated at the gene transcription level by some hormonal factors such as gibberellins (Cheikh & Brenner, 1992), and by the 'fine' regulatory metabolite controls. Leaf metabolism in group III is strongly expressed by high accumulation of starch, compared to a substantial sucrose and low starch accumulation in illuminated leaves of classes I and II. SPS in non-photosynthetic organs, such as potato tubers, seems to be regulated by both non-light dependent protein phosphorylation and by metabolites (Reimholz et al., 1994). In a co-ordinated effect, light also modulates the transcription and post-translational modification of the fructose-1,6-bisphosphatase isoenzymes both in the chloroplast and the cytoplasm. This increase in activity of this enzyme is supportive in accelerating the carbohydrate flux towards the synthesis of sucrose (and starch) and in groups I and II plants described above, it is complementary to the effect of light on SPS activity (Khayat et al., 1993).

At the level of metabolite control, SPS is allosterically regulated by inorganic phosphate which inhibits the reaction and is activated by glucose-6-phosphate. The glucose-6-phosphate/P_i ratio in the cell has, therefore, an important effect on the level of catalysis by the enzyme (Doehlert & Huber, 1985; Stitt et al., 1988). Thus, in a situation when P_i in the cell is sequestered and hexose phosphates are elevated, SPS is activated, accelerating sucrose synthesis. Such conditions also favor the activation of SPS due to protein dephosphorylation as described earlier in this section. However, because of the complex metabolic effects of P_i deprivation on overall cellular metabolism, the outcome of this condition on the rate of sucrose synthesis in vivo can not be predicted in a simple way (Usuda & Shimogawara, 1991, 1993; Crafts-Brandner, 1992; Theodorou & Plaxton, 1993; Cakmak et al., 1994; Rychter & Randall, 1994; Huber et al., 1994a). Surprisingly, in many studies, it was found that P_i deprivation, while diminishing overall growth of photosynthetic tissue, did not reduce (or even increased) sucrose and starch accumulation in sink non-photosynthetic storage sites, such as roots and grain. This suggests that the dramatic metabolic changes which occur during phosphate deprivation (e.g. reduction of overall CO_2 assimilation and respiration, reduced ATP and UDP-glucose levels, reduced level of protein phosphorylation, reduced gluconeogenesis and slowed down glycolysis) allowed for the stimulation of alternate glycolysis by-passing, ATP and P_i independent reactions, particularly those provided by phosphoenol pyruvate phosphatase, phosphoenol pyruvate kinase and pyrophosphate: fructose 6-phosphate 1-phosphotransferase (PFP) (Theodorou & Plaxton, 1993). Maintaining a high rate of sucrose

synthesizing activity, rather than sucrose cleavage reactions, under such conditions will be supportive to the overall metabolic re-routing which requires the maintenance of a supply of pyrophosphate. Deprivation of inorganic phosphate has also a direct effect as a transcription level signal that increases SPS synthesis (Usuda & Shimogawara, 1993; Huber et al., 1994a). Similar signaling effect for P_i was demonstrated for the synthesis of the regulatory α-subunit protein of PFP (Theodorou et al., 1992), and of vacuolar acid phosphatase (VspB) storage proteins (Sadka et al., 1994), two enzymes whose activities are closely intertwined with sucrose metabolism.

Another factor that will determine the rate of sucrose phosphate synthesis is the availability of the substrates and their concentration in the cytoplasm. In this case, the indirect regulatory effect of fructose-2,6-P_2 on the rate of sucrose synthesis has been greatly emphasized. Fructose-2,6-P_2 is considered to have a major role in regulating carbohydrate fluxes and partitioning in plant tissues. As a cytosolic metabolite, it is a powerful inhibitor of fructose-1,6-bisphosphatase, thus it can modulate the rate by which hexose-phosphate substrates are being provided for the synthesis of sucrose. Indeed, many studies have clearly indicated an inverse correlation between physiological levels of fructose-2,6-P_2 in the tissue and that of sucrose synthesis. Factors that could affect fructose-2,6-P_2 metabolism and turnover, e.g. during light and dark periods, will thus also influence the rate of sucrose synthesis and carbon partitioning (Stitt, 1990; Neuhaus et al., 1990; Huber et al., 1990). A specific site for fructose-2,6-P_2 control is its strong stimulation of PFP (pyrophosphate : fructose-6-phosphate 1-phosphotransferase) activity. The intensity of this reaction may affect the strength of hexose flow to or from sucrose particularly in non-photosynthetic tissues. Direct experimental support for these conjectural assumptions was found in the study of transformed tobacco plants which could express the mammalian liver 6-phosphofructo-2-kinase gene (Kruger & Scott, 1994; Scott et al., 1995). Both during photosynthesis and dark periods these plants displayed increased production and higher level (50–230%) of intracellular fructose-2,6-P_2. During light periods, sucrose synthesis was strongly reduced whereas accumulation of starch moderately increased. During dark periods, these transgenic leaves containing high fructose-2,6-P_2 levels had a lower degree of starch turnover and mobilization and a significant sequestration of P_i in the form of various phosphate esters, particularly as glycerate-3-phosphate. Such an increase in the 3-phosphoglycerate/P_i ratio may result in the stimulation of ADP-glucose pyrophosphorylase and starch biosynthesis. In contrast with these conclusions, a recent study by Paul *et al.* (1995) found that transgenic tobacco seedlings which expressed only 1–3% of wild-type PFP did not differ much from wild-type plant in their carbohydrate partitioning patterns and in the level of sucrose synthesis. It may be that the correlation of fructose-2,6-P_2 with the rate of sucrose synthesis is not necessarily explained through its effect on PFP activity alone. In summary, the experimental data, though indirectly, support the premise that fructose-2,6-P_2 levels contribute to the modulation of carbon partitioning, particularly by affecting those reactions involved in sucrose synthesis. However, the highly variable quantitative aspects of this 'long range' modulation in different plants makes it difficult to clearly define its molecular basis.

(ii)   Sucrose-6′-phosphate phosphatase

Sucrose-6′-phosphate, the product of the SPS reaction, is rapidly hydrolyzed by sucrose-6′-phosphate phosphatase (SPP), and therefore does not accumulate in plant tissue. In addition, this molecule was found to be devoid of any role in the regulation of SPS activity (Krause & Stitt, 1992). SPP is found at an ample level in the cytoplasm, together with and much higher than SPS. It is an enzyme which depends on Mn^{2+} or Mg^{2+} for its activity but only very limited information is available about its detailed molecular properties (Avigad, 1982; Hawker *et al.*, 1987; Echeverria & Salerno, 1993). The study of SPP is limited by the difficulty in providing sufficient quantities of sucrose-6′-phosphate for use as the substrate. A recent development of a chemical procedure for its synthesis may help advance the research of this enzyme (Kim & Behrman, 1995).

A suggestion that SPP participates in the group transport mechanism responsible for sucrose transport has not been substantiated experimentally. Evidence that the combined action of SPS and SPP is the dominant mechanism for net sucrose synthesis was shown in an experiment where the maize SPS gene was attached to various promoters and expressed in transgenic tomato plants. These transformed plants were thus endowed with SPS permitting production of sucrose at levels much higher than those in the wild type. In one case, when using the 35S (cabbage mosaic virus) promoter, the SPS was constitutively expressed in the transgenic tomato resulting in a large increase in the rate of photo-synthesis and of sucrose production indicating that the endogeneous level of SPP activity was adequate. The disaccharide accumulated to levels which resulted in up to 100% increase in the biomass (Galtier *et al.*, 1993; Foyer & Ferrario, 1994).

(iii)   Sucrose synthase

Sucrose synthase (SS) is a highly characterized enzyme. The enzyme usually resides in the cytoplasm of both photosynthetic and non-photosynthetic cells including the vascular bundles of various plants (Ho, 1988; Tomlinson *et al.*, 1991; Geigenberger *et al.*, 1993; Nolte & Koch, 1993). This location may indicate that SS participates in sucrose translocation but such a role has not been conclusively proven. The active SS enzyme is a tetramer built of a monomeric unit of about 90 kDa. Small variations in this size were found for different isoenzymes between species, or even between those present in the same organism. SS isoenzymes were purified, for example, from tomato (Sun *et al.*, 1992; Sowokinos & Varus, 1992), peach fruit (Moriguchi & Yamaki, 1988), soybean nodules (Morell & Copeland, 1985), *Vicia faba* cotyledons (Ross & Davies, 1992), rice (Wang *et al.*, 1992) and carrot (Sebkova *et al.*, 1995). In monocotyledons, such as the developing maize kernels, two SS isoenzymes encoded by two separate genes designated *Shrunken*-1 (or *Sh*-1) and *Sus*-1, or alternatively as *SS*-1 and *SS*-2. In the leaf, particularly in young growing tissue, the SS-2 enzyme dominates whereas SS-1 is predominantly high in starch sink storage tissues. SS-1 subunit is a 96 kDa, whereas SS-2 is a 90 kDa protein. There are almost no kinetic differences between these two enzymes, but significant variations in patterns of their control at the gene transcription level, which could be correlated to the structure of the individual genes were identified (Echt & Chourey, 1985; Heinlein & Starlinger, 1989; Nguyen-Quoc *et al.*, 1990; Maas *et al.*, 1991; Koch *et al.*, 1992; Heim *et al.*, 1993). A SS monomer was identified as the nodulin-100 protein that accumulates in soybean nodules (Zammit & Copeland, 1993).

Gene structures and cDNA sequences for several SS isoenzymes were determined. Among these are the SS-1 of maize (Huang *et al.*, 1994), tomato (Wang F. *et al.*, 1993b), rice RSS-1 isoenzyme (Werr *et al.*, 1985; Yu *et al.*, 1992; Wang M. *et al.*, 1992), potato (Salanoubat & Balliard, 1987), wheat (Marana *et al.*, 1988), *Arabidopsis* (Chopra *et al.*, 1992) and carrot (Sebkova *et al.*, 1995). The structural enzyme for most of these preparations is a protein of about 805 amino acids.

There is close homology (50–70%) between the different SS species when determined immuno-chemically or when comparing the amino acid sequences, but less than 25% with SPS. Differences in the gene flanking promoter regions probably account for variations in the transcription of SS isoenzymes as a response to various environmental or metabolic stimuli.

A noteworthy fact is that anaerobiosis was found to stimulate expression of SS-1 gene in maize or RSS1 in rice (McCarty *et al.*, 1986; Springer *et al.*, 1986; McElfresh & Chourey, 1988; Ricard *et al.*, 1991; Xue *et al.*, 1991). It was determined that oxygen inhibited enzyme transcription (mRNA synthesis) rather than the translation reactions. This is compatible with the observation that oxygen in general has some control on gene expression (Pahl & Baeuerie, 1994). It can be assumed that under normal aerobic conditions, SS synthesis (at least the SS-1 isoenzyme) is not fully expressed in the plant growing at normal atmos-pheric conditions, but variation in oxygen level imposed by extreme stress may have a detrimental effect on the pattern and rate of sucrose metabo-lism by regulating the level of the SS reaction. For example, in germinating seeds which may often be under temporary anoxia or 'flood stress', sucrose from the scutellum that enters the embryo, will be effectively cleaved by high levels of SS to provide intermediates for glycolysis and for cell growth (Kennedy *et al.*, 1992).

Another critical factor that controls SS gene expression is sucrose itself. This disaccharide serves as a strong inducer for this gene. There is a distinct difference of response to sucrose between the different isomeric genes studied in maize and rice. The SS-1 enzyme increased 10-fold in the presence of high concentrations of carbohydrate, but this did not occur with *Sh*-1, for which suc-rose could act as a repressor (Maas *et al.*, 1990; Koch *et al.*, 1992; Karrer & Rodriguez, 1992; Heim *et al.*, 1993). The promoter region of this enzyme, similar to other sucrose inducible genes, has a specific sucrose response element which, by a mechanism not yet deciphered, promotes gene transcription (see discussion of this topic further on in this chapter).

4.2.2 Sucrose mobilization

(a) *The reactions*

There are two reactions that initiate the cleavage of the sucrose molecule to provide intermediates which enter the metabolic pathways of the cell and

to furnish the energy and substrates required for sustaining viability and growth (Fig. 4.3). One reaction, common and abundant in plant tissues, is hydrolysis to glucose and fructose, catalyzed by invertase isoenzymes. The second reaction is provided by sucrose synthase which in the presence of UDP converts sucrose into UDP-glucose and fructose. The free hexoses can temporarily accumulate or are phosphorylated into hexose 6-phosphates by ATP-dependent hexokinases and fructokinases (Kruger, 1990; Gardner *et al.*, 1992; Renz *et al.*, 1993). The UDP-glucose provides a substrate that can be used directly or after modification, and transported to other cellular compartments for the synthesis of cellulose, callose, pectin, starch and many other glycosides as well as providing glucose-1-phosphate for respiration. In addition to the established property of the 'classical' reaction of SS with UDP, it can, though, at a much slower rate, produce ADP-glucose with ADP. Lim *et al.* (1992) found that a SS isoenzyme, more specific to ADP than UDP as the acceptor, is present in developing rice grain and is responsible for significant ADP-glucose formation in the cyto-

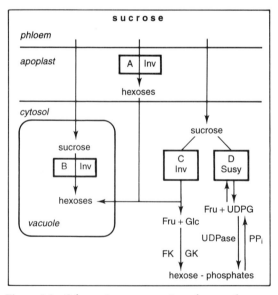

Figure 4.3 Schematic representation of routes for sucrose cleavage. Inv, invertase; A, apoplastic ('cell-wall') 'acid' enzyme; B, vacuolar, 'acid' enzyme; C, cytosolic, 'neutral' enzyme; D, sucrose synthase (after Zrenner *et al.*, 1995). Sucrose and the monosaccharide are able to pass into or out of the vacuole and the hexose phosphates in the cytosol are used by various biosynthetic and other metabolic pathways, such as providing for the re-synthesis of sucrose phosphate (in the cytoplasm) or the formation of starch (in the amyloplasts).

plasm. This finding by cloning and genetic analysis has yet to be substantiated.

Originally, the role of SS as a sucrose cleaving enzyme, was based mostly on its kinetic parameters, predominantly on its very high K_m for sucrose (between 50 and 200 mM compared to less than 0.03–0.50 mM for the enzyme's three other substrates) and the fact that the K_{eq} of the reaction is close to 1.0. SS is also not known to be modulated by metabolites, and the direction of the reaction it catalyzes is exclusively dependent on the concentration of the reactants. In addition, some tissues, particularly storage organs, developing fruit or germinating seeds, have active sucrose utilization but very low invertase activities (Feingold & Avigad, 1980; Avigad, 1982). Subsequent research of many species firmly established the fact that at specific physiological states SS provides a key reaction for sucrose cleavage specifically in the cytoplasm. Many experimental data based on correlating enzyme levels to the rate of sucrose utilization, specific ^{14}C-labeling patterns between sucrose and other metabolites and cellular constituents particularly starch, pulse–chase measurements tracing sucrose turnover, and other types of kinetic analysis of rates of the carbohydrate flux catalyzed by SS, are all in support of a central role of SS in sucrose cleavage (see for example Echeverria & Humphreys, 1984a; Edwards & apRees, 1986; Hargreaves & apRees, 1988; Kruger, 1990; Geigenberger & Stitt, 1993; Wang et al., 1994).

In order to understand sucrose mobilization, it has to be noticed that SS action is confined to the cytoplasm, whereas invertase activity can be found both in the cytoplasm and also in the vacuole and the apoplast. Sucrose can pass through the membrane barrier between these three compartments, but does not appear inside the plastids. The levels of various invertase isoenzymes and sucrose synthase isoenzyme may vary appreciably during various physiological states of the tissue, and there may also be significant species differences in the expression of these sucrose-cleaving activities. Consequently, the quantitative values which apportion the different contributions of invertase and SS found in a set of experiments for one individual plant or organ cannot be simply applied without careful examination to predict patterns of sucrose mobilization in a different species or strain.

(b) Invertase

(i) The enzymes

Invertases (β-D-fructofuranosidase) are present in multiple forms in most plant tissues, and there are numerous studies describing these enzyme entities (for review on early literature see Avigad, 1982; apRees, 1984; Hawker, 1985). Invertase hydrolyzes sucrose to its monosaccharide constituents, but at a much slower rate it can also remove terminal β-fructosyl residues from short-chain oligosaccharides, such as raffinose and kestoses. Long-chain fructans (see later in this chapter) are poor substrates for this enzyme. In addition, invertases can carry out transfructosylation reactions between its substrate molecules resulting in formation of oligosaccharides such as kestotrioses when sucrose serves as both the donor and acceptor. Production of these oligosaccharides is directly related to initial substrate concentration, and since they are subsequently hydrolyzed by invertase, their presence in the system is usually transitory. Two classes of invertase differentiated by their pH optima are present in plant tissues. An alkaline invertase, with maximal activity at pH 7.0, is localized in the cytoplasm, and isoforms of an acid invertase, a glycoprotein with a pH optimum at about 5.0, have at least two subcellular locations. One site is in the vacuole (tonoplast) and another, considered to be 'extracellular', is in the apoplast and a significant portion is attached to the cell wall. This latter type, named also 'free space', 'insoluble' or 'particulate' invertase, can be solubilized by exposure to high salt concentrations. The invertase isoenzymes which are localized at different subcellular sites are most probably encoded by separate genes.

High levels of acid invertase are often characteristic of growing and differentiating tissues, for example in growing stems, leaves and seedlings, during seed germination and in developing sink organs such as seeds, kernels, roots and fruits. The amount of the soluble acid invertase, however, tends to diminish in mature, filled sink organs. Less dramatic variations in the activity of the cytoplasmic neutral invertase have been found at different developmental and physiological phases. In general, the maximal activity of invertase that can be expressed in various plant tissue is extremely variable in different plant species and within the species in different genetic variants. It is, therefore, difficult to propose a general scheme which will universally describe the exact role of all invertase isoenzymes. Evaluation of their position and function in maintaining the flow of carbohydrates in a particular plant should be individualized for each specific case.

Acid invertase has a K_m for sucrose of about 5 mM and its purified preparations obtained from many sources have molecular sizes of 50–70 kDa.

Smaller cross-reactive peptides in the 20–30 kDa range were often isolated during the purification which probably indicates proteolytic fragmentation of the enzyme. It is a glycoprotein which has three N-linked oligosaccharide chains. A very strong immunochemical cross-reaction was found between enzymes from different species and between isoenzymes which are encoded by separate genes in the same plant. Selected examples describing detailed structures and characterization of this enzyme are as follows. In the tomato (Klann et al., 1992; Ohyama et al., 1992; Konno et al., 1993) the major monomer of the vacuolar enzyme is 52 kDa and that of the cell wall enzyme is a 68 kDa peptide. In the carrot, two isoenzymes one a 68 kDa vacuolar invertase and one a 63 kDa for the cell wall enzyme were identified. The nucleotide sequences for these three isoenzymes have close (48–68%) identity (Sturm & Chrispeels, 1990; Unger et al., 1992, 1994). The acid invertase of the potato is a 58 kDa protein (Bracho & Whitaker, 1990; Burch et al., 1992; Hedley et al., 1993). This subunit is converted into a 48 kDa protein under special experimental conditions which may dissociate it from a complexing inhibitor, or shorten the enzyme by other unidentified mechanisms (Burch et al., 1994, Pressey, 1994). In the tobacco plant, the cell wall-associated invertase is a 63 kDa protein (Weil & Rausch, 1990). The mung bean seedling invertase is a protein of about 70 kDa which is a heterodimer of 38 and 30 kDa subunits. The cDNA of this enzyme contains a leader sequence corresponding to 101 amino acids which is absent from the mature protein (Arai et al., 1992). The barley enzyme has three acid invertase isoenzymes, the dominant species is a 64 kDa protein which forms a dimer (Oberland et al., 1993), whereas in rice seedlings, the active enzyme is a dimer of a 46 kDa protein (Isla et al., 1995).

In comparison with the abundant knowledge on acid invertases, the neutral invertase has been purified and characterized in only a limited number of cases such as citrus leaves (Schaffer, 1986), soybean hypocotyls (Chen & Black, 1992) and chicory leaves (Van den Ende & VanLaere, 1995). The enzyme has a molecular size of about 60 kDa which aggregate into a tetrameric active structure. This protein is not a glycoprotein and its K_m for sucrose is close to 20 mM. The alkaline invertase, similar to acid invertase, is sensitive to inhibition by high glucose and fructose concentrations, by pyridoxal phosphate and by heavy metal cations. It is also inhibited by Tris buffer, a trait which may interfere with its assay.

(ii)   Metabolic role of invertase

Invertase isoenzymes are potentially to be found in all compartments which may contain sucrose as a metabolite whether as a transient or a stored molecule. In the cytosol, sucrose synthase provides an important alternative to invertase action and sometimes even becomes the dominant obligatory route of sucrose cleavage. The degree of expression of each of the invertase isoenzymes can show large variation according to the physiological and developmental stage of the particular organ examined and may vary in different species of plants. The molecular basis for the regulatory patterns of invertase activity is little understood.

A description of several selected number of studies which illustrate a role for invertase in overall carbohydrate fluxes is presented in the following examples. In the ripening fruits of strains of cultivated tomatoes, sucrose accumulates because of a very low ability to produce acid invertase, whereas other strains which are strong acid invertase producers accumulate glucose and fructose (Miron & Schaffer, 1991; Yelle et al., 1991; Stommel, 1992; Klann et al., 1993). It has been suggested that when sucrose accumulates in fruit with diminished vacuolar invertase, its level is dominantly determined by a striking increase in the activity of SPS (Dali et al., 1992). Sucrose synthase activity is the major reaction responsible for sucrose utilization in these senescent fruit. Tomato hybrids which incorporated the acid invertase genes produce the enzyme both in the apoplast and in the vacuole and accumulate hexoses rather than sucrose. In a related experiment, introduction of the cDNA of acid invertase in the antisense direction into an invertase-producing tomato, resulted in a significant increase of sucrose accumulation (Ohyama et al., 1995). In the tomato, similar to other plants, mechanical wounding or fungal infection strongly induced increased expression of acid invertase (Sturm & Chrispeels, 1990; Benhamou et al., 1991; Ohyama et al., 1995). The physiological value of this phenomenon is not clear, but it may be related to the profound changes elicited under such conditions by the oligosaccharine regulators.

Transgenic tobacco plants which express yeast invertase were used to evaluate the relative role of different subcellular compartments in relation to sucrose metabolism. These transformed plants were of stunted growth, had distinct morphological changes in the leaves, a reduced rate of photosynthesis and increased respiration. Compared to the normal non-transformed plants, accumulation of starch, sucrose and reducing hexoses had greatly increased in the illuminated leaves. Starch levels did not diminish even after a prolonged dark period indicating deficiencies in its mobilization into sucrose and in sugar transport. Some variation in the distribution and level of accumulated sugars was noticed depending on whether the inserted yeast invertase was localized in the cell wall, the cytosol or vacuolar compartments (Stitt *et al.*, 1990; Sonnewald *et al.*, 1991, Heineke *et al.*, 1994a,b).

Similar to the creation of transgenic tobacco plants, yeast invertase has been expressed in transformed potato plant (Heineke *et al.*, 1992). In this case there was a large accumulation of glucose and fructose and elimination of sucrose from the apoplast in the leaves. Consequently, the osmolarity of the cell sap was elevated, tubers which were smaller in size and number contained higher levels of monosaccharide and less starch, all of which reflected the decreased level of sucrose that was available for transport. It is interesting that in the mature and stored excised and stored potato tubers, there was a sharp decline in the activity of sucrose synthase, and acid invertase activity was retained as the dominating catalyst responsible for sucrose cleavage (Ross & Davies, 1992). This suggests that the presence of sucrose itself contributes to the regulation of both sucrose synthase and invertase gene expression. In another experiment (Zrenner *et al.*, 1995), tubers of transgenic potato plants which expressed antisense RNA for sucrose synthase had low levels of this enzyme, but expressed a large increase of both alkaline and acid invertase activity resulting in high concentrations of glucose and fructose. These 'antisense' tubers also had accumulated a low level of starch suggesting that sucrose synthase activity is a key determinant in regulating carbohydrate sink strength in the tubers.

The indispensable need for invertase to allow normal tissue development is seen in the seed miniature-1 (*mn*) mutant of maize. The homozygote of this mutant lost the ability to produce invertase in the basal endosperm cells, resulting in an inability to sustain normal development of kernels after pollination (Miller & Chourey, 1992). This finding supports the concept that endosperm invertase constitutes a key rate-limiting regulatory step which provide hexoses for the biosynthesis of sucrose and starch in the developing kernel (Doehlert & Felker, 1987).

An evaluation of the role of invertase in leaves (Huber, 1989) concluded that high activity of vacuolar invertase in such species as soybean and tobacco prevents sucrose accumulation within the vacuole. The hexoses produced are metabolically re-utilized in the cytoplasm, including some cycling back into sucrose synthesis. In comparison, other plants, such as pea, broad bean and spinach have a very low level of acid invertase. Consequently, accumulation of sucrose in the leaves is significant. In a starchless mutant of *Nicotiana*, more carbon is partitioned into sucrose during photosynthesis to levels which exceed the capacity for its export from the leaves. This sucrose is hydrolyzed by the vacuolar invertase, resulting in a 5–10-fold increase of free hexose accumulation (Huber & Hanson, 1992).

(iii)   Invertase inhibitors

The presence of endogenous proteins which specifically inhibit acid invertase was established for several sink deposits such as potato tuber and beets (Avigad, 1982). The inhibitor isolated from potato tubers, tomato fruit and from *Nicotiana tabacum* was identified as a 18 kDa protein which forms a complex with acid invertase most efficiently at pH 4.5 (Bracho & Whitaker, 1990; Weil *et al.*, 1994; Pressey, 1994). Sucrose at low mM concentrations and divalent cations strongly inhibit this association. It is very likely that invertase isolated from cell homogenates may, in part, be in a complex with the inhibitor protein. Dissociation is induced by high ionic strength, elevated pH, presence of sucrose and by divalent cations. It is not known whether this inhibitor system has any *in vivo* regulatory role in sucrose metabolism.

4.2.3   Sucrose as a regulator

Sucrose itself has also a role in controlling the expression of several important enzymes and other proteins (Williams *et al.*, 1992). Pontis (1977) strongly advocated such a role based on various preliminary data and study of whole plant or tissue metabolism. Echeverria & Humphreys (1984b) noticed that sucrose itself had an inhibitory effect on the activity of SPS and sucrose level in maize scutellum, but the molecular basis for this effect at the time was not clear. Interactions affecting SS synthesis by sucrose were discovered later as discussed above. Among recently described studies, sucrose was found to enhance the expression of a wound-inducible potato proteinase inhibitor II, antagonistically to its inhibition by salicylic acid. The sucrose response element on this gene has been identified (Johnson & Ryan, 1990; Jefferson *et al.*, 1990; Kim *et al.*, 1992). Sucrose also stimulated, antagonistically to auxin, the expression of the Vsp storage protein in soybean

(DeWald & Mullet, 1994) and the expression of the patatin genes, the main storage proteins in potato tubers (Wenzler et al., 1989; Kim et al., 1994). Additional examples where sucrose induced gene expression are the β-amylase and sporamine in the sweet potato (Hattori & Nakamura, 1988; Hattori et al., 1990; Takeda et al., 1995), and the light-dependent expression of phytochrome-1 and NADPH-photochlorophyllide oxidoreductase in rape cells (Harter et al., 1993).

Another interesting example for a transcriptional control by sucrose was described for transgenic Arabidopsis plants transformed by the CHS-A gene for chalcone synthase, an enzyme catalyzing a key step in the synthesis of anthocyanin. The presence of sugars, particularly sucrose, was required to allow expression of this gene and the formation of flower anthocyanin (Tsukaya et al., 1991). Two homologous nucleotide sequences identified as 'sucrose boxes' which were also identified in several other genes responding to stimulation by sucrose were defined.

In addition to the direct allosteric effect of P_i on the activity of key enzymes thus altering fluxes of carbohydrate metabolism as discussed earlier, this anion was found to repress gene transcription when it is induced by sucrose. This was shown for the formation of the Vsp mRNA, transcribing a gene encoding the production of vegetative storage glycoprotein with phosphatase activity in the soybean vacuoles. Sucrose modulates transcription of Vsp through interaction with a specific sugar responsive domain downstream in the promoter region. Compared to the low level of Vsp mRNA production in the presence of P_i, its abundance was stimulated about 8-fold by addition of sucrose or by the elimination of phosphate, but it increased about 80-fold when sucrose was present in absence of phosphate altogether (Sadka et al., 1994). It is suggested that such antagonistic play between P_i and sucrose may modulate expression of other genes that have a similar specific sugar responsive domain in their promoter (Fig. 4.4).

Compared to the enhancement effects, sugars such as glucose and sucrose can repress the expression of enzymes such as was found for the rbcS photosynthetic elements in Arabidopsis and in maize mesophyll (Sheen, 1990; Cheng et al., 1992; Krapp et al., 1993; Yang & Sheen 1994), and α-amylase in rice (Krapp et al., 1993; Yu et al., 1991). The mechanism of such gene repression by carbohydrates is directly related to the pool of free glucose such as when sucrose was hydrolyzed by invertase. This expression also involves the hexokinase molecule in a role of a 'glucose sensor' (Yang & Sheen, 1994). This phenomenon

is completely distinct from the direct involvement of sucrose as a stimulant of gene transcription described above.

A close co-ordination between the rate of NO_3^- reduction and sucrose level in plants has been observed in many studies (Sechley et al., 1992; Champigny & Foyer, 1992; Huber et al., 1994a,b; Foyer & Ferrario, 1994). Sucrose displayed a prominent ability to up-regulate the expression of nitrate reductase which is usually induced by light. Thus, sucrose replacing light could elicit the synthesis of nitrate reductase mRNA, and lead to the expression of active reductase in dark-adapted leaves (Cheng et al., 1992). In contrast, NO_3^- enhances protein phosphorylation, for example SPS and PEP-carboxylase. It is suggested that the outcome of these complementary regulatory actions by NO_3^- and by sucrose will contribute to a decrease in carbohydrate partitioning towards sucrose, while increasing production of NH_3^+ and the flow of carbon into intermediates diverted into the synthesis of amino acids. Understanding the events which occur at this key juncture between the processes of CO_2 and NO_3^- assimilation is of great importance for our ability to genetically manipulate an increase in plant growth and biomass accumulation for the benefit of humans.

In conclusion, the foregoing discussion clearly indicates that the role of sucrose as a regulator of cellular metabolism is extremely central and potent. The examples for which this effect was detected involve proteins of extremely diverse function and place in metabolic pathways, and this list emerged at random in studies not primarily designed to explore carbohydrate metabolism. It is expected that future research will discover other genes that have the 'sucrose box' and unravel the molecular mechanism for their induction by the disaccharide.

As a broad evaluation of the biological value of sucrose as a regulator of gene expression, it seems advantageous to the organism that when sucrose production is intense and its presence is abundant, it promotes (together with other cellular factors) synthesis of various enzymes and storage proteins that can utilize the ample supply of substrate needed for their production. At first it is coordinated with nitrate metabolism to provide for the synthesis of amino acids, and subsequently it culminates with an increased presence of specific enzymes which enhance provision of energy, the biosynthesis of cellular components and growth. The sucrose effect may be synergistic or antagonistic to the input by other signals, such as P_i concentrations, auxins and gibberellins and also maybe to oxygen levels.

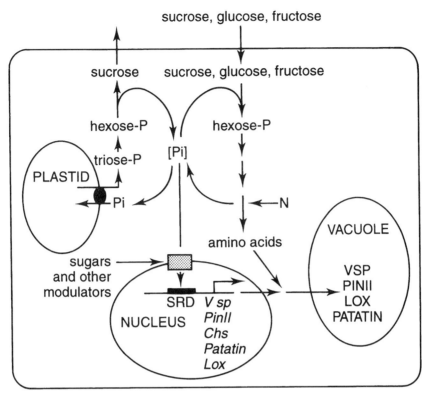

Figure 4.4 A scheme showing possible connections between sucrose, phosphate and expression of sucrose-modulated genes for vacuolar proteins. The action of sugar, phosphate and other modulators (such as auxin, oxygen) is mediated via a sugar responsive domain (or box), SRD, in the gene promoter. Intensive synthesis of sucrose during photosynthesis or the addition of external sucrose, glucose or fructose to the cells causes an increase in hexose phosphates, and reduction of cytoplasmic P_i (see Roby *et al.*, 1987), a condition that enhances gene expression. Utilization of hexose phosphates for biosynthetic reactions and in energy yielding metabolic reactions, would release P_i to be recycled for further carbon utilization. Excess carbon is stored as starch and sucrose, and in some plants also as fructans, and a supply of nitrogen is required to promote synthesis of amino acids. The genes shown are representative of some of those found to be sensitive to regulation by sucrose (from Sadka *et al.*, 1994).

4.2.4 Sucrose transport

(a) *The pathway*

Sucrose, a primary product of photosynthetic CO_2 fixation is transported through the sieve elements of the phloem to all the non-photosynthetic parts of the plant. It is estimated that 80–85% of the organic content of the phloem sap is sucrose. A large body of information about this process has been published and many ideas and hypotheses about the physicochemical and cellular basis of this process have been proposed. Only recently, with the isolation of sucrose transporters, a clear understanding of sucrose transport is emerging.

The long distance transport of sucrose is thought to occur by a passive mass flow mechanism between the source (photosynthetic tissue) and the sink organs. This movement of sugar is to be driven by an active uptake of sucrose into the conducting complex, the phloem (loading), and the exit of sucrose out of the sieve tubes in the receiving organ (unloading). Numerous studies of sucrose transport carried out in whole plants, in protoplasts and other isolated vesicles have indicated that this active process is maintained by involvement of several membrane-associated steps. The data are strongly supportive of the presence of specific membrane sucrose carriers whose substrate specificity, susceptibility to inhibitors and kinetic parameters could be accurately defined. Mechanistically, these carriers were identified as a H^+-sucrose symport systems (Hitz & Giaquinta, 1987; Lemoine *et al.*, 1988; Hitz *et al.*, 1986; Hecht *et al.*, 1992). These are energy-consuming steps which require the membrane-associated enzymic hydrolysis of a pyrophosphoryl linkage in ATP and/or in pyrophosphate. The proton-sucrose co-transporter in the membranes' bilayer are specific sites of passage that are located between other points of sugar progress by diffusion along the route of sucrose transport. The controlled activity of these transporters propels the movement of sucrose forward, often against concentration gradients, all the way from the symplast of the photosynthetic mesophyll through the phloem down to the remote organs that need carbohydrate for growth and for storage. Many recent reviews should be consulted for detailed evaluations of sucrose transport, for example: Hitz & Giaquinta, 1987; Humphreys, 1988; Getz, 1991; Bush, 1993; Sauer & Tanner, 1993, and Sauer *et al.*, 1994.

Some elements of sucrose transport in plants are presented in Fig. 4.5 and the following points about individual steps of the reaction should be considered.

1. Sucrose is transported down a concentration gradient throughout the cytoplasm of the mesophyll cells (symplast) via the interconnecting plasmodesmata pores.
2. Sucrose from the symplast is passively transported ('unloaded') by simple diffusion or by facilitated diffusion into the intercellular space (apoplast) which surrounds the symplast.
3. Sucrose is actively transported from the apoplast into the cells of the phloem structure (companion cells, sieve elements). This transport (apoplastic unloading) is carried out against a large concentration gradient, and it

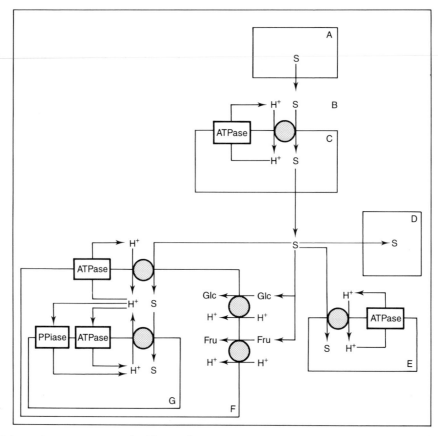

Figure 4.5   Schematic representation of pathways for sucrose (S) transport. A, symplast of mesophyll cells; B, apoplast; C, sieve elements of the phloem system; D, target young cells in growing leaves and roots; E, target cells in developing fruit; F, target cells in sink and storage organs, and G, tonoplast. Shaded circles indicate H^+-symport transporters. The ion translocases are H^+-ATPase and H^+-Pyrophosphatase. Some passage of sucrose through the transplast membrane occurs by diffusion.

is coupled with proton transport. This sucrose–proton co-transport (symport) is driven by membrane translocator-ATPase activity (a proton pump) which maintains an electrochemical proton gradient across the membrane. This controlled transport requires a one to one ratio of sugar to proton as an obligatory stoichiometry, and it is inhibited by many typical uncouplers of electron transport chains.

4. The plasmalemma and the vacuolar transport proton translocating ATPase are probably two distinct proteins. The H^+-sucrose symporters in the vacuole and the plasmalemma may also be two distinct proteins. Similar to other sugar-H^+ symport systems, the exact molecular mechanism that links the proton translocator to that of sucrose is not known, but several reasonable hypotheses for this reaction have been proposed. The direction of sucrose movement can be the same as that of the proton (co-transport) or may pass in the opposite direction (antiport) as determined by the direction of the proton pumping electrochemical gradient (Bush, 1989). In addition to ATPase activity, the action of two tonoplast pyrophosphatases is also considered to be an important contributor to H^+ gradients across the vacuolar membrane and thus could also participate in promoting sugar-H^+ symport function (Rea et al., 1992; Leigh et al., 1992).

5. Invertase activities in the cytoplasm and in the apoplast (often in a membrane-bound form) have been sometimes directly implicated as an obligatory step in sucrose transport (unloading). Experimental evidence, mostly based on the level and distribution of this invertase activity in relation to rates of sucrose transport, and particularly the fact that transport of sucrose does not involve sucrose cleavage (as seen by using unsymmetrically ^{14}C-labeled sucrose as the substrate) suggests that invertase does affect sucrose transport only indirectly by lowering sucrose presence, but not as a component of the transporter complex itself. Invertase reaction will provide glucose and fructose which will pass the membrane by independent H^+-glucose symport transporters (and probably also by a fructose transporter). Some investigators suggest that the cleavage of sucrose by invertase just prior to transport could serve to prevent backflow of this disaccharide into the sieve tubes (Fieuw & Willenbrink, 1990; Thom & Maretzki, 1992; Sauer et al., 1994).

6. A proposal advocating the presence of a UDP-glucose group translocator mechanism for sucrose transport in the tonoplast of sugar cane and beets has been proposed (Thom & Maretzki, 1985; Maretzki & Thom, 1987). A description of this transport system which culminates in a vacuolar synthesis of sucrose from the transported UDP-glucose and fructose is still occasionally presented in the literature as a viable idea. However, work from several laboratories has shown that it was based on experimental artifacts (Preisser & Komor, 1988, 1991; Niemietz & Hawker, 1988).

(b) Sucrose carriers

A more direct characterization of sucrose transport is evolving from recent work related to the isolation of specific sucrose binding proteins from plant plasmalemma and from tonoplast membranes. Strong evidence, such as inhibition of sucrose transport in intact membrane vesicles after interaction with antibodies specific to these proteins, functional expression of the transporter in yeast, kinetics of binding and substrate specificity, justifies the identification of this protein as a 'sucrose carrier' that facilitates a H^+-sugar symport. Future information about these sugar carrier proteins is expected to be obtained from the study of mutants and of transgenic plants. Findings in such studies will significantly add to our understanding of the mechanism of sugar transport in general including the specific transport of sucrose.

Among the sucrose carriers (SUT-1) isolated from the plasma and tonoplast membranes are the 42 kDa (which may be a truncated form of a 55 kDa) protein from sugar beet (Lemoine et al., 1989; Gallet et al., 1992; Li et al., 1992b; Getz et al., 1993), a 55 kDa protein (pS21) from spinach (Riesmeier et al., 1992), a 55 kDa protein from potato (Riesmeier et al., 1993b) and a 63 kDa soybean protein (Grimes et al., 1992). Two sucrose transporters SUC1 (54.9 kDa, which was isolated from transgenic yeast as a 45 kDa polypeptide) and SUC2 (54.5 kDa) were identified in Arabidopsis thaliana (Sauer & Stolz, 1994) and Plantago major (Gahrtz et al., 1994). These carriers are highly specific to sucrose; they are thiol(-SH)-sensitive proteins and sequences of those sucrose transporters obtained from different sources have a high degree of homology (65–85%) but lower homology to hexose-H^+ transporters. When functionally expressed in yeast, they had a K_m for sucrose of about 0.5–1.0 mM, pH optimum of 4.5–5.0, and were inhibited by maltose, but not by raffinose and also inhibited by known uncouplers of cellular transport such as *m*-chlorophenylhydrazone (CCCP). In conclusion, the experimental evidence supports the identification of these sucrose carriers as proton symporters.

High levels of both SUC2 and SUC1 gene expressions were found to happen in mature leaves and particularly in the phloem but to be low in sink tissues where the monosaccharide transporters dominate. The sucrose transport proteins are associated with the external leaflet of the membrane bilayer but have some variable pH optimum activity profile. Maximal activity for SUC2 is at pH 5 and it precipitously loses its activity at higher pH values. In comparison, SUC1 shows a broad optimum activity between pH 5.0 and 6.5. This property, as well as other considerations, may indicate that multiforms of sucrose transporters are differently situated in membranes to facilitate unloading of sucrose from the mesophyll cells, for phloem loading and for phloem unloading. It can be argued that catalysis of phloem loading and unloading by differently regulated sucrose carriers will provide a more physiologically adaptable transport system (Reismeier *et al.*, 1993a; Sauer & Stolz, 1994; Overvoorde & Grimes, 1994).

4.2.5 Sucrose derivatives

Many sucrose esters, particularly O-acylated in the glucosyl group, were detected in several plant tissues as components of the trichomes glycolipid fraction or the cuticular waxes of the leaves. These groups of esters are considered to be flavor precursors and an interest in this trait led to their discovery. Esters studied in most detail are those in the tobacco leaves (Wahlberg *et al.*, 1986; Garegg *et al.*, 1988; Arrendale *et al.*, 1990; Matsuzaki *et al.*, 1991). A variety of acyl groups could be found in these structures, e.g. acetyl, methylpentanoyl, methylbutyryl, methylhexanoyl and others. Similar types of sucrose esters were subsequently found in other plants such as in the glandular trichome of *Solanum berthaultii* (King *et al.*, 1987) and the *Lycopersicon hirsutum* (King *et al.*, 1990). It has been estimated that these esters comprise 5–10% of the dry weight of mature tobacco leaves and almost half the trichome exudate. It was suggested that these molecules are involved in plant–pest interactions. Acyl-CoA derivatives are most probably the acyl donor substrates used for the ester biosynthesis from the sucrose acceptor (Kandra & Wagner, 1990).

Other sucrose derivatives such as galloylsucrose from rhubarb (Kashiwada *et al.*, 1988), feruloyl sucrose from *Heloniopsis* (Liliaceae) (Nakano *et al.*, 1986) and cumaroyl sucrose from *Prunus* (Yoshinari *et al.*, 1990) are examples which indicate that a wide variety of structural metabolites built on the sucrose molecule are present in plant tissue, a group about which there is very meager information.

A derivative of sucrose is agrocinopine-A which is one of the opines induced by infective *Agrobacterium tumefaciens* bacteria carrying a tumor inducing (Ti) plasmid. This plasmid induces the plant host into an uncontrolled growth, production and excretion of opines, a mixed group of compounds used by the bacteria for growth. A common structure is a mannityl opine, derived from the condensation of glucose and glutamic acid (Savka & Farrand, 1992). Agrocinopin-A, on the other hand, is a phosphate diester between the 4-OH of the fructosyl residue of sucrose and the 2-OH of L-arabinose (Ryder *et al.*, 1984).

4.3. α,α-Trehalose

4.3.1 Biology

The non-reducing disaccharide trehalose (2) (α-D-glucopyranosyl 1-α-D-glucopyranoside) is widely distributed in nature (Elbein, 1974; Avigad, 1982). In addition to being a major metabolite in insect hemolymph, as well as in some crustaceans and nematodes, it is found in many bacteria, actinomycetes, fungi, yeast, mosses, ferns and algae. The fern *Selaginella* is a major source for commercial trehalose. Only in rare cases has the presence of trehalose been definitely identified in spermatophytes and higher vascular plants (Bianchi *et al.*, 1993; Drennan *et al.*, 1993). Trehalose is a major component present in the 'manna' exudate which appears during the summer season on several Middle Eastern desert plants, such as *Tamarix* and *Hammada*. The production of this honeydew gum is an outcome of the excretory activity of aphid colonies which infest these plants (Avigad, 1982).

(2)
α,α-Trehalose
α-D-Glucopyranosyl α-D-Glucopyranoside

Traditionally, trehalose was considered as a storage or reserve carbohydrate, often found next to starch (glycogen), and sometimes also with sucrose and polyol sugars, and tended to accumulate in high amounts in spores, cysts and sporangia of many organisms. The concentration of trehalose in stationary vegetative yeast can reach levels of 5–10% of its dry weight and much higher in spores. In addition to it serving as energy storage, trehalose is viewed as having a major role in providing protection to cell viability during exposure to episodes of stress such as desiccation, increased osmotic pressure in a halophilic environment and during high and low temperature shock (Van Laere, 1989; Wiemken, 1990). This role is similar and complementary to that of other solutes such as proline and various polyols (Delauney & Verma, 1993). Some noteworthy findings underlying the enzymatic and molecular basis for these physiological adaptations as related to trehalose biochemistry and the regulation of the cellular carbohydrate metabolism in general, have been described in several organisms, particularly in yeast.

Being a common metabolite of fungi, the presence of trehalose in plant tissue preparations may strongly indicate a fungal infection such as the presence of ergot, smut or rust pathogens. In other biological habitats, trehalose metabolism is intimately involved in the symbiotic relationship between host and parasite of various organisms such as mycorrhizae and vascular plants, and rhizobia in the nodules of the legumes (Mellor, 1992). Large quantities of trehalose are found as a by-product in fermentation liquors of agricultural products. The source of the disaccharide is the yeast and fungi that promoted the fermentation process (Hull et al., 1995).

4.3.2 Synthesis

Two catalytic steps are responsible for trehalose biosynthesis. The first is catalyzed by UDP-glucose:D-glucose-6-phosphate 1-α-D-glucosyltransferase (or trehalose-6-phosphate synthase, TPS):

$$\text{UDP-glucose} + \text{Glucose-6-phosphate}$$
$$\rightarrow \alpha,\alpha\text{-Trehalose-6-Phosphate} + \text{UDP}$$

This transglucosylation is practically irreversible under physiological conditions. The second reaction, catalyzed by α,α-trehalose-6-phosphate phosphatase (TPP) releases free trehalose and orthophosphate:

$$\alpha,\alpha\text{-Trehalose-6-phosphate} + \text{H}_2\text{O}$$
$$\rightarrow \alpha,\alpha\text{-Trehalose} + \text{P}_i$$

The two enzymes from yeast as well as from other micro-organisms were cloned and sequenced. They are cytosolic proteins which tend to assemble as a complex generally called 'Trehalose synthase'. This complex exists as a trimeric protein coded by (a) the gene *TPS1* (also identified as GGS1/TPS1) which encodes the production of the catalytic 56 kDa protein subunit responsible for TPS; (b) *TPS2* gene, which prescribes the production of a 102 kDa protein having the TPP activity, and (c) a *TPS3* (or *TSL1*) gene which encodes the production of a 123 kDa 'regulatory' protein, whose role has not been yet defined. It is obvious that the rate of trehalose synthesis is generally dependent on the flux and supply of carbohydrate intermediates in the cell, and will be in competition for UDP-glucose and glucose-6-phosphate which are also being utilized by other biosynthetic systems, such as for glycogen, sucrose and cell wall polymers.

Based on the study of many mutants of the trehalose synthase system, a complex regulatory phenomenon that relates trehalose biosynthesis to patterns of glucose utilization and cellular viability, under adverse environmental conditions has been described for several micro-organisms (Blazquez et al., 1994; Thevelein & Hohmann, 1995). For example, a direct correlation between thermotolerance (usually in the temperature range 35–60°C) and trehalose accumulation has been observed in yeast, fungi and bacterial spores (Neves et al., 1991; Arguelles, 1994; DeVirgilo et al., 1994; Hottiger et al., 1994; VanDijck et al., 1995). In addition, the TPS1 and TPS2 proteins themselves are considered to be heat shock proteins. Mutants that are defective in these genes and have lost the ability to accumulate trehalose, become heat sensitive. In comparison, a mutation in the *NTH* gene (responsible for the formation of a trehalose hydrolase, the key mobilization enzyme for trehalose utilization), resulted in cells which accumulate and store trehalose longer and can sustain prolonged periods of cellular thermotolerance. In addition to being linked to thermotolerance, trehalose accumulation was also found to be directly correlated with the ability of cells to withstand extreme changes in osmotic pressure in a halophilic environment or during desiccation. Similar to these observations made for yeast and fungi, *Escherichia coli* mutants defective in the synthesis of trehalose were also found to be severely impaired in their osmotic tolerance response. The molecular basis of this process has been described in detail (Boos *et al.*, 1990; Strom & Kaasen, 1993).

As a compound that may reach a significant level of intracellular accumulation, it should be noted that trehalose is an excellent solute which even in very high concentrations stabilizes proteins against denaturation at elevated temperatures, and does not interfere with the catalytic activities of many enzymes. As a highly stable, non-reducing disaccharide, it is inert (similar to sucrose) with respect to the formation of Schiff-base adducts with proteins, a process that leads to the production of modified protein structures. Such non-enzymatic interaction ('glycation') actively occurs with reducing sugars such as glucose, fructose or hexose 6-phosphates, making them, in comparison, much less suitable for long intracellular storage and for serving as protectors of cellular viability during periods of dormancy and stress (Labuza *et al.*, 1993; Sun & Leopold, 1995).

An additional observation is that the *TPS2* gene product is probably identical to that of the *PFK3* gene which encodes the production of the particle-bound form of the phosphofructokinase II complex in yeast (Sur *et al.*, 1994). Mutants in this gene cannot accumulate trehalose, are temperature sensitive and have lost the ability to sporulate. This finding indicates even further the close molecular linkage between patterns of glucose uptake, glycolysis and trehalose metabolism and how these processes relate to the cellular responses during environmental stress.

4.3.3   Trehalose catabolism

In most organisms, particularly as studied in fungi and yeasts, two different hydrolases (α,α-trehalase) specific for trehalose are known (Thevelein, 1988; McBride & Ensign, 1990; Carrilo *et al.*, 1995): a cytosolic neutral trehalase (pH optimum 7.0) and a separate acid trehalase (pH optimum 4.5) which is a heavily glycosylated glycoprotein present in the vacuoles or in the periplasmic space. The catalytic activity of neutral trehalase (*NTH1* gene) is increased by a 3,5-cAMP-dependent protein phosphorylation. Consequently, for example, physiological conditions which terminate dormancy by the induction of germination and growth, result in the rapid hydrolysis of stored trehalose providing glucose for cellular metabolism. This event will occur

simultaneously with the activation of glycogenolysis and glycolysis which are also triggered by 3,5-cAMP protein phosphorylation reactions. As indicated earlier, mutations deficient in *NTH1* and other genes which are responsible for 3,5-cAMP-dependent protein phosphorylations, may lead to increased trehalose accumulation and with it an elevated resistance to heat shock element which encodes the formation of heat shock proteins. It is also possible that similar to *TPS2* gene, the *NTH-1* gene includes a sequence defined as a heat shock element which encodes the formation of heat shock protein domains (Mitenbühler & Holzer, 1991; Nwaka *et al.*, 1995).

The physiological role of the acid trehalase is not clear. Since it is compartmentalized outside the cytosol, its contact with cytosolic trehalose is expected to occur only during periods of cell membrane damage that may happen during stress periods such as carbon or nitrogen starvation. In fungi, it provides a mechanism by which extracellular trehalose taken up (e.g. by pinocytosis or endocytosis) is utilized for growth. Similarly, in *E. coli*, the periplasmic acid trehalase which is induced by osmotic stress and starvation provides a mechanism by which extracellular trehalose is metabolized and utilized for the synthesis of intracellular trehalose and for growth. Detailed descriptions of the complex pathway of trehalose metabolism, transport and their relevancy to stress regulation of gene expression in bacteria are available (Strom & Kaasen, 1993; Kaasen *et al.*, 1994).

4.3.4   Trehalose in plant symbiotic associations

Trehalose is a metabolite found at significant levels in legume nodules, usually next to sucrose. The trehalose is produced by a large number of the bacterioids, such as *Bradyrhizobium, Rhizobium* and *Azotobacter* species (Mellor, 1992; Müller *et al.*, 1994a). Trehalose is also synthesized by the mycorrhizal symbionts of vesicular plants (Martin *et al.*, 1988; Mellor, 1992; Schubert *et al.*, 1992; Shachar-Hill *et al.*, 1995). In both cases trehalose synthesis and degradation play a role in the process of reciprocal carbon mobilization that exists between the host and the symbiont and the dynamics of this metabolic interaction are strongly modulated by many environmental and physiological factors which affect both partners.

In the rhizobia, trehalose is synthesized by the trehalose synthase of the bacterioid whereas the monosaccharide substrates are provided by the host. Conversely, the disaccharide in the bacterioid is hydrolyzed by a trehalase found at high levels in the peribacterial space and it is most probably produced by the plant host (Salminen & Streeter, 1986; Kinnback & Werner, 1991; Müller *et al.*, 1992). In addition to trehalase, activities of trehalose phosphorylase that convert trehalose into β-glucose-1-phosphate, and also of phosphotrehalase that convert trehalose-6-phosphate into glucose and glucose-6-phosphate have been detected in the bacterioid cytosol. The physiological role of these two enzymes is not clear. It is of interest that a dramatic increase in trehalose accumulation occurs in rhizobia grown in condition of high osmotic pressure (Breedveld *et al.*, 1993). Also, increased levels of nitrate in the medium cause a reduction in the amount of trehalose present in the bacterioid (Müller *et al.*, 1994b). It seems that the symbiont accumulates trehalose to provide for its own needs of building up an energy reserve and at the same time benefits the organism with an osmotic stress protector. Supply of carbon for the synthesis of trehalose depends on the plant host metabolism and photosynthesis. Conversely, the trehalose segregated in the symbiont becomes metabolically utilizable by the plant host only after its hydrolysis to glucose. Trehalose itself cannot usually permeate through plasma membranes, and was found to be toxic when used as a growth substance for many plant cells which lack trehalase (Müller *et al.*, 1992; 1994a).

4.4 OTHER OLIGOSACCHARIDES

Two classes of oligosaccharides are found in plants: those which are synthesized *in vivo*, termed primary oligosaccharides, and those which are produced as a result of carbohydrate degradation, termed secondary oligosaccharides. Members of the former class are discussed in this section. Sucrose is discussed in section 4.2. Glucose-containing oligosaccharides such as maltose [O-α-glucopyranosyl-(1 → 4)-glucopyranose] and its higher homologs are generally the products of starch degradation in tissues; however, in some species, short-term photosynthesis in the presence of $^{14}CO_2$ has resulted in the formation of maltose (Kandler & Hopf, 1980). Isomaltose [O-α-glucopyranosyl-(1 → 6)-glucopyranose] and cellobiose [O-β-glucopyranosyl-(1 → 4)-glucopyranose] do not seem to be primary oligosaccharides. Melibiose [O-α-galactopyranosyl-(1 → 6)-glucopyranose] is found in the exudates, nectaries and tissues of various plants. In *Aconitum napellus* and various other plants grown at higher altitudes,

this sugar accumulates in sufficient quantities and is also translocated. Its physiological importance is, however, unclear (Dey, 1980a).

Planteobiose [O-α-galactopyranosyl-(1 → 6)-fructofuranose] was isolated in a free form from the seeds of *Nicotiana tabacum* and various species of *Plantago*. Some minor disaccharides which contained α-galactosyl groups have also been detected in plant tissues, for example, O-α-galactopyranosyl-1 → 3)-L-arabinopyranose, O-α-galactopyranosyl-α-galactopyranoside and O-α-galactopyranosyl-(1 → 4)-galactopyranose. Common primary storage oligosaccharides are the galactose derivatives of either polyols or sucrose.

4.4.1 Oligosaccharides of polyols

Both acyclic (alditols) and cyclic (cyclitols) polyols occur in plants, mostly in combined forms (mainly as glycosides Fig. 4.6). Acyclic members, glucitol, mannitol and glycerol may occur in free form, for example, glucitol is translocated via the phloem of some plants, and have physiological importance in providing winter-hardiness. Glycerol commonly occurs as part of lipids and also as glycosyl derivatives. Inositols are the most widely occurring cyclitol; *myo*-inositol is physiologically the most important member and plays a diverse role in plant metabolism (Avigad, 1982; Bieleski, 1982; Loewus & Dickinson, 1982).

Two storage galactosylglycerols, floridoside (2-O-α-galactosylglycerol; **3**) and isofloridoside (1-O-galactosylglycerol; **4**) have been well studied in red and brown algae. They play important roles in osmoregulation. Their levels increase with hypersalinity and decrease with hyposalinity; under the latter condition high-molecular-weight cell components are synthesized (Dey, 1985).

The alditol galactoside, clusianose (1-O-α-galactosyl hamamelitol; **5**) occurs in a number of plants and confers cold-resistance to them. The compound accumulates in the leaves at high levels and is not known to be translocated to other parts of the plant. The biosynthetic pathway probably involves transfer of galactosyl group of a nucleotide galactoside from the pool of photosynthetic intermediates to free hamamelitol:

NDP-D-Galactose + hamamelitol

→ clusianose + NDP

Among the cyclitol galactosides, galactinol [1L-1-O-(-α-D-galactosyl)-*myo*-inositol; **6**] is the most

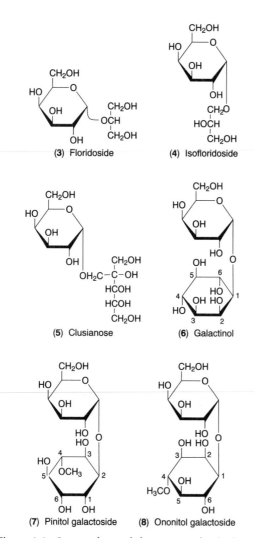

(3) Floridoside

(4) Isofloridoside

(5) Clusianose

(6) Galactinol

(7) Pinitol galactoside

(8) Ononitol galactoside

Figure 4.6　Some galactosyl derivatives of polyols.

functions as the galactosyl donor, instead of UDP-galactose, for the biosynthesis of raffinose oligosaccharides. There is no substantial evidence to show that galactinol is translocated to other parts of the plant; this aspect remains controversial. In addition to leaves, galactinol is also synthesized in seeds, roots and rhizomes. In leaves galactinol synthesis occurs predominantly in intermediary cells (minor vein companion cells) which are characterized by the presence of numerous plasmodesmata. There is also evidence of synthesis in mesophyll cells (Beebe & Turgeon, 1992). The branched plasmodesmata interconnect the intermediary cells with the cells of bundle sheath.

The enzyme responsible for the biosynthesis of galactinol is galactinol synthase (UDP-galactose: *myo*-inositol galactosyl transferase, GS) and it requires Mn^{2+} for its activity (Liu *et al.*, 1995):

UDP-Galactose + *myo*-inositol

$$\xrightarrow{\ Mn^{2+}\ } galactinol + UDP$$

The enzyme is extravacuolar in tubers. The activity of GS correlates positively with the levels of the raffinose family of oligosaccharides in leaves and seeds. It probably regulates the levels of the storage oligosaccharides in specified parts of plants (Castillo *et al.*, 1990; Madore, 1991; Keller, 1992a; Mitchell *et al.*, 1992). The synthase is an SH-enzyme and is inhibited by uridine nucleotides and UDP-glucose. Therefore, it can be regulated via redox reactions and by the photosynthetic pool of metabolites. A phosphorylation/dephosphorylation mechanism has also been suspected in its regulation but not proven. As galactinol can occur in various parts of a plant, GS is likely to exist in multiple molecular forms. Higher homologs of galactinol with up to four galactose residues attached to each other via O-α-(1,6)-linkages have also been reported to occur in plants. It is, however, not known whether one or different enzymes catalyze the sequential transfer of galactose to synthesize the higher homologs of galactinol.

Two other inositol-derived galactosides have also been detected in plants. Pinitol galactoside [1D-2-O-(-α-D-galactosyl)-4-O-methyl-*chiro*-inositol; 7] occurs in the seeds of a number of legumes and coexists with galactinol in some (Dey, 1985). It was suggested that like galactinol, pinitol galactoside may also act as α-galactosyl donor for the biosynthesis of galactose-containing compounds.

well-studied and characterized member. It was first isolated from sugar beet juice (Brown & Serro, 1953) and to-date it is known to occur in various organs of a large number of plants which also possess the raffinose family of oligosaccharides (Dey, 1985). Its synthesis at high levels was demonstrated in leaves subjected to photosynthesis in the presence of $^{14}CO_2$ and the radioactive incorporation detected in the compound was far greater than in raffinose and stachyose. The galactose residue of galactinol was labeled much faster and at a higher level than *myo*-inositol. From the kinetic studies of ^{14}C incorporation it became apparent that a rapid turnover of galactinol pool takes place during photosynthesis and it is then utilized for the synthesis of the raffinose family of oligosaccharides. Thus, galactinol

Ononitol galactoside [1D-1-O-(α-D-galactosyl)-4-O-methyl-myo-inositol; 8] was isolated from adzuki beans (*Vigna angularis*). Its presence has also been reported in a large number of plant species from over 20 different families. The usefulness of this compound as a chemotaxonomic marker was suggested, but its physiological role, either as galactosyl donor or a reserve material, is unclear. Other cyclitol glycosides have also been reported to occur in plants (Avigad, 1982), for example, lilioside (glucosyl glycerol) (Kaneda *et al.*, 1984), arabinosyl inositol (Sakata *et al.*, 1989) and fagopyritol (galactosyl inositol) (Brenac *et al.*, 1995).

4.4.2   Sucrose oligosaccharides

It is well established that sucrose is the most abundant soluble disaccharide in the plant kingdom. It is the major transportable form of carbohydrate synthesized in the leaves during photosynthesis. It also forms the nucleus for the synthesis of a variety of oligosaccharides. These are mainly galactose substituted derivatives of sucrose in which α-galactosidic linkages have been formed at different carbon atoms of the glucose and fructose moieties (Dey, 1990). Thus, a range of trisaccharides, and tetrasaccharides (Fig. 4.7) are formed (Table 4.1). Higher homologs can also be formed by multiple addition of α-galactosyl groups to an existing galactose moiety. Glucosyl and fructosyl groups can also be linked to sucrose, for example, β-(1 → 6)-linked glucosyl group at the glucose moiety of sucrose yielding gentianose, β-(2 → 6)-linked fructosyl group at the glucose moiety yielding neokestose, β-(2 → 6)-linked fructosyl group at the fructose moiety yielding kestose, and β-(2 →1)-linked fructosyl group at the fructose moiety yielding isokestose. These products are the potential precursors of fructose-containing polymer biosynthesis (see section 4.5).

(a)   *Trisaccharides*

Of the eight possible α-galactosides of sucrose as shown in Table 4.1, all but two have been found to occur in plants as primary oligosaccharides. However, only three members, raffinose (**9**), umbelliferose (**12**) and planteose (**13**), have been studied in detail. Although the trisaccharide with α-galactosyl group at C-4 of glucose moiety in sucrose has not been reported, a tetrasaccharide containing this structure (**21**) has been identified.

Raffinose (**9**) occurs widely in higher plants mainly in leaves, stems and storage organs (seeds, roots and rhizomes). Its level in the leaves is low, however it is translocated and accumulates in higher concentrations in the storage organs during plant development. As seeds mature and lose water, there is a concomitant increase in raffinose concentration. This trisaccharide is often found in the plant tissue together with its higher homologs and galactinol. The relative ratio of the individual galactosyl oligosaccharides may vary in plant species and their organs (Dey, 1980a, 1985). The kinetics of the synthesis of oligosaccharides are typical of reserve substances and radioactive labeling experiments suggest that synthesis occurs by the transfer of galactosyl group from a galactose-containing intermediate (derived from the photosynthetic pool) to sucrose or galactosyl sucrose derivatives.

The galactose-containing intermediate was initially suspected to be UDP-galactose, the hexose being enzymically transferred to sucrose yielding raffinose. However, this pathway was disputed and galactinol (**6**) was found to be the galactosyl donor. The enzyme involved was galactinol: sucrose 6-α-D-galactosyltransferase (raffinose synthase). The wide occurrence of galactinol in plant organs that possess raffinose and related oligosaccharides provides a good argument in favor of it being the galactosyl donor. The transferase exists in two forms in *Vicia faba* seeds and catalyzes the following reactions:

Galactinol + sucrose → raffinose + *myo*-inositol

Raffinose + [^{14}C]sucrose

$$\rightleftharpoons [^{14}C]raffinose + sucrose$$

Both forms were devoid of α-galactosidase and stachyose (**17**) synthesizing activities.

In storage organs raffinose may either be utilized as a precursor for the synthesis of higher homologous oligosaccharides or be hydrolyzed by α-galactosidases and invertases (Dey, 1980a, 1985; Avigad, 1982; Dey & Del Campillo, 1984). In maturing pea seeds the level of raffinose at an early stage is higher than its higher homologs, however, levels of the latter rapidly increase as maturation progresses. During seed germination raffinose and its higher homologs are degraded to galactose and sucrose. The fructosyl group of raffinose can be cleaved by invertase to yield melibiose (see Fig. 4.8; **25**) but the reaction is slow. These sugars are further metabolized by the growing seedling. Some resynthesis of the oligosaccharides also occurs during seed germination. Raffinose and its family

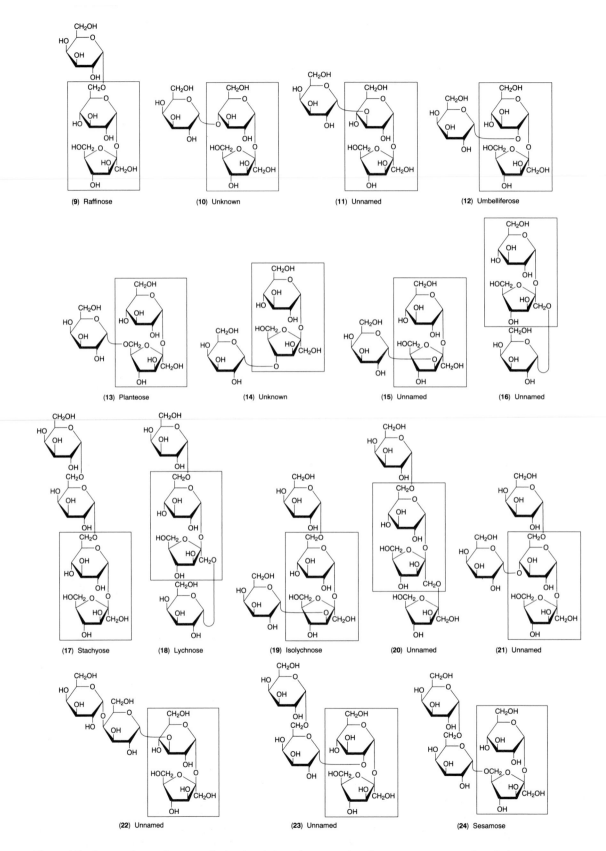

Figure 4.7 Some galactosyl sucrose derivatives (tri- and tetrasaccharides). Sucrose unit is identified within the box.

Table 4.1 O-α-Galactosyl sucrose derivatives (see Fig. 4.7 for structures)

α-Galactosyl linkage with glucose/fructose moiety of sucrose	Trisaccharides	Tetrasaccharide	
C-6 of Glucose	Raffinose (9)	Stachyose (17)	
		Lychnose (18)	
		Isolychnose (19)	
		Unnamed (20)	
C-4 of Glucose	Not Known (10)	Unnamed (21)	
C-3 of Glucose	Unnamed (11)	Unnamed (22)	
C-2 of Glucose	Umbelliferose (12)	Unnamed (23)	
C-6 of Fructose	Planteose (13)	Sesamose (24)	
C-4 of Fructose	Not Known (14)	Not known	
C-3 of Fructose	Unnamed (15)	See Isolychnose (19)	
C-1 of Fructose	Unnamed (16)	See Lychnose (18)	

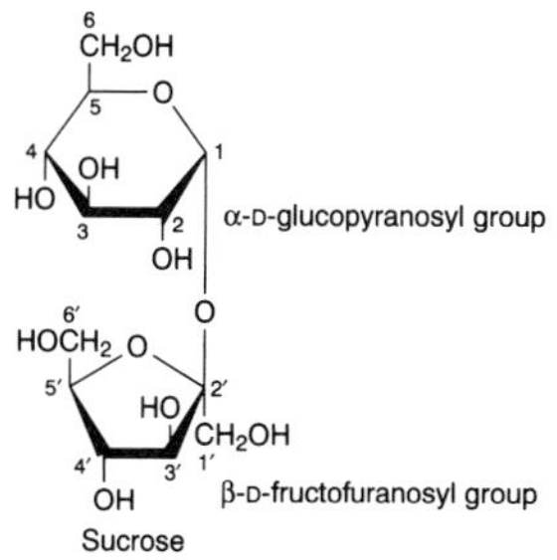

of oligosaccharides are important sugar translocates in several plant species such as, cucurbits. The accumulation of raffinose is often linked to frost-hardiness of plants and their levels can be regulated by both temperature and photoperiod (Kandler & Hopf, 1980).

Umbelliferose (12) is an isomer of raffinose and is ubiquitous among umbellifers especially in storage organs (roots, fruits), leaves and stems. Studies on foliar photosynthetic assimilation of $^{14}CO_2$ and incorporation of ^{14}C into carbohydrates show nearly equal distribution of radioactivity into umbelliferose and sucrose. However, the percentage label in sucrose increases in distant parts and organs of the plant as compared to umbelliferose. Thus, kinetic studies of the distribution of label in constituent monosaccharides of umbelliferose suggest a biosynthetic process that involves transfer of an α-galactosyl group of an activated galactosyl donor to sucrose molecule. The following pathway of synthesis is established using an enzyme preparation from the leaves of *Aegopodium podagaria*;

UDP-[^{14}C]galactose + sucrose

$\rightarrow$ [^{14}C]umbelliferose + UDP

In the mature leaves, photosynthetic $^{14}CO_2$ fixation did not result in the formation of [^{14}C]galactinol yet [^{14}C]umbelliferose was formed. It is of note that no raffinose was detected in these leaves. However, in young leaves galactinol becomes labeled but in this case raffinose is also labeled in addition to umbelliferose. Thus, galactinol is not a galactosyl donor for the synthesis of umbelliferose. The enzyme catalyzing the above reaction,

umbelliferose synthase, has a pH optimum of ~7.5 and is stable at −20°C. Unlike raffinose synthase, this enzyme is specific for UDP-galactose as the galactosyl donor and is unable to catalyze an exchange reaction between umbelliferose and [^{14}C]sucrose.

The main function of umbelliferose is as a reserve carbohydrate. It is utilized by the action of α-galactosidase which yields galactose and sucrose. Invertase can cleave the fructosyl residue of umbelliferose but the reaction is slow and a negligible amount of isomelibiose is liberated.

Planteose (13) is another isomer of raffinose and was first identified in *Plantago* seeds. A number of plants such as, Gentianales, Lamiales, Hippuridales, Loasales, Oleales, Scrophulariales and Solanales contain this trisaccharide which seems to indicate its chemotaxonomic significance. In seeds, planteose is deposited during maturation. In some species sucrose, raffinose and stachyose are found in high amounts in the roots and stems while seeds contain the highest concentration of planteose. Thus, planteose is synthesized *de novo* in seeds. Galactose from UDP-galactose is transferred to sucrose and the reaction is catalyzed by planteose synthase (Dey, 1980b; Hopf *et al.*, 1984a)

UDP-galactose + sucrose $\rightarrow$ planteose + UDP

The most likely function of planteose in seeds is as storage oligosaccharide which is utilized during germination with primary cleavage of the galactosyl group by α-galactosidase.

Of the possible isomers of raffinose shown in Table 4.1, **10** and **14** have not been detected in plant species. However, **11**, **15** and **16** are known

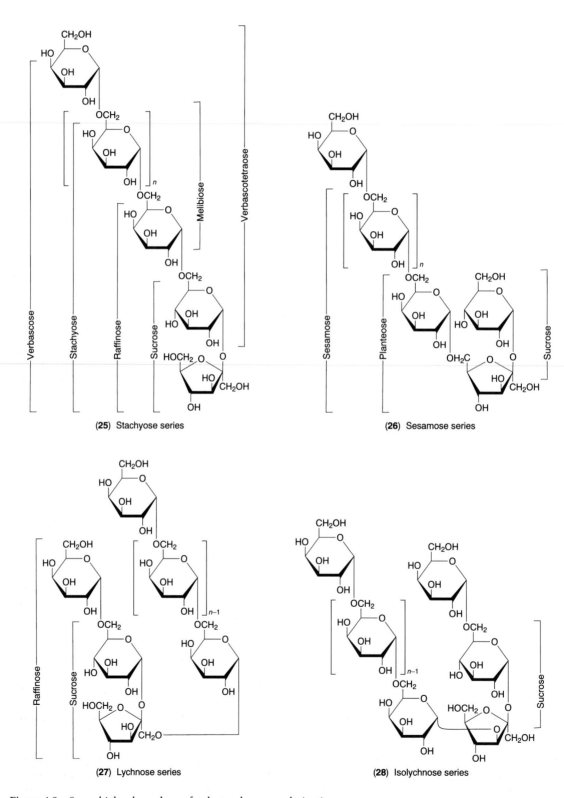

Figure 4.8   Some higher homologs of galactosyl sucrose derivatives.

to occur: **11** is found in the seeds of *Lolium* and *Festuca* (Gramineae); **15** and **16** are found in the roots of *Silena inflata* (Dey, 1985).

(b) *Tetrasaccharides*

One of the abundant tetrasaccharides in plants is stachyose (**17**). It is a higher homolog of raffinose (Table 4.1). First isolated from the rhizomes of *Stachys tuberifera*, this sugar was found to coexist with raffinose and other related oligosaccharides in various organs of a large variety of plant species (Dey, 1985). Stachyose is recognized as a major storage and transport sugar in woody plants, cucurbits and legumes.

The kinetics of ^{14}C labeling of oligosaccharides during photosynthetic fixation of $^{14}CO_2$ in leaves of *Catalpa* and *Buddleia* demonstrated that stachyose is synthesized at the expense of the pool of raffinose and galactinol. As raffinose synthase is unable to synthesize stachyose, a specific enzyme must exist in the plant source for the synthesis of this tetrasaccharide. The enzyme was isolated from several sources and found to catalyze the following reaction (Dey, 1985; Holthaus & Schmitz, 1991):

$$[^{14}C]\text{Galactinol} + \text{raffinose}$$
$$\rightarrow [^{14}C]\text{stachyose} + myo\text{-inositol}$$

UDP-galactose does not act as galactosyl donor. The transfer reaction is much higher ($\sim$20-fold) than hydrolysis of galactinol. Although the transfer reaction is freely reversible, the synthesis of stachyose is favored (Dey, 1990). The enzyme is unable to synthesize raffinose, verbascose or higher homologous oligosaccharides that are generally found in the storage organs. The enzyme also catalyzes the following exchange reaction:

$$\text{Stachyose} + [^{14}C]\text{raffinose}$$
$$\rightleftharpoons [^{14}C]\text{stachyose} + \text{raffinose}$$

Immunolocalization of galactinol synthase and ^{14}C-labeling studies have shown that the complete pathway of stachyose synthesis is present in both mesophyll and intermediary cells of *Cucurbita pepo* leaves (Beebe & Turgeon, 1992). However, intermediary cells are the predominant sites of raffinose and stachyose synthesis (Turgeon & Gowan, 1992; Flora & Madore, 1993). In stachyose-transporting plants a symplastic pathway of phloem loading seems to be operative (Turgeon & Gowan, 1990; Turgeon & Beebe,

1991; Flora & Madore, 1993, Turgeon *et al.*, 1993). During growth of the leaf to maturity, stachyose is translocated to all parts of the plant except to other mature leaves. In the young leaves and shoots the sugar is readily metabolized; in stem and roots, the metabolism is slower. A major portion of the translocate passes from the nodal region to the upper and lower parts of the plant but a small portion is retained by the petiole. The import of stachyose into the phloem of immature petioles is followed by its radial movement into the surrounding tissues (Dey, 1985). The immature leaves are not able to synthesize this sugar and the specific galactosyl transferase is absent from the tissues. The young leaves contain a unique alkaline α-galactosidase which efficiently hydrolyzes stachyose and raffinose for further metabolism. As young leaves mature, synthesis and export of stachyose begins and import decreases; the fully mature leaf is a net exporter. The synthesis of stachyose occurs even in senescent and partly yellowed leaves.

In storage organs, for example tubers of *Stachys sieboldii*, vacuoles are the site of storage of stachyose. The mechanisms of uptake of this oligosaccharide and sucrose are quite similar. A steep concentration gradient of stachyose from cytosol to vacuole exists and the transport is anticipated to proceed actively in a proton-antiport manner (Keller, 1992b; Greuterrt & Keller, 1993). In roots and seeds, stachyose may be stored as such or can be transformed to other α-galactosyl oligosaccharides. Thus, the enzyme-catalyzed exchange reactions may prove to be of importance.

In seeds, stachyose is metabolized during germination. In winter-hardy plants, stachyose and related oligosaccharides confer frost-hardiness; there is a seasonal variation of their pool size (Kandler & Hopf, 1980). In *C. pepo* the export of these sugars from leaves is completely inhibited at 5°C. Galactosyl oligosaccharides are also becoming the focus of research in relation to desiccation and other environmental stresses in seeds, roots and leaves (Blackman *et al.*, 1992; Bucker & Guderian, 1993; Nichols *et al.*, 1993; Turgeon *et al.*, 1993; Wiemken & Incichen, 1993; Bachman *et al.*, 1994; Lin & Huang, 1994; Steadman *et al.*, 1996). At low water concentration, instead of crystallizing, the sugars in solution undergo a transition to a highly 'viscous glass' state in which the aqueous component has an infinitely small diffusional freedom (Leopold *et al.*, 1994). This tends to stabilize and preserve the activities of proteins and vital membrane structures within the tissue.

The tetrasaccharides lychnose (**18**) and isolychnose (**19**) are isomers of stachyose and have been found in the vegetative storage organs and leaves of plants belonging to the family Caryophyllaceae. During foliar photoassimilation of $^{14}CO_2$, lychnose, isolychnose, raffinose and its family of oligosaccharides become labeled. Galactinol, sucrose and raffinose are labeled at an early stage but lychnose and isolychnose acquire label after longer photosynthetic periods. This indicates that raffinose is probably the precursor of the tetrasaccharides. In the photoassimilation experiments the labeling of these sugars increased even at a stage when their net synthesis was constant. One way of explaining this observation is via an exchange reaction:

Lychnose/isolychnose + [^{14}C]raffinose

$\rightleftharpoons$ [^{14}C]lychnose/isolychnose + raffinose

In *in vitro* experiments using preparations from leaves of *Cerastium arvense*, the synthesis of raffinose from sucrose and galactinol could be demonstrated; however, no synthesis of lychnose/ isolychnose occurred from raffinose and galactinol or UDP-galactose. The mechanism of the initial synthesis of lychnose/isolychnose is now known to take place via specific transferase-catalyzed attachment of galactosyl residue from a raffinose molecule to the fructose moiety of raffinose (Hopf *et al.*, 1984b):

Raffinose + raffinose

$\rightarrow$ lychnose/isolychnose + sucrose

It is, however, not clear whether the same enzyme catalyzes the synthesis of both oligosaccharides.

Lychnose and isolychnose can be degraded by α-galactosidase to yield monogalactosyl sucrose derivatives or ultimately sucrose and galactose as final products. In ^{14}C labeling experiments with an 18 h 'chase' following photosynthesis for 3 h in $^{14}CO_2$, two isoplanteoses, compounds **15** and **16** respectively, were detected. These must have emerged as a result of α-galactosidase-catalyzed cleavage of the galactosyl group attached to the glucose moiety of isolychnose and lychnose, respectively. The main function of the oligosaccharides is as reserve carbohydrates in parallel with the raffinose family of oligosaccharides.

Sesamose (**24**) is the higher homolog of the trisaccharide planteose (**13**) with an additional α-galactosyl group attached to C-6 of the galactose moiety. Analogous to the raffinose family of oligosaccharides sesamose co-exists in plants

which contain planteose. Sesamose has been identified in maturing seeds of *Sesamum indicum* and using seed extract as the enzyme source, its *in vitro* synthesis was demonstrated with UDP-galactose rather than galactinol as the galactosyl donor (Dey, 1985).

Several other tetrasaccharides with α-galactosyl residues in their overall structure, e.g. **20**, **21**, **22**, **23** and **25**, may exist as secondary oligosaccharides in plants. Verbascotetraose (**25**) can be derived from verbascose (a pentasaccharide) (**25**) after invertase-catalyzed removal of the fructosyl group. Verbascotetraose also occurs in a free form in several plant species (Dey, 1985). The tetrasaccharides **20** and **21** are present in cotton seeds and several *Triticum* species in association with the raffinose family of oligosaccharides. Compound **23**, is the higher homolog of umbelliferose (**12**) and compound **22** was isolated from *Festuca rubra*.

(c) *Higher homologous oligosaccharides*

These oligosaccharides are formed by subsequent linking of new α-galactosyl groups to the existing α-galactosyl moieties of the tetrasaccharides described in the preceding section (Fig. 4.8). By far the most widely distributed oligosaccharides in plants are the homologs of stachyose (**25**). Verbascose and ajugose are the penta- and hexasaccharides, respectively. These oligosaccharides co-exist with raffinose and stachyose in most leguminous plants, the highest concentration being in the storage organs (Dey, 1985, 1990).

The synthesis of verbascose and higher homologs in leaves was demonstrated during photoassimulation of $^{14}CO_2$ (Kandler & Hopf, 1980; Bachman & Keller, 1995). *In vitro* experiments using an enzyme preparation from the mature seeds of *Vicia faba* demonstrated transfer of galactosyl group of galactinol to stachyose yielding verbascose. When raffinose was the galactosyl acceptor, stachyose is formed. There is a mutual inhibition of the synthesis when the enzyme preparation is incubated with galactinol in the presence of both acceptors. The synthesis of verbascose is inhibited to a greater extent by raffinose. However, the enzyme from *Phaseolus vulgaris* which synthesized stachyose is unable to synthesize verbascose. It seems that the *V. faba* preparation contains two specific transferases. The pathway of ajugose (the homologous hexasaccharide) synthesis also involves galactinol as the galactosyl donor. The enzyme preparations from *Pisum sativum* and *V. sativa* are able to synthesize this sugar when verbascose is used as galactosyl acceptor. *V. faba* enzyme is unable to catalyze the reaction. A general pathway for the biosynthesis of the stachyose series of oligosaccharides is shown in Fig. 4.9. It is presumed that a specific transferase catalyzes the synthesis of each member of the series. The function of these oligosaccharides *in vivo* is mainly as storage carbohydrates, similar to raffinose and stachyose. Their concentration in leaves is low compared to raffinose and stachyose, and thus they are relatively unimportant as translocates or in providing frost-hardiness to the plant.

The penta- and hexasaccharides of the sesamose series (Fig. 4.8; **26**) have been isolated from the seeds of *Sesamum indicum*. These oligosaccharids remain little studied as they are not widely distributed among plants. The physiological state of seeds seems to determine their relative concentration; only the mature seeds contain the higher homologs at detectable concentrations. The biosynthetic pathway requires UDP-galactose as the galactosyl donor. Whereas the enzyme preparation from 8-week-old seeds (following anthesis) of *S. indicum* was unable to synthesize the pentasaccharide, the extract from 10-week-old seeds gave positive results (Dey, 1980b). The degradation of the oligosaccharides *in vivo* require α-galactosidase and invertase; the levels of these enzymes generally increase during seed germination.

The oligosaccharides of lychnose (Fig. 4.8; **27**) and isolychnose (**28**) series occur in a number of caryophyllous plants; they often coexist in the same plant source. In *Lychnis dioica*, the lychnose series occur mainly during the autumn season, whereas members of the isolychnose series preponderate in the spring season.

Oligosaccharides of both series containing 8–12 galactosyl residues have been separated. Foliar photoassimilation of $^{14}CO_2$ yielded labeled members of both series only after prolonged photosynthesis ($\sim$18 h), whereas raffinose was labeled at an earlier stage. There could be a threshold level requirement of raffinose prior to synthesis of the oligosaccharides.

4.5 FRUCTANS

Fructans, next to starch and sucrose, are probably the most abundant metabolizable storage carbohydrates found in plants. The fructan family of polymers accumulates in roots, tubers or bulbs of numerous plants among the Alliaceae, Liliaceae and Compositae as well as in other angiosperm families, in species such as dahlia, tulip, onion, asparagus, chicory, agave and Jerusalem

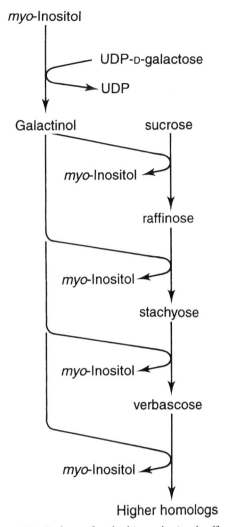

Figure 4.9 Pathway for the biosynthesis of raffinose and its higher homologs.

artichoke. Fructans are also found in seeds of the Gramineae particularly at early stages of grain development. Of special interest and of great economic and nutritional value is the presence of fructans in large quantities in stems and leaf sheaths of pasture and fodder grasses, particularly among the Gramineae, but also in edible weeds of other families such as the Compositae and Liliaceae. Fructan oligo- and polysaccharides are also formed by a large number of fungi and bacteria.

Knowledge of fructan biochemistry, physiology and ecology has impressively expanded in recent years and several detailed reviews on the subject are recommended for further reading (Meier & Reid, 1982; Nelson & Smith, 1986; Pollock &

Chatterton, 1988; Smouter & Simpson, 1989; Kühbach & Schnyder, 1989; Pontis, 1989; Smeekens *et al.*, 1991; Fuchs, 1991; Hidaka *et al.*, 1991; Pollock & Cairns, 1992; John, 1992; Suzuki & Chatterton, 1993).

4.5.1 Structures

Fructans are linear or branched polymers in which one or more of the β-fructofuranosyl-fructose linkages constitute the predominant unit. In plants, the group includes oligosaccharides with a degree of polymerization (DP) of 3–6 as well as longer chains which can reach a DP of about 50 in the inulin group and somewhat longer (DP of about 200) in the case of the levan type of polysaccharides. In comparison, the fructans produced by bacteria are usually very large polysaccharides reaching a molecular weight of 10^5–10^7. An α-glucosyl residue in the sucrose configuration is present as the end group of newly synthesized fructan molecules. In plant cells, fructans accumulate in the vacuole and are usually found as a mixture of chains of varying lengths, depending on the physiological state of the tissue. Sucrose and galactosyl-sucrose oligosaccharides are often present together with fructans in the same compartment.

Fructans are classified into the following basic structures.

Inulins have exclusively the β-fructofuranosyl 2:1-fructose linkage (29).

Levans have mostly the β-fructofuranosyl 2:6-fructose linkage (30). The name levan is usually used for the bacterial fructans which are often branched molecules also containing some β-fructofuranosyl 2:1-fructose linkages. Phlein is a name used to describe the predominantly linear

levan structures which are present in plants. Graminan is used to describe plant fructans which are highly branched, containing both 2:1 and 2:6 type β-fructofuranosyl-fructose linkages. Examples of some branched structures are those of the wheat phlean (31) and the *Agave vera cruz* fructan (32).

Kesto-*n*-oses denote short chain fructans that contain a sucrose unit (33–37), when *n* indicates the DP of the molecules, e.g. kestotriose. The three trisaccharides in this group are 1-kestose (or 1-kestotriose); 6-kestose (or 6-kestotriose) and neokestose (6_G-kestotriose, once also called isokestose). The kestotrioses can be lengthened by one or more fructosyl units providing a family of fructan oligosaccharides with different structures according to the nature of the fructosidic bonds formed. Thus, the two common tetrasaccharides found are nystose (or 1,1-kestotetraose), a linear inulin type structure, and bifurcose (or 1&6-kestotetraose) which is a branched structure. A system of nomenclature based on the names indicated above has been proposed for the identification of longer fructan oligosaccharide chains (Lewis, 1993; Suzuki & Chatterton, 1993).

4.5.2 Synthesis

The process of fructan synthesis in plant tissues is catalyzed by a multifunctional enzyme system originally proposed by Edelman & Jefford (1968) for fructans of the inulin type. Although not completely satisfactory as the singular path for fructan biosynthesis, it has been identified, with slight variations to exist in many species, such as the Jerusalem artichoke, barley, asparagus and others. The enzymes responsible for the

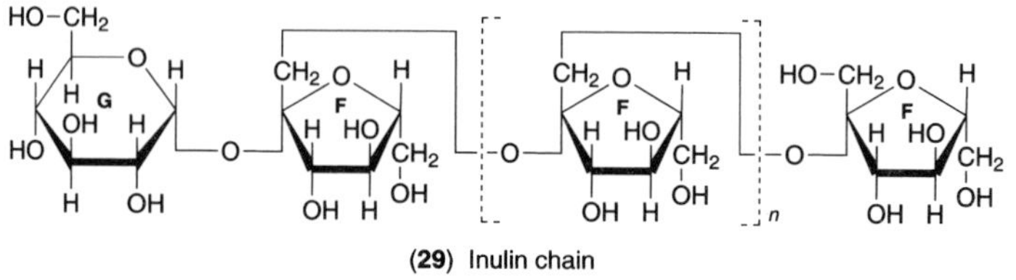

(29) Inulin chain

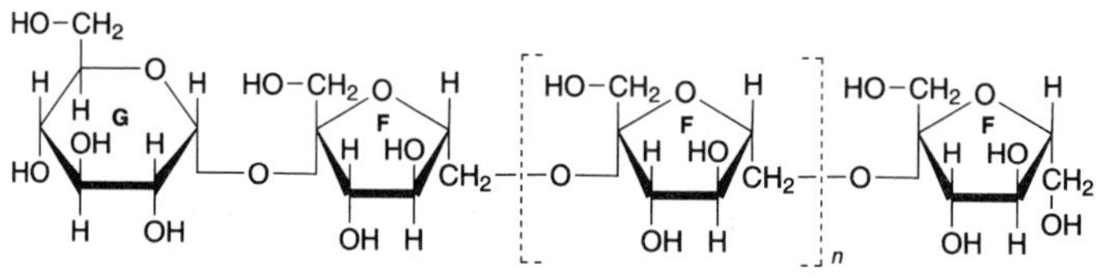

(30) Levan (Phlein) chain

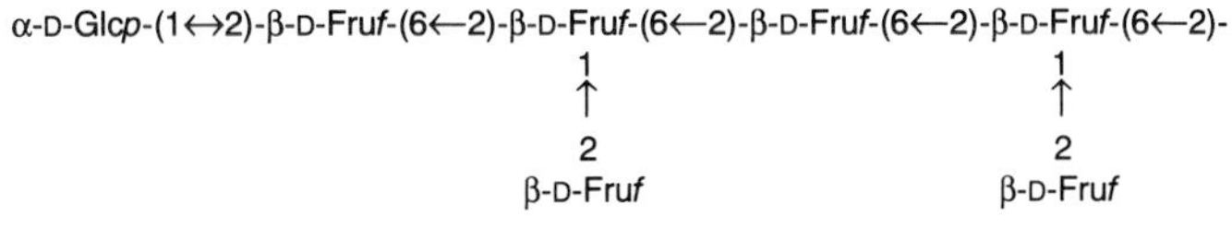

α-D-Glcp-(1↔2)-β-D-Fruf-(6←2)-β-D-Fruf-(6←2)-β-D-Fruf-(6←2)-β-D-Fruf-(6←2)-

1
↑
2
β-D-Fruf

1
↑
2
β-D-Fruf

(31)

β-D-Fruf-(2→1)-β-D-Fruf-(2→1)-β-D-Fruf-(2→1)-β-D-Fruf-(2↔1)-β-D-Glcp

6
↑
2
β-D-Fruf
6
↑
2
β-D-Fruf]₅

]₉

6
↑
2
β-D-Fruf]₂

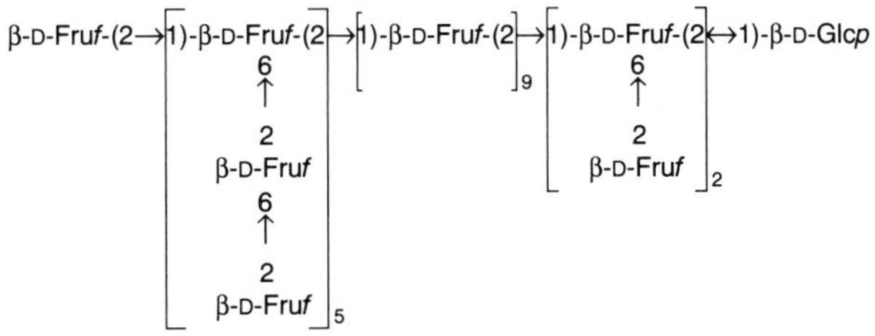

(32)

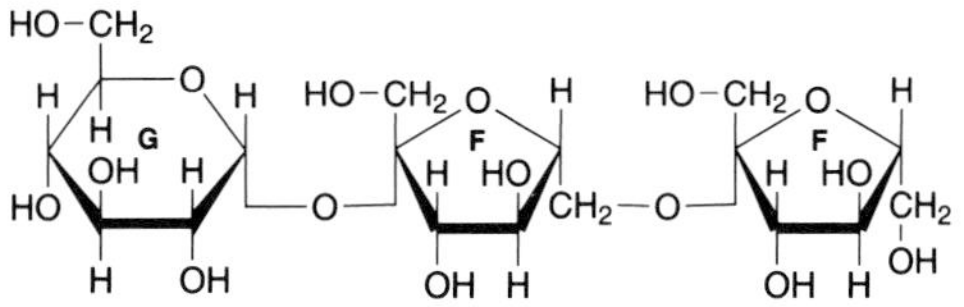

(34) 6-Kestose (6-Kestotriose)

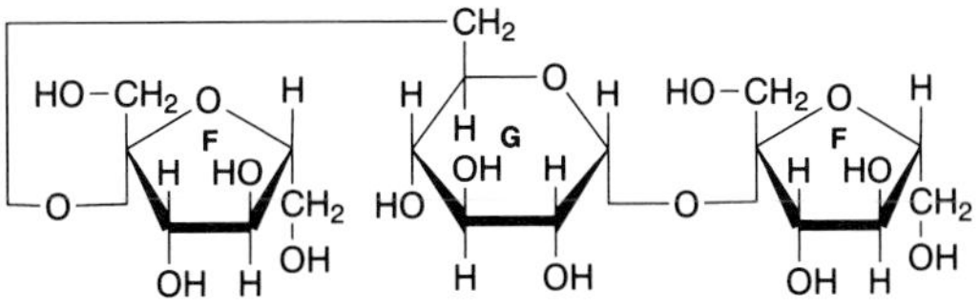

(35) Neokestose (6$_G$-Kestotriose)

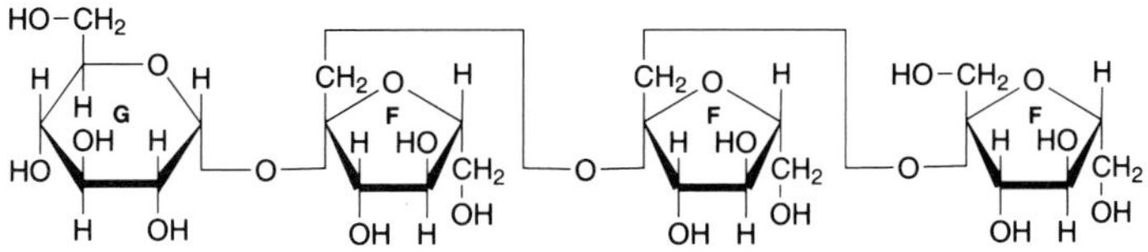

(36) Nystose (1,1-Kestotetraose)

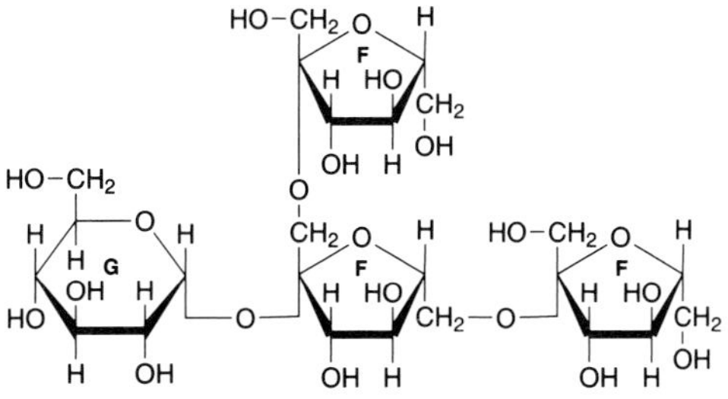

(37) Bifurcose (1&6-Kestotetraose)

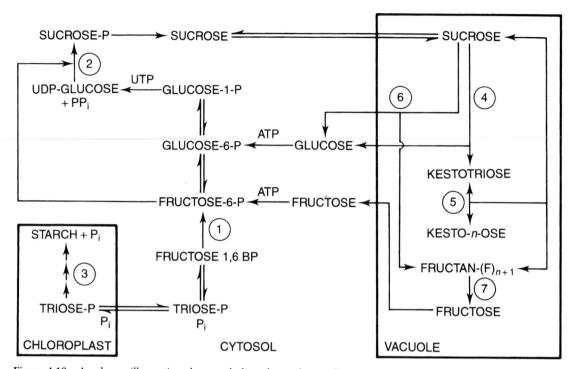

Figure 4.10 A scheme illustrating the metabolic relationship and compartmentation of the three major storage carbohydrates in leaves (modified after Housley & Pollock, 1993). Numbers indicate key reactions important in the regulation of carbohydrate flux. (1) fructose 1,6-bisophosphatase, inhibited by fructose-2,6-bisphosphate; (2) sucrose phosphate synthase (SPS), regulated by glucose-6-phosphate, P_i and by protein phosphorylation; (3) ADP-glucose pyrophosphorylase in the chloroplast, activated by 3PGA and inhibited by P_i; (4) SST, and (5) FFT, in the vacuole, controlled by gene expression and sensitive to sucrose level; (6) SFT, sensitive to level of sucrose; (7) FEH, regulated by gene expression and inhibited by high levels of fructose and sucrose. Sucrose levels are also modulated by the activity of acid invertase in the vacuole, and by a neutral invertase and the activity of sucrose synthase in the cytoplasm. The controlled process of starch degradation is an additional source of hexose that may replenish the pool of sucrose. Similar pathways for fructans occur in tuber storage tissues.

trans-fructosylation reactions which result in fructan synthesis are localized in the vacuole, the site of fructan accumulation (Fig. 4.10).

In the first step, a sucrose:sucrose fructosyltransferase (SST) catalyzes the transfer of a fructosyl between two molecules of sucrose irreversibly to form 1-kestose and glucose:

Sucrose + sucrose → 1-kestotriose + glucose

G-F + G-F   G-F-F + G

The optimum pH for this enzyme is about 5.0 and its K_m for sucrose is very high (above 0.1 M), consequently, the rate of the reaction in the normal physiological ranges of sucrose availability will be a major determinant for fructan synthesis.

The 1-kestose produced by SST can serve as the acceptor for single terminal fructosyl residue

transferred from other fructan chains being the donors in a reaction catalyzed by fructan:fructan fructosyltransferase (FFT):

$$G\text{-}(F)_m + G\text{-}(F)_n \rightleftharpoons G\text{-}(F)_{m+1} + G\text{-}(F)_{n-1}$$

Fructan oligosaccharides from the triose level and longer can be fructosyl donors. The optimum pH for FFT is 6.0–7.0, noticeably higher than for SST. In the FFT reversible chain elongation reaction, sucrose cannot serve as a fructosyl donor, but it can serve as an effective acceptor, resulting in the formation of 1-kestose, 6-kestose or neokestose. Similarly, any one of these neokestotrioses can be further lengthened by the additional, sequential, FFT-mediated fructosyl transfers from donor fructan oligosaccharides. Overall, FFT is defined as the enzyme that leads to a disproportionate distribution of fructosyls between chains of oligosaccharides, a process initiated by the 1-kestose as the primary substrate. FFT by itself does not lead to any net synthesis of fructan. During periods of intense fructan synthesis and accumulation, the ample supply of 1-kestose provided by SST will be transformed into longer fructan chains by repeated FFT catalyzed transfructosylations. In contrast, during periods of fructan utilization and sometimes because of the effect of certain environmental causes, FFT activity will contribute to reducing the average length of the stored polymers. The use of specifically ^{14}C-labeled substrates as donors or acceptors in the SST and FFT systems can greatly facilitate the understanding of fructan metabolism and provide a mechanism for the preparation of differentially labeled fructosides. For example, incubation of $[^{14}C]$-sucrose and $[^{12}C]$-1-kestose with FFT will result in transfer of the terminal $[^{12}C]$fructosyl to yield a $[^{14}C]$-1-kestose. The fact that sucrose serves as an acceptor partially explains its strong inhibitory activity on the rate of chain elongations mediated by FFT.

The synthesis of fructans by SST and FFT was established primarily for the inulin type structures, but experimental evidence suggests that a similar process is responsible to the synthesis of the $\beta2:6$ levan-type linkages (Shiomi, 1989, 1992). The diverse structural profile of fructans that can be isolated from plants indicates that more than one FFT isomer may exist (Chatterton et al., 1993a,b). It has been suggested that in addition to an 1-FFT that produces the 1^F-inulin type linkage, isomeric FFT entities with variable substrate specificities may be responsible for the production of the 6^F (6-kestose, levan) or also the 6^G (neokestose) products. Although such diverse patterns of corresponding transfructosylation reactions have been analyzed in crude plant tissue preparations, purification and a detailed characterization of these individual FFT isoenzymes have had limited success. Some diversity may also exist for the SST system to account for the synthesis of $\beta2:1$ or $\beta2:6$ linkages. Simmen et al. (1993), Chatterton et al. (1993a,b), Livingston et al. (1994) and Penson & Cairns (1994) observed that in addition to 1-SST which is the established 'universal' enzyme that synthesizes 1-kestose, a separate 6-SST isoform that converts sucrose into 6-kestose and bifurcose is present in various grasses, but the relative distribution of 1-SST and 6-SST varies considerably between species.

When discussing the 'classical' Edelman and Jefford's scheme for fructan synthesis, it should

be emphasized that it is still considered to be a deficient model for several reasons. There has been only a limited success in purifying and characterizing the enzymes involved (see for example, Jeong & Housley, 1992; Van Den Ende & Van Laere, 1993; Lüscher et al., 1993); the SST/FFT system tested in vitro could in most cases produce from sucrose only a low yield of short oligosaccharides but did not lead to the net formation of long chain fructans such as inulin; the pattern of oligosaccharide produced in vitro was often different from that found in the intact tissue; the rate of SST activity depends on elevated sucrose concentrations, the same condition that may inhibit chain elongation by FFT; the presence in plants of such a large diversity of fructan structures with respect to degree of branching and size calls for a larger number of specific enzymes and type of reaction to account for their synthesis; also, many enzyme preparations studied were contaminated by invertase, an enzyme that on its own may carry out trisaccharide formation from sucrose. These and several other considerations led Cairns (1993) to consider the SST/FFT system, as traditionally presented, to be inadequate in explaining the net fructan biosynthesis in plants. This unclarity for the plant fructans is in contrast to the detailed defined characterization of the bacterial levansucrase (sucrose:2,6-β-D-fructan 6-β-D-fructosyltransferase), an enzyme which by direct repetitive transfructosylation reactions from sucrose as the single donor leads to the mass production of macromolecular levan. Several recent studies (Cairns & Ashton, 1993, 1994; Penson & Cairns, 1994; Lüscher & Nelson, 1995) have shown that a partially purified, sucrose-stabilized enzyme fraction identified as FSA (fructan synthetic activity) could be obtained from leaves of several grasses. This FSA is different in its properties from the SST/FFT pair, and could produce a more impressive yield of fructan oligosaccharides when presented with high sucrose concentration as the substrate. The Agave enzyme described by Nandra & Bhatia (1980) has somewhat similar activity. For the synthesis of levan type (graminan/phlein) oligosaccharides, Duchateau et al. (1995) found that a glycoprotein enzyme purified from barley leaves, a sucrose:fructan 6-β-fructosyltransferase (6SFT) can use sucrose as the fructosyl donor to synthesize oligofructans. A variety of oligofructans, such as 1-kestotriose produced by the 6-SST or the 1-SST reactions, serve as the initial acceptor (Fig. 4.10).

A sequence of reactions for synthesis of levan type oligosaccharide chains is:

1. G-F   +   G-F $\xrightarrow{\text{1-SST}}$ G-F-F   +   G
   Sucrose   Sucrose   1-kestotriose   Glucose

2. G-F-F + G-F $\xrightarrow{\text{6-SFT}}$ G-F-F   +   G
                                              (with branch F above middle F)
                                   Bifurcose

3. G-F-F + G-F $\xrightarrow{\text{6-SFT}}$ G-F-F   +   G
   (branch F)   (branch F-F)
                              (1 & 6,6) kestopentaose

4. G-F-F + (G-F)$_n$ $\xrightarrow{\text{6-SFT}}$ G-F-F   +   nG
   (branch F-F)   (branch F-F-(F)$_n$)
                                   Phlein

The pool of fructan oligosaccharides (DP 3–24) found in extracts from an individual species is usually large. Structural determination of just the fructans of DP 3–6 have indicated about 14 different oligosaccharides that could generically be formed by the sequential extension of kestotriose structures by both 6-SFT and 1-SFT (Fig. 4.11). It is plausible that synthesis of these different series in the same plant is determined by variation in the relative affinity of the enzymes to the fructosyl acceptors. This preference will dictate which series of the fructans formed will be dominant in the particular plant examined. The SFT reaction is reminiscent of the bacterial levansucrose reaction, but the final products obtained here are oligosaccharides of DP 3–20 and not a polysaccharide. If further substantiated, these studies suggest that various combinations of 1- and 6-SST activities with sucrose : fructan 1- and 6-fructosyltransferase (SFT) activities are responsible for net biosynthesis of the fructans. FFT, on the other hand, should be considered only as an enzyme for fructan processing. In conclusion, it is obvious that the enzymology and molecular biology of fructan synthesis in plants is a complex system that is far from being clearly characterized.

4.5.3 Fructan depolymerization

Chain shortening of fructans can be facilitated by the action of FFT, but this disproportionating reaction will not lead to a net loss of fructosyl bonds (Simpson & Bonnett, 1993). The key reaction that mobilizes stored fructan back into the metabolic flux of the cell is catalyzed by a fructan exohydrolase (FEH), a vacuolar enzyme with a pH 5.0 optimum. In this FEH reaction, the terminal β-fructosyl residue is released as free fructose. A stepwise repetition of this hydrolysis will result in a complete chain truncation down to sucrose The fructose produced re-enters cell metabolism and subsequently can also be used for the resynthesis of sucrose in the cytosol. Several isoforms of FEH may exist, the one characterized best is 1-FEH which hydrolyzes the β-2 : 1 inulin linkage. Additional exohydrolase(s) are most probably involved in cleaving β-2 : 6 end groups in fructans. 1-FEH does not effectively hydrolyze sucrose and is distinct from acid invertase activity which is also localized in the vacuole but can only poorly hydrolyze long-chain fructan oligosaccharides. Studies in vitro indicated that high fructose and sucrose concentrations exert some inhibitory effect on 1-FEH but it is not clear whether this has any physiological significance. The fact that invertase can effectively hydrolyze short-chain fructans such as kestoses, has to be taken into account when interpreting overall pattern of fructan metabolism in the sink tissues (Bonnett & Simpson, 1993).

4.5.4 Patterns of fructan metabolism

There is a very large body of information describing the patterns of fructan accumulation and utilization in various plants and the different physiological, developmental and environmental factors that affect it (see literature reviews cited above). Kinetically, the basic biochemical determinant for the synthesis of fructans is the availability of sucrose for the SST and the subsequent SFT reactions. Hence, during periods in which abundant sucrose is produced by CO_2 fixation during photosynthesis, the rate of 1-kestose production and the subsequent elongation by transfructosylation reactions will maintain an intense fructan oligosaccharides synthesis, culminating in their massive accumulation. When the supply of sucrose is diminished or stopped, net accumulation will cease, and as fructan depolymerization will intensify, the pre-stored polymers will be depleted. The fructose liberated will be transported into the cytosol and diverted into a variety of metabolic pathways. With some degree of variation, many examples illustrating such patterns of fructan metabolism in different plants in vivo have been analyzed and described. However, correlation of these changes with the action of specific enzymes is far from being satisfactory (Cairns, 1993). It is

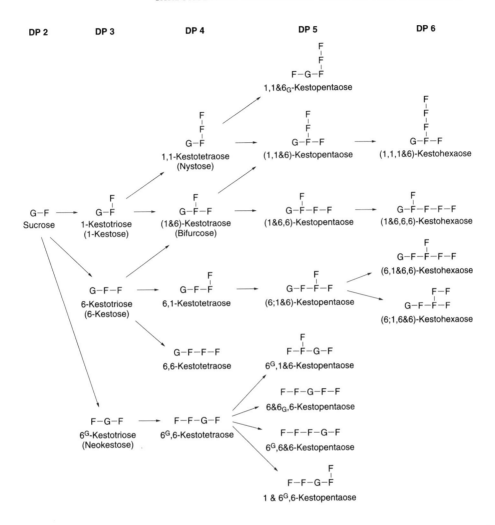

Figure 4.11 Composite chart showing probable relationship between fructan oligomers of up to DP 6 according to structural analysis in *Asparagus officinalis* roots (Shiomi, 1989, 1993), oat (*Avena* sp.) leaves (Livingston *et al.*, 1993), and cheatgrass (*Bromus tectorum*) (Chatterton *et al.*, 1993a). The predominant line in the grasses is the 1 & 6-kesto-*n*-oses, and in the Asparagus the 6^G-kesto-*n*-oses.

clear that the fructan and sucrose pools in the cells are intimately linked metabolically and are found in a continuous state of equilibrium and turnover as expressed by fructosyl exchange reactions (Sims *et al.*, 1993). This exchange occurs both when the net level of the stored fructans is increasing and also during periods of intense depletion of these stores. Hypotheses relating to the regulatory mechanisms of fructan and sucrose metabolism have been proposed (Pollock & Cairns, 1992; Housley & Pollock, 1993). There is also a dynamic relationship between the fructan and the starch storage pools linked through sucrose, a molecule which provides the substrate for the syntheses of the two polymers (Schnyder, 1993). An interesting example was described for *shx* barley which is

severely deficient in its ability to synthesize starch. In this mutant, excess carbohydrate produced by photosynthesis and not deposited as starch is diverted via sucrose to bring about an increased synthesis of fructan (Tyynla *et al.*, 1995). In another study, potato plants that normally do not produce fructans were genetically modifed to express a bacterial β-fructosyltransferase (levansucrase) which transforms sucrose into a high-molecular-weight levan. The transgenic plant accumulated enormous amounts of this bacterial fructan in all tissues, particularly in the leaves (to about 30% of dry weight) while the level of starch was reduced. Also in the microtubers, starch accumulation was significantly reduced while fructan was deposited. This modified plant did not show abnormalities in sucrose metabolism, except the fact that it was effectively diverted from starch to fructan synthesis. It is noteworthy that the levan that accumulated here is a dead-end metabolite since no enzyme capable of hydrolyzing these huge bacterial polymers is present (Van der Meer et al., 1994).

Not much is known about direct molecular mechanisms that control fructan synthesis and degradation. Various preliminary observations indicate that the enzymes responsible for fructan metabolism are subject to several types of controls both allosteric, which inhibit enzyme activity (such as the inhibition of FEH and SST by sucrose as indicated above), and more significantly at the level of gene transcription. SST, for example, turns over rapidly, a process particularly noticed in the dark. Conversely, enhancement of SST activity due to increased transcription is induced by high level of sucrose and during light periods. This replenishment of SST overcomes the losses resulting from the rapid protein degradation (Obenland et al., 1991; Winters et al., 1994). It was also noticed that K^+ at physiological concentrations (K_i 122 mM) inhibits SST activity, an effect that may have physiological significance (Chevalier & Rupp, 1993). It is important to remember that the level of acid invertase activity in the vacuole may inversely affect fructan synthesis primarily because of its ability to hydrolyze sucrose and thereby deprive SST of its substrate. Invertase also cleaves kestotrioses, and thus may compete with FFT. Consequently, it is difficult to devise a simple scheme that explains the regulation, functional co-existence and co-ordination between SST, FFT, FEH and invertase which are present together in the same compartment. Related to this point, it has been observed in many cases that during periods of active fructan synthesis and accumulation, the level of invertase activity in the vacuole is extremely low or absent (Smeekens et al., 1991; Housley & Pollock, 1993; Simmen et al., 1993). It is most likely that the dominant mechanism for regulating these competing vacuolar enzymes, all of which are most likely glycoproteins, reside at the level of gene transcription as well as in post-translational processing and modifications which contribute to their compartmentation, activation and stability.

Findings which strongly indicate that control of FEH at the gene level is a key step in setting the pace of fructan mobilization (Simpson & Bonnett, 1993) were demonstrated in the study of petal expansion of the day lily (Bieleski, 1993). In this case, a large amount of carbohydrates, close to 90% in the form of fructan oligosaccharides and some sucrose, accumulate in the developing flower bud. At about 24 h before opening, a very rapid and complete degradation of the fructan to fructose and glucose occurs causing a 3–5-fold increase in cell sap osmolarity and the petals dramatically open. Flower senescence then follows when the sugar pool in the petals is completely depleted. It has been shown here that fructan degradation is triggered by a sudden increase in the activity of FEH. A similar process, though attributed to an increased invertase activity, was observed to occur during growth of the carnation petals (Woodson & Wang, 1987).

An interesting and important aspect of fructan biochemistry is its relationship to cold acclimatization, a phenomenon particularly observed in cereals and grasses (see, for example, Kühbach & Schnyder, 1989; Pollock & Cairns, 1992). During prolonged exposure to cold temperatures (3–15°C), the amount of fructan which accumulates mostly as short and medium-length oligosaccharides, increases significantly, up to 20–30% of dry weight. It seems that fructans serve as an ancillary carbohydrate sink during cold seasons. Under these cold weather conditions, sucrose continues to be synthesized during the light periods, but its translocation into other sink deposits (such as starch) which already may be full, is slowed down. In addition, as the need for carbohydrate substrates to provide for biosynthetic process in the cell is limited, fructan is stored in the leaves and stems at high levels because of limited growth. When temperatures are elevated and intensive growth resumes, an increased rate of fructan turnover, often with a decrease in the size of the fructan pool will occur. A more direct role for fructan as a cryoprotectant for survival of hardy grasses during winter and also as molecules which contribute to osmotic regulation during other types of environmental stress has been suggested (Pollock & Chatterton, 1988; Pontis, 1989; Suzuki & Chatterton, 1993).

During cold acclimatization, it has been observed that the average chain length of fructans in tubers or in leaves is reduced. For example, a drop in average to DP 17 from DP 31 for inulin in the Jerusalem artichoke tuber occurred when stored in low temperature (Liu et al., 1994). It is likely that the activity of FFT is the key catalyst that promoted this disproportionation. It should be noted that understanding of the cellular factors that control the level of accumulated fructan and its DP is of considerable importance with regard to the commercial production of these polysaccharides such as inulin from the Jerusalem artichoke and from chicory (John, 1992).

4.6 STARCH

4.6.1 Structure and occurrence

Next to sucrose, starch is the principal reserve carbohydrate in plants. Its site of deposition is completely separate from that of sucrose, and it is localized in the plastids, both the chloroplasts in leaves and the amyloplasts in non-photosynthetic tissues. There is an active dynamic relationship of flow of glucosyls between sucrose and starch, but being on different sides of a membrane, the

exchange between the two is not direct and involves several enzymic steps.

Starch is an α-glucan. Two types of starch molecules are present together in plants. Amylose, a molecule that tends to precipitate in a crystalline form from cold water solutions, contains up to several thousand α-glucosyl units linked almost exclusively in $\alpha(1 \rightarrow 4)$ linkage. Contrary to the classical concept, most native amylose molecules contain a small number (1–5) of $\alpha(1 \rightarrow 6)$ branches usually clustered close to the reducing head of the molecule (Manners, 1985). The population of

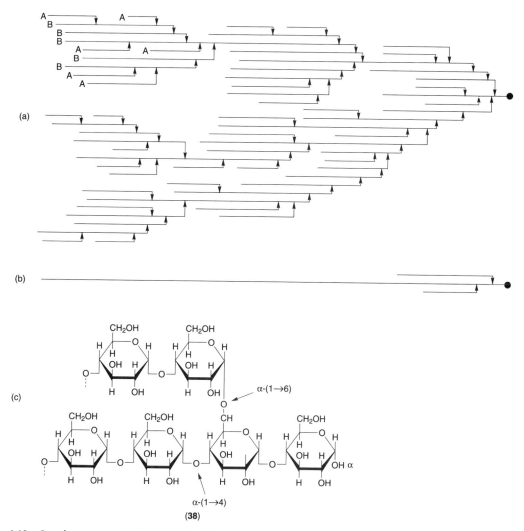

Figure 4.12   Starch structure. (a) Model of the 'cluster' structure of amylopectin. A, unbranched 'external' $\alpha(1 \rightarrow 4)$ chains; B, variable length chains that carry A or B chains as $\alpha(1 \rightarrow 6)$ branches, and ——●, the single initiation chain containing the reducing end group (the chain originally linked to an amino acid residue on the protein primer identified as tyrosine for glycogenin or aginine for amylogenin). (b) Model for an amylose molecule which contains a very few number of branches. (c) Structural representation of the branched α-glucan chain of starch.

amylose molecules is heterogeneous and can vary broadly. In most cereals, for example, a range of amylose with an average DP of 200–4000 is present; in potatoes, there are molecules with an average DP of 8000–10 000. The second type (or component) of starch is amylopectin, a branched molecule which forms stable gel-like solutions and may contain up to 10^6 glucosyl residues with an average chain length of 20–24 $\alpha(1 \rightarrow 4)$-linked glucosyl units that are linked by $\alpha(1 \rightarrow 6)$ branch linkages. Thus, 4–5% of the linkages in amylopectin molecules represent branch points. Between 1000 and 3000 of these chains are linked together to form the amylopectin molecule. Many models describing the structure of amylopectin have been proposed over the years. The current accepted model (Fig. 4.12; structure (38)) based on both chemical and enzymatic analysis is the cluster type and it has the following elements (Manners, 1985; Takeda et al., 1993; Ong et al., 1994): a heterologous group of A chains, which are the exterior ones in the molecule, and do not carry any branch on them, has an average DP of 12–16. The A chains comprise 40–60% of the total glucosyls in the molecule. A group of B chains are those which carry one or more branches mostly of A chains. B chains are divided into three or four groups according to length: long B chains with DP 70–110 (about 1–3% of the total molecule); medium B chains with DP 45–60 (about 10–15% of the total molecule), and short B chains of DP 21–24 (about 25–35% of the total molecule). A single C chain is the one that has the reducing 'head' of the molecule, or the site of linkage to the protein primer. The ratio of the number of A : B chains is 1.0–1.8. The average interior chain length between branch points is 5–8 glucosyl units. The dimensions described vary somewhat between species and strains. Models based on this information show that domains of branch sites appear concentrated in a defined laminated arrangement, with some periodicity, hence the use of the term cluster model. The region containing the high density of branch linkages is amorphous, and the side chains are packed as double helices in parallel arrays to provide crystalline lamellae (Imberty et al., 1991). The degree of hydrolysis by β-amylase which removes maltose units from the free $\alpha(1 \rightarrow 4)$ glucosyl chains, roughly represents the degree of branching of the molecule. Thus, amylose is degraded 80–90% by β-amylolysis and the degradation will reach 100% when the branch residues are removed. Most plant amylopectins have a β-amylolysis limit of 55–60% whereas glycogen, which is a much more branched glucan has a β-amylolysis limit of about

(39)
Maltose
4-O--α-D-Glucopyranosyl-D-glucopyranose

(40)
Isomaltose
6-O--α-D-Glucopyranosyl-D-glucopyranose

42%. The disaccharide structures which represent the $\alpha(1 \rightarrow 4)$ and the $\alpha(1 \rightarrow 6)$ glucosyl residues are maltose (39) and isomaltose (40) respectively.

Starch usually accumulates in granules which differ in shape and size for different plant species. The structure of the granules is not well defined in the chloroplasts, but they have a compact and distinct shape when present in storage tissues. The different properties of the starch granules are predominantly determined by the relative presence of amylose and amylopectin, by the length and branching of the amylose chains by the small variations in length of chains and degree of branching in amylopectin, and by the presence of a variety of minor components such as the existence of some glucosyl residues which are phosphorylated, various mineral ions, lipids and some proteins. Enzymes involved in starch synthesis and degradation are often found to be strongly attached to the starch granule. The granule, however, is relatively resistant to amylolysis but has a high susceptibility to attack by α-amylase. The ratio of amylose to amylopectin is in the range 1 : 5–7, a value that is different and fairly constant for individual species. This ratio can vary slightly in the same plant during different conditions and phases of growth. 'Waxy' starch granules are those which have very little or no amylose. The arrangement of amylopectin component is the one that most strongly affects the granule structure and crystallinity (Manners, 1989).

Starch accumulation in the chloroplast occurs during periods of photosynthesis. When this transient store is rapidly filled, a larger portion of photoassimilated carbons will flow out of the plastids towards an enhanced sucrose synthesis.

During the dark period, chloroplast starch is rapidly broken down and its products are exported into the cytoplasm where a major portion of them is converted into sucrose. Generally, an active flow and exchange of carbon between starch and sucrose exists at all times.

In the amyloplasts of non-photosynthesizing reserve tissues such as in seeds, tubers and roots, starch has a low rate of turnover, it accumulates and is stored for long periods, it is then rapidly mobilized and metabolized when new growth is initiated (for a large selection of general reviews on starch biochemistry, see Preiss & Levi, 1980; Steup, 1988, 1990; James et al., 1985; Guilbot & Mercier, 1985; Morrison & Karkalas, 1990; Preiss, 1988, 1991a,b; Preiss et al., 1991; Beck & Ziegler, 1989; Akazawa, 1991; Stark et al., 1992; Duffus, 1992; Okita, 1992; Emes & Tobin, 1993).

4.6.2   Starch biosynthesis

The enzymic reactions leading from ADP-glucose to the production of starch are similar in both the chloroplasts and the amyloplasts (apReese & Entwistle, 1989; Okita, 1992; Preiss, 1991a; Nakata & Okita, 1995; Nelson & Pan, 1995). The pyrophosphorylase reaction leading to ADP-glucose synthesis is a major pacemaker for starch biosynthesis and it is being controlled by the level of 3-phosphoglyceric acid (3PGA) and P_i. Various mutants such as in maize, pea and potato which have a defective genetically modified ADP-glucose pyrophosphorylase exhibited severe deficiencies in the rate and ability to synthesize starch. Also, experimental conditions that drastically altered the cellular concentration of 3PGA or P_i had a strong effect on starch synthesis.

In the leaves during photosynthesis, allocation of triose phosphate is a strictly regulated process. Part of it is channeled into starch synthesis in the chloroplast itself and another portion is exported as 3PGA to provide for the synthesis of sucrose in the cytoplasm. Leaves that lack chloroplast phosphoglucomutase or phosphoglucoisomerase activities lost their ability to synthesize starch in these plastids, but could continue to intensively synthesize sucrose.

During the dark period, starch is degraded providing more triose phosphate which is exported from the chloroplast to secure the continuous synthesis of sucrose in the cytoplasm. The 3PGA/P_i translocator in the plastid membrane is the vehicle which maintains the regulated flow of organic carbon between the chloroplast (starch)

and cytoplasm (sucrose). The transfer of triose phosphate molecules that is facilitated by this translocator explains the rapid exchange of carbon (as can be seen by using a ^{14}C label) that occurs between starch and sucrose at all times when there is a net synthesis and accumulation of starch, and also during periods of starch depletion. In the amyloplasts of non-photosynthetic storage tissue dependence on the 3PGA/P_i translocator for the transport of carbon between the plastid and the cytoplasm is quantitatively of negligible importance. Transported sucrose, which arrives at the storage organ provides hexoses and hexose phosphates, the products of invertase, sucrose synthase and hexokinase activities. These metabolites, particularly glucose-6-phosphate and to a lesser degree glucose-1-phosphate, are transported into the amyloplasts by a 'modified' 3PGA/phosphate translocator, and there immediately used for the synthesis of ADP-glucose and starch, by-passing a need for gluconeogenesis which is required when starting the pathway with triose phosphates (Flügge & Heldt, 1991; Neuhaus et al., 1993; Tetlow et al., 1994). When starch in the amyloplasts is being depolymerized, products of its degradation, such as maltose, glucose and glucose-1-phosphate, are transported back into the cytoplasm, and much of it is reincorporated into sucrose. Consequently the routes that lead to interchange of the carbon skeleton of glucosyl between starch and sucrose in the leaves are distinctly different from those in non-photosynthetic organs. In order to build hexoses in the leaves, almost all of these products have to go through a triose phosphate level and gluconeogenesis both in the plastid and the cytoplasm resulting in extensive distribution of carbons in the glucosyls of the two storage sugars (Viola et al., 1991). In contrast, in non-photosynthetic storage tissues, the carbon skeleton of hexoses is mobilized between starch and sucrose with only a small degree of dilution by the triose phosphate pool, and therefore with very little redistribution of carbons. The fact that amyloplast has very little or no fructose-1,6-bisphosphatase is a major cause for this metabolic outcome.

Another pathway for starch synthesis, which was claimed to exist for both amyloplasts and chloroplasts, starts directly with ADP-glucose molecules transported intact from the cytoplasm (Pozueta-Romero et al., 1991; Tetlow et al., 1994). The ADP-glucose in this case is produced by reversal of sucrose synthase reaction, starting with sucrose and ADP. However, in many studies particularly with mutants, it was clearly established that starch synthesis is always very susceptible to inhibition of the plastid ADP-glucose pyrophosphorylase activity. These observations indicate that ADP-glucose supplied by the cytoplasmic sucrose synthase reaction is, at best, a minor source for starch biosynthesis in the plastid and if this enzyme is deficient, cytoplasmic ADP-glucose cannot salvage the diminished starch synthesis.

Other factors that differentially affect starch synthesis in various organs are related to enzyme heterogeneity. As will be described later on, many of the enzymes associated with starch synthesis and degradation are found in more than one form, usually as genetically distinct products. These isoenzymes that have the same catalytic activity may appear in variable distribution in different cells or vary slightly in their kinetic properties, but they may be differently regulated at the level of gene expression and at post-translational processing. The fine structure of the starch end product with regard to its chain length, degree of branching, molecular size and association into a granular structure will be determined by the properties of the synthesizing isoenzymes present at the site of the glucan's formation.

About 0.5% of the glucosyls in amylopectin that accumulates in tuber crops such as the potato is phosphorylated. Approximately 80% of the phosphate is linked to C-6 and most of the rest to C-3. The esters can amount to a significant amount of the phosphorus stored in the tuber and it also affects the physical properties of the starch. Complete enzymic depolymerization of such starch will require participation of a phosphatase activity to remove phosphate esters than could impede the action of phosphorylase and α-glucosidase. The mechanism responsible for this phosphorylation is not known, but it is clear that it proceeds concurrently with the process of starch synthesis and not after completion of the amylopectin molecule biosynthesis (Nielsen et al., 1994).

As has been mentioned earlier, the donor of glucosyls for the production of starch is ADP-glucose. Formation of this molecule is catalyzed by ADP-glucose pyrophosphorylase catalyzing the key regulatory reaction for the synthesis of starch as discussed in greater detail later.

$$ATP + \alpha\text{-D-glucosyl-1-phosphate}$$
$$\rightleftharpoons ADP\text{-glucose} + PP_i$$

In the chloroplasts, hydrolysis of pyrophosphate by pyrophosphatase will support the flow of the reaction towards ADP-glucose production.

The $\alpha(1 \rightarrow 4)$ glucosyl chains of starch are synthesized by starch synthase which carry out the reaction:

$$(Glucosyl)_n + ADP\text{-glucose}$$
$$\rightarrow (Glucosyl)_{n+1} + ADP$$

An additional enzyme, branching enzyme, is responsible for the formation of the $\alpha(1 \rightarrow 6)$ branches:

$$\alpha1\text{-}4 \text{ Glucan} \rightarrow \alpha1\text{-}6 \text{ branched } \alpha1\text{-}4 \text{ Glucan}$$

In passing, it should be noted that starch phosphorylase, an enzyme involved in depolymerization of the α-glucan, may still have at special circumstances some minor contribution for adding α-glucosyls to an amylose chain when α-D-glucose-1-phosphate is the donor (Nelson & Pan, 1995). Fig. 4.13 illustrates the major route for starch synthesis and a more detailed discussion of these individual steps is given below.

(a) *ADP-glucose pyrophosphorylase (ADPGPP)*

This enzyme is found in the plastids of all tissues containing starch. Studies with enzymes from spinach, maize and potato and many other species and plants, established that the active protein is a tetramer of about 230 kDa, composed of four immunologically distinct subunits: two 55 kDa and two 60 kDa. The maize *Shrunken*-2 (*Sh*-2) gene codes for the 60 kDa subunit, and the *brittle*-2 (*br*-2) gene codes the 55 kDa subunit. Molecular sizes of the two subunits vary slightly in different plant species. The rice enzyme, on the other hand, is a tetramer composed of 50 kDa subunits (Nakamura & Kawaguchi, 1992). By studying various mutant strains, it was established that the presence of the larger subunit in the tetrameric holoenzyme is needed for expression of maximal catalytic activity. A mutated enzyme which contains only the small subunits could still bind substrates and allosteric effectors, but had only a very low catalytic ability to produce ADP-glucose resulting in a significantly reduced starch production in the seeds. Other mutants analyzed in *Arabidopsis*, pea and potato which were deficient in ADPGPP all had a server reduction in starch biosynthesis (Lin et al., 1988; Bhave et al., 1990; Hylton & Smith, 1992). Smith-White & Preiss (1992) concluded on the basis of sequence analysis that the distribution of the ADPGPP subunits in tissues vary markedly and could be divided

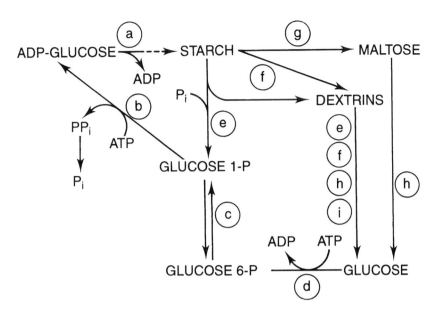

Figure 4.13   Schematic representation of pathways of starch metabolism. The dextrins comprise a structurally heterogeneous group, depending on the enzymes responsible for their formation. Enzymes indicated: a, starch synthase and branching enzyme; b, ADP-glucose pyrophosphorylase; c, phosphoglucoisomerase; d, hexokinase; e, starch phosphorylase; f, α-amylase; g, β-amylase; h, α-glucosidase, and i, debranching enzymes. Participation of UDP-glucose:amylogenin glucosyltransferase is required for the initiation reaction preceding steps a.

according to the presence of at least three groups: (1) a small subunit that is present in all cells; (2) a large subunit mostly found in non-photosynthetic cells, and (3) a different large subunit that is found mostly in photosynthetic cells. Heterogenic combinations of small and large subunits in different cell types will form tetrameric enzymes with variable kinetic and regulatory responses. It is difficult to predict from electrophoretic profiles of a purified enzyme preparation which of the closely sized subunits are present in an isolated tetramer. They may be a mixture of several combinations made from the genetically distinct subunits of the three groups, and sometimes also from partial proteolytically degraded subunits (Kleczkowski et al., 1991, 1993). Müller-Röber et al. (1994) found that in tobacco plants which were transformed with the regulatory element of the gene from the large ADPGPP subunit of potato tubers, the enzyme was functional in most non-photosynthetic tissue, but not in leaf mesophyll. These experiments, as well as a similar study defining a co-ordinated expression of the large and small subunits in the potato plant by Nakata & Okita (1995) suggest the existence of specific regulatory mechanisms for this enzyme which are different in photoassimilate exporters vs. importers tissues. In another experiment, transgenic potato plants which were transformed with the small ADPGPP gene in the antisense orientation almost ceased to

synthesize starch both in leaves and in tubers (Müller-Röber et al., 1992). In addition to various morphological changes noticed for these plants, most of the photoassimilate accumulated as sucrose rather than in starch representing not only an increased flow of carbon into the synthesis of the disaccharide, but also an increased synthesis of sucrose phosphate synthase. However, at the same time, the synthesis of the protein storage patatin in these tubers was markedly reduced. This observation is contrary to what was expected from other studies claiming that the high level of sucrose can induce the synthesis of this protein (see discussion earlier in this chapter).

The molecular biology of ADPGPP has been studied and reviewed in great detail (Kleczkowski et al., 1991; Preiss et al., 1991; Smith-White & Preiss, 1992; Stark et al., 1992). cDNA clones for the two ADPGPP subunits from several sources were prepared and sequenced, indicating the presence of a wide degree of homology between different species. Among those sequenced are the maize Sh-2 gene (Shaw & Hannah, 1992), a large subunit of barley (Villand et al., 1992), and potato tubers (du Jardin & Berhin, 1991). There is a close homology in sequences of enzyme domains which are responsible for binding of substrates and allosteric effectors.

A most important property of ADPGPP is its allosteric regulation by 3PGA which is an activator (maximum activation of about 25–50-fold at 2–5 mM) and by inorganic phosphate (P_i) which is an inhibitor (50% inhibition at 2–10 mM). Rates of ADP-glucose synthesis, and hence starch, as well as partitioning of photosynthetic into sucrose in the cytoplasm, are, therefore, directly correlated to the $3PGA/P_i$ ratio in the plastid. In addition to 3PGA, other phosphate esters such as fructose-6-phosphate can also act as weak effectors. It is important to notice the difference in the pattern of the $3PGA/P_i$ effect on starch synthesis between photosynthetic and non-photosynthetic cells. This effect is particularly apparent in the chloroplasts where the 3PGA is produced abundantly during the light period. In plastids of non-photosynthetic storage tissues, on the other hand, there is at best only a low level of 3PGA and the bulk of carbohydrate for starch synthesis arrives into the stroma predominantly as glucose-6-phosphate rather than as triose-P. Amyloplasts lack fructose-1,6-bisphosphatase, thus rendering them unable effectively to produce hexose from triose phosphate (Entwistle & apRees, 1990; Smith & Denyer, 1992). From all these considerations, it can be assumed that in the amyloplasts fluctuations in P_i levels will have a more dramatic influence on the ADPGPP activity, and consequently, on starch synthesis.

The regulatory effect of $3PGA/P_i$ on ADPGPP activity is commonly found for enzymes of many species, but as expected for diverse genomes, it is not universal. In barley endosperm, the large 60 kDa subunit is a protein with a much faster (half-life in order of minutes) turnover, then the small 51 kDa subunit. The holoenzymes constructed from partially digested large subunits and the native small subunits had similar kinetic properties, but its activity was less affected by the $3PGA/P_i$ ratio when compared to the unprocessed native tetramer (Kleczkowski et al., 1993). This observation, as well as those made for other mutants discussed earlier, underscores the fact that the regulation of starch synthesis by the allosteric effects on ADPGPP activity has a quantitatively and qualitatively variable dimension in different strains and tissues and even in the same tissue at different times. It can also be concluded that the rate of ADP-glucose synthesis in the amyloplast is modulated by regulatory mechanisms which are less precisely defined than those for the chloroplasts.

Pyridoxal phosphate is also a strong activator of ADPGPP (maximal stimulation of about 6-fold by 20 mM) by its ability to bind a lysyl residue at the 3PGA activator site (Preiss et al., 1991). As it is effective only at such high concentrations, it is not expected to be physiologically important.

(b)   Initiation primer

Whenever there are sufficient numbers of pre-existing $\alpha(1 \rightarrow 4)$ glucan chain oligosaccharides present even as short as maltotriose, they can be used as acceptors for the growth of the amylose structure by the starch synthase reaction. However, to initiate the synthesis of starch molecules by in vivo transglucosylation from ADP-glucose (starch synthase reaction) from ground zero in total absence of malto-oligosaccharide chains there is a need for the presence of an alternate specific primer molecule. The identity of this intracellular primary primer that serves as the initiation site for synthesis of the macromolecule was for a long time a mystery. The first glucose residue in the mature polysaccharide was always represented as the 'reducing head' of the molecule, but it was questionable whether this is the natural state of the polysaccharide in vivo. Most starch (and glycogen) molecules isolated from tissues have a 'reducing head', but this could have been formed by earlier cleavage in the tissue or during the isolation process or by both. Early experiments (see Moreno et al., 1987 for literature) identified the presence of a protein acceptor to whom a glucosyl unit from UDP-glucose could be attached covalently in an in vitro system to provide an initial primer for growth of the starch molecule. The study of this primer in animal tissue has yielded much detail about this initiation process and a brief description is of interest. A 37 kDa protein called glycogenin is abundantly present in glycogen-forming tissues and it is glycosylated at a single tyrosyl residue by a glucosyltransferase reaction from UDP-glucose (Cao et al., 1993a). The monoglucosylated glycogenin, while forming a complex with the very complex heteromeric glycogen synthase molecule, carries out a self-glucosylation reaction using UDP-glucose as the donor and extending the tyrosine-attached $\alpha(1 \rightarrow 4)$ glucan chain to a length of about eight glucosyl units. Only at this point is an effective primer for glycogen synthase formed, and the glycogen molecule from now on continues to grow by the action of this enzyme (Whelan, 1986; Smythe & Cohen, 1989). Consequent to these reactions, all newly synthesized glycogen molecules are covalently anchored in glycogenin and indeed it was found that most glycogen isolated from mammalian tissue is still attached to the protein. It is obligatory that the glycogenin-dependent initiation of glycogen synthesis by glycogen synthase should have an attached malto-octose chain made by the self-glycosylating glycogenin itself. Digesting this chain with glycogen phosphorylase or α-amylase renders the glycogenin incompetent to serve as a primer (Cao *et al.*, 1993b; Skurat *et al.*, 1993). This effect suggests the existence of a regulatory mechanism which controls initiation of glycogen synthesis when glycogenolysis is stimulated. It is interesting that glycogenin can use UDP-xylose as a substrate for self-glycosylation instead of UDP-glucose. The glycogen synthesized with such a xylosylated primer, therefore, has a small number of xylosyl residues in its structure (Roden *et al.*, 1994).

The study of the equivalent protein primer in plants has not advanced as it has in animal tissue. However, the information at hand suggests a similar system. In the potato tuber, a glycogenin-like 38 kDa protein serves as the initiator for starch synthesis. Also here, similar to mammalian tissues, the first glucosylation step detected in an *in vitro* system was catalyzed by a UDP-glucose : protein transglucosylase reaction followed by the subsequent steps of transglucosylations from UDP-glucose to lengthen the glucan primer chain until it can be an acceptor for glucosyls transferred from ADP-glucose by the starch synthase reaction (Ardila & Tandecarz, 1992). A re-evaluation of the role of UDP-glucose as the first glucosyl donor is required, since this substrate is not considered to be produced in the plastids. An autocatalytic self-glucosylating protein using UDP-glucose as the substrate, was recently isolated from sweet corn and cloned. Named amylogenin, it functionally resembles glycogenin. However, amylogenin, incorporates a β-glucosyl by linking it to an arginine residue (rather than to a tyrosyl) in the protein (Singh *et al.*, 1995). A more detailed characterization of the starch initiation primer system in plants should now be expected.

(c) Starch synthase (ADP-glucose: 1,4-α-D-glucan 4-D-glucosyltransferase)

This enzyme which exists in two major forms called the granule associated and the soluble species is responsible for the 'side by side' synthesis of amylose and amylopectin. They represent two genetically distinct products which differ in substrate affinities. The granule-bound enzyme is associated with the synthesis of amylose. In maize, producers of 'waxy starch' which have the *wx* genotype have none or very low levels of this enzyme and produce almost exclusively a highly branched amylopectin. A similar mutant of potato (*amf*) lost the activity of the granule-bound synthase also resulting in the synthesis of amylose free starch. The Waxy (Wx) locus product is a

starch synthase molecule of 60 kDa (Taira *et al.*, 1991; Sivak *et al.*, 1993). Dry *et al.* (1992) claim that there are two isoforms of this enzyme (I and II) in pea embryo and in the potato tuber. cDNA for both the isoenzymes was sequence. One of the isoforms (II) is ubiquitous in every plant organ, whereas the other is more limited in its distribution. Whether these two isoenzymes have a differential role in the synthesis of amylose and also in elongating the $\alpha(1 \rightarrow 4)$ chains in amylopectin during growth of the starch granule is not clear. The information available on the Wx enzyme suggests that regulation of its transcription determines the amylose content in the endosperm.

Inhibition of the expression of the granule-bound synthase with an antisense RNA led to the production of amylose-free starch in the potato (Visser *et al.*, 1991; Kulpers *et al.*, 1994a,b). It is noteworthy that in contrast to the soluble starch synthase, the granule-associated enzyme can, though at a poor rate, use not only ADP-glucose as the substrate, but also UDP-glucose. It is not clear if this ability has any physiological value particularly since UDP-glucose is synthesized in the cytosol.

The soluble starch synthase which is responsible for adding the $\alpha 1 : 4$ glucosyls to growing amylopectin was purified from pea embryos and found to consist of two isoforms. The relatively higher susceptibility of this species of starch synthase to high temperatures (30–45°C) is considered to be the prime cause for reduced starch production in cereal grains if they are exposed to such heat levels during the filling stage (Jenner *et al.*, 1993; Denyer *et al.*, 1994). The major component of this starch synthase is a 77 kDa protein and a second, immunologically cross-reactive protein is a 60 kDa molecule. The kinetic properties of these two are very similar (Denyer & Smith, 1992; Smith & Denyer, 1992). Isomeric forms of this soluble enzyme were also detected in other plant tissues such as maize (Mu *et al.*, 1994), and barley where one isoenzyme was found to be missing in the *shx* mutant (Tyynela & Schulman, 1993). The cDNA for three isomeric forms of the soluble starch synthase in rice has been cloned and characterized (Baba *et al.*, 1993), a study which opens the way for better understanding the regulation and functional role of these individual isoenzymes.

The nature of the starch synthase itself and not necessarily only contributions of the branching enzyme is responsible for the structure of the final starch product (Shewmaker *et al.*, 1994). In these experiments, transgenic potato tubers which expressed the glycogen synthase by *Escherichia coli* produced a very highly branched amylopectin and starch granules with physical and chemical properties close to that of cereal starches rather than those of potato starch. Since branching in these transgenic tubers was carried out by the potato's Q enzymes, the nature of the product indicates that the bacterial glycogen synthase, which also uses ADP-glucose as the substrate, somehow had a diverting effect in altering the elongation/branching ratio during the progress of amylopectin synthesis.

It is accepted as a rule by most investigators that starch synthase is exclusively localized in the plasmid. However, Tacke *et al.* (1991) claimed that some of the enzyme in spinach leaves could be identified as extrachloroplastic. If this finding is substantiated, it will be of interest to determine its function in that location. It is not clear where the glucan-synthesizing enzymes are located in Rhodophyta, where the starch granules are present in the cytoplasm (Yu *et al.*, 1993). It should be noted that most starch degradation enzymes in green plant cells are also present in the cytosol with no clear physiological role.

(d)   *Branching (Q) enzyme [1,4-α-glucan: 1,4- α-D glucan 6-α-D-(1,4-α-D-glucano)-transferase]*

This enzyme catalyzes transfer of an oligoglucan segment of several glucosyls from a linear α1 : 4-glucan oligosaccharide (maltodextrin) of 30–50 DP long and linking it to a C-6 of a α-glucosyl residue in another α1 : 4 oligoglucan chain serving as the acceptor. The chain length of a maltooligosaccharide transferred is usually considered to be 11–14 glucosyls, but other studies indicate a longer chain, averaging 20–25 glucosyls. The enzyme is basically a protein of about 85 kDa closely associated with the soluble starch synthase form which is responsible for the synthesis of amylopectin. Indeed, mutants whose branching enzyme level is low, such as the *ae* (amylose extender) or *du* (dull) of maize, produce an abundance of amylose (70%) in the endosperm, compared with the wild type that contains 70% amylopectin. In a sugary-1 (*Su*-1) mutant, the amylopectin formed is branched at a higher degree, resembling glycogen. As found for maize, potato, sweet potato and rice, the enzyme is usually found in two or three isoforms, with a variable distribution in different organs (Singh & Preiss, 1985; James *et al.*, 1985; Nakamura *et al.*, 1992; Yamanuchi & Nakamura, 1993; Nakayama & Nakamura, 1994). All isoforms transfer a glucan with an average chain length of 21–24,

but the isoenzymes have differences in primer affinity. Some prefer longer chains as acceptors and others shorter chains, consequently, it was suggested that they form different degrees and densities of branching which contributes to creating the 'cluster' amylopectin structure (Mizuno *et al.*, 1993; Takeda, *et al.*, 1993; Guan *et al.*, 1994). When the two branching enzyme genes from maize were cloned in a branching enzyme deficient *E. coli* mutant, the average chain length of the amylopectin-like products synthesized was different for the two clones (Guan *et al.*, 1995). The isoforms also exhibit variable patterns of expression during embryo development (Burton *et al.*, 1995). It is interesting that the potato branching enzyme is very close in structure to the large subunit of ADPGPP which may indicate some close functional and regulatory association between these two enzymes (Kossmann *et al.*, 1991). This assumption is further supported by a study of the cDNA sequence of the rice Q-enzyme I isozyme (Nakamura & Yamanuchi, 1992).

4.6.3   Starch mobilization

The pathways of starch degradation in the photosynthetic and the non-photosynthetic storage tissues are different in the dynamics of their fluxes and their regulation, even though qualitatively they both involve similar enzyme reactions. In the amyloplasts of mature storage tissues such as seeds and tubers, enzymes that degrade starch are very low or completely dormant, being inactive, missing altogether, physically sequestered or inhibited, or inactive because of a combination of several of these reasons. When germination and plant growth take place, activity levels of these catabolic enzymes increase dramatically. This includes the glucan phosphorylase as well as the several specific α-glucanohydrolases which often appear in several isoactive species. Hormonal factors, such as gibberellins (GA$_3$), participate in the induction of hydrolase synthesis, particularly that studied in detail in the case of α-amylases. Other environmental factors, not always clearly identified, also participate in this regulation. The patterns of induction of a specific enzymic activity may differ markedly for different isoenzymes of the same catalytic activity, consequently modulating the pace of starch degradation in different species in a multitude of ways. The major end products of this starch mobilization in the amyloplast are maltose, glucose and glucose-1-phosphate. Maltose also is eventually broken down to glucose by α-glucosidase both in the amyloplast and in the cytoplasm. The glucose and glucose-1-phosphate enter the general metabolic pool of carbohydrate, and are used for cell respiration and for the biosynthesis of cellular matter (Fig. 4.13).

In photosynthetic tissue, on the other hand, starch degradation is a process that exists to some extent all the time in balance with starch biosynthesis. Proof for this situation is the fact that there is a continuous dynamic exchange of ^{14}C between the two major photoassimilates, starch in the plastid and sucrose in the cytoplasm, even during periods when intense net synthesis of starch occurs. The rate of these opposing reactions, or starch turnover, may vary markedly at different times. During the diurnal cycle, while intensive starch synthesis and accumulation occurs, there is seemingly no starch degradation if this evaluation is based on measurement of the net starch increase alone. During the dark period, starch in the plastid is rapidly depleted because of an increased activity of the degradative enzymes and a decreased supply of 3-carbon intermediates available to replace the depleted glucosyls. The final products of starch depolymerization, glucose and glucose-1-phosphate, are converted into triosephosphate in the chloroplast and in this form transported to the cytoplasm. In contrast to the amyloplasts only a minute amount of hexose phosphates move in and out of the chloroplast, but some glucose and probably maltose can exit into the cytoplasm.

A puzzling situation is that all the degradative enzymes for starch are also present in the cytoplasm and are often at levels even higher than those found in the plastid. The function of these cytoplasmic enzymes while their α-glucan substrates (with the exception of some maltose) are exclusively located in the plastid with its own subset of some or all of these degradative enzymes has not yet been determined (Beck & Ziegler, 1989).

(a) α-Amylase (1,4-α-glucan 4-glucanohydrolase)

This Ca^{2+}-binding, endohydrolase family cleaves $\alpha(1 \rightarrow 4)$ glucosidic linkages in starch in a random manner (Akazawa et al., 1988; Fincher, 1989; MacGregor, 1993; Søgaard et al., 1993). Amylose will yield primarily maltose, some glucose, some maltotriose which will be slow to hydrolyze and a trace of isomaltose. Amylopectin will yield a mixture of short-chain oligosaccharides (DP 2-12), and their distribution will depend on the structural profile of the branched substrate. The digest includes maltose, maltotriose and other malto-oligosaccharide structures which have the α1:6 glucosyls corresponding to the original branch linkages in the amylopectin. α-Amylase can attack not only soluble starch, but also when the glucan is present in the granule. In general, particularly in the non-photosynthetic storage organs, α-amylase is considered to initiate the starch depolymerization reactions. The random fragmentation of the large starch molecule will at first yield medium size length α-glucan chains (limit dextrins and malto-oligosaccharides) which are substrates for further cleavage by phosphorylase and by other starch hydrolyzing enzymes.

α-Amylase participation in the mobilization of starch is most notable in germinating seeds. It is synthesized in the scutellum and aleurone tissue and secreted into the endosperm. It is also active in pods of maturing fruit to support growth of the seed. Patterns of α-amylase induction in different stages of growth in organs such as the aleurone, scutellum and the embryo axis have been studied in many systems (Akazawa et al., 1988; Sanwo & DeMasson, 1992; Minamikawa et al., 1992).

Multiple forms of α-amylase were isolated from plant tissues. Several genes for α-amylase isoenzymes are present in the same organism, for example seven genes in barley, one of the most detailed studied. Some of these isomeric enzymes were identified by their different isoelectric point (pI) a reflection of differences in amino acid sequences in the native protein but also the possible outcome of a post-translational carboxypeptidase digestion (Svensson & Søgaard, 1992). The DNA sequences for many of the polymorphic α-amylase genes from various plants have been determined (close to 20 by now) and classified into two main classes (AMY1 and AMY2) and three subfamilies organized in a phylogenetic relationship according to sequence homologies which also translate into differences in structural features, induction of transcription, substrate specificity, activation by Ca^{2+} and inhibition by a protein inhibitor (Huang et al., 1992).

Resting seeds usually do not contain or have only very low levels of α-amylase. The enzyme is intensively produced during germination and precursor protein species have been identified. In the genome of the same plant, expression of the different α-amylase genes during seed germination and development occurs in a complex pattern. Many studies, particularly of germinating rice and barley seeds, suggest interaction between metabolic signals such as carbohydrate intermediates, hormones and Ca^{2+} to modulate expression of α-amylase activity in various organs of the developing seedling. The most effective hormone to induce α-amylase synthesis at the transcriptional level in the aleurone layer of the seed is gibberellic acid (GA$_3$) (Beck & Ziegler, 1989; Thomas & Rodriguez, 1994). Three GA$_3$ *cis* elements responsive domains ('boxes') in the promoter region of an α-amylase gene in cereal seeds have been identified as having the motifs: TAACAAA, TATCCAC/T and CCTTTT (Itoh *et al.*, 1995). The process of α-amylase synthesis, secretion and post-translational modification is typical of secretory glycoprotein (Fincher, 1989; Sticher & Jones, 1992). It is of interest that most of the α-amylase is found as an extrachloroplastic protein and comparatively, only varying degrees of a lower level of this enzyme activity are located in the chloroplasts (Ghiena *et al.*, 1993).

Protein inhibitors of α-amylase are present in the seeds of many cereals and in storage vacuoles of cotyledonary cells of some legume seeds (Beck & Ziegler, 1989; Iguti & Lajolo, 1991; Santino *et al.*, 1992). Some of these inhibitors were characterized in great detail and they were found to inhibit α-amylases from many biological sources. The inhibitor from beans, for example, is a glycoprotein of 15–20 kDa with an inactive precursor of 28 kDa when present in the endoplasmic reticulum. The barley inhibitor appears as three 15 kDa isoproteins one of which also exists in a glycolysated form. The inhibitor can exert inhibitory effects on α-amylases when in a monomeric, homodimeric, homotetrameric and heterotetrameric combination; however, these various forms have large variations in inhibitor effectivity (Morallejo *et al.*, 1993; Marci *et al.*, 1993). The physiological role of these inhibitors is not clear. They are usually deposited and accumulate in the developing endosperm and gradually disappear during germination. There must be a value for protecting stored starch against endogenous amylase activity at some stages of development, and one hypothesis suggests that these inhibitors are to provide seed resistance to digestion by insects whose own α-amylase activity is obligatory for starch digestion.

(b) β-Amylase (1,4-α-D-glucan maltohydrolase)

This enzyme is an exohydrolase that removes maltose residues from the non-reducing end of $\alpha(1 \rightarrow 4)$ amylose chains. Progress of β-amylolysis is halted at 1–3 glucosyl residues away from a branch $\alpha(1 \rightarrow 6)$ linkage. The best substrates for β-amylase are long malto-oligosaccharide chains in amylose and such as those produced first by the partial α-amylolysis of starch. Very short oligosaccharides, such as maltotetraose and maltotriose, are very poor substrates for this enzyme.

The β-amylase is present as an extrachloroplastic glycoprotein in the leaves and also in non-photosynthetic tissues. A major portion of the enzyme in both leaves and storage organs is stored in the vacuoles, a site which does not contain starch.

Activity of the enzyme in germinating cereal seeds increases parallel to that of α-amylase and is found in several isoforms, for example three in wheat (Ziegler *et al.*, 1994) but only one in maize (Lauriere *et al.*, 1992). Sucrose and polygalacturonic acid are strong inducers of β-amylase production in sweet potato (Takeda *et al.*, 1995). In cereals, such as barley, β-amylase is present in 'latent' or 'bound' and 'free' forms. Proteolytic cleavage generates the free and most active form. The enzyme protein also needs its sulfhydryl group to be kept in a reduced form in order to retain its catalytic activity. It is interesting that cereal mutants deficient in β-amylase, as well as several wild strains which lack this enzyme can still germinate normally, indicating that β-amylolysis is not an obligatory process for growth of the seedling, and activities of α-glucanohydrolases and phosphorylase are adequate to promote starch degradation effectively. It has, therefore, been argued that β-amylase has a biological value more as a storage protein than as an enzyme (Beck & Ziegler, 1989). The β-amylase has been purified as a kDa 56 protein from various sources and its cDNA sequenced (Kreis *et al.*, 1987; Monroe *et al.*, 1991).

(c) α1:4 Glucan phosphorylase (1,4-α-D-glucan : orthophosphate α-D-glucosyltransferase)

This enzyme which has been detected in all species that use starch for storage catalyzes the reversible reaction:

$$\alpha1 : 4(\text{Glucan})_n + P_i$$
$$\rightleftharpoons \alpha\text{-D-glucosyl-1-phosphate} + \alpha1 : 4(\text{Glucan})_{n-1}$$

The K_{eq} for this reaction is about 3, and under most physiological conditions the reaction will proceed in the direction of glucan depolymerization. Historically, before the discovery of starch synthase (by L.F. Leloir in 1961), phosphorylase was considered to be responsible for net amylose synthesis. The enzyme is a pyridoxal-P containing protein and it is active as a dimer of about 210 kDa. Contrary to the complex regulatory properties of the mammalian glycogen phosphorylase, the enzyme in plants seems to be a much 'simpler' protein. It is not known to be regulated by phosphorylation and dephosphorylation and is not significantly affected by allosteric regulators, such as activation by ATP and hexose 6-phosphate. ADP-glucose and UDP-glucose were found to have an inhibitory effect on the enzyme from some sources and this inhibition is very potent for the algal enzyme (Yu & Pedersen, 1991). Regulation of starch phosphorylase isoenzymes in plant tissue is primarily based on the level of gene expression. Specific details defining the signals and response patterns for this control mechanism are poorly defined. In several species, e.g. peas, banana, maize and spinach, at least two phosphorylase isoenzymes exist. The dominant form (II) is found in the plastids and has a size of 105 kDa for its monomer. It may be accompanied by a slightly longer precursor protein. Another isomer (I) of 90 kDa is predominantly a cytosolic enzyme (van Berkel *et al.*, 1991). In photosynthetic algae, the phosphorylase is exclusively located in the pyrenoid and not in the stroma of the chloroplasts (Yu *et al.*, 1994). It is surprising that the level of the plastid phosphorylase is very high during seed development and low during germination, the opposite situation expected from the state of the starch stores in those conditions. A similar presence of isoenzymes is found in the potato where in both leaves and tubers, type L form is localized in the plastids whereas type H phosphorylase is in the cytoplasm. The H isoenzyme is a molecule 78 amino acids shorter than the L isomer. The H type has low affinity for branched α-oligoglucans (e.g. glycogen, amylopectin) as substrates whereas for the L type this affinity is high. cDNA cloning indicates heterogeneity between the L isoenzyme isolated from leaves and tubers. The two are about 84% homologous but have distinct sequence differences in some defined domains and when compared to sequences of phosphorylase isoenzyme from other species (Sonnewald *et al.*, 1995; Liu *et al.*, 1995).

(d) D-enzyme (4-α-glucanotransferase)

This 'disproportionating' enzyme was first found in the plastids of potato tubers and subsequently in other plant tissues (Preiss & Levi, 1980; Liu & Preiss, 1988). It catalyzes the reversible condensation of donor and acceptor 4-α-D-oligoglucan chains. Free glucose can also serve as an acceptor:

$$\alpha 1 : 4\text{-}(\text{Glc})_m + \alpha 1 : 4\text{-}(\text{Glc})_n$$
$$\rightarrow \alpha 1 : 4\text{-}(\text{Glc})_{(m+n-1)} + \text{Glc}$$

Short-chain malto-oligosaccharides are the best substrates, but even large starch molecules can function as donors. Maltose is the smallest unit transferred when maltotriose is the substrate. The position of this reaction in the overall scheme of starch breakdown is not clear. One hypothesis is that it helps to 'rebuild' longer $\alpha(1 \rightarrow 4)$-glucan chains from short fragments (DP 3–5) provided by the action of α-glucosidases making them better suited for further degradation by phosphorylase. Also, it can modify the length of the A type terminal chains in the amylopectin structure, increasing their susceptibility to cleavage by phosphorylase as well as to the branching enzyme.

A cDNA clone encoding the enzyme was isolated from potato tubers and identified as a 59.5 kDa protein of 54 amino acids. The cloned enzyme had a N-terminal transit peptide which may be involved in enzyme targeting to the plastids. The D-enzyme is not known to be regulated by protein phosphorylation nor modulated by phosphorylated metabolites. It is present in all tissues that have starch. Surprisingly, it is induced (mRNA production) strongly by light and by sucrose, conditions that promote starch synthesis and accumulation. This raises the question of whether D-enzyme in the potato is indeed part of the starch degradation process or it fulfills some other role for disproportionation which is part of starch turnover and synthesis (Takaha *et al.*, 1993).

(e) α-Glucosidase (α-D-glucoside hydrolase)

This enzyme is an exohydrolase which sequentially removes terminal α-glucosyl residues from dextrins of a broad range of chain lengths as short as disaccharides up to macromolecular starch whether present in solution or though less effectively, in the granule. It has been considered to provide a mechanism for the hydrolysis of maltose formed as the final product of starch hydrolysis by α and β-amylases (Preiss & Levi, 1980). The enzyme is found in all starch-storing tissues. It has a broad substrate specificity and the $\alpha(1 \rightarrow 4)$ glucosidic linkage is not the only one cleaved. The $\alpha(1 \rightarrow 6)$-glucosyl linkage, as well as many other α-glucosides with a variety of aglycons can be substrates. Consequently, to some degree, the α-glucosidase can fulfill the needs of a debranching enzyme since it can remove the $\alpha(1 \rightarrow 6)$-glucosyl 'stubs' in branched limit dextrins and in short-chain malto-oligosaccharides produced by the action of amylases or by starch phosphorylase. Several isoforms of the enzyme have been identified which may reside in both the cytoplasm and in the plastids (Sun *et al.*, 1995). The plant enzyme resembles the powerful enzyme amyloglucosidase (or glucoamylase) from *Aspergillus niger* which is extensively used in starch research and in biotechnology.

(f) Debranching enzyme (α-dextrin 6-glucanohydrolase)

A large number of debranching activities have been detected in starch storing and metabolizing tissues. These activities were defined and named as separate entities according to variations in substrate specificity creating a confusing classification (Preiss & Levi, 1980; Beck & Ziegler, 1989). For example, 'R-enzyme' is a debrancher that can act on amylopectin; a 'pullulanase' degrades phosphorylase-limit dextrin, and an 'isoamylase' has the broadest specificity, but does not attack glycogen. Doehlert & Knutson (1991) have tried to introduce some better definition and characterization to classify the debranching enzymes. They identified in maize endosperm two species of debranchers. One, an 'isoamylase' which hydrolyzes $\alpha(1 \rightarrow 6)$-branch linkages in amylopectin and in phosphorylase or β-amylase limit dextrins but does not hydrolyze pullulan (a fungal linear α-glucan with about 60% $\alpha(1 \rightarrow 4)$ and 40% $\alpha(1 \rightarrow 6)$ linkages) and has a pH optimum of 7.0. A second debrancher enzyme is 'pullulanase' or 'limit dextrinase' with a pH optimum at 5.5, which hydrolyzes pullulan and β-amylase limit dextrins. The limit dextrinase of barley malt appears as two or three isoenzymes of 104 kDa and is very sensitive to oxidation of its sulfhydryl groups (Sissons *et al.*, 1992). There is limited information on the molecular biology, regulation and physiological dynamics of the debranching enzyme. Its synthesis and level in starch metabolizing tissues is often parallel to that of starch phosphorylase and it increases significantly during seed germination.

(g) α-1,4-Glucan lyase

This enzyme catalyzes the degradation of maltose, malto-oligosaccharides and starch to produce 1,5-anhydrofructose by an eliminative cleavage of the terminal non-reducing $\alpha(1 \rightarrow 4)$-glucosyl residues (Fig. 4.14). This lyase was found in a species of red seaweeds (e.g. *Gracilariopsis lemaneiformis*). The purified enzyme is a 111 kDa protein with a broad pH optimum of 3.0–7.0. It is located exclusively in the stroma of the chloroplasts and not next to cytosolic floridean starch granules which are present in the alga (Yu *et al.*, 1993; Yu & Pedersen, 1993). The biological role of this glucan lyase has yet to be defined and the metabolic fate of the product 1,5-anhydrofructose has to be established.

(h) α-Glucosyltransferases

Most glycohydrolases exhibit an ability to catalyze transglycosylation reactions from a glycosidic donor to an alcohol group, e.g. in another sugar, as the acceptor. The intensity of such a reaction is directly related to substrate concentration and the accumulation of transfer product during this process is usually transitory. Generally, it seems that finding an *in vitro* α-glucosyltransferase activity in plant tissue extracts should not be simply credited to the presence of a specific and distinct enzymic entity, but actually to reactions catalyzed by any one, or a mixture, of the well-defined glucosidases characterized above.

4.6.4 Sucrose–starch interrelationship

The movement of organic carbon to and from starch and sucrose and between these two glycosides constitutes a major route for photoassimilate

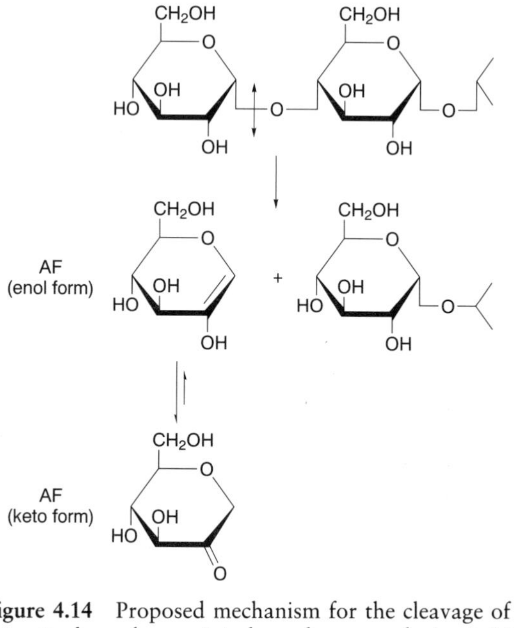

Figure 4.14 Proposed mechanism for the cleavage of 1,4 α-D-glucan by a α-1,4-glucan lyase, producing 1,5-D-anhydrofructose (AF).

mobilization and compartmentation in plants. Several basic aspects of the metabolic interconnection between these two components and their regulation have been pointed out in earlier discussions above. Because of the importance of these processes and in the general scheme of carbohydrate metabolism and also in order to reflect on the large research effort invested in studying them, it is of value to re-capitulate an overview of the current concept on the biochemistry of the sucrose–starch relationship (Fig. 4.15).

At all times during the diurnal cycle in photosynthetic tissues, there is a need to maintain the continuous synthesis of sucrose whose role is to be exported from the leaf to provide the carbon source for respiration and growth of the whole plant. During the light period apportioning of carbon towards sucrose production in the cytoplasm or towards starch synthesis in the chloroplast occurs at the triose phosphate level. Tight regulation of the ADPGPP activity by 3PGA/P_i level in the chloroplasts is expressed in the plastid and it is influenced by metabolic events that occur both in the plastid and in the cytoplasm. These events are those that modulate the level of 3PGA and free P_i in these two compartments. The 3PGA/P_i translocator in the chloroplast membrane provides the conduit for these two metabolites to pass to and from the plastid. Only a very low level of hexose monophosphates through the chloroplast membrane occurs. During the dark periods, degradation of the chloroplast starch by phosphorylase will produce glucose-1-phosphate which will continue to supply triose phosphate for export and for sucrose synthesis. The glucose and maltose produced by amylolysis will also reach the cytoplasm; Some directly, but mostly via conversion into triose-3-phosphate. At this metabolic juncture, it has to be considered that the synthesis of sucrose itself is a highly controlled process. This regulation occurs both at the level of control by metabolite such as the concentrations of fructose-2,6-P_2, fructose- and glucose-6-phosphates and P_i, and by protein phosphorylation/dephosphorylation (sometimes light-dependent), and by various other signals that regulate gene expression. Several examples showing how the strength of the sucrose synthesizing machinery decisively influences the degree of carbon apportionment between sucrose and starch in the leaves have been described earlier.

In non-photosynthetic sink organs, the arriving sucrose may accumulate to varying degrees mostly in the vacuole. In order to be used for starch synthesis in the amyloplast, the disaccharide is cleaved by invertase (vacuolar or cytoplasmic)

and/or by reversal of the sucrose synthase reaction. The relative importance of these two reactions is different in different species and sink organs. Glucose and fructose are phosphorylated by specific kinases, enter the amyloplasts and via ADP-glucose are converted into starch. The UDP-glucose which is produced in the cytoplasm by the sucrose synthase reaction is split into glucose-1-phosphate and UTP, the driving force for this reaction is the pyrophosphate provided by the pyrophosphate : fructose 6-phosphate 1-phosphotransferase reaction. When starch in the amyloplasts is mobilized, glucose and glucose-1-phosphate may be converted back into sucrose when they reach the cytoplasm. For further discussion, only a selection of limited examples from the vast literature on this topic will be described.

Studies of the potato plant have provided outstanding information about the sucrose–starch metabolic relationship. Reduction in the level of ADPGPP in the plant caused by mutation, by other genetic manipulation or by inhibition of its activity will dramatically reduce the accumulation of starch in all the plastids both in the leaves and in the tubers and at the same time increase the amount of sucrose produced and stored (Preiss, 1991a; Müller-Röber et al., 1992; Nakata & Okita, 1995). Antisense repression of the triose phosphate translocator resulted in a dramatic decrease in partitioning of carbon towards sucrose production and almost tripled (up to 80% or more of the CO_2 assimilated) the production of starch during the light period (Riesmeier et al., 1993a,b; Heinecke et al., 1994a,b). Here, this synthesis became the outlet for the bulk of triose phosphate produced during photosynthesis and the increased level of ADPGPP activity. The export of assimilate during dark periods in these plants was as intense or even higher than that observed for the wild-type organism under the same conditions. This effect was in part because the level of triose phosphate coming from starch degradation through glucose-1-phosphate in the dark is only a fraction (about 10%) of the concentration of triose phosphate present during the light period. Under those circumstances, an increased relative importance is given to the usually low level hexose transporter that transfers glucose out of the chloroplast as the product of starch hydrolysis. Other interesting studies on the dynamics of sucrose–starch interrelationship in leaves of *Reilla helicophylla* (Witt, 1989), rye (Otto & Feierabend, 1989), *Nicotiana sylvestris* (Hanson & McHale, 1988), sugar beet (Li et al., 1992) and *Arabidopsis thaliana* (Sicher & Kremer, 1992) have been described.

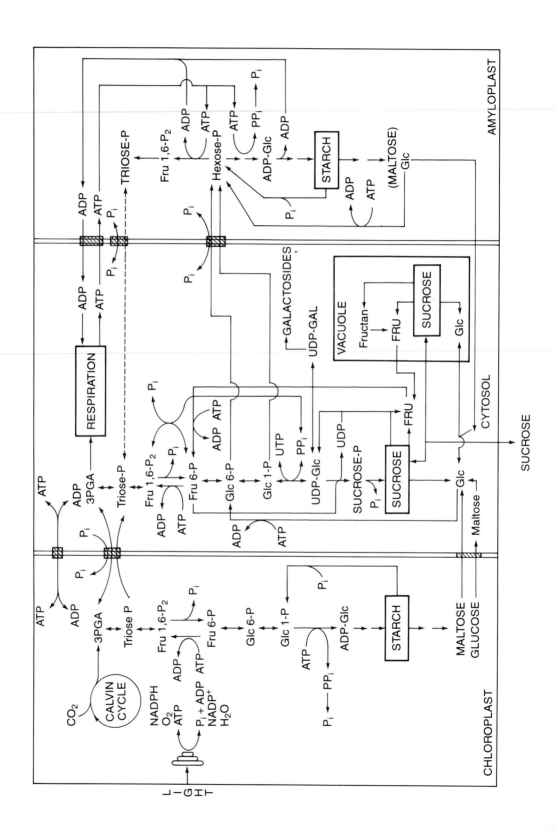

In the potato tuber, it was clearly established that sucrose is rapidly turning over and the reversible sucrose synthase reaction provides the major step that cleaves sucrose coming into the organ from the phloem. The pace of this reaction will depend on the rate through which the products, UDP-glucose and fructose are withdrawn from the system. A very high level of the cytoplasmic enzyme UDP-glucose pyrophosphorylase is almost a prerequisite to promote enhancement of the sucrose synthase reverse reaction (Morrell & apRees, 1986; Geigenberger & Stitt, 1993; Merlo et al., 1993; Spychalla et al., 1994). In support of those reactions are the observations that transgenic potato plants with a decreased expression of pyrophosphate:fructose-6-phosphate 1-phosphotransferase had tubers with a significant lower level of starch and some increase in sucrose (Hajirezaei et al., 1994).

Starch–sucrose transformations in the potato tubers are also strongly affected by the sharp decline in ADPGPP activity in detached tubers during storage. Consequently, starch synthesis is strongly inhibited and sucrose synthesis is greatly stimulated. These effects are the result of a hexose phosphate activated sucrose synthase reaction now moving in the synthesizing direction using UDP-glucose and fructose as the substrates (Reimholz et al., 1994; Geigenberger et al., 1994). This event in detached stored tubers is part of the reason for sucrose accumulation at the expense of starch degradation which under these conditions is also accompanied by strong inhibition of glycolysis (Bredemeijer et al., 1991).

A central role of sucrose synthase in sucrose–starch interconversions as established for the potato tuber was also found in many other species, such as the developing endosperm of maize, pea and wheat (Doehlert et al., 1988; Tyson & apRees, 1989; Duffus, 1992; Smith & Denyer, 1992). Although there are quantitative differences between species, the general mechanism for sucrose–starch interconversion in non-photosynthetic tissue is similar to the model described for potato tuber. These fluxes of carbohydrate metabolism have also been studied in many cases of fruit and seed development. For example, reports on tomato (Galtier et al., 1993; Wang et al., 1994), banana (Hubbard et al., 1990; Hill & apRees, 1994), Ricinus communis seedlings (Geigenberger & Stitt, 1991), pea seedlings (Hargreaves & apRees, 1988; Smith et al., 1989; Hill & Smith, 1991) and Araucaria araucana seeds (Cardemil & Varner, 1984) should be consulted for further details to evaluate the variability of metabolic processes related to sucrose and starch compartmentation.

4.7 OTHER RESERVE POLYSACCHARIDES

In addition to the abundant starch and fructose polymers, plants produce a number of other polysaccharides which can be categorized as 'reserve' material. These compounds accumulate during times of growth and development, stored away for an interim period of dormancy, then intensively utilized to furnish metabolizable monosaccharides for a new growth cycle. It should be pointed out that in many cases the demarcation line between 'structural' polysaccharides and 'reserve' polysaccharides is not very definite. Many polymers which are considered to be typical 'reserve' material, for example in seeds, appear as very similar molecules which are considered to be 'cell wall' material when isolated from other tissues. Even from the same tissue, one fraction of a heterogeneous population of a polysaccharide is considered as 'reserve' whereas other fractions are defined as cell-wall matter. Most structural cell wall polysaccharides are subject to a process of depolymerization and extensive hydrolysis associated with tissue growth and the monosaccharides produced by this process, by what is called the salvage reaction, re-enter the general pool of cellular metabolites. Thus, even these small

Figure 4.15 A simplified composite scheme showing the relationship between sucrose and starch metabolism in the cytoplasm and in the plastids, both the chloroplasts and the amyloplasts. Major transport of carbon from the chloroplasts is via the 3PGA/P_i translocator, whereas to the amyloplast it is dominated by glucose-6-phosphate/P_i translocation, and by a low level transport of glucose-1-phosphate. In some cases it has been claimed that the adenylate translocator also supports transport of ADP-glucose transport which is produced by reversal of the sucrose synthase reaction with ADP as the acceptor. The key regulatory reaction which modulates sucrose synthesis by metabolite control are the fructose-1,6-bisphosphatase and sucrose phosphate synthase, and for starch, the activity of the ADP-glucose pyrophosphorylase. It is now recognized that the redundancy of many of the enzymes responsible for sucrose and starch synthesis and degradation, the selective differential induction patterns of gene transcription of these isoenzymes, different processes of post-translational modification of various enzymes, combined with metabolite control and kinetic consideration, will all result in a very flexible and adaptive system of metabolic fluxes and carbon partitioning along the routes shown in this scheme.

portions of cell wall elements at specific transitory physiological states, can serve as a 'reserve' material.

The most abundant information available about reserve polysaccharides is their chemical structure and distribution. Biochemical knowledge about these polymers is mostly related to their enzymic depolymerization, and only limited information is available on the enzymology of their biosynthesis. The best studied group is that of the galactomannans and the reason for this interest is the great economic importance of these materials which are extensively used as stabilizers in the food and chemical industries.

For detailed reviews and sources of information on this group of polysaccharides see Meier & Reid (1982), Manners & Sturgeon (1982), Stephen (1983), Painter (1983) and Reid (1985). The work of Dea & Morrison (1975), Dey (1978), Matheson (1990), Reid et al. (1992), Hegnauer and Grayer-Barkmeijer (1993), Ganter et al. (1993), Nonogaki et al. (1995) and Kontos and Spyropoulos (1995) present detailed discussions of the galactomannan group of reserve polysaccharides and the enzymes involved in their depolymerization.

4.7.1 Mannans

This heterogeneous family of polysaccharides is subdivided into four basic structural groups. In addition to being well-defined stored reserve material, mannan-based polysaccharides, particularly galactomannans, are also found as ubiquitous components of plant cell walls (Carpita & Gibeau, 1993).

(a) Pure mannans

A linear chain of 100–2000 residues of $\beta(1 \rightarrow 4)$-D-mannopyranosyls constitutes the basic structure of this group.

$$-4 \, \text{Man} \, \beta(1 \rightarrow 4) \, \text{Man} \, \beta(1 \rightarrow 4)$$
$$\text{Man} \, \beta(1 \rightarrow 4) \, \text{Man} \, \beta(1 \rightarrow 4) \, \text{Man}-$$

$$(41)$$

Mannans are the dominant reserve material in the seed endosperm of species of the Palmae; the best known are ivory 'nut' and date kernels. It was also isolated from seeds of Umbelliferae species and from coffee beans. Some mannan preparations have a small amount (<2%) of branch glucosyl or galactosyl residues. The mannans depolymerize readily during seed germination by the action of endo- and exo-β-mannanases. The mannose

released is metabolized via conversion into fructose and much of it is channeled into the synthesis of sucrose. Little is known about mannan biosynthesis in which GDP-mannose is the ultimate glycosyl donor.

(b) Glucomannans

Similar in structure to the β-mannan, glucomannans contain a variable proportion (6–50%) of β-glucosyls which replace some β-mannosyl residues along the chain:

$$-4 \, \text{Glc} \, \beta(1 \rightarrow 4) \, \text{Man} \, \beta(1 \rightarrow 4) \, \text{Glc} \, \beta(1 \rightarrow 4)$$
$$\text{Man} \, \beta(1 \rightarrow 4) \, \text{Man} \, \beta(1 \rightarrow 4) \, \text{Glc}-$$

$$(42)$$

Some glucomannans have a small number (2–5%) of α-galactosyls which are linked as $\alpha(1 \rightarrow 6)$ monomeric branches to the backbone chain. Some of the residues in the polysaccharide (about 4–8%) are 6-O-acetylated. These polymers are present as storage material in many monocotyledons especially in bulbs and tubers of Liliaceae, Amaryllidaceae and Iridaceae, and are often found with some starch. Chain length can vary from DP 30 to 2000 in bulbs of various species of Lillium, up to 6000 in tubers of Amorphophalus konjac (konjac mannan) and 12 000 in Aloe vera leaves. Some general patterns of synthesis of glucomannan and its depolymerization in Lillium, Asparagus and orchids have been described. The enzymology of the degradation process involves the action of end- and exo-β-mannosidases and β-glucosidase and requires first the action of an esterase to remove the O-acetyl group and an α-galactosidase to hydrolyze branch galactosyls. All these activities are required in order to facilitate a complete hydrolysis to monosaccharides.

(c) Galactomannans

This is the largest group, the most abundant and best studied group of mannan-based reserve polysaccharides. Galactomannans are also commonly identified as 'cell wall' substances in numerous plants.

The storage galactomannans are especially abundant in the endosperm of leguminous seeds. Their basic structure is the linear chain of $\beta(1 \rightarrow 4)$-D-mannosyl units, which are substituted by single $\alpha(1 \rightarrow 6)$-D-galactopyranosyl residues as branches dispersed randomly along the chain (43). The ratio of Man/Gal can vary from 1:1 to 4:1 and is species specific. Cold-water solubility of the polysaccharide increases significantly with the increased number of galactosyl branches. Some of the best known galactomannans are the locust bean 'gum' from *Ceratonia siliqua* (about 25% galactosyl branches) and guaran from *Cyamopsis tetragonolobus* (about 65% galactosyl branches). These two polymers are abundantly used as stabilizers, gelling agents and thickeners in the food and chemical industries. The molecular size of galactomannan can vary greatly. For example, DP of guaran is up to 20 000, whereas the polysaccharide from *Sophora japonica* seeds had only 30–50 hexosyl units.

Mobilization of galactomannans during seed germination involves the actions of α-galactosidase, endo β-mannanase and β-mannosidase, whose patterns of activity and characterization has been studied in some detail. The biosynthesis of galactomannan was studied in various species, but particularly in fenugreek (*Trigonella foenum-graecum*) which produces a polymer with 96% galactosyl substitutions. These experiments (Reid *et al.*, 1992) identified membrane-bound specific glycosyltransferases that use GDP-mannose, UDP-glucose and UDP-galactase as glycosyl donors and which are responsible for the construction of the galacto- and glucomannan. Detailed characterization of the individual enzymes involved has yet to be carried out.

(d) *Galactoglucomannans*

These polysaccharides are β-glucomannan chains which have $\alpha(1 \rightarrow 6)$ galactosyl branches linked to the β-mannosyl backbone. The density of the galactosyl branches and the overall hexose composition can vary, for example, for the galactoglucomannan from *Cercis siliquastrum* seeds, the Glc:Gal:Man ratio is 2:1:11, and for the polymer from asparagus tuber, the ratio is 1:4:5.

4.7.2 Xyloglucans (amyloid)

This group of polysaccharides based on a structure of a $\beta(1 \rightarrow 4)$-glucan cellulose type chain has some short branch units of a single α-D-xylosyl or disaccharide branches in which the D-xylosyl is topped off by an additional β-D-galactosyl (**44**). These polysaccharides when in the category of 'reserve' matter, are called 'amyloid' because of their tendency to stain blue with the iodine reagent for amylose.

The amyloid is a galactoxyloglucan branched polysaccharide (Glc:Xyl:Gal ratio 3:2:1 with some species variations). It is present in some starchless or starch-poor exalbuminous seed kernels of many species of Leguminosae with fleshy cotyledons. The best characterized are those from *Tamarindus* and *Tropaeolum* with a suggested structure of a $\beta(1 \rightarrow 4)$-linked glucopyranosyl chain to which side chains of β-D-galactopyranosyl-$(1 \rightarrow 2)$-α-D-xylopyranosyl and/or of individual α-D-xylopyranosyls are linked to 6-OH of the cellulose type backbone. The usually insoluble cellulose type polysaccharide is rendered water soluble by these branch substitutions. This galactoxyloglucan structure represents one class of a large group of xyloglucans associated with cell wall, which also contain a number of $\alpha(1 \rightarrow 2)$-L-fucopyranosyl residues capping the galactosyl-xylosyl disaccharide branches (Joseleau *et al.*, 1992, 1994). During growth, xyloglucan fragments obtained from these polysaccharides are found among the group of oligosaccharins which serve as strong growth elicitors in plants (Albersheim *et al.*, 1994; Joseleau *et al.*, 1994).

Not much is known about amyloid biochemistry as a reserve substance. It is mobilized during germination (Edwards *et al.*, 1985) and its synthesis in soybean membrane vesicles used UDP-glucose and UDP-xylose as substrates (Hayashi *et al.*, 1988). Most of the detailed investigations of xyloglucan degradation were directed at defining the association of this polysaccharide with cellulose microfibrils as a key structural element of plant cell walls (Carpita & Gibeau, 1993; Fry, 1995) and not related to the 'amyloid' reserve fraction. Based on the similarity of chemical structure, one can expect almost identical patterns of biochemical processes to be responsible for the synthesis and depolymerization of both the amyloid and the cell wall-associated xyloglucans. A complete advanced degradation of the oligosaccharide fragments down to the monosaccharide constituents is expected to be the same whether it originated from a cell wall complex or from apoplastic or vacuolar storage sites.

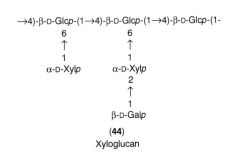

(**44**)

Xyloglucan

α-D-Gal*p*
1
↓
6
α-D-Gal*p*
1
↓
6
→4)-β-D-Man*p*-(1→4)-β-D-Man*p*-(1→4)-β-D-Man*p*-(1→4)-β-D-Man*p*-(1-

(**43**)

Galactomaman

4.7.3 α-Glucan

(a) Floridean starch

Starch is the ubiquitous α-glucan stored in plants. A structural variant of amylopectin which has distinctly different chemical properties is the floridean starch found as a reserve polysaccharide in red algae of the Rhodophyceae, particularly studied in *Dilsea edalis*. Floridean starch is a highly branched α-glucan with an average chain length of DP 12–18 and a β-amylolysis degree of 40%. Like starch, the main chain has α(1 → 4) linkages, the branch linkages are α(1 → 6), but in addition the molecule contains some α(1 → 3) glucosyls (about 3%). Floridean starch accumulates in the cytoplasm and not in plastids of the algae cells. Its synthesis depends on both ADP-glucose and UDP-glucose as substrates and probably requires participation of a specific enzyme that is responsible for the synthesis of the α(1 → 3) linkages. Its degradation involves the action of α-amylase, debrancher enzyme, phosphorylase, and a α(1 → 4)-glucan lyase, all detected in red algae extracts (Yu *et al.*, 1993).

(b) Nigeran

This group of linear α-glucans which have alternate α(1 → 3)- and α(1 → 4)-glucopyranosyl linkages (in a ratio as variable as 1 : 0.2–5) is produced by many molds and fungi, therefore, sometimes called mycodextran.

A soluble polysaccharide, isolichenan, where the ratio of α(1 → 3) to α(1 → 4) linkages can vary between about 3 and 5.2, is present in large amounts in many lichens such as the Iceland moss *Cetraria islandica* (Barreto-Bergter, 1983; Gorin *et al.*, 1988), but whether it is a reserve material or predominantly a structural polymer is not clear. Partial hydrolysis of nigeran yields the disaccharides maltose and nigerose (**45**).

4.7.4 β-glucans

Many aspects of β-glucan synthesis from UDP-glucose and degradation by a group of endo- and exo-β-glucanase are summarized by Mullins (1990), Meins *et al.* (1992), Vogali-Lange *et al.* (1994), Hoj & Fincher (1995), Fry (1995) and Alonso *et al.* (1995). This intensively studied group of glucans in higher plants is considered to be a component of cell wall and usually not counted as a 'reserve' material. However, its rate of turnover

(45)
3-*O*-α-D-Glucopyranosyl-D-glucopyranose
Nigerose

during the process of cell wall metabolism during periods of intensive growth and development indicates that to some modest extent products of β-glucan hydrolysis can serve as a source of metabolizable substrates.

(a) β1,3-D-Glucans

A β(1 → 3)-glucan whose bulk is better defined as 'reserve' is the laminaran which is found in many algae, particularly in Phaeophytae (brown seaweeds) such as *Laminaria hyperborea* and in Chlorophytae (green seaweeds). The main β(1 → 3) chain of laminaran has a very small number of β(1 → 6) glucosyl branches. Partial acid or enzymic hydrolysis of the polymer yields the disaccharide laminaribiose (**46**). This polysaccharide is also the main food reserve in diatoms, making it one of the most abundant polysaccharides in nature (Painter, 1983). Laminaran is a linear α(1 → 3)-D-glucopyranose chain with a DP of about 30–200 units and a terminal 1-*O*-mannitol residue. Enzymes that depolymerize laminaran are the same multicomponent system of β-glucanases that hydrolyze all β(1 → 3) glucans. These enzymes, abundant in nature are often called 'laminarases' when laminaran is used as the substrate. Another type of a reserve β(1 → 3)-D-glucan is paramylon which accumulates in a granular form in *Euglenophyte*. Paramylon is synthesized from UDP-glucose and its degradation of paramylon is carried out by a series of endo- and exo-β-glucosidases as well as by a laminaran phosphorylase and by a laminaribiose phosphorylase which yield α-D-glucose-1-phosphate as the product.

(46)
Laminaribiose
3-*O*-β-D-Glucopyranosyl-D-glucopyranose

(b) β(1 → 6) Glucan

This is a polysaccharide called pustulan found in some lichens, such as *Umbilicus pustulata* and the edible *Gyrophora esculenta* (Gorin *et al.*, 1988).

(c) β(1 → 3)(1 → 4) Glucan

The $\beta(1 \rightarrow 3)/\beta(1 \rightarrow 4)$-glucans are abundantly present in plant cell walls. They are found in most cereal kernels and consequently in flour together with the major commercial sources of starch. They constitute 3–5% of the dry weight of the kernel whose total comprises about 72–80% starch, 1–2% fructan, 4–8% various 'hemicellulose' β-glucans, 6–11% protein, and some other minor substances. Like most other cell wall polysaccharides, these β-glucans are structural and not 'reserve' material in the pure sense. Occasionally, however, because of their location in cereal grain and being the subject of enzymic depolymerization, these β-glucans may have a minor role as a reserve substance.

REFERENCES

Akazawa, T. (1991) *Plant Mol. Biol. Rep.* 9, 145–165.

Akazawa, T., Mitsui, T. & Hayashi, M. (1988) In *Biochemistry of Plants* Vol. 14, (J. Preiss, ed.), pp. 465–492. Academic Press, New York.

Albersheim, P., An, J., Freshour, G., Fuller, M.S., Guillen, R., Ham, K.S., Hahn, M.G., Huang, J., O'Neill, M., Whitcombe, A., Williams, M.V., York, W.S. & Darvill, A. (1994) *Trans. Biochem. Soc.* 22, 374–378.

Alonso, E., Niebel, F. de C., Obregon, P., Gheysen, G., Inze, D., Van Montagu, M. & Castresana, C. (1995) *Plant J.* 7, 309–320.

apRees, T. (1984) In *Storage Carbohydrates in Vascular Plants* (D.H. Lewis, ed.), pp. 53–73. Cambridge University Press, London.

apRees, T. (1988) In *Biochemistry of Plants* Vol. 14, (J. Preiss, ed.), pp. 1–33. Academic Press, New York.

apRees, T. & Entwistle, G. (1989) In *Physiology, Biochemistry and Genetics of Nongreen Plastids* (C.D. Boyer, J.C. Shannon & R.C. Hardison, eds), pp. 46–62. American Society Plant Physiologists, Rockville, MD.

Arai, M., Mori, H. & Imaseki, H. (1992) *Plant Cell Physiol.* 33, 245–252.

Ardila, F.J. & Tandecarz, J.S. (1992) *Plant Physiol.* 99, 1342–1347.

Argüelles, J.C. (1994) *FEBS Lett.* 350, 266–270.

Arrendale, R.F., Severson, R.F., Sisson, V.A., Costello, C.E., Laery, J.A., Himmelsbach, D.S. & VanHalbeek, H. (1990) *J. Agr. Food Chem.* 38, 75–85.

Avigad, G. (1982) In *Encyclopedia of Plant Physiology* (F.A. Loewus & W. Tanner, eds), New Series, Vol. 13A, Plant Carbohydrates I, pp. 217–347. Springer Verlag, Berlin.

Baba, T., Nishihara, M., Mizuno, K., Kawasaki, T., Shimada, H., Kobayashi, E., Ohnishi, S., Tanaka, K.I. & Arai, Y. (1993) *Plant Physiol.* 103, 565–573.

Bachman, M. & Keller, F. (1995) *Plant Physiol*, 109, 991–998.

Bachman, M., Matile, P. & Keller, F. (1994) *Plant Physiol.* 105, 1335–1345.

Barreto-Bergter, E. (1983) *Adv. Carb. Chem. Biochem.* 41, 67–103.

Beck, E. & Ziegler, P. (1989) *Annu. Rev. Plant Physiol. Plant Mol. Biol.* 40, 95–117.

Beebe, D.U. & Turgeon, R. (1992) *Planta* 188, 354–361.

Benhamou, N., Grenier, J. & Chrispeels, M.J. (1991) *Plant Physiol.* 97, 739–750.

Bhave, M.R., Lawrence, S., Barton, C. & Hannah, I.C. (1990) *Plant Cell* 2, 581–588.

Bianchi, G., Gamba, A., Limiroli, R., Pozzi, N., Elster, R., Salamini, F. & Bartels, D. (1993) *Physiol. Plant* 87, 223–226.

Bieleski, R.L. (1982) In *Encyclopedia of Plant Physiology.* New Series Vol. 13A, (F.A. Loewus & W. Tanner, eds), pp. 158–192. Springer-Verlag, Berlin and New York.

Bieleski, R.L. (1993) *Plant Physiol.* 103, 213–219.

Blackman, S.A., Obendorf, R.L. & Leopold, A.C. (1992) *Plant Physiol.* 100, 225–230.

Blazquez, M.A., Stucka, R., Feldman, H. & Gancedo, C. (1994) *J. Bacteriol.* 176, 3895–3902.

Bonnett, G.D. & Simpson, R.J. (1993) *New Phytol.* 123, 443–451.

Boos, W., Ehmann, U., Forkl, H., Klein, W., Rimmele, M. & Postma, P. (1990) *J. Bacteriol.* 172, 3450–3461.

Bracho, G.E. & Whitaker, J.R. (1990) *Plant Physiol.* 92, 381–385; 386–394.

Bredemeijer, G.M.M., Burg, H.C.J., Claassen, P.A.M. & Stiekema, W.J. (1991) *J. Plant Physiol.* 138, 129–135.

Breedweld, M.W., Dijkema, C., Zevenhuizen, L.P.T.M. & Zehnder, J.B. (1993) *J. Gen. Microbiol.* 139, 3157–3163.

Brenac, P., Horbowicz, M. & Obendorf, R.L. (1995) *Plant Physiol.* 108, 5–19.

Brown, R.J. & Serro, F.R. (1953) *J. Am. Chem. Soc.* 75, 1040–1042.

Bruneau, J.M., Worrell, A.C., Cambou, B., Lando, P. & Voelker, T.A. (1991) *Plant Physiol.* 96, 473–478.

Bucker, J. & Guderian, R. (1993) *Plant Physiol.* 141, 634–636.

Burch, L.R., Davies, H.V., Cuthbert, E.M., Machray, G.C., Hedley, P.H. & Waugh, R. (1992) *Phytochemistry* 31, 1901–1904.

Burch, I.R., Davies, H.V., Ross, H.A., Machray, G.C., Hedley, P. & Waugh, R. (1994) *Phytochemistry* 35, 579–582.

Burton, R.A., Bewley, J.D., Smith, A.M., Bhattacharyye, M.K., Tatge, H., Ring, S., Bull, V., Hamilton, D.V. & Martin, C. (1995) *Plant J.* 7, 3–15.

Bush, D.R. (1989) *Plant Physiol.* **89**, 1318–1323.

Bush, D. (1993) *Annu. Rev. Plant Physiol. Plant Mol. Biol.* **44**, 513–542.

Cairns, A.J. (1993) *New Phytol.* **121**, 15–24.

Cairns, A.J. & Ashton, J.E. (1993) *New Phytol.* **124**, 381–388.

Cairns, A.J. & Ashton, J.E. (1994) *New Phytol.* **126**, 3–10.

Cakmak, I., Hengeler, C. & Marshner, H. (1994) *J. Exp. Bot.* **45**, 1245–1250.

Cao, Y., Mahrenholz, A.M., DePaoli-Roach, A.A. & Roach, P.J. (1993a) *J. Biol. Chem.* **268**, 14687–14693.

Cao, Y., Skurat, A.J., DePaoli-Roach, A.A. & Roach, P.J. (1993b) *J. Biol. Chem.* **268**, 21717–21721.

Cardemil, L. & Varner, J.E. (1984) *Plant Physiol.* **76**, 1047–1054.

Carpita, N.C. & Gibeau, D.M. (1993) *Plant J.* **3**, 1–30.

Carrilo, D., Vicente-Soler, J., Fernandez, J., Soto, T., Cansado, J. & Gacto, M. (1995) *Microbiology* **141**, 679–686.

Castillo, E.M., De Lumen, B.O., Reyes, P.S. & De Lumen, H.Z. (1990) *J. Agric. Food Chem.* **38**, 361–365.

Champigny, M.L. & Foyer, C. (1992) *Plant Physiol.* **100**, 7–12.

Chatterton, N.J., Harrison, P.A., Thornley, W.R. & Bennett, J.H. (1993a) *New Phytol.* **124**, 389–396.

Chatterton, N.J., Harrison, P.A., Thornley, W.R. & Bennett, J.H. (1993b) *J. Plant Physiol.* **142**, 552–556.

Cheikh, N. & Brenner, M.L. (1992) *Plant Physiol.* **100**, 1230–1237.

Chen, J.Q. & Black, C.C. (1992) *Arch. Biochem. Biophys.* **295**, 61–69.

Cheng, C.L., Acedo, G.N., Christinsin, M. & Conkling, M.A. (1992) *Proc. Natl. Acad. Sci. U.S.A.* **89**, 1861–1864.

Chevalier, P.M. & Rupp, R.A. (1993) *Plant Physiol.* **101**, 589–594.

Chopra, S., Del-favero, J., Dolferus, R. & Jacobs, M. (1992) *Plant Mol. Biol.* **18**, 131–134.

Crafts-Brandner, S.J. (1992) *Plant Physiol.* **98**, 1133–1138.

Dali, N., Michaud, D. & Yelle, S. (1992) *Plant Physiol.* **99**, 434–438.

Dea, I.C.M. & Morrison, A. (1975) *Adv. Carbohydr. Chem. Biochem.* **35**, 341–376.

Delauney, A.J. & Verma, D.P.S. (1993) *Plant J.* **4**, 215–223.

Denyer, K. & Smith, A.M. (1992) *Planta* **186**, 609–617.

Denyer, K., Hylton, C.M. & Smith, A.M. (1994) *Am. J. Plant Physiol.* **21**, 783–789.

DeVirgilio, C., Hottiger, T., Dominguez, J., Boller, T. & Wiemken, A. (1994) *Eur. J. Biochem.* **219**, 179–186.

DeWald, D.B. & Mullet, J.E. (1994) *Plant Physiol.* **104**, 439–444.

Dey, P.M. (1978) *Adv. Carbohydr. Chem. Biochem.* **35**, 341–373.

Dey, P.M. (1980a) *Adv. Carbohydr. Chem. Biochem.* **37**, 283–372.

Dey, P.M. (1980b) *FEBS Lett.* **114**, 153–156.

Dey, P.M. (1985) In *Biochemistry of Storage Carbohydrates in Green Plants* (P.M. Dey & R.A. Dixon, eds), pp. 53–129. Academic Press, London, New York.

Dey, P.M. (1990) In *Methods in Plant Biochemistry* (P.M. Dey, ed.), pp. 189–218. Academic Press, London, New York.

Dey, P.M. & Del Campillo, E. (1984) *Adv. Enzymol. Mol. Biol.* **56**, 141–249.

Doehlert, D.C. & Huber, S.C. (1985) *Biochim. Biophys. Acta* **830**, 267–273.

Doehlert, D.C. & Felker, F.C. (1987) *Physiol. Plant* **70**, 51–57.

Doehlert, D.C. & Knutson, C.A. (1991) *J. Plant Physiol.* **138**, 566–572.

Doehlert, D.C., Kuo, T.M. & Feker, F.C. (1988) *Plant Physiol.* **86**, 1013–1019.

Drennan, P.M., Smith, M.T., Goldsworthy, D. & vanStaden, J. (1993) *Plant Physiol.* **141**, 493–496.

Dry, I., Smith, A., Edwards, A., Bhattacharyya, M., Dunn, P. & Martin, C. (1992) *Plant J.* **2**, 193–202.

Duchateau, N., Bortlik, K., Simmen, U., Wiemken, A. & Bancal, P. (1995) *Plant Physiol.* **107**, 1249–1255.

Duffus, C.M. (1992) *Trans. Biochem. Soc.* **20**, 13–18.

du Jardin, P. & Berhin, A. (1991) *Plant Mol. Biol.* **16**, 349–351.

Echeverria, E. & Humphreys, T. (1984a) *Phytochemistry* **23**, 2173–2178.

Echeverria, E. & Humphreys, T. (1984b) *Phytochemistry* **23**, 2727–2731.

Echeverria, E. & Salerno, G. (1993) *Physiol. Plant.* **88**, 434–438.

Echt, C.S. & Chourey, P.S. (1985) *Plant Physiol.* **79**, 530–536.

Edelman, J. & Jefford, T.G. (1968) *New Phytol.* **67**, 517–531.

Edwards, J. & apRees, T. (1986) *Phytochemistry* **25**, 2027–2039.

Edwards, M., Dea, I.C.M., Bulpin, P.V. & Reid, J.S.G. (1985) *Planta* **163**, 133–140.

Elbein, A.D. (1974) *Adv. Carbohydr. Chem. Biochem.* **30**, 227–256.

Emes, M.J. & Tobin, A.K. (1993) *Int. Rev. Cytol.* **145**, 149–216.

Entwistle, G. & apRees, T. (1990) *Biochem. J.* **271**, 467–472.

Feingold, D.S. & Avigad, G. (1980) In *Biochemistry of Plants* Vol. 3, (J. Preiss, ed.), pp. 101–170. Academic Press, New York.

Felix, G., Regenass, M., Spanu, P. & Boller, T. (1994) *Proc. Natl. Acad. Sci. U.S.A.* **91**, 952–956.

Fieuw, S. & Willenbrink, J. (1990) *J. Plant Physiol.* **137**, 216–223.

Fincher, G.B. (1989) *Annu. Rev. Plant Physiol. Plant Mol. Biol.* **40**, 305–346.

Flora, L.L. & Madore, M.A. (1993) *Planta* **189**, 484–490.

Flügge, U.I. & Heldt, H.W. (1991) *Annu. Rev. Plant Physiol. Plant Mol. Biol.* **42**, 129–144.

Foyer, C.H. & Ferrario, S. (1994) *Trans. Biochem. Soc.* **22**, 909–915.

Fry, S.C. (1995) *Annu. Rev. Plant Physiol. Plant Mol. Biol.* **46**, 497–520.

Fuchs, A. (1991) *Biochem. Soc. Trans.* **19**, 555–560.

Gahrtz, M., Stolz, J. & Sauer, N. (1994) *Plant J.* **6**, 697–706.

Gallet, O., Lemoine, R., Gaillerd, C., Larsson, C. & Delrot, S. (1992) *Plant Physiol.* **98**, 17–23.

Galtier, N., Foyer, C.H., Huber, J., Voelker, T.A. & Huber, S. (1993) *Plant Physiol.* **101**, 535–543.

Ganter, J.L.S.M., Zawadzki-Baggio, S.F., Leitner, S.C.S., Sierakowski, M.R. & Reicher, F. (1993) *J. Carbohydr. Chem.* **12**, 753–756.

Gardner, A., Davies, H.V. & Burch, L.R. (1992) *Plant Physiol.* **100**, 178–183.

Garegg, P.J., Oscarson, S. & Ritzen, H. (1988) *Carbohydr. Res.* **181**, 89–96.

Geigenberger, P. & Stitt, M. (1991) *Planta* **185**, 81–90.

Geigenberger, P. & Stitt, M. (1993) *Planta* **189**, 329–339.

Geigenberger, P., Langenberger, S., Wilke, I., Heineke, D., Heldt, H.W. & Stitt, M. (1993) *Planta* **190**, 446–453.

Geigenberger, P., Merlo, L., Reinholz, R. & Stitt, M. (1994) *Planta* **193**, 486–493.

Getz, H.P. (1991) *Planta* **185**, 261–268.

Getz, H.P., Grosdande, J., Kurkdjian, A., Lelievre, F., Muretzki, A. & Guern, J. (1993) *Plant Physiol.* **102**, 751–760.

Ghiena, C., Schultz, M. & Schnabl, H. (1993) *Plant Physiol.* **101**, 73–79.

Gorin, P.A.J., Baron, M. & Iacomini, M. (1988) In *Handbook of Lichenology* Vol. 3, (M. Galun, ed.), pp. 9–23. CRC Press, Boca Raton.

Greuterrr, H. & Keller, F. (1993) *Plant Physiol.* **101**, 1317–1322.

Grimes, H.D., Overvoorde, P.J., Ripp, K., Franceschi, V.R. & Hitz, W.D. (1992) *Plant Cell* **4**, 1561–1574.

Guan, H.P., Baba, T. & Preiss, J. (1994) *Plant Physiol.* **104**, 1449–1453.

Guan, H., Kuriki, T., Sivar, M. & Preiss, J. (1995) *Proc. Natl. Acad. Sci. U.S.A.* **92**, 964–967.

Guilbot, A. & Mercier, C. (1985). In 'The Polysacchrides' (G.O. Aspinall, ed.), Vol. 3, pp. 209–282. Academic Press, Orlando.

Hajirezaei, M., Sonnewald, U., Viola, R., Carlisle, S., Dennis, O. & Stitt, M. (1994) *Planta* **192**, 16–30.

Hanson, K.R. & McHale, N.A. (1988) *Plant Physiol.* **88**, 838–844.

Hargreaves, J.A. & apRees, T. (1988) *Phytochemistry* **27**, 1621–1625.

Harter, K., Talke-Messerer, C., Barz, W. & Schefer, E. (1993) *Plant J.* **4**, 507–516.

Harter, K., Frohnmeyer, H., Kircher, S., Kunkel, T., Mühlbauer, S. & Schäfer, E. (1994) *Proc. Natl. Acad. Sci. U.S.A.* **91**, 5038–5042.

Hattori, T. & Nakamura, K. (1988) *Plant Mol. Biol.* **11**, 417–426.

Hattori, T., Nakagawa, S. & Nakamura, K. (1990) *Plant Mol. Biol.* **14**, 595–604.

Hawker, J.S. (1985) In *Biochemistry of Storage Carbohydrates in Green Plants* (P.M. Dey & R.A. Dixon, eds), pp. 1–51. Academic Press, London.

Hawker, J., Smith, G.M., Phillips, H. & Wiskich, J.T. (1987) *Plant Physiol.* **84**, 1281–1285.

Hayashi, T., Koyanna, T. & Matsuda, K. (1988) *Plant Physiol.* **87**, 341–345.

Hecht, R., Slone, J.H., Buckhout, T.J., Hitz, W.D. & VanDerWoude, W.J. (1992) *Plant Physiol.* **99**, 439–444.

Hedley, P.E., Machray, G.C., Davies, H.V., Burch, L. & Wangh, R. (1993) *Plant Mol. Biol.* **22**, 917–922.

Hegnauer, R. & Grayer-Berkmeijer, R.J. (1993) *Phytochemistry* **34**, 3–16.

Heim, U., Weber, H., Baumlein, H. & Wobus, U. (1993) *Planta* **191**, 394–401.

Heineke, D., Sonnewald, U., Büssis, D., Günter, G., Leidreiter, K., Wilke, I., Raschke, K., Willmitzer, L. & Heldt, H.W. (1992) *Plant Physiol.* **100**, 301–308.

Heineke, D., Kruse, A., Flügge, U.I., Frommer, W.B., Riesmeier, J.W., Willmitzer, L. & Heldt, H.W. (1994a) *Planta* **193**, 174–186.

Heineke, D., Wildenberger, K., Sonnewald, U., Willmitzer, L. & Heldt, H.W. (1994b) *Planta* **194**, 29–33.

Henlein, M. & Starlinger, P. (1989) *Mol. Gen. Genet.* **215**, 441–446.

Hidaka, H., Hirayama, M. & Yamada, K. (1991) *J. Carbohydr. Chem.* **10**, 509–522.

Hill, S.H. & apRees, T. (1994) *Planta* **192**, 52–60.

Hitz, W.D. & Giaquinta, G.R.J. (1987) *BioAssays* **6**, 217–221.

Hitz, W.D., Card, P.J. & Ripp, K.G. (1986) *J. Biol. Chem.* **261**, 11986–11991.

Ho, I.C. (1988) *Annu. Rev. Plant Physiol.* **39**, 355–378.

Hoj, P.B. & Fincher, G.B. (1995) *Plant J.* **7**, 367–379.

Holthaus, U. & Schmitz, K. (1991) *Planta* **184**, 525–531.

Hopf, H., Gruber, G., Zinn, A. & Kandler, O. (1984b) *Planta* **162**, 282–288.

Hopf, H., Spanfelnar, M. & Kandler, O. (1984a) *Z. Pflanzenphysiol.* **114**, 485–492.

Horbowicz, M. & Obendorf, R.L. (1994) *Seed Sci. Res.* **4**, 385–405.

Hottiger, T., DeVirgilio, C., Hall, N., Boller, T. & Wiemken, A. (1994) *Eur. J. Biochem.* **219**, 187–193.

Housley, T.L. & Pollock, C.J. (1993) In *Science and Technology of Fructans* (M. Suzuki & N.J. Chatterton, eds), pp. 191–227. CRC Press, Boca Raton.

Huang, N., Stebbins, G.L. & Rodriquez, R.L. (1992) *Proc. Natl. Acad. Sci. U.S.A.* **89**, 7526–7530.

Huang, X.F., Quoc, B.N., Chourey, P.S. & Yelle, S. (1994) *Plant Physiol.* **104**, 293–294.

Hubbard, N.L., Pharr, G.M. & Huber, S.C. (1990) *Plant Physiol.* **94**, 201–208.

Huber, J.L.A. & Huber, S.C. (1992a) *Biochem. J.* **283**, 877–882.

Huber, S.C. (1989) *Plant Physiol.* **91**, 656–662.

Huber, S.C. & Huber, J.L. (1991) *Plant Cell Physiol.* **32**, 319–326.

Huber, S.C. & Hanson, K.R. (1992) *Plant Physiol.* **99**, 1449–1454.

Huber, S.C. & Huber, J.L. (1992b) *Plant Physiol.* **99**, 1275–1278.

Huber, S.C., Nielsen, T.H., Huber, J.L.A. & Pharr, D.M. (1989) *Plant Cell Physiol.* **30**, 277–285.

Huber, S.C., Huber, J.A. & Hanson, K. (1990) In *Perspectives in Biochemical and Genetic Regulation of Photosynthesis* (I. Zelitch, ed.), pp. 85–101. Wiley-Liss, New York.

Huber, S.C., Huber, J.L. & Kaiser, W.M. (1994a) *Physiol. Plant.* 92, 302–310.

Huber, S.C., Huber, J.L. & McMichael, R.W. Jr (1994b) *Int. Rev. Cytol.* 149, 47–98.

Hull, S.R., Gray, J.S.S., Koerner, T.A.W. & Montgomery, R. (1995) *Carbohydr. Res.* 266, 147–152.

Humphreys, T.E. (1988) In *Solute Transport in Plant Cells and Tissues* (D.A. Baker & J.L. Hall, eds), pp. 305–345. Longmans-Wiley, Essex.

Hylton, C. & Smith, A.M. (1992) *Plant Physiol.* 99, 1626–1635.

Iguti, A.M. & Lajolo, F.M. (1991) *J. Agr. Food Chem.* 39, 2131–2136.

Imberty, A., Buleon, A., Tran, V. & Perez, S. (1991) *Starch* 43, 375–384.

Isla, M.I., Salerno, G., Pontis, H., Vattuone, M.A. & Sampietro, A.R. (1995) *Phytochemistry* 38, 321–325.

Itoh, K., Yamaguchi, J., Huang, N., Rodriguez, R.L., Akazawa, T. & Shimamato, K. (1995) *Plant Physiol.* 107, 25–31.

James, D.W., Preiss, J. & Elbein, A.D. (1985) In *The Polysaccharides* Vol. 3, (G.O. Aspinall, ed.), pp. 107–208. Academic Press, Orlando.

Jefferson, R., Goldsbrough, A. & Bevan, M. (1990) *Plant Mol. Biol.* 14, 995–1008.

Jenner, C.F., Siwek, K. & Hawker, J.S. (1993) *Aust. J. Plant Physiol.* 20, 329–335.

Jeong, B.R. & Housley, T. L. (1992) *Plant Physiol.* 100, 199–204.

John, P. (1992) *Biosynthesis of the Major Crop Products.* John Wiley, Chichester.

Johnson, R. & Ryan, C.A. (1990) *Plant Mol. Biol.* 14, 527–536.

Joseleau, J.P., Cartier, N., Chambat, G., Faik, A. & Ruel, K. (1992) *Biochimie* 74, 81–88.

Joseleau, J.P., Chambat, G., Cortelazzo, A., Faik, A. & Ruel, K. (1994) *Trans. Biochem. Soc.* 22, 403–407.

Kaasen, I., McDougall, J. & Strom, A.R. (1994) *Gene* 145, 9–15.

Kalt-Torres, W., Kerr, P.S. & Huber, S.C. (1987) *Physiol. Plant* 70, 653–658.

Kandler, O. & Hopf, H. (1980) In *Biochemistry of Plants: A Comprehensive Treatise.* Academic Press, New York.

Kandra, L. & Wagner, G.J. (1990) *Plant Physiol.* 94, 906–912.

Kaneda, M., Kobayashi, K., Nishida, K. & Katsuta, S. (1984) *Phytochemistry* 23, 795–798.

Karrer, E.F. & Rodriguez, R.L. (1992) *Plant J.* 2, 517–523.

Kashiwada, Y., Nomaka, G. & Nishioka, I. (1988) *Phytochemistry* 27, 1469–1472.

Keller, F. (1992a) *Plant Physiol.* 99, 1251–1253.

Keller, F. (1992b) *Plant Physiol.* 98, 442–445.

Kennedy, R.A., Rumpha, M.E. & Fox, T.C. (1992) *Plant Physiol.* 100, 1–6.

Khayat, E., Harn, C. & Daie, J. (1993) *Plant Physiol.* 101, 57–64.

Kim, K.B. & Behrman, E.J. (1995) *Carbohydr. Res.* 270, 71–75.

Kim, S.R., Kim, Y., Costa, M. & An, G. (1992) *Plant Physiol.* 98, 1479–1483.

Kim, S.Y., May, G.D. & Park, W.D. (1994) *Plant Mol. Biol.* 26, 603–615.

King, R.R., Singh, R.P. & Calhoun, L.A. (1987) *Carbohydr. Res.* 166, 113–121.

King, R.R., Calhoun, L.A., Singh, R.P. & Boucher, A. (1990). *Phytochemistry* 29, 2115–2118.

Kinnback, A. & Werner, D. (1991) *Plant Sci.* 77, 47–55.

Klann, E., Yelle, S. & Bennett, A.B. (1992) *Plant Physiol.* 99, 351–353.

Klann, E.M., Chetelat, R.T. & Bennett, A.B. (1993) *Plant Physiol.* 103, 863–870.

Kleczkowski, L.A., Villand, P., Lönneborg, A., Olsen, O.A. & Lüthi, E. (1991) *Z. Naturforsch.* 46c, 605–612.

Kleczkowski, L., Villand, P., Lüthi, E., Olsen, O.A. & Preiss, J. (1993) *Plant Physiol.* 101, 175–186.

Klein, R.R., Crafts-Brandner, S.J. & Salvucci, M.E. (1993) *Planta* 190, 498–510.

Koch, K.E., Nolte, K.D., Duke, E.R., McCarty, D.R. & Avigue, W.T. (1992) *Plant Cell* 4, 59–69.

Konno, Y., Vedvick, T., Fitzmaurice, L. & Mirkov, T.E. (1993) *J. Plant Physiol.* 141, 385–392.

Kossmann, J., Visser, R.G.F., Müller-Röber, B., Willmitzer, L. & Sonnewald, U. (1991) *Mol. Gen. Genet.* 230, 39–44.

Kontos, F. & Spyropoulos, C.G. (1995) *J. Exp. Bot.* 46, 577–583.

Krapp, A., Hoffmann, B., Shafer, C. & Stitt, M. (1993) *Plant J.* 3, 817–828.

Krause, K.P. & Stitt, M. (1992) *Phytochemistry* 31, 1143–1146.

Kreis, M., Williamson, M., Buxton, B., Pywell, J., Hejgaard, J. & Svendsen, I. (1987) *Eur. J. Biochem.* 169, 517–525.

Kruger, N.J. (1990) In *Plant Physiology, Biochemistry and Molecular Biology* (D.T. Dennis & D.M. Turpin, eds), pp. 59–76. Longmans, Harlow.

Kruger, N.J. & Scott, P. (1994) *Trans. Biochem. Soc.* 22, 904–909.

Kühbach, W. & Schnyder, H. (eds) (1989) *Fructan Special Issue. J. Plant Physiol.* 134, 121–260.

Kulpers, A.G.J., Jacobsen, E. & Visser, R.G.F. (1994a) *Plant Cell* 6, 43–52.

Kulpers, A.G.J., Soppe, W.J.J., Jacobsen, E. & Visser, R.G.F. (1994b) *Plant Mol. Biol.* 26, 1759–1773.

Labuza, T.P., Reineccius, G.A., Monnier, V.M., O'Brien, J. & Baynes, J.W. (1994) *Maillard Reactions in Chemistry, Food and Health.* The Royal Society of Chemistry, London.

Lauriere, C., Doyen, C., Thevenot, C. & Daussant, J. (1992) *Plant Physiol.* 100, 887–893.

Leigh, R.A., Pope, A.J., Jennings, I.R. & Sanders, D. (1992) *Plant Physiol.* 100, 1698–1705.

Lemoine, R., Daie, J. & Wyse, R. (1988) *Plant Physiol.* 86, 575–580.

Lemoine, R., Derlot, S., Gallet, O. & Larsson, C. (1989) *Biochim. Biophys. Acta* **978**, 65–71.

Leopold, A.C., Sun, W.Q. & Bernal, -Lugo, I. (1994) *Seed Sci. Res.* **4**, 267–274.

Lewis, D.H. (1993) *New Phytol.* **124**, 583–594.

Li, B., Geiger, D.R. & Shieh, W.J. (1992) *Plant Physiol.* **99**, 1393–1399.

Li, Z.S., Gellet, O., Gallard, C., Lemoine, R. & Delrot, S. (1992b) *Biochim. Biophys. Acta* **1103**, 259–267.

Lim, P.Y., Perata, P., Pozueta-Romero, J., Akazawa, T. & Yamaguchi, J. (1992) *Biosci. Biotech. Biochem.* **56**, 695–696.

Lin, C.T., Lin, M.T., Chou, H.Y., Lee, P.D. & Su, J.C. (1995) *Plant Physiol.* **107**, 277–278.

Lin, T.P. & Huang, N.H. (1994) *J. Exp. Bot.* **45**, 1289–1294.

Lin, T.P. & Preiss, J. (1988) *Plant Physiol.* **86**, 260–265.

Lin, T.P., Caspar, T., Somerville, C. & Preiss, J. (1988) *Plant Physiol.* **86**, 1131–1135.

Liu, J.J., Odegard, W. & Lumen, B.O. (1995) *Plant Physiol.* **109**, 505–511.

Liu, J., Waterhouse, A.L. & Chatterton, N.J. (1994) *J. Carbohydr. Chem.* **13**, 859–872.

Livingston, D.F., Chatterton, N.J. & Harrison, P.A. (1993) *New Phytol.* **123**, 725–734.

Livingston, U.P., Knievel, D.P. & Gildow, F.E. (1994) *New Phytol.* **127**, 27–36.

Loewus, F.A. & Dickinson, D.B. (1982) In *Encyclopedia of Plant Physiology*, New Series Vol. 13A (F.A. Loewus & W. Tanner, eds), pp. 193–216. Springer-Verlag, Berlin, New York.

Lüscher, M. & Nelson, C.J. (1995) *Plant Physiol.* **107**, 1419–1425.

Lüscher, M., Frehner, M. & Nösberger, J. (1993) *New Phytol.* **123**, 437–442; 717–724.

Maas, C., Shaal, S. & Werr, W. (1990) *EMBO J.* **9**, 3447–3452.

Maas, C., Laufs, J., Grant, S., Korfhage, C. & Werr, W. (1991) *Plant Mol. Biol.* **16**, 199–207.

MacGregor, A. (1993) *Starch* **45**, 232–237.

Madore, M.A. (1991) In *Recent Advances in Phloem Transport and Assimilate Compartmentation* (J.L. Bonnemain, S. Delrot, W.J. Lucas & J. Dainty, eds), pp. 23–26. Auest Editions, Nantes.

Manners, D.J. (1985) *Cereal Foods World* **30**, 461–467.

Manners, D.J. (1989) *Carbohydr. Polymers* **11**, 87–112.

Manners, D.J. & Sturgeon, R.J. (1982) In *Encyclopedia of Plant Physiology* (F.A. Loewus & W. Tanner, eds), New Series, Vol. 13A, pp. 472–514. Springer Verlag, Berlin.

Marana, C., Garcia-Olmedo, F. & Carbonero, P. (1988) *Gene* **63**, 253–260.

Marci, L.I., MacGregor, A.W., Schroeder, S.W. & Bazin, S.L. (1993) *J. Cereal Sci.* **18**, 103–106.

Maretzki, A. & Thom, M. (1987) *Plant Physiol.* **83**, 235–237.

Martin, F., Ramstedt, M., Soderhall, K. & Canet, D. (1988) *Plant Physiol.* **86**, 935–940.

Matheson, N.K. (1990) In *Methods in Plant Biochemistry* Vol. 2, (P.M. Dey & J.B. Harborne, eds), pp. 371–413. Academic Press, London.

Matsuzaki, T., Shinozaki, Y., Suhara, S., Tobita, T., Shigematsu, H. & Koiwai, A. (1991) *Agr. Biol. Chem.* **55**, 1417–1419.

McBride, M.J. & Ensign, J.C. (1990) *J. Bacteriol.* **172**, 3637–3643.

McCarty, M., Shaw, J.R. & Hannah, L.C. (1986) *Proc. Natl. Acad. Sci. U.S.A.* **83**, 9099–9103.

McElfresh, K.C. & Chourey, P.S. (1988) *Plant Physiol.* **87**, 542–546.

Meier, H. & Reid, J.S.G. (1982) In *Encyclopedia of Plant Physiology* (F.A. Loewus & W. Tanner, eds), New Series, Vol. 13A, pp. 418–471. Springer Verlag, Berlin.

Meins, Jr F., Neuhaus, J.M., Sperisen, C. & Ryals, J. (1992) In *Genes Involved in Plant Defense* (T. Boller & F. Meins, Jr, eds), pp. 245–282. Springer, Berlin.

Mellor, R.B. (1992) *Symbiosis* **12**, 113–129.

Merlo, L., Geigenberger, P., Hajirezaei, M. & Stitt, M. (1993) *J. Plant Physiol.* **142**, 392–402.

Miller, M.E. & Chourey, P.S. (1992) *Plant Cell* **4**, 297–305.

Minamikawa, T., Yamanuchi, D., Wade, S. & Takeuchi, H. (1992) *Plant Cell Physiol.* **33**, 253–258.

Miron, D. & Schaffer, A.A. (1991) *Plant Physiol.* **95**, 623–627.

Mitchell, D.E., Gadus, V. & Madore, M.A. (1992) *Plant Physiol.* **99**, 959–965.

Mittenbühler, K. & Holzer, H. (1991) *Arch. Microbiol.* **155**, 217–220.

Mizuno, K., Kawasaki, T., Shimada, H., Satoh, H., Kobayashi, E., Okumura, S., Arai, Y. & Baba, J. (1993) *J. Biol. Chem.* **268**, 19084–19091.

Monroe, J.D., Salminen, M.D. & Preiss, J. (1991) *Plant Physiol.* **97**, 1599–1601.

Morallejo, M., Garcia-Casado, G., Sanchez-Monge, R., Lopez-Otin, C., Molina-Cano, J.L., Romagosa, I. & Salcedo, C. (1993) *J. Cereal Sci.* **17**, 107–113.

Morell, S. & apRees, T. (1986) *Phytochemistry* **25**, 1579–1585.

Morell, M. & Copeland, L. (1985) *Plant Physiol.* **78**, 149–154.

Moreno, S., Cardini, C.E. & Tandecarz, J.S. (1987) *Eur. J. Biochem.* **182**, 609–614.

Moriguchi, T. & Yamaki, S. (1988) *Plant Cell Physiol.* **29**, 1361–1366.

Morrison, W.R. & Karkalas, J. (1990) In *Methods in Plant Biochemistry*, Vol. 2 (P.M. Dey & J.B. Harborne, eds), pp. 323–352. Academic Press, London.

Mu, C., Harn, C., Ko, Y.T., Singletary, G.W., Keeling, P.L. & Wasserman, B.P. (1994) *Plant J.* **6**, 151–159.

Müller, J., Staehelin, C., Mellor, K.B., Boller, T. & Wiemken, A. (1992) *J. Plant Physiol.* **140**, 8–13.

Müller, J., Xie, Z.P., Mellor, R.B., Boller, T. & Wiemken, A. (1994a) *Physiol. Plant* **90**, 86–92.

Müller, J., Xie, Z.P., Staehelin, C., Boller, T. & Wiemken, A. (1994b) *J. Plant Physiol.* **143**, 153–160.

Müller-Röber, B., Sonnewald, U. & Willmitzer, L. (1992) *EMBO J.* **11**, 1229–1238.

Müller-Röber, B., LeCognata, U., Sonnewald, U. & Willmitzer, L. (1994) *Plant Cell* 6, 601–612.

Mullins, J.T. (1990) *Physiol. Plant* 78, 309–314.

Nakamura, Y. & Kawaguchi, K. (1992) *Physiol. Plant* 85, 336–342.

Nakamura, Y. & Yamanuchi, H. (1992) *Plant Physiol.* 99, 1265–1266.

Nakamura, Y., Takeichi, T., Kawaguchi, K. & Yamanuchi, H. (1992) *Physiol. Plant.* 84, 329–335.

Nakano, K., Murakami, K., Takaishi, Y. & Tomimatsu, T. (1986). *Chem. Pharm. Bull.* 34, 5005–5010.

Nakano, K., Murakami, K., Takaishi, Y. & Tomimatsu, T. (1988) *Chem. Pharm. Bull.* 34, 5003–5010.

Nakano, K., Omura, M. & Fukui, T. (1989) *J. Biochem.* 106, 528–532.

Nakata, P.A. & Okita, T.W. (1995) *Plant Physiol.* 108, 361–368.

Nakayama, S. & Nakamura, Y. (1994) *Physiol. Plant.* 91, 763–769.

Nandra, K.S. & Bhatia, I.S. (1980) *Phytochemistry* 19, 965–966.

Nelson, C.V. & Smith, D. (1986) *Curr. Top. Plant Biochem. Physiol.* 5, 1–16.

Nelson, O. & Pan, D. (1995) *Annu. Rev. Plant Physiol. Plant Mol. Biol.* 46, 475–496.

Neuhaus, H.E., Quick, W.P., Siegel, G. & Stitt, M. (1990) *Planta* 181, 583–592.

Neuhaus, H.E., Heinrichs, G. & Scheibe, R. (1993) *Plant Physiol.* 101, 573–578.

Neves, M., Jorge, J.A., Francois, J.M. & Terenzi, H.F. (1991) *FEBS Lett.* 283, 19–22.

Nguyen-Quoc, B., Krivitzky, M., Huber, S.C. & Lecharny, A. (1990) *Plant Physiol.* 94, 516–523.

Nichols, M.B., Bancal, M.O., Foley, M.E. & Volenee, J. (1993) *Plant Physiol.* 88, 225–228.

Nielsen, T.H., Wischmann, B., Enevoldsen, K. & Møller, B.L. (1994) *Plant Physiol.* 105, 111–117.

Niemietz, C. & Hawker, J.S. (1988) *Aust. J. Plant Physiol.* 15, 359–366.

Nolte, K.D. & Koch, K. (1993) *Plant Physiol.* 101, 899–905.

Nonogaki, H., Nomaguchi, M. & Morobashi, Y. (1995) *Physiol. Plant.* 94, 328–334.

Nwaka, S., Mechler, B., Destruelle, M. & Holzer, H. (1995) *FEBS Lett.* 360, 286–290.

Obenland, D.M., Simmen, U., Boller, T. & Wiemken, A. (1991). *Plant Physiol.* 97, 811–813.

Obenland, D.M., Simmen, U., Boller, T. & Wiemken, A. (1993) *Plant Physiol.* 101, 1331–1339.

Ohyama, A., Hirai, M. & Nishimura, S. (1992) *Jpn. J. Genet.* 67, 491–492.

Ohyama, A., Ito, H., Sato, T., Nishimura, S., Imai, T. & Hirai, M. (1995) *Plant Cell Physiol.* 36, 369–376.

Okita, T.W. (1992) *Plant Physiol.* 100, 560–564.

Ong, M.H., Jumel, K., Tokarczuk, P.F., Blanshard, J.M.V. & Harding, J.E. (1994) *Carbohydr. Res.* 260, 99–117.

Otto, S. & Feierabend (1989) *Physiol. Plant.* 76, 65–73.

Overvoorde, P.J. & Grimes, H.D. (1994) *J. Biol. Chem.* 269, 15154–15161.

Pahl, H.L. & Baeuerie, P.A. (1994) *BioEssays* 16, 497–502.

Painter, T.J. (1983) In *The Polysaccharides*, Vol. 2, (G.O. Aspinall, ed.), pp. 195–285. Academic Press, New York.

Paul, M., Sonnewald, U., Hajirezaei, M., Dennis, D. & Stitt, M. (1995) *Planta* 196, 277–283.

Penson, S.P. & Cairns, A.J. (1994) *New Phytol.* 128, 395–402.

Pollock, C.J. & Cairns, A.J. (1992) *Annu. Rev. Plant Physiol. Mol. Biol.* 42, 77–101.

Pollock, C.J. & Chatterton, N.J. (1988) In *The Biochemistry of Plants*, Vol. 14 (J. Preiss, ed.), pp. 109–139. Academic Press, San Diego.

Pontis, H.G. (1977) In *International Review of Biochemistry*, Vol. 13 (D.H. Northcote, ed.), pp. 79–117. Butterworth, London.

Pontis, H.G. (1989) *J. Plant Physiol.* 134, 148–150.

Pozueta-Romero, J., Ardila, F. & Akazawa, T. (1991) *Plant Physiol.* 97, 1565–1572.

Preiss, J. (1988) In *The Biochemistry of Plants*, Vol. 14 (J. Preiss, ed.), pp. 181–254. Academic Press, San Diego.

Preiss, J. (1991a). In *Biology and Molecular Biology of Starch Synthesis and Regulation* (B.J. Miflin, ed.), Oxford Surveys of Plant Molecular and Cellular Biology, Vol. 7, pp. 59–114. Oxford University Press, Oxford.

Preiss, J. (1991b) *Plant Mol. Cell Biol.* 7, 59–114.

Preiss, J. & Levi, C. (1980) In *The Biochemistry of Plants* Vol. 3 (J. Preiss, ed.), pp. 371–423. Academic Press, New York.

Preiss, J., Bell, K., Smith-White, B., Iglesies, A., Kakefuda, G. & Li, L. (1991) *Trans. Biochem. Soc.* 19, 539–547.

Preisser, J. & Komor, E. (1988) *Plant Physiol.* 88, 259–265.

Preisser, J. & Komor, E. (1991) *Planta* 186, 109–114.

Pressey, R. (1994) *Phytochemistry* 36, 543–546.

Quy, L.V. & Champigny, M.L. (1992) *Plant Physiol.* 99, 344–347.

Rea, P.A., Kim, Y., Sarafian, V., Poole, R.J., Davies, J.M. & Sanders, D. (1992) *Trends Biochem. Sci.* 17, 348–352.

Reid, J.S.G. (1985) In *Biochemistry of Storage Carbohydrates in Green Plants* (P.M. Dey & R.A. Dixon, eds), pp. 265–288. Academic Press, London.

Reid, J.S.G., Edwards, M.E., Gidley, M.J. & Clark, A.H. (1992) *Trans. Biochem. Soc.* 20, 23–26.

Reimholz, R., Geigenberger, P. & Stitt, M. (1994) *Planta* 192, 480–488.

Renz, A., Merlo, L. & Stitt, M. (1993) *Planta* 190, 156–165.

Ricard, B., Rivoal, J., Spiter, A. & Pradet, A. (1991) *Plant Physiol.* 95, 669–674.

Riesmeier, J.W., Willmitzer, L. & Frommer, W.B. (1992) *EMBO J.* 11, 4705–4713.

Riesmeier, J.W., Hirner, B. & Frommer, W.B. (1993a) *The Plant Cell* 5, 1591–1590.

Riesmeier, J.W., Flügge, U.I., Schultz, B., Heineke, D., Heldt, H.W., Willmitzer, L. & Frommer, W.B. (1993b) *Proc. Natl. Acad. Sci. U.S.A.* 90, 6160–6164.

Roby, C., Martin, J.B., Bligny, R. & Donce, R. (1987) *J. Biol. Chem.* 262, 5000–5007.

Roden, L., Ananth, S., Campbell, P., Manzella, S. & Meezan, E. (1994) *J. Biol. Chem.* **269**, 11509–11513.

Ross, H.A. & Davies, H.V. (1992) *Plant Physiol.* **100**, 1008–1013.

Rychter, A.M. & Randall, D.D. (1994) *Physiol. Plant.* **91**, 383–388.

Ryder, M.H., Tate, M.E. & Jones, G.P. (1984) *J. Biol. Chem.* **259**, 4704–4710.

Sadka, A., DeWald, D.B., May, G.D., Park, W.D. & Mullet, J.E. (1994) *Plant Cell* **6**, 737–749.

Sakata, K., Yamaguchi, H., Yagi, A. & Ina, K. (1989) *Agr. Biol. Chem.* **53**, 2975–2979.

Salanoubat, M. & Balliard, G. (1987) *Gene* **60**, 47–56.

Salminen, S.O. & Streeter, J.E. (1986) *Plant Physiol.* **81**, 538–541.

Salvucci, M.E. & Klein, R.R. (1993) *Plant Physiol.* **102**, 529–536.

Santino, A., Daminati, M.G., Vitale, A. & Bollini, R. (1992) *Physiol. Plant.* **85**, 425–432.

Sanwo, M.M. & DeMasson, D.A. (1992) *Plant Physiol.* **99**, 1184–1192.

Sauer, N. & Tanner, W. (1993) *Bot. Acta* **106**, 277–286.

Sauer, N. & Stolz, J. (1994) *Plant J.* **6**, 67–77.

Sauer, N., Baier, K., Gahrz, R., Stolz, J. & Truernit, E. (1994) *Plant Mol. Biol.* **26**, 1671–1679.

Savka, M.A. & Farrand, S.K. (1992) *Plant Physiol.* **90**, 784–789.

Schaffer, A.A. (1986) *Phytochemistry* **25**, 2275–2277.

Schnyder, H. (1993) *New Phytol.* **123**, 233–245.

Schubert, A., Wyss, P. & Wiemken, A. (1992) *J. Plant Physiol.* **140**, 41–45.

Scott, P., Lange, A.J., Pilkis, S.J. & Kruger, N.J. (1995) *Plant J.* **7**, 461–469.

Sebkova, V., Unger, C., Hardegger, M. & Sturm, A. (1995) *Plant Physiol.* **108**, 75–83.

Sechley, K.A., Yamaga, T. & Oaks, A. (1992) *Int. Rev. Cytol.* **134**, 85–163.

Shachar-Hill, Y., Pfeffer, P.E., Douds, D., Osman, S.F., Doner, L.W. & Ratcliff, R.G. (1995) *Plant Physiol.* **108**, 7–15.

Shaw, J.R. & Hannah, L.C. (1992) *Plant Physiol.* **98**, 1214–1216.

Sheen, J. (1990) *Plant Cell* **2**, 1027–1038.

Shewmaker, C.K., Boyer, C.D., Wiesenborn, D.P., Thompson, D.B., Boersig, M.R., Oaks, J.V. & Stalker, D.M. (1994) *Plant Physiol.* **104**, 1159–1166.

Shiomi, N. (1989) *J. Plant Physiol.* **134**, 151–155.

Shiomi, N. (1992) *New Phytol.* **122**, 421–432.

Shiomi, N. (1993) *New Phytol.* **123**, 265–270.

Sicher, R.C. & Kremer, D.F. (1992) *Physiol. Plant.* **85**, 446–452.

Siegel, G. & Stitt, M. (1990) *Plant Sci.* **66**, 205–210.

Siegel, G., Mackintosh, C. & Stitt, M. (1990) *FEBS Lett.* **270**, 198–202.

Simmen, V., Oberland, D., Boller, T. & Wiemken, A. (1993) *Plant Physiol.* **101**, 459–468.

Simpson, R.S. & Bonnett, G.D. (1993) *New Phytol.* **123**, 455–469.

Sims, I.M., Horgan, R. and Pollock, C.J. (1993) *New Phytol.* **123**, 25–29.

Singh, B.K. & Preiss, J. (1985) *Plant Physiol.* **78**, 34–40.

Singh, D.G., Lomako, J., Lomako, W.M., Whelan, W.J., Meyer, H.E., Serwe, M. & Metzger, J.W. (1995) *FEBS Lett.* **376**, 61–64.

Sissons, M.J., Lance, R.C.M. & Sparrow, D.H.B. (1992) *J. Cereal Sci.* **16**, 107–128.

Sivak, M.N., Wagner, M. & Preiss, J. (1993) *Plant Physiol.* **103**, 1355–1358.

Skurat, A.V., Cao, Y. & Roach, P.J. (1993) *J. Biol. Chem.* **268**, 14701–14707.

Smeekens, S., Angenene, G., Ebskamp, E. & Weisbeek, P. (1991) *Trans. Biochem. Soc.* **19**, 565–569.

Smith, A.M. & Denyer, K. (1992) *New Phytol.* **122**, 21–33.

Smith, A.M., Bettey, M. & Bedford, I.D. (1989) *Plant Physiol.* **89**, 1279–1284.

Smith-White, B. & Preiss, J. (1992) *J. Mol. Evol.* **34**, 449–464.

Smouter, H. & Simpson, R.J. (1989) *New Phytol.* **111**, 359–368.

Smythe, C. & Cohen, P. (1989) *Eur. J. Biochem.* **200**, 625–631.

Søgaard, M., Abe, J., Martin-Eauclaire, M.F. & Svensson, B. (1993) *Carbohydr. Polymers* **21**, 137–146.

Sonnewald, U., Brauer, M., von Schaewen, A., Stitt, M. & Willmitzer, L. (1991) *Plant J.* **1**, 95–106.

Sonnewald, U., Basner, A., Greve, B. & Sterys, M. (1995) *Plant Mol. Biol.* **27**, 567–576.

Sowokinos, J.R. & Varus, J. L. (1992) *J. Plant Physiol.* **139**, 672–679.

Springer, B., Werr, W., Starlinger, P., Bennett, D.C., Zokolica, M. & Freeling, M. (1986) *Mol. Gen. Genet.* **205**, 461–648.

Spychalla, J.P., Scheffler, B.E., Sawakinos, J.R. & Beven, M.W. (1994) *J. Plant Physiol.* **144**, 444–453.

Stark, D.M., Timmerman, K.P., Barry, G.F., Preiss, J. & Kishore, G.M. (1992) *Science* **258**, 287–292.

Stark, J.R. & Lynn, A. (1992) *Trans. Biochem. Soc.* **20**, 7–12.

Steadman, K.J., Pritchard, H.W. & Dey, P.M. (1996) *Ann. Bot.* **77**, 667–674.

Stephen, A.M. (1983) In *The Polysaccharides*, Vol. 2 (G.O. Aspinall, ed.), pp. 97–193. Academic Press, New York.

Steup, M. (1988) In *The Biochemistry of Plants*, Vol. 14 (J. Preiss, ed.), pp. 255–296. Academic Press, San Diego.

Steup, M. (1990) In *Methods in Plant Biochemistry*, Vol. 3 (P.J. Lea, ed.), pp. 103–128. Academic Press, London.

Sticher, L. & Jones, R.L. (1992) *Plant Physiol.* **98**, 1080–1086.

Stitt, M. (1990) *Annu. Rev. Plant Physiol. Plant Mol. Biol.* **41**, 153–185.

Stitt, M., Wilke, I., Feil, R. & Heldt, H.W. (1988) *Planta* **174**, 217–230.

Stitt, M., von Schaewen, A. & Willmitzer, L. (1990) *Planta* **183**, 40–50.

Stommel, J.R. (1992) *Plant Physiol.* **99**, 324–328.

Strom, A.R. & Kaasen, I. (1993) *Mol. Microbiol.* **8**, 205–210.

Sturm, A. & Chrispeels, M.J. (1990) *Plant Cell* **2**, 1107–1119.

Sun, J., Loboda, T., Sung, S.J.S. & Black, C.C. (1992) *Plant Physiol.* **98**, 1163–1169.

Sun, W.Q. & Leopold, A.C. (1995) *Physiol. Plant.* **94**, 94–104.

Sun, Z., Duke, S.H. & Henson, C.A. (1995) *Plant Physiol.* **108**, 211–217.

Sur, I.P., Lobo, Z. & Maitra, P.K. (1994) *Yeast* **10**, 199–209.

Suzuki, M. & Chatterton, N.J. (1993) *Science and Technology of Fructose*, CRC Press, Boca Raton.

Svensson, B. & Søgaard, M. (1992) *Trans. Biochem. Soc.* **20**, 34–42.

Tacke, M., Yang, Y. & Sterys, M. (1991) *Planta* **185**, 220–226.

Taira, T., Uematsu, M., Nakano, Y. & Morikawa, T. (1991) *Biochem. Genet.* **29**, 301–311.

Takaha, T., Yanase, M., Okada, S. & Smith, S.M. (1993) *J. Biol. Chem.* **268**, 1391–1396.

Takeda, Y. & Preiss, J. (1993) *Carbohydr. Res.* **240**, 265–275.

Takeda, Y., Guan, H.P. & Preiss, J. (1993) *Carbohydr. Res.* **240**, 253–263.

Takeda, S., Kowyama, Y., Takeuchi, Y., Matsuoka, K., Nishimura, M. & Nakamura, K. (1995) *Plant Cell Physiol.* **36**, 321–333.

Tetlow, I.J., Blissett, K.J. & Emes, M.J. (1994) *Planta* **194**, 454–460.

Theodorou, M.E., Cornel, F.A. Duff, S.M.G. & Flaxton, W.C. (1992) *J. Biol. Chem.* **267**, 21901–21905.

Theodorou, M.E. & Plaxton, W.C. (1993) *Plant Physiol.* **101**, 339–344.

Thevelein, J.M. (1988) *Exp. Mycol.* **12**, 1–12.

Thevelein, J.M. & Hohmann, S. (1995) *Trends Biochem. Sci.* **20**, 3–10.

Thom, M. & Maretzki, A. (1985) *Proc. Natl. Acad. Sci. U.S.A.* **82**, 4697–4701.

Thom, M. & Maretzki, A. (1992) *J. Plant Physiol.* **139**, 555–559.

Thomas, B.R. & Rodriguez, R.L. (1994) *Plant Physiol.* **106**, 1235–1239.

Tomlinson, P.T., Duke, E.R., Nolte, K.D. & Koch, K.E. (1991) *Plant Physiol.* **97**, 1249–1252.

Tsukaya, H., Ohshima, T., Naito, S., China, M. & Komeda, Y. (1991) *Plant Physiol.* **97**, 1414–1421.

Turgeon, R. & Beebe, D.U. (1991) *Plant Physiol.* **96**, 349–354.

Turgeon, R. & Gowan, E. (1990) *Plant Physiol.* **94**, 1244–1249.

Turgeon, R. & Gowan, E. (1992) *Planta* **187**, 388–394.

Turgeon, R., Beebe, D.U. & Gowan, E. (1993) *Plant Physiol.* **191**, 446–456.

Tyson, R.H. & apRees, T. (1989) *Plant Sci.* **59**, 71–76.

Tyynela, J. & Schulman, A.H. (1993) *Physiol. Plant.* **89**, 835–841.

Tyynela, J., Stitt, M. & Lonneborg, A., Smeckens, S. & Schulman, A.H. (1995) *Physiol. Plant.* **93**, 77–84.

Unger, C., Hofsteenge, J. & Sturm, A. (1992) *Eur. J. Biochem.* **204**, 915–921.

Unger, C., Hardegger, M., Lienhard, S. & Sturm, K. (1994) *Plant Physiol.* **104**, 1351–1357.

Usuda, H. & Shimogawera, K. (1991) *Cell Physiol.* **32**, 499–504.

Usuda, H. & Shimogawera, K. (1993) *Plant Cell Physiol.* **34**, 767–770.

van Berkel, J., Conrads-Strauch, J. & Steup, M. (1991) *Planta* **185**, 432–439.

Van den Ende, W. & Van Laere, A. (1995) *Physiol. Plant.* **93**, 291–298.

Van den Ende, W. & Van Laere, A. (1993) *New Phytol.* **123**, 31–37.

Van der Meer, I.M., Ebskamp, J.M., Visser, R.G.F., Weisbeek, P.J. & Smeekens, S.C.M. (1994) *Plant Cell* **6**, 561–570.

Van Dijck, P., Colavizza, D., Smet, P. & Thevelein, J.M. (1995) *Appl. Env. Microbiol.* **61**, 109–115.

Van Laere, A. (1989) *FEMS Microbiol. Rev.* **63**, 201–210.

Villand, P., Olsen, O.A., Kilian, A. & Kleczkowski, L.A. (1992) *Plant Physiol.* **100**, 1617–1618.

Viola, R., Davies, H.V. & Chuteck, A.R. (1991) *Planta* **183**, 203–208.

Visser, R.G.F., Samhorst, I., Kuipersd, G.J., Ronys, N.J., Feenstra, W.J. & Jacobsen, E. (1991) *Mol. Gen. Genet.* **225**, 289–296.

Vogeli-Lange, R., Frundt, C., Hart, C.M., Beffa, R., Nagy, F. & Meins, F. (1994) *Plant J.* **5**, 273–278.

Wagner, K.G. & Backer, A.I. (1992) *Int. Rev. Cytol.* **134**, 1–84.

Wahlberg, I., Walsh, E.B., Fossblom, I., Oscarson, S., Enzell, C.R., Ryhage, R. & Isaksson, R. (1986) *Acta Chem. Scand.* **1340**, 724–730.

Walker, J.L. & Huber, S.C. (1989a) *Plant Physiol.* **89**, 518–524.

Walker, J.L. & Huber, S.C. (1989b) *Planta* **177**, 116–120.

Wang, A.Y., Yu, W.P., Juang, R.H., Huang, J.W., Sung, H.Y. & Su, J.C. (1992) *Plant Mol. Biol.* **18**, 1191–1194.

Wang, M.B., Boulter, D. & Gatehouse, A. (1992). *Plant. Mol. Biol.* **19**, 881–885.

Wang, F., Sanz, A., Brenner, M.L. & Smith, A. (1993a) *Plant Physiol.* **101**, 321–327.

Wang, F., Smith, A.G. & Brenner, M.L. (1993b) *Plant Physiol.* **103**, 1463–1464.

Wang, F., Smith, A.G. & Brenner, M.L. (1994) *Plant Physiol.* **104**, 535–540.

Wang, Z.Y., Zheng, F.G., Shan, G.Z., Gao, J.P., Snusted, P., Li, M.G., Zhang, J.L. & Hong, M.M. (1995) *Plant J.* **7**, 613–622.

Weil, M. & Rausch, T. (1990) *Plant Physiol.* **94**, 1575–1581.

Weil, M., Kransgrill, S., Schuster, A. & Rausch, T. (1994) *Planta* **193**, 438–445.

Weiner, H., McMichael, R.W. & Huber, S.C. (1992) *Plant Physiol.* **99**, 1435–1442.

Weiner, H., Huber, S.C., Weiner, H. & Stitt, M. (1993) *Curr. Top. Plant Biochem. Physiol.* **12**, 111–112.

Wenzler, H., Migneri, G., Fisher, L. & Park, W. (1989) *Plant Mol. Biol.* **13**, 347–354.

Werr, W., Frommer, W.B., Maas, C. & Starlinger, P. (1985) *EMBO J.* **4**, 1375–1380.

Whelan, W.J. (1986) *Bioessays* **5**, 136–140.

Wiemken, A. (1990) *Antonie van Leeuwenhoek* **58**, 209–217.

Wiemken, V. & Incichen, K. (1993) *Planta* **191**, 387–392.

Williams, J.H.H., Winters, A.L. & Farrar, J.F. (1992) In *Molecular Biochemical and Physiological Aspects of Plant Respiration* (H. Lambers & I.H.W. Van der Plas, eds), pp. 1–7. Academic Press, The Hague.

Winter, H., Robinson, D.G. & Heldt, H.W. (1994) *Planta* **193**, 530–535.

Winters, A.L., Williams, J.H.H., Thomas, D.S. & Pollock, C.J. (1994) *New Phytol.* **128**, 591–600.

Witt, H.J. (1989) *J. Plant Physiol.* **135**, 99–104.

Woodson, W.R. & Wang, H. (1987) *Physiol. Plant.* **71**, 224–228.

Worrell, A.C., Bruneau, J-M., Summerfeld, K., Boersig, M. & Voelker, T.A. (1991) *Plant Cell* **3**, 1121–1130.

Xue, Z., Larsen, K. & Jochimsen, B.U. (1991) *Plant Mol. Biol.* **16**, 899–906.

Yamanuchi, H. & Nakamura, Y. (1993) *Plant Cell Physiol.* **33**, 985–991.

Yang, J.C. & Sheen, J. (1994) *Plant Cell* **6**, 1665–1697.

Yelle, S., Chetelat, R.T., Dorais, M., DeVerna, J.W. & Bennett, A.B. (1991) *Plant Physiol.* **95**, 1025–1035.

Yoshinari, K., Sashida, Y., Mimaki, Y. & Shimomura, H. (1990) *Chem. Pharm. Bull.* **38**, 415–417.

Yu, S. & Pedersen, M. (1991) *Physiol. Plant.* **81**, 149–155.

Yu, S. & Pedersen, M. (1993) *Planta* **191**, 137–142.

Yu, S., Kenne, I. & Pedersen, M. (1993) *Biochim. Biophys. Acta* **1156**, 313–320.

Yu, S., Marcussen, J. & Pedersen, M. (1994) *Planta* **193**, 307–311.

Yu, S.M., Kuo, Y.H., Shen, G., Shen, Y.G. & Liu, L.F. (1991) *J. Biol. Chem.* **266**, 21131–21137.

Yu, W.P., Wang, A.Y., Juang, R.H., Jung, H.Y. & Su, J.C. (1992) *Plant Mol. Biol.* **18**, 139–142.

Zammit, A. & Copeland, L. (1993) *Aust. J. Plant Physiol.* **20**, 25–32.

Ziegler, P., Daussant, J. & Loos, K. (1994) *J. Exp. Bot.* **48**, 1147–1155.

Zrenner, R., Salanoubat, M., Willmitzer, L. & Sonnewald, U. (1995) *Plant J.* **7**, 97–107.

FURTHER READING

Aspinall, G.O. (ed.) (1985) *The Polysaccharides*, Vols. 2 and 3. Academic Press, Orlando.

Baker, D.A. & Hall, J.L. (eds) (1988) *Solute Transport in Plant Cells and Tissues*. Longmans-Wiley, Essex.

Baker, D.A. & Milburn, J.A. (1989) *Transport of Photoassimilates*. Wiley, New York.

Bewley, D.J. & Black, M. (1994) *Seeds: Physiology of Development and Germination*. Plenum Press, New York.

Bonnemain, J.L., Delrot, S., Lucas, W.J. & Dainty, J. (1991) *Recent Advances in Phloem Transport and Assimilate Compartmentation*. Auest Editions, Nantes.

Bush, D.R. (1993) Proton coupled sugar and amino acid transporters in plants. *Annu. Rev. Plant Physiol. Plant Mol. Biol.* **44**, 513–554.

Catley, B.J. (ed.) (1992) Biochemistry of plant polysaccharides. *Trans. Biochem. Soc.* **20**, 7–53.

Dennis, D.T. & Turpin, D.M. (eds) (1990) *Plant Physiology, Biochemistry and Molecular Biology*. Longmans, Harlow.

Dey, P.M. & Dixon, R.A. (eds) (1985) *Biochemistry of Storage Carbohydrates in Green Plants*. Academic Press, London.

Dey, P.M. & Harborne, J.B. (eds) (1990) *Methods in Plant Biochemistry*, Vol. 2, Carbohydrates. Academic Press, London.

Emes, M.J. & Tobin, A.K. (1993) Control of metabolism and development in higher plant plastids. *Int. Rev. Cytol.* **145**, 149–216.

Flügge, U.I. & Heldt, H.W. (1991) Metabolite translocators of the chloroplast envelope. *Annu. Rev. Plant Physiol. Mol. Biol.* **42**, 129–144.

Fry, S.C. (1995) Polysaccharide modifying enzymes in the plant cell wall. *Annu. Rev. Plant Physiol. Mol. Biol.* **46**, 497–520.

Geiger, D.R. & Servaites, J.C. (1994) Diurnal regulation of photosynthetic carbon metabolism in C_3 plants. *Annu. Rev. Plant Physiol. Mol. Biol.* **45**, 235–256.

Huber, S.C., Huber, J.L. & McMichael, R.W. Jr (1994) Control of plant enzyme activity by reversible protein phosphorylation. *Int. Rev. Cytol.* **149**, 47–98.

John, P. (ed.) (1991) Redesigning crop products for biotechnology: Starch and fructan. *Trans. Biochem. Soc.* **19**, 539–572.

John, P. (1992) *Biosynthesis of the Major Crop Products*. Wiley, Chichester.

Kandler, O. & Hopf, H. (1982) In *Encyclopedia of Plant Physiology*. New Series (F.A. Loewus & W. Tanner, eds), Vol. 13A, pp. 348–383. Springer-Verlag, Berlin, New York.

Keller, F. (1989) *J. Plant Physiol.* **134**, 141–147.

Lewis, D.H. (1984) *Storage Carbohydrate in Vascular Plants*. Cambridge University Press, London.

Loewus, F.A. & Tanner, W. (eds) (1982) *Encyclopedia of Plant Physiology, Carbohydrates*, New Series, Vol. 13A. Springer Verlag, Berlin.

Miflin, B.J. (ed.) (1991) *Biology and Molecular Biology of Starch Synthesis and Regulation*. Oxford University Press, Oxford.

Nelson, O. & Pan, D. (1995) Starch synthesis in maize endosperm. *Annu. Rev. Plant Physiol. Mol. Biol.* **46**, 475–496.

Pollock, C.J., Farr, J.F. & Gordon, A.J. (1992) *Carbon Partitioning Within and Between Organisms*. Bios Scientific Publishers, Oxford.

Preiss, J. (ed.) (1980, 1988) *The Biochemistry of Plants, Carbohydrates*, Vols. 3 and 14 (P.K. Sumpf & E.E. Conn, chief eds), Academic Press, New York.

Rea, P.A. & Poole, R.J. (1993) Vacuolar H^+-translocating pyrophosphatases. *Annu. Rev. Plant Physiol. Mol. Biol.* **44**, 157–180.

Stitt, M. (1990) Fructose 2,6-bisphosphate as a regulatory molecule in plants. *Annu. Rev. Plant Physiol. Mol. Biol.* **41**, 153–185.

Stitt, M. & Sonnewald, U. (1995) Regulation of metabolism in transgenic plants. *Annu. Rev. Plant Physiol. Mol. Biol.* **46**, 341–368.

Suzuki, M. & Chatterton, N.J. (eds) (1993) *Science and Technology of Fructans*. CRC Press, Boca Raton.

Wagner, K.G. & Backer, A.I. (1992) Dynamics of nucleotides in plants studied on a cellular basis. *Int. Rev. Cytol.* **134**, 1–84.

Zelitch, I. (ed.) (1990) *Perspectives in Biochemical and Genetic Regulation of Photosynthesis*. Wiley-Liss, New York.

5 Carbohydrate Metabolism: Structural Carbohydrates

J.S. Grant Reid

5.1 INTRODUCTION

Apart from water, structural carbohydrates are the main chemical constituents of most plant tissues and most plant cells. This is because carbohydrates form the bulk of the plant cell's supporting structure – *the cell wall or extracellular matrix*. Consequently plant structural carbohydrates together form the most abundant natural compounds available on earth. They are clearly our most important renewable natural resource, and will be used increasingly both as a source of energy and as raw materials for industrial processes.

5.2 THE PLANT CELL WALL OR EXTRACELLULAR MATRIX

5.2.1 Significance of the cell wall

With very few exceptions, plant cells are enclosed within a wall. The wall has mechanical strength, and defines the shape and size of the cell. Under the light microscope, the walls separating the cells in a plant tissue are usually clearly visible. The walls of adjacent cells meet at a dividing-line known as the *middle lamella*, which can be distinguished with the electron microscope (Fig. 5.1a,b). There is strong cell–cell adherence across the middle lamella, since the individual cells

within a plant tissue do not normally fall apart when the tissue is placed under mechanical stress. Indeed the mechanical strength of a plant tissue is a function of the properties of the walls of cells within the tissue.

The walls of plant cells of different types differ greatly in appearance. Young cells, which still retain the capacity to divide and/or elongate, invariably have a very thin cell wall (0.1–1 μm in cross-section). This is the *primary cell wall* (Fig. 5.1a,b). Primary cell walls are of fundamental importance in the process of cell expansion. It is evident that primary cell walls must yield to allow a cell to grow. It is equally clear that cell growth is accompanied by an increase in the total area of the wall. Thus if, as is observed, wall thickness is maintained during cell expansion, growth must be accompanied by intensive biosynthesis of cell wall constituents. It must be also accompanied by the rearrangement of those molecular interactions which confer rigidity on the cell wall. There is convincing evidence, derived mainly from measurements of cell mechanical properties and from measurements of cell turgor pressure, that cell expansion is in fact *controlled* by metabolic events within the primary cell wall (Taiz, 1984).

Young plant tissues, the cells of which have only thin primary walls, are relatively soft. Their rigidity is maintained by cell *turgor*, that is

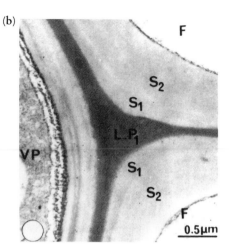

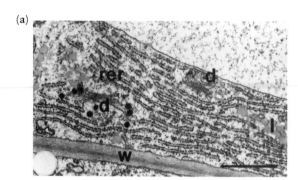

Figure 5.1 Electron micrographs of plant cell walls. (a) Micrograph of developing lupin cotyledons, showing primary cell walls (W). The middle lamella is the central, darker-staining line. Bar represents 1 μm. Photo reproduced from Parker (1984), with permission. (b) Micrograph showing contact between two fiber cells (F) and a parenchyma cell (VP) in linden wood. The layered secondary walls (S₁, S₂) of the fiber cells are visible. L.P₁, middle lamella plus primary cell wall. Photo reproduced from Vian *et al.* (1986), with permission.

hydrostatic pressure exerted by the protoplast on the wall and resisted by the latter. If such tissues are subjected to water-stress to the extent that turgor pressure cannot be maintained, *wilting* of the tissue occurs.

When plant cells lose the capacity to grow and divide, they may differentiate into cells of different types, some of which have cell walls which are very thick. Such thickenings take the form of deposits laid down inside the primary cell wall, that is between the primary cell wall and the plasma membrane of the cell. The thickenings, which are usually many times the thickness of the primary cell wall, are termed *secondary*, and the whole wall internal to the primary cell wall is termed the *secondary cell wall* (Fig. 5.1b). Cells with secondary wall-thickenings are very important in conferring rigidity on plant tissues, independently of cell turgor. Sclerenchyma cells (fiber cells or stone-cells) have this function, whereas tracheids and xylem elements both strengthen tissues and function in water-conduction. The walls of all of the above cell types are not only rigid, but extremely hard. Hardness is conferred by the deposition throughout the cell wall, after the completion of secondary thickening, of the phenolic polymer *lignin*. Lignin deposition (*lignification*) is usually followed by cell death and the disappearance of the cytoplasmic contents. It is noteworthy that the secondary wall thickenings are not always uniform over the cell surface. In xylem elements, for example, the secondary thickenings take the form of elaborate patterns (spirals, reticular etc.), implying precise targeting

of carbohydrate deposits to particular areas of the wall during its biosynthesis. It is significant also that the spiral and reticular thickenings of the water-conducting cells are exactly those best-suited to strengthening a tubular vessel subjected to radial stress (in this case lower hydrostatic pressure inside than outside). In contrast to primary cells walls, walls which have undergone lignification are relatively inert metabolically.

Although 'cell wall' is the most commonly used expression for the carbohydrate-rich structure surrounding the plant cell, research workers have sought an alternative descriptor lacking the idea of inertia and immovability associated with the word 'wall'. This is because, in most cells, the 'wall' is very much an active metabolic compartment, despite being external to the plasma membrane. Thus, in recent years there has been a tendency to use the more general term 'extracellular matrix'. In this chapter both expressions are used interchangeably.

5.2.2 Cell wall architecture

When viewed using various electron microscopic protocols, the walls of most plant cells can be shown to consist of distinctive rodlike or fibrillar structures embedded in seemingly amorphous material. The visible structures are *microfibrils of cellulose* (Fig. 5.2) and they are largely crystalline. The microfibrils are embedded in material which is generally termed the *matrix* of the cell wall. The cellulose microfibrils are not normally arranged at

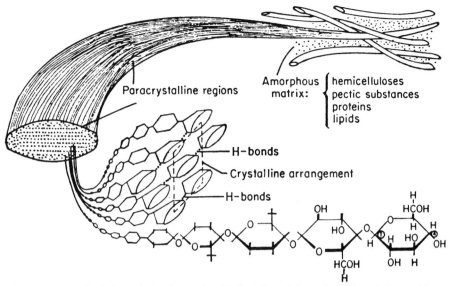

Figure 5.2 Arrangement of cellulose chains in a microfibril. Adapted from Roelofsen (1965), with permission.

random, implying a high degree of control over their biosynthesis and/or deposition.

A classic example of microfibril orientation is provided by the wood fiber or tracheid. In a typical tracheid, several wall layers are distinguishable, all with different microfibril orientations, namely a thin primary cell wall and a thick, three-layered secondary cell wall. In the primary cell wall, the cellulose microfibrils are laid down in an apparently disordered network. By contrast, in the three

layers of the secondary wall the microfibrils are highly ordered. In the outer layer (S_1) the microfibrils are orientated at a fairly shallow angle to the cell axis. In the middle layer (S_2) they run at a very shallow angle to the cell axis, and in the inner layer (S_3) they are at a very steep angle to the cell axis (Fig. 5.3).

In recent years, much light has been shed on the phenomenon of microfibril orientation in cell walls by the observation that virtually all cell walls are

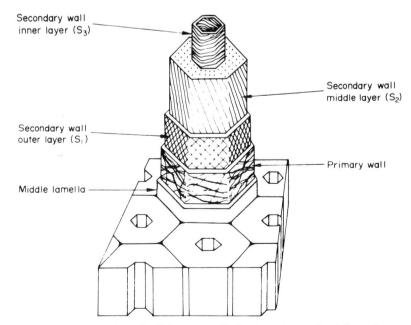

Figure 5.3 A diagrammatic representation of cellulose microfibril orientation in the different layers of the cell wall of a wood fiber. Reproduced from Preston (1974), with permission.

polylamellate, that is they have very fine lamellae (very much finer than the wall layers illustrated in Fig. 5.3) in which the cellulose microfibrils have different orientations. In many such walls the microfibril orientation in successive lamellae can be described in terms of a *helicoid*. That is to say, successive wall lamellae are deposited in which the microfibrils are parallel, but orientated at a fixed angle to those in the previous lamella (Neville & Levy, 1985; Roland *et al.*, 1987). The arced structures visible in many electron micrographs of cell walls (Fig. 5.4) result from helicoidal organization of the walls. The arcs are optical illusions caused by the helicoidal arrangement, but are diagnostic of it and were used in the first place to deduce its existence.

The helicoidal wall introduces the interesting concept that there is some mechanism within the living plant cell which changes microfibril orientation at regular intervals, always in the same direction, like a biological clock the hands of which do not revolve smoothly but in distinct 'ticks' (Vian & Roland, 1987). It also prompts several interesting questions. What is the time interval between the deposition of successive

lamellae? How are wall organizations as illustrated in Fig. 5.3 related to the helicoid? What cytoplasmic factors control microfibril orientation? In the case of the mung bean primary cell wall a partial answer has been obtained to the first question by temporarily interrupting wall deposition, making sections and observing a break in the pattern of arcs as illustrated in Fig. 5.4. Since it can be deduced that two complete arcs in such a pattern represent a turn of 180° in microfibril orientation the time taken for a complete turn of the hands of the 'ticking' biological clock can be estimated. This interval was approximately 6 h (Vian & Roland, 1987). The relationship between finely polylamellate helicoidal walls and walls with massively thick layers has been elegantly clarified by the observation (Roland *et al.*, 1987) that the transition between the layers of the secondary cell wall is not a sharp one. Between the massive, uniform layers there are brief helicoidal interruptions (Fig. 5.5). Thus it may be deduced that the biological clock determining the time intervals between successive lamellae does not always tick regularly. It can remain stopped over a long period to allow the deposition of a thick wall layer with microfibrils running in a given direction, can then be restarted to allow microfibril reorientation to occur and then stop once more. The mechanism of the clock will almost certainly involve the cytoskeleton of the cell, since the arrangement of microtubular elements in the outer layers of the cell cytoplasm have repeatedly been reported to reflect the orientations of cellulose microfibrils in the wall (Lloyd, 1984). The establishment of cause-and-effect relationships between external or internal stimuli, cytoplasmic activities and microfibrillar orientation will require much ingenious and exciting research.

5.2.3 Structural components of cell walls

The classification of cell wall components into microfibrillar and matrix elements is ultrastructure-based. However, even before cel-lulose microfibrils could be viewed in the microscope, a classification based on chemical extraction was taking shape. Lignified secondary walls are extraordinarily resistant to extraction, being permeated by the highly cross-linked, hydro-phobic, highly stable lignin polymer. Accordingly the fractionation of the carbohydrate components of such walls must be preceded by a chemical treatment designed to degrade the phenolic polymer selectively, to give fragments capable of being dissolved from the wall.

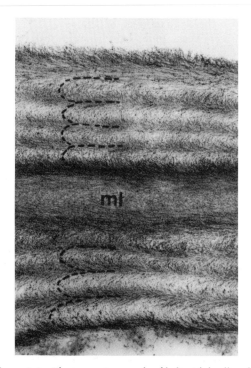

Figure 5.4 Electron micrograph of helicoidal cell walls. The arcs (dotted lines) are optical illusions arising from the helicoidal arrangement of cellulose microfibrils. Walls of adjacent cells in aerenchyma tissue of *Papyrus*. ml, middle lamella. (×30 000) Reproduced from Roland *et al.* (1987), with permission.

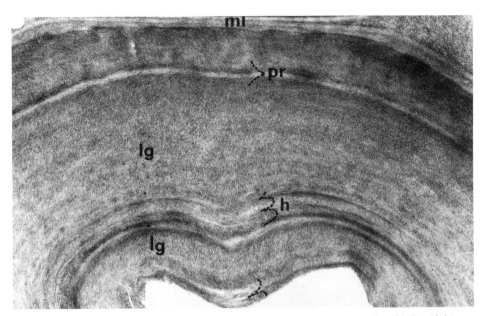

Figure 5.5 Electron micrograph of a cell wall with secondary thickening, showing brief helicoidal interruptions (h, dotted lines) between thick uniform layers with longitudinally orientated microfibrils (lg). ml, middle lamella. pr, reversion point. Sclerenchyma cell from the stem of *Aristolochia clematitis*. (×21 000) Reproduced from Roland *et al.* (1987), with permission.

Examples of such processes (*delignifications*) are acid chlorite and acid sulfite treatments: both oxidize the phenolic rings and convert the lignin into water-soluble derivative. Acid sulfite treatment is part of the paper-making process. To isolate and study the structures of the carbohydrates of highly lignified tissues, such as woods and crop straws, it is virtually essential to delignify the tissue, to give an almost lignin-free starting material, known as 'holocellulose'. Non-lignified cell walls are not normally subjected to delignification.

On the basis of extractability, three principal groups of plant cell wall carbohydrates can be distinguished: the pectins, the hemicelluloses and cellulose. The *pectins* are extracted using aqueous solutions containing substances, such as ammonium oxalate or EDTA, which are capable of chelating Ca^{2+} or other divalent metal cations. The *hemicelluloses* are more difficult to remove from the cell wall material, requiring fairly concentrated solutions of sodium or potassium hydroxide. The residue after extraction of the pectins and the hemicelluloses is rich in cellulose, and is often termed *alpha-cellulose*. This term, which originated in the paper-making industry, is somewhat confusing since the main component of the *alpha*-cellulose fraction is cellulose, a polymer composed entirely of *beta*-linked glucose residues! To avoid such confusion the term alkali-insoluble residue will be used throughout this section.

Fortunately, the independently developed microscopic and extraction-based classifications of the carbohydrate components of the cell wall are compatible one with the other. Microscopic examination of tissues undergoing traditional extraction confirms that the pectic and hemicellulose fractions derive from the matrix of the cell wall, whereas the alkali-insoluble residue clearly includes the microfibrillar component.

The pectin fraction, the hemicellulose fraction and alkali-insoluble residue also differ from each other in their molecular structures, which are discussed in section 5.3. Although it is intuitively obvious that the microfibrillar component of the cell wall will differ in its biological function from the matrix components, it is becoming increasingly apparent that the pectin and hemicellulose fractions may also have different roles to play in wall functionality. Current functional wall models are discussed in section 5.4.

5.3 STRUCTURES AND INTERACTIONS OF PLANT CELL WALL POLYSACCHARIDES

This section outlines the main structural features of the macromolecular components of the pectin and hemicellulose fractions of the wall, as well as of cellulose, the polysaccharide which makes up the bulk of the alkali-insoluble residue.

It should be noted that there is no such thing as 'the typical cell wall'. Wall material isolated from tissues dominated by cells with thick secondary walls (e.g. wood) is very rich in both cellulose and the hemicellulose fraction, but contains very little or no pectin. On the other hand cell wall material isolated from the flesh of many fruits is particularly rich in the pectin fraction. In general the structures of the hemicellulose polymers present in primary cell walls are different from those in secondary cell walls. Cell walls (primary or secondary) of plants from different major taxonomic groupings can also differ greatly with respect to the relative proportions and the macromolecular compositions of their hemicellulose and pectin fractions.

Also the molecular structures of the polymeric carbohydrates of plant cell walls are often obtained after their isolation from the cell wall, and that isolation process may itself have brought about structural changes, possibly including unsuspected ones. For example, delignification will oxidize not only lignin but also phenolic substituents attached to carbohydrate polymers. Also alkaline extraction procedures will hydrolyze ester linkages. This may result in the loss of acetyl substituents attached to the hydroxyl groups of carbohydrate polymers, or in the breakage of any ester-linkages cross-linking different macromolecular components of the cell wall. Accidental hydrolysis or other degradation during extraction is also capable of removing labile features from a polymer, or of decreasing its molecular weight.

Despite the above cautionary words, it *is* possible to draw general conclusions concerning the structures of plant cell wall polymers, and this is attempted below. Nonetheless modern cytochemical studies, particularly on primary cell walls, have revealed differences in cell wall molecular composition between different cells in a tissue and even between different areas of the wall of a given cell (Knox *et al.*, 1990). Such differences are potentially of great relevance to the understanding of cellular differentiation and recognition.

5.3.1 Cellulose

From the structural viewpoint cellulose is probably the best-understood of all the carbohydrates of the plant cell wall (Nevell & Zeronian, 1985). It is easily obtained in a relatively pure state by exhaustive extraction with alkali of the hemicellulosic fraction of the cell wall, and can be subjected to structural determination by standard chemical methods. Cellulose obtained in this way from most tissues yields over 90% D-glucose on acid hydrolysis, the remainder being small amounts of other sugars, notably xylose, galactose and mannose. Since some tissues, such as cotton hairs, yield cellulose which gives essentially 100% glucose on hydrolysis it is usually assumed that the non-glucose sugars arise from incomplete removal of hemicellulose polymers.

Structural analyses of cellulose show that the primary structure of the molecule is very simple: a long, monotonous, linear sequence of D-glucose residues joined together by $\beta(1 \rightarrow 4)$ glycosyl linkages (Fig. 5.6). The average *degree of polymerization* (number of glucose residues in a chain) of cellulose from secondary cell wall sources is high, in the region of 10 000, whereas the limited number of studies that have been carried out on primary cell wall celluloses indicate a lower value (c. 2000–6000). When considering the molecular weights of carbohydrate polymers it must, of course, be borne in mind that the molecular weight is an *average* value. Ideally the measured average value will reflect the peak of the natural distribution of chain lengths. Often it will also reflect accidental degradation during isolation and preparation of the sample for analysis by a physico-chemical technique. No simple comparative procedure such as sodium dodecyl sulfate (SDS)-gel electrophoresis (used for proteins) is available for carbohydrate polymers.

In the cell wall and after purification free of the pectin and hemicellulose fractions, cellulose exists as rod-like, highly insoluble fibrils. In the cell wall they are known as *microfibrils*. Viewed using the electron microscope, the microfibrils are tubular structures, usually elliptical in cross-section with diameters ranging from about 5 to 30 nm (Figs. 5.2 and 5.6b). Much, although not all, of the cellulose within the microfibril exhibits crystalline order when subjected to X-ray crystallographic analysis. The remainder is amorphous or 'paracrystalline' (Fig. 5.2). Within the individual chains, the glucose units (in their energetically preferred chair conformations) are arranged in a spiral or helical pattern in which each residue is rotated at 180° relative to the preceding and following ones (Fig. 5.2). This arrangement can also be predicted using minimum-energy calculations, and is confirmed by X-ray crystallography. Two forms of crystalline cellulose, which may be differentiated by X-ray crystallography, are known: cellulose I and cellulose II. In its native state cellulose always exists in the cellulose I form. However, if native cellulose is dissolved and recrystallized (regenerated), or even simply treated with strong alkali

(a)

$$\cdots\cdots\text{Glc}1\overset{\beta}{\rightarrow}4\text{Glc}1\overset{\beta}{\rightarrow}4\text{Glc}1\overset{\beta}{\rightarrow}4\text{Glc}1\overset{\beta}{\rightarrow}4\text{Glc}1\overset{\beta}{\rightarrow}4\text{Glc}1\rightarrow\cdots\cdots$$

(b)

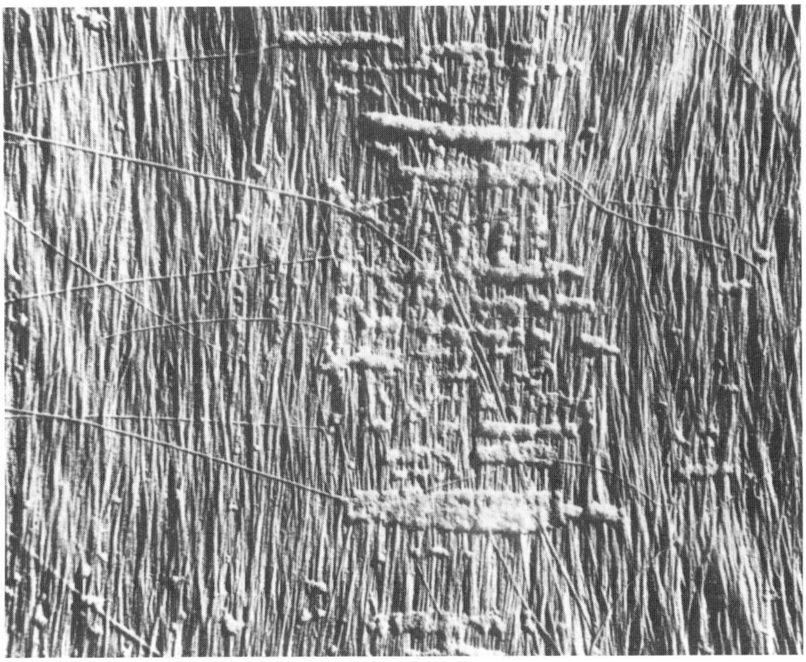

Figure 5.6 Cellulose structure and organization. (a) Chemical structure. (b) Electron micrograph showing microfibrils of cellulose. Reproduced from Roelofsen (1965), with permission.

without dissolution (mercerization) it is converted into cellulose II. Cellulose II is the thermodynamically stable form of cellulose, and cannot be reconverted into cellulose I. The existence of native cellulose in the less thermodynamically stable (metastable) form is a strong indication that the biosynthesis of cellulose involves a process of orientation of the newly formed $(1 \rightarrow 4)$-β-D-glucan chains which results in their crystallization in the less thermodynamically stable cellulose I form (Haigler, 1985). There is cytochemical evidence (Chanzy & Henrissat, 1985) that in cellulose I the parallel glucan chains all run in the same direction, whereas the X-ray evidence has been interpreted to indicate that the glucan chains in cellulose II are arranged in an alternating fashion. However, there is still some controversy over this point, and the molecular arrangements differentiating cellulose I and cellulose II cannot be held to be fully understood (Delmer, 1987).

Cellulose is present almost universally in the cell walls of higher plants, constituting the microfibrillar phase. It accounts for about 30–40% by weight of the cell walls of woody tissues which are dominated by secondary cell walls. If lignin is neglected this is over 50% of the wall carbohydrates. In primary cell walls cellulose accounts for only about 20% of the wall.

5.3.2 Hemicelluloses

The definition of 'hemicellulose' often causes communication difficulties between plant scientists. To some, the 'hemicellulose fraction' is the material removed from cell wall material by alkaline extraction. To others a 'hemicellulose' is a cell wall polymer with a particular type of molecular structure and perhaps a particular notional function within the cell wall. In fact, if the hemicellulose polymers are differentiated clearly from the pectic polymers, which may be largely removed from cell wall material by prior extraction with neutral solutions of divalent metal chelators, then the two usages of the word 'hemicellulose' are compatible. After removal of the pectic polymers, alkaline extraction of cell walls does yield a range of carbohydrate polymers with structural features in common, and which may well play a role in wall architecture which is distinct from that of both cellulose and pectin. In this chapter the alkali-soluble post-pectin wall fraction is the 'hemicellulose fraction', and each of the individual polymers obtained by further purification of the hemicellulose fraction is 'a hemicellulose'.

(a)   Hemicelluloses from secondary cell walls

Much of the early work (up to the 1970s) on the structures of the carbohydrate polymers in the hemicellulose fraction of the cell wall was done with reference to readily available bulk plant tissues such as woods, and the straws of crop grasses (Timell, 1964, 1965; Wilkie, 1979). Both, especially the woods, have a high proportion of lignified cells with thick secondary cell walls. Consequently, any polymers extracted with alkali from such tissues will inevitably reflect the composition of the secondary cell wall. This subsection summarizes briefly the structures of the principal polymers isolated from hemicellulose fractions prepared from wood tissues and grass straws. It should be noted that the great majority of species studied are commercially important and originate in the temperate zone of the northern hemisphere. Thus comparisons will be drawn between the hemicelluloses of the woods (softwoods) of temperate conifers (gymnosperms), the woods (hardwoods) of temperate broad-leaved trees (angiosperms) and the straws of temperate grasses (Gramineae).

(i)   The hemicellulose fractions obtained from softwoods

These can generally be fractionated to obtain relatively pure samples of two polymer types, best described as galactoglucomannan and an arabino-4-O-methylglucuronoarabinoxylan (Timell, 1965).

Galactoglucomannan, which is present in greater amounts, is an essentially linear molecule. It has a backbone chain made up of D-glucopyranose and D-mannopyranose residues which are joined together by $\beta(1 \rightarrow 4)$-linkages. Thus the backbone recalls that of cellulose, but differs from it both in the presence of mannosyl residues and in its much shorter length (average chain-length around 100 sugar residues). The ratio of mannosyl to glucosyl residues in the backbone (Man/Glc ratio) is generally about (but not exactly) 3, and the distribution of the two residues may be statistically random, although this has not been rigorously proved. The glucomannan backbone carries short branches consisting of single D-galactopyranosyl residues attached by $\alpha(1 \rightarrow 6)$-linkages to mannose residues in the backbone. About 2–4% of the backbone residues carry galactosyl substituents (Fig. 5.7), and again the statistical distribution of these substituents has not been ascertained. In its native state, galactoglucomannan is almost certainly acetylated. When extraction procedures which avoid the use of alkali are used, the removal of galactoglucomannan from the cell wall is inefficient, but the polysaccharide obtained is partially acetylated, apparently on mannosyl residues. This indicates that the galactoglucomannan, like several of the other cell wall matrix polysaccharides mentioned below, may be acetylated in its native state.

The *arabino-4-O-methylglucuronoxylan* is an essentially linear polymer with a backbone consisting of D-xylose (a pentose) residues in the pyranose ring-form joined together by $\beta(1 \rightarrow 4)$-linkages. Again the backbone carries short side chains consisting of single sugar residues. Two side-chain types are present, residues of 4-O-methyl-D-glucuronic acid in the pyranose ring form linked $\alpha(1 \rightarrow 2)$ to backbone xylose, and residues of L-arabinose in the furanose ring form linked $\alpha(1 \rightarrow 3)$ to backbone xylose (Fig. 5.8a). The ratio of 4-O-methylglucuronic acid to xylose is about $1:5$–10, and the ratio of arabinose to xylose is very variable. This may reflect varying losses of arabinose during isolation of the polysaccharide, since arabinose residues in the furanose form are very easily hydrolyzed, even with mild acid. Little is known about the statistical distributions of the two substituents.

(ii)   The hemicellulose fractions obtained from hardwoods

These fractions contain a pair of hemicellulose polymers which are structurally similar to, but not identical with those described above. The principal

```
        ß        ß        ß        ß        ß        ß        ß        ß        ß
----Man1→4Man1→4Man1→4Glc1→4Glc1→4Man1→4Man1→4Glc1→4Man1→4Man1→ ---
        6                                 6
       ↑α                                ↑α
     Gal1                              Gal1
```

Figure 5.7   Structural features of galactoglucomannans.

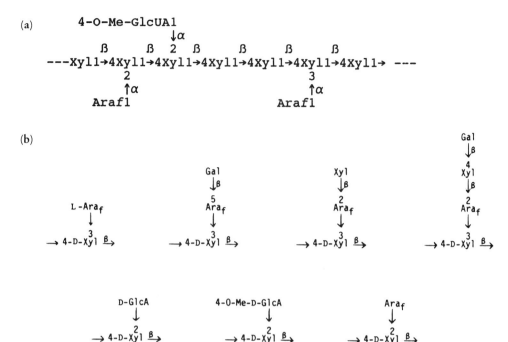

Figure 5.8 Structural features of xylans. (a) Main structural features of arabino-4-O-methyl-glucuronoxylans from softwoods. (b) Examples of side chains present in xylans from other sources. Reproduced, with permission, McNeil *et al.* (1984).

hemicellulose polymer is a 4-O-methylglucurono-xylan, and it is accompanied by lesser amounts of a glucomannan (Timell, 1964). This is in contrast to the softwoods where the major hemicellulose is the galactoglucomannan and the minor one is the arabino-4-O-methylglucuronoxylan.

The backbone of the hardwood 4-O-methylglu-curonoarabinoxylan is again a $\beta(1 \rightarrow 4)$-linked chain of D-xylopyranose residues, and it is substituted by single residues of 4-O-methyl-D-glucuronic acid, in the pyranose ring form, attached $\alpha(1 \rightarrow 2)$ to backbone xylose. The ratio of xylose to 4-O-methylglucuronic acid residues in the polymer is about $10:1$, and the statistical distribution of the uronic acid residues is unknown. Like the galactoglucomannan of the softwoods, this hemicellulose is partially acetylated in its native state, the acetyl ester groups being attached to positions 2 and/or 3 of xylose. On average, the molecule contains 150–200 sugar residues.

The glucomannan hemicellulose of hardwoods is similar in structure to the galactoglucomannan from the softwoods (Fig. 5.7). The glucose and mannose residues are both in the pyranose ring form, and constitute a $\beta(1 \rightarrow 4)$-linked backbone. The ratio Man/Glc is about 2, and the two types of residue may be randomly distributed. Galactose substituents, linked as in the galactoglucomannans are either infrequent or absent altogether.

(iii)   Hemicellulose fractions obtained from the straws of crop grasses

These fractions yield predominantly hemicelluloses based on the $\beta(1 \rightarrow 4)$-linked D-xylan backbone, that is they have the same backbone as the 4-O-methylglucurono- and arabino-4-O-methyl-glucuronoxylans of woods. The grass xylans, however, appear to carry a greater variety of short side chains, including α-L-arabinofuranose attached to C-2 or C-3 of xylose, α-D-glucurono- and α-4-O-methyl-D-glucuronopyranose attached to C-2 of xylose and more complex side chains containing galactose or xylose. Some other side chains associated with grass xylans are shown in Fig. 5.8b.

All the hemicelluloses described above share the common structural features of a linear $\beta(1 \rightarrow 4)$-linked backbone of conformationally related sugar residues in the pyranose ring form. In each taxonomic grouping, at least the predominant hemicellulose has a backbone which is laterally substituted by short branches, either neutral, or acidic as in the case of the 4-O-methylglucuronic acid and glucuronic acid side chains of the xylans. This structural pattern may be of significance for the macromolecular interactions which almost certainly occur within the secondary cell wall.

(b) Hemicelluloses from primary cell walls

Structural information on the hemicellulose polymers of the secondary cell wall was gathered over
several decades. The accumulation of sufficient
bulk plant tissue for large-scale hemicellulose
extractions presented no real problem, and sufficient quantities of purified hemicelluloses for
conventional structural analyses (methylation analysis, periodate oxidation, partial hydrolysis etc.)
could be obtained. By contrast tissues with cell
walls which are almost exclusively primary
(growing tips, elongation zones etc.) could not
easily be harvested in bulk and in any case yielded
a very much smaller proportion of wall material.
Thus, up to the mid 1970s, there was little or no
direct experimental evidence of the chemical
structures of the polymers present in the hemicellulose fractions of primary cell walls. Primary
cell wall analysis was virtually restricted to
analyses of sugar residue composition (Meier &
Wilkie, 1959; Thornber & Northcote, 1962).

In the mid 1970s the structural analysis of
primary cell walls became practically feasible.
Rapid, efficient protocols for the total methylation
of very small amounts of carbohydrate polymers
were developed, as well as new separation and
identification procedures for partially methylated
alditol acetates. Accordingly it became possible to
carry out methylation analysis on milligram
amounts of purified polysaccharides or even
intact cell walls. Studies on plant pathogenic
micro-organisms were beginning to yield enzymes
capable of hydrolyzing specific glycosyl linkages
within the polymers of the cell wall. Suspension
cultures of plant cells, which have walls which, at
least superficially, resemble primary cell walls
were recognized as possible model systems for
the investigation of primary cell wall structure.
Most importantly, pioneering work, notably that
carried out at the University of Colorado by
P. Albersheim and colleagues, brought all of these
new approaches to bear on the problem of the
molecular structure of the plant primary cell wall
(Talmadge et al., 1973; Bauer et al., 1973; Keegstra
et al., 1973). The primary cell wall is now known
to contain cellulose, pectin and hemicellulose
polymers some of which are quite distinct in
structure from the hemicelluloses of the secondary
wall. Many primary cell walls also contain
significant quantities of structural protein, and
small amounts of phenolic materials, some bound
covalently to the carbohydrate polymers (McNeil
et al., 1984; Fry, 1988).

Structural studies have been carried out on the
primary cell walls of a relatively large number of
plant species, most of them angiosperms of some
economic importance in temperate countries. The
available data indicate two distinct types of
primary cell walls, which differ in the compositions of their matrix components. The primary cell
walls of dicotyledonous plants and those monocotyledonous plants that are not grasses have cell
walls which are rich in both pectin and hemicellulose. Such primary cell walls have been
described as type 1 (Carpita & Gibeaut, 1993).
The primary cell walls of the grasses (Gramineae),
termed type 2 primary cell walls (Carpita &
Gibeaut, 1993) contain very little pectin, and
proportionally more polymers which, on the
basis of their alkali-solubility can be classified as
hemicellulose. Furthermore, the relative proportions and the molecular structures of the hemicelluloses in walls of the two types also differ
(McNeil et al., 1984).

(i) Type 1 primary cell walls

The hemicellulose fractions from type 1 primary
cell walls normally contain xyloglucan as the
principal component, alongside a smaller amount
of a xylan.

The xyloglucan (Hayashi, 1989), which
accounts for about 20% of the dry weight of the
cell wall, is remarkable in the degree of precision
and order in its structure. It is best considered as a
backbone with successive layers of substitution.
The backbone is a chain of $\beta(1 \rightarrow 4)$-linked
D-glucopyranosyl residues identical to that of
cellulose, but considerably shorter. The backbone
carries single xylopyranosyl substituents which are
joined to it by $\alpha(1 \rightarrow 6)$-linkages. Most of these
substituents (apparently all of them in some cases)
are arranged in a very regular fashion along the
backbone: three consecutive glucosyl residues
carry substituents while the fourth does not
(Fig. 5.9). Thus most of the xyloglucan chain may
be considered to be composed of a series of
Glc_4Xyl_3 repeating units (Fig. 5.9). The subunits
may be further substituted. Of the three side-chain
xylose units on the Glc_4Xyl_3 repeating unit, the
two farthest from the non-reducing end of the
subunit may carry D-galactopyranosyl residues
linked $\beta(1 \rightarrow 2)$ to xylose (Fig. 5.9). Of these,
one may bear a further substituent, namely an
L-fucopyranose residue linked $\alpha(1 \rightarrow 2)$ to galactose (Fig. 5.9). The precise arrangement of
substituents along the backbone of the xyloglucan
molecule indicates a high degree of metabolic
control over the biosynthetic process. There is
evidence that xyloglucan may be partially acetylated in its native state (York et al., 1988).

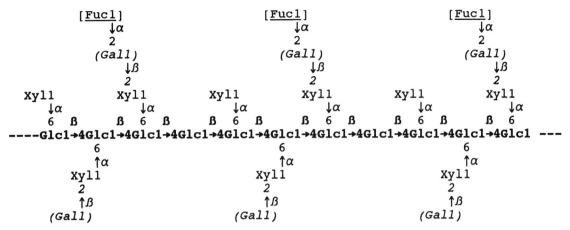

Figure 5.9 Structural features of primary cell wall xyloglucans. Note the concentric layers of substitution and the Glc₄Xyl₃ based structural subunits (see text). The glucan backbone is in bold type, and the xylose substituents in normal type. (*Gal*), possible galactose substituent. [*Fuc*], possible fucose substituent.

The existence of the structural subunit based on Glc$_4$Xyl$_3$ was detected by hydrolysis of xyloglucans using a purified *endo*-$\beta(1 \rightarrow 4)$-D-glucanase, which hydrolyzed the backbone chain at the glycosyl linkage following the unsubstituted glucosyl residue (Fig. 5.9), thus releasing the subunit oligosaccharides (Hayashi & Maclachlan, 1984). This illustrates the value of enzymes of known specificity in the investigation of plant cell wall polysaccharide structures.

Also present in the hemicellulose fraction from type 1 primary cell walls is a hemicellulose which is structurally very similar to the xylan-type hemicelluloses of secondary walls. It has a $\beta(1 \rightarrow 4)$-linked D-xylan backbone which carries a range of different substituents attached to C-2 and/or C-3 of backbone xylose.

(ii) Type 2 primary cell walls

The hemicellulose fractions from type 2 primary cell walls are very rich in polysaccharides of the xylan type. They are accompanied by smaller amounts of a xyloglucan, and a mixed-linkage $\beta(1 \rightarrow 3, 1 \rightarrow 4)$-glucan.

Although the xylans of primary cell walls have been studied to a lesser degree than those from secondary walls it is clear that the former are the more complex in their molecular structures. A $\beta(1 \rightarrow 4)$-linked D-xylan backbone is fairly heavily decorated by a variety of short side chains, the most important of which are illustrated in Fig. 5.8(b). The site of attachment of the side chains is to C-2 and/or C-3 of the backbone xylose residues. The acidic nature of the majority of the side chains is perhaps of significance, since pectin (see section

5.3.3), which provides most of the acidic functions in the type 2 primary cell wall, is present only in very low amounts in the type 2 primary wall.

Although the primary cell walls of the grasses do contain xyloglucans in small amounts, these polysaccharides are apparently less regular in their substitution pattern than the xyloglucans in type 1 primary cell walls. The degree of xylose substitution of the glucan backbone is lower, and there are fewer of the distinct Glc$_4$Xyl$_3$-based structural units which dominate the structures of the xyloglucans present in type 1 cell walls. Some of the xylose residues are galactose-substituted, but very few, if any, of these carry a fucosyl substituent (Hayashi, 1989).

A characteristic hemicellulose of the type 2 primary cell wall is the mixed-linkage $\beta(1 \rightarrow 3, 1 \rightarrow 4)$-glucan (Stone & Clarke, 1992). The backbone is linear, in the sense that it carries no side branches. However, two linkage-types, $(1 \rightarrow 3)$ and $(1 \rightarrow 4)$, are present in the backbone, and the overall shape or conformation of the molecule depends on their relative proportions and distribution (Fig. 5.10). In fact the arrangement of the two linkage types within the molecule shows a high degree of order. In the first place the relative proportions of the two types of linkages within the molecule are apparently constant (ca. 2.2). Secondly, the distribution of the two linkage types within the molecule is non-random. The $(1 \rightarrow 4)$-linkages do not occur singly, but in blocks. The blocks of $(1 \rightarrow 4)$-linkages are separated by $(1 \rightarrow 3)$-linkages, which only occur singly. The molecule is therefore a series of $\beta(1 \rightarrow 4)$-linked structural domains or blocks separated by single $\beta(1 \rightarrow 3)$-linkages. Most of the structural blocks

$$
\overset{\beta}{}\quad\overset{\beta}{}\quad\overset{\beta}{}\quad\overset{\beta}{}\quad\overset{\beta}{}\quad\overset{\beta}{}\quad\overset{\beta}{}\quad\overset{\beta}{}\quad\overset{\beta}{}\quad\overset{\beta}{}
$$

---4Glc1→4Glc1→3Glc1→4Glc1→4Glc1→3Glc1→4Glc1→4Glc1→4Glc1→3Glc1→4--

Figure 5.10 Structural features of mixed-linkage $\beta(1 \rightarrow 3, 1 \rightarrow 4)$-glucan.

contain two or three $(1 \rightarrow 4)$-linked glucose residues, but smaller proportions of longer blocks are present (Stone & Clarke, 1992).

As in the case of xyloglucans the structural regularity of the mixed-linkage β-glucan molecule was detected with the help of an enzyme which hydrolyzed the molecule in a specific and predictable way. In this instance the enzyme was an unusually specific β-glucanase ('lichenase') from the bacterium *Bacillus subtilis*. This enzyme hydrolyzes $\beta(1 \rightarrow 4)$-glucosyl linkages, but only if they immediately follow a $\beta(1 \rightarrow 3)$-glucosyl link. Thus digestion of a $\beta(1 \rightarrow 3, 1 \rightarrow 4)$-glucan with the *B. subtilis* lichenase releases β-linked gluco-oligosaccharides all of which are $(1 \rightarrow 4)$-linked except for the linkage nearest to the reducing end, which is $\beta(1 \rightarrow 3)$ (Fig. 5.10). Each originated from a structural block, and the number of sugar residues in each is equal to the number of $\beta(1 \rightarrow 4)$-linked glucose residues in the block. By analyzing the relative proportions of all the saccharides released, the relative numbers of structural blocks of different sizes within the glucan molecule may be determined.

5.3.3 Pectin

The definition of 'pectin' is almost as problematic as that of 'hemicellulose'. To the food manufacturer (or consumer) pectin is a natural fruit polysaccharide which is used, in jams for example, because of its ability to gel in the presence of high concentrations of sugar. The commercially important pectins originate in the cell walls of some fruits (citrus fruits and apples), the primary cell walls of which are particularly rich in them. In fact pectin appears to be present universally in primary cell walls, and it is a major constituent of primary cell walls of type 1. It is also present in the middle lamella between cells of all types.

The term 'pectin' encompasses a complex group of polysaccharides, some of which may be structural domains of larger, more complex molecules. This is, however, not certain. In addition to 'pectin' the terms 'the pectic substances' and 'the pectic polysaccharides' are in common use. In this chapter 'pectin' or 'the pectin fraction' will be used to describe the entire chelator-soluble polysaccharide fraction from a given type of cell wall. The term 'pectic polysaccharide' will be used for a single structural entity derived from or present in the pectin fraction, whether it exists as an independent macromolecule or as a major structural domain of a larger one.

Pectin is acidic, containing a high proportion of D-galacturonic acid residues, most of which are present in a linear backbone. In some pectic polysaccharides the backbone consists almost entirely of D-galacturonic acid residues, in the pyranose ring form, joined together by $\alpha(1 \rightarrow 4)$-linkages. The carboxylic acid groups of some of the galacturonic acid residues are esterified with methanol (Fig. 5.11). Pectic polysaccharides which have this structure, predominantly or exclusively, are known as '*homogalacturonans*'. The term is possibly misleading, because the chain seldom consists exclusively of galacturonic acid and methylgalacturonate residues. A number of L-rhamnose (a 6-deoxyhexose) residues are normally interspersed within the chain. The linkage from D-galacturonic acid to L-rhamnose is $\alpha(1 \rightarrow 2)$, and the linkage from rhamnose to the following galacturonic acid is $\alpha(1 \rightarrow 4)$ (Fig. 5.11). Other pectic polysaccharides have a much higher proportion of rhamnose residues in the backbone, often to the extent of an alternating arrangement of galacturonic acid and rhamnose residues. Such rhamnose-rich pectic polysaccharides are called '*rhamnogalacturonans*' (Fig. 5.12a) (McNeil *et al.*, 1984). The rhamnose residues in rhamnogalacturonans, and probably also the homogalacturonans, can serve as anchorage points for side chains attached to the backbone. Consequently, because of the variable frequency of the rhamnose residues, there are structures within pectin that are much more highly branched than others. The highly branched regions are often referred to as the '*hairy regions*' (de Vries *et al.*, 1982), and the less highly branched regions as the '*smooth regions*'.

$$
\overset{\alpha}{}\qquad\overset{\alpha}{}\qquad\overset{\alpha}{}\qquad\overset{\alpha}{}\qquad\overset{\alpha}{}\qquad\overset{\alpha}{}
$$

---*Me*GalUA1→4*Me*GalUA1→4*Me*GalUA1→2Rha1→4*Me*GalUA1→4*Me*GalUA1→4*Me*GalUA1---

Figure 5.11 Structural features of the homogalacturonan component of pectin. *Me*GalUA, methyl ester of galacturonic acid.

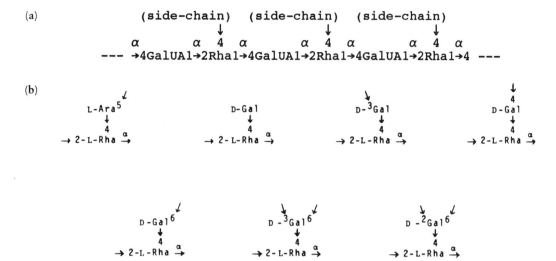

(a)

```
        (side-chain)  (side-chain)  (side-chain)
             ↓             ↓             ↓
      α     α   4  α      α   4  α      α   4  α
--- →4GalUA1→2Rha1→4GalUA1→2Rha1→4GalUA1→2Rha1→4 ---
```

(b)

```
            ↙                                    ↘             ↓
                                                             4
      L-Ara⁵              D-Gal            D-³Gal          D-Gal
        ↓                   ↓                 ↓               ↓
        4                   4                 4               4
  → 2-L-Rha →ᵃ        → 2-L-Rha →ᵃ      → 2-L-Rha →ᵃ     → 2-L-Rha →ᵃ


            ↙                    ↘  ↙                 ↘  ↙
      D-Gal⁶             D-³Gal⁶             D-²Gal⁶
        ↓                   ↓                   ↓
        4                   4                   4
  → 2-L-Rha →ᵃ        → 2-L-Rha →ᵃ        → 2-L-Rha →ᵃ
```

Figure 5.12 Structural features of the rhamnogalacturonan I component of pectin. (a) Backbone, with possible sites of attachment of side chains. (b) Structural features of some of the side chains of rhamnogalacturonan I. Reproduced, with permission, McNeil *et al*. (1984).

Rhamnogalacturonan structures of the type illustrated (Fig. 5.12a) are called rhamnogalacturonan I (or RGI) to differentiate them from a very different type of rhamnose and galacturonic rich molecule cell wall molecule, known as rhamnogalacturonan II (RGII), the structure of which is mentioned later. RGI carries side chains consisting mainly of neutral sugar residues. Not all side-chain structures have yet been identified and/or characterized fully; some are illustrated in Fig. 5.12(b).

```
              Araf1
               ↓α
        α   3  α         α
a)  ---Araf1→5Araf1→5Araf1→5---


          ß       ß       ß
b)  ---Gal1→4Gal1→4Gal1→4---


          ß       ß       ß
c)  ---Gal1→4Gal1→4Gal1→4---
              3
              ↑α
            Araf1
              5
              ↑α
            Araf1
```

Figure 5.13 Structural features of the arabinan, galactan and arabinogalactan components of pectin. (a) Arabinan; (b) galactan: (c) arabinogalactan.

In addition to the acidic polysaccharides based on residues of D-galacturonic acid and L-rhamnose, pectins from different plant sources contain variable proportions of neutral polysaccharides. On the basis of their large molecular size, these components must be differentiated from the relatively short side chains of RGI. The neutral pectic components were formerly believed to be distinctly separate polysaccharides. It is now thought that, in the cell wall, they are anchored to one or other of the acidic pectic polysaccharides as large side chains. Three main structural types are known: the arabinans, the arabinogalactans and the galactans (McNeil *et al*., 1984).

Pectic arabinans are highly branched molecules consisting of L-arabinose (a pentose) residues, mainly in the furanose ring form. A core of $\alpha(1 \rightarrow 5)$-linked residues carries $\alpha(1 \rightarrow 3)$- and $\alpha(1 \rightarrow 2)$-linked arabinofuranosyl side chains. The arabinogalactans are branched polysaccharides containing some galactose residues which are terminal, non-reducing, and others which carry substituents on the 3- and/or 6-position. The arabinose residues are predominantly in the furanose ring-form, and carry substituents at the 3- or 5-position. The galactans are essentially linear molecules composed of a backbone chain of $\beta(1 \rightarrow 4)$-linked D-galactose residues, in the pyranose ring form. The chain carries a few side branches consisting of single L-arabinose residues (Fig. 5.13).

Rhamnogalacturonan II (RGII) (McNeil *et al*., 1984) is a relatively small, but highly complex,

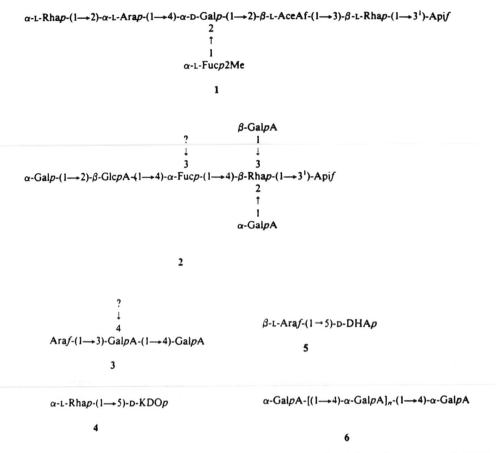

Figure 5.14 Some structural features of rhamnogalacturonan II. Reproduced from Stevenson *et al.* (1988), with permission.

polysaccharide which appears to be present universally in plant primary cell walls. It is normally present in the pectic fraction, but does not appear to be part of a larger molecular complex. Some of the structural features of RGII are shown in Fig. 5.14. The presence in the primary cell wall of a molecule of such complexity and apparent structural precision is intriguing. It may well play a role in cell recognition and/or signaling.

5.4 SUPRAMOLECULAR INTERACTIONS OF STRUCTURAL POLYSACCHARIDES IN CELL WALLS

Plant cell walls have, in the past, been likened to steel-reinforced concrete. It is useful to compare the cellulose microfibrils in the wall to the steel elements in reinforced concrete. The microfibrils do possess high tensile strength, superior to that of steel, and they are generally arranged in mechanically relevant orientations. To compare the matrix of the cell wall to concrete is less appropriate. The matrix components clearly do not constitute a hard, inert mass, and their specific molecular interactions almost certainly contribute in an important, although by no means fully understood, way to the mechanical properties of the cell wall.

The reinforced concrete analogy is perhaps just acceptable in the case of cells with thick, secondary cell walls which have undergone lignification. Lignification hardens the wall matrix, effectively fixing any polysaccharide–polysaccharide interactions. On the other hand, primary cell walls are not at all concrete-like. Primary cell walls are strong and rigid to resist the outward pressure exerted by the plant protoplast, yet they are also capable of expansion. Primary walls allow cell growth by undergoing controlled, usually directional, extension. Thus their carbohydrate components are biosynthesized and deposited continuously, to keep pace with the increase in wall surface area. Carbohydrate components already present within the wall undergo rearrangement, both passively as a result of growth-related changes in cell dimensions, and actively to permit and even to control the amount, the rate and the direction of growth. Since the realization that cell growth is dependent on metabolic processes going on within the wall, there has been an urgent research interest in determining which wall components undergo modification, and how such modifications in turn influence the mechanical properties of the cell wall. It has also become clear that changes in the mechanical properties of primary cell walls accompany many developmental processes, notably fruit-ripening, and that the texture and quality of most of our vegetable foods is intimately bound-up with the mechanical properties of the plant primary cell wall. Cell growth, fruit ripening and texture are three important parameters over which agronomists would like to exercise control in crop plants. Yet the successful manipulation of these parameters in a systematic manner will depend upon research progress in two related areas. A primary requirement is an understanding of how individual cell wall components interact to determine the mechanical properties of the cell wall. Also the eventual structural modification within the plant of individual cell wall components requires an understanding, at the molecular level, of the metabolic pathways which bring about their biosynthesis and metabolic turnover. Current ideas on the supramolecular interactions of the carbohydrate components of the primary cell wall are introduced below. Biosynthesis and turnover are covered in the following sections.

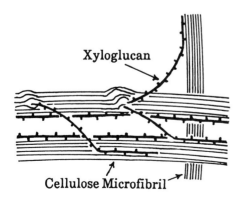

Figure 5.15 Potential linkages between xyloglucan and cellulose. Reproduced, with permission, Hayashi (1989).

5.4.1 Cellulose–hemicellulose interactions

There is no evidence for covalent linkages between cellulose, the microfibrillar component of the cell wall, and any of the carbohydrate or other components of the matrix. On the other hand, there is good evidence that certain of the hemicelluloses of the matrix undergo strong non-covalent binding to cellulose. The best-documented example of this is the interaction between xyloglucan, the main hemicellulose polysaccharide of the type 1 primary cell wall, and cellulose.

The total extraction of xyloglucan from cell walls is very difficult, requiring very high concentrations of alkali, or other reagents capable of disrupting hydrogen bonds. Once isolated, xyloglucan binds specifically to cellulose *in vitro* (Hayashi *et al.*, 1987), and can be removed from the surface of the cellulose microfibrils only by treatments which disrupt hydrogen bonds.

There can be little doubt that at least some of the xyloglucan present in the primary cell wall of type 1 is bound to cellulose. It is possible, given the known distance between cellulose microfibrils within the wall and the estimated length of a xyloglucan molecule, that a single xyloglucan molecule could bind to more that one cellulose microfibril, effectively tethering microfibrils together (Fig. 5.15). Micrographs of onion cell walls show clear intermicrofibril bridges (McCann *et al.*, 1990), which are visible after the removal of pectin but not after the extraction of xyloglucan (Fig. 5.16). Thus it is possible that the hydrogen-bonding interaction between cellulose and xyloglucan is the basis of a network within the native primary cell wall, known as the *xyloglucan–cellulose network*. There is insufficient xyloglucan in the type 2 primary cell walls of grasses for a xyloglucan–cellulose network to be important. Hemicelluloses present in the type 2 primary cell wall (mixed-linkage glucan and xylan) have, however, been shown to hydrogen-bond to cellulose, although less strongly than xyloglucan. It is possible that a network involving the bridging of cellulose microfibrils by these molecules exists in the type 2 wall.

5.4.2 Pectin interactions

There is little direct evidence that the pectin in the plant primary cell wall is covalently linked to hemicellulose or to cellulose, or that it is strongly hydrogen-bonded to other components. On the

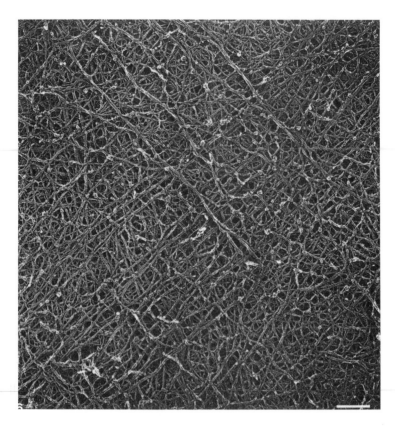

Figure 5.16   Electron micrograph of onion cell walls. Fast-freeze-deep-etch rotary-shadowed replica technique, following removal of most of the pectin from the wall. Note the thin bridges between thicker cellulose microfibrils. Bar = 200 nm. Reproduced from McCann *et al.* (1990), with permission.

other hand pectin undergoes self-interaction to the extent that it may itself form the basis of a network within the primary cell wall.

It is well known that many pectins, *in vitro*, form gels in the presence of calcium ions (Jarvis, 1984). The formation of a gel by an otherwise soluble polymer implies the formation of intermolecular *junction zones* (Fig. 5.17), and there is general agreement that in the case of pectin these take the form of an interaction between the negatively charged homogalacturonan domains (or smooth regions) and the divalent, positively charged calcium ions. There is good evidence that calcium-pectin crosslinks exist also in the type 1 primary cell wall. Calcium levels in the wall are high enough to support such interactions, treatment of cell walls with calcium chelators results in loss of wall strength, and histochemical studies with antibodies indicate that such structures are present within the wall

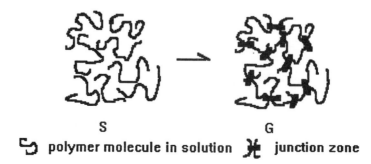

S   G

🖑 **polymer molecule in solution**   ⚓ **junction zone**

Figure 5.17   Simplified representation of the formation of a gel network, following the formation of junction zones between polymer molecules in solution. S, polymer molecules in solution. G, gel network.

(Liners *et al.*, 1992). For these reasons it is supposed that at least some of the pectin within the primary cell wall exists as a crosslinked network, *the pectin network*. Other types of crosslinkages are known to be possible within the pectin network, for example ester crosslinks and crosslinkages between the phenolic substituents which are known to be attached in low amounts to pectin. These will not be discussed here since direct evidence for them at the present time is scant.

The type 2 primary cell walls of the grasses contain very little pectin, and do not contain an equivalent molecular complex.

5.4.3 Primary cell wall models

Much of the pioneering work on the structures of the carbohydrate components of the primary cell wall was done in the laboratory of Peter Albersheim and coworkers at the University of Colorado, and it was that group which put forward the first tentative function model of the primary cell wall (the type 1 wall as defined here) (Keegstra *et al.*, 1973). The Albersheim model (Fig. 5.18) was the first of its kind and the clear

inspiration for more recent versions, it incorporated a framework of cellulose microfibrils linked to a xyloglucan–pectin–protein matrix by the xyloglucan–cellulose hydrogen-bonding interaction described above. The xyloglucan, pectin and protein components were depicted as being linked together via xyloglucan–pectin and pectin–protein covalent linkages (Fig. 5.18). No such covalent linkages have been generally confirmed (but see, Qi *et al.*, 1995) and more recent cell wall models are based on the hypothesis of *independent networks* within the cell wall. The type 1 cell wall is envisaged to be composed of the cellulose–xyloglucan network, the pectin network, plus in some cases a protein network, all contributing independently to the overall mechanical and other properties of the cell wall. The principal network within the type 2 cell wall may be considered to be cellulose–xylan. The networks within the type 1 cell wall were differentiated clearly by Carpita & Gibeaut (1993) (Fig. 5.19), although the representation is not to scale. The model by McCann & Roberts (1991) (Fig. 5.20) for the type 1 cell wall of onion has the cellulose microfibrils and the remaining matrix-filled spaces between them, scaled correctly relative to the overall thickness of the primary wall.

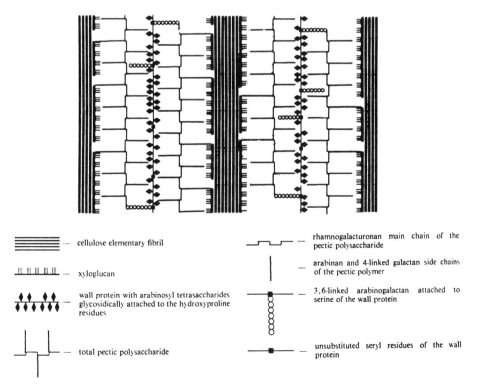

cellulose elementary fibril — === —

xyloglucan — ∥∥∥∥∥∥ —

wall protein with arabinosyl tetrasaccharides glycosidically attached to the hydroxyproline residues — ♦♦♦♦♦♦ —

total pectic polysaccharide

rhamnogalacturonan main chain of the pectic polysaccharide

arabinan and 4-linked galactan side chains of the pectic polymer

3,6-linked arabinogalactan attached to serine of the wall protein

unsubstituted seryl residues of the wall protein

Figure 5.18 The Albersheim model for polymer interactions in the plant primary cell wall. From Keegstra *et al.* (1973), with permission.

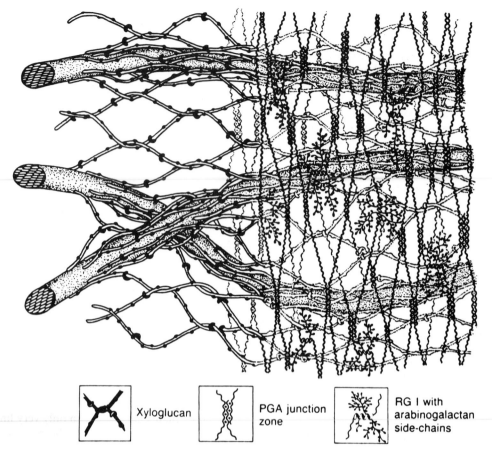

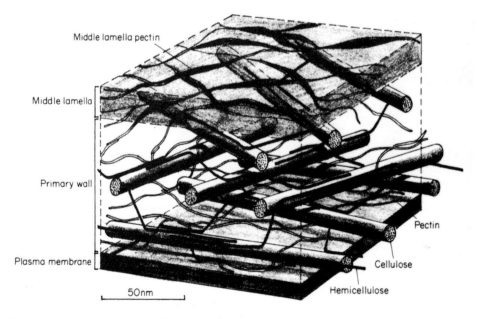

Figure 5.19 Diagrammatic interpretation of polysaccharide interactions in the type 1 primary cell wall. Reproduced from Carpita & Gibeaut (1993), with permission.

Figure 5.20 Schematic representation of how wall polymers might be spatially arranged in the primary cell wall of onion (McCann & Roberts, 1991). In this model the sizes and spacings of the polymers are based on direct measurements of native walls. Reproduced with permission.

Confronted with some of the current detailed models (Figs 5.19, 5.20), the reader might be excused for thinking that the biochemistry of the plant primary cell wall is understood. It is not! Although the various networks mentioned above are probably present in the wall, we know relatively little about their relative contributions to overall wall properties. Although the types of structural domains present in hemicelluloses and pectin are known, we do not know how they contribute individually to the establishment and maintenance of networks, or influence wall properties. Nor do we understand fully the molecular alterations to the wall networks which accompany and control the process of cell expansion. These problems will be approached more closely when it is possible to alter the structures of hemicelluloses and pectin within the wall and observe the mechanical and developmental consequences of the structural changes. To achieve this it is desirable, if not essential, to gain an understanding of the pathways of biosynthesis and metabolic turnover of individual cell wall components. These topics are discussed in sections 5.5 and 5.6.

5.5 BIOSYNTHESIS OF STRUCTURAL POLYSACCHARIDES

The biosynthesis of structural cell wall carbohydrates must be seen in the overall context of the carbohydrate metabolism of the cell. The photosynthetic process (Chapter 2), and the processes of breakdown of starch and other carbohydrates (Chapter 4) are relevant. It is these processes which generate sucrose, the main transported carbohydrate in higher plants. The pathways of sucrose catabolism (Chapter 4) are relevant. They are the main energy-generating process in most plant cells,

and they are responsible for the generation of phosphorylated sugar intermediates which are in turn required for the synthesis of the sugar nucleotides (Chapter 12) which serve as direct precursors for the formation of cell wall polysaccharides. In this section, only the synthesis of polysaccharides from sugar nucleotide precursors will be discussed. It should, however, be remembered that the pathways of sugar nucleotide formation are important in the regulation of intracellular sugar nucleotide levels, which may themselves participate in the regulation of cell wall polysaccharide biosynthesis.

The sugar nucleotides which have been demonstrated to be, or are believed to be, precursors of cellulose, hemicellulose and pectin are listed in Table 5.1. The current state of knowledge of the biosynthetic processes themselves is summarized below.

5.5.1 Cellulose

Attempts to demonstrate cellulose biosynthesis (Delmer, 1987) *in vitro* using enzyme preparations from higher plants have met with only very limited success, although the subject has been researched intensively for over 25 years. This is surprising, since the molecular structure of cellulose (linear $\beta(1 \rightarrow 4)$-glucan, section 5.3.1) is relatively simple, and cellulose biosynthesis is a major metabolic process in almost all plant cells. By contrast, a great deal is known about the mechanism of cellulose biosynthesis in the bacterium *Acetobacter xylinum*. The cellulose of *A. xylinum* is not part of the cell wall of the bacterium. It is synthesized by the bacterial cell and secreted into the culture medium in the form of long microfibrils of cellulose I (Brown *et al.*, 1983).

Table 5.1 Some sugar nucleotides thought to be involved in hemicellulose and pectin biosynthesis

Nucleotide precursor	Polysaccharide product
GDP-fucose	Xyloglucan
GDP-mannose	Glucomannan, galactomannan[a]
GDP-glucose	Glucomannan
UDP-arabinose	Arabinoxylan
UDP-galactose	Pectin, galactomannan[a]
UDP-galacturonic acid	Pectin
UDP-glucose	$\beta(1 \rightarrow 3)(1 \rightarrow 4)$-glucan, callose, xyloglucan
UDP-glucuronic acid	Glucuronoxylan
UDP-xylose	Xylans, xyloglucan

[a] Cell wall storage polysaccharide. See section 5.5.2c.

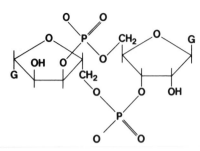

Figure 5.21 Structure of cyclic diguanylic acid. G, guanine. Reproduced, with permission, from Delmer (1987).

In *A. xylinum*, the high-energy glucose donor for cellulose biosynthesis is UDP-glucose (UDP-Glc). The enzyme complex responsible for the synthesis (*cellulose synthase*) is tightly bound to the bacterial cell membrane. Careful detergent treatment of the membranes results in their dispersion and the formation of a solution containing protein/detergent/lipid micelles which retain low cellulose synthase activity. A high level of activity is restored by adding the effector molecule *cyclic diguanylic acid* (Fig. 5.21), which was isolated from *A. xylinum* extracts and is believed to be an activator of the enzyme complex *in vivo* (Ross *et al.*, 1987).

The cellulose synthase of *A. xylinum* was partially purified by a technique known as *product entrapment*, which has proved useful for the purification of soluble or detergent-solubilized enzymes which catalyze the formation of an insoluble product. When the insoluble cellulose fibers formed by the detergent-solubilized *A. xylinum* cellulose synthase are collected by centrifugation, the synthetic enzymes remain associated with the cellulose and can be partially separated from contaminating proteins (Fig. 5.22) (Lin & Brown, 1989).

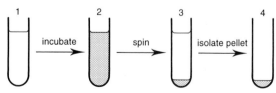

1. Detergent-solubilised enzyme preparation plus NDP-sugar substrate
2. Suspension of insoluble polysaccharide product
3. Product pelleted by centrifugation
4. Isolated pellet may contain polysaccharide synthase enzyme(s) specifically bound to product

Figure 5.22 Diagrammatic representation of product entrapment.

The catalytic subunit of *A. xylinum* cellulase synthase was identified positively by a further technique called *photoaffinity labeling*, which has been used to identify a variety of membrane-associated enzymes or receptors (Lin *et al.*, 1990). A radiolabeled, photoreactive analog of a substrate or other ligand is mixed at low concentration with the membrane preparation, and the mixture is irradiated with ultraviolet light to convert the photoactivable reagent into a chemical species so reactive that it will bind covalently to any molecule in its vicinity. Reagent molecules tightly bound to a protein will react with it, thus radiolabeling it and permitting its localization on SDS-gels by autoradiography. Non-bound reagent molecules react with water. The use of 5'-azido-UDP-glucose (Fig. 5.23), a UDP-Glc photoanalog, in experiments of this type allowed the positive identification of the catalytic subunit of *A. xylinum* cellulose synthase, eventually permitting the cloning and characterization of the genomic sequence encoding the protein (Saxena *et al.*, 1990).

There is considerable ultrastructural evidence that plant cellulose synthesis occurs at the plasma membrane of the cell, and that rosette-like structures on the plasma membrane may be the catalytic centers (Fig. 5.24). Plasma membrane preparations, or indeed any mixed membrane preparations from higher plants, catalyze at best only extremely small amounts of $\beta(1 \rightarrow 4)$-linked glucan from UDP-glucose, which may possess some of the properties of cellulose (Okuda *et al.*, 1993). Much glucan is formed, but it is mainly $\beta(1 \rightarrow 3)$-linked. $\beta(1 \rightarrow 3)$-Linked glucan is not a normal constituent of the plant cell wall, but it is formed by many plant cells as a response to injury, and is usually called 'callose'. Thus any breakage of the plant cell membrane 'switches' its biosynthetic machinery to the formation of wound callose. The essential completeness of the switch-over, and other circumstantial evidence, has fostered the idea that perhaps the enzymes responsible for the synthesis of callose and cellulose are one and the same, and that cell breakage triggers a change in the transfer specificity from $(1 \rightarrow 4)$ to $(1 \rightarrow 3)$, mediated possibly by the elimination of the membrane potential and changes in ion-concentration following cell disruption. Should this be true, current efforts in several laboratories throughout the world to identify the enzymes involved in callose synthesis would have enhanced significance.

An obvious approach to the identification of the catalytic subunit of higher plant cellulose synthase is to probe plant cDNA libraries using the

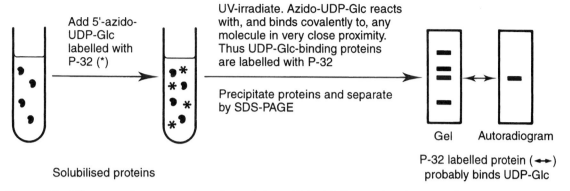

Figure 5.23 Diagrammatic representation of use of photoaffinity labeling to identify a UDP-glucose binding protein.

sequence of the *A. xylinum* cellulose synthase catalytic subunit. At the present time, however, no success has been reported, indicating that the two sequences may not be highly homologous. Nor has the addition of cyclic diguanylic acid to plant membrane preparations been reported to enhance cellulose synthesis *in vitro*. Thus, at the time of writing, the biosynthesis of the most abundant pure substance on earth remains something of an enigma.

5.5.2 Hemicelluloses and pectin

Although the matrix polysaccharides are more complex structurally than cellulose, attempts to obtain their biosynthesis *in vitro* from sugar nucleotide precursors have been accompanied by some success. The biosynthesis *in vitro* of several hemicelluloses, notably glucomannan (section 5.3.2a), xylan (sections 5.3.2a and b), xyloglucan (section 5.3.2b) and mixed linkage $\beta(1 \to 3, 1 \to 4)$-glucan (section 5.3.2b) has been reported. Of the pectin structural domains, only galactan

(section 5.3.3) biosynthesis has been demonstrated convincingly.

The enzymes catalyzing the formation of matrix polysaccharides from sugar nucleotide precursors are tightly bound to membranes. The subcellular origin of the active membranes has not always been determined, but when it has, Golgi membranes have always been indicated. This is in contrast to cellulose and callose biosynthesis, which are plasma membrane associated.

To detect the biosynthesis of cell wall matrix polysaccharides *in vitro*, tissues are normally homogenized, and partially purified membrane preparations are obtained by centrifugation protocols. The membrane preparations are incubated with appropriate sugar nucleotide precursors, labeled (usually with ^{14}C) in the sugar moiety, and labeled high-molecular-weight material is separated from non-incorporated precursor and from any low-molecular-weight labeled products (Fig. 5.25). In experiments such as this, the identification of the high-molecular-weight labeled material is of the utmost importance. For example, in an experiment to investigate the incorporation

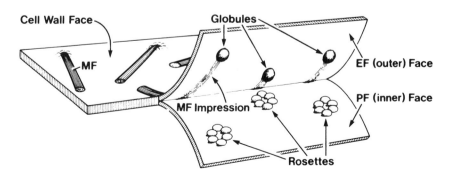

Figure 5.24 Stylized drawing of structures revealed by freeze-fracture of plasma membranes of certain algae and of higher plants. The 'rosettes' are believed to be terminal complexes associated with the ends of growing cellulose microfibrils (MF). Reproduced, with permission, from Delmer (1987).

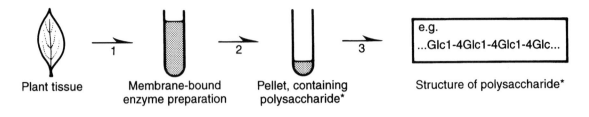

1. Homogenisation followed by differential centrifugation
2. Incubation with sugar nucleotide labelled with C-14(*) in the sugar part (e.g. UDP-Glc*). Precipitation of labelled polysaccharide, and removal of remaining labelled nucleotide by centrifugation and washing pellet
3. Structural studies on labelled polysaccharide (methylation analysis or enzymatic analysis)

Figure 5.25   Schematic representation of methods used to demonstrate cell wall polysaccharide biosynthesis *in vitro.*

of label from UDP-[^{14}C]glucose, several different hemicelluloses (glucomannan, xyloglucan, mixed-linkage glucan), as well as callose or even cellulose are all possible products. The interpretation of some of the earlier work on matrix polysaccharide biosynthesis is difficult, because it was carried out before techniques for the characterization of very small amounts of polysaccharides became available. In recent work, labeled polysaccharides formed *in vitro* have increasingly been subjected to full methylation analysis or to fragmentation using enzymes of known substrate specificity. Glucomannan (a secondary cell wall hemicellulose (section 5.3.2a)) and mixed-linkage glucan (a hemicellulose of type 2 primary cell walls (section 5.3.2b)) biosynthesis are used below as examples of the modern approach to product characterization.

(a)   *Glucomannan biosynthesis*

The biosynthesis of glucomannan, the principal hemicellulose of the secondary cell walls of coniferous woods, provides a good example of the importance of characterizing polymeric products formed *in vitro* from sugar nucleotide precursors (Dalessandro *et al.,* 1986). Membrane preparations from cambial cells (i.e. cells actively differentiating into wood fibers with secondary walls) of pine catalyzed the incorporation of label from GDP-[^{14}C]mannose into a labeled polymeric product which, on acid hydrolysis, released not only labeled mannose, but also labeled glucose. Similarly when labeled GDP-mannose was the precursor a polymeric product was formed which yielded both labeled glucose and mannose on hydrolysis. On methylation analysis the product clearly contained glucose and mannose residues linked together in a linear fashion by (1 → 4)-linkages as in glucomannan. Thus the catalytic membranes from pine cambium must contain an enzyme capable of interconverting GDP-mannose and GDP-glucose (a GDP-mannose-2-epimerase) and glycosyltransferases capable of catalyzing the transfer of mannose and glucose residues from GDP-mannose or GDP-glucose to elongate a growing glucomannan chain (Fig. 5.26). The membrane-bound enzymes have not yet been isolated and purified, and little is known about the control of the ratio mannose/glucose *in vivo.*

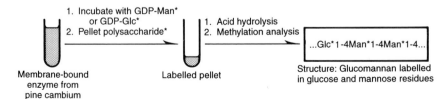

Figure 5.26   Schematic representation of methods used to demonstrate secondary cell wall glucomannan biosynthesis *in vitro.* (Dalessandro *et al.,* 1986.)

(b)   Mixed-linkage β-glucan biosynthesis

The biosynthesis of mixed-linkage $\beta(1 \to 3, 1 \to 4)$-glucan, a main hemicellulose of type 2 primary cell walls (section 5.3.2b), provides an excellent example of the value of enzymes in probing the structures of small amounts of labeled polysaccharides formed *in vitro*. Membranes isolated from tissues or cell cultures of plant species with type 2 primary walls catalyze the incorporation of label from UDP-[^{14}C]glucose into polymeric material. As mentioned above in the context of cellulose biosynthesis, the plasma membranes of ruptured plant cells have the ability to catalyze the synthesis of $\beta(1 \to 3)$-glucan (callose) from UDP-glucose. Thus any mixed-linkage glucan product would inevitably be contaminated by callose, possibly in large amounts, and other glucan products are possible. Methylation analysis might provide information on the relative numbers of $(1 \to 3)$- and $(1 \to 4)$-linkages in the mixture, but assignment of them to individual $\beta(1 \to 3)$- and $\beta(1 \to 3, 1 \to 4)$-glucan polymers would be impossible. By contrast, the enzyme lichenase (section 5.3.2b), which hydrolyzes a $\beta(1 \to 4)$-glucosyl linkage only if it immediately follows a $\beta(1 \to 3)$-glucosyl linkage, gave a means of identifying and analyzing $\beta(1 \to 3, 1 \to 4)$-glucan even in the presence of large amounts of other β-glucans. Treatment of labeled *in vitro* polysaccharide with this enzyme hydrolyzes $\beta(1 \to 3, 1 \to 4)$-glucan selectively, converting it into oligosaccharides with the general structure $[\text{Glc}(1 \to 4)]_n\text{Glc}(1 \to 3)\text{Glc}$, where $n \geq 1$ (section 5.3.2b). Accordingly the use of this enzyme has not only allowed the quantitative estimation of $\beta(1 \to 3, 1 \to 4)$-glucan in the presence of other products, but also the calculation of the relative numbers of $\beta(1 \to 4)$-linked 'blocks' of different lengths in the labeled polysaccharide product (Fig. 5.27) (Cook *et al.*, 1980; Henry & Stone, 1982; Becker *et al.*, 1994).

In this way, the biosynthesis of $\beta(1 \to 3, 1 \to 4)$-glucan *in vitro* from UDP-glucose has been demonstrated conclusively. Furthermore it has been demonstrated that the ratio of trisaccharide to tetrasaccharide structural blocks within the polymer formed *in vitro* is constant, and equal to that in the native cell wall glucan, irrespective of the experimental conditions under which the *in vitro* synthesis is carried out (Becker *et al.*, 1994). This indicates that the structure of the polymer is not regulated via intracellular substrate or cofactor levels, but by the specificity of the membrane-bound glycosyltransferase enzyme(s). It is possible that a single transferase is present, catalyzing the formation of both linkage-types. Equally there may be two glucosyltransferases, each forming a single linkage-type, but very tightly coupled functionally. The answer will no doubt become clear when current attempts to purify the transferase(s) succeed.

(c)   Regulation of hemicellulose and pectin biosynthesis

Until the pathways of hemicellulose and pectin biosynthesis have been defined fully, there can be no definitive description of the regulation of these processes. However knowledge of the biosynthesis of certain polymers has advanced to the point where indicators as to the possible control mechanisms can be obtained.

As mentioned above, the molecular fine structure of the product of $\beta(1 \to 3, 1 \to 4)$-glucan biosynthesis *in vitro* from UDP-glucose cannot be altered by changing the conditions of synthesis, indicating that control of the structure of the product lies exclusively at the level of the membrane-bound glycosyltransferases. Tight control of polymer structure at the transferase level has also been demonstrated in the case of xyloglucan biosynthesis, where the formation of the glucan backbone cannot continue without simultaneous formation of xylosyl side chains

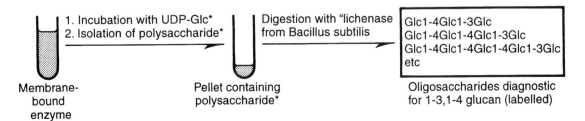

Figure 5.27   Schematic representation of methods used to demonstrate $\beta(1 \to 3)(1 \to 4)$-glucan biosynthesis *in vitro*.

Table 5.2 Examples of cell wall storage polysaccharides in seeds

Polysacccharide	Typical source
Galactomannan	Endosperms, mainly of leguminous seeds
Glucomannan	Endosperms, e.g. of *Asparagus*, *Endymion*, *Iris*
Mannan	Endosperms of palm seeds (e.g. date), Endosperms of Umbelliferae
Xyloglucan	Cotyledons of several species, e.g. tamarind, nasturtium (*Tropaeolum*)
Galactan	Cotyledons of some lupins
$\beta(1 \rightarrow 3)(1 \rightarrow 4)$-glucan	Endosperms of cereal grasses

(Fig. 5.9) (Hayashi & Matsuda, 1981; Gordon & Maclachlan, 1989). Little information is available currently on the formation of the galactosyl substituents attached to xylose, but the outer fucosyl substituents (attached to galactose) are apparently formed independently of the other structural features, by the action of a fucosyltransferase which catalyzes the transfer of fucosyl residues from GDP-fucose to pre-formed galactosylated xyloglucan polymer (Hanna *et al.*, 1991).

Insufficient information is available on the factors regulating the frequency of occurrence of the different sugar residues in the majority of cell wall matrix carbohydrates. These are either linear mixed polymers (glucomannans, rhamnogalacturonans) with two or more different types of sugar residue in the main chain, or consist of a backbone which is decorated by short side chains (galactoglucomannans, 4-O-methylglucuronoxylans, arabinoxylans etc.). In neither group is there a regular repeating pattern as in the xylosyl-glucose core of the xyloglucans, and it has been suggested repeatedly that structural features may be distributed at random. Studies on *galactomannan* biosynthesis, however, indicate strongly that this is not the case, and that precise statistical rules may govern the distribution of substituents in such molecules.

The galactomannans are matrix polysaccharides (hemicelluloses) present in large amounts in the endosperm cell walls of leguminous seeds (Reid, 1985a). Leguminous endosperm cell walls are specialized both in structure and composition. A thin primary cell wall, of apparently normal composition, is present plus a very thick inner wall layer composed almost exclusively of galactomannan (no cellulose, apparently no other polysaccharide components). The galactomannan is laid down during the development of the seed, and is mobilized after germination, acting as a carbohydrate reserve for the developing seedling. Galactomannan is an example of a *cell wall storage polysaccharide* – a seed cell wall polysaccharide which serves as a storage molecule (Reid, 1985b). Cell wall storage polysaccharides are all similar in structure to hemicelluloses or pectin structural features present in the normal primary or secondary wall, from which they almost certainly arose in the course of evolution. Other examples of cell wall storage polysaccharides are xyloglucan, glucomannan, mixed-linkage $\beta(1 \rightarrow 3, 1 \rightarrow 4)$-glucan and arabinoxylan (Table 5.2).

The galactomannans have a linear $\beta(1 \rightarrow 4)$-linked D-mannan backbone, which carries single-unit D-galactosyl side chains linked $\alpha(1 \rightarrow 6)$. They, therefore, bear a strong structural resemblance to the galactoglucomannans of secondary cell walls (Fig. 5.28; compare Fig. 5.7). Although all galactomannans conform to the structural pattern in Fig. 5.28, the galactomannans from the seeds of leguminous species are galactose-substituted to different degrees. Mannose/galactose (Man/Gal) ratios range from just over 1 (high-galactose) to about 3.5 (low-galactose). By studying galactomannan biosynthesis in three leguminous species, fenugreek, guar and senna, which form high-, medium- and low-galactose galactomannans, respectively, it has been possible to identify the enzymes responsible for galactomannan biosynthesis and to pinpoint the biochemical

```
               Gal1          Gal1          Gal1
                 ↓α            ↓α            ↓α
        ß      ß 6  ß      ß 6  ß      ß 6  ß      ß      ß
----Man1→4Man1→4Man1→4Man1→4Man1→4Man1→4Man1→4Man1→4Man1→4Man1→ ---
          6                          6             6             6
          ↑α                         ↑α            ↑α            ↑α
        Gal1                       Gal1          Gal1          Gal1
```

Figure 5.28 Structural features of galactomannans.

levels at which the Man/Gal ratio is controlled (Edwards *et al.*, 1989, 1992; Reid *et al.*, 1995).

Two mechanisms are involved, the biosynthetic process, and a post-depositional modification. In fenugreek and guar, the Man/Gal ratios (1.1 and 1.6, respectively) are determined by the biosynthetic mechanism alone. On the other hand in senna, the Man/Gal ratio (3.3) is established by the controlled removal of galactose by α-galactosidase action from a primary biosynthetic product (Man/Gal ratio = 2.3) during late seed development (Edwards *et al.*, 1992). Post-depositional modification may be of much wider importance in the control of the structures of other cell wall polysaccharides than has been supposed hitherto. In the case of pectin, the removal of methyl ester groups by pectin methylesterase action is known to occur.

Two tightly membrane-bound enzymes are involved in the biosynthesis of galactomannans. A GDP-mannose-dependent β(1 → 4)-D-mannosyltransferase (or mannan synthase) is responsible for elongating the mannan backbone towards the non-reducing end by catalyzing the transfer of mannosyl residues one at a time from GDP-mannose. A highly specific UDP-galactose-dependent α-D-galactosyltransferase catalyzes the transfer of a galactosyl residue to an acceptor mannose residue at or near the non-reducing end of the growing D-mannan chain (Fig. 5.29). *In vitro*, in the presence of labeled GDP-mannose alone (no UDP-galactose), the product is labeled β(1 → 4)-D-mannan. In the presence of GDP-mannose and UDP-galactose, the product is a galactomannan. In the presence of labeled UDP-galactose alone, no labeled galactomannan is formed. Some degree of control of the Man/Gal ratio of the *in vitro* galactomannan can be obtained by varying the *in vitro* concentrations of the two sugar nucleotides; by lowering the GDP-mannose concentration at constant (saturating) UDP-galactose, higher degrees of galactose substitution may be obtained. Thus the formation of the galactosyl side chains in galactomannans is strictly dependent on simultaneous elongation of the mannan chain, whereas the mannan chain can be elongated independently of the formation of side chains (Edwards *et al.*, 1989).

In the investigation of galactomannan biosynthesis, as in the case of β(1 → 3, 1 → 4)-glucan, enzymatic hydrolysis was crucial to the identification and characterization of labeled *in vitro* products of biosynthesis. The enzyme in this case was an endo-β(1 → 4)-D-mannanase from *Aspergillus niger*. This enzyme recognizes a sequence of five mannosyl residues in a mannan chain, hydrolyzing between residues 3 and 4 numbered from the non-reducing end (Fig. 5.30) (McCleary & Matheson, 1983). Galactosyl substitution of the mannan chain in galactomannans affects the action of the enzyme in a very well-defined fashion, substituents at residues 2 and/or 4 of the recognition sequence preventing hydrolysis. Thus galactomannans with higher degrees of substitution are, in general, hydrolyzed to a lesser extent than galactomannans with lower degrees of substitution. However, the pattern of substitution of the galactomannan also influences both the degree of hydrolysis, and the quantitative distribution of galactomannan oligosaccharide fragments released in the enzyme digests.

Galactomannan biosynthesis by the experimental model depicted in Fig. 5.29 has been computer-simulated with the inbuilt assumption of a *second-order Markov-chain*. That is to say, it has been assumed that the probability of obtaining galactose-substitution at the acceptor mannosyl residue (labeled a in Fig. 5.29) is influenced by the pre-existing state of substitution at only the nearest and the second-nearest neighbor mannosyl residues (labeled 1 and 2 in Fig. 5.29). On this assumption

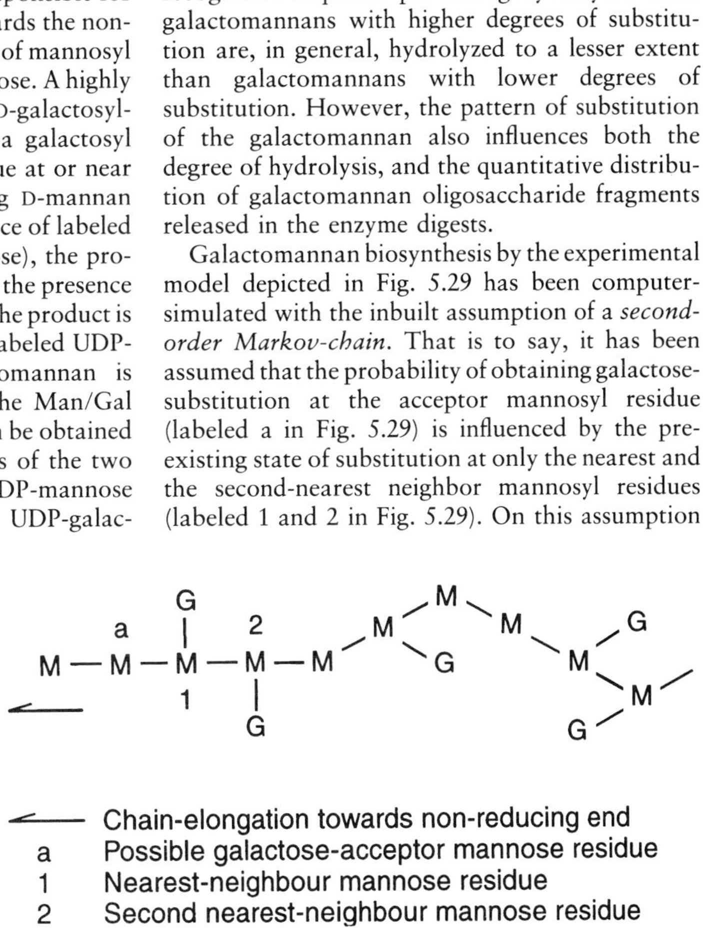

Figure 5.29 Enzyme interaction in galactomannan biosynthesis. A GDP-mannose-dependent mannan synthase catalyzes the elongation of the mannan backbone towards the non-reducing end (arrow). A UDP-galactose-dependent galactosyltransferase catalyzes the transfer of a galactosyl residue to an acceptor mannose residue (a) at or near the end of the growing mannan chain. 1,2, nearest and second-nearest neighbor mannosyl residues to the acceptor residue (see text).

— Chain-elongation towards non-reducing end
a — Possible galactose-acceptor mannose residue
1 — Nearest-neighbour mannose residue
2 — Second nearest-neighbour mannose residue

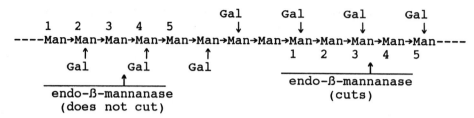

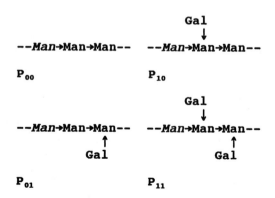

Figure 5.30   Action on galactomannans of an endo-$\beta(1 \rightarrow 4)$-D-mannanase from *Aspergillus niger* (McCleary & Matheson, 1983).

there are four independent probabilities of obtaining galactose substitution at the acceptor mannose: P_{00}, P_{10}, P_{01} and P_{11} (Fig. 5.31). Any numerical set of four probabilities will then suffice for the computer-generation of a hypothetical galactomannan with a uniquely described distribution of galactose residues along the mannan backbone and a defined Man/Gal ratio. The action of the *A. niger* mannanase has also been computer-simulated, so that the input of a probability tetrad into the computer will generate not only a unique galactomannan structure but the relative amounts of the various manno- and galactomanno-oligosaccharides which would be released on the exhaustive hydrolysis of that galactomannan using the *A. niger* mannanase. Fragmentation, using the *A. niger* mannanase, of labeled galactomannans generates the relative amounts of labeled manno- and galactomanno-oligosaccharides as experimental data. Thus by inputting an experimental data set derived from the enzymatic fragmentation of a labeled galactomannan biosynthesized *in vitro*, the computer algorithm will generate a unique probability tetrad. This was done for galactomannans, covering a wide range of Man/Gal ratios formed *in vitro* using the enzymes from the high-, medium- and low-galactose species. To allow comparison of probability-sets derived from galactomannans with different Man/Gal ratios all probability sets were scaled linearly, with P_{00} being ascribed the arbitrary value 1.00. When this was done it was observed that all probability sets from a single species were closely similar, yet quite different from those obtained from the other two species. This confirmed the correctness of the second-order Markov-chain model, and demonstrated clearly that the specificities of the galactomannan-forming transferase enzymes were different in the three species. The transfer specificities, defined mathematically in the probability sets, may each be considered to define a maximum Man/Gal ratio for a galactomannan synthesized according to the transfer rules thus defined. This is the Man/Gal ratio of the galactomannan formed when the highest of the four probabilities within a set is set to 1.00 and the other three are rescaled correspondingly in a linear fashion. When this was done for the fenugreek, guar and senna consensus probability sets, the maximum galactose-content predicted for each species was either identical with or only very slightly higher than that of the primary product of the biosynthetic machinery. This was remarkable, indicating not only that the distribution of substituents on the main chain of the galactomannan, but also the total amount of substitution is determined by the specificities of the membrane-bound transferase enzymes (Reid *et al.*, 1995).

Although galactomannans are not present generally in cell walls, most hemicelluloses have a non-regular distribution either of different residues within a backbone or substituents along a linear backbone. It seems likely that the application of Markov-chain statistics in these cases also, will reveal the operation of precisely defined statistical rules generated by the specificities of highly organized membrane-bound transferase complexes.

Figure 5.31   Illustration of the four possible states of substitution at the nearest and second-nearest neighbor mannosyl residues to the galactosylacceptor mannosyl residue (italicized). P_{00}, P_{10}, P_{01} and P_{11} are the corresponding probabilities of obtaining galactose substitution at the acceptor mannosyl residue.

5.6 METABOLIC TURNOVER OF STRUCTURAL COMPONENTS

The plant cell wall is an extracellular matrix and, as such, might be expected to undergo only slow metabolic turnover. It is unlikely that the lignified walls of non-living cells such as vessel elements and fibers are turned over. On the other hand, living cells might be expected to possess the metabolic machinery to rearrange the molecular structures of their (mainly primary) cell walls. There is ample evidence that they do. The cell wall storage polysaccharides of seeds, for example (section 5.5.2c) are completely mobilized after germination, and the enzymatic mechanisms involved have in several cases been described in detail (Reid, 1985b). This section highlights two incidences where the enzymatic modification of a particular cell wall matrix component is associated with, and believed to control, developmental events, namely xyloglucan turnover in the context of cell elongation, and pectin turnover in the context of fruit ripening. It draws the reader's attention finally to the steadily increasing evidence that oligosaccharide fragments, originating almost certainly from the hydrolysis *in vivo* of cell wall polysaccharides, are powerful regulatory and signaling molecules in plants.

5.6.1 Xyloglucan turnover and cell growth

The process of cell wall expansion (Taiz, 1984) is one in which the wall of the cell extends in response to the internal pressure (turgor) of the cell's contents. Although turgor pressure is necessary for growth to occur, there is much evidence that, in higher plants at least, cell expansion is controlled by metabolic events occurring within the cell wall itself. In other words, cell expansion occurs as a result of cellular control over wall *extensibility*. The changeover from a non-extensible to an extensible cell wall is often called '*loosening*'. The biochemical mechanism of loosening (which must, of course, be reversible) has occupied cell wall biochemists for some years, and there is a great deal of circumstantial evidence that, in type 1 primary cell walls at any rate, xyloglucan modification may be involved.

Xyloglucan has been associated with cell expansion following the demonstration that it is strongly hydrogen-bonded to cellulose (section 5.3.2b), and especially since the realization that a xyloglucan–cellulose network may exist within the primary cell wall (section 5.4.3). Initially it was suggested that wall loosening might be brought about by the controlled weakening of the hydrogen bonds between cellulose and xyloglucan, thus allowing relative movement of the cellulose microfibrils. Change in wall pH was suggested as a possible agent to induce weakening of the hydrogen bonds, since the growth-promoting substance auxin is known to bring about wall acidification by stimulating the active transport of hydrogen ions out of the cytoplasm. In the absence of evidence for the direct weakening of cellulose–xyloglucan binding by naturally occurring pH changes within the wall (Valent & Albersheim, 1974), attention turned to the possibility that glycosyl linkages within xyloglucan molecules might be broken, possibly reversibly by transglycosylase action (Albersheim, 1975).

Xyloglucan was further implicated in auxin-induced growth by the observation that auxin-induced elongation of stem sections of peas and other dicotyledonous plants was accompanied by xyloglucan turnover and solubilization (Labavitch & Ray, 1974). Xyloglucan solubilization occurred as a response to auxin even when growth was prevented by lowering cell turgor using osmotically active substances. Auxin treatment also increased the levels of the enzyme *endo-β(1 → 4)-D-glucanase* (Fan & Maclachlan, 1966) (often incorrectly called 'cellulase'). On purification (Byrne *et al.*, 1975), this enzyme was shown to hydrolyze xyloglucan in preference to cellulose (Hayashi *et al.*, 1984), the final products of xyloglucan hydrolysis *in vitro* being the Glc_4Xyl_3-based structural subunits depicted in Fig. 5.9. The possible involvement of the *endo-β(1 → 4)-D-glucanase* in the control of growth was further promoted by the observation that one of the structural subunits of xyloglucan, the nonasaccharide $Glc_4Xyl_3GalFuc$ (Fig. 5.9) was able, at nanomolar concentrations, to inhibit auxin-induced extension of pea stem sections (York *et al.*, 1984). Was this a natural feedback control mechanism?

It was recognized that *endo-β(1 → 4)-D-glucanase* action alone could not control cell expansion, since wall-loosening is reversible, and the glucanase was capable of breaking glycosyl linkages but not re-forming them. More recently an enzyme has been discovered which breaks internal linkages in the xyloglucan backbone reversibly. This enzyme, xyloglucan *endo-transglycosylase* (or XET), catalyzes the breakage of the intersubunit glycosyl linkage in xyloglucans, freeing the portion of the polymer towards the reducing end and transferring the portion of the molecule towards the non-reducing end onto the non-reducing end of a xyloglucan or xyloglucan oligosaccharide acceptor. The first XET to be purified was from germinated nasturtium seeds (Edwards *et al.*, 1986; Fanutti *et al.*, 1993) where it is involved in the post-germinative depolymerization of the cell wall storage xyloglucan localized in the cotyledons. This enzyme is also able to hydrolyze xyloglucans, hydrolysis predominating over transglycosylation only when the concentration of suitable chain acceptors is very low (Fig. 5.32). XET activity has been detected in a wide variety of plant tissues (Fry *et al.*, 1992), and has been purified from elongating bean hypocotyls (Nishitani & Tominaga, 1992). In contrast to the nasturtium seed XET, the bean enzyme did not appear to have any hydrolytic action. The amino acid sequences of the nasturtium (de Silva *et al.*, 1993) and bean (Okazawa *et al.*, 1993) enzymes (deduced from cDNA clones) are closely similar.

The action of XET is exactly that which might be predicted for an enzyme involved in loosening a xyloglucan–cellulose network, and this may be its role in the plant. There is no evidence yet, however, that it is involved in elongation growth. An *Arabidopsis* gene, the product of which is, on the basis of its deduced amino acid sequence, almost certainly an XET, is expressed mainly in areas where cells are dividing rather than expanding (Medford *et al.*, 1991). Furthermore, extracts containing high XET activities did not accelerate the elongation of frozen-thawed plant stem segments placed under tension (McQueen-Mason *et al.*, 1993). On the other hand, two proteins have been isolated from elongating cucumber stem segments which have exactly this property (McQueen-Mason *et al.*, 1992). Named *expansins*, they promote the extension of tissue segments under tension, but have no apparent hydrolytic activity against cell wall polysaccharide substrates. Their action is clearly catalytic, with an optimum around pH 3.5–4.5, but there is still controversy surrounding the nature of their action. The observation that the expansins catalyze the extension of paper under tension (McQueen-Mason & Cosgrove, 1993) suggests that they may well loosen non-covalent linkages.

Is there a xyloglucan–cellulose network within the type 1 primary cell wall? Does the loosening of the xyloglucan–cellulose network control elongation growth? Are *endo*-glucanases, XET and expansins involved? Answers to these questions will clearly be obtained very soon.

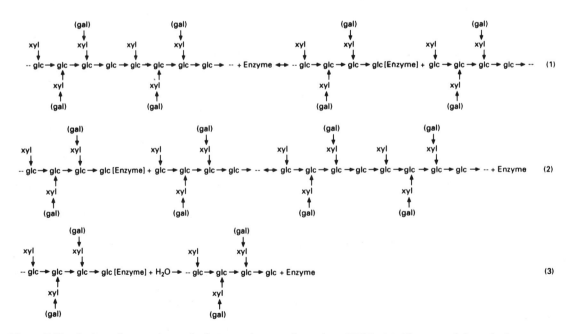

Figure 5.32 Action of nasturtium xyloglucan *endo*-transglycosylase (XET). (1) Cleavage of the xyloglucan substrate, forming a donor–enzyme complex and releasing a (potential acceptor) xyloglucan fragment. (2) Transfer of donor from enzyme complex to an acceptor xyloglucan chain-end. (3) Decomposition of donor–enzyme complex by hydrolysis (transfer of donor fragment to water). (1) + (2) = transglycosylation. (1) + (3) = hydrolysis of xyloglucan. Hydrolysis occurs only when donor concentrations are very low. The substrate illustrated is a seed xyloglucan, which does not have fucosyl side chains. (gal) = possible galactose-substitution. Reproduced from Fanutti *et al.* (1993), with permission.

5.6.2   Pectin metabolism in fruit ripening

The ripening of a fruit is a highly complex developmental sequence, involving wide-ranging metabolic changes (Brady, 1987). For example, chlorophyll biosynthesis may cease and the levels of other pigments may increase, respiration rates may increase dramatically, acid levels may decrease and sugars increase, and the flesh of the fruit will usually soften. Since softening is a change in the mechanical properties of the fruit tissue, and mechanical properties are strongly associated with the cell wall, it is scarcely surprising that there is a considerable literature on changes in the structures of cell wall components associated with ripening (Fischer & Bennett, 1991). The cell wall changes accompanying tomato fruit ripening are particularly well documented, and are highlighted below.

The cell walls of many fruit tissues, including those of tomato, are particularly rich in pectin, and most work on ripening-associated cell wall changes has concentrated on this component. In the case of tomato, light microscopic evidence has suggested strongly that primary cell wall and middle lamella pectin were extensively degraded during ripening. Furthermore, the visible changes in wall ultrastructure could be mimicked to a great extent by treatment of the tissues of unripe fruit with purified $endo$-$\alpha(1 \rightarrow 4)$-D-polygalacturonase (or pectinase) (Crookes & Grierson, 1983). The effect of the pectinase is to hydrolyze the pectin backbone, mainly within the homogalacturonan domains, thus causing extensive depolymerization. It could effectively disintegrate any pectin network within the wall. In tomato and some other fruits, ripening is accompanied by a large increase in the activity within the tissue of a plant $endo$-polygalacturonase, presumed to be responsible for pectin depolymerization and, by implication, involved in the softening process. An increase in the synthesis of three polygalacturonase isoenzymes, all products of one gene (Bird $et\ al.$, 1988) was responsible for most of the ripening-associated increase in polygalacturonase activity, and the increased synthesis was ascribed to increased genetic transcription of it (Brady $et\ al.$, 1982; Grierson $et\ al.$, 1985).

To determine whether or not the softening of tomatoes could be attributed entirely to the action of this enzyme on cell wall pectin, the suppression of its activity by genetic modification was attempted. Transgenic tomato plants with single or multiple copies of the polygalacturonase cDNA inserted in the antisense orientation were constructed, and in many such plants the expression of the major polyglacturonase isoenzyme was suppressed by over 90%. In these plants, fruit softening was delayed and modified, but not eliminated, indicating the participation of other enzymes, and probably the importance of other cell wall structural components in the softening process. A transgenetic antisense-$endo$-polygalacturonase tomato, which is claimed to soften more slowly and to retain its flavor longer than other tomatoes, has been marketed recently in the USA.

There is naturally considerable commercial impetus behind research which is capable of yielding fruits with improved textural qualities, and current activity in this area is intensive.

5.6.3   Bioactive cell wall polysaccharide fragments

Although the polysaccharide components of plant cell walls are best described as 'structural', it must be borne in mind that their biological roles are as diverse as those of the plant cell wall itself. The cell walls, apart from being 'structural', provide a continuous extracellular pathway (the apoplast) within a plant for the movement of water and ions. The hydrophilic properties of the matrix polysaccharides are clearly important in apoplastic transport, and the ability of pectin to chelate calcium may also be relevant. Also, in seeds, cell wall polysaccharides may regulate tissue mechanical properties and also serve as substrate reserves.

In all plant tissues the cell wall is a physical barrier to pathogenic attack, but recent evidence indicates a more active role for cell wall polysaccharides in the defense mechanism. Oligosaccharide fragments, derived apparently from the action of pathogen-derived hydrolytic enzymes, are powerful $elicitors$ of defense responses in plant tissues, such as the biosynthesis of phytoalexins (antimicrobial toxins, Chapter 13) and hypersensitive cell death (a strategy to limit infection) (Ryan & Farmer, 1991). Thus cell wall-degrading enzymes produced by the pathogen to effect a passage through the plant cell wall have the added effect of releasing oligosaccharide fragments which are capable of transmitting a signal to initiate defense mechanisms in neighboring plant cells. The best-documented $plant$ cell wall-derived elicitors (there is another group, originating in the fungal cell wall as a result of hydrolysis of it by plant-derived enzymes), are short sequences of $\alpha(1 \rightarrow 4)$-linked D-galacturonic acid residues. They are presumably generated by the action of $endo$-polygalacturonases and/or $endo$-pectate lyases secreted by the pathogen on the homogalacturonan domains of pectin. $Endo$-pectate lyase cleaves glycosyl linkages, not by hydrolysis, but by 1,2-elimination, producing oligosaccharides with an unsaturated sugar derivative at the non-reducing end. Both the normal and the unsaturated oligosaccharides are active elicitors. The response is chain-length dependent, the optimum being at about 10–15 sugar residues depending on the plant and the response. In this instance, a 'structural' plant cell wall polysaccharide carries encoded information.

A further example of cellular signaling by oligosaccharides derived from 'structural' cell wall polysaccharides was alluded to in the context of xyloglucan metabolism and cell expansion (section 5.6.1). Since the initial observation that the xyloglucan oligosaccharide $Glc_4Xyl_3GalFuc$ (Fig. 5.9) inhibits auxin-induced elongation of pea stem segments (York et al., 1984), other effects of xyloglucan oligosaccharides on plant development have been documented (Darvill et al., 1992; Aldington & Fry, 1993). The structural requirements for the inhibition of auxin-induced growth in pea stem segments have been investigated, and the structural motif $Glc_2GalFuc$ at the reducing terminus of the molecule singled out as essential to the effect (McDougall & Fry, 1989). Other oligosaccharides lacking this motif, however, were shown to be weak promoters of growth (e.g. Glc_4Xyl_3Gal, Fig. 5.9) in the same system (McDougall & Fry, 1990).

It is not known for certain whether or not xyloglucan oligosaccharides are regulatory molecules in vivo, although they have been detected in culture filtrates of plant cells. If they have a regulatory role, and if they are generated only during the turnover of xyloglucan in the cell wall, then it might be expected that the pathways forming and removing them would be highly regulated. Two enzymes capable of forming the Glc_4Xyl_3-based oligosaccharides from xyloglucan are known: endo-$\beta(1 \rightarrow 4)$-D-glucanase and XET (section 5.6.1). An α-L-fucosidase capable of removing the outer fucose residues from

xyloglucan oligosaccharides has been characterized (Augur et al., 1993), as has a β-galactosidase which catalyzes the removal of the subsequent layer of galactosyl residues (Edwards et al., 1988). The further degradation of the remaining Glc_4Xyl_3 core depends on the co-operative interaction of two enzymes. A xyloglucan-specific α-D-xylosidase (O'Neill et al., 1989; Fanutti et al., 1991) catalyzes the removal of only the xylosyl residue attached to the non-reducing terminal glucose, and cannot act again until this has been removed by the action of β-glucosidase. In all, three rounds each of α-xylosidase and β-glucosidase hydrolysis are required for the total hydrolysis of the oligosaccharide to glucose and xylose (Fig. 5.33).

The term 'oligosaccharin' has been coined for those cell wall-derived oligosaccharides which, when applied externally, modify plant development or metabolism (Darvill et al., 1992; Aldington & Fry, 1993). No oligosaccharin receptors have yet been described, and their status as regulatory molecules in vivo is still uncertain.

5.7 CONCLUSIONS

It was mentioned at the beginning of this chapter that plant structural polysaccharides constitute the most important renewable resource on earth, and that their direct use as industrial raw materials is inevitable. It should now be clear that they are of economic and commercial importance for other reasons. This chapter has highlighted current efforts to clarify the links between the chemical structures of plant cell wall carbohydrates, their interactions, cell wall ultrastructural organization, and tissue mechanical and textural properties. The practical goal of exercising control over the mechanical properties of crop plants and the textures of fruits and vegetables adds purpose to the excitement of current research in the area of plant 'structural' carbohydrates.

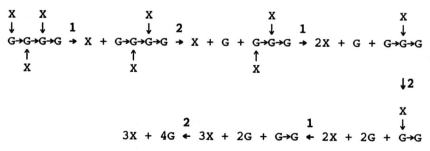

Figure 5.33 Co-operative action of hydrolytic enzymes in the final stages of xyloglucan oligosaccharide degradation. 1, action of xyloglucan oligosaccharide-specific α-xylosidase. 2, action of β-glucosidase.

REFERENCES

Albersheim, P. (1975) *Sci. Am.* **232**, 81–95.

Aldington, S. & Fry, S.C. (1993) *Adv. Bot. Res.* **19**, 1–101.

Augur, C., Benhamou, N., Darvill, A. & Albersheim, P. (1993) *Plant J.* **3**, 415–426.

Bauer, W.D., Talmadge, K.W., Keegstra, K. & Albersheim, P. (1973) *Plant Physiol.* **51**, 174–184.

Becker, M., Vincent, C. & Reid, J.S.G. (1995) *Planta* **195**, 331–338.

Bird, C.R., Smith, C.J.S., Ray, J.A., Moureau, P., Bevan, M.J., Birds, A.S., Hughes, S., Morris, P.C., Grierson, D. & Schuch, W. (1988) *Plant Mol. Biol.* **11**, 6541–662.

Brady, C.J. (1987) *Annu. Rev. Plant Physiol.* **38**, 155–178.

Brady, C.J., MacAlpine, G., McGlasson, W.B. & Ueda, Y. (1982) *Aust. J. Plant Physiol.* **9**, 171–178.

Brown, R.M. Jr, Haigler, C.H., Suttie, J., White, A.R., Roberts, E., Smith, C., Itoh, T. & Cooper, K. (1983) *J. Appl. Polymer Sci.: Appl. Polymer Symp.* **37**, 33–78.

Byrne, H., Christou, N.V., Verma, D.P.S. & Maclachlan, G.A. (1975) *J. Biol. Chem.* **250**, 1012–1018.

Carpita, N.C. & Gibeaut, D.M. (1993) *Plant J.* **3**, 1–30.

Chanzy, H. & Henrissat, B. (1985) *FEBS Lett.* **184**, 285–288.

Cook, J.A., Fincher, G.B., Keller, F. & Stone, B.A. (1980) In *Mechanisms of Saccharide Polymerization and Depolymerization* (J.J. Marshall, ed.), pp. 301–315. Academic Press, New York.

Crookes, P.R. & Grierson, D. (1983) *Plant Physiol.* **72**, 1088–1093.

Dalessandro, G., Piro, G. & Northcote, D.H. (1986) *Planta* **169**, 564–574.

Darvill, A., Augur, C., Bergmann, C., Carlson, R.W., Cheong, J.-J., Eberhard, S., Hahn, M.G., Ló, V.M., Marfa, V., Meyer, B., Mohnen, D., O'Neill, M.A., Spiro, M.D., van Halbeek, H., York, W.S. & Albersheim, P. (1992) *Glycobiology* **2**, 181–198.

Delmer, D.P. (1987) *Annu. Rev. Plant Physiol.* **38**, 259–290.

de Silva, J., Jarman, C.D., Arrowsmith, D.A., Stronach, M.S., Chengappa, S., Sidebottom, C. & Reid, J.S.G. (1993) *Plant J.* **3**, 701–711.

de Vries, J.A., Rombouts, F.M., Voragen, A.G.J. & Pilnik, W. (1982) *Carbohydr. Polym.* **2**, 25–33.

Edwards, M., Dea, I.C.M., Bulpin, P.V. & Reid, J.S.G. (1986) *J. Biol. Chem.* **261**, 9489–9494.

Edwards, M., Bowman, Y.J.L., Dea, I.C.M. & Reid, J.S.G. (1988) *J. Biol. Chem.* **263**, 4333–4337.

Edwards, M., Bulpin, P.V., Dea, I.C.M. & Reid, J.S.G. (1989) *Planta* **178**, 41–51.

Edwards, M., Scott, C., Gidley, M.J. & Reid, J.S.G. (1992) *Planta* **187**, 67–74.

Fan, D-F. & Maclachlan, G.A. (1966) *Can. J. Bot.* **44**, 1025–1034.

Fanutti, C., Gidley, M.J. & Reid, J.S.G. (1991) *Planta* **184**, 137–147.

Fanutti, C., Gidley, M.J. & Reid, J.S.G. (1993) *Plant J.* **3**, 691–700.

Fischer, R.L. & Bennett, A.B. (1991) *Annu. Rev. Plant Physiol. Plant Mol. Biol.* **42**, 675–703.

Fry, S.C. (1988) *The Growing Plant Cell Wall.* Longman, Harlow, 333 pp.

Fry, S.C., Smith, R.C., Renwick, K.F., Martin, D.J., Hodge, S.K. & Matthews, K.J. (1992) *Biochem. J.* **282**, 821–828.

Gordon, R. & Maclachlan, G. (1989) *Plant Physiol.* **91**, 373–378.

Grierson, D., Slater, A., Spiers, J. & Tucker, G.A. (1985) *Planta* **163**, 263–271.

Haigler, C.H. (1985) In *Cellulose Chemistry and its Applications* (R.P. Nevell & S.H. Zeronian, eds), pp. 30–83. Horwood, Chichester.

Hanna, R., Brummell, D.A., Camirand, A., Hensel, A., Russell, E.F. & Maclachlan, G.A. (1991) *Arch. Biochem. Biophys.* **290**, 7–13.

Hayashi, T. (1989) *Annu. Rev. Plant Physiol. Plant Mol. Biol.* **40**, 139–168.

Hayashi, T. & Maclachlan, G. (1984) *Plant Physiol.* **75**, 596–604.

Hayashi, T. & Matsuda, K. (1981) *J. Biol. Chem.* **256**, 11117–11122.

Hayashi, T., Marsden, M.P.F. & Delmer, D.P. (1987) *Plant Physiol.* **83**, 384–389.

Hayashi, Y., Wong, Y-S. & Maclachlan, G. (1984) *Plant Physiol.* **75**, 605–610.

Henry, R.J. & Stone, B.A. (1982) *Plant Physiol.* **69**, 632–636.

Jarvis, M.C. (1984) *Plant Cell Env.* **7**, 153–164.

Keegstra, K., Talmadge, K.W., Bauer, W.D. & Albersheim, P. (1973) *Plant Physiol.* **51**, 188–196.

Knox, J.P., Linstead, P.J., King, J., Cooper, C. & Roberts, K. (1990) *Planta* **181**, 512–521.

Labavitch, J.M. & Ray, P.M. (1974) *Plant Physiol.* **53**, 669–673.

Lin, F.C. & Brown, R.M. Jr (1989) In *Cellulose and Wood – Chemistry and Technology* (C. Schuerch, ed.), pp. 473–492. Wiley, New York.

Lin, F.C., Brown, R.M., Drake, R.R. & Haley, B.E. (1990) *J. Biol. Chem.* **265**, 4782–4784.

Liners, F., Thibault, J.F. & van Cutsem, P. (1992) *Plant Physiol.* **99**, 1099–1104.

Lloyd, C.W. (1984) *Int. Rev. Cytol.* **86**, 1–51.

McCann, M.C. & Roberts, K. (1991) In *The Cytoskeletal Basis of Plant Growth and Form* (C.W. Lloyd, ed.), pp. 109–129. Academic Press, London.

McCann, M.C., Wells, B. & Roberts, K. (1990) *J. Cell Sci.* **96**, 323–334.

McCleary, B.V. & Matheson, N.K. (1983) *Carbohydr. Res.* **119**, 191–219.

McDougall, G.J. & Fry, S.C. (1989) *Plant Physiol.* **89**, 883–887.

McDougall, G.J. & Fry, S.C. (1990) *Plant Physiol.* **93**, 1042–1048.

McNeil, M., Darvill, A.G., Fry, S.C. & Albersheim, P. (1984) *Annu. Rev. Biochem.* **53**, 625–663.

McQueen-Mason, S.J. & Cosgrove, D.J. (1993) *Plant Physiol.* **102**, 122 (Meeting Abstract).

McQueen-Mason, S.J., Durachko, D.M. & Cosgrove, D.J. (1992) *Plant Cell* **4**, 1425–1433.

McQueen-Mason, S.J., Fry, S.C., Durachko, D.M. & Cosgrove, D.J. (1993) *Planta* 190, 327–331.

Medford, J.I., Elmer, J.S. & Klee, H.J. (1991) *Plant Cell* 3, 359–370.

Meier, H. & Wilkie, K.C.B. (1959) *Holzforschung* 13, 177–182.

Nevell, R.P. & Zeronian, S.H. (1985) *Cellulose Chemistry and its Applications*. Horwood, Chichester.

Neville, A.C. & Levy, S. (1985) In *Biochemistry of Plant Cell Walls* (C.T. Brett & J.S. Hillman, eds), pp. 99–124. Society for Experimental Biology Seminar Series 28, Cambridge University Press, Cambridge.

Nishitani, K. & Tominaga, R. (1992) *J. Biol. Chem.* 267, 21058–21064.

Okazawa, K., Sato, Y., Nakagawa, T., Asada, K., Kato, I., Tomita, E. & Nishitani, K. (1993) *J. Biol. Chem.* 268, 25364–25368.

Okuda, K., Li, L., Kudlicka, K., Kuga, S. & Brown, R.M. Jr (1993) *Plant Physiol.* 101, 1131–1142.

O'Neill, R.A., Albersheim, P. & Darvill, A.G. (1989) *J. Biol. Chem.* 264, 20430–20437.

Parker, M. (1984) *Protoplasma* 120, 224–232.

Preston, R.D. (1974) In *Dynamic Aspects of Plant Ultrastructure* (A.W. Robards, ed.), pp. 256–309. McGraw-Hill, London.

Qi, X., Behrens, B.X., West, P.R. & Mort, A.J. (1995) *Plant Physiol.* 108, 1691–1701.

Reid, J.S.G. (1985a) In *Biochemistry of Storage Carbohydrates in Green Plants* (P.M. Dey & R.A. Dixon, eds), pp. 265–288. Academic Press, London.

Reid, J.S.G. (1985b) *Adv. Bot. Res.* 11, 125–155.

Reid, J.S.G., Edwards, M., Gidley, M.J. & Clark, A.H. (1995) *Planta* 195, 489–495.

Roelofsen, P.A. (1965) *Adv. Bot. Res.* 2, 69–149.

Roland, J.C., Reis, D., Vian, B., Satiat-Jeunemaitre, B. & Mosiniak, M. (1987) *Protoplasma* 140, 75–91.

Ross, P., Weinhouse, H., Aloni, Y., Michaeli, D. & Weinberger-Ohana, P. (1987) *Nature* 325, 279–281.

Ryan, C.A. & Farmer, E.A. (1991) *Annu. Rev. Plant Physiol. Plant Mol. Biol.* 42, 651–674.

Saxena, I.M., Lin, F.C. & Brown, R.M. Jr (1990) *Plant Mol. Biol.* 15, 673–683.

Stevenson, T.T., Darvill, A.G. & Albersheim, P. (1988) *Carbohydr. Res.* 182, 207–226.

Stone, B.A. & Clarke, A.E. (1992) *Chemistry and Biology of (1→3)-β-glucans*. La Trobe University Press, Victoria, Australia.

Taiz, L. (1984) *Annu. Rev. Plant Physiol.* 35, 585–657.

Talmadge, K.W., Keegstra, K., Bauer, W.D. & Albersheim, P. (1973) *Plant Physiol.* 51, 158–173.

Thornber, J.P. & Northcote, D.H. (1962) *Biochem. J.* 82, 340–346.

Timell, T.E. (1964) *Adv. Carbohydr. Chem.* 19, 247–302.

Timell, T.E. (1965) *Adv. Carbohydr. Chem.* 20, 409–483.

Valent, B.S. & Albersheim, P. (1974) *Plant Physiol.* 54, 105–108.

Vian, B. & Roland, J.C. (1987) *New Phytol.* 105, 345–357.

Vian, B., Reis, D., Mosiniak, M. & Roland, J.C. (1986) *Protoplasma* 131, 185–199.

York, W.S., Darvill, A.G. & Albersheim, P. (1984) *Plant Physiol.* 75, 295–297.

York, W.S., Oates, J.E., van Halbeek, H., Darvill, A.G., Albersheim, P., Tiller, P.R. & Dell, A. (1988) *Carbohydr. Res.* 173, 113–132.

Wilkie, K.C.B. (1979) *Adv. Carbohydr. Chem. Biochem.* 36, 215–264.

6 Plant Lipid Metabolism

John L. Harwood

Interest in plant lipids developed firstly from the observation in the 1930s that animals fed a diet devoid of lipid developed a series of symptoms collectively called essential fatty acid (EFA) deficiency. It transpired later that the 'essential' structure required was provided in the diet by plant lipids and was characterized by polyunsaturated fatty acids with double bonds 6 and 9 carbons from the methyl terminus. Such acids cannot be synthesized by mammals but are needed as precursors of eicosanoids and may play other vital roles in healthy individuals.

Since that time, interest in plant lipids has increased and there are now many other areas of industrial application. For example, efforts are being made to understand the regulation of seed oil quality in order to increase the usefulness of individual crop products by molecular biological manipulation. Moreover, plant membrane lipids have implications for the action of various growth regulators and are targets for several important classes of herbicides. In addition, plant lipids are being used by industry for detergent, nylon, and cosmetic manufacture, as highly stable lubricants and, recently, as a renewable source of fuel.

6.1 THE CHEMISTRY OF PLANT LIPIDS

6.1.1 Plant fatty acids

Most lipids in plants are acyl lipids; that is, the main membrane lipids are glycosylglycerides or phosphoglycerides whereas the principal storage material is triacylglycerol. All of these molecules have fatty acids esterified to the glycerol backbone.

Although over 300 different fatty acids have been isolated in plants, only a few are commonly used by these organisms in their storage or membrane lipids (Table 6.1). Saturated acids are almost invariably even carbon numbered because of their route of synthesis (section 6.2). Again, because of the nature of the fatty acid synthase of most plants, 16- and 18-carbon acids (palmitic and stearic, respectively) predominate. Most plants contain significant, though minor, amounts of myristic acid and some important crops contain large quantities of medium-chain fatty acids such as capric or lauric (Table 6.2). The latter acids are important in the detergent and cosmetic industries as well as providing easily absorbed lipid for patients with digestive disorders. On the other hand, a fat with a high stearate content is cocoa butter (Table 6.2) where such a composition is an important determinant of the unique properties of chocolate.

Because plants are poikilotherm (organisms which cannot regulate their own temperature) they have to contain membrane lipids with melting properties such that they are fluid under environmental conditions (section 6.7.3). In effect this means that such lipids will contain a large percentage of unsaturated fatty acids because the introduction of a *cis*-double bond in an acyl chain has a dramatic effect on the transition temperature (Tc) of such an acid (e.g. stearic acid Tc = 70°C, oleic acid Tc = 16°C). Oleic acid is the most

PLANT BIOCHEMISTRY
ISBN 0-12-214674-3

Table 6.1 Fatty acids found abundantly in plants

	Common name	Systematic name	Remarks
Saturated	Lauric acid (12:0)	Dodecanoic acid	Usually found in small amounts
	Myristic acid (14:0)	Tetradecanoic acid	Usually found in small amounts
	Palmitic acid (16:0)	Hexadecanoic acid	The most common saturated acid
	Stearic acid (18:0)	Octadecanoic acid	Usually found in small amounts
Unsaturated	Oleic acid (*cis* 9-18:1)	*cis* 9-Octadecenoic acid	The most common monounsaturated acid
	Linoleic acid (*cis,cis* 9,12-18:2)	*cis,cis* 9,12-Octadecadienoic acid	The most common diunsaturated acid. Abundant in membrane lipids
	α-Linolenic acid (all *cis* 9,12,15-18:3)	all *cis* 9,12,15-Octadecatrienoic acid	The most abundant plant fatty acid. Exceptionally high in leaves

common monounsaturated fatty acid whereas the polyunsaturated fatty acids, linoleic and α-linolenic acids, are found extensively especially as components of membrane lipids. In virtually all plant unsaturated fatty acids the double bond is in the *cis*-configuration and, for polyunsaturated fatty acids, the individual double bonds are methylene-interrupted (i.e. three carbons apart).

A notable fact about the fatty acid content of different plant tissues is the constancy of leaf composition where most of the lipids are components of membranes (Table 6.3). This is in striking contrast to the composition of seed oils and fats where, not only are the percentages of individual fatty acids very variable, but unusual components may be major constituents (Table 6.2). This undoubtedly relates to the functions of lipids as membrane or storage molecules. In membranes it is extremely important for the structure and properties to be carefully maintained so a tight regulation of lipid composition is maintained. By contrast, for energy provision the important properties are that the lipid can be easily stored and broken down, as required, by degradative enzymes. The structural requirements for these two properties are wide and, hence, an extensive variety of fatty acids can be tolerated.

Table 6.2 The fatty acid composition of some important oil crops

	Fatty acid composition (% total)					
	16:0	18:0	18:1	18:2	18:3	Other[a]
Avocado	20	1	60	18	0	1
Castor bean	1	1	3	4	trace	91
Cocoa butter	27	34	35	3	trace	1
Coconut	9	2	7	2	0	80
Corn (maize)	14	2	30	50	2	2
Linseed	6	3	17	14	60	trace
Olive	12	2	72	11	1	2
Palm	42	4	43	8	trace	3
Palm Kernel	9	3	15	2	0	71
Rape	4	1	54	23	10	8
Soybean	11	4	22	53	8	2
Sunflower	6	5	19	68	trace	2

[a]For castor bean 90% of the fatty acids in the storage oil are ricinoleic (12-hydroxyoctadecenoic) acid, for coconut oil medium chain acids of 8–14 carbons represent 79% and in palm kernel oil medium chain acids represent 71%. In both coconut and palm kernel oil, lauric acid is the dominant constituent (about 45% total).

Table 6.3 The fatty acid composition of some plant leaves

Plant	Fatty acid composition (% total)				
	16:0	18:1	18:2	α-18:3	Others
Barley	13	6	6	64	11
Pea	12	2	25	53	8
Rape[a]	13	4	16	50	17
Soybean	12	7	14	56	11
Spinach[a]	13	5	14	54	14
Wheat	18	5	15	55	7

[a]These species are '16:3-plants' which have a distinctive pathway of lipid metabolism which involves the desaturation of palmitate to hexadecatrienoate (16:3) within the chloroplast.

6.1.2 The main plant membrane lipids are glycosylglycerides and phosphoglycerides

Almost all plant membrane lipids are based on the trihydric alcohol glycerol. In the chloroplast membranes of photosynthetic tissues (as well as the plastids of non-photosynthetic tissues) the glycosylglycerides are the major components (Table 6.4). The only significant phospholipid is phosphatidylglycerol. Thus, in marked contrast to virtually every other membrane in Nature, chloroplasts contain protein–glycolipid structures. Three glycosylglycerides are important and their structures are shown in Fig. 6.1. The two galactolipids carry no charge although their carbohydrate head groups are polar. These molecules conform to the usual characteristic of membrane lipids in that the diacylglycerol portion is hydrophobic and orientates itself such that it is located in the interior of the membrane, while the head group is hydrophilic and is on the surface of the membrane. The third glycosylglyceride, the plant sulfolipid, carries a full negative charge at physiological pH. Therefore, in chloroplasts it is likely to associate with positively charged protein groups and/or be shielded by water-soluble cations.

Plants contain the same types of phosphoglycerides (Fig. 6.2) as animals but in somewhat different proportions. Phosphatidylglycerol, which is only a minor component of animals, is a major lipid mainly by virtue of its location in photosynthetic membranes. In the extrachloroplastid membranes,

phosphatidylcholine is the main lipid with phosphatidylethanolamine also important (Table 6.5). Phosphatidylinositol is also a significant lipid and its phosphorylated derivatives (phosphatidylinositol 4-phosphate and phosphatidylinositol 4,5-*bis* phosphate) are found in plasma membranes where they function in signal transduction (section 6.5.3). Phosphatidylserine is a minor component of most membranes whereas diphosphatidylglycerol (cardiolipin) is located in the inner mitochondrial membrane.

Although phosphoglycerides and glycosylglycerides are, by far, the most important membrane lipids, a large number of minor lipids will often also be found. Thus sterols (and their glucoside- or acylglucoside-derivatives) and sphingolipids may be significant in the plasma membrane, whereas plasmalogens or other ether-linked lipids may be minor components in many plant membranes.

Because the *sn*-1 and *sn*-2 positions of the glycerol backbone are usually esterified with fatty acids, and these are a complex mixture in plants, then each lipid class will contain various combinations of fatty acyl groups. This phenomenon is known as the 'molecular species' of a lipid. The proportion of individual molecular species in an individual membrane lipid is carefully regulated and, moreover, the positional distribution of particular fatty acids is subject to control during synthesis and turnover (see section 6.5.1). A striking illustration of the way in which fatty acid composition is regulated is shown in Table 6.4 for the four main chloroplast lipids. Thus,

Table 6.4 The lipid composition of chloroplast membranes

Plant lipid	% total lipid	Fatty acid composition (% total)					
		16:0	16:3	18:1	18:2	18:3	Other[a]
Pea							
MGDG	42	4	–	1	3	90	2
DGDG	31	9	–	3	7	78	3
SQDG	11	32	–	2	5	58	3
PG	12	18	–	2	11	38	31
Spinach							
MGDG	41	trace	25	1	2	72	trace
DGDG	29	3	5	2	2	87	1
SQDG	16	39	–	1	7	53	trace
PG	12	11	–	2	4	47	36

[a]Includes 27% *trans*-hexadecenoate in the phosphatidylglycerol of pea and 32% *trans*-hexadecenoate in the phosphatidylglycerol of spinach.
Abbreviations: MGDG, monogalactosyldiacylglycerol; DGDG, digalactosyldiacylglycerol; SQDG, sulfoquinovosyldiacylglycerol; PG, phosphatidylglycerol.

Common name	Structural and chemical name

Monogalactosyl
diacylglycerol
(MGDG)

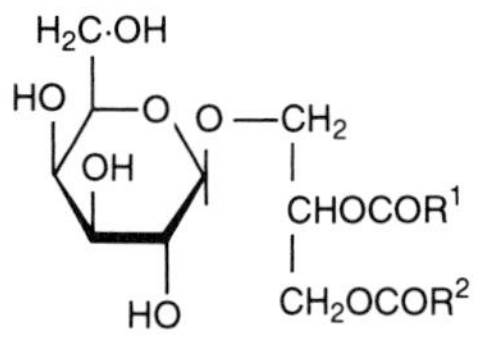

1,2-diacyl-[β-D-galactopyranosyl-(1'→3)]-sn-glycerol

Digalactosyl
diacylglycerol
(DGDG)

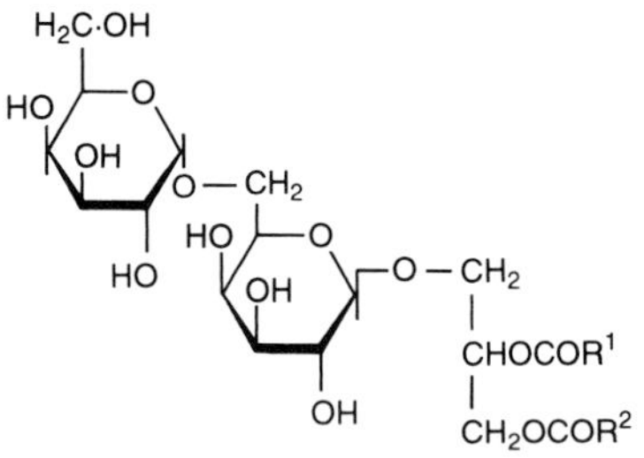

1,2-diacyl-[α-D-galactopyranosyl-(1'→6')-
β-D-galactopyranosyl-(1'→3)]-sn-glycerol

Plant sulpholipid
(sulphoquinovosyl-
diacylglycerol;
(SQDG)

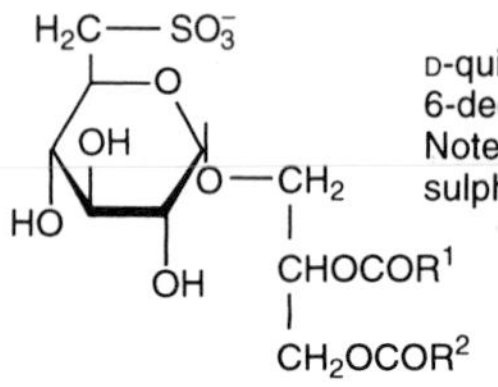

D-quinovose is
6-deoxy-D-glucose.
Note the carbon-
sulphur bond

1,2-diacyl-[6-sulpho-α-D-quinovopyranosyl-(1' → 3)]-sn-glycerol

Figure 6.1 The structure of plant glycosylglycerides.

monogalactosyl diacylglycerol is highly unsaturated and, in the case of the '16:3 plant' spinach is the predominant location of hexadecatrienoate. Digalactosyldiacylglycerol contains the same fatty acids as its monogalactosyl counterpart but is always more saturated whereas the sulfolipid contains mainly palmitate and linolenate. Chloroplast phosphatidylglycerol has the most distinctive fatty acid pattern with an unusual component, trans-Δ3-hexadecenoate (located exclusively at the sn-2 position). This fatty acid is never found in any other membrane lipid – a striking testament to the careful control of lipid metabolism in plants.

6.1.3 Plants store energy as triacylglycerol

When plants use lipids as an energy store they almost invariably accumulate triacylglycerol. This lipid is located in oil bodies of seeds but may form a heterogeneous mixture of droplets in the partly decomposed tissues of fruits such as those of avocado, olive or palm. Small quantities of its metabolic precursor, diacylglycerol (Fig. 6.3) are also usually present, together with minor degradative products such as non-esterified fatty acids. By contrast to the usual situation, the desert plant, jojoba, stores wax esters (Fig. 6.3) which are useful alternatives to sperm whale oil and are used extensively in cosmetic and shampoo products.

6.1.4 Sometimes unusual fatty acids are present in lipid stores

As discussed earlier, the majority of fatty acids in plant tissues are restricted to only a few types (Table 6.1). Indeed, the membranes of plants have a very carefully regulated balance of lipid classes and their acyl constituents. These are typified by the data in Tables 6.3–6.5. However, in some cases

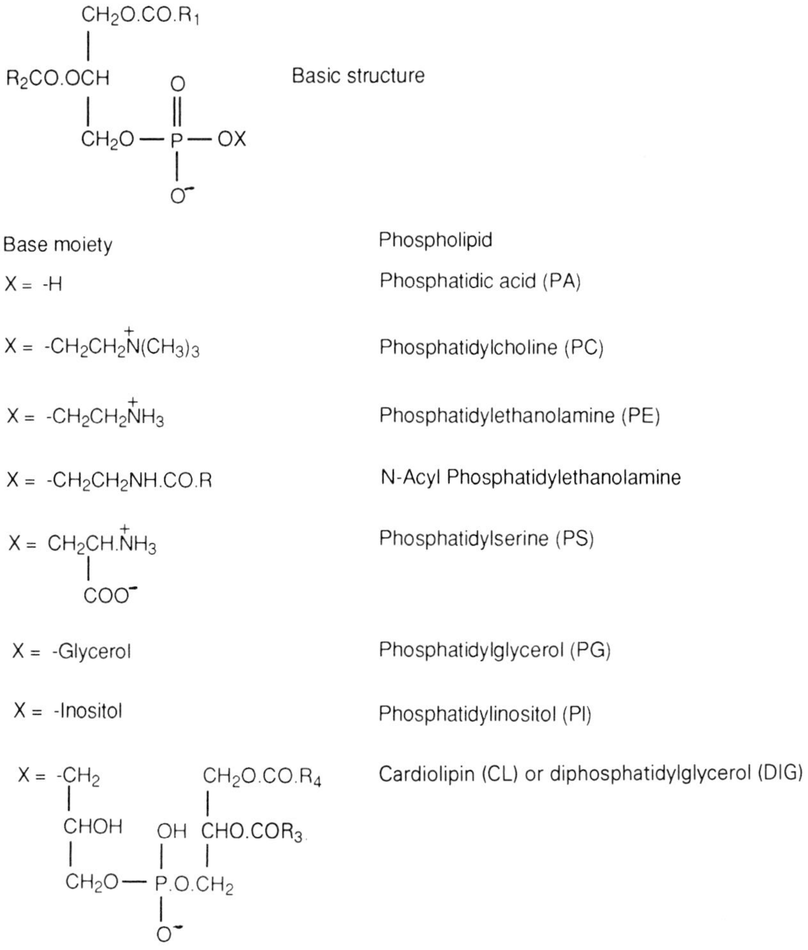

```
        CH2O.CO.R1
         |
R2CO.OCH        O
         |      ||
        CH2O — P — OX        Basic structure
                |
                O⁻
```

Base moiety	Phospholipid
X = -H	Phosphatidic acid (PA)
X = -CH₂CH₂N⁺(CH₃)₃	Phosphatidylcholine (PC)
X = -CH₂CH₂N⁺H₃	Phosphatidylethanolamine (PE)
X = -CH₂CH₂NH.CO.R	N-Acyl Phosphatidylethanolamine
X = CH₂CH.N⁺H₃ COO⁻	Phosphatidylserine (PS)
X = -Glycerol	Phosphatidylglycerol (PG)
X = -Inositol	Phosphatidylinositol (PI)

```
X = -CH2        CH2O.CO.R4
     |           |
    CHOH   OH  CHO.COR3
     |      |   |
    CH2O — P.O.CH2
            |
            O⁻
```
Cardiolipin (CL) or diphosphatidylglycerol (DIG)

Figure 6.2 The structure of important plant phosphoglycerides.

Table 6.5 The phosphoglyceride composition of some extrachloroplast membranes

Membrane	% total phosphoglycerides					
	DPG	PC	PE	PG	PI	PS
Castor bean endosperm						
Outer miochondrial	0	51	39	trace	10	
Inner mitochondrial	7	34	58	1	2	
Peroxisome	trace	49	31	trace	6	0
Glyoxysome	trace	51	27	3	9	1
Potato tuber						
Inner mitochondrial	19	33	33	5	7	
Peroxisome	–	61	20	15	4	
Microsome	1	45	33	2	16	

Phosphoglycerides represent more than 90% of the total lipids of all of these membrane fractions. The microsomal fraction contains small membrane fragments and is typically enriched in endoplasmic reticulum.

Abbreviations: DPG, diphosphatidylglycerol; PC, phosphatidylcholine; PE, phosphatidylethanolamine; PG, phosphatidylglycerol; PI, phosphatidylinositol; PS, phosphatidylserine.

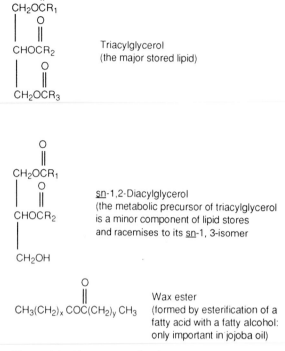

Figure 6.3 Plant storage lipids.

plants accumulate significant amounts of unusual fatty acids in their storage oils. Some examples are given in Table 6.6. Many of these 'unusual' acids have important industrial uses and there is increasing interest in their production as sources of raw materials.

6.1.5 Lipid analysis is highly sensitive and sophisticated

Plant lipid analysis is not just an academic exercise – it is important for the evaluation of crops in agriculture, in estimating nutritional qualities and even adulteration in food. (A particularly horrendous case of the latter occurred in the early 1980s when some unscrupulous individuals bought rapeseed oil designed for industrial purposes and containing aniline, tried to remove the latter and then passed the oil off as the much more expensive olive oil. Some 600 people died as a result of consuming derivatives from the aniline and over 15 000 people suffered side effects with various degrees of severity.)

Lipids must first be extracted from plant tissues and mixtures of polar and non-polar solvents (e.g. methanol and chloroform) are frequently used.

Table 6.6 Some unusual fatty acids present in plant seed oils

	Name	Notes
Cyclic fatty acid $\overset{\displaystyle CH_2}{\overset{\displaystyle /\ \backslash}{CH_3(CH_2)_7C=C(CH_2)_7COOH}}$	Sterculic	From plants of the Malvales family
$-(CH_2)_{12}COOH$	Chaulmoogric	Used for leprosy treatment
Epoxy fatty acid $\overset{\displaystyle O}{\overset{\displaystyle /\ \backslash}{CH_3(CH_2)_4CH-CHCH_2CH=CH(CH_2)_7COOH}}$	Vernolic	Epoxy derivative of oleate
Trans-fatty acid $CH_3(CH_2)_{11}CH=CHCH_2COOH$	*Trans*-hexadecenoic	Characteristic of phosphatidylglycerol in chloroplasts
Cis-fatty acids $CH_3(CH_2)_{10}CH=CH(CH_2)_4COOH$	Petroselinic	Used as a source of precursors for detergent and nylon manufacture
$CH_3(CH_2)_7CH=CH(CH_2)_{11}COOH$	Erucic	Present in old varieties of rape. Trierucin is a useful lubricant
Hydroxy-fatty acid $CH_3(CH_2)_5\underset{\underset{\displaystyle OH}{\mid}}{CH}CH_2CH=CH(CH_2)_7COOH$	Ricinoleic	Represents <90% of the acids of castor bean oil. Triricinolein used as a lubricant

These mixtures allow good penetration of the tissues as well as good solubility characteristics for all major lipid classes. Some plant degradative enzymes are active in organic solvents and it may be necessary to inactivate these before the analytical process begins. Steam treatment or hot isopropanol are two possible methods.

Once extraction has taken place, analysis will usually continue by removal of non-lipid contaminants and various forms of chromatography in order to separate lipid classes (Fig. 6.4). In general, lipids do not have chromophores and usually have to be detected by staining or other derivatization. The best reagents for this purpose on thin-layer chromatography are non-destructive chemicals (e.g. 1-anilino-4-naphthosulfonic acid) which reveal lipids by fluorescence under UV light. Because the sprays are non-destructive further analysis is then possible. A recent advance in high performance liquid chromatography is the development of an evaporative mass detector which converts the effluent from a column into a fine spray which scatters light in proportion to the mass of lipid. This is relatively sensitive; $1\,\mu g$ of lipid is easily measured and gives broadly similar results with most lipid types. For acyl lipids (and some other types such as sterols) it is possible to make volatile derivatives (e.g. fatty acid methyl esters) and analyze them by very sensitive gas-liquid chromatography (GLC). The latter method not only quantifies lipids but also gives information on the structure of the acyl moieties. Identification is particularly clear if GLC is combined with mass-spectrometry.

For more sophisticated analysis it may be necessary to carry out further tests. Three examples are indicated in Fig. 6.4. Because acyl lipids contain mixtures of fatty acids attached to the glycerol backbone, an individual lipid class does not contain a single chemical moiety. The various combinations of fatty acids (molecular species) can be separated from each other by hydrophobic chromatography (e.g. on HPLC) where the longer chains are held back more on

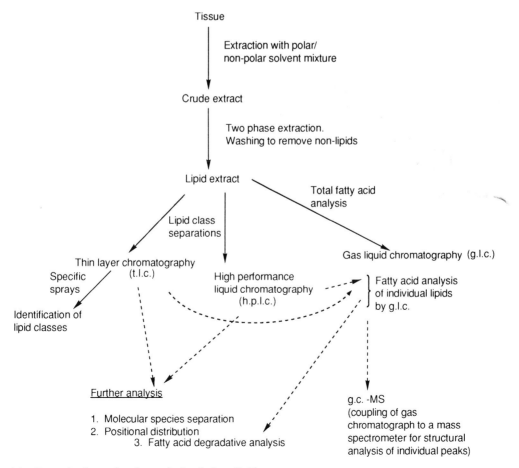

Figure 6.4 General scheme for the analysis of plant lipids.

the column due to the hydrophobicity or by silver nitrate chromatography when separation is on the basis of double-bond content. Silver nitrate chromatography can also be used to differentiate between *trans*- and *cis*-double bonds and will even resolve positional isomers. A second test uses enzymes to hydrolyze fatty acids at the different carbons of the glycerol backbone. Thus, for example, phospholipase A$_2$ (see section 6.5.1) will hydrolyze the *sn*-2 positions of phospholipids, thus allowing the fatty acids at that position to be analyzed separately from those at the *sn*-1 position. A third test is used in order to delineate the double-bond position(s) in fatty acids. Such an analysis is needed, for example, to differentiate oleic acid (*cis* Δ9-18:1) from vaccenic (*cis* Δ11-18:1) or petroselenic (*cis* Δ6-18:1) acids, all of which may be eluted at the same position off a GLC column. Oxidative attack is used to split the hydrocarbon chains at the position of the double bond with the production of two fragments – a dicarboxylic acid from the carboxyl fragment and a new monocarboxylic acid from the methyl end. Ozone and potassium permanganate are typical oxidative reagents employed.

6.2 FATTY ACID BIOSYNTHESIS

6.2.1 *De novo* synthesis involves two enzyme complexes

The ultimate source of carbon for lipid synthesis comes from photosynthesis. In seeds and oil-accumulating fruits, this photosynthate has to be transported from leaf tissue via sucrose (or mannose). The terminal enzyme of glycolysis can be regarded as pyruvate dehydrogenase/decarboxylase (PDH) which yields acetyl-CoA. PDH is usually present in plastids which is the site of *de novo* fatty acid synthesis although in some plants, such as spinach, mitochondrial PDH may act to produce acetyl-CoA which is hydrolyzed to acetate. The unionized acetate (i.e. acetic acid) is

capable of crossing membranes easily and can be taken into plastids where it is activated again to acetyl-CoA.

The first of the two enzymes involved in *de novo* fatty acid formation is acetyl-CoA carboxylase. This is a type 1 biotin-containing enzyme. It has two activities (Fig. 6.5) which, together with the biotin carboxyl carrier protein (BCCP), are usually contained in a single high molecular mass (220–240 kDa) multifunctional protein.* The first partial reaction involves the ATP-driven carboxylation of the biotin moiety. The reaction is believed to proceed via a carbamyl phosphate intermediate. The biotin is attached to protein in the BCCP via a flexible linkage (involving proline) which allows it to interact also with the second active site, that of the carboxyl transferase. The carboxyltransferase catalyzes the carboxylation of acetyl-CoA to form malonyl-CoA. It is this second partial reaction that is specific for the enzyme as opposed to other carboxylases which also share the biotin carboxylase.

In many situations, acetyl-CoA carboxylase is believed to control the flux of carbon into lipids. However, the mechanism of regulation is unclear. Unlike in animals tricarboxylic acids (such as citrate) and phosphorylation/dephosphorylation do not appear to be involved in its control. In maturing oil seeds, acetyl-CoA carboxylase is made in increasing quantities as oil accumulation begins (see Fig. 6.9). The vital nature of acetyl-CoA carboxylase is revealed by the herbicidal action of the 'grass-specific' herbicides, aryloxyphenoxy propionates and cyclohexanediones. These chemicals kill grass species, such as barley or rye-grass, by inhibiting acetyl-CoA carboxylase without affecting the enzyme in other monocotyledons or in dicotyledons.

Once malonyl-CoA has been generated, it can be used as the source of two-carbon addition units for fatty acid biosynthesis. The plant fatty acid synthetase is a type II dissociable complex in which the proteins catalyzing the partial reactions can each be isolated and purified. The individual

$$ATP + CO_2 + BCCP \xrightleftharpoons{\text{biotin carboxylase}} CO_2\text{–}BCCP + ADP + Pi$$

$$CO_2\text{–}BCCP + Acetyl\text{–}CoA \xrightleftharpoons{\text{Carboxyl transferase}} Malonyl\text{–}CoA + BCCP$$

Figure 6.5 Action of acetyl-CoA carboxylase. BCCP, biotin carboxyl carrier protein.

*In dicotyledons, the chloroplast acetyl-CoA carboxylase is a multienzyme complex consisting of three separate proteins. All plants have a second, multifunctional, isoform in the cytosol.

reactions involved in the addition process are depicted in Fig. 6.6. Originally it was assumed that both acetyl-CoA (the original primer molecule) and malonyl-CoA underwent acyl transfer with acyl carrier protein (ACP) to generate acetyl-ACP and malonyl-ACP, respectively. However, the physiological activity of acetyl-ACP is now in doubt since a condensing enzyme is present in plants which can condense malonyl-ACP directly with acetyl-CoA (short-chain condensing enzyme or β-ketoacyl-ACP synthase III; KAS III). Condensation of a fatty acyl primer with malonyl-ACP yields a β-ketoacyl-ACP which is reduced, dehydrated and reduced again to yield a new fatty acyl-ACP, some two carbons longer than the original primer. The product acyl-ACP then undergoes a condensation to initiate a new cycle (Fig. 6.6). The first reductase is β-ketoacyl-ACP reductase which can be present in isoforms. Usually NADPH is the source of reductant, but some isoforms have activity with NADH. The dehydrase is specific for the D(-)β-hydroxyacyl-ACP and yields a trans-2-acyl-ACP which is reduced by the second

reductase, enoyl-ACP reductase. The latter enzyme is also present in isoforms and both NADH and NADPH forms have been purified.

Apart from the short-chain condensing enzyme, two other β-ketoacyl-ACP synthetases are present in plants. In most plants the short-chain condensing enzyme (KAS III) can use acetyl- and, possibly, butyryl-primers. Further condensation is catalyzed by β-ketoacyl-ACP I (KAS I) which uses primers of up to 14-carbons and, hence, is responsible for palmitoyl-ACP formation. β-ketoacyl-ACP II (KAS II) can then condense palmitoyl-ACP with malonyl-ACP to give the final product of fatty acid synthetase, stearoyl-ACP. The three condensing enzymes were originally discovered by the use of inhibitors and are similar to the condensing enzymes of Escherichia coli fatty acid synthetase which is also a dissociable type II complex.

All the individual components of fatty acid synthetase are coded by nuclear DNA but are active in the chloroplast (plastid) stroma. Thus, they are synthesized with a signal peptide which

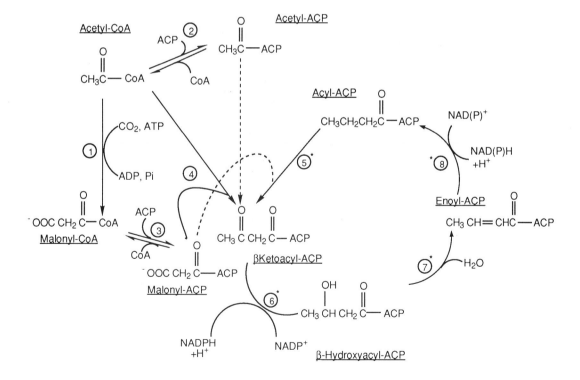

Figure 6.6 The reactions of plant fatty acid synthetase. ACP, acyl carrier protein. 1, Acetyl-CoA carboxylase; 2, acetyl-CoA : ACP transacylase (probably not important in most plants); 3, malonyl-CoA : ACP transacylase; 4, short-chain condensing enzyme (KAS III) (important for the initial condensation); 5, β-ketoacyl-ACP synthetase I (continues condensation up to C14-ACP), β-ketoacyl-ACP synthetase II (condenses palmitoyl-ACP with malonyl-ACP); 6, β-ketoacyl-ACP reductase; 7, β-hydroxyacyl-ACP dehydrase; 8, enoyl-ACP reductase. *Substrates are reacted successively around the cycle 7 times to give palmitoyl-ACP or 8 times to yield stearoyl-ACP.

allows movement across the envelope membranes of the organelle. The process has been studied most extensively with acyl carrier protein which, as with bacterial ACP, is a small molecular weight, acidic protein. Considerable sequence homology is observed between different plant ACPs and with those from bacteria and, even, the appropriate domain of the multifunctional animal fatty acid synthetase.

The end products of fatty acid synthetase are nearly always a mixture of palmitate and stearate The ratio of these products is dependent on the activity of β-ketoacyl-ACP synthetase II in comparison to other competing reactions such as the transfer of palmitate from palmitoyl-ACP into complex lipids or its hydrolysis by a thioesterase. Plants with relatively high β-ketoacyl-ACP synthetase II, such as oilseed rape, produce a high proportion of 18-carbon acids.

In some plants, such as coconut or *Cuphea*, medium-chain fatty acids (C_8–C_{12}) accumulate. For some years the mechanism of (premature) chain termination was unclear. However, a specific medium chain thioesterase was discovered in the Californian Bay plant which also produces medium-chain products. Similar thioesterases have now been found in agriculturally important oil crops such as *Cuphea* so that the mechanism of medium-chain fatty acid production appears to be similar to that involved in milk lipid production by the mammary glands of mammals.

6.2.2 An important modification of fatty acids is aerobic desaturation

Tables 6.2 and 6.3 show that the majority of plant fatty acids are unsaturated molecules. These are formed from the saturated products of *de novo* fatty acid synthesis, namely palmitic and stearic acids. The overall process of fatty acid desaturation is aerobic and involves a cyanide-sensitive terminal desaturase protein. However, other features of the various desaturases, such as their positional specificity, the substrates used and the details of the associated electron transport system may differ.

As mentioned in section 6.2.1, the major product of fatty acid synthetase in plants is stearoyl-ACP. This compound is the substrate for the first important plant fatty acid desaturase (stearoyl-ACP $\Delta9$-desaturase) which produces oleoyl-ACP as product (Fig. 6.7). The enzyme is soluble and present in the stroma of plastids. It has such high activity that stearate rarely accumulates

in plants but is nearly all converted into oleate. The enzyme has been purified from several sources and cDNAs corresponding to the genes from these plants cloned. Active desaturase protein has been produced upon expression of safflower cDNA in *E. coli* and of castor bean cDNA in yeast. All the $\Delta9$-desaturases showed strong conservation of polypeptide sequence but little sequence identity with the stearoyl-CoA $\Delta9$-desaturase from animals or to the $\Delta12$-desaturase from cyanobacteria.

Oleoyl-ACP is a good substrate for either the stromal thioesterase (which hydrolyzes it to free oleic acid) or the acyltransferases which are responsible for acylating glycerol 3-phosphate (see section 6.3.2). If oleic acid is produced, this can be esterified again, this time to oleoyl-CoA, by an acyl-CoA synthetase on the envelope membrane (Fig. 6.7). Acyl-CoAs, including oleoyl-CoA, are substrates for acyltransferases on the endoplasmic reticulum and there are extra ways in which oleate can become incorporated into the major endoplasmic reticulum lipid, phosphatidylcholine (section 6.4.2).

The desaturase which converts oleate into the polyunsaturated fatty acid linoleate acts at the $\Delta12$-position and uses a complex lipid substrate; that is, the desaturation occurs *in situ* while the oleate substrate is part of a membrane lipid. Phosphatidylcholine is the major substrate in plants and the desaturation process results in a local change in the fluidity of the membrane, since linoleate has a lower melting point than oleate. Some progress has been made in our understanding of the $\Delta12$-desaturase recently. It seems that the endoplasmic reticulum enzyme of safflower uses electrons from NADH which are passed on to the terminal desaturase protein via cytochrome b_5. In chloroplasts the $\Delta12$-desaturase is present in the envelope membrane and probably uses a variety of complex lipid substrates. Certainly in cyanobacteria (blue–green algae) which do not contain phosphatidylcholine, $\Delta12$-desaturation appears to take place on all of the four membrane lipids (MGDG, DGDG, SQDG and PG). In two species of cyanobacteria, the gene coding for the $\Delta12$-desaturase has been transferred to strains lacking the capacity to make linoleate. Expression of the $\Delta12$-desaturase protein not only permitted the strains to make linoleate but also made them tolerant to cold temperatures (as expected from the melting properties of linoleate compared to oleate).

The most abundant plant (and, indeed, world) fatty acid is α-linolenate. This is made by the action of a $\Delta15$-desaturase on linoleate. In most

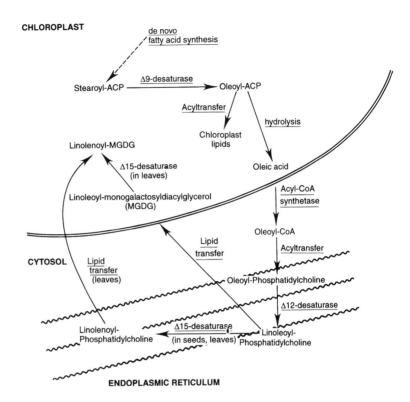

Figure 6.7 Major pathways for the desaturation of fatty acids.

leaves the major substrate for this desaturase is the chloroplast lipid, monogalactosyldiacylglycerol (MGDG). Alternatively, in some leaves phosphatidylcholine can act as a substrate for some of the desaturation of linoleate and in seeds (such as linseed which produces significant linoleate) this phospholipid is a major substrate. In order for MGDG to function as a substrate it may be necessary to transfer lipid from the endoplasmic reticulum back to the chloroplast. This is probably catalyzed by the back reaction of choline phosphotransferase (Fig. 6.13 and section 6.4.2). Because of difficulties in purifying the Δ15-desaturase, little is known about its properties. It is subject to inhibition by certain pyridazinone herbicides (see section 6.7.4).

Another plant fatty acid desaturase which uses a complex lipid substrate is the enzyme producing the unusual *trans*-Δ3-hexadecenoic acid (see Table 6.4). As might be expected for a fatty acid which is only present at the *sn*-2 position of phosphatidylglycerol, the substrate for the enzyme is phosphatidylglycerol containing palmitate at that position. There are many other desaturases present in different plants, some of which are indicated in Fig. 6.7. All of them are aerobic desaturases and the anaerobic pathway found in *E. coli* is not used by plants at all.

Plants differ in how much desaturation occurs exclusively in their chloroplasts. This is due to the relative activity of chloroplast phosphatidate phosphohydrolase which may make diacylglycrol available from the plastid (or not as the case may be). (See also section 6.4.2 and Fig. 6.7.)

6.2.3 Elongation of preformed fatty acids takes place on the endoplasmic reticulum

Although the main products of fatty acid synthetase are 16- and 18-carbon acids, longer-chain products are produced in substantial amounts. Many seed oils contain very long-chain (>C18) fatty acids, particularly monounsaturated compounds. A good example is the original type of oilseed rape which contained about 60% erucic acid (*cis* Δ13-22 : 1). The enzymes responsible for making erucic acid are elongases which add two-carbon units from malonyl-CoA to the original precursor, oleic acid. Since the new carbon is added to the carboxyl end of the fatty acyl chain, the double-bond numbering changes in relation to the carboxyl terminus (Fig. 6.8). Acyl-CoAs are used for the elongase enzymes and NADPH is the preferred reductant.

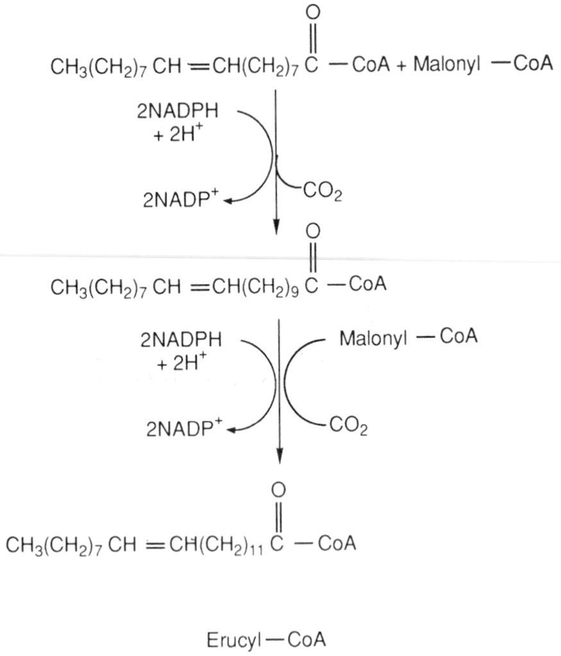

$$CH_3(CH_2)_7 CH =CH(CH_2)_7 \overset{\overset{\displaystyle O}{\parallel}}{C} -CoA + Malonyl -CoA$$

2NADPH + 2H⁺ → 2NADP⁺ + CO₂

$$CH_3(CH_2)_7 CH =CH(CH_2)_9 \overset{\overset{\displaystyle O}{\parallel}}{C} -CoA$$

2NADPH + 2H⁺ ← Malonyl — CoA ; 2NADP⁺ ← CO₂

$$CH_3(CH_2)_7 CH =CH(CH_2)_{11} \overset{\overset{\displaystyle O}{\parallel}}{C} -CoA$$

Erucyl—CoA

Figure 6.8 The elongation of oleoyl-CoA to form erucic acid.

Most of the elongation of fatty acids takes place in order to provide precursors for surface lipids (section 6.6) such as waxes, cutin and suberin. The reactions involve saturated fatty acids and are localized on the endoplasmic reticulum or the Golgi membranes for the longer chain products. Again, NADPH, malonyl-CoA and acyl-CoA substrates are used. Each elongase complex seems to contain a condensing activity, two reductases and a dehydrase. Thus, the sequence of partial reactions involved in two-carbon addition is comparable to those of *de novo* synthesis, and some of the proteins catalyzing these partial reactions have been solubilized and partly purified. Some herbicides, such as thiocarbamates, which affect wax formation, seem to inhibit fatty acid elongation and this may explain their action on surface lipid formation.

6.3 TRIACYLGLYCEROL SYNTHESIS

6.3.1 Accumulation of storage lipid

When plants need to store energy in the form of lipid they do so only at particular, and carefully regulated, times in their development. The seeds of plants mature in three distinct phases as shown in Fig. 6.9. In the first stage cell division occurs but

there is little accumulation of reserves. In the second stage a rapid deposition of storage material (lipid, carbohydrate etc.) takes place. For plant seeds where unusual fatty acids are abundant it is at this second stage where their synthesis is localized rather than being present at all developmental stages. In the third stage, desiccation takes place and there is little further synthesis of food reserves.

As discussed in section 6.2.1, acetyl-CoA carboxylase activity is often considered to control the overall carbon flux to lipids and, indeed, the activity of this enzyme in developing seeds (Fig. 6.9) is consistent with such a role.

The storage oil accumulates in oil bodies although, initially, the triacylglycerol may be produced as naked droplets. The oil bodies are surrounded by a half-unit membrane of phospholipid in which specific proteins, the oleosins, are located. Oleosins seem to function in preventing the coalescence of oil bodies and may also act as anchors for the lipases involved in the degradation of lipid stores during germination (section 6.4). The oleosins from different seeds seem to have a very well conserved amino acid sequence which gives

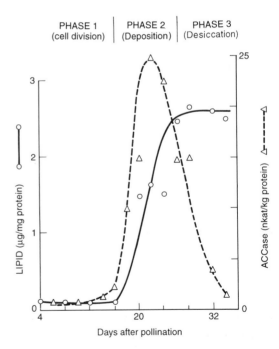

Figure 6.9 The deposition of triacylglycerol in developing seeds of oilseed rape: correlation with acetyl-CoA activity. (Redrawn from data in Turnham, E. & Northcote, D.N. (1983) *Biochem. J.* 212, 223–229, with permission.)

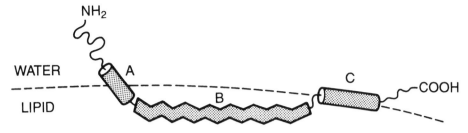

Figure 6.10 Possible arrangement of the three domains of an oleosin protein at the surface of an oil body. Section A, N-terminal proximal α-helical domain that may help the binding of the lipase active during germination. Section B, a central 70-residue proline-rich hydrophobic domain. Due to its hydrophobic nature it is probably immersed in the triacylglycerol of the storage oil droplet. Section C, an amphipathic α-helix with charged amino acid side chains in contact with the aqueous phase and uncharged side chains on the lipid side. (From D.J. Murphy (1990) *Prog. Lipid Res.* **29**, 299–324, with permission.)

three distinct regions to the proteins (Fig. 6.10). These regions are very reminiscent of mammalian apoprotein associated with the low density serum lipoprotein. The N-terminal end has an α-helical region penetrating into the cytosol whereas a central proline-rich hydrophobic domain is believed to reside in the half-unit membrane or, perhaps, even in the triacylglycerol core of the oil body. The C-terminal has an α-helical arrangement with positively charged amino acids which seem to arrange themselves around the helix such that they can interact ionically with the negative charges on the phospholipids of the half-unit membrane.

6.3.2 Triacylglycerols are synthesized by the Kennedy pathway

In 1961 the American biochemist, Kennedy, elucidated the pathway for the synthesis of tri-acylglycerols in animals. The same route is used by plants and consists of four reactions (Fig. 6.11). Glycerol-3-phosphate (derived by reduction of the dihydroxyacetone phosphate intermediate of glycolysis) is the initial acceptor. It is acylated, first at the *sn*-1 position and then at the *sn*-2 position by fatty acyl-CoA acyltransferases. These enzymes are located on the endoplasmic reticulum.

The first enzyme (glycerol 3-phosphate acyltransferase) prefers to use saturated (palmitic) acids in most plants but has a fairly broad substrate specificity. As a result of its action, palmitate is much enriched at the *sn*-1 position compared to the *sn*-2 position. However, in situations where there is little palmitoyl-CoA available, other fatty acids are easily used. Thus, in olive fruits (where oleoyl-CoA is the dominant acyl-CoA) oleate is the most abundant fatty acid at the *sn*-1 position of triacylglycerol. In contrast, the

second acylation is highly specific for oleoyl-CoA and saturated acids are virtually excluded from the *sn*-2 position.

The two acylations of glycerol 3-phosphate produce the phospholipid, phosphatidic acid. In cells this is present in salt forms and, therefore, the name phosphatidate is a better term. It is hydrolyzed by phosphatidate phosphohydrolase to yield diacylglycerol which is used by phospholipid and glycosylglyceride synthesis (section 6.4) in addition to storage lipids. Finally, in the only reaction specific to triacylglycerol formation, diacylglycerol is acylated by a third fatty acid using the enzyme diacylglycerol acyltransferase. The latter has a broad substrate specificity in most plants and the fatty acid composition of the *sn*-3 positions of triacylglycerol is usually dependent on which acyl groups are available from the acyl-CoA pool.

In animals, triacylglycerol deposition is under very strict metabolic control and takes place virtually whenever excess calories are available. In plants, however, such control seems unnecessary since triacylglycerol synthesis is governed by temporal (Fig. 6.9) rather than metabolic factors. Thus, it is not surprising that no single reaction in the Kennedy pathway seems to exert much flux control, although diacylglycerol acyltransferase may be barely adequate when triacylglycerol synthesis is high. Instead, the overall rate of lipid deposition is probably dependent on the supply of acyl-CoAs and, therefore, ultimately on the activity of acetyl-CoA carboxylase (Fig. 6.9).

Two aspects of storage oil quality are worth mentioning. First, in those plants where highly unsaturated oils accumulate (e.g. safflower, linseed: Table 6.2) the diacylglycerol from the Kennedy pathway may be incorporated into

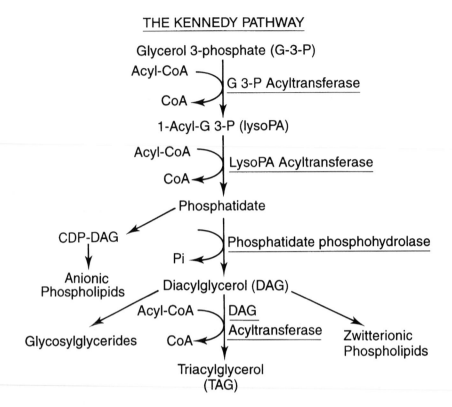

Abbreviations: G 3-P = glycerol 3-phosphate; PA = phosphatidic acid;
DAG = diacylglycerol; TAG = triacylglycerol

Figure 6.11   The Kennedy Pathway for triacylglycerol synthesis (and other lipid formation).

phosphatidylcholine via the back reaction of the cholinephosphotransferase. The oleate at the *sn*-2 positions will then be a substrate for the Δ12-desaturase (section 6.2.3 and Fig. 6.7) and a linoleate-enriched diacylglycerol can then be returned to the Kennedy pathway. For oils which are rich in oleate but poor in polyunsaturates, such as olive (Table 6.2), the reaction of the cholinephosphotransferase during oil accumulation is minimal. A second aspect of stored oil quality is that the acyltransferases of any particular crop seem to have special structural features which enable them to deal with unusual fatty acids as necessary. Thus, whereas the acyltransferases of castor bean are well able to use the hydroxy-fatty acid, ricinoleic, the same substrate cannot be used by the acyltransferases of, say, oilseed rape. Another interesting feature is that unusual fatty acids are excluded from membrane lipids (where they could impair function) even though the diacylglycerol intermediate of the Kennedy pathway is also used for membrane lipid synthesis. It is not known how

this is achieved although the use of specific pools of diacylglycerols for membrane or storage lipid synthesis is one possible explanation. Alternatively, it may be that, since triacylglycerol accumulation is confined to a period of minimal membrane biogenesis, the opportunity for damaging modification of membrane lipid composition is reduced considerably.

6.4   MEMBRANE LIPID BIOGENESIS

6.4.1   Synthesis of glycosylglycerides, the main membrane lipids of chloroplasts

Not surprisingly, the three chloroplast glycosylglycerides (MGDG, DGDG and SQDG) are made within that organelle. The pathways involved are shown in Fig. 6.12. Monogalactosyldiacylglycerol is made on the chloroplast envelope using diacylglycerol generated by phos-

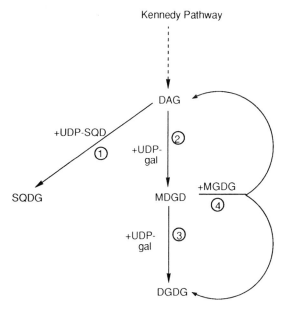

Kennedy Pathway

Figure 6.12   The synthesis of glycosylglycerides. DAG, diacylglycerol; MGDG, monogalactosyldiacylglycerol; DGDG, digalactosyldiacylglycerol; SQDG, sulfoquinovosyldiacylglycerol. Reaction 1, UDP-SQD : DAG sulfoquinovosyltransferase; reaction 2, UDP-galactose : DAG galactosyltransferase; reaction 3, UDP-galactose : MGDG galactosyltransferase; reaction 4, galactolipid : galactolipid galactosyltransferase.

phatidate phosphohydrolase (either in the chloroplast envelope or from the endoplasmic reticulum (see section 6.3.2) and UDP-galactose from the cytosol. The enzyme involved is called a galactosyltransferase. For digalactosyldiacylglycerol (DGDG) there are two possible pathways. Either MGDG is subject to a second galactosyltransferase using UDP-galactose or, alternatively, two molecules of MGDG undergo an interlipid transfer reaction with the formation of DGDG and diacylglycerol. This latter reaction (catalyzed by galactolipid : galactolipid galactosyltransferase) is believed to be the more important in those plants which have been examined.

The plant sulfolipid (sulfoquinovosyldiacylglycerol : SQDG) is made by a reaction analogous to that for MGDG (Fig. 6.12), except that the water-soluble substrate is UDP-sulfoquinovose. The origin of the sulfoquinovose itself has not been unequivocally demonstrated. It may originate from UDP-galactose via a keto-gluco-5-ene intermediate which can accept sulphite. The reactions would be analogous to those for deoxysugar formation.

6.4.2   Phosphoglyceride synthesis

Almost all phosphorus-containing lipids in plants are phosphoglycerides and only their synthesis will be considered in this chapter. Phosphoglyceride formation can be described under three headings – formation of zwitterionic lipids, formation of anionic lipids and conversions of one phospholipid into another.

Zwitterionic lipids are those containing both negative and positive groups but no net charge. At physiological pH, phosphatidylcholine and phosphatidylethanolamine would be such lipids. They are the major constituents of non-chloroplast plant membranes. Both are synthesized by similar three-reaction pathways (Fig. 6.13). First, a kinase enzyme uses ATP to phosphorylate the base (choline or ethanolamine). Separate choline and ethanolamine kinases have been purified from soybean seeds and these, and other studies, have defined some of their characteristics. In the second step, the phosphorylated bases are transferred to a cytidine nucleotide to yield CDP-base intermediates. The enzymes involved (again separate for cholinephosphate and ethanolaminephosphate) are called cytidylyltransferases. Evidence has shown that, in peas, the cholinephosphate cytidylyltransferase is rate-limiting for the overall pathway to phosphatidylcholine. The enzyme activity appears to be regulated by allosteric activation of pre-existing enzyme and also by new protein synthesis. The final enzymes of the pathway transfer base phosphates from CDP-bases to diacylglycerol. The reactions are freely reversible and, as discussed earlier (section 6.3), allow ready exchange of diacylglycerol from the phosphatidylcholine pool. The enzymes involved in this final transfer are cholinephosphotransferase and ethanolaminephosphotransferase, respectively.

Instead of being made via CDP-bases, the anionic (negatively charged) phosphoglycerides are synthesized using CDP-diacylglycerol (Fig. 6.13). Again phosphatidate plays a central role but instead of generating diacylglycerol, it is involved in another cytidylyltransferase reaction with CTP. The CDP-diacylglycerol product then reacts directly with inositol to give phosphatidylinositol or with glycerol 3-phosphate to yield phosphatidylglycerol-3-phosphate which is rapidly dephosphorylated to give phosphatidylglycerol. The latter is also used to synthesize diphosphatidylglycerol (cardiolipin) which is a major lipid of mitochondria where it is localized in the inner membrane.

The reaction involved in phosphatidylinositol formation appears to be most active in the

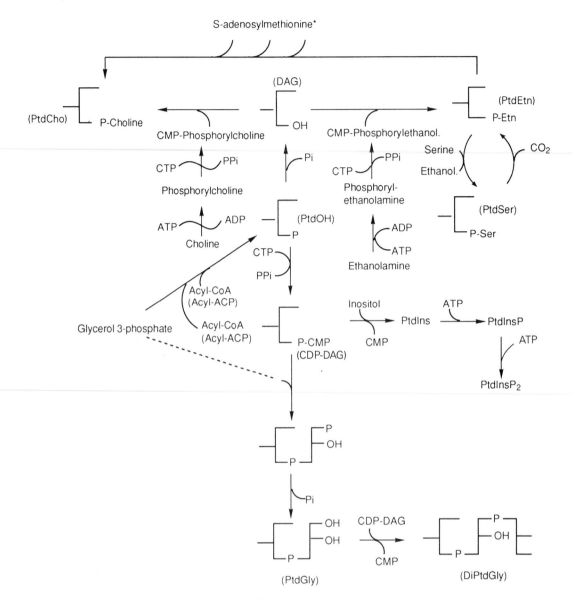

Figure 6.13   Pathways for phospholipid formation in plants. *The direct methylation of phosphatidylethanolamine to phosphatidylcholine may not occur in plants (see Fig. 6.14 for alternative pathways to phosphatidylcholine).

endoplasmic reticulum. However, further phosphorylation of this phosphoglyceride appears to be in the plasma membrane. This is in keeping with the function of phosphatidylinositol 4,5-*bis*-phosphate in signal transduction (section 6.5.3). In contrast, phosphatidylglycerol is made mainly in chloroplasts, where it is the only major phospholipid. A small amount of synthesis occurs also in mitochondria where it is used predominantly as a precursor of diphosphatidylglycerol.

Once a given phospholipid has been made there are some possibilities to convert it into another phospholipid. (We have already seen the reactions for synthesis of inositol phospholipids and the production of cardiolipin from phosphatidylglycerol). Phosphatidylserine is made by a Ca^{2+}-requiring base-exchange reaction using another phospholipid (often phosphatidylethanolamine) and serine (Fig. 6.13). In addition, there is some evidence that phosphatidylethanolamine can be successively methylated using *S*-adenosylmethionine to yield phosphatidylcholine. As an alternative, in a few plant tissues, ethanolamine is methylated in the form

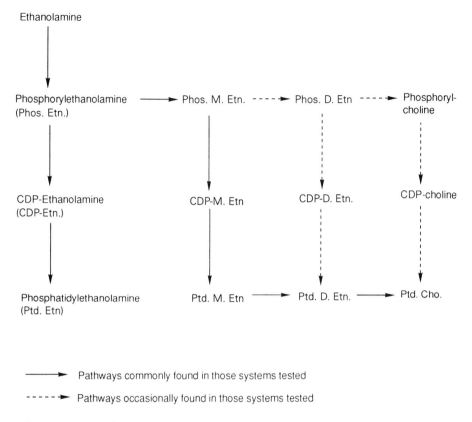

Pathways commonly found in those systems tested

- - - - - ► Pathways occasionally found in those systems tested

Figure 6.14 Alternative ways of generating phosphatidylcholine by methylation reactions. Phos. M. Etn., phosphorylmethylethanolamine; Phos. D. Etn., phosphoryldimethylethanolamine; Ptd. M. Etn., phosphatidylmethylethanolamine; Ptd. D. Etn., phosphatidyldimethylethanolamine; Ptd. Cho., phosphatidylcholine.

of ethanolaminephosphate or CDP-ethanolamine and methylation of phosphatidylethanolamine does not take place directly (Fig. 6.14). In addition, as shown in Fig. 6.13, decarboxylation of phosphatidylserine can yield phosphatidylethanolamine. This reaction has been studied well in micro-organisms and reported in plants – although its importance in the latter has not been fully elucidated.

6.4.3 Lipid transfer proteins

When lipids are made in one part of the cell but are destined for membranes in another, the problem of their movement must be addressed. At present, we do not know exactly how this takes place. Certainly, although some lipids (e.g. acyl-CoAs) can equilibrate across aqueous compartments quite easily, free diffusion does not seem a tenable hypothesis. One idea is that the membrane systems are all interconnected and, because of the relatively rapid movement of lipids

laterally in membranes (section 6.7.1), it is possible that this could be a mode of transport. Certainly, connections have been observed in electron micrographs between the endoplasmic reticulum and other organelles. However, there are still places where no such connections have been observed and one is left to explain the massive transport from such places as the chloroplast envelope to the thylakoids during leaf development. A more tenable idea seems to be the use of lipid transport proteins. These are molecules which act in the same way as albumin transports fatty acids in mammalian blood. That is, they have lipid binding sites which may be quite specific for different classes of molecules. A considerable number of such lipid binding proteins have been purified and characterized. They have been used to allow movement of phospholipids or glycosylglycerides between membrane systems *in vitro* and are presumed to play such a role *in vivo*. Several of the proteins have been sequenced and work is now underway on their molecular biology.

6.5 LIPID CATABOLISM AND TURNOVER

6.5.1 Membrane turnover and breakdown

Basically, lipids can be said to have three roles in Nature. They act as energy stores, and as constituents of membranes, and they can be involved in the control of metabolism, e.g. as second messengers or growth regulators.

The phospholipids and glycosylglycerides of plant membranes are not static molecules but undergo a significant turnover. This 'turnover' (the speed with which individual molecules are synthesized and then broken down) (often measured as a $t_{1/2}$ value when 50% of a given lipid class will have been turned over) can either take place in part of the molecule or for the whole molecule.

For membrane lipids the fatty acyl moieties are often removed and replaced much faster than the whole molecule. This may be in connection with metabolism (e.g. during fatty acid desaturation) or as a response to a particular need. For example, during cold acclimation there may be a need to 'retailor' the molecular species of a particular lipid class, i.e. to change the combination of fatty acids present in order to lower the melting point of the lipid and keep the membrane fluid.

For phospholipids, hydrolysis can take place as a result of the activity of phospholipases (Fig. 6.15) or acyl hydrolases. The latter are present at high activity in many plant tissues and are capable of hydrolyzing any acyl lipid. They do, however, have a substrate specificity and usually are particularly active towards monogalactosyldiacylglycerol. Under most circumstances, their activity is carefully controlled (probably by compartmentation) but, if the tissue is damaged mechanically or by frost then they gain access to the membrane lipid substrates. Thus, for example, if potatoes are homogenized and an attempt is made to isolate membrane fractions then these are often partly digested (even during their isolation) due to acyl hydrolase activity. The acyl hydrolases are very resistant to heat and frequently work well at temperatures of 60 or 70°C. Because they act on acyl chains which are fluid at low temperatures, they may also be active in frozen tissues. The active acyl hydrolases of tissues such as Brussels sprouts makes it necessary to boil these vegetables briefly before storing them frozen.

Phospholipase A enzymes remove a fatty acid from an intact phospholipid molecule. Phospholipase A_1 acts at the *sn*-1 position whereas A_2 acts at the *sn*-2 group. Phospholipase A_2 has an important role in allowing remodeling of phospholipids and specific enzymes may permit newly synthesized fatty acids to leave their parent phospholipid after formation. For example, the unusual hydroxy fatty acid ricinoleic is made at the *sn*-2 position of phosphatidylcholine. After it has been synthesized it is rapidly hydrolyzed by a specific phospholipase A_2 which makes it available for incorporation in triacylglycerol stores and, at the same time, prevents it being retained as part of a membrane lipid.

Phospholipase C removes the base-phosphate moiety leaving a diacylglycerol product. It is uncertain how important such enzymes are in plants but, certainly, a specific enzyme is involved in the turnover of inositol lipids (section 6.5.3). On the other hand, plants are often very good sources of phospholipase D enzymes which can remove the base moiety. This reaction could be involved in the general production of phospholipids but there is little evidence for this role *in vivo*. For example, the phospholipase D-catalyzed synthesis of phosphatidylglycerol produces a racemic mixture instead of the head-group glycerol being in the opposite configuration to the glycerol backbone (such as occurs naturally). Thus phospholipase D is one of those plant enzymes (like urease) which is in search of a physiological role.

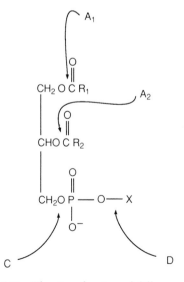

Figure 6.15 The site of action of different phospholipases. Note that no reliable report of phospholipase B has been made for plants.

6.5.2 Degradation of lipid stores

As emphasized in section 6.1, the lipid stores of higher plants are almost invariably triacylglycerol. Such lipid is stored in discrete bodies (section 6.4.2) which pack the interior of oil-rich seeds. When seeds such as castor bean germinate, lipases are activated and bind to the surface of the oil bodies possibly through specific anchoring sites on the oleosins. Hydrolysis of the triacylglycerols then takes place at the primary hydroxyls (*sn*-1 and *sn*-3 positions). The 2-monoacylglycerol product will naturally racemize to the more stable 1-monoacylglycerol which will then be a substrate for lipase attack even if the enzyme could not hydrolyze the secondary position originally. A large number of different lipases has now been isolated from plants and their substrate specificities and other enzymic characteristics determined. They are often very stable to heat, organic solvents and other extreme conditions. Some of them have important industrial implications and are either used in organic syntheses or cause food spoilage. In the latter regard are enzymes which cause rancidity in olive oil or deterioration of whole flour products. On the other hand, the ability of lipases to catalyze many

esterification and other synthetic reactions in low water media allows them to be used industrially for the modification of edible oils and even to catalyze peptide production from amino acids.

Once fatty acids have been released from lipid stores (or membranes) they can be subjected to oxidative attack usually by the β-oxidation cycle. Oxidation of fatty acids is discussed in section 6.5.4.

6.5.3 Plant lipids as second messengers

It is well established in animals that phosphatidylinositol-4,5-*bis*phosphate plays a key role in metabolic control. Such a function also seems to pertain to plants and the key features of the process are outlined in Fig. 6.16. In brief, an agonist arrives on the outside of a cell where it binds to a specific receptor. The binding leads to a conformational change in an adjacent G-protein which, in turn, activates a specific phospholipase C. The latter only acts on phosphatidylinositol-4,5-*bis*phosphate and hydrolyzes the lipid to yield a membrane-localized diacylglycerol and a water-soluble inositol-1,4,5-*tris*phosphate (InsP$_3$). These two products both act as second messengers. Diacylglycerol stimulates a protein kinase C

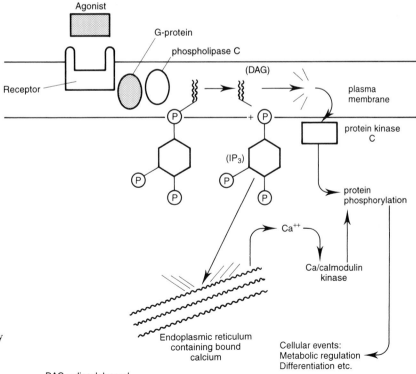

Figure 6.16 Mechanism by which the hydrolysis of phosphatidylinositol-4,5-*bis*phosphate can control metabolism.

DAG = diacylglycerol
IP$_3$ = Inositol-1,4,5-*tris*phosphate

whereas InsP$_3$ releases intracellular calcium stores which stimulate the Ca^{2+}/calmodulin-stimulated protein kinases. The different types of protein kinase phosphorylate different target proteins and so regulate metabolism.

Work on the metabolism of phosphatidylinositol-4,5-bisphosphate in plants has lagged behind that in animals but a role for the lipid at least in salt stress and flowering has been suggested. Certainly, all the necessary enzymes are in place and the turnover of inositol lipids in relation to calcium signaling has been well proven.

Other lipid-mediated phenomena have been shown in plants but the main such area involves oxidized (lipoxygenized) fatty acids and is discussed below.

6.5.4 Fatty acid oxidation

Fatty acids can be subject to α-, β-, ω-oxidations, in-chain hydroxylation (oxidation) and lipoxygenase attack.

α-Oxidation involves oxidation of the carbon adjacent to the terminal carboxyl function (i.e. the α-carbon). The proposed pathway is shown in Fig. 6.17. The process has been studied in pea leaves and in peanuts. Molecular oxygen is a requirement and, in the peanut system, a hydrogen peroxide-generating system is involved. Enzymes which catalyze the reduction of peroxides, such as glutathione peroxidase, reduce α-oxidation and increase the production of D-hydroxypalmitate. This pointed to the existence of 2-hydroperoxy-

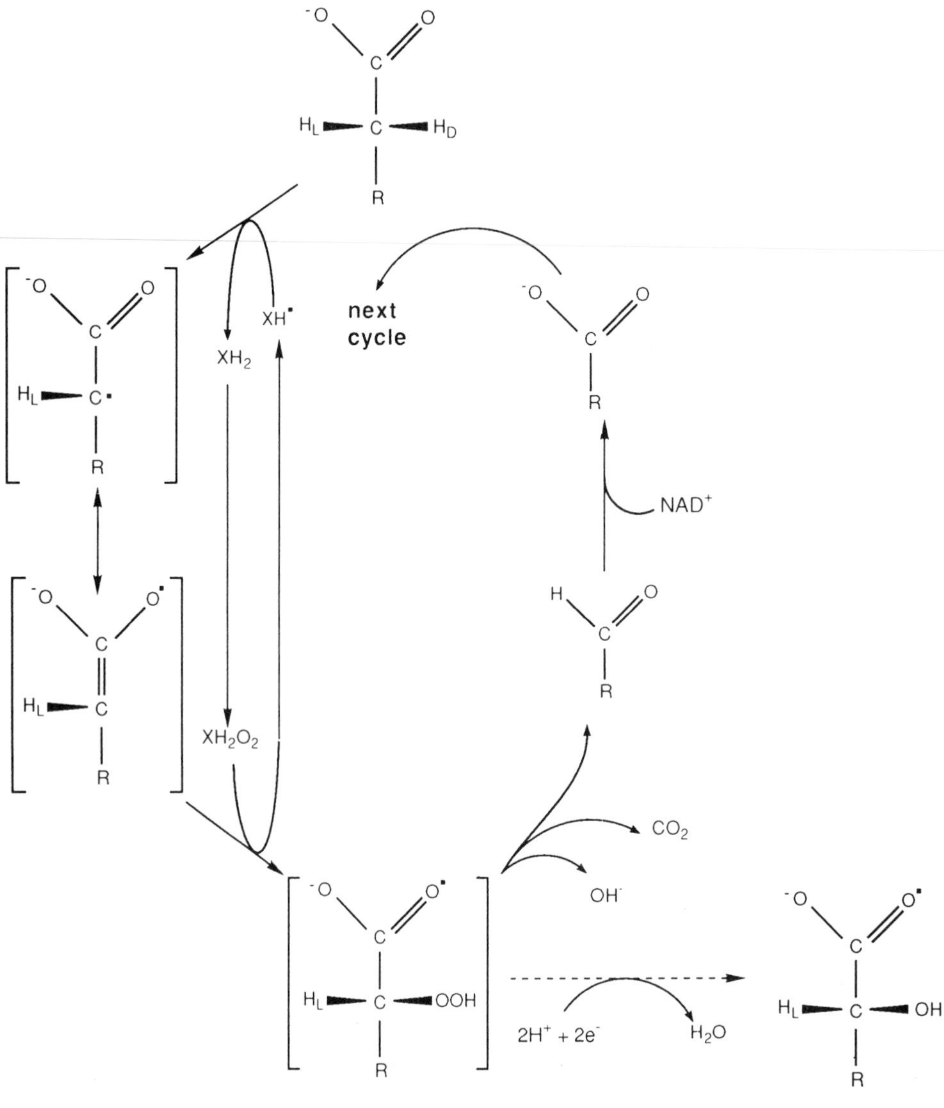

Figure 6.17 Proposed α-oxidation pathway in plants.

palmitate as an intermediate. Accumulation of hydroxy fatty acids in experiments was due to intermediates being channeled into a dead-end pathway. After the hydroperoxy intermediate, CO_2 is lost and a fatty aldehyde produced. This is oxidized by a NAD^+-requiring system to yield a fatty acid one carbon less than the original (palmitate) substrate. α-Oxidation in plants is important for metabolism of compounds where methyl side chains prevent normal β-oxidation. Phytol, from the side chain of chlorophylls, is a particular example. The oxidation process may be soluble or associated with the endoplasmic reticulum (depending on the plant tissue).

β-Oxidation of fatty acids is the major way in which the energy from lipid stores can be released.

In plants the enzymes of β-oxidation have been found to be localized in microbodies and show some distinction from those used in animal mitochondrial β-oxidation. Fatty acyl-CoAs are used as starting material but no involvement of carnitine has been demonstrated for their catabolism by microbodies. By analogy with animal systems, non-esterified fatty acids are converted into their CoA thiolesters by use of an acyl-CoA synthetase associated with the microbody membrane (Fig. 6.18). Both non-esterified fatty acids and acyl-CoAs appear to be able to cross the microbody membrane freely.

The reactions of β-oxidation are shown in Fig. 6.18. Initial oxidation of acyl-CoA is catalyzed by acyl-CoA oxidase in a reaction which generates

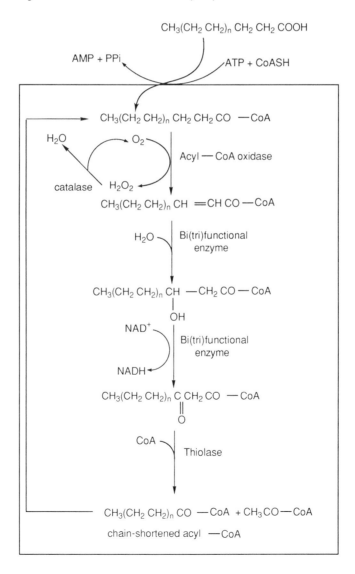

Figure 6.18 β-Oxidation in plant microbodies.

hydrogen peroxide. The latter is rapidly destroyed by peroxisomal catalase activity. The next two reactions (enoyl-CoA hydratase, 3-hydroxyacyl-CoA dehydrogenase) are catalyzed by a bifunctional protein and result in the generation of NADH. Finally, a thiolase cleaves acetyl-CoA from the acyl chain with the formation of another acyl-CoA which is two-carbons shorter than the original substrate. The latter substrate is then recycled so that the overall process can begin again.

Of course, not all fatty acids catabolized by microbodies are even carbon numbered, saturated acids. When odd-chain substrates are used, propionyl-CoA is the end product. This propionate is processed further in plants by a pathway quite distinct from that in animals. Two dehydrogenations are involved, together with a hydration. In addition, the thiolester is moved from one end of the molecule to the other in order to permit a final decarboxylation and co-production of acetyl-CoA (Fig. 6.19).

Most plant fatty acids are cis-unsaturated and contain $\Delta 9$-double bonds (see section 6.1.2). These two facts, in themselves, present problems for β-oxidation because both the position and the configuration of the double bonds are inappropriate for the trans-$\Delta 2$-intermediate in β-oxidation (Fig. 6.18). To overcome this, a ^3cis-^2trans-enoyl-CoA isomerase is present as a third activity of

the tri(bi)functional enzyme. For fatty acids such as linoleate, a 2,4-dienyl-CoA reductase is present and the oxidation sequence is illustrated in Fig. 6.20.

The need for β-oxidation in germinating seeds is transient and work by Beevers and his colleagues, in particular, revealed details of the process in castor beans. As the seeds imbibe water the special microbodies are produced in which all the enzymes for β-oxidation and the subsequent glyoxysome cycle (which processes acetyl-CoA) are present. A lipase also becomes active which liberates fatty acids from the triacylglycerol stored in the oil bodies. Within a short period (about 5–6 days) all the triacylglycerol stores have been hydrolyzed with the bulk of the carbon converted into carbohydrate for further metabolism (Fig. 6.21). In animals, acetyl-CoA liberated by β-oxidation enters the citric acid (Krebs) cycle in which two carbons are released as CO_2. Thus, it is not possible for animals to convert fatty acids into glucose and other carbohydrates. By contrast, it is vital for seeds to be able to use stored lipids as a source of carbohydrate before photosynthesis becomes active in the young leaves. Therefore, the glyoxylate cycle (in which the two decarboxylation reactions of the citric acid cycle are missing) is used (Fig. 6.22). The glyoxysomal enzymes are relatively unimportant (or missing) in

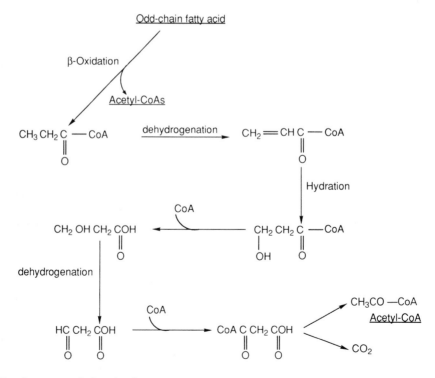

Figure 6.19   Propionate metabolism in plants.

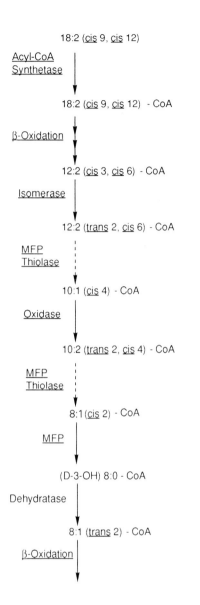

18:2 (<u>cis</u> 9, <u>cis</u> 12)

<u>Acyl-CoA</u>
<u>Synthetase</u>

18:2 (<u>cis</u> 9, <u>cis</u> 12) - CoA

<u>β-Oxidation</u>

12:2 (<u>cis</u> 3, <u>cis</u> 6) - CoA

<u>Isomerase</u>

12:2 (<u>trans</u> 2, <u>cis</u> 6) - CoA

<u>MFP</u>
<u>Thiolase</u>

10:1 (<u>cis</u> 4) - CoA

<u>Oxidase</u>

10:2 (<u>trans</u> 2, <u>cis</u> 4) - CoA

<u>MFP</u>
<u>Thiolase</u>

8:1 (<u>cis</u> 2) - CoA

<u>MFP</u>

(D-3-OH) 8:0 - CoA

<u>Dehydratase</u>

8:1 (<u>trans</u> 2) - CoA

<u>β-Oxidation</u>

Figure 6. 20 β-Oxidation of linoleic acid by plant peroxisomes. MFP, multifunctional protein.

the microbodies (peroxisomes) of leaves where β-oxidation functions in the overall turnover of fatty acids and to provide an additional source of acetyl-CoA for general metabolism.

6.5.5 Plant lipoxygenases and their function

Lipoxygenase is a term applied to a group of enzymes which catalyze the oxygenation, by molecular oxygen, of fatty acids containing a *cis*, *cis*-1,4-pentadiene system to produce conjugated hydroperoxydiene derivatives (Fig. 6.23).

Lipoxygenase activity is widespread in the plant kingdom, often in very high amounts. The enzymes are particularly important in food plants where they catabolize polyunsaturated fatty acids to produce derivatives with characteristic tastes and flavors. They are also used for bleaching natural pigments such as wheat flour carotenoids or alfalfa chlorophyll. Leguminous plants, such as soya bean, are often good sources of lipoxygenases but the absence of measured activity in certain plant tissues does not mean that the enzyme is absent, since there are often inhibitors present.

A number of lipoxygenases have been purified from different plants and enzymes with both acid and alkaline pH optima found. Because lipoxygenases lack cofactors other than non-heme iron, the number of inhibitors is small. Usually, the enzymes are most active towards linoleic acid although other acids with a double bond at C-13 (*n*-6) were also active.

Aerobic lipoxygenase reactions give rise to a conjugated *cis*, *trans*-pentadienyl hydroperoxide which has its *trans* bond next to the hydroperoxide. The enzymes differ in the position of the hydroperoxide group which for linoleic acid may be C-9 or C-13. In addition, lipoxygenases also catalyze an anaerobic reaction between the product of the aerobic reaction and linoleic acid. A complex mixture of chemicals is produced which includes decomposition products of the 13-hydroperoxide and dimeric products.

Lipoxygenases will also catalyze co-oxidation reactions. These are used to assay the enzymes and in commercial applications. An example of the latter is the addition of soya bean or broad bean flours (both rich in lipoxygenase activity) to wheat flour in order to bleach pigments for white bread production. Enzymes from different sources differ in their co-oxidation ability which probably depends on the lifetime of the free-radical intermediates.

The hydroperoxide products of lipoxygenase action are not found in healthy plant cells because they are rapidly destroyed by enzymatic processes in order to protect cells from membrane or protein damage. A full discussion of all the possible products is beyond the scope of this chapter but some products have important physiological functions (Fig. 6.24). Thus, volatile aldehydes account for the characteristic smell of many fruits and vegetables, e.g. cucumber. The product of hydroperoxide dehydrase, 12-oxophytodienoic acid, is metabolized to the plant hormone, jasmonic acid. The latter compound (and/or its methyl ester) has often been cited to cause growth inhibition, abscission, senescence and other effects. Alterna-

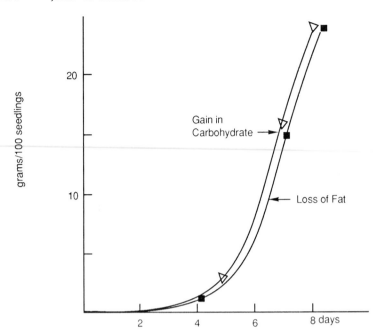

Figure 6.21 Conversion of fat into carbohydrate during the germination of castor bean seeds. From H. Beevers (1980) In *The Biochemistry of Plants* (P.K. Stumpf & E.E. Conn, eds), Vol. 4, pp. 117–131. Academic Press, New York, with permission.

tively, wounding appears to activate the hydro-peroxide lyase pathway to give 12-oxo-10(E)-dodecenoic acid, which has been reported to have wound-healing properties. In addition, certain lipoxygenase pathway products have been found to have anti-fungal properties.

6.6   CUTINS, SUBERINS AND WAXES

6.6.1   Introduction to surface layers

All plant surfaces are covered in a protective layer, which serves the dual function of preventing water loss and also stopping microbial entry into plant tissues. The nature of the surface layer is different for the various plant parts with the leaf covering formed of cutin and wax whereas roots (and wounds) contain suberin. Actually the chemical compositions and structures of wax, cutin and suberin are imprecise and differ not only from species to species but also with development age. However, generalizations may be made and these are detailed below. Both cutin and suberin contain extensive polymeric structures but wax comprises a looser mixture of aliphatic compounds.

6.6.2   The chemistry of surface coverings

Cuticular waxes can be readily extracted by washing intact tissue with suitable organic solvents such as hexane or chloroform. Waxes are also found associated with suberin in barks, tubers, roots or wounds as well as in the seeds of certain plants, notably jojoba (whose wax esters serve as a substitute for sperm whale oil).

Hydrocarbons (alkanes) are ubiquitous in plant waxes and usually make up a high proportion of the total components. Odd-chain alkanes predominate because of their pathway of synthesis. Monoesters of long-chain alcohols and fatty acids are also found in virtually every plant examined (Table 6.7). In addition, very long-chain alcohols (mainly C_{22}–C_{34}) and very long-chain saturated fatty acids are usual components. However, the proportion of these two chemicals can vary from a few per cent to about half of the total wax. Of the other components in wax, ketones and β-diketones may be major constituents of some waxes.

Cutin is composed chiefly of a C_{16}-family and/or a C_{18} family of monomers. The hydroxy groups of monomers such as 16-hydroxypalmitic acid or 10,16-dihydroxypalmitic acid then allow ester links to be formed and so build up a polyester structure.

The major aliphatic components of suberin (20–50% of suberin-enriched samples) are ω-hydroxy acids, the corresponding dicarboxylic acids, very long chain (>C_{18}) acids and similarly long alcohols. There are also substantial amounts of phenolic compounds of which the major component is *p*-coumaric acid (see Chapter 10). The polymeric structure of suberin is not well characterized but has been suggested to be based on a phenolic core to which dicarboxylic acids and some hydroxy fatty acids can be attached and, themselves, linked in ester bonds.

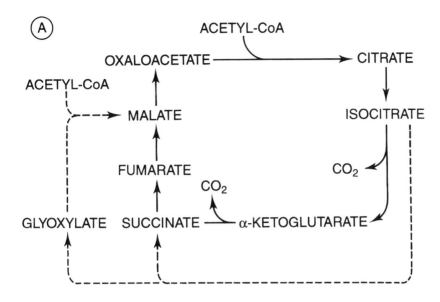

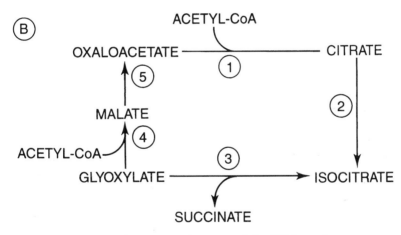

(A) the Glyoxylate Cycle as a bypass of the TCA cycle
(B) The Glyoxylate Cycle as it functions in the glyoxysome, showing the production of succinate from 2 mol of acetyl-CoA. The five steps constituting the cycle are catalyzed by the following enzymes: (1) citrate synthetase, (2) aconitase, (3) isocitrate lyase, (4) malate synthetase, (5) malate dehydrogenase.

Figure 6.22 The glyoxylate cycle for conversion of acetyl-CoA carbon into carbohydrate. From H. Beevers (1980) In *The Biochemistry of Plants* (P.K. Stumpf & E.E. Conn, eds), Vol. 4, pp. 117–131. Academic Press, New York, with permission.

6.6.3 The biosynthesis of wax, cutin and suberin

The pathways for the formation of the major monomers of wax, cutin and suberin are shown in Fig. 6.25. Fatty acid synthesis (section 6.2.1) provides palmitate and stearate, the latter of which can be desaturated to oleate. Palmitate and oleate then form the main precursors of both cutin and suberin. In all cases ω-hydroxylation is the first step. The ω-hydroxypalmitate or ω-hydroxyoleate produced can then be converted

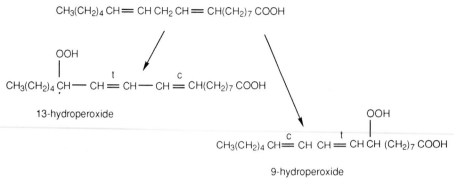

Figure 6.23 Lipoxygenase attack on linoleic acid.

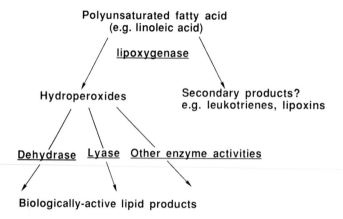

Figure 6.24 Metabolism of polyunsaturated fatty acids to give biologically active lipid products.

into dicarboxylic acids for suberin production or to in-chain hydroxy acids for cutin formation. Some details are given in Fig. 6.25 but see Kolattukudy (1980, 1987) for more discussion.

The aliphatic components of waxes are formed after the elongation of stearate (see section 6.2.3). Very long-chain fatty acids can be reduced via an aldehyde intermediate to fatty alcohols or, alternatively, decarboxylated to alkanes. The fatty alcohols and acids can also combine together to form simple wax esters.

To form the polymeric structures of cutin and suberin needs transacylation reactions between free hydroxyl groups of cutin and an incoming hydroxy fatty acid or between the hydroxyl group of the phenol core of suberin and incoming dicarboxylic acids. Little is known, however, about the enzymes involved or their characteristics.

Table 6.7 Typical components of waxes, cutins and suberins

	Waxes	Cutins	Suberins
Alkanes	Major component	–	–
Ketones	Minor usually	–	–
β-Diketones	Sometimes major	–	–
Esters	Minor, common	–	–
VLC alcohols	Common	Rare, minor	Common, major
VLC acids	Common	Rare, minor	Common, major
Dicarboxylic acids	Rare	Minor	Major
In-chain subst. acids	Rare	Major	Minor
Phenolics	–	Low	High

Abbreviation: VLC, very long chain ($>C_{18}$).

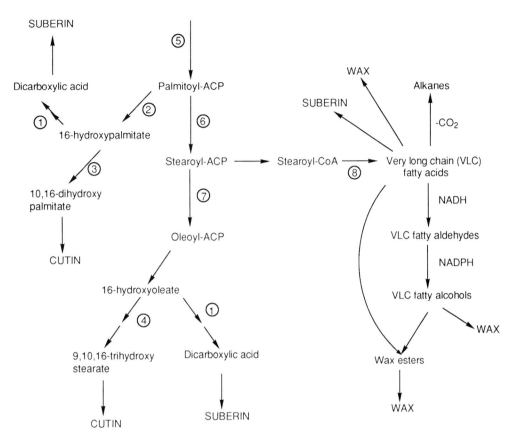

Figure 6.25 Overall scheme for the synthesis of components used in plant surface coverings. 1, Two dehydrogenase steps via an ω-oxo acid intermediate; 2, ω-hydroxylation; 3, in-chain hydroxylation; 4, epoxidation of the double bond, followed by hydration; 5, fatty acid synthetase; 6, the β-ketoacyl-ACP synthetase II step of fatty acid synthetase; 7, stearoyl-ACP Δ9-desaturase; 8, fatty acid elongation.

6.7 PLANT MEMBRANES

6.7.1 Introduction to membrane structure

The overall structure for membranes which is now accepted generally is that proposed by Singer and Nicholson (Fig. 6.26). The 'fluid mosaic' model envisages a basic structure consisting of a lipid bilayer in which proteins may be embedded as well as attached to the outer surface. Because of their amphipathic (both water-loving and water-hating) properties, most membrane lipids naturally form bilayers in aqueous media so that their acyl chains are shielded from the aqueous environment. This arrangement leaves the polar head groups (phosphobase or phosphopolyhydroxyl in phospholipids or sugar(s) in glycosylglycerides) to form appropriate interactions with extrinsic proteins, ions or other molecules.

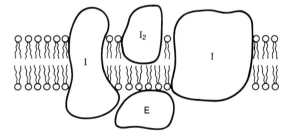

Figure 6.26 Simplified depiction of a fluid-mosaic membrane structure. The lipid bilayer composed mainly of phosphoglycerides or, in chloroplasts, glycosylglycerides forms the backbone of the structure. The acyl chains are above their Tc and, therefore, are fluid for a normal functional membrane. I, integral proteins which may span the membrane or, as in I_2, are only embedded in one half of the bilayer. Extrinsic proteins (E) are bound to the surface of the membrane through ionic interactions.

For a membrane to function efficiently it is thought that the vast majority of the acyl chains must be above their gel to liquid transition temperature (Tc) and, therefore, be fluid. This requirement has important implications for frost sensitivity/resistance in plants (section 6.7.3). 'Fluid' membranes are needed for efficient functioning of membrane proteins since it not only permits appropriate movement of substrates and products to enzymes but also permits lateral (and transverse) movement of proteins used in transport, electron transport and light-harvesting. Thus, the proteins of transport systems which act through pore-opening mechanisms need to be able to undergo conformational changes while those which act as carriers have to cross the hydrophobic interior of the bilayer.

Once it was realized that intrinsic proteins were present in membranes, according to the Singer/Nicholson model, then interest turned to examination of the two halves of biological membranes. Freeze-fracture and freeze-etching techniques in electron microscopy had already revealed that membranes showed sidedness with regard to their proteins and the same question was now asked with regard to lipid distribution. In order to examine sidedness, it is necessary to be able to prepare pure samples of particular membranes and then to have probes available which cannot penetrate the membrane vesicles analyzed. The general principle of the methods is shown in Fig. 6.27. Intact membrane vesicles are probed first to determine which lipids are present in the outer half of the bilayer. Then the vesicles are disrupted

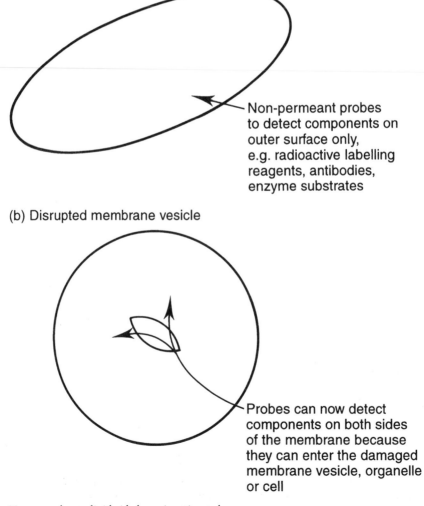

(a) Intact membrane vesicle

Non-permeant probes to detect components on outer surface only, e.g. radioactive labelling reagents, antibodies, enzyme substrates

(b) Disrupted membrane vesicle

Probes can now detect components on both sides of the membrane because they can enter the damaged membrane vesicle, organelle or cell

Figure 6.27 How membrane lipid sidedness is estimated.

to allow the probes to also interact with the inner surface. Some lipids may be inaccessible to the probes (e.g. in a tight complex with intrinsic proteins) and these will be unlabeled even in disrupted vesicles. When these techniques were applied to plant membranes it was found that, as in other organisms, a sided distribution of lipids is found (Fig. 6.28). Presumably, this sidedness has important functional significance.

By spin-labeling the acyl chains of membrane lipids it has been found that there is considerable movement of the hydrocarbon chain. Thus, there is free rotation about the individual carbons of the acyl chains (except where constrained by a double bond) and flexing takes place at a rate of about once every 10^{-9} s. Moreover, the whole lipid is usually able to move freely in a transverse direction (i.e. in the plane of the membrane). For model membranes, each lipid molecule can move sideways once every 10^{-7} s. This rapid movement, of course, allows lipids such as plastoquinone to function efficiently as mobile electron carriers in photosynthesis. In contrast to the ease of sideways movement, lipids are constrained from crossing

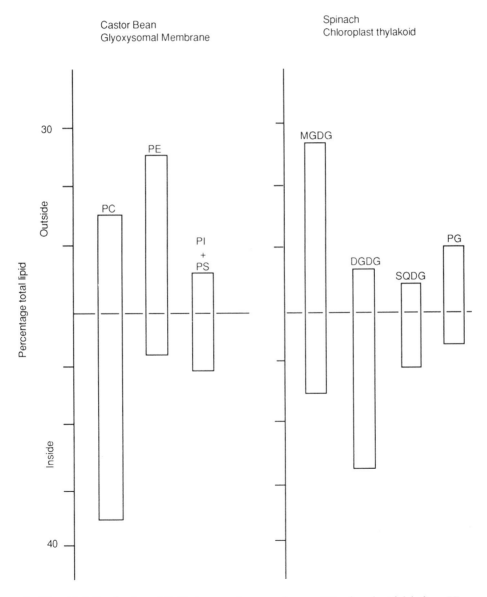

Figure 6.28 The sided distribution of lipids in two plant membranes. PC, phosphatidylcholine; PE, phosphatidylethanolamine; PI, phosphatidylinositol; PS, phosphatidylserine; MGDG, monogalactosyldiacylglycerol; DGDG, digalactosyldiacylglycerol; SQDG, sulfoquinovosyldiacylglycerol; PG, phosphatidylglycerol.

from one side of the bilayer to the other (a process known as flip-flop) by the hydrophilic nature of their head groups which are unsuited to crossing the hydrophobic membrane interior. Flip-flop rates in model membranes have been estimated to take place once every 10^5 s or a billion, billion times slower than sideways movement! Thus, the sided distribution of membrane lipids is preserved.

6.7.2 Functional aspects of plant membranes

Membranes in any eukaryotic organism have many features in common. First, they provide a permeability barrier to many molecules thus allowing cells and their organelles to maintain unique environments in which their metabolic and other processes can take place. Because membranes are impermeable to large and/or water-soluble molecules, then specific transport processes are needed to allow such substances to move across membranes when needed.

Second, membranes provide the site for enzymatic and other activities of intrinsic proteins. For enzymes dealing with hydrophobic molecules, such as those engaged in many of the reactions of lipids, presence in the membrane gives them just the right milieu for activity. Furthermore, for reactions which have a sided requirement, the membrane allows the proteins to be arranged specifically with regard to one face or other of the membrane. Thus, the photosynthetic complexes (photosystem 1, photosystem 2, cytochrome b_6/f complex) are able to function with the simultaneous generation of a proton gradient across the membrane which is so vital for subsequent photophosphorylation.

Third, membranes may be used for packaging materials and, sometimes, in secretory processes. The half-unit membranes of oil bodies have been mentioned in section 6.3.1. Another example in plants are the salt-secreting glands in the leaves of *Tamarix* which allow this organism to tolerate high salinity. In addition, many plant cells have well-developed smooth endoplasmic reticulum as Golgi bodies which is indicative of active synthesis of substances for secretion or export. Some lipid molecules are examples of such substances and the production of very long chain fatty acids (by elongation) for surface layer production (see section 6.6) could be cited.

Finally, membranes can contain receptors which serve to mediate the influence of external signals. Receptors for auxins and phytochrome have been quite well characterized and, clearly, allow these plant growth regulators to exert their effects. Another example would be the role the membrane plays in the production of second messengers by the hydrolysis of phosphatidylinositol 4,5-*bis*phosphate (section 6.5.3).

6.7.3 Environmental aspects of membranes with particular reference to temperature

The lipid composition and/or metabolism of plants has been shown to be altered by a number of environmental influences. These include light, temperature, atmospheric pollutants, the salts available in soils and xenobiotics such as pesticides.

The most obvious influence of light is through its stimulation of photosynthetic membrane function. Thus, light is needed for chloroplast development and, because of that, the lipid composition of leaves will undergo significant change. The most obvious alteration is, of course, the production of chlorophyll (i.e. leaf greening) but the acyl lipid composition also changes (Table 6.8). Prominent amongst these changes is the increase in α-linolenate content and the appearance of the unique *trans*-Δ3-hexadecenoate..

Because light stimulates photosynthesis (and the attendant production of ATP and NADPH) it also stimulates fatty acid (and lipid) synthesis. Acetyl-CoA carboxylase is activated; probably through the changes in stromal conditions (Table 6.9). Thus, measurements of fatty acid synthesis using radiolabeled precursors show that formation of lipid is about 20 times as fast in leaves in the light as in the dark.

There is much interest, from an agricultural standpoint, in the ability of certain plant species to be resistant to cold (not necessarily freezing) temperatures. If crops which are now sensitive to cold could be made resistant, then the range of countries in which they could be grown would be

Table 6.8 Alterations in broad bean fatty acids seen on growth in light compared to etiolated tissues

| | Fatty acids (% total) | | | | | | |
	16:0	16:1[a]	18:0	18:1	18:2	18:3	Other
Light-grown	12	7	3	3	14	56	5
Dark-grown	17	1	5	2	34	39	2

[a] 4% as *trans*-Δ3-hexadecenoate which is not present in the dark.

Table 6.9 Possible ways in which chloroplast acetyl-CoA carboxylase is stimulated by light

Parameter	In vivo change during photosynthesis		Activity change in vitro[a]		Additive fold change
	From	To	From	To	
pH	7.1	8.0	50	155	3.1×
Mg^{2+}	2 mM	5 mM	70	165	7.3×
ATP	0.5–0.8 mM	0.8–1.4 mM	80	135	12.3×
ADP	0.6–1.0 mM	0.3–0.6 mM	65	125	23.6×

[a]Activity expressed as nmol/min/mg protein.
From Harwood (1988), with permission.

increased. Therefore, biochemists have been researching on the mechanisms by which plants adapt to or can already cope with low (0–10°C) temperatures.

There are a number of ways in which modification of membrane lipid composition can contribute to a more 'fluid' hydrocarbon core or, alternatively, to a bilayer which is fluid to lower environmental temperatures. Changes which have been noted for plants or algae are listed in Table 6.10. Molecular species re-modeling could involve either the wholesale creation of new combinations of fatty acids (Fig. 6.29) or alternatively a swapping of the positions of two existing acids or the glycerol backbone (Fig. 6.30). Because no new fatty acids have to be synthesized, these changes are often regarded as an emergency response to a sudden lowering of environmental temperature.

The importance of molecular species composition in determining temperature sensitivity is

Table 6.10 Lipid changes which can affect membrane fluidity and which have been observed in plants

1. Molecular species re-modeling.
2. Changes in unsaturation.
3. Alterations in the proportions of lipid classes.
4. Increases in lipid to protein ratios at lower temperatures.

emphasized by studies showing that the impairment of photosynthesis on chilling in some plants but not in others is due to the proportions of certain species of phosphatidylglycerol. When plants contained greater than 20% of their total thylakoid phosphatidylglycerol as either the dipalmitoyl-(Tc = 42°C) or the 1-palmitoyl, 2-trans-Δ3-hexadecenoyl combination (Tc = 32°C) then they were susceptible to chilling damage (Table 6.11). This was presumed to be due to the fact that a significant amount of the total thylakoid lipid would phase separate under these conditions and

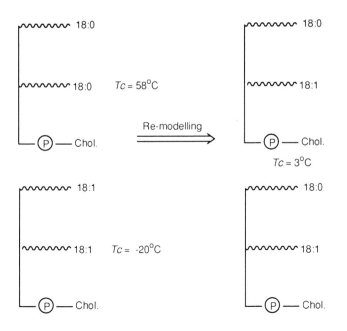

Figure 6.29 How the creation of new molecular species of membrane lipids could prevent phase separation which is due to the transition of some molecules to the gel phase.

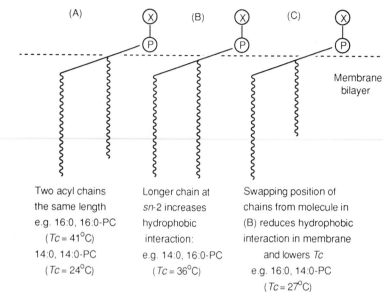

(A)
(B)
(C)
X
P
Membrane bilayer
Two acyl chains the same length e.g. 16:0, 16:0-PC (Tc = 41°C) 14:0, 14:0-PC (Tc = 24°C)
Longer chain at sn-2 increases hydrophobic interaction: e.g. 14:0, 16:0-PC (Tc = 36°C)
Swapping position of chains from molecule in (B) reduces hydrophobic interaction in membrane and lowers Tc e.g. 16:0, 14:0-PC (Tc = 27°C)

Figure 6.30 Because the glycerol backbone of membrane lipids is not parallel to the membrane surface, swapping acyl positions can affect fluidity.

Table 6.11 Correlation of chilling sensitivity in different plants with their content of 16 : 0/16 : 0 – and 16 : 0/*trans* 16 : 1 – molecular species of phosphatidylglycerol

Chilling-sensitive plants	%[a]	Chilling-resistant plants	%
Sweet potato	65	Lettuce	17
Squash	64	Spinach	6
Taro	62	Oat	5
Rice	44	Cabbage	4
Tobacco	39	Dandelion	4
Maize	37	Wheat	3

[a]Values as mol. % of total phosphatidylglycerol.

so cause a loss of membrane function. No other thylakoid lipid (see Table 6.4) seemed to be involved in chilling resistance/sensitivity.

Plants, such as squash, which produce large amounts of the above ('saturated') molecular species of phosphatidylglycerol seem to contain glycerol-3-phosphate acyltransferase enzymes which has a substrate specificity for palmitate in addition to oleate (Fig. 6.31). When the genes for such enzymes are transferred into moderately sensitive plants (tobacco) they make them more chilling-sensitive and when the genes for acyltransferases from resistant plants such as *Arabidopsis* are similarly transferred, then tobacco becomes more temperature-resistant (Table 6.12).

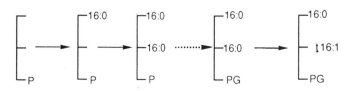

Pathway A
P
18:1
P
18:1
16:0
18:1
16:0
PG
18:2
16:0
PG
18:3
16:0
PG
Pathway B
P
16:0
P
16:0
16:0
P
16:0
16:0
PG
16:0
t16:1
PG

Figure 6.31 The production of disaturated species of phosphatidylglycerol in temperature-sensitive plants depends on the specificity of their glycerol 3-phosphate acyltransferases. Pathway A is dominant in chilling resistant plants and produces phosphatidylglycerol molecules of high fluidity (low melting point). Pathways A and B are of comparable activities in chilling-sensitive species and result in an appreciable proportion of the phosphatidylglycerol having a Tc above 32°C (see Table 6.11).

Table 6.12 Genetic manipulation of tobacco plants to change the composition of phosphatidylglycerol and increase chilling tolerance

	PG Fatty acid composition (% total)			Photosynthesis[a]	
	16:0	16:1, *t*	C_{18} unsat.	Before	After chilling
Squash (SQ)	54	22	18	–	–
Arabidopsis (ARA)	29	29	40	–	–
Tobacco	20	46	32	–	–
Transgenic tobacco with plasmid alone	23	44	32	0.016	0.010
pSQ[b]	52	29	12	0.024	0.004
pARA	24	38	36	0.017	0.017

[a]Photosynthesis was measured as O_2 evolution in ml min^{-1} $10\,cm^{-2}$ leaf area. Plants were grown at 27°C, chilled to 1°C for 4 h and then returned to 27°C. The photosynthetic rates 5 h later are shown.
[b]pSQ corresponds to the plasmid containing the gene for the squash glycerol 3-phosphate acyltransferase and pARA the gene for the *Arabidopsis* enzyme.
For more details refer to Murata *et al.* (1992) *Nature* 356, 710–713.

A very common change which is seen in membrane lipids with a lowering of environmental temperature is an increase in unsaturation. Introduction of a *cis* double bond into a saturated chain can make a tremendous difference to its melting properties. Thus stearic acid has a Tc of 58°C whereas oleic acid has a Tc of 14°C. The addition of a second double bond lowers the melting point further (linoleic acid Tc = −5°C) although not as much as for the first double bond. Thus, an increase in membrane lipid unsaturation makes very good sense in terms of an adaptive response.

The mechanism by which fatty acid desaturation is increased by a lowering of environmental temperature has been studied in a number of organisms. Three main theories have been proposed (Table 6.13). The available evidence favors a dual influence with membrane fluidity activating existing desaturase complexes as well as the synthesis of new desaturase protein occurring through gene transcription.

Table 6.13 How fatty acid desaturation may be increased at low temperaures

1. Supply of substrate increased (oxygen is a required substrate for aerobic desaturases and its solubility in water is increased at low temperatures).
2. Activation of pre-existing desaturase (the desaturases are usually membrane-bound enzymes and low temperatures may cause changes in membrane fluidity that activate such complexes).
3. New desaturase protein is made (low temperatures are known to increase protein synthesis in a number of cases – probably by increasing gene transcription).

The importance of fatty acid unsaturation in aiding chilling resistance has been demonstrated clearly in gene transfer experiments using cyanobacteria. A gene which coded for a Δ-12 desaturase (which converted, for example, oleate into linoleate) was transferred into mutants of *Synechocystis* which lacked the activity as well as into an organism which did not make polyunsaturated fatty acids. In both cases, the increase in membrane polyunsaturated fatty acids increased membrane fluidity and made the organisms much more resistant to low temperature damage (Table 6.14).

Other lipid changes are also seen with alterations in environmental temperature. Increases in the relative amounts of phosphatidylcholine are often found. If the ratio of this phospholipid to phosphatidylethanolamine is increased in a given membrane then this might be expected to be beneficial in terms of fluidity. This is because the Tc for phosphatidylcholine is about 13°C lower than phosphatidylethanolamine for equivalent molecular species. Of course, conversion of phosphatidylethanolamine into phosphatidylcholine is easily possible through methylation (Fig. 6.13) and this could form a mechanism for adaptation. A similar conversion of monogalactosyldiacylglycerol into digalactosyldiacylglycerol (Fig. 6.12) would also increase membrane fluidity due to the lower Tc of the digalactosyl-lipid. Changes in the ratio of these two galactolipids with altered environmental temperatures have also been seen in a number of cases.

Finally, the fluidity of a membrane can be influenced by the relative proportions of lipids and proteins. For low temperature growth, an increased lipid content is usually seen and this

Table 6.14   Genetic modification of cyanobacteria to allow Δ12-desaturation and the formation of linoleic acid

	Fatty acids (%)			Loss in photosynthetic activity at 5°C (%)
	Sat.	Monounsat.	*cis*, 9,12-Polyunsat.	
Wild type *Synechocystis*	59	8	29	–
Mutant	60	32	trace	–
Transformant (with 12-desat. gene)	61	9	27	–
Wild type *Anacystis*	54	42	–	55
Transformant	52	31	11	10

For more details refer to the article by Wada *et al.* (1990) Nature **347**, 200–203.

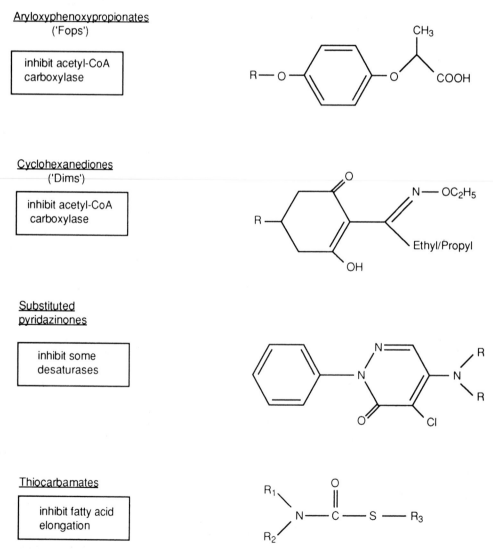

Figure 6.32   Basic structure of some herbicide classes affecting fatty acid synthesis. For the aryloxyphenoxypropionates the R group is usually a benzene or pyridine ring with halogen substituents. For substituted pyridazones the R group can represent hydrogens or methyl groups and the benzene ring may be substituted. In thiocarbamates R₁ and R₂ are usually alkyl side chains (ethyl, propyl, isopropyl) and R₃ may be ethyl, propyl or a chloride-substituted alkyl group.

appears to aid fluidity. Of course, such gross changes in membrane composition cannot be accomplished easily and require gross membrane biogenesis rather than modification of existing molecules.

6.7.4 Herbicides which act by affecting lipid metabolism

There are three main classes of herbicides which have a primary action in altering lipid synthesis (Fig. 6.32). Substituted pyridazinones have three main actions: they may (1) inhibit carotenoid formation, (2) alter fatty acid desaturation and (3) interfere with electron transport. Particularly prone among the fatty acid desaturases appears to be the $\Delta 15$-desaturase that converts linoleate into α-linolenate. However, the most herbicidal action of many pyridazinones is their inhibition of carotene synthesis (usually at the level of phytoene desaturase). The lack of carotenoids allows extensive photodestruction of chlorophyll during photosynthesis and has led to the popular description for such compounds as 'bleaching herbicides'.

A second class of herbicides are the thiocarbamates (Fig. 6.32). Many of these compounds are known to interfere with wax formation. The exact mechanism of action is still a matter of research but it is believed that the thiocarbamates are activated to their sulfoxide derivatives which are the actual herbicidal compounds. The sulfoxide derivatives of thiocarbamate herbicides inhibit fatty acid elongation, thus preventing the production of very long chain fatty acids which are the precursors of many wax components (see Fig. 6.25).

A third group of compounds represent recently discovered 'grass-specific' herbicides. Two main chemical structures are represented – the aryloxyphenoxypropionates and the cyclohexanediones. These compounds act by inhibiting acetyl-CoA carboxylase in sensitive grasses (Poaceae) but not in other monocotyledons or in dicotyledons. Interestingly, the mechanism of selectivity is due to differences in the structure of the target enzyme, acetyl-CoA carboxylase, and is not due to differential uptake, transport or metabolism of the herbicide. This is a very unusual basis for selectivity for a herbicide. The aryloxyphenoxypropionates and cyclohexanediones inhibit the carboxyltransferase partial reaction (see section 6.2.1) and have mutually exclusive binding characteristics. They are some of the most active herbicides known and appear to have low toxicity towards other organisms.

FURTHER READING

Browse, J. & Somerville, C.R. (1991) Glycerolipid synthesis: biochemistry and regulation. *Annu. Rev. Plant Physiol.* **42**, 467–506.

Cherif, A. *et al.* (eds) (1992) Various specialised articles in *Metabolism, Structure and Utilization of Plant Lipids*, Centre National Pedagogique, Tunis.

Gerhardt, B. (1990) Fatty acid degradation in plants. *Prog. Lipid Res.* **31**, 417–446.

Gunstone, F.G., Harwood, J.L. & Padley, F.B. (eds) (1994) *The Lipid Handbook*, 2nd edn. Chapman & Hall, London.

Gurr, M.I. & Harwood, J.L. (1991) *Lipid Biochemistry*, 4th edn. Chapman & Hall, London.

Hamilton, R.J. (ed.) (1995) *Wakes: Chemistry, Molecular Biology and Functions*. The Oily Press, Dundee.

Harwood, J.L. (1980) Plant acyl lipids: structure, distribution and analysis. In *The Biochemistry of Plants* (P.K. Stumpf & E.E. Conn, eds), Vol. 4, pp. 1–55. Academic Press, New York.

Harwood, J.L. (1988) Fatty acid metabolism. *Annu. Rev. Plant Physiol.* **39**, 101–138.

Harwood, J.L. (1991) Lipid synthesis. In *Target Sites for Herbicide Action* (R.C. Kirkwood, ed.), pp. 57–94. Plenum Press, New York.

Harwood, J.L. (1996) Recent advances in the biosynthesis of plant fatty acids. *Biochim. Biophys. Acta* **1301**, 7–56.

Harwood, J.L. & Bowyer, J.R. (eds) (1990) *Methods in Plant Biochemistry*, Vol. 4, Lipids, membranes and aspects of photobiology. Academic Press, London.

Hetherington, A.M. & Drobak, B.K. (1992) Inositol-containing lipids in higher plants. *Prog. Lipid Res.* **31**, 53–64.

Huang, A.H.C. (1987) Lipases. In *The Biochemistry of Plants* (P.K. Stumpf & E.E. Conn, eds), Vol. 9, pp. 91–120. Academic Press, New York.

Huang, A.H.C. (1992) Oil bodies and oleosins in seeds. *Annu. Rev. Plant Physiol.* **43**, 177–200.

Joyard, J. & Douce, R. (1987) Galactolipid synthesis. In *The Biochemistry of Plants* (P.K. Stumpf & E.E. Conn, eds), Vol. 9, pp. 215–274. Academic Press, New York.

Kader, J.C. (1990) Intracellular transfer of phospholipids, galactolipids and fatty acids in plant cells. *Subcell. Biochem.* **16**, 69–111.

Kader, J.C. & Mazliak, P. (eds) (1995) *Plant Lipid Metabolism*. Kluwer, Dordrecht.

Kolattukudy, P.E. (1980) Cutin, suberin and waxes. In *The Biochemistry of Plants* (P.K. Stumpf & E.E. Conn, eds), Vol. 4, pp. 571–646. Academic Press, New York.

Kolattukudy, P.E. (1987) Lipid-derived defensive polymers and waxes and their role in plant-microbe interaction. In *The Biochemistry of Plants* (P.K. Stumpf & E.E. Conn, eds), Vol. 9, pp. 291–314. Academic Press, New York.

Moore, T.S. (1982) Phospholipid biosynthesis. *Annu. Rev. Plant Physiol.* **33**, 235–259.

Mudd, J.B. (1980) Phospholipid biosynthesis. In *The Biochemistry of Plants* (P.K. Stumpf & E.E. Conn, eds), Vol. 4, pp. 250–282. Academic Press, New York.

Mudd, J.B. & Kleppinger-Sparace, K.F. (1987) Sulpholipids. In *The Biochemistry of Plants* (P.K. Stumpf & E.E. Conn, eds), Vol. 9, pp. 275–290. Academic Press, New York.

Murphy, D.J. (1990) Storage lipid bodies in plants and other organisms. *Prog. Lipid Res.* **29**, 299–324.

Murphy, D.J. (ed.) (1993) *Designer Oil Crops*. VCH, Weinheim.

Ohlrogge, J.B., Browse, J. & Somerville, C.R. (1991) The genetics of plant lipids. *Biochim. Biophys. Acta* **1082**, 1–26.

Pollard, M.R. (1987) Analysis and structure determination of acyl lipids. In *The Biochemistry of Plants* (P.K. Stumpf & E.E. Conn, eds), Vol. 9, pp. 1–30. Academic Press, New York.

Quinn, P.J. & Harwood, J.L. (eds) (1990) Various articles in *Plant Lipid Biochemistry, Structure and Utilization*, Portland Press, London.

Quinn, P.J. & Williams, W.P. (1990) Structure and dynamics of plant membranes. In *Methods in Plant Biochemistry*, Vol. 4 (J.L. Harwood & J.R. Bowyer, eds), pp. 297–340. Academic Press, London.

Siedow, J.N. (1991) Plant lipoxygenase: structure and function. *Annu. Rev. Plant Physiol.* **42**, 145–188.

Slabas, A.R. & Fawcett, T. (1992) The biochemistry and molecular biology of plant lipid biosynthesis. *Plant Mol. Biol.* **19**, 169–191.

Stymne, S. & Stobart, A.R. (1987) Triacylglycerol synthesis. In *The Biochemistry of Plants* (P.K. Stumpf & E.E. Conn, eds), Vol. 9, pp. 175–214. Academic Press New York.

Vance, D.E. & Vance, J. (1991) Various chapters on membranes. In *Biochemistry of Lipids, Lipoproteins and Membranes*. Elsevier, Amsterdam.

Vick, B.A. & Zimmerman, D.C. (1987) Oxidative systems for modification of fatty acids: the lipoxygenase pathway. In *The Biochemistry of Plants* (P.K. Stumpf & E.E. Conn, eds), Vol. 9, pp. 54–90. Academic Press, New York.

Walton, T.E. (1990) Waxes, cutin and suberin. In *Methods in Plant Biochemistry*, Vol. 4 (J.L. Harwood & J.R. Bowyer, eds), pp. 105–158. Academic Press, London.

7 Primary Nitrogen Metabolism

Peter J. Lea

7.1 INTRODUCTION

Plants play a vitally important role in the conversion of inorganic into organic nitrogen. The major source of nitrogen to all organisms is that present as the gas dinitrogen in the atmosphere (N_2). The industrial fixation of nitrogen is carried out by the Haber–Bosch process, which takes place at high temperatures (400–600°C) and pressures (100–200 atmospheres):

$$N_2 + 3H_2 \rightarrow 2NH_3$$

The majority of this fixed nitrogen is used as a fertilizer for crop plants. The release of nitrate into rivers, lakes and oceans following the application of fertilizer has frequently been a cause of public concern, although the precise origin of the nitrate is still a topic of considerable argument (Goulding & Poulton, 1992). Nitrogen may also be fixed biologically by a range of prokaryotic organisms, and this process is discussed in detail in section 7.2. N_2 may be released back to the atmosphere following the action of denitrifying bacteria in the soil (von Berg *et al.*, 1992):

$$NO_3^- \rightarrow NO_2^- \rightarrow NO \rightarrow N_2O \rightarrow N_2$$

N_2O (dinitrogen oxide) may also be released into the atmosphere where it could play a major role in the breakdown of the ozone layer (Mengel, 1992). A schematic diagram of the cycling of nitrogen is shown in Fig. 7.1. Although nitrogen may only comprise 0.5–5% of the dry weight of plant tissues, it is a constituent of vitally important compounds, e.g. amino acids, proteins and nucleic acids.

Humans are not able to use any forms of inorganic nitrogen as a nitrogen source, so organic nitrogen mainly as proteins must be an important constituent of the diet. Humans are also unable to synthesize nine amino acids, which are termed essential. The following sections of this chapter will follow the flow of nitrogen from the atmosphere into all the amino acids found in the proteins of higher plants.

7.2 NITROGEN FIXATION

7.2.1 The organisms involved

A range of prokaryotic organisms have the capacity to fix nitrogen, these have been discussed in detail by Gallon & Chaplin (1987) and Sprent & Sprent (1990). As can be seen from Table 7.1, the ability to fix nitrogen can be carried out at different concentrations of ambient oxygen. The additional complexity of carrying out nitrogen

PLANT BIOCHEMISTRY
ISBN 0-12-214674-3

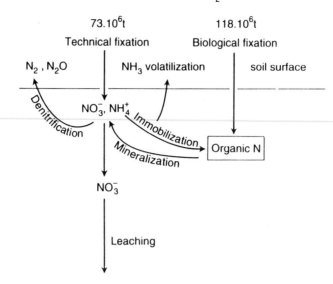

Figure 7.1 Principal processes in global nitrogen turnover (Mengel, 1992).

fixation in the cyanobacteria that are producing oxygen from photosynthesis will be discussed in section 7.2.5. Although strictly speaking *Azospirillum* species are free living, they are frequently found in association with the roots of grasses, where they fix nitrogen microaerobically.

Probably the most important form of agricultural nitrogen fixation is carried out by the rhizobium–legume symbiosis in the root nodule (Table 7.2; see section 7.2.4). However, some rhizobium, e.g. *Azorhizobium, Caulinodans* and *Photorhizobium thompsonia* form stem nodules. A number of non-leguminous plants also produce nitrogen-fixing nodules, but the symbiotic

prokaryote is an actinomycete *Frankia*. Cyanobacteria can also form nitrogen-fixing symbiotic relationships with a variety of organisms including the coral leaf sponge *Siphonochalina* (Sprent & Sprent, 1990: Rai, 1990)

7.2.2 Nitrogenase

(a) *Structure*

Nitrogenase carries out the following reaction:

$$N_2 + 8H^+ + 8e^- \rightarrow 2NH_3 + H_2$$

Each electron transfer is accompanied by the hydrolysis of two ATP molecules, and therefore the reaction also includes:

$$16MgATP \rightarrow 16MgADP + 16P_i$$

The majority of studies on nitrogenase were originally carried out on the enzyme isolated from *Klebsiella pneumoniae*, which was shown to contain molybdenum (Mo). The soil bacterium *Azotobacter chroococcum* normally has a Mo-containing nitrogenase but also has the capacity to synthesis a vanadium (V)-containing enzyme if Mo is not available. The closely related species *Azotobacter vinelandii* has, in addition to both Mo and V-nitrogenases, a third iron-containing nitrogenase which functions only when the Mo and V levels are very low (Smith & Eady, 1992).

Table 7.1 Free living nitrogen fixing organisms

Archaebacteria	
Methanogens	*Methanococcus volate*
Eubacteria	
Heterotrophs	
Anaerobes	*Clostridium pasteurianum*
Facultative anaerobes/	*Klebsiella pneumoniae*
Microaerobes	*Azotobacter vinelandii*
Aerobes	*Azospirillum lipoferum*
Autotrophs	
Chemotrophic bacteria	*Thiobacillus ferrooxidans*
Photosynthetic bacteria	*Rhodospirillum rubrum*
Cyanobacteria	
Unicellular	*Gloeothece* spp.
Filamentous	*Oscillatoria* spp.
Heterocystous	*Anabaena, Nostoc* spp.

Table 7.2 Symbiotic nitrogen fixing organisms

Name	Host
Rhizobiaceae	Legumes and *Parasponia*
Azorhizobium	
Bradyrhizobium	
Photorhizobium	
Rhizobium	
Sinorhizobium	
Actinomycetales	
Frankia	
Cyanobacteria	*Gunnera* (angiosperm)
	Macrozamia (gymnosperm)
	Azolla (pteridophyte)
	Blasia (bryophyte)
	Rhizalenia (diatom)
	Lichens
	Siphonochaliana (sponge)

All nitrogenase enzymes contain an Fe protein that is made up of two oxygen sensitive subunits with molecular weights of approximately 62 kDa and contain 4Fe and $4S^{2-}$ atoms per dimer. The iron–sulfur center has a midpoint potential of -350 mV. When MgATP is bound, a conformational change of the Fe protein takes place and the midpoint potential becomes 100 mV more negative. Mechanistic studies have demonstrated that the Fe protein is a specific MgATP-dependent electron donor to the molybdenum-containing protein (Smith, 1990).

The molybdenum-containing part of the nitrogenase enzyme is termed the MoFe protein, and the structure is outlined in Table 7.3. The iron–molybdenum cofactor (FeMoCo) which is derived from homocitrate is bound to the α-subunit of the MoFe protein at a cleft formed from two regions of the polypeptide at Cys-275, Gln-191 and His-195 (Scott *et al.*, 1990). The structures of the VFe protein and the third nitrogenase protein of *A. vinelandii* are shown in Table 7.3.

Detailed structures of the Fe and MoFe proteins have recently been obtained by X-ray crystallography (Georgiadis *et al.*, 1992; Kim & Rees, 1992) and have been discussed by Eady & Smith (1992).

(b) *Mechanism of action*

The most comprehensive description of the mechanism of nitrogenase has been developed by Thornley & Lowe (1985). First the reduced Fe–protein–MgATP complex binds to the MoFe protein. Then the electron is transferred with concomitant hydrolysis of MgATP. Fig. 7.2 shows the eight consecutive steps of the reduction of N_2, each of which corresponds to one electron transfer from the Fe protein to the Mo–Fe protein. For simplicity the Mo–Fe protein is shown as E, with a subscript denoting the number of electrons that have been transferred to it. Following the evolution of hydrogen after the transfer of three electrons, nitrogen is then reduced by the sequential addition of electrons and protons until two molecules of ammonia are released. A probable intermediate in this reduction is the hydrazido complex $E=N-NH_2$; it is likely that it is this intermediate that releases hydrazine on acid or base quenching of the reaction.

Table 7.3 Comparisons of the structures of MoFe, VFe and third nitrogenase proteins

Property	MoFe *K. pneumoniae*	VFe		Third Nitrogenase of *A. vinelandii*	
		A. chroococcum	*A. vinelandii*	Fast species	Slow species
Native molecular weight	220 kDa	210 kDa	200 kDa	158 kDa	216 kDa
Subunit structure	$\alpha_2\beta_2$	$\alpha_2\beta_2\delta_2$	$\alpha_2\beta_2(\delta_2)$	$\alpha_1\beta_2$	$\alpha_2\beta_2\delta_2$
Subunit molecular weight	2×50 kDa	2×50 kDa	2×52 kDa	2×50 kDa	2×50 kDa
	2×60 kDa	2×55 kDa	2×55 kDa	1×58 kDa	2×58 kDa
+		2×13 kDa			
Metal and S^{2-} content (mol/mol)					
V	0.05	2 ± 0.3	0.7 ± 3	0.015	0.01
Mo	2	0.06	0.05	0.015	0.085
Fe	32 ± 3	21 ± 1	9 ± 2	11	24
S^{2-}		19 ± 0.2	21 ± 1	9	18

Data derived from Table 1, Smith & Eady (1992).

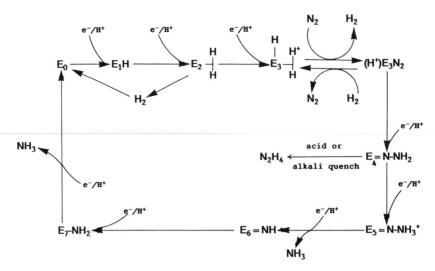

Figure 7.2 The catalytic cycle for the reduction of nitrogen by Mo nitrogenase of *Klebsiella pneumoniae* (Smith & Eady, 1992).

In addition to the reduction of nitrogen, nitrogenase can also reduce a variety of other substrates (Table 7.4). These substrates have in common, their small size and in many cases the possession of a triple bond. Notable among these substrates is acetylene (ethyne), the conversion of which to ethylene, is the basis of the most commonly used assay for nitrogen fixation. The ethylene formed by nitrogenase can be readily assayed using a simple gas chromatographic method. The more accurate measurements of nitrogen fixation following the incorporation of $^{15}N_2$ into organic material using a mass spectrometer is expensive and very time consuming. However, data obtained using the acetylene reduction assay should be treated with caution and the inexperienced investigator is recommended to read the work of Minchin *et al.* (1986) and Denison *et al.* (1992).

Both the Fe protein and the MoFe protein of nitrogenase are inactivated by O_2, with typical half decay times in air of 45 s and 10 min, respectively. For this reason all manipulations during the isolation and assay of the enzyme *in vitro*, have to be carried out under anaerobic conditions. In the case of the MoFe protein, O_2 causes a change in the oxidation state of the Fe centers, which after prolonged exposure results in the loss of Fe, Mo and S^{2-} from the protein. The components of the Mo-independent nitrogenases have been reported to be even more sensitive to O_2, with the half decay time of the VFe protein being only 45 s. Due to this oxygen sensitivity, nitrogen fixation organisms have developed a number of strategies to allow nitrogen fixation to take place aerobically; these mechanisms have recently been described in detail by Gallon (1992).

Table 7.4 Additional substrates used by the enzyme nitrogenase

Substrate		Products
Azide	$(N{\equiv}N{-}N^-)$	N_2, N_2H_4, NH_3
Nitrous oxide	$(N{\equiv}N{-}O)$	N_2
Cyanide	$(C{\equiv}N^-)$	CH_4, NH_3, CH_3NH_2
Alkyl cyanides	$(R{-}C{\equiv}N)$	$R{-}CH_3$, NH_3
Cyanamide	$(N{\equiv}CNH_2)$	CH_4, NH_3, CH_3NH_2
Acetylene	$(HC{\equiv}CH)$	$H_2C{=}CH_2$
Alkynes	$(R{-}C{\equiv}CH)$	$R{-}CH{=}CH_2$
Proton	(H^+)	H_2

7.2.3 Genetics

The genetics of nitrogen fixation has been extensively studied in *K. pneumoniae*. There are 20 genes (*nif*) involved in the assembly of nitrogenase (Fig. 7.3). The synthesis of nitrogenase is repressed by the presence of combined nitrogen (detected by the intracellular ratio of glutamine:2-oxoglutarate) and by O_2. In *K. pneumoniae* the general nitrogen regulation system (*ntr*) is also involved. The product of the *ntrA* gene is a sigma factor (σ^{54}) that is required for the recognition of the unusual promoters of the *ntr* and *nif* genes.

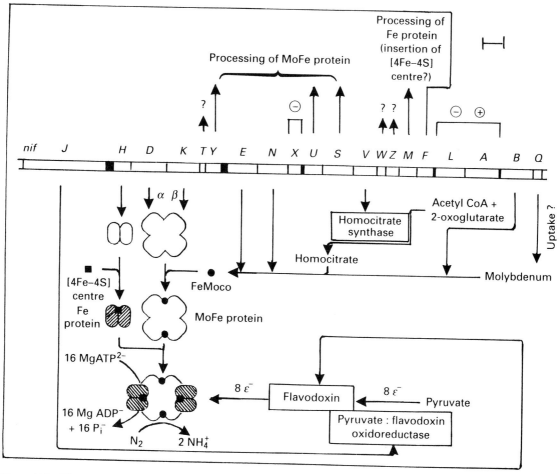

Figure 7.3 The nitrogen fixation (*nif*) regulon of *Klebsiella pneumoniae* showing the roles of the gene products in nitrogen fixation. Genes designated (+) and (−) are regulatory genes (Gallon, 1992).

Since a full description of this complex system is beyond the scope of this chapter, the reader should refer to Buck (1990) and Gallon (1992). The additional genes required in rhizobium for the formation of the legume root nodule are described in the next section.

7.2.4 The rhizobium–legume symbiosis

(a) Nodule development

The process begins with the infection of a root hair by a rhizobium bacterium. The first recognizable reaction of the plant is the curling of the root hair. A cell wall is then broken down and the bacterium is able to enter. An infection thread develops that penetrates the inner cortex of the root. The infected cortex cells increase in size and divide to form a sphere surrounded by uninfected cells and

an outer fibrous layer. Within the infected cells the rhizobia divide and enlarge to form bacteroids, which are separated from the plant cytoplasm by the peribacteroid membrane (Fig. 7.4). This process has been described in detail by Sprent & Sprent (1990).

(b) Rhizobium nodulation genes

It is now thought that specific compounds exuded from the roots are able to switch on nodulating (*nod*) genes in the rhizobium bacteria. These compounds are flavones and isoflavones (Rolfe, 1988). The nodulating genes in *R. leguminosarum* are shown in Fig. 7.5 (Economou *et al.*, 1990). The gene *nod D*, appears to be directly under the influence of flavones or isoflavones and to regulate the other genes of the *nod* complex, including the host specific *nod E* and *nod F* genes. Some rhizobia, e.g. *R. meliloti*, have three copies of

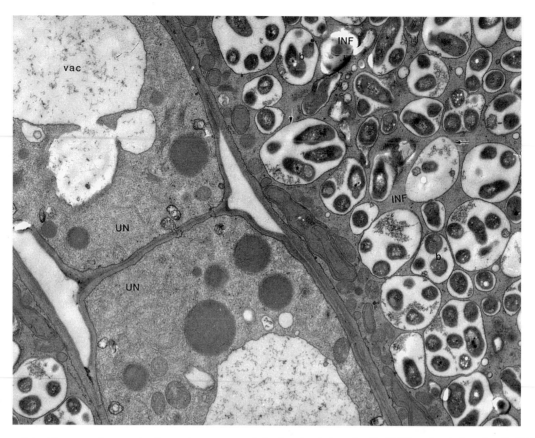

Figure 7.4 Electron micrograph of a soybean root nodule showing infected cells (INF) and uninfected cells (UN). The bacteroids (b) are enclosed within the peribacteroid membrane (arrow). Note the absence of vacuoles (vac) in the infected cells. Photographs by courtesy of Dr F.W. Wagner, University of Nebraska; reproduced by permission from Gallon & Chaplin, 1987.

nod D which may vary in their flavonoid recognition and in this way may regulate the host range, or other factors such as the number of nodules formed, or the time between inoculation and nodule formation. Each operon of the *nod* genes is preceded by a conserved DNA sequence the 'nod box', which is presumably regulated by the product of the *nod D* gene (Fisher & Long, 1992).

The development of nodules is elicited by specific lipo-oligosaccharide signal molecules released by the rhizobium bacteria. These molecules consist of a chito-oligosaccharide backbone of different chain length that contain at the non-reducing end a long-chain unsaturated fatty acid (Downie, 1991; Schultze *et al.*, 1992). The genes *nod ABC* are highly conserved in rhizobium and the corresponding proteins have been shown to induce root deformation, branching and cortical plant cell division (Dudley *et al.*, 1987; Banfalvi & Kondorosi, 1989). Transgenic plants containing the *nod A* and *nod B* genes have been shown to exhibit unusual growth abnormalities. It has recently been confirmed that the three genes are involved in the synthesis of acetylated glucosamine oligosaccharide signal molecules (John *et al.*, 1993).

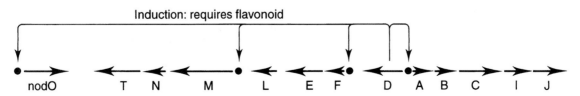

Figure 7.5 The nodulation genes of *Rhizobium leguminosarum* (Economou *et al.*, 1990).

(c) *Plant nodulin genes*

Once the appropriate species of rhizobium infects the roots of the correct host, the plant responds by sequentially expressing a series of genes which are required for the formation and efficient functioning of the nodule. These genes are termed nodulins, which can be divided into early nodulins that are involved in the infection process and root nodule morphogenesis and late nodulins that are expressed following the release of the rhizobium bacteria from the infection threads, but prior to commencement of nitrogen fixation (Verma & Miao, 1992; Govers & Bisseling, 1992).

Amino acid sequence analysis of the early nodulating gene (ENOD2) has revealed that it resembles other plant cell wall proteins and contains two repeating pentapeptides Pro-Pro-Glu-Tyr-Glu and Pro-Pro-His-Glu-Lys (Van der Weil *et al.*, 1990). Although there is no direct evidence that the protein is further glycosylated, it is likely that ENOD2 is a hydroxyproline-rich glycoprotein. Since the nodule inner cortex has been suggested to be an oxygen diffusion barrier (Denison *et al.*, 1992), the presence of the putative cell wall protein ENOD2 may contribute to the morphology of the nodule parenchyma and thus indirectly to the establishment of the oxygen barrier.

The ENOD12 gene codes for a similar protein to that encoded by the ENOD2 gene, but it is located in a different cell type. ENOD12 mRNA is produced in the invasion zone, the site where infection thread growth occurs and where bacteria are released from the infection threads. The ENOD12 gene is unusual in that it is also expressed in the stem and flower; the precise mechanism of the regulation of the gene is still not clear (Govers & Bisseling, 1992). Amino acid sequence analysis of other early nodulins has suggested that ENOD3 and ENOD14 might encode proteins involved in metal transport and that ENOD5 may be a protein in the peribacteroid membrane.

There are three major functions of the plant cells in a rapidly nitrogen fixing root nodule.

1. To provide a rapid flux of oxygen to support rhizobial respiration at a concentration low enough to prevent the deactivation of nitrogenase.
2. To supply carbohydrates, usually in the form of C_4 acids to support the rapid rates of rhizobial respiration.
3. To rapidly remove the ammonia synthesized by the rhizobial bacteroids to prevent the inhibition of nitrogenase.

Many of the later nodulin genes encode proteins that support these functions.

Leghemoglobin is the most abundant (up to 20%) protein in root nodules. Leghemoglobin is localized in the cytoplasm of the infected plant cells and not inside the peribactoid membrane. This evidence suggests that leghemoglobin controls the oxygen flux in the plant cytoplasm but not inside the rhizobium bacteroids. The globin is encoded by the host plant, whereas the heme group is primarily synthesized by the bacterium and transported into the host cytoplasm. There is now evidence that the heme precursor δ-aminolevulinic acid is synthesized in the plant cell, whereas the rest of the heme group is assembled by the bacteria (Sangwan & O'Brian, 1991). The globin protein is encoded by a family of genes in soybean, where at least four genes have been identified, *Lba*, *Lbc1*, *Lbc2* and *Lbc3* (Lee *et al.*, 1983).

The ammonia excreted by the bacteroids following the action of nitrogenase is assimilated in the plant cytoplasm by the enzyme glutamine synthetase. Several nodule specific glutamine synthetase isoenzymes have been identified and these will be discussed in section 7.4.1. Following the synthesis of glutamine, the fixed nitrogen is transported to the leaves as amides or ureides. Nodulin-35, which is the second most abundant soluble protein in the nodules of soybean has been shown to be a subunit of uricase II which catalyzes the conversion of uric acid into allantoin (Susuki & Verma, 1991). The enzyme is present in the peroxisomes of uninfected cells in the nodule.

Sucrose is the main source of fixed carbon transported from the leaves to the roots and root nodules. The first enzyme involved in sucrose breakdown is the inappropriately named sucrose synthase, which converts sucrose into UDP-glucose and fructose. The subunits of sucrose synthase have been identified as nodulin 100 of soybean. Free heme causes the sucrose synthase enzyme to dissociate and this inhibition may be involved in the regulation of carbohydrate metabolism during nodule senescence (Thummler & Verma, 1987). A number of nodulins are also thought to be involved in the structure and function of the peribacterioid membrane; these have been discussed in detail by Verma & Miao (1992).

7.2.5 The Cyanobacteria

(a) *Heterocyst structure*

The cyanobacteria (formerly known as the blue–green algae) have the capacity to obtain the energy and reducing power required for nitrogen fixation from light. Their simultaneous use of both atmospheric carbon and nitrogen has enabled them to colonize environments that are unfavorable to other organisms. They solve the problem of oxygen inactivation of nitrogenase by separating photosynthetic oxygen evolution from nitrogen fixation by space or time.

When growing in a medium containing ammonia, the cyanobacteria grow as filaments of undifferentiated vegetative cells. Upon transfer to a nitrogen free medium, 5–10% of the cells undergo differentiation to form thick-walled heterocysts, that are distributed evenly along the cyanobacterial filament. After differentiation, the heterocysts are not able to divide. The heterocyst cell wall consists of three layers: an outer polysaccharide fibrous layer, a homogeneous central layer and an inner glycolipid layer (Wolk, 1982). Mutants deficient in heterocyst glycolipids or polysaccharides lack the capacity to fix nitrogen aerobically, but maintain the capacity to do so in the absence of oxygen (Murry & Wolk, 1989). The permeability of heterocysts is such that nitrogen may enter at a rate sufficient to allow nitrogenase to operate but at the same time limit the entry of oxygen (Walsby, 1985). It is possible that the major regulation of gaseous and metabolite exchange takes place through the polar body or 'plug' that links the heterocyst to the vegetative cells.

(b) Heterocyst metabolism

The heterocyst lacks photosystem II activity which ensures that no oxygen is liberated in the location of the nitrogenase enzyme. The presence of photosystem I ensures that sufficient ATP can be generated by cyclic photophosphorylation. The heterocysts also lack Rubisco and the Calvin cycle enzymes so that CO_2 fixation is limited to the vegetative cells. Thus carbohydrate has to be transported into the heterocysts, probably in the form of maltose. Glucose is oxidized via the pentose phosphate pathway to yield NADPH in the reaction catalyzed by glucose-6-phosphate dehydrogenase (see Fig. 7.6).

The ammonia formed by nitrogenase is immediately assimilated by glutamine synthetase, and the majority is exported to the vegetative cells in the form of glutamine. Glutamate which is formed by glutamate synthase is returned to the heterocyst to assimilate another molecule of ammonia (Hager et al., 1983). Full details of the glutamate synthase cycle are given in section 7.4. The system described is thus a rather elegant division of labor, in which CO_2 assimilation and nitrogen fixation are spatially separated but at the same time intimately linked.

In the unicellular cyanobacterium Gloeothece, nitrogen fixation takes place at night and photosynthetic CO_2 assimilation and O_2 evolution occur during the day. In Gloeothece, nitrogenase

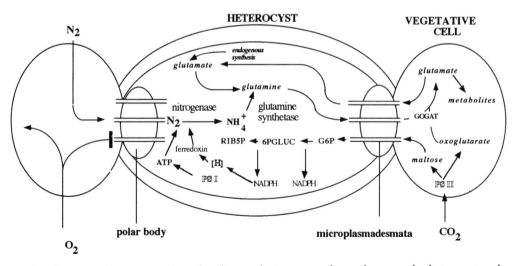

Figure 7.6 Diagrammatic representation of carbon and nitrogen exchange between the heterocyst and vegetative cells. Carbon compounds are transported from the vegetative cells into the nitrogen fixing heterocyst in order to supply the necessary NADPH via the oxidative pentose pathway. The ammonia produced by nitrogen fixation is assimilated by glutamine synthetase and the majority of glutamine transported to the vegetative cell where it is converted into glutamate by glutamate synthase (GOGAT). The heterocyst and vegetative cell thus demonstrate an early evolved form of cellular specialization (Smith & Gallon, 1993).

synthesis commences two hours before entering the dark period. After the onset of darkness, both synthesis and activity of nitrogenase are supported by ATP, reductant and carbon skeletons formed from the metabolism of stored glucan reserves, accumulated during the light period. The newly fixed ammonia is rapidly assimilated into amino acids. A fuller description of nitrogen fixation in the cyanobacterium is given by Gallon & Chaplin (1988) and Gallon (1992).

(c) *Regulation of heterocyst formation*

As mentioned previously heterocysts occur singly and at regular intervals along the cyanobacterial filament. This has led to the suggestion that vegetative cells produce an inhibitor of differentiation that requires activation by a co-inhibitor produced by heterocysts (Adams, 1992). The molecular nature of the inhibitor and co-inhibitor are unknown, but vegetative cells contain a factor (VFI) that binds to *nifH* and also to *xisA*, a gene involved in heterocyst differentiation (Chastain *et al.*, 1990). In addition, a gene designated *hetR* is involved in the early stages of heterocyst differentiation in *Anabaena* PCC 7120. Cyanobacteria carrying mutations in this gene do not produce heterocysts, but when present in multicopies, *hetR* allows heterocyst formation even in the presence of fixed nitrogen (Haselkorn, 1992).

In *Anabaena* PCC 7120, a major rearrangement of the *nif* gene region occurs during heterocyst differentiation. During this rearrangement, two regions of DNA are excised from the chromosome, thus producing a functional *nif* regulon. Although transcription of the newly formed operon *nifHDK* closely follows the rearrangement of the *nif* region, the two events are independent of each other. No rearrangement of the *nif* genes have been reported in non-heterocystous cyanobacteria, as *nifHDK* are already clustered. A full description of the regulation of nitrogen fixation in the cyanobacteria is given by Elhai & Wolk (1990) and Haselkorn (1992).

7.3 NITRATE UPTAKE AND REDUCTION

7.3.1 Nitrate uptake

It is necessary at this point to summarize our current knowledge of root nitrate uptake (Larsson *et al.*, 1992; Oaks & Long, 1992; Rufty *et al.*, 1992; Ullrich, 1992), although information on the proteins involved in the transport process is still

not available. In barley, root nitrate uptake is carried out by three different systems (Siddiqi *et al.*, 1990; Aslam *et al.*, 1992). At high external nitrate concentrations, a low affinity system operates that appears to be constitutive and unregulated. At low external concentrations, two high-affinity systems appear to operate. One (K_m for nitrate, $7 \mu M$) is constitutive and the second (K_m for nitrate, $15-34 \mu M$) is induced by nitrate (Lee & Drew, 1986; Aslam *et al.*, 1992).

Following uptake, nitrate may be immediately reduced and converted into amino acids in the root as described in sections 7.3 and 7.4. Alternatively the nitrate may be temporarily stored in the root vacuole or transported to the leaves for subsequent reduction. The proportion of nitrate reduced in the root or leaves is variable and depends on the concentration of external nitrate, developmental age and plant species (Andrews, 1986). If nitrate reduction and assimilation takes place in the root, higher rates of root respiration are required (Bloom *et al.*, 1992).

The level of expression of the inducible high affinity nitrate transport system is subject to negative feedback regulation dependent on the nitrogen status of the plant (Siddiqi *et al.*, 1989). It has been suggested that the feedback regulation is mediated by a product of ammonia assimilation (Cooper & Clarkson, 1989; Lee *et al.*, 1992). However, King *et al.* (1993) have compared the uptake of $^{13}NO_3^-$ in wild type barley and mutants totally lacking nitrate reductase (see section 7.3.2). They concluded that nitrate itself was able to regulate its own uptake without the requirement of further metabolism. Nitrite may also be taken up by barley roots and is able to induce the nitrate uptake system, suggesting that nitrate and nitrite are transported by the same process (Aslam *et al.*, 1993).

7.3.2 Nitrate reductase

Nitrate is reduced to nitrite by the enzyme nitrate reductase which catalyzes the following reaction:

$$NO_3^- + 2H^+ + 2e^- \rightarrow NO_2^- + H_2O$$

The reducing power for nitrate reduction is normally supplied as NADH; however, in some plants, a bispecific enzyme that is able to use either NADH or NADPH as a substrate is also present. In soybean three isoenzymes of nitrate reductase have been isolated: constitutive NADH and NAD(P)H forms and the more extensively studied inducible NADH-dependent form (Streit *et al.*,

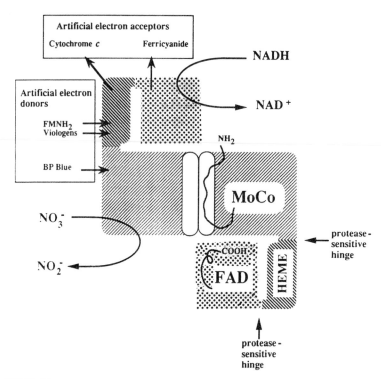

Figure 7.7 Hypothetical model for higher plant nitrate reductase. The main structural features are shown on the right hand subunit. The three catalytic domains containing their respective prosthetic groups, the amino terminal domain (possibly involved in the interaction between subunits) and the carboxy terminus of the polypeptide chain (Pelsy & Caboche, 1992).

1987). Nitrate reduction can take place in the root depending on the plant species, developmental age or nitrate supply (Andrews, 1986). In general, as the external nitrate concentration increases, the proportion that is transported to the shoot for reduction increases. The intracellular location of nitrate reductase has been the subject of intense study and considerable discussion. The majority of investigations have demonstrated that the enzyme is present in the cytoplasm. Studies that indicate that the enzyme is loosely bound to the plasma or chloroplast membrane, have in general, been discounted (Askerlund et al., 1991).

(a) Structure

NADH-dependent nitrate reductase comprises two identical subunits of 100–115 kDa. The redox cofactors FAD, heme (cytochrome b_5) and a molybdenum cofactor (MoCo) are bound to each monomer in a 1:1:1 ratio. Electrons are passed from NADH to nitrate as shown:

$$NADH \rightarrow FAD \rightarrow Cytochrome\, b_5$$

$$\rightarrow MoCo \rightarrow Nitrate$$

The monomer is not active by itself and the two subunits are linked by a disulfide bond, which can be reduced without loss of activity (Hyde et al., 1989). The amino acid sequences of eight plant enzymes have now been obtained and are discussed in detail by Rouze & Caboche (1992). Three catalytic domains have been identified by comparison of the sequence with other proteins of known function (Fig. 7.7). The N-terminal half of the enzyme of 430–460 amino acids is the MoCo domain, determined by homology with sulfite oxidase. The central 81–83 amino acids contain the heme domain, which is a member of the cytochrome b_5 superfamily. The 270–275 amino acids in the C-terminal sequence, comprise the FAD domain. The three domains are linked to one another by two hydrophilic hinge regions, which are very susceptible to proteolytic cleavage and probably account for the early problems in isolating the enzyme in an intact and active state.

The individual domains are able to carry out a number of partial reactions, e.g. the heme domain can act as a NADH-dependent cytochrome c reductase. Although these partial reactions have been studied in detail it is unlikely that they have any physiological role. In addition nitrate reductase can also convert chlorate into the toxic metabolite chlorite. This reaction is the basis of herbicidal activity and has been used extensively for the isolation of mutants that lack nitrate reductase activity (Kleinhofs & Warner, 1990).

(b) Genetics and molecular biology

A vast array of higher plant mutants lacking nitrate reductase activity have been characterized. The plants used include barley, tobacco, tomato and *Arabidopsis thaliana*; this subject area has been reviewed in detail by Kleinhofs & Warner (1992), Pelsy & Caboche (1992) and Wray & Kinghorn (1989).

A range of mutants that lack both nitrate reductase and xanthine dehydrogenase activity are probably defective in the ability to synthesize the MoCo factor. Five loci have been characterized in barley (*Nar2 – Nar6*), three in *A. thaliana* (B25, B73 and G1) and six in *Nicotiana plumbaginifolia* (*CnxA – CnxF*).

In *A. thaliana*, the analysis of chlorate-resistant mutants has shown that two genes *Chl2* and *Chl3* are involved in the biosynthesis of the nitrate reductase apoenzyme. In barley, the *Nar1* locus has been identified as the NADH-specific nitrate reductase structural gene. None of the *Nar1* mutants are completely deficient in nitrate reductase activity. This is due to the expression of a second gene *Nar7* that is presumed to code for the NAD(P)H nitrate reductase apoenzyme. Double mutants of barley, *Nar1, Nar7*, totally lack the ability to use nitrate as a nitrogen source. In *N. plumbaginifolia* only one functional *Nia* gene is present that codes for the nitrate reductase apoenzyme. However, four classes of *nia* mutant have been isolated that refer to the catalytic domain affected by the mutation.

Nia genes have been cloned and sequenced from a range of plants, including tobacco, tomato, rice, barley, *A. thaliana* and the alga *Chlamydomonas reinhardtii* (Rouze & Caboche, 1992). Southern blot analysis has shown that in tobacco there is one *Nia* locus per haploid genome, whereas in *A. thaliana* two sequences have been detected. A more complex system exists in barley where one to three copies have been detected and in rice where six to eight copies per haploid genome have been identified, depending on the subspecies (Kleinhofs *et al.*, 1988). In general three introns of variable size have been detected, which are located at strictly conserved positions (Hoff *et al.*, 1991).

Rouze & Caboche (1992) have compared the amino acid sequences of the nitrate reductase protein from eight higher plants with those isolated from algae and fungi. The sequences were shown to be very similar, suggesting that a similar catalytic mechanism can be applied to the enzymes so far studied (Solomonson & Barber, 1990).

7.3.3 Nitrite reductase

Nitrite is reduced to ammonia by the enzyme nitrite reductase, which catalyzes the following reaction:

$$NO_2^- + 8H^+ + 6e^- \rightarrow NH_4^+ + H_2O$$

The enzyme is present in the leaves and roots where it is located totally in the chloroplast or plastid. The reducing power is obtained from reduced ferredoxin directly from the light reactions in the leaf or from carbohydrate metabolism in the root (Bowsher *et al.*, 1993).

(a) Structure

Ferredoxin-dependent nitrite reductase comprises one subunit of molecular weight 60–64 kDa. The polypeptide contains a siroheme prosthetic group and a 4Fe–4S cluster at the active site (Siegel & Wilkerson, 1989). It was originally thought that the enzyme contained an additional subunit of 24–25 kDa that was involved in the binding of ferredoxin, but this suggestion has now been disproved (Hilliard *et al.*, 1991).

(b) Genetics and molecular biology

In contrast to the large numbers of plant mutants lacking various activities of nitrate reductase, until recently no mutants lacking the enzyme have been isolated. This may be due in part to the toxic nature of nitrite, which would have a deleterious effect on metabolism. A mutant of barley has been isolated that lacks nitrite reductase activity in the leaves and roots and can only be maintained by growth on ammonia or glutamine. The mutation is in a single nuclear gene *Nir1* and the plants have been shown to contain elevated levels of nitrate reductase activity (Duncanson *et al.*, 1993). Vaucheret *et al.* (1992) have expressed antisense nitrite reductase mRNA in tobacco plants. One transformed plant was shown to lack nitrite reductase activity and accumulated nitrite. The plant had a greatly reduced level of ammonia, glutamine and protein in the leaf tissue.

Nitrite reductase cDNA clones have been characterized and sequenced from spinach (Back *et al.*, 1988), maize and tobacco (Vaucheret *et al.*, 1992) and the birch tree *Betula pendula* (Friemann *et al.*, 1992).

7.3.4 The regulation of nitrate assimilation

It has been known for over 30 years that nitrate reductase activity is induced by the presence of nitrate and light (Hageman & Flesher, 1960). The induction of nitrate reductase was one of the first enzymes in plants shown to be caused by *de novo* protein synthesis (Zielke & Filner, 1971). This early work has been reviewed by Beevers & Hageman (1980) and Kleinhofs & Warner (1990).

With the onset of molecular biological techniques, it is now possible to follow the induction of nitrate reductase activity at the gene level. Nitrate supply leads to an increase in the steady-state level of mRNA encoding nitrate reductase in a variety of plant species and tissues (Crawford *et al.*, 1986; Cheng *et al.*, 1991; Friemann *et al.*, 1991). This in turn, results in nitrate reductase protein synthesis and increased activity, shortly after nitrate supply to the roots and after a lag in the leaves (Melzer *et al.*, 1989).

Similarly it has been shown that nitrite reductase mRNA increases following nitrate supply in a range of species (Gupta & Beevers, 1985; Back *et al.*, 1988; Faure *et al.*, 1991). The 5′ promoter sequence of the spinach nitrite reductase gene has been shown to confer nitrate-inducible glucuronidase (GUS) activity in transgenic tobacco, confirming that nitrate controls transcription of the nitrite reductase gene (Back *et al.*, 1991; Rastogi *et al.*, 1993). Although nitrate is able to co-regulate the expression of the nitrate and nitrite reductase, there is no evidence at the present time, of the existence of a specific nitrate regulatory gene, using a similar mechanism to that described in fungi (Burger *et al.*, 1991).

Light is also involved in the increase in mRNA levels of both nitrate and nitrite reductase in the presence of nitrate. It has been suggested that this enhanced light effect is mediated by phytochrome (Gowri & Campbell, 1989; Melzer *et al.*, 1989; Schuster & Mohr, 1990). In light-grown plants treated with nitrate, the levels of nitrate and nitrite reductase mRNAs, proteins and enzyme activities slowly decrease after two days in the dark (Deng *et al.*, 1990; Bowsher *et al.*, 1991). After return to light there is a rapid increase in the levels of nitrate and nitrite reductase mRNAs and enzyme activities (Deng *et al.*, 1990; Cheng *et al.*, 1991).

Diurnal variations in nitrate reductase activity have been known for some time (Hagemann *et al.*, 1961; Lillo, 1984). More recently strong oscillations in the levels of nitrate reductase mRNA have been observed, with a major peak being detected at the end of the dark period (Galangou *et al.*, 1988). The circadian rhythm in the level of nitrate reductase mRNA is maintained in continuous light, but disappears slowly in the dark (Deng *et al.*, 1990) (Fig. 7.8). Oscillations in the levels of nitrite reductase mRNA in phase with nitrate reductase have also been detected (Bowsher *et al.*, 1991; Faure *et al.*, 1991).

Very recently an overall scheme for the regulation of nitrate and nitrite reductase has been proposed by Vincentz *et al.*, (1993). It has been shown that the addition of fructose, glucose and sucrose to dark-adapted tobacco leaves resulted in the induction of nitrate reductase expression in the dark. On the other hand glutamate or glutamine (but not asparagine) down-regulate the expression of nitrate reductase in low light. It had previously been shown that glutamine concentrations correlated in the opposite phase with the expression of nitrate reductase mRNA (Deng *et al.*, 1991; Fig. 7.8).

The formation of nitrite reductase mRNA was not stimulated by soluble carbohydrates in the dark, but required light for the stimulation of transcription. Repression of nitrite reductase mRNA synthesis was, however, detected to a lesser extent in the presence of glutamine, glutamate or asparagine. It would, therefore, appear that the induction of the transcription of

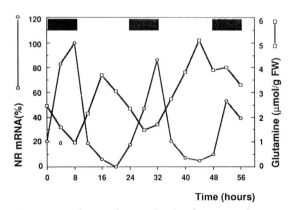

Figure 7.8   The steady-state levels of nitrate reductase mRNA and glutamine in tobacco leaves follow a circadian rhythm in opposite phase. After a final dark period (0–8 h) plants were kept in continuous light. The effective last night of the light/dark regime is shown by a black rectangle. The grey rectangles indicate the positions of the night periods if the light/dark regime had been maintained (Deng *et al.*, 1991).

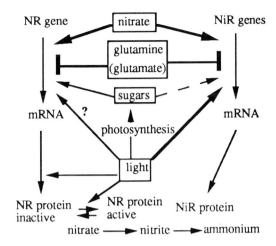

Figure 7.9 Scheme for the regulation of nitrate and nitrite reductase genes in tobacco, nitrate, light and sugars are inducers, glutamine and glutamate are repressors (Vincentz *et al.*, 1993).

the nitrate and nitrite reductase genes in tobacco leaves is regulated by the balance of soluble sugars and amino acids, as shown in Fig. 7.9 (Vincentz *et al.*, 1993).

It has been shown that the activity of nitrate reductase can be modulated *in vivo*. The activity of the enzyme was shown to decline under conditions when the rate of photosynthetic CO_2 assimilation was low (Kaiser & Brendle-Behnisch, 1991). Similarly a rapid fall in the activity of nitrate reductase was determined following the transfer from light to dark (Reins & Heldt, 1992). Kaiser & Spill (1991) were able to show that the enzyme could be inactivated *in vitro* by ATP and Mg^{2+} and reactivated in the presence of AMP and Mg^{2+}. It has now been suggested that the nitrate reductase activity in leaves can be regulated *in vivo* in a similar manner to sucrose phosphate synthase (Huber *et al.*, 1992a). The enzyme is subject to a covalent modification by phosphorylation and inactivation by ATP. Removal of the phosphate groups by phosphatase causes a reactivation of the nitrate reductase (Huber *et al.*, 1992b).

7.4 AMMONIA ASSIMILATION

It has been shown in previous sections that ammonia is generated via nitrogen fixation and nitrate reduction. A small amount may also be taken up from the soil under acidic conditions, particularly by trees (Stewart *et al.*, 1992). These three processes are termed primary assimilation.

Ammonia may also be synthesized in large amounts by a number of secondary metabolic reactions:

1. The conversion of glycine into serine in the photorespiratory carbon and nitrogen cycle (see Chapter 2),
2. The catabolism of the transport compounds, asparagine, arginine and ureides,
3. During normal amino acid metabolism, e.g. conversion of cystathionine into homocysteine in methionine biosynthesis and threonine into 2-oxobutyrate in isoleucine biosynthesis. The enzyme phenylalanine ammonia lyase produces ammonia in the conversion of phenylalanine into cinnamate, the first reaction involved in the synthesis of lignin a major constituent of secondary cell walls,
4. Proteins are frequently hydrolyzed during the germination of seedlings or following leaf senescence. Prior to the synthesis of transport compounds, ammonia is liberated through the operation of glutamate dehydrogenase:

$$Glutamate + NAD^+ + H_2O$$
$$\rightarrow 2\text{-Oxoglutarate} + NH_3 + NADH + H^+$$

The enzyme has also been shown to metabolize glutamate when the level of 2-oxoglutarate in the tricarboxylic acid cycle is low due to a shortage of available carbohydrate (Robinson *et al.*, 1992).

Two enzymes glutamine synthetase and glutamate synthase operate in tandem to form the glutamate synthase cycle of ammonia assimilation (Fig. 7.10).

7.4.1 Glutamine synthetase

There is now a substantial body of evidence that clearly demonstrates that glutamine synthetase (GS) is the sole port of entry into amino acids in higher plants (Lea *et al.*, 1990, 1992; Lea, 1991). GS catalyzes the ATP-dependent conversion of glutamate into glutamine:

$$Glutamate + ammonia + ATP$$
$$\rightarrow Glutamine + AMP + P_i$$

The enzyme is an octameric protein with a native molecular weight of 350–400 kDa and has a very high affinity for ammonia ($K_m = 3$–$5\,\mu M$). Early investigations suggested that there were two isoenzymes of GS present in plants, one located

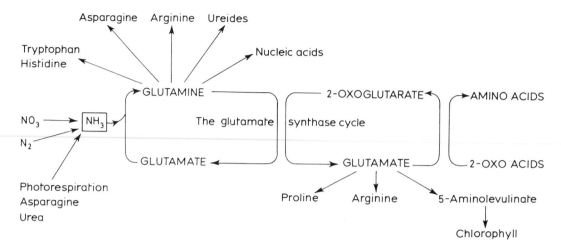

Figure 7.10 The assimilation of ammonia in higher plants via the glutamine synthetase/glutamate synthase cycle.

in the cytoplasm and one in the chloroplast or plastid (McNally *et al.*, 1983). More recent molecular studies have suggested that this hypothesis is an oversimplification. The enzyme has been studied in detail in *Phaseolus vulgaris* and *Pisum sativum* and a brief summary of the data obtained with these two plants will be discussed below.

(a) *Phaseolus vulgaris*

Five genes coding for GS have been identified in *P. vulgaris* (Fig. 7.11) (Forde & Cullimore, 1989). The *gln-α* *gln-β* and *gln-γ* genes encode the cytosolic a, *β* and *γ* polypeptides which are located in the cytoplasm. The DNA sequences show that these three polypeptides have molecular weights of approximately 39 kDa and are 85–89%

identical. The fourth expressed GS gene *gln δ* encodes a polypeptide of 47 kDa, which has extensions to the amino and carboxy termini of 57 and 16 amino acids, when compared to the three cytosolic GS polypeptides. The N-terminal extension encodes a chloroplast targetting sequence, which is cleaved during transport into the organelle. The resulting chloroplast located GS polypeptide, termed *δ* is about 42 kDa and assembles into an octamer of identical subunits. In contrast, the three cytosolic polypeptides are able to assemble into a range of isoenzymes containing varying proportions of the *α*, *β* and *γ* polypeptides (Bennett & Cullimore, 1989; Cai & Wong, 1989). The fifth GS gene, *gln-ε* has been identified in a genomic clone, but no evidence for the expression of the gene has yet been obtained.

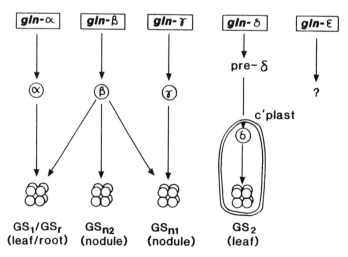

Figure 7.11 The genetic control of glutamine synthetase isoenzyme in nodules, roots and leaves of *Phaseolus vulgaris* (Forde & Cullimore, 1989).

(i)   gln-α

This is the least characterized of the four expressed genes, although a high level of mRNA has been demonstrated in dry seeds. During germination, the mRNA remains the most abundant in the cotyledons, radicles and plumules for up to 2 days. After this time, the mRNA declines to barely detectable levels (Cock *et al.*, 1991).

(ii)   gln-β

The mRNA for this gene has been detected in all organs of *P. vulgaris*. In roots, gln-β mRNA is the major form of GS mRNA and is induced during the early stages or radicle emergence (Ortega *et al.*, 1986). A similar gene in soybean has been shown to be induced by ammonia, (Miao *et al.*, 1991), but this is not the case in *P. vulgaris*. During nodulation, the level of gln-β mRNA remains constant. Experiments with GUS fusion in *Lotus corniculatus*, indicated that the promoter was expressed in both the cortical and infected regions of the nodule, but as the nodules matured, expression was restricted to the vascular system (Forde *et al.*, 1989). The gln-β gene is also expressed in leaves, but only at the early stages of development (Cock *et al.*, 1991).

(iii)   gln-γ

The gln-γ GS gene is highly expressed in the nitrogen-fixing nodule but not in roots and leaves. Low levels of expression have also been detected in the stems, petioles and cotyledons of germinating seeds (Bennett *et al.*, 1989). Experiments with GUS fusions in *L. corniculatus* has revealed that the gln-γ promoter is only expressed in the infected cells of the central nodule tissue (Forde *et al.*, 1989). It has been proposed that the release of the *Rhizobium* bacteria may lead to the production of factors specific to the infected cells, which are required for the expression of the gln-γ gene in nodules. In soybean it has been proposed that the increased expression of plant GS genes in nodules is caused by ammonia generated by nitrogen fixation (Hirel *et al.*, 1987; Miao *et al.*, 1991). This mechanism has not been confirmed in *P. vulgaris* (Cock *et al.*, 1990).

(iv)   gln-δ

During leaf development, the chloroplast gln-δ gene is expressed 2 days after germination, and after 8 days becomes the major GS transcript. The gene is only expressed at a very low level in dark-grown plants and on transfer to light there is a massive induction of gln-δ mRNA synthesis (Cock *et al.*, 1990). The gene is also expressed in the roots and root nodules, but only at a low level. A full description of the regulation of the GS genes in *P. vulgaris* has been published by Cullimore *et al.* (1992). A summary of the induction of the four GS genes during the development of the nitrogen fixing nodule and the primary leaf are shown in Fig. 7.12 (A, B).

(b)   Pisum sativum

Three genes, termed *GS1, GS3A* and *GS3B* have been shown to code for cytosolic GS in *P. sativum*. In roots, *GS1* is the predominant isoform, although it is also expressed in nodules, *GS3A* and *GS3B* are highly expressed in nodules and also in the cotyledons of germinating seeds (Tingey *et al.*, 1987). Whereas the *GS3A* and *GS3B* genes have near identical nucleic acid sequences, *GS3A* is expressed at higher levels (Walker & Coruzzi, 1989). The chloroplast nuclear gene *GS2* encodes the chloroplast enzyme of GS, which contains a transit peptide (Tingey *et al.*, 1988).

Light causes a dramatic induction in the level of *GS2* mRNA in pea leaves. Experiments with etiolated plants demonstrated that the accumulation of *GS2* mRNA was only partially due to a phytochrome-mediated response. Similar investigations have also been carried out in tobacco and tomato (Becker *et al.*, 1992). The presence of mature chloroplasts in dark-adapted pea plants was shown to accelerate the induction of *GS2* mRNA synthesis following transfer of the plants to white light (Edwards & Coruzzi, 1989). Levels of the *GS2* mRNA were found to be reduced in the leaves of pea plants that had been grown in elevated CO_2, conditions that would suppress the generation of ammonia from the photorespiratory pathway (see section 7.4.3). Although similar results were obtained by Cock *et al.* (1991) in *P. vulgaris*, there was no increase in the level of chloroplastic GS mRNA following the transfer of plants from elevated CO_2 to air. The role of photorespiratory ammonia release in the regulation of GS is still therefore in doubt. The regulation of the synthesis of GS in peas has been reviewed by McGrath & Coruzzi (1991).

In an attempt to obtain more information on the control of the different *P. sativum* GS genes, Coruzzi and her colleagues have used GUS expression in transgenic tobacco plants as an elegant detection method. The promoter for the

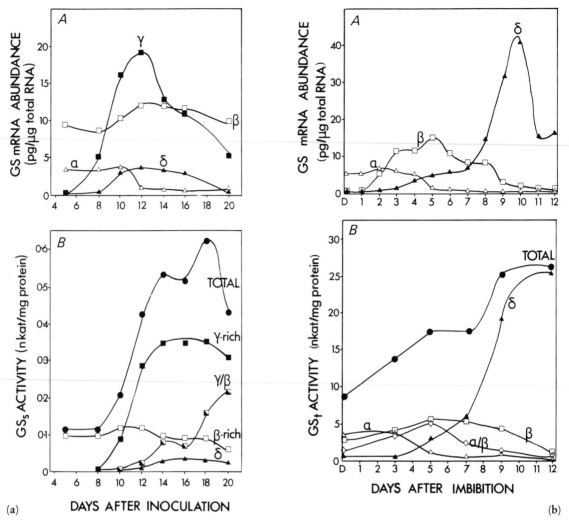

Figure 7.12 Changes in the level of glutamine synthetase mRNA and isoenzymes during the development of the nitrogen fixing nodule (a) and leaf (b) of *Phaseolus vulgaris* (adapted from Cock *et al.*, 1990 and Cullimore *et al.*, 1992).

GS2 chloroplast gene was shown to direct GUS expression in photosynthetically active cells, e.g. pallisade and spongy parenchyma of the leaf, collenchyma of the stem and photosynthetic cotyledons. In contrast, the promoter for cytosolic *GS3A* gene directed expression of GUS in the vascular system of the cotyledons, leaves, stems and roots (Edwards *et al.*, 1990). The expression of the *GS3A* promoter was later shown to be confined to the phloem cells. The promotor conferred a similar pattern of expression in transgenic alfalfa, where expression was also observed, at high levels, in root nodules (Brears *et al.*, 1991).

The evidence that cytosolic and chloroplastic GS are expressed in different cell types, suggests that they play different roles in nitrogen metabolism. The expression of *GS3A* in the phloem cells would indicate that it plays a major role in the synthesis of glutamine required for nitrogen transport, this would be particularly important in the seedling cotyledons and the nitrogen fixing nodule. The role of the chloroplast GS in photorespiration is discussed in section 7.4.3.

7.4.2 Glutamate synthase

This enzyme catalyzes the reductant dependent conversion of glutamine and 2-oxoglutarate to yield two molecules of glutamate. The two different forms of glutamate synthase are present in higher plants, one utilizes NADH as a source of reductant and the other ferredoxin:

Glutamine + 2-oxoglutarate

+ ferredoxin (reduced)

→ 2-glutamate + ferredoxin (oxidized)

(a)   *Ferredoxin-dependent enzyme*

This enzyme, which was first detected in pea leaves (Lea & Miflin, 1974), is an iron–sulfur flavoprotein and may represent up to 1% of the protein content of leaves (Marquez *et al.*, 1988). The enzyme is a large monomeric protein with a molecular weight of 140–160 kDa. Tissue fractionation studies have shown that ferredoxin-dependent glutamate synthase is localized in the chloroplasts of leaves (Wallsgrove *et al.*, 1982). Using immuno-gold antibody labeling techniques, Botella *et al.* (1988) were able to show that the enzyme was localized solely in the chloroplast stroma of mesophyll xylem parenchyma. The activity of the enzyme increases during leaf development in the light (Suzuki *et al.*, 1987), a response which may be phytochrome dependent (Hecht *et al.*, 1988).

Despite initial difficulties, data are now available concerning the molecular biology of the enzyme. The first report of the sequence of a gene encoding ferredoxin-dependent glutamate synthase was made by Sakakibara *et al.* (1991). The cDNA was shown to encode a polypeptide of 1616 amino acids, including a chloroplast transit peptide of 97 amino acids. The enzyme was shown to exhibit a strong similarity to the *E. coli* NADPH-dependent glutamate synthase protein with 42% of the amino acids residues being identical (Oliver *et al.*, 1987). The enzyme also contained a short region similar to a potential FMN-binding region of yeast flavocytochrome b_2. Two partial sequences of the gene and protein isolated from tobacco (Zehnacker *et al.*, 1992) and barley (Avilla *et al.*, 1993) have confirmed the strong similarity with the *E. coli* enzyme.

In maize, mRNA transcription was shown to increase after illumination of etiolated leaves (Sakakibara *et al.*, 1991). Similar results were obtained in tobacco, and it was proposed that in tomato, the response was mediated by phytochrome. Levels of glutamate synthase mRNA in illuminated leaves of tobacco were considerably lower than expected, when compared to chloroplastic GS (Zehnacker *et al.*, 1992).

(b)   *NADH-dependent enzyme*

In green leaves the activity of the NADH-dependent enzyme is low in comparison to ferredoxin-dependent activity (Wallsgrove *et al.*, 1982; Hecht *et al.*, 1988). Enzyme activity has been detected in a range of non-green tissues, e.g. roots, cotyledons and tissue culture cells (Lea *et al.*, 1992). The NADH-dependent enzyme appears to play a major role in the ammonia assimilation in nitrogen fixing nodules (Robertson *et al.*, 1975; Awonaike *et al.*, 1981; Anderson *et al.*, 1989; Chen *et al.*, 1990). The enzyme is a monomer and has a high molecular weight in the region of 200–225 kDa (Anderson *et al.*, 1989; Chen & Cullimore, 1989). Antisera raised against alfalfa root nodule glutamate synthase did not detect the presence of a similar protein in roots or leaves (Anderson *et al.*, 1989).

A 7.2 kb cDNA clone encoding a 240 kDa NADH-dependent glutamate synthase protein has been isolated from alfalfa root nodules (Gregerson *et al.*, 1993). The protein shares significant sequence identity with maize ferredoxin glutamate synthase and with both the large and small subunit of *E. coli* NADPH-glutamate synthase. The expression of NADH-glutamate synthase mRNA, protein and enzyme activity was developmentally regulated in the root nodules. A dramatic increase in gene expression occurred coincidentally with the onset of nitrogen fixation in the bacteroid. Using both *Rhizobium* and plant mutants, Gregerson *et al.* (1993) concluded that a signal is present in effective nodules at the onset of nitrogen fixation, that is required for the induction of mRNA synthesis.

7.4.3   Mutants lacking enzymes of ammonia assimilation

The metabolism of phosphoglycolate in the photorespiratory nitrogen cycle in C_3 plants is described in detail in Chapter 2; CO_2 is liberated following the conversion of glycine into serine in the mitochondria (Oliver *et al.*, 1990). Ammonia is also liberated in this process at the same rate as CO_2, but is immediately reassimilated via the glutamate synthase cycle (Lea *et al.*, 1990, 1992). The rate of ammonia production during photorespiration has been calculated to be ten times the rate of nitrate assimilation (Keys *et al.*, 1978).

Following an initial suggestion by Somerville & Ogren (1979), it has been possible to isolate mutant plant lines that are deficient in the key enzymes of the photorespiratory nitrogen cycle (Somerville,

1986). Such mutants have been selected on the basis that they grow normally under non-photorespiratory conditions (0.7% CO_2), but show severe stress symptoms and chlorotic lesions on exposure to air.

Mutants of barley deficient in chloroplastic GS have been isolated by two laboratories (Blackwell et al., 1987; Wallsgrove et al., 1987). The mutants exhibit remarkably similar properties to plants treated with the GS inhibitors methionine sulfoximine and phosphinothricin (Lea, 1991). These are characterized by a massive accumulation of ammonia and a reduction in the rate of photosynthetic CO_2 assimilation following exposure to air. The maximum rate of ammonia evolution has been calculated to be 40% of the rate of CO_2 assimilation, and can be taken as a direct measurement of the rate of photorespiration.

Freeman et al. (1990) were able to identify three classes of barley mutants lacking chloroplastic GS_2 activity: class I in which the absence of GS_2 protein was correlated with low levels of mRNA; class II which had normal or increased levels of GS_2 mRNA but very little protein; and class III, which had significant amounts of mRNA and protein. Mutant lines of plant species, other than barley, lacking GS_2 have not been isolated.

Mutants lacking the chloroplastic ferredoxin-dependent enzyme have been isolated from Arabidopsis thaliana (Somerville & Ogren, 1980), barley (Kendall et al., 1986; Blackwell et al., 1988) and pea (Lea et al., 1992). After exposure to air, the leaves of the deficient mutants accumulated large quantities of glutamine and small amounts of ammonia; the levels of all other amino acids were greatly reduced. Ten minutes after the transfer to air, the rate of photosynthetic CO_2 assimilation of the mutant leaves had fallen to 10% of the rate of the wild type plant. This loss of photosynthetic capacity was much faster than that seen in the GS-deficient mutants and could be reversed by prefeeding alanine or asparagine.

Western blot analysis of four of the barley mutants lacking ferredoxin-dependent glutamate synthase indicated that three of the mutants had no detectable cross-reacting material of molecular weight 154 kDa, whereas a fourth (RPr 84/42) had a detectable, but greatly reduced amount of enzyme protein (Avilla et al., 1993). Using a 1.3 kb cDNA fragment specifying the amino terminal portion of barley ferredoxin glutamate synthase, Avilla et al. (1993) were able to determine the mRNA levels in the barley mutants. In two of the mutants (RPr 82/9 and RPr 84/82) the probe hybridized to a mRNA similar in mobility (5.7kb) and abundance to that of the wild type. The mutants RPr 82/1 and RPr 84/82, however,

exhibited two RNA hybridizing species, one of which was larger than 5.7kb and one smaller. The latter two therefore represent a fourth type of mutant (class IV), in which normal mRNA is replaced by RNA(s) of a different size.

The use of mutants lacking the enzymes GS and glutamate synthase have confirmed that ammonia is assimilated as shown in Fig. 7.10 and not via the enzyme glutamate dehydrogenase (Lea et al., 1992; Robinson et al., 1992). Mutants deficient in both enzymes of ammonia assimilation have been obtained by normal genetic crossing. Such mutants lacking both chloroplastic GS and ferredoxin-dependent glutamate synthase are still capable of growing in elevated CO_2. The double mutants contain low levels of amino acids and are very sensitive to exposure to air (Blackwell et al., 1988).

7.5   ASPARAGINE METABOLISM

Asparagine is the major transport and storage compound found in all plants (Lea & Miflin, 1980; Sieciechowicz et al., 1988; Rabe, 1990) and may also accumulate under stress conditions.

7.5.1   Asparagine synthesis

The amide group of glutamine can be transferred directly to aspartate in an enzyme reaction catalyzed by asparagine synthetase:

Glutamine + aspartate + ATP

$\rightarrow$ asparagine + glutamate + ADP + PP_i

The cotyledons of germinating seeds have proved to be a major source of asparagine synthetase activity and the enzyme has been studied in lupins (Rognes, 1975, 1980; Lea & Fowden, 1975) soybean (Streeter, 1973) and Vigna (Kern & Chrispeels, 1978). In maize roots, the enzyme is also able to use ammonia as a substrate which may be of physiological importance in this tissue (Oaks & Ross, 1984). The enzyme also plays an important role in root nodules (Reynolds et al., 1982). In leaves, the enzyme has proved very difficult to assay, probably due to the presence of inhibitors (Joy et al., 1983; Joy & Ireland, 1990). Tsai & Coruzzi (1990) have been able to clone asparagine synthetase from pea using the human gene as a probe. Two classes of asparagine synthetase cDNAs (AS1 and AS2) that encoded homologous but distinct polypeptides with molecular weights of 66.3 and 65.6 kDa were isolated; glutamine binding sites (Met-Cys-Gly-Ile) were detected at the amino terminus of both AS1 and AS2.

Northern blot analysis revealed that the level of AS1 mRNA increased in the dark in the leaves of both etiolated seedlings and mature pea plants, in agreement with physiological experiments on the synthesis of asparagine (Joy et al., 1983). The light repression of AS1 mRNA synthesis was found to be a phytochrome-mediated response. Both AS1 and AS2 mRNA were shown to accumulate in germinating cotyledons and nitrogen fixing nodules (Tsai & Coruzzi, 1990). A similar gene has been isolated from soybean and expressed in E. coli, with the aim of obtaining a large quantity of the asparagine synthetase protein (Hughes & Matthews, 1993).

Asparagine was the first amino acid to be discovered in plants, being isolated from Asparagus sativus in 1806. Asparagus spears are an important commercial crop, but there is a rapid deterioration following harvest and the levels of asparagine increase markedly in 24 hours (King et al., 1990). A cDNA clone encoding asparagine synthetase has now been isolated from asparagus spear tips (Davies & King, 1993). Levels of mRNA began to increase in the tips 2 h after harvest and in other sections of the spear after 4 h. Unlike the pea asparagine synthetase genes, mRNA levels were not affected by light treatment and only low levels of transcript were detected in the fleshy roots, ferns or spears before harvest (Davies & King, 1993).

7.5.2 Asparagine breakdown

(a) Asparaginase

In lupins, 80% of the asparagine entering the maturing fruit is metabolized, initially in the testa (seed coat). The nitrogen liberated from the amide group of asparagine has been detected in the endosperm fluid in the form of ammonia, alanine and glutamine. At later stages of maturation, the developing lupin cotyledon is able to metabolize asparagine directly (Atkins et al., 1975; Pate, 1989). Similarly asparagine has been shown to be rapidly metabolized in the seed coats of maturing pea seeds giving rise to high concentrations of ammonia in the endosperm fluid (Murray, 1992). Asparaginase catalyzes the deamination of asparagine:

$$\text{Asparagine} + H_2O \rightarrow \text{Aspartate} + NH_3$$

The enzyme has been purified from Lupinus polyphyllus where it was shown to have a molecular weight of 72 kDa with subunits of 38 kDa (Lea et al., 1978). Antisera raised against the purified asparaginase cross-reacted with a protein of similar characteristics isolated from the cotyledons and leaves of L. polyphyllus, but no reaction was detected with extracts from the roots and nodules following Western blot analysis (Sodek & Lea, 1993). The enzyme has also been purified from a related species L. arboreus which has a molecular weight of 75 kDa and subunits of 14–19 kDa (Lough et al., 1992a). The discrepancies in the molecular weight of the putative subunit have now been clarified by Lough et al. (1992b) who have isolated a 1.2 kb cDNA clone from L. arboreus seeds that codes for a truncated asparaginase protein of 32.8 kDa, indicating that the enzyme comprises two subunits. A genomic sequence encoding the enzyme has now been obtained, which contains three introns and a 5' flanking region with a sequence associated with seed specific expression (Dickson et al., 1992).

Following the identification of asparaginase in lupin seeds, Sodek et al. (1980) isolated a second form of the enzyme from developing seeds that required potassium ions for activity. A potassium-dependent asparaginase was also detected in the developing seeds of Vicia faba, Phaseolus multiflorus, Zea mays and Hordeum vulgare. The existence of asparaginase activity in the presence of potassium ions in the testa and cotyledons of pea have been confirmed by Murray & Kennedy (1980) and the enzyme activity has been shown to be reduced by the presence of glutamine in soybean cotyledons (Tonin & Sodek, 1990). The two forms of asparaginase in seeds have been shown to be immunologically distinct (Lea et al., 1984; Lough et al., 1992a). A potassium-dependent asparaginase has been isolated from the young expanding leaves of pea with a molecular weight of 58 kDa. The enzyme activity has been shown to undergo a complex diurnal variation that is regulated by light (Sieceichowicz et al., 1988).

(b) Asparagine transamination

A series of detailed ^{15}N-labeling experiments have shown that the major route of asparagine metabolism in mature pea leaves was via transamination (Ta & Joy, 1985; Ta et al., 1985). It was argued that asparagine made a small but significant contribution to the synthesis of glycine by the transamination of glyoxylate during the process of photorespiration (Ta & Joy, 1986).

Asparagine aminotransferase activity has been detected in high levels in pea leaves, where it is located in the peroxisome (Ireland & Joy, 1983a), but the enzyme is either absent or present in low activity in other tissues (Walton & Woolhouse, 1983). The enzyme has been purified from pea leaves and it was proposed that one protein catalyzed both serine : glyoxylate aminotransferase and asparagine : glyoxylate aminotransferase activity (Ireland & Joy, 1983b). Confirmation that the two activities were carried out by the same protein (and coded for by one gene) was obtained by the isolation of a mutant barley plant that was blocked in the photorespiratory nitrogen cycle (see section 7.4.3) and lacked both serine : glyoxylate and asparagine : glyoxylate aminotransferase activity (Murray et al., 1987).

7.6 AMINOTRANSFERASES

Aminotransferases (also known as transaminases) catalyze the transfer of the amino group from the 2 position of the amino acid to a 2-oxo acid to yield a new oxo acid and amino acid. Glutamate is a major amino donor and is involved in two important reactions:

Aspartate Aminotransferase

Glutamate + Oxaloacetate

$$\rightleftharpoons 2\text{-Oxoglutarate} + \text{Aspartate}$$

Alanine Aminotransferase

Glutamate + Pyruvate

$$\rightleftharpoons 2\text{-Oxoglutarate} + \text{Alanine}$$

Both reactions liberate 2-oxoglutarate, which may then return to the glutamate synthase cycle. Aminotransferases have been detected in plants that can synthesize all the protein amino acids (except proline), provided the correct 2-oxo acid precursor is available. Nitrogen can thus be readily distributed from glutamate, via aspartate and alanine to all amino acids (Givan, 1980; Ireland & Joy, 1985). The reversibility of the reaction allows for sudden changes in demand for key amino acids, although in specific pathways, e.g. the synthesis of glycine in photorespiration, certain enzymes tend only to operate in one direction. Since it is not possible to describe the large variety of aminotransferases that have been detected in higher plants, this chapter will concentrate on one enzyme only.

7.6.1 Aspartate aminotransferase

As well as the normal 'housekeeping' role in the formation of aspartate required for protein synthesis and the synthesis of the aspartate family of amino acids, this enzyme also has three other major functions:

1. The rapid transfer of amino groups from glutamate via aspartate to asparagine in nitrogen fixing root nodules (Vance & Gantt, 1992). The 2-oxoglutarate molecule cycles to collect more amino groups in the glutamate synthase reaction, and oxaloacetate is synthesized by the action of PEP carboxylase.
2. The malate–oxaloacetate shuttle is often used to transfer reducing power from the mitochondria and chloroplasts to the cytoplasm via the enzyme malate dehydrogenase (Kromer & Heldt, 1991). Due to the inherent instability of oxaloacetate, the oxoacid may also be transaminated to aspartate to facilitate transport (Ireland & Joy, 1985).
3. Following the formation of oxaloacetate, in the PEP carboxylase reaction in C_4 plants, aspartate is transported from the mesophyll cells to the bundle sheath cells in NAD-ME type plants. Full details of C_4 photosynthesis are given in Chapter 2.

Due to the multiple roles of aspartate aminotransferase, distinct isoenzymes have been identified in the mitochondria, chloroplasts, peroxisomes and cytoplasm of higher plants, which appear to be under separate genetic control (Manganaris & Alston, 1988).

In the alfalfa root nodule, two forms of aspartate aminotransferase have been identified, designated AAT-1 and AAT-2. The alfalfa AAT-1 has a native molecular weight of 84 kDa with two subunits of 42 kDa. AAT-2 is slightly smaller with a native molecular weight of 80 kDa and two subunits of 40 kDa. Antibodies raised against the two enzymes indicate that they are immunologically distinct (Griffith & Vance, 1989; Farnham et al., 1990). During the development of nitrogen fixing root nodules, AAT-2 is the predominant isoenzyme, whereas AAT-1 is the major form in roots (Egli et al., 1989). In C_4 plants two forms of aspartate aminotransferase have been characterized in *Panicum* and *Eleusine* species, that display physical characteristics and immunological properties similar to the alfalfa enzymes (Numazawa et al., 1989; Taniguchi & Sugiyama, 1990). A cytosolic isoenzyme is localized in the mesophyll cells and a mitochondrial enzyme is present in the bundle sheath cells (Taniguchi et al., 1992).

A cDNA clone encoding the alfalfa AAT-2 has now been isolated; the deduced amino acid sequence is 53 and 47% identical to the animal mitochondrial and cytoplasmic enzyme respectively. A N-terminal sequence of 59 amino acids suggests that the enzyme is localized in the plastid. The expression of AAT-2 mRNA in nodules was severalfold higher than the level detected in roots or leaves. Northern and Western blot analysis showed there was a sevenfold increase in AAT-2 mRNA and a comparable increase in enzyme protein during nodule development (Gantt et al., 1992).

A full length cDNA clone encoding aspartate aminotransferase has also been identified in carrot. The deduced amino acid sequence has a 52 and 53% identity with the sequences of the mouse cytoplasmic and mitochondrial enzymes, respectively. Northern blot analysis of the mRNA levels indicated that the gene was expressed throughout the cell culture up to 7 days and was highly expressed in roots but not in leaves (Turano et al., 1992). cDNA clones encoding the cytoplasmic and mitochondrial enzymes in the C_4 plant Panicum miliaceum have also been recently isolated (Taniguchi et al., 1992).

7.7 THE ASPARTATE FAMILY

The biosynthesis of lysine, threonine, methionine and isoleucine is shown in Fig. 7.13 (Bryan, 1990).

Experiments with ^{14}C-labeled aspartate have shown that chloroplasts are capable of synthesizing lysine, threonine, isoleucine and homocysteine in light-driven reactions, which can therefore be termed photosynthetic (Mills et al., 1980). Subsequent subcellular localization studies have shown that all the enzymes required are present in the chloroplast (Wallsgrove et al., 1983). The final methylation step involved in the conversion of homocysteine into methionine has been shown to take place in the cytoplasm. It is no coincidence that all the essential amino acids required by animals in their diet are synthesized by plants in the chloroplast. Animals do, however, have the capacity to convert homocysteine into methionine.

7.7.1 Aspartate kinase

The enzyme aspartate kinase (Fig. 7.13, 1) catalyzes the ATP-dependent phosphorylation of the β-carboxyl group of aspartate. Three isoenzymes of aspartate kinase have been isolated from barley and their presence confirmed by genetic analysis (Rognes et al., 1983). The first isoenzyme (AKI) is subject to feedback inhibition by threonine alone and is only present in low levels in rapidly growing tissue. The other two isoenzymes (AKII and AKIII) are similar to each other and are both inhibited by lysine. Methionine alone has no direct effect on any of the aspartate kinase isoenzymes but is able to regulate the flow of carbon through the activated

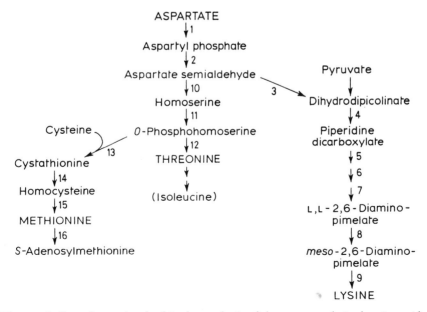

Figure 7.13 The metabolic pathways involved in the synthesis of the aspartate-derived amino acids lysine, threonine, methionine and isoleucine in higher plants. The enzymes 1–16 are described in the text.

form, S-adenosylmethionine. AKII and AKIII are both subject to synergistic inhibition in the presence of S-adenosylmethionine and lysine (Rognes *et al.*, 1980). In carrot, three forms of aspartate kinase have also been identified; one inhibited by lysine, one inhibited by threonine and one partially inhibited by both end product amino acids (Wilson *et al.*, 1991).

(a) *Threonine-sensitive enzyme*

Initial studies (Aarnes & Rognes, 1974) suggested that the threonine-sensitive aspartate kinase enzyme may exist as a bifunctional protein in combination with threonine-sensitive homoserine dehydrogenase, as has been shown in *E. coli* (Cohen, 1983).

In a series of studies on threonine-sensitive homoserine dehydrogenase isolated from carrot tissue culture cells, Wilson *et al.* (1991) were able to show that the enzyme copurified with aspartate kinase. The amino acid sequence of the 85 kDa polypeptide indicated that both homoserine dehydrogenase and aspartate kinase activities were present on the same protein. A similar copurification of the two enzyme activities was also determined in maize tissue culture cells (Azevedo *et al.*, 1992a). Confirmation of the existence of the bifunctional aspartate kinase–homoserine dehydrogenase enzyme in carrot has been obtained by Weisemann & Matthews (1993). Two overlapping cDNA clones encoding the enzyme have been isolated. Together the cDNA clones delineate a 3089 bp long sequence encompassing an open reading frame encoding 921 amino acids, including the mature protein and a long transit peptide. The deduced amino acid sequence has high homology with the two isoenzymes of threonine-sensitive aspartate kinase–homoserine dehydrogenase found in *E. coli*.

(b) *Lysine-sensitive enzyme*

There have been three major attempts to purify lysine-sensitive aspartate kinase from higher plants. Due to the low specific activity of the enzyme, this work has been confined to tissue culture cells (Relton *et al.*, 1988; Dotson *et al.*, 1989; Azevedo *et al.*, 1992b). As yet, the enzyme has not been purified to homogeneity and the gene encoding the protein has not been isolated. Attempts to determine the native molecular weight of the enzyme have encountered difficulties. In carrot tissue cultures, values of 100–225 kDa were obtained (Relton *et al.*, 1988), in maize cultures, values varied between 104–225 kDa (Dotson *et al.*,

1989) and 139–150 kDa (Azevedo *et al.*, 1992b). A detailed kinetic analysis of the mechanism of lysine inhibition of the maize enzyme has been carried out by Dotson *et al.* (1990a).

Aspartyl phosphate is connected to aspartate semialdehyde by aspartate semialdehyde dehydrogenase (Fig. 7.13; 2). A single form of the enzyme has been isolated from maize tissue cultures. The enzyme does not appear to be regulated by end product amino acids (Myers & Gengenback, 1982).

7.7.2 Lysine synthesis

The first branch in the pathway leads to lysine. Dihydrodipicolinate is formed from the combination of pyruvate and aspartate semialdehyde catabolized by dihydrodipicolinate synthase (Fig. 7.13, 3). The molecular weight of the native enzyme is in the range 115 kDa for spinach (Wallsgrove & Mazelis, 1981), 123 kDa for wheat (Kumpaisal *et al.*, 1987), 130 kDa for maize (Frisch *et al.*, 1991) and 127 kDa for pea (Dereppe *et al.*, 1992). Frisch *et al.* (1991) proposed that the maize enzyme was a homotetramer with a subunit molecular weight of 38 kDa. However, Dereppe *et al.* showed that the pea enzyme was a trimer with three subunits of molecular weight 43 kDa. Kampaisal *et al.* (1987) suggested a ping-pong mechanism for wheat dihydrodipicolinate synthase in which pyruvate binds to the enzyme first, followed by the release of water before aspartate semialdehyde binds to the activated enzyme complex. The data obtained with the maize enzyme were consistent with this mechanism. All the enzymes so far examined have been shown to be extremely sensitive to feedback inhibition by lysine, with I_{50} values in the order of 20–50 μM. cDNA clones encoding dihydrodipicolinate synthase have been isolated from wheat (Kaneko *et al.*, 1990) and maize (Frisch *et al.*, 1991). The maize sequence indicated the presence of a 54 amino acid transit peptide and a subunit molecular weight of 36 kDa, consistent with the chloroplast localization of the enzyme (Wallsgrove & Mazelis, 1980).

Dihydrodipicolinate reductase (Fig. 7.13, 4) has been isolated from maize kernels (Tyagi *et al.*, 1983). The enzyme was initially detected by the ability to catalyze diaminopimelate synthesis from aspartate and pyruvate in extracts of an *E. coli* mutant lacking the reductase. Further evidence of the presence of the enzyme was based on the demonstration of competitive inhibition by dipicolinate. The penultimate step in the synthesis of lysine in bacteria is the epimerization of L-diaminopimelate to *meso*-diaminopimelate (Fig. 7.13, 8). Evidence for the presence of this enzyme in maize has been presented by Tyagi *et al.* (1982).

Meso-diaminopimelate decarboxylase (Fig. 7.13, 9) catalyzes the final step in the synthesis of lysine (Kelland *et al.*, 1985) and has been shown to be localized solely in the chloroplast (Mazelis *et al.*, 1976). The molecular weight of the enzyme in higher plants varies between 75 kDa for wheat germ (Mazelis & Crevelling, 1978) and 85 kDa for maize endosperm (Sodek, 1978). Lysine at concentrations up to 1 mM has little effect on enzyme activities, although 20 mM inhibited the enzyme by 60%. It is unlikely that inhibition at this high concentration would indicate any physiological feedback regulation.

Three enzymes in the center of the pathway of lysine synthesis (5, 6, 7) have not been detected in higher plants (Bryan, 1990). Wenko *et al.* (1985) reported evidence of the presence of a single enzyme that could synthesize *meso*-diaminopimelate directly from piperidine dicarboxylate via a reductive amination step. The presence of this putative enzyme, *meso*-diaminopimelate dehydrogenase has been confirmed in a few bacterial species (White, 1983) but not in higher plants.

7.7.3 Threonine synthesis

Homoserine dehydrogenase (Fig. 7.13, 10) catalyzes the reductant dependent conversion of aspartate semialdehyde into homoserine. Distinct threonine-sensitive and threonine-resistant isoenzymes have been isolated in higher plants, the former being localized in the chloroplast and the latter in the cytoplasm (Sainis *et al.*, 1981). The isoenzyme isolated from maize has been shown to differ in size, subunit composition (Walter *et al.*, 1979) and antigenic properties (Krishnaswamy & Bryan, 1986). The threonine-sensitive maize enzyme can exist in at least four ligand-induced conformational states, two of which are dimeric and two of which are tetrameric (Krishnaswamy & Bryan, 1986).

In carrot, homoserine dehydrogenase activity reversibly converts between a threonine insensitive form in the presence of K^+ and a threonine sensitive form in the presence of threonine (Matthews *et al.*, 1989; Turano *et al.*, 1990). As discussed previously, threonine-sensitive homoserine dehydrogenase and aspartate kinase from carrot were shown to copurify and to be different activities of the same enzyme protein (Wilson *et al.*, 1991). The full amino acid and nucleic acid sequence of the aspartate kinase–homoserine dehydrogenase has now been determined (Weisemann & Matthews, 1993).

Homoserine is converted into O-phosphohomoserine by homoserine kinase (Fig. 7.13, 11). The enzyme isolated from radish was shown to be inhibited by isoleucine, threonine and S-adenosylmethionine (Baum *et al.*, 1983). Pea homoserine kinase is also subject to inhibition by several amino acids including S-adenosylmethionine (Thoen *et al.*, 1978a; Muhitch & Wilson, 1983). In contrast, Aarnes (1976) was unable to identify significant effects of amino acids on the activity of the enzyme isolated from barley.

Riesmeier *et al.* (1993) have purified homoserine kinase from wheat germ to homogeneity. The enzyme had a native molecular weight of 75 kDa and a subunit molecular weight of 36 kDa. These data conflict with previous reported molecular weights in the range 120–240 kDa (Thoen *et al.*, 1978a), suggesting that the enzyme can form aggregates. Threonine, methionine, valine and isoleucine were found to have no inhibitory effect on the enzyme at 10 mM. S-adenosylmethionine was shown to inhibit the wheat germ enzyme by 26% at 10 mM, although it is unlikely that this effect would be of any physiological importance. It is possible that plants contain regulated and unregulated forms of homoserine kinase in an analogous way to homoserine dehydrogenase; however, the presence of distinct isoenzymes has not been confirmed (Muhitch & Wilson, 1983). O-phosphohomoserine is converted into threonine in an internal rearrangement, catalyzed by threonine synthase (Fig. 7.13, 12). The enzyme, which is localized in the chloroplast, has been partially purified from radish (Madison & Thompson, 1976), pea (Thoen *et al.*, 1978b), barley (Aarnes, 1978) and *Lemma* (Giovanelli *et al.*, 1984). The enzyme has a virtual absolute requirement for S-adenosylmethionine as an allosteric activator. Half maximal activation is obtained at 40–200 μM S-adenosylmethionine. Threonine synthase activity is not inhibited by any potential feedback inhibitors, but the enzyme is repressed by methionine supplementation of the growth medium (Giovanelli *et al.*, 1984; Rognes *et al.*, 1986).

7.7.4 Methionine synthesis

Detailed ^{35}S and ^{14}C labeling experiments have indicated that the first committed step in the synthesis of methionine is the formation of cystathionine (Giovanelli *et al.*, 1985a, 1989a, b). Cystathionine-γ-synthase (Fig. 7.13, 13) catalyzes the combination of O-phosphohomoserine and cysteine to yield cystathionine and inorganic phosphate. The enzyme has not been extensively purified from higher plants, but has been isolated from sugar beet (Madison & Thompson, 1976) and barley (Aarnes, 1980). In *Lemna paucicostata* the enzyme has been shown to be sensitive to a range of potent inhibitors including propargylglycine (Thompson *et al.*, 1982a). Cystathionine-γ-synthase appears to be present in excess in *L. paucicostata* as the enzyme can be inhibited by up to 84% by propargylglycine, before there is a detectable decrease in the rate of methionine synthesis (Thompson *et al.*, 1982a).

Although cystathionine-γ-synthase activity *in vitro* is not regulated by any potential feedback inhibitors, the level of the enzyme activity *in vivo* is regulated in *L. paucicostata*. The addition of $2 \mu M$ methionine decreased the enzyme activity to 15% of the control value. Conversely, the addition of lysine and threonine which would reduce the flow of carbon into methionine, increased the *in vivo* activity of cystathionine-γ-synthase three-fold (Thompson *et al.*, 1982b). Similar results have been obtained with barley seedlings grown in sterile culture (Rognes *et al.*, 1986).

Cystathionine is metabolized to homocysteine, pyruvate and ammonia by cystathionine-β-lyase (Fig. 7.13, 14). Mutants of tobacco lacking the enzyme activity require homocysteine or methionine for growth (Negrutiu *et al.*, 1985). The enzyme is present in both the chloroplast and cytoplasm of higher plants (Wallsgrove *et al.*, 1983). The enzyme was initially isolated and purified from spinach (Giovanelli & Mudd, 1971; Giovanelli, 1987).

More recently the enzyme has been purified to apparent homogeneity from spinach leaves. The enzyme has a mature molecular weight of 210 kDa with a subunit molecular weight of 53 kDa (Staton & Mazelis, 1991). Cystathionine-β-lyase has also been purified to homogeneity from the tissue culture cells of *Echinochloa colonum* with a specific activity ten times that of the spinach enzyme. The mature molecular weight of the enzyme is 160 kDa with a subunit molecular weight of 41 kDa (W. Turner and P.J. Lea, unpublished results).

The characteristic flavors and odors of a number of important vegetables are due to the enzymic degradation of non-protein sulfur-containing amino acids. These enzymes also have the capacity to cleave a C–S bond to form a sulfur-containing product, pyruvate and ammonia by a mechanism analogous to that of cystathionine-β-lyase. These enzymes have been characterized in garlic, onion and leek (Won & Mazelis, 1989; Nock & Mazelis, 1989).

The final enzyme in the pathway catalyzes the S-methyltetrahydropteroyltriglutamate-dependent methylation of homocysteine. The enzyme methionine synthase (Fig. 7.13, 15) has been detected in a small number of plants including pea leaves (Shah & Cossins, 1970) and carrot root (Fedec & Cossins, 1976). Two forms of the enzyme have been partially purified from pea seeds (Dodd & Cossins, 1970). In *E. coli*, methionine synthase is a cobalamin-dependent enzyme with a molecular weight of 133 kDa. The *metH* gene encoding the enzyme has been isolated and sequenced (Banerjee *et al.*, 1989).

S-Adenosylmethionine is used as an important methyl donor for a range of methylation reactions in plants including the synthesis of derivatives of ethanolamine, pectins, chlorophyll, lipids and nucleic acids (Mudd & Datko, 1986; Cossins, 1987). The derivative is formed by the combination of methionine and ATP catalyzed by S-adenosylmethionine synthetase (16). Three isoenzymes of S-adenosylmethionine synthetase have been isolated from wheat embryos. One of the isoenzymes has been purified to homogeneity and shown to be a dimer with a subunit molecular weight of 84 kDa (Mathur *et al.*, 1991).

The gene (*sam-1*) coding for S-adenosylmethionine synthetase has been isolated from *A. thaliana*. High levels of mRNA were expressed in the vascular tissues of the root and stem. Peleman *et al.* (1989) suggested that the *sam-1* gene may code for the enzyme involved in lignification. The molecular weight of subunits of the *A. thaliana* and carnation (Larsen & Woodson, 1991) enzymes were calculated to be 43 kDa, considerably lower than the value determined for the wheat germ enzyme.

By a complex series of analyses, Giovanelli *et al.* (1985b) were able to follow the metabolism of the methyl carbon atom, sulfur atom and C_4 group of methionine in *L. paucicostata*. The results showed that the rate of synthesis of methionine from homocysteine is about five times faster than the rate of incorporation of sulfate into homocysteine (Fig. 7.14). This can be accounted for by the high rate of turnover of S-adenosylmethionine and the recycling of S-adenosylhomocysteine back to methionine.

In another series of experiments, Giovanelli *et al.* (1983) showed that the sulfur atom and methyl carbon atom of methionine are recycled independently of the C_4 skeleton. They proposed that S-adenosylmethionine is decarboxylated to yield a C_3 fragment that is used in the synthesis of polyamines (Tiburcio *et al.*, 1990). The remainder of the S-adenosylmethionine molecule,

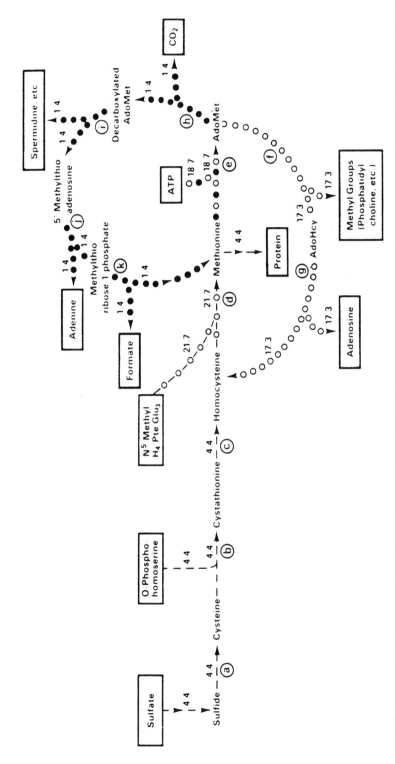

Figure 7.14 Estimates of the fluxes associated with the metabolism of methionine in *Lemna paucicostata*. Fluxes are expressed as nmol per colony doubling. AdoMet = S-adenosylmethionine; AdoHcy = S-adenosylhomocysteine; H₄PteGlu₃ = tetrahydrofolate (Giovanelli *et al.*, 1985).

S-methylthioadenosine is then metabolized, releasing adenine and formate and finally yielding methionine. In this novel pathway the original methylthio group is retained and the C_4 moiety in the regenerated methionine is derived from the ribosyl moiety of ATP (Fig. 7.14).

7.7.5 Regulation of the aspartate pathway (of amino acid biosynthesis)

A simple scheme of the overall regulation is shown in Fig. 7.15. The various mechanisms known to operate *in vitro* can be summarized as follows:

1. *Lysine* inhibits two isoenzymes of aspartate kinase and the first enzyme unique to its own synthesis, dihydrodipicolinate synthase.
2. *Threonine* inhibits one isoenzyme of aspartate kinase and one isoenzyme of homoserine dehydrogenase.
3. *Methionine* appears to operate through the activated form *S*-adenosylmethionine. Two isoenzymes of aspartate kinase are synergistically inhibited in the presence of lysine. *S*-Adenosylmethionine is able to activate threonine synthase. There is also evidence that methionine and/or *S*-adenosylmethionine can repress the synthesis of cystathioine-γ-synthase.
4. *Isoleucine* inhibits the first enzyme unique to its own synthesis, threonine dehydratase (see section 7.8.1).

Giovanelli *et al.* (1989a,b) examined in detail the regulation of the aspartate pathway in *Lemna paucicostala* and concluded, after much discussion, that there was little regulation of the pathway at the position of aspartate kinase. They also proposed that there was no regulation

of threonine synthesis along the whole pathway, suggesting that the inhibition of homoserine dehydrogenase and the activation of threonine synthase did not operate *in vivo*. The elegant regulatory scheme outlined by Giovanelli *et al.* (1989a,b), requires further confirmation by the use of transgenic and mutant plants (section 7.7.6).

7.7.6 The isolation of mutants and transgenic plants with altered regulation

Initial studies by Green & Phillips (1974) indicated that the growth of germinating embryos of a range of cereal crops could be inhibited by the presence of lysine and threonine. The inhibition of growth could be alleviated by the presence of low concentrations of methionine or homoserine. An examination of Fig. 7.15 would suggest that methionine synthesis is prevented by the complete inhibition of all the aspartate kinase isoenzymes and also of threonine-sensitive homoserine dehydrogenase.

Maize tissue culture cell lines have been isolated following azide mutagenesis, that are resistant to the combined action of lysine and threonine. Three of the cell cultures gave rise to genetically stable plant lines (Hibberd & Green, 1982; Diedrick *et al.*, 1990). Similar mutant selection procedures have been successful in barley (Bright *et al.*, 1982) and tobacco (Frankard *et al.*, 1991). The mutant lines isolated from all three plant species were shown to contain elevated levels of soluble threonine.

In barley leaves, three forms of aspartate kinase activity have been detected following ion exchange chromatography (Bright *et al.*, 1982). In the barley mutant R3202 resistant to lysine plus threonine,

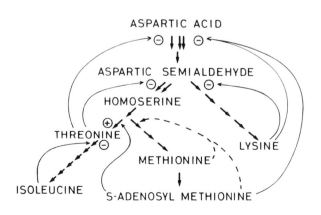

Figure 7.15 Regulation of biosynthesis of the aspartate-derived amino acids (−), enzyme inhibition; (+), enzyme activation; − − − enzyme repression.

the AKII peak was not inhibited by lysine even at the high concentration of 10 mM, whereas the AKIII enzyme peak was still inhibited by lysine. Conversely in the mutant R3004, AKIII activity was less sensitive to lysine, with a maximum inhibition detected at 3.5 mM lysine compared to 0.36 mM for wild-type barley (Rognes et al., 1983). Genetic analysis of the two mutants suggested that resistance to lysine plus threonine was controlled by two dominant genes termed It1 and It2 which are probably the structural genes for AKII and AKIII (Bright et al., 1982). A double mutant derived from a cross between R3202 (It1) and R3004 (It2) has been constructed which contained AKII insensitive to lysine inhibition and AKIII with a decreased sensitivity to lysine (Arruda et al., 1984). A full description of a range of barley mutants resistant to lysine and threonine has been given by Bright et al. (1984).

In maize, two separate genes (termed Ask1 and Ask2) have also been shown to confer resistance to lysine and threonine and the overproduction of threonine (Diedrick et al., 1990). Aspartate kinase purified from the Ask1 and Ask2 mutant lines was less sensitive to inhibition by lysine. Wild-type aspartate kinase activity was inhibited 50% by 10 μM lysine. In contrast, approximately 760 μM lysine was required to inhibit 50% of the aspartate kinase activity of the homozygous mutant Ask2/Ask2 and 25 μM lysine was required for 50% inhibition of the heterozygous Ask1/+ enzyme. Dotson et al. (1990b) suggested that maize aspartate kinase could be a heteromeric enzyme, consisting of two lysine sensitive polypeptides encoded by Ask1 and Ask2.

A number of research groups have attempted to select mutant lines that are resistant to the lysine analogue S-(2-aminoethyl)-L-cysteine (AEC) on the grounds that the mutant plants may overproduce lysine. The majority of these experiments have produced plants with a reduced uptake of AEC (Bright et al., 1979; Matthews et al., 1980). However, one mutant of Nicotiana sylvestris regenerated from protoplast culture, was resistant to AEC (RAEC-1) and was shown to accumulate lysine. Only 50% of the dihydrodipicolinate synthase activity was inhibited by lysine (Negrutiu et al., 1984).

In an elegent series of experiments, Frankard et al. (1992) crossed the lysine overproducing mutant (15-fold, RAEC-1) with a threonine overproducing mutant (60-fold, RLT-70) resistant to lysine plus threonine (Frankard et al., 1991). The sensitivities of aspartate kinase and dihydrodipicolinate synthase to feedback inhibition are shown in Table 7.5. The level of soluble lysine in the

Table 7.5 Comparison of the inhibition of aspartate kinase (AK) and dihydrodipicolinate synthase (DHDPS) activities in mutant and wild type tobacco

Genotype	AK	DHDPS
Wild type	80	100
RAEC-1 X RLT 70		
Double heterozygote	40	50
RAEC-1		
Heterozygote	80	50
Homozygote	80	0
RLT 70		
Heterozygote	40	100
Homozygote	0	100

Data expressed as percentage of total activity. Lysine added at 10 mM. From Frankard et al. (1992).

leaves of the double mutant was 20-fold higher than the wild type and sometimes rose to 50% of the total soluble amino acid pool. Unfortunately the accumulation of lysine in the leaves caused an abnormal phenotype characterized by a reduced leaf blade surface, absence of stem elongation and sterility. Surprisingly the level of soluble threonine in the double mutant was lower or the same as in the wild type.

Following on almost 20 years of mutant selection, Shaul and Galili have attempted to construct transgenic plants that overproduce lysine and threonine. Dihydrodipicolinate synthase in E. coli is considerably less sensitive to feedback inhibition by lysine ($I_{50} = 1$ mM) than the plant enzyme ($I_{50} = 20 \mu$M). Shaul & Galili (1992a) constructed transgenic tobacco plants containing the E. coli dapA gene coding for dihydrodipicolinate synthase in the cytoplasm and in the chloroplast. The increase in lysine concentration (up to a maximum of 40-fold), was shown to be correlated with the level of dihydrodipicolinate synthase activity. Compartmentalization of the enzyme inside the chloroplast was found to be essential for lysine overproduction. The highest lysine overproducing plants exhibited abnormal characteristics including the possession of leaves of a mottled yellow/green colour and a lack of apical dominance. Similar results with transgenic tobacco plants containing the E. coli dapA gene have been published by Glassman (1992).

In a second series of experiments Shaul & Galili (1992b) used the mutant E. coli lysC gene that coded for aspartate kinase insensitive to lysine (Boy et al., 1979). Increases in the soluble level of threonine varied between 20- and 80-fold in transgenic plants containing the enzyme expressed in the chloroplast. Small increases in the levels of lysine and isoleucine was also detected. In contrast to the experiments with dihydrodipicolinate synthase, a chloroplast localization of aspartate kinase was important but not essential for threonine overproduction.

It is clear from the experiments with mutants and transgenic plants that it is possible to produce plants with high levels of soluble lysine and threonine but not methionine. Such plants may themselves be of a greater nutritive value, but the abnormal growth defects are an obvious drawback. It will be necessary in the future to include genes coding for modified storage proteins that are rich in the essential amino acids lysine and threonine.

7.8 THE BRANCHED CHAIN AMINO ACIDS

Although strictly speaking isoleucine is derived from the aspartate derived amino acid threonine, I shall consider the synthesis of the three branched chain amino acids isoleucine, leucine and valine together (Fig. 7.16).

7.8.1 Threonine dehydratase

This enzyme is also known as threonine deaminase and catalyzes the first committed step of isoleucine synthesis, forming 2-ketobutyrate and ammonia. Confirmation of the key role of the enzyme has been obtained by Negrutiu *et al.* (1985) who

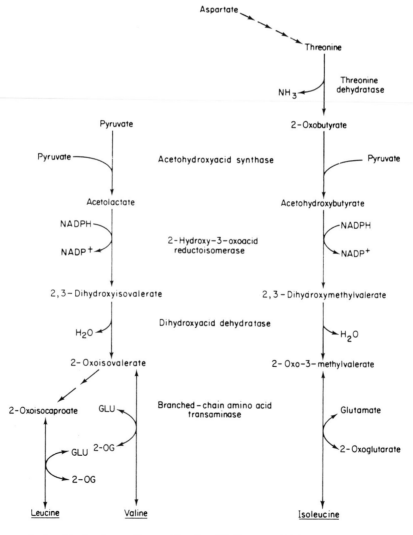

Figure 7.16   The synthesis of isoleucine, valine and leucine (Wallsgrove, 1990).

isolated isoleucine-requiring mutants of tobacco that lacked the enzyme activity. Two forms of the enzyme have been isolated from maize, both of which are totally inhibited by low concentrations of the end product isoleucine (Kirchner & Bryan, 1985). In tomato, an isoleucine-sensitive form of threonine dehydratase was shown to predominate in young leaves, but as the leaves aged, an isoleucine insensitive form of the enzyme was shown to be present at high levels. The mature molecular weight of the isoleucine sensitive form was shown to be 370 kDa whereas the insensitive form was 200 kDa. Szamosi *et al.* (1992) proposed that the former enzyme was involved in isoleucine synthesis and the latter in threonine breakdown.

Threonine dehydratase is one of the most abundant soluble proteins in tomato flowers. The gene coding for the 55 kDa polypeptide has been sequenced and shown to contain a plastid transit peptide and eight introns. The expression of threonine dehydratase mRNA was shown to be 50-fold higher in sepals and 500-fold higher in the rest of the flower than in leaves and roots. The reason for this unusual overexpression in floral organs is not clear at the present time (Samach *et al.*, 1991).

7.8.2 Acetohydroxy acid synthase

The pathways leading to the synthesis of isoleucine, leucine and valine are considered to be biochemically parallel, being catalyzed by enzymes possessing dual substrate specificities. Acetohydroxy acid synthase can either catalyze the synthesis of acetohydroxybutyrate or acetolactate (Fig. 7.16). Control of the pathway is achieved by feedback inhibition of acetohydroxy acid synthase by leucine and valine. Leucine in combination with valine has been shown to inhibit growth due to a lack of isoleucine (Miflin & Cave, 1972; Borstlap, 1981).

The apparently innocent looking enzyme acetohydroxy acid synthase has been the subject of intense investigation, since it was shown to be the site of action of two major classes of herbicides, the sulfonylureas and imidazolinones (Wittenbach *et al.*, 1992). Full details of the mechanism of action of the herbicides have been described (Schloss *et al.*, 1988; Schloss, 1992).

Genomic clones of acetohydroxy acid synthase have been isolated from *A. thaliana* and tobacco using a heterologous hybridization probe derived from yeast. The genes proved to be approximately 85% homologous at the amino acid level and neither contained introns (Mazur *et al.*, 1987). In tobacco, two genes (designated *SurA* and *SurB*) coding for acetohydroxy acid synthase have been shown to have 95% sequence homology in the protein coding regions (Lee *et al.*, 1988).

In the allotetraploid species *Brassica napus*, five acetohydroxy acid synthase genes have been identified, of which four have been cloned and sequenced (Rutledge *et al.*, 1991). The AHAS1 and AHAS3 genes are constitutively expressed in a wide range of somatic and reproductive tissues, suggesting that they encode a fundamental class of enzymes which are vital for growth and development. AHAS2 mRNA accumulation was predominantly in mature ovules and in the extra-embryogenic tissues of developing seeds, suggesting a specific function in seedling development (Ouellet *et al.*, 1992).

In situ hybridization studies in tobacco showed that the acetohydroxy acid synthase mRNA was expressed ubiquitously, with the greatest accumulation being in metabolically active cell types in the root, stem and flowers. There was a co-ordinated regulation of expression of the two genes *SurA* and *SurB* suggesting that these may be functionally redundant and are retained as a result of the tetraploid nature of *Nicotiana tabacum* (Keeler *et al.*, 1993).

There has been considerable interest in obtaining plants that are resistant to the toxic effect of the imidazolinone and sulfonylurea herbicides (Bright, 1992). Mutant lines of tobacco (Chaleff & Ray, 1984) *A. thaliana* (Haughn & Somerville, 1986), soybean (Sebastien & Chaleff, 1987), canola (Swanson *et al.*, 1989), *Datura innoxia* (Rathinasabapathi & King, 1991) and maize (Anderson & Georgeson, 1989; Bright, 1992) have been isolated that show varying degrees of resistance to the herbicide inhibitors of acetohydroxy acid synthase.

The transformation of the *A. thaliana* gene (*csr-1*) coding for the mutant acetohydroxy acid synthase (Haughn & Somerville, 1986) has been shown to confer resistance in tobacco to the sulfonylurea herbicides (Haughn *et al.*, 1988; Charest *et al.*, 1990; Odell *et al.*, 1990). Recently Tourneur *et al.* (1993) have used a P70 promoter and the *csr-1* mutant gene of *A. thaliana* to produce transgenic tobacco with a 1500-fold increase in resistance to chlorsulfuron. The activity of acetohydroxy acid synthase in the transgenic plants was increased 12-fold and the plants were resistant to the external supply of valine. However, there was no increase in the soluble level of valine, leucine or isoleucine indicating that the enzyme is not the rate-limiting step in the pathway.

7.8.3 Acetohydroxy acid isomeroreductase

The enzyme catalyzes a two-step reaction, an alkyl migration and a NADPH-dependent reduction to yield 2,3-dihydroxy-3-isovalerate (substrate acetolactate) or 2,3-dihydroxy-3-methylvalerate (substrate acetohydroxybutyrate). Three forms of the enzyme have been isolated from spinach, each of which have a native molecular weight of 114 kDa and are made up of two subunits of identical molecular weight (57 kDa) but different isoelectric points (Dumas et al., 1989).

A full length cDNA clone encoding the enzyme has been isolated from spinach which contains a 72 amino acid transit peptide. The amino acid sequence shows only a 23% homology to the yeast and E. coli enzymes (Dumas et al., 1991). The gene encoding acetohydroxy acid isomeroreductase has been overexpressed in E. coli and large quantities of the enzyme used for a detailed kinetic analysis. The enzyme was shown to have a preference for 2-acetohydroxybutyrate over acetolactate and to operate at high ratios of NADPH/NADP$^+$, conditions which are normally present in illuminated chloroplasts (Dumas et al., 1992).

7.8.4 Dihydroxyacid dehydratase

This enzyme has been purified to homogeneity from spinach leaves, and is a dimer with subunits of molecular weight of 63 kDa. Dihydroxyacid dehydratase carries out the dehydration and tautomerization of the 2,3-dihydroxycarboxylic acid to the corresponding 2-keto acid. The enzyme contains a [2Fe–2S] cluster which is a novel finding for enzymes of the hydratase class (Flint & Emptage, 1988). A mutant of Datura innoxia requiring isoleucine and valine for growth has been shown to lack dihydroxyacid dehydratase (Wallsgrove et al., 1986).

7.8.5 Isopropylmalate dehydrogenase

The gene encoding this enzyme has been isolated from Brassica napus chloroplast DNA by complementation of a yeast leu 2 mutant. The cDNA encodes a 52 kDa protein which has a putative chloroplast transit peptide. Analysis of the sequence of the isopropylmalate dehydrogenase protein, suggested that the enzyme is more similar to bacterial than fungal proteins (Ellerstrom et al., 1992). The gene encoding the enzyme has also been isolated from potato (Jackson et al., 1993).

A preliminary report has indicated that the final three enzymes of leucine synthesis can be detected in spinach chloroplast extracts (Hagelstein et al., 1993).

7.8.6 Branched-chain amino acid aminotransferases

These enzymes carry out the final reactions in the synthesis of leucine, isoleucine and valine. Two forms of leucine aminotransferase have been isolated from barley (Aarnes, 1981) and soybean (Parthre et al., 1987). In barley, the two forms have the same molecular weight (95 kDa) whereas in soybean molecular weights of 68 and 93 kDa were determined. In barley, using glutamate as the amino donor, the relative rates of oxo acid utilization were oxisocaproate > oxomethylvalerate > oxoisovalerate (Wallsgrove, 1990).

7.9 THE BIOSYNTHESIS OF PROLINE AND ARGININE

The synthesis of proline and arginine have long been considered to be related pathways (Thompson, 1980). Proline may either be synthesized directly from glutamate or via the transamination of ornithine (Fig. 7.17).

7.9.1 Proline

As well as being a major constitutent of protein, in particular in cereal seeds, proline is also thought to

Figure 7.17 The synthesis of proline (Verma et al., 1992). P5CS, Δ'-pyrroline-5-carboxylate synthase; P5CR, Δ'-pyrroline-5-carboxylate reductase; OAT, ornithine aminotransferase.

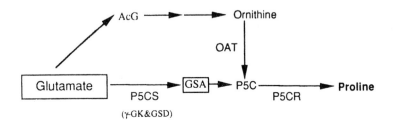

act as an osmotic protectant in plants that have been subjected to either drought or high salt stress (Hanson & Hitz, 1982; McCue & Hanson, 1990; Delauney & Verma, 1993). Tomato cells culture in 400 mM NaCl or 25% polyethylene glycol can accumulate proline up to concentrations 300-fold higher than in unstressed conditions (Handa et al., 1986). Kinetic studies using ^{15}N-labeling showed that the accumulation of proline was due to a 10-fold increase in the rate of synthesis and a concomitant reduction in the rate of degradation (Rhodes et al., 1986). Proline metabolism has also been shown to play an important role in the process of nitrogen fixation in the legume root nodule (Kohl et al., 1988, 1991).

For many years the conversion of glutamate into glutamate semialdehyde was assumed to be the initial pathway of proline synthesis in higher plants although the enzyme activity has not been unequivocally demonstrated in vivo. However, Hu et al. (1992) have isolated a cDNA clone from Vigna aconitifolia nodules that encodes a bifunctional enzyme Δ'-pyrroline-5-carboxylate synthetase (P5CS) with both γ-glutamyl kinase and glutamate semialdehyde dehydrogenase activity. The P5CS protein was shown to have two distinct domains that had 55.3% and 57.9% overall similarity to the E. coli ProB and ProA proteins (Deutch et al., 1984).

The plant P5CS enzyme protein was isolated from the complemented E. coli ProBA mutant and was shown to be 50% inhibited by 5 mM proline; however, the wild type E. coli enzyme was 50-fold more sensitive to end-product inhibition. Northern blot analysis indicated that the P5CS gene was expressed at a greater level in the leaves than the roots and nodules of V. aconitifolia. However, the level of mRNA in the roots was shown to be increased by the addition of 200 mM NaCl to the plants (Hu et al., 1992).

Δ'-pyrroline-5-carboxylate reductase (P5CR) has been purified from barley and shown to increase in activity following drought stress (Kruger et al., 1986; Argandona & Pahlich, 1991). The gene encoding the enzyme has been isolated from soybean nodules by the complementation of the E. coli ProC mutant (Delauney & Verma, 1990). Levels of P5CR mRNA are elevated in salt stressed plants (Delauney & Verma, 1990; Williamson & Slocum, 1992). The enzyme has been purified to homogeneity by expression of the soybean nodule cDNA in E. coli and shown to have a subunit molecular weight of 29 kDa. Subcellular fractionation and Western blot analysis were used to localize P5CR in transgenic tobacco plants

synthesizing P5CS constitutively. Enzyme activity was present in the cytoplasm of the root, nodule and leaf but 15% of the activity was also detected in the chloroplast fraction (Szoke et al., 1992). As P5CR has previously been detected in the chloroplasts of pea leaves (Rayapati et al., 1989), it would appear that at least two isoenzymes exist and that proline may be synthesized in different subcellular compartments in the leaf and root. Transgenic tobacco plants containing a 50-fold elevated level of P5CR activity did not contain increased concentrations of proline (Szoke et al., 1992), confirming that the enzyme is not the rate-limiting step in proline synthesis (LaRosa et al., 1991).

It has been proposed that proline may be synthesized from glutamate via ornithine, an intermediate in the synthesis of arginine (Thompson, 1980; Stewart, 1981). Ornithine may be transaminated via either the α or δ amino groups yielding P2C or P5C, respectively. Delauney et al. (1993) have isolated a cDNA encoding ornithine aminotransferase (OAT) by trans-complementation of E. coli ProBA mutants in the presence of ornithine. Sequence data suggested that the Vigna aconitifolia OAT gene had a significant homology to other δ-aminotransferases from mammals and yeast, indicating that P5C is the likely product of the enzyme reaction.

Based on the levels of P5CS and OAT mRNA in seedlings, Delauney et al. (1993) concluded that both enzymes were involved in proline synthesis under normal physiological conditions. However, under conditions of salt stress or nitrogen starvation, levels of OAT mRNA decreased, whereas those of P5CS increased. These data (based on the regulation of gene expression only) suggest that the direct conversion of glutamate into proline is the predominant pathway under stress conditions.

7.9.2 Arginine

Arginine is a major form of storage nitrogen in plants, where it may form 40% of the nitrogen in seed protein and 50–90% of the soluble nitrogen in fruit trees, grape vines and flower bulbs (Micallef & Shelp, 1989a). Ornithine and arginine also act as precursors of the important secondary metabolites, polyamines (Tiburcio et al., 1990). In addition, citrulline may act as a nitrogen transport compound in non-legume nitrogen fixing symbioses (Pate et al., 1988; Sellstedt & Atkins, 1991).

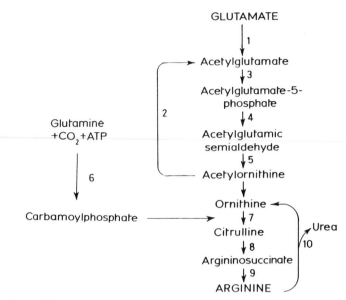

Figure 7.18   The synthesis of arginine. The enzymes 1–10 are described in the text.

(a)   Pathway of synthesis

Arginine is synthesized from glutamate via ornithine and citrulline (Fig. 7.18) (Shargool et al., 1988). Inhibition studies have shown that in soybean cell suspension cultures the acetyl group is rapidly returned from N-acetylornithine to glutamate via ornithine acetyltransferase (Fig. 7.18, 2), (Shargool & Jain, 1985). N-acetylglutamate kinase (3) has been purified to homogeneity from pea cotyledons and shown to exist in dimeric and tetrameric forms. Arginine was found to strongly inhibit the enzyme activity, this inhibition was relieved by N-acetylglutamate, which activated the kinase enzyme (McKay & Shargool, 1981).

In an elegant series of tracer studies, Micallef & Shelp (1989b) examined the metabolism of [ureido-^{14}C]citrulline and [1-^{14}C]ornithine in the developing cotyledons of soybean. They were able to confirm the pathway of the conversion of ornithine into arginine as shown in Fig. 7.18. Further metabolism of ornithine requires carbamoylphosphate synthetase (6), an enzyme that is involved in both arginine and pyrimidine biosynthesis. The enzyme is subject to feedback inhibition by UMP which is relieved by ornithine (Kolloffel & Verkerk, 1982). Multiple forms of ornithine carbamoyltransferase (7) have been detected in higher plants, some of which may be involved in polyamine biosynthesis (Martin et al., 1983; Baker et al., 1983). Arginosuccinate synthetase (8) has been shown to be regulated by

energy change, with arginine acting as a modifier of this regulation (Shargool, 1973).

(b)   Subcellular localization of enzymes

The first two enzymes in the pathway N-acetylglutamate synthase (1) and N-acetylglutamate kinase (3) are located in the cytoplasm of soybean tissue culture cells (Jain et al., 1987), and pea leaves (Taylor & Stewart, 1981). Similarly, the last two enzymes of the pathway arginosuccinate synthetase (8) and arginosuccinate lyase (9) are also present in the cytoplasm. However, the three enzymes in the center of the pathway, ornithine acetyltransferase (2), carbamoyl phosphate synthetase (6) and ornithine carbamoyltransferase (7), have been shown to be localized in either plastids or chloroplasts (Taylor & Stewart, 1981; Shargool et al., 1978a,b; DeRuiter & Kolloffel, 1985).

In germinating legume cotyledons arginine is degraded to ornithine and urea by arginase (10) in the mitochondria (De Ruiter & Kolloffel, 1983). In a recent very interesting observation, Ludwig (1993) has shown that Arabidopsis thaliana chloroplasts are able to metabolize arginine. The author was able to demonstrate that the chloroplasts contained arginine iminohydratase that yielded citrulline and ammonia. The citrulline may be further metabolized to ornithine and carbamoyl phosphate by ornithine carbamoyltransferase. The carbamoyl phosphate can then be broken down to bicarbonate and ammonia by carbamate kinase. Ludwig (1993) proposed that the interconversion of ornithine–citrulline–arginine may be a novel mechanism of transporting ammonia and CO_2 from the mitochondria to the chloroplast.

7.10 SULFUR AMINO ACIDS

7.10.1 Cysteine

Sulfate is reduced to sulfide in a series of reactions involving ATP and reduced ferredoxin in the chloroplast (Anderson, 1990), the full details of which are beyond the scope of this chapter. Cysteine synthase [O-acetylserine (thiol) lyase] catalyzes the following reaction:

$$O\text{-Acetylserine} + S^{2-} \rightarrow Cysteine + Acetate$$

The enzyme is also responsible for the synthesis of a number of heterocyclic β-substituted alanines, some of which are important plant secondary products (Ikegami et al., 1990, 1993).

Multiple forms of cysteine synthase have been found in Phaseolus (Bertagnolli & Wedding, 1977), Brassica napus leaves (Nakamura & Tamura, 1989) and spinach leaves (Lunn et al., 1990). In a detailed study of cauliflower bud protoplasts, Rolland et al. (1992) demonstrated the presence of four forms by ion exchange chromatography. Two isoenzymes were located in purified proplastids, one in the mitochondria and one in the cytoplasm. The chloroplast cysteine synthase has been purified to homogeneity from spinach. The enzyme consists of two identical subunits of molecular weight 35 kDa containing one molecule of pyridoxal phosphate per subunit. On gel filtration chromatography, part of the enzyme activity eluted in association with serine acetyltransferase activity at a position corresponding to a molecular weight of 310 kDa (Droux et al., 1992). Antisera raised against the chloroplast enzyme showed very low cross-reactivity with the cytoplasmic and mitochondrial enzymes.

A cDNA clone encoding the cytoplasmic form of cysteine synthase has been isolated from spinach with a molecular weight of 34 kDa (Saito et al., 1992). A comparison of the amino acid sequence indicated a 53% identity with the E. coli enzyme. The cDNA contained no apparent transit sequence and constitutive expression of mRNA was detected in roots and leaves.

A cDNA clone encoding the complete sequence of the chloroplast enzyme has been isolated by Rolland et al. (1993). The 35 kDa polypeptide also contained a 52 amino acid transit peptide. A good primary sequence homology (50%) was demonstrated between the spinach chloroplast cysteine synthase and the A isoenzyme of E. coli. Southern blot analysis indicated that the chloroplast enzyme is encoded by multiple genes in spinach. The gene encoding the chloroplast cysteine synthase has also been isolated from spinach by Saito et al. (1993). Northern blot analysis indicated that the mRNA was expressed at a higher level in the leaves than the roots. A cDNA clone encoding cysteine synthase from Capsicum annuum chromoplasts has been shown to be up-regulated during fruit development (Romer et al., 1992).

7.10.2 Glutathione

The tripeptide glutathione: Glutamate – Cysteine – Glycine is considered to be the major storage and transport compound containing reduced sulfur (Rennenberg, 1982; Rennenberg & Lamoureux, 1990; Schupp et al., 1992). Glutathione may exist in the reduced form of GSH or in the oxidized state as a dimer GSSG, where two molecules are linked by a disulfide bond. It has also been proposed that the tripeptide may play a key role in the defence of plants against various environmental stresses, e.g. cold, heat, drought, high light, fungal attack and herbicides (Smith et al., 1989; Alscher, 1989). Phytochelatins are polymers of γ-glutamylcysteine, which have been shown to be involved in the chelation of certain toxic metals (Rauser, 1990; Steffens, 1990).

(a) Synthesis

γ-Glutamylcysteine synthetase catalyzes the first ATP-dependent step involved in glutathione synthesis. The enzyme has been partially purified from cell suspension cultures of Nicotiana tabacum by the determination of γ-glutamylcysteine as the monobromobimane derivative (Hell & Bergmann, 1990). Buthionine sulfoximine and methionine sulfoximine were inhibitors of the enzyme as had previously been reported for γ-glutamylcysteine synthase from mammalian tissues (Orlowski & Meister, 1971). The enzyme was completely inhibited by glutathione ($K_i = 0.42$ mM). The data indicated that the rate of glutathione synthesis in vivo may be substantially influenced by the concentration of cysteine and glutamate and may be further feedback regulated by the end product glutathione (Hell & Bergmann, 1990).

The second enzyme in the pathway, glutathionine synthetase, catalyzes the ATP-dependent attachment of glycine to the C-terminal group of γ-glutamylcysteine. Enzyme activity in relatively crude preparations has been detected in spinach leaves (Law & Halliwell, 1986), legume seedlings (Macnicol, 1987) and tobacco suspension cultures (Hell & Bergmann, 1988). Glutathione synthetase in pea and tobacco may also use γ-glutamyl-α-aminobutyrate as a substrate. In plants that contain homoglutathione, the enzyme shows a greater affinity for β-alanine, than glycine (Klapheck et al., 1988).

(b) Subcellular localization

Glutathione is present in the chloroplasts of leaves at high concentrations (1–6 mM) although it may also be present in the cytoplasm (Foyer & Halliwell, 1976; Smith et al., 1985; Bielawski & Joy, 1986; Klapheck et al., 1987). In pea leaves, 72% of the γ-glutamyl cysteine synthetase and 47% of glutathione synthetase were localized in the chloroplasts. In spinach leaves the distribution in the chloroplast was 61% and 58%, respectively (Hell & Bergmann, 1990). The data in leaves suggested that there are two major sites of synthesis of glutathione. However, in maize roots, almost 50% of the γ-glutamylcysteine synthetase activity was detected in the root plastids of 4-day-old seedlings, whereas less than 10% of glutathione synthetase was determined in the organelle (Ruegsegger & Brunold, 1993).

7.11   HISTIDINE

Histidine has been described as the 'Cinderella' of plant amino acids (Miflin, 1980) with little work being carried out on the route of synthesis in plants, although the pathway has been established in bacteria (Bryan, 1990). Mutant cell lines have been isolated that require histidine for growth, with enzyme deficiencies proposed in imidazole glycerol phosphate dehydratase or histidinol phosphate aminotransferase (Gebhardt et al., 1983; Shimamoto & King, 1983; Negrutiu et al., 1985).

The final enzyme in the pathway, histidinol dehydrogenase, catalyzes a four-electron dehydrogenase reaction:

$$\text{Histidinol} + \text{NAD}^+ \rightarrow \text{Histidinal} + \text{NADH}$$

$$\text{Histidinal} + \text{NAD}^+ \rightarrow \text{Histidine} + \text{NADH}$$

The enzyme was initially detected in 10 different plant species including asparagus, cabbage and wheat germ (Wong & Mazelis, 1981). Recently histidinol dehydrogenase has been purified to homogeneity from cabbage, a procedure that commenced with 10 kg of leaf material (Nagai & Scheidegger, 1991). The enzyme is a dimer with a subunit molecular weight of 52 kDa. At least five different forms of the enzyme could be separated by isoelectric focusing.

A full length cDNA encoding histidinol dehydrogenase has been isolated from cabbage and shown to contain a putative 31 amino acid chloroplast transit peptide. The predicted protein sequence was 51% identical to the yeast enzyme and 49% identical to the E. coli enzyme (Nagai et al., 1991). The cDNA encoding the enzyme has been expressed in 519 cells using the baculovirus expression vector system, and a simple one-step purification procedure carried out to obtain large quantities of pure enzyme (Nagai et al., 1992).

ACKNOWLEDGEMENTS

Work on nitrogen metabolism in plants has been funded by the AFRC and SERC over the last 7 years. I am extremely grateful to Annie Brotheridge for her tireless help and support in typing this manuscript.

REFERENCES

Aarnes, H. (1976) Plant Sci. Lett. 7, 187–194.
Aarnes, H. (1978) Planta 40, 185–192.
Aarnes, H. (1980) Plant Sci. Lett. 19, 81–89.
Aarnes, H. (1981) Z. Pflanzenphysiol. 102, 81–89.
Aarnes, H. & Rognes, S.E. (1974) Phytochemistry 13, 2717–2724.
Adams, D.G. (1992) In Prokaryote Structure and Function : A New Prospective (S. Mohan, C. Dow & J.A. Cole, eds), pp. 341–384. Cambridge University Press, Cambridge.
Alscher, R.G. (1989) Physiol. Plant. 77, 457–464.
Anderson, H.W. (1990) In The Biochemistry of Plants (B.J. Miflin & P.J. Lea, eds), Vol. 16, pp. 327–381. Academic Press, San Diego.
Anderson, M.P., Vance, C.P., Heichel, G.H. & Miller, S.S. (1989) Plant Physiol. 90, 351–358.
Anderson, P.C. & Georgeson, M. (1989) Genome 31, 994–999.
Andrews, M. (1986) Plant Cell Environ. 9, 511–519.
Argandona, V. & Pahlich, E. (1991) Phytochemistry 30, 1093–1094.
Arruda, P., Bright, S.W.J., Kueh, J.S.H., Lea, P.J. & Rognes, S.E. (1984) Plant Physiol. 76, 442–446.
Askerlund, P., Laurent, P., Nakagawa, H. & Kadar, J-C. (1991) Plant Physiol. 95, 6–13.
Aslam, M., Travis, R.L. & Huffaker, R.C. (1992) Plant Physiol. 99, 1124–1133.

Aslam, M., Travis, R.L. & Huffaker, R.C. (1993) *Plant Physiol.* **102**, 811–819.

Atkins, C.A., Pate, J.S. & Sharkey, P.J. (1975) *Plant Physiol.* **56**, 807–812.

Avilla, C., Marquez, A.J., Pajuelo, P., Cannell, M.E., Wallsgrove, R.M. & Forde, B.G. (1993) *Planta* **189**, 475–483.

Awonaike, K.O., Lea, P.J. & Miflin, B.J. (1981) *Plant Sci. Lett.* **23**, 189–195.

Azevedo, R.A., Smith, R.J. & Lea, P.J. (1992a) *Phytochemistry* **31**, 3731–3734.

Azevedo, R.A., Blackwell, R.D., Smith, R.J. & Lea, P.J. (1992b) *Phytochemistry* **31**, 3725–3730.

Back, E., Burkhart, W., Moyer, M., Privalle, L. & Rothstein, S. (1988) *Mol. Gen. Genet.* **212**, 20–26.

Back, E., Dunne, W., Schneiderbauer, A., deFramond, A., Rastogi, R. & Rothstein, S.J. (1991) *Plant Mol. Biol.* **17**, 9–18.

Baker, S.R., Jones, L.H. & Yon, R.J. (1983) *Phytochemistry* **22**, 2167–2169.

Banerjee, R.V., Johnston, N.L., Sobeski, J.K., Datta, P. & Matthews, R.G. (1989) *J. Biol. Chem.* **264**, 13888–13893.

Banfalvi, Z. & Kondorosi, A. (1989) *Plant Mol. Biol.* **13**, 1–12.

Baum, H.J., Madison, J.T. & Thompson, J.F. (1983) *Phytochemistry* **22**, 2409–2412.

Becker, T.W., Caboche, M., Carrayol, E. & Hirel, B. (1992) *Plant Mol. Biol.* **19**, 367–379.

Beevers, L. & Hageman, R.H. (1980) In *The Biochemistry of Plants* (B.J. Miflin, ed.), Vol. 5, pp. 115–168. Academic Press, New York.

Bennett, M.J. & Cullimore, J.V. (1989) *Planta* **179**, 433–440.

Bennett, M.J., Lightfoot, D.A. & Cullimore, J.V. (1989) *Plant Mol. Biol.* **12**, 553–565.

Bertagnolli, B.L. & Wedding, R.T. (1977) *Plant Physiol.* **60**, 115–121.

Bielawski, W. & Joy, K.W. (1986) *Planta* **169**, 267–272.

Blackwell, R.D., Murray, A.J.S. & Lea, P.J. (1987) *J. Exp. Bot.* **38**, 1799–1809.

Blackwell, R.D., Murray, A.J.S., Lea, P.J. & Joy, K.W. (1988) *J. Exp. Bot.* **39**, 845–858.

Bloom, A.J., Sikrapanna, S.S. & Warner, R.L. (1992) *Plant Physiol.* **99**, 1294–1301.

Borstlap, A.C. (1981) *Planta* **151**, 314–319.

Botella, J.R., Verbelen, J.P. & Valpuesta, V. (1988) *Plant Physiol.* **87**, 255–257.

Bowsher, C.G., Long, D.M., Oaks, A. & Rothstein, S.J. (1991) *Plant Physiol.* **95**, 281–285.

Bowsher, C.G., Hucklesby, D.P. & Emes, M.J. (1993) *Plant J.* **3**, 463–467.

Boy. E., Borne, F. & Patte, J.C. (1979). *Biochimie* **61**, 1151–1160.

Brears, T., Walker, E.L. & Coruzzi, G.M. (1991) *Plant J.* **1**, 235–244.

Bright, S.W.J. (1992) In *Biosynthesis and Molecular Regulation of Amino Acids in Plants* (B.K. Singh, H.E. Flores & J.C. Shannon, eds), pp. 184–194. American Society of Plant Physiologists, Maryland.

Bright, S.W.J., Norbury, P.B. & Miflin, B.J. (1979) *Theor. Appl. Genet.* **55**, 1–4.

Bright, S.W.J., Kueh, J.S.H., Franklin, J., Rognes, S.E. & Miflin, B.J. (1982) *Nature* **299**, 278–279.

Bright, S.W.J., Lea, P.J., Arruda, P., Hall, N.P., Kendall, A.C., Keys, A.J., Kueh, J.S.H., Parker, M.L. Rognes, S.E., Turner, J.C., Wallsgrove, R.M. & Miflin, B.J. (1984) In *Genetic Manipulation of Plants and its Application to Agriculture* (P.J. Lea, ed.), Vol. 23, pp. 141–169. Oxford University Press, Oxford.

Bryan, J.K. (1990) In *The Biochemistry of Plants* (B.J. Miflin & P.J. Lea, eds), Vol. 16, pp. 161–195. Academic Press, San Diego.

Buck, M. (1990) In *Nitrogen Fixation: Achievements and Objectives* (P.M. Gresshoff, E.L. Roth, G. Stacey & W.E. Newton, eds), pp. 451–457. Chapman & Hall, New York.

Burger, G., Tilbur, J. & Scazzocchio, C. (1991) *Mol. Cell Biol.* **11**, 795–802.

Cai, X. & Wong, P.P. (1989) *Plant Physiol.* **91**, 1056–1062.

Chaleff, R.S. & Ray, T.B. (1984) *Science* **223**, 1148–1151.

Charest, P.J., Hattori, J., DeMoor, J., Iyer, V.N. & Miki, B.L. (1990) *Plant Cell Rep.* **8**, 643–646.

Chastain, C.J., Brusca, J.S., Ramasubrammian, T.S., Wei, T-F. & Golden, J.W. (1990) *J. Bacteriol.* **172**, 5044–5051.

Chen, F.L. & Cullimore, J.V. (1989) *Planta* **179**, 441–447.

Chen, F.L., Bennett, M.J. & Cullimore, J.V. (1990) *J. Exp. Bot.* **41**, 1215–1221.

Cheng, C-L., Acedo, G.N., Dewdney, J., Goodman, H.M. & Conkling, M.A. (1991) *Plant Physiol.* **96**, 275–279.

Cock, J.M., Mould, R.M., Bennett, M.J. & Cullimore, J.V. (1990) *Plant Mol. Biol.* **14**, 549–560.

Cock, J.M., Brock, I.W., Watson, A.T., Swarup, R., Morby, A.P. & Cullimore, J.V. (1991) *Plant Mol. Biol.* **17**, 761–771.

Cohen, G.N. (1983). In *Amino Acids: Biosynthesis and Genetic Regulation* (K.M. Herrmann & R.L. Somerville, eds), pp. 147–171. Addison-Wesley, Reading, MA.

Cooper, H.D. & Clarkson, D.T. (1989) *J. Exp. Bot.* **40**, 753–762.

Cossins, E.A. (1987) In *The Biochemistry of Plants* (D.D. Davies, ed.), Vol. 2, pp. 366–418. Academic Press, New York.

Crawford, N.M., Campbell, W.H. & Davis, R.W. (1986) *Proc. Natl. Acad. Sci. U.S.A.* **83**, 8073–8076.

Cullimore, J.V., Cock, J.M., Daniell, T.J., Swarup, R. & Bennett, M.J. (1992) In *Inducible Plant Proteins* (J.L. Wray, ed.), pp. 79–95. Cambridge University Press, Cambridge.

Davies, K.M. & King, G.A. (1993) *Plant Physiol.* **102**, 1337–1340.

Delauney, A. & Verma, D.P.S. (1990) *Mol. Gen. Genet.* **221**, 299–305.

Delauney, A.J. & Verma, D.P. (1993) *Plant J.* **4**, 215–223.

Delauney, A.J., Hu, C.A., Kavi Kishor, P.B. & Verma, D.P.S. (1993) *J. Biol. Chem.* **268**, 18673–18678.

Deng, M.D., Moureaux, T., Leydecker, M.T. & Caboche, M. (1990) *Planta* **180**, 257–261.

Deng. M.D., Moureaux, T., Cherel, I., Boutin, J.P. & Caboche, M. (1991) *Plant Physiol. Biochem.* **29**, 239–247.

Denison, R.F., Witty, J.F. & Minchin, F.R. (1992) *Plant Physiol.* **100**, 1863–1868.

Dereppe, C., Bold, G., Ghisalba, O., Ebert, E. & Schar, H.P. (1992) *Plant Physiol.* **98**, 813–821.

DeRuiter, H. & Kolloffel, C. (1983) *Plant Physiol.* **73**, 525–528.

DeRuiter, H. & Kolloffel, C. (1985) *Plant Physiol.* **77**, 695–699.

Deutch, A.H., Rushlow, K.E. & Smith, R.J. (1984) *Nucleic Acids Res.* **12**, 6337–6355.

Dickson, J.M.J.J., Vince, E., Grant, M.R., Smith, L.A., Rodber, K.A., Farnden, K.J.F. & Reynolds, P.H.S. (1992) *Plant Mol. Biol.* **20**, 333.

Diedrick, T.J., Frisch, D.A. & Gengenbach, B.G. (1990) *Theor. Appl. Genet.* **79**, 209–215.

Dodd, W.A. & Cossins, E.A. (1970) *Biochim. Biophys. Acta.* **201**, 461–470.

Dotson, S.B., Somers, D.A. & Gengenbach, B.G. (1989) *Plant Physiol.* **91**, 1602–1608.

Dotson, S.B., Somers, D.A. & Gengenbach, B.G. (1990a) *Plant Physiol.* **93**, 98–104.

Dotson, S.B., Frisch, D.A., Somers, D.A. & Gengenbach, B.G. (1990b) *Planta* **182**, 546–552.

Downie, J.A. (1991) *Curr. Biol.* **1**, 382–384.

Droux, M., Martin, J., Sajus, P. & Douce, R. (1992) *Arch. Biochem. Biophys.* **295**, 379–390.

Dudley, M.E., Jacobs, T.W. & Long, S.R. (1987) *Planta* **171**, 289–301.

Dumas, R., Job, D., Ortholand, J-Y., Emeric, G., Greiner, A. & Douce, R. (1992) *Biochem J.* **288**, 865–874.

Dumas, R., Joyard, J. & Douce, R. (1989) *Biochem. J.* **262**, 971–976.

Dumas, R., Lebrun, M. & Douce, R. (1991) *Biochem. J.* **277**, 469–475.

Duncannon, E., Gilkes, A.F., Kirk, D.W., Sherman, A. & Wray, J.L. (1993) *Mol. Gen. Genet.* **236**, 275–282.

Eady, R.R. & Smith, B.E. (1992) *Curr. Biol.* **2**, 637–639.

Economou, A., Hamilton, W.D.O., Johnston, A.W.B. & Downie, J.A. (1990) *EMBO J.* **9**, 349–354.

Edwards, J.W. & Coruzzi, G.M. (1989) *Plant Cell* **1**, 241–248.

Edwards, J.W., Walker, E.L. & Coruzzi, G.M. (1990) *Proc. Natl. Acad. Sci. U.S.A.* **87**, 3459–3463.

Egli, M.A., Griffith, S.M., Miller, S.S., Anderson, M.A. & Vance, C.P. (1989) *Plant Physiol.* **91**, 898–904.

Elhai, J. & Wolk, C.P. (1990) *EMBO J.* **9**, 3379–3388.

Ellerstrom, M., Josefsson, L-G., Rask, L. & Ronne, H. (1992) *Plant Mol. Biol.* **18**, 557–566.

Farnham, M.W., Griffith, S.M., Vance, C.P. & Miller, S.S. (1990) *Plant Physiol.* **94**, 1634–1640.

Faure, J.D., Vincentz, M., Kronenberger, J. & Caboche, M. (1991) *Plant J.* **1**, 107–113.

Fedec, P. & Cossins, E.A. (1976) *Phytochemistry* **15**, 1819–1823.

Fisher, R.F. & Long, S.R. (1992) *Nature* **357**, 655–660.

Flint, D.H. & Emptage, M.H. (1988) *J. Biol. Chem.* **263**, 3558–3563.

Forde, B.G. & Cullimore, J.V. (1989) In *Oxford Surveys of Plant Molecular and Cell Biology* (B.J. Miflin, ed.), Vol. 6, pp. 247–296. Oxford University Press, Oxford.

Forde, B.G., Day, H.M., Turton, J.F., Shen, W.J., Cullimore, J.V. & Oliver, J.E. (1989) *Plant Cell* **1**, 391–401.

Foyer, C.H. & Halliwell, B. (1976) *Planta* **133**, 21–25.

Frankard, V., Ghislain, M. & Jacobs, M. (1991) *Theor. Appl. Genet.* **82**, 273–282.

Frankard, V., Ghislain, M. & Jacobs, M. (1992) *Plant Physiol.* **99**, 1285–1293.

Freeman, J., Marquez, A.J., Wallsgrove, R.M., Saarelainen, R. & Forde, B.G. (1990) *Plant Mol. Biol.* **14**, 297–311.

Friemann, A., Brinkmann, K. & Hachtel, W. (1991) *Mol. Gen. Genet.* **227**, 97–105.

Friemann, A., Brinkmann, K. & Hachtel, W. (1992) *Mol. Gen. Genet.* **231**, 411–416.

Frisch, D.A., Gengenbach, B.G., Tommey, A.M., Sellner, J.M., Somers, D.A. & Myers, D.E. (1991) *Plant Physiol.* **96**, 444–452.

Galangau, F., Daniel-Vedele, F., Moureaux, T., Dorbe, M.F., Leydecker, M.T. & Caboche, M. (1988) *Plant Physiol.* **88**, 383–388.

Gallon, J.R. (1992) *New Phytol.* **122**, 571–609.

Gallon. J.R. & Chaplin, A.E. (1987) *An Introduction to Nitrogen Fixation.* Cassell, London.

Gallon, J.R. & Chaplin, A.E. (1988) In *Biochemistry of the Algae and Cyanobacteria* (L.J. Rogers & J.R. Gallon, eds), pp. 147–173. Oxford University Press, Oxford.

Gantt, J.S., Larson, R.L., Farnham, M.W., Pathirana, S.M., Miller, S.S. & Vance, C.P. (1992) *Plant Physiol.* **98**, 868–878.

Gebhardt, C., Shimamoto, K., Lazar, G., Schnebli, V. & King, P.J. (1983) *Planta* **159**, 18–24.

Georgiadis, M.M., Komiya, H., Chakrabarti, P., Woo, D., Kornuc, J.J. & Rees, D.C. (1992) *Science* **257**, 1653–1659.

Giovanelli, J. (1987) *Methods Enzymol.* **143**, 443–449.

Giovanelli, J. & Mudd, S.H. (1971) *Biochim. Biophys. Acta* **227**, 654–670.

Giovanelli, J., Datko, A.H., Mudd, S.H. & Thompson, G.A. (1983) *Plant Physiol.* **71**, 319–326.

Giovanelli, J., Mudd, S.H. & Datko, A.H. (1985a) *Plant Physiol.* **77**, 450–455.

Giovanelli, J., Mudd, S.H. & Datko, A.H. (1985b) *Plant Physiol.* **78**, 555–560.

Giovanelli, J., Mudd, S.H. & Datko, A.H. (1989a) *Plant Physiol.* **90**, 1577–1583.

Giovanelli, J., Mudd, S.H. & Datko, A.H. (1989b) *Plant Physiol.* **90**, 1584–1599.

Giovanelli, J., Veluthambi, K., Thomson, G.A., Mudd, S.H. & Datko, A.H. (1984) *Plant Physiol.* **76**, 285–292.

Givan, C.V. (1980) In *The Biochemistry of Plants* (B.J. Miflin, ed.), Vol. 5, pp. 329–357. Academic Press, New York.

Glassman, K.F. (1992) In *Biosynthesis and Molecular Regulation of Amino Acids in Plants* (B.K. Singh, H.E. Flores & J.C. Shannon, eds), pp. 217–228. American Society of Plant Physiologists, Maryland.

Goulding, K. & Poulton, P. (1992) *Chem. Brit.* **28**, 1100–1102.

Govers, F. & Bisseling, T. (1992) In *Nitrogen Metabolism of Plants* (K. Mengel & D.J. Pilbeam, eds), pp. 31–37. Clarendon Press, Oxford.

Gowri, G. & Campbell, W.H. (1989) *Plant Physiol.* **90**, 792–798.

Green, C.E. & Phillips, R.L. (1974) *Crop. Sci.* **14**, 827–830.

Gregerson, R.G., Miller, S.S., Twary, S.N., Gantt, J.S. & Vance, C.P. (1993) *Plant Cell.* **5**, 215–226.

Griffith, S.M. & Vance, C.P. (1989) *Plant Physiol.* **90**, 1622–1629.

Gupta, S.C. & Beevers, L. (1985) *Planta* **166**, 89–95.

Hagelstein, P., Klein, M. & Schultz, G. (1993) *Plant Physiol.* **102**, 961S.

Hageman, R.H. & Flesher, D. (1960) *Plant Physiol.* **35**, 700–708.

Hageman, R.H. Flesher, D. & Gitter, A. (1961) *Crop Sci.* **1**, 201–204.

Hager, K.P., Danneberg, G. & Bothe, H. (1983) *FEMS Microbiol. Lett.* **17**, 179–183.

Handa, S., Handa, A.K., Hasegawa, P.W. & Bressan, R.A. (1986) *Plant Physiol.* **80**, 938–945.

Hanson, A.D. & Hitz, W.D. (1982) *Annu. Rev. Plant Physiol.* **33**, 163–103.

Haselkorn, R. (1992). *Annu. Rev. Genet.* **26**, 113.

Haughn, G.W. & Somerville, C.R. (1986) *Mol. Gen. Genet.* **204**, 430–434.

Haughn, G.W., Smith, J., Mazur, B. & Somerville, C.F. (1988) *Mol. Gen. Genet.* **211**, 266–271.

Hecht, U., Oelmuller, R., Schmidt, S. & Mohr, H. (1988) *Planta* **175**, 130–138.

Hell, R. & Bergmann, L. (1988) *Physiol. Plant.* **72**, 70–76.

Hell, R. & Bergmann, L. (1990) *Planta* **180**, 603–612.

Hibberd, K.A. & Green, C.E. (1982) *Proc. Natl. Acad. Sci. U.S.A.* **79**, 559–563.

Hilliard, N.P., Hirasawa, M., Knaff, D.B. & Shaw, R.B. (1991) *Arch. Biochem. Biophys.* **291**, 195–199.

Hirel, B., Bouet, C., King, B., Layzell, D., Jacobs, F. & Verma, D.P.S. (1987) *EMBO J.* **6**, 1167–1171.

Hoff, T., Stummann, B.M. & Henningssen, K.W. (1991) *Physiol. Plant.* **82**, 197–204.

Hu, C.A., Delauney, A.J. & Verma, D.P.S. (1992) *Proc. Natl. Acad. Sci. U.S.A.* **89**, 9354–9358.

Huber, J.L., Huber S.C., Campbell, W.H. & Redinbaugh, M.G. (1992a) *Arch. Biochem. Biophys.* **296**, 58–65.

Huber, J.L., Huber, S.C., Campbell, W.H. & Redinbaugh, M.G. (1992b) *Plant Physiol.* **100**, 706–712.

Hughes, C.A. & Matthews, B.F. (1993) *Plant Physiol.* **102**, 82S

Hyde, G.E., Wilberding, J.A., Meyer, A.L., Campbell, E.R. & Campbell, W.H. (1989) *Plant Mol. Biol.* **13**, 233–246.

Ikegami, F., Mizuno, M., Kihara, M. & Murakoshi, I. (1990) *Phytochemistry* **29**, 3461–3465.

Ikegami, F., Ongena, G., Sakai, R., Itagaki, S., Kobori, M., Ishikawa, T., Kuo, T-H., Lambein, F. & Murakoshi, I. (1993) *Phytochemistry* **33**, 93–98.

Ireland, R.J. & Joy, K.W. (1983a) *Plant Physiol.* **72**, 1127–1129.

Ireland, R.J. & Joy, K.W. (1983b) *Arch. Biochem. Biophys.* **223**, 291–296.

Ireland, R.J. & Joy, K.W. (1985) In *Transaminations* (P. Christen & D.E. Metzler, eds), pp. 376–383. John Wiley, New York.

Jackson, S.D., Sonnewold, V. & Willmetzer, L. (1993) *Mol. Gen. Genet.* **236**, 309–311.

Jain, J.C., Shargool, P.D. & Chung, S. (1987) *Plant Sci.* **51**, 17–21.

John, M., Rohrig, H., Schmidt, J., Wienebe, U. & Schell, J. (1993) *Proc. Natl. Acad. Sci. U.S.A.* **90**, 625–629.

Joy, K.W. & Ireland, R.J. (1990) In *Methods in Plant Biochemistry* (P.J. Lea, ed.), Vol. 3, pp. 288–295. Academic Press, London.

Joy, K.W., Ireland, R.J. & Lea, P.J. (1983) *Plant Physiol.* **73**, 165–168.

Kaiser, W.M. & Brendle-Behnisch, E. (1991) *Plant Physiol.* **96**, 363–367.

Kaiser, W.M. & Spill, D. (1991) *Plant Physiol.* **96**, 368–375.

Kaneko, T., Hashimoto, T., Kumpaisal, R. & Yamada, Y. (1990) *J. Biol. Chem.* **265**, 17451–17455.

Keeler, S.J., Sanders, P., Smith, J.K. & Mazur, B.J. (1993) *Plant Physiol.* **102**, 1009–1018.

Kelland, J.G., Palcic, M.M., Pickard, M.A. & Verderas, J.C. (1985) *Biochemistry* **24**, 3263–3267.

Kendall, A.C., Wallsgrove, R.M., Hall, N.P., Turner, J.C. & Lea P.J. (1986) *Planta* **168**, 316–323.

Kern, R. & Chrispeels, M.J. (1978) *Plant Physiol.* **62**, 815–819.

Keys, A.J., Bird, I.F., Cornelius, M.J., Lea, P.J., Wallsgrove, R.J. & Miflin, B.J. (1978) *Nature* **275**, 741–743.

Kim, J. & Rees, D.C. (1992) *Science* **257**, 1677–1682.

King, B.J., Siddiqi, M.Y., Thomas, T.R., Warner, R.L. & Glass, A.D.M. (1993) *Plant Physiol.* **102**, 1279–1286.

King, G.A., Woodland, D.C., Irving, D.E. & Borst, W.M. (1990) *Physiol. Plant.* **80**, 393–400.

Kirchner, S.C. & Bryan, J.K. (1985) *Plant Physiol.* **77**, 109S.

Klapheck, S., Latus, C. & Bergmann, L. (1987) *Plant Physiol.* **131**, 123–131.

Klapheck, S., Zopes, H., Levels, H.G. & Bergmann, L. (1988) *Physiol. Plant.* **74**, 733–739.

Kleinhofs, A. & Warner, R.L. (1990) In *The Biochemistry of Plants* (B.J. Miflin & P.J. Lea, eds), Vol. 16, pp. 89–120. Academic Press, San Diego.

Kleinhofs, A. & Warner, R.L. (1992) In *Barley : Genetics Biochemistry, Molecular Biology and Biotechnology* (P.R. Stewry, ed.), pp. 209–230. C.A.B. International, Wallingford.

Kleinhofs, A., Warner, R.L., Hamat, H.B., Juricek, M., Huang, C. & Schnorr, K. (1988) *Curr. Top. Plant Biochem. Physiol.* **7**, 35–42.

Kohl, D.H., Schubert, K.R., Carter, M.B., Hagendorn, C.H. & Shearer, G. (1988) *Proc. Natl. Acad. Sci. U.S.A.* **85**, 2036–2040.

Kohl, D.H., Kennelly, E.J., Zhu, Y., Schubert, K.R. & Shearer, G. (1991) *J. Exp. Bot.* **42**, 831–837.

Kolloffel, C. & Verkerk, B.P. (1982) *Plant Physiol.* **69**, 143–145.

Krishnaswamy, S. & Bryan, J.K. (1986) *Arch. Biochem. Biophys.* **246**, 250–262.

Kromer, S. & Heldt, H.W. (1991) *Biochim. Biophys. Acta* **1057**, 42–50.

Kronenberger, J., Lapingle, A., Caboche, M. & Vaucheret, H. (1993) *Mol. Gen. Genet.* **236**, 203–208.

Kruger, R., Jager, H.J., Hintz, M. & Pahlich, E. (1986) *Plant Physiol.* **80**, 142–144.

Kumpaisal, R., Hashimoto, T. & Yamada, Y. (1987) *Plant Physiol.* **85**, 145–151.

LaRosa, P.C., Rhodes, D., Rhodes, J.C., Bressan, R. & Csonka, L.N. (1991) *Plant Physiol.* **96**, 245–250.

Larsen, P.B. & Woodson, W.T. (1991) *Plant Physiol.* **96**, 997–999.

Larsson, C.M., Mattsson, M., Duarte, P., Samuelson, M., Ohlen, E., Oscarson, P., Ingermarsson, B., Larsson, M. & Lundborg, T. (1992) In *Nitrogen Metabolism in Plants* (K. Mengel & D.J. Pilbeam, eds), pp. 71–90. Clarendon Press, Oxford.

Law, M.Y. & Halliwell, B. (1986) *Plant Sci.* **43**, 185–191.

Lea, P.J. (1991) In *Topics in Photosynthesis* (N.R. Baker, ed.), Vol. 10, pp. 267–298. Elsevier, Amsterdam.

Lea, P.J. & Fowden, L. (1975) *Proc. R. Soc. London B.* **192**, 13–26.

Lea, P.J. & Miflin, B.J. (1974) *Nature* **251**, 614–616.

Lea, P.J. & Miflin, B.J. (1980) In *The Biochemistry of Plants* (B.J. Miflin, ed.), Vol. 5. pp. 569–607. Academic Press, New York.

Lea, P.J., Blackwell, R.D. & Joy, K.W. (1992) In *Nitrogen Metabolism in Plants* (K. Mengel & D.H. Pilbeam, eds), pp. 153–186. Claredon Press, Oxford.

Lea, P.J., Festenstein, G.N., Hughes, J.S. & Miflin, B.J. (1984) *Phytochemistry* **23**, 511–516.

Lea, P.J. Fowden, L. & Miflin, B.J. (1978) *Phytochemistry* **17**, 217–222.

Lea, P.J., Robinson, S.A. & Stewart, G.R. (1990) In *The Biochemistry of Plants* (B.J. Miflin & P.J. Lea, eds), Vol. 16, pp. 121–159. Academic Press, San Diego.

Lee, J.S., Brown, G.G. & Verma, D.P.S. (1983) *Nucleic Acids Res.* **11**, 5541–5553.

Lee, K.Y., Townsend, J., Tepperman, J., Black, M., Chui, C-F., Mazur, B., Dunsmuir, P. & Bedbrook, J. (1988) *EMBO J.* **7**, 1241–1248.

Lee, R.B. & Drew, M.C. (1986) *J. Exp. Bot.* **37**, 1768–1779.

Lee, R.B., Purves, J.V., Ratcliffe, R.G. & Saker, L.R. (1992) *J. Exp. Bot.* **43**, 1385–1396.

Lillo, C. (1984) *Physiol. Plant.* **61**, 219–223.

Lough, T.J., Chang, K.S., Carne, A., Monk, B.C., Reynolds, P.H.S. & Farnden, K.J.F. (1992a) *Phytochemistry* **31**, 1519–1527.

Lough, T.J., Reddington, B.D., Grant, M.R., Hill, D.F., Reynolds, P.H.S. & Farnden, K.J.F. (1992b) *Plant Mol. Biol.* **19**, 391–399.

Ludwig, R.A. (1993) *Plant Physiol.* **101**, 429–434.

Lunn, J.E., Droux, M., Martin, J. & Douce, R. (1990) *Plant Physiol.* **94**, 1345–1352.

Macnicol, P.K. (1987) *Plant Sci.* **53**, 229–235.

Madison, J.T. & Thompson, J.F. (1976) *Biochem. Biophys. Res. Commun.* **71**, 684–691.

Manganaris, A.G. & Alston, F.H. (1988) *Theor. Appl. Genet.* **76**, 449–454.

Marquez, A.J., Avila, C., Forde, B.G. & Wallsgrove, R.M. (1988) *Plant Physiol. Biochem.* **26**, 645–651.

Martin, F., Hirel, B. & Gadal, P. (1983) *Z. Pflanzenphysiol.* **111**, 413–422.

Mathur, M., Saluja, D. & Sachar, R.C. (1991) *Biochim. Biophys. Acta* **1078**, 161–170.

Matthews, B.F., Shye, S.C.H. & Widholm, J.M. (1980) *Z. Pflanzenphysiol.* **96**, 453–463.

Matthews, B.F., Farrer, M.J. & Gray, A.C. (1989) *Plant Physiol.* **91**, 1569–1574.

Mazelis, M. & Crevelling, R.K. (1978) *J. Food Biochem.* **2**, 29–37.

Mazelis, M., Miflin, B.J. & Pratt, H.M. (1976) *FEBS Lett.* **66**, 197–200.

Mazur, B.J., Chui, C.K. & Smith, J.K. (1987) *Plant Physiol.* **85**, 1110–1117.

McCue, K.F. & Hanson, A.D. (1990) *Trends Biotechnol.* **8**, 359–362.

McGrath, R.B. & Coruzzi, G.M. (1991) *Plant J.* **1**, 275–280.

McKay, G. & Shargool, P.D. (1981) *Biochem. J.* **195**, 71–81.

McNally, S.G., Hirel, B., Gadal, P., Mann, A.F. & Stewart, G.R. (1983) *Plant Physiol.* **72**, 22–25.

Melzer, J.M., Kleinhofs, A. & Warner, R.L. (1989) *Mol. Gen. Genet.* **217**, 341–346.

Mengel, K. (1992) In *Nitrogen Metabolism in Plants* (K. Mengel & D.J. Pilbeam, eds), pp. 1–15. Clarendon Press, Oxford.

Miao, G.H., Hirel, B., Marsolier, M.C., Ridge, R.W. & Verma, D.P.S. (1991) *Plant Cell* **3**, 11–22.

Micallef, B.J. & Shelp, B.J. (1989a) *Plant Physiol.* **90**, 624–630.

Micallef, B.J. & Shelp, B.J. (1989b) *Plant Physiol.* **90**, 631–634.

Miflin, B.J. (1980) In *The Biochemistry of Plants* (B.J. Miflin, ed.), Vol. 5, pp. 569–607. Academic Press, New York.

Miflin, B.J. & Cave, P.R. (1972) *Plant Physiol.* **54**, 550–555.

Mills, W.R., Lea, P.J. & Miflin, B.J. (1980) *Plant Physiol.* **65**, 1166–1172.

Minchin, F.R. Sheehy, J.E. & Witty, J.F. (1986) *J. Exp. Bot.* **37**, 1581–1591.

Mudd, S.H. & Datko, A.H. (1986) *Plant Physiol.* **81**, 103–114.

Muhitch, M.J. & Wilson, K.G. (1983) *Z. Pflanzenphysiol.* **110**, 39–46.

Murray, A.J.S., Blackwell, R.D., Joy, K.W. & Lea, P.J. (1987) *Planta* **172**, 106–112.

Murray, D.R. (1992) *New Phytol.* **120**, 259–268.

Murray D.R. & Kennedy, I.R. (1980) *Plant Physiol.* **66**, 782–785.

Murry, M.A. & Wolk, C.P. (1989) *Arch. Microbiol.* **151**, 469–474.

Myers, P.D. & Gengenbach, B.G. (1982) *Plant Physiol.* **69**, 128S.

Nagai, A. & Scheidegger, A. (1991) *Arch. Biochem. Biophys.* **284**, 127–132.

Nagai, A., Ward, E., Beck, J., Tada, S., Chang, J.Y., Scheidegger, A. & Ryals, J. (1991) *Proc. Natl. Acad. Sci. U.S.A.* **88**, 4133–4137.

Nagai, A., Suzuki, K., Ward, K., Moyer, M., Hashimoto, M., Mano, J., Ohta, D. & Scheidegger, A. (1992) *Arch. Biochem. Biophys.* **295**, 235–239.

Nakamura, K. & Tamura, G. (1989) *Agric. Biol. Chem.* **53**, 2537–2538.

Negrutiu, I., Cattoir-Reynaerts, A., Verbruggen, I. & Jacobs, M. (1984) *Theor. Appl. Genet.* **68**, 11–20.

Negrutiu, I., DeBrouwer, D., Dirks, R. & Jacobs, M. (1985) *Mol. Gen. Genet.* **199**, 330–337.

Nock, L.P. & Mazelis, M. (1989) *Phytochemistry* **28**, 729–731.

Numazawa, T., Yamada, S., Hase, T. & Sugiyama, T. (1989) *Arch. Biochem. Biophys.* **270**, 313–319.

Oaks, A. & Long, D.M. (1992) In *Nitrogen Metabolism in Plants* (K. Mengel & D.J. Pilbeam, eds), pp. 91–102. Clarendon Press, Oxford.

Oaks, A. & Ross, D.W. (1984) *Can. J. Bot.* **65**, 68–73.

Odell, J.T., Caimi, P.J., Yadav, N.S. & Mauvais, C.J. (1990) *Physiol Plant.* **94**, 1647–1654.

Oliver, D.J., Neuberger, M., Bourguignon, J. & Douce, R. (1990) *Physiol. Plant.* **80**, 487–491.

Oliver, G., Gosset, G., Sanchez-Pescador, R., Lazoya, E., Ku, L.M., Flores, N., Becerril, B., Valle, F. & Bolivar, F. (1987) *Gene* **60**, 1–11.

Orlowski, M. & Meister, A. (1971) *J. Biol. Chem.* **246**, 7095–7105.

Ortega, J.L., Campos, F., Sanchez, F. & Lara, M. (1986) *Plant Physiol.* **80**, 1051–1054.

Ouellet, T., Rutledge, R.G. & Miki, B.L. (1992) *Plant J.* **2**, 321–330.

Parthre, V., Singh, A.K., Viswanathan, P.N. & Sane, P.V. (1987) *Phytochemistry* **26**, 2913–2917.

Pate, J.S. (1989) In *Plant Nitrogen Metabolism* (J.E. Poulson, J.T. Romeo & E.E. Conn, eds), pp. 65–115. Plenum Press, New York.

Pate, J.S., Lindblad, P. & Atkins, C.A. (1988) *Planta* **176**, 461–471.

Peleman, J., Boerjan, W., Engler, G., Seurinck, J., Botterman, J., Alliotte, T., Van Montagu, M. & Inze, D. (1989) *Plant Cell* **1**, 81–93.

Pelsy, F. & Caboche, M. (1992) *Adv. Genet.* **30**, 1–40.

Rabe, E. (1990) *J. Hort. Sci.* **65**, 231–243.

Rai, A.W. (1990) In *A Handbook of Symbiotic Cyanobacteria*. CRC Press, Boca Raton.

Rastogi, R., Back, E., Schneiderbauer, A., Bowsker, C.G., Moffatt, B. & Rothstein, S.J. (1993) *Plant J.* **4**, 317–326.

Rathinasabapathi, B. & King, J. (1991) *Plant Physiol.* **96**, 255–261.

Rauser, W.E. (1990) *Annu. Rev. Biochem.* **59**, 61–86.

Rayapati, P.J., Stewart, C.R. & Hack, E. (1989) *Plant Physiol.* **91**, 581–586.

Relton, J.M., Bonner, P.L.R., Wallsgrove, R.M. & Lea, P.J. (1988) *Biochim. Biophys. Acta* **953**, 48–60.

Rennenberg, H. (1982) *Phytochemistry* **21**, 2771–2781.

Rennenberg, H. & Lamoureux, G.L. (1990) In *Sulphur Nutrition and Sulphur Assimilation in Higher Plants* (H. Rennenberg, C.H. Brunold, L.J. DeKok & I. Stulen, eds), pp. 53–66. SBP Academic, The Hague.

Reynolds, P.H.S., Blevins, D.G., Boland, M.J., Schubert, K.R. & Randall, D.D. (1982) *Physiol. Plant.* **55**, 255–260.

Rhodes, D., Handa, S. & Bressan, R.A. (1986) *Plant Physiol.* **82**, 890–893.

Riens, B. & Heldt, H.W. (1992) *Plant Physiol.* **98**, 573–577.

Riesmeier, J., Klonus, A.K & Pohlenz, H.D. (1993) *Phytochemistry* **32**, 581–584.

Robertson, J.G., Warburton, M.P. & Farnden, K.J.F. (1975) *FEBS Lett.* **55**, 33–37.

Robinson, S.A., Stewart, G.R. & Phillips, R. (1992) *Plant Physiol.* **98**, 1190–1195.

Rognes, S.E. (1975) *Phytochemistry* **14**, 1975–1982.

Rognes, S.E. (1980) *Phytochemistry* **19**, 2287–2292.

Rognes, S.E. Lea, P.J. & Miflin, B.J. (1980) *Nature* **287**, 357–359.

Rognes, S.E., Bright, S.W.J. & Miflin, B.J. (1983) *Planta* **151**, 32–38.

Rognes, S.E., Wallsgrove, R.M., Kueh, J.S.H. & Bright, S.W.J. (1986) *Plant Sci.* **43**, 45–50.

Rolfe, B.G. (1988) *Bio Factors* **1**, 3–10.

Rolland, N., Droux, M. & Douce, R. (1992) *Plant Physiol.* **98**, 927–935.

Rolland, N., Droux, M., LeBrun, M. & Douce, R. (1993) *Arch Biochem. Biophys.* **300**, 213–222.

Romer, S., d'Harlingue, A., Camara, B., Schantz, R. & Kuntz, M. (1992) *J. Biol. Chem.* **267**, 17966–17970.

Rouze, P. & Caboche, M. (1992) In *Inducible Plant Proteins* (J.L. Wray, ed.), pp. 45–77. Cambridge University Press, Cambridge.

Ruegsegger, A. & Brunold, C. (1993) *Plant Physiol.* **101**, 561–566.

Rufty, T.W., Volk, R.J. & Glass, A.D.M. (1992) In *Nitrogen Metabolism of Plants* (K. Mengel & D.J. Pilbeam, eds), pp. 103–120. Clarendon Press, Oxford.

Rutledge, R.G., Ouellet, T., Hattori, J. & Miki, B.L. (1991) *Mol. Gen. Genet.* **229**, 31–40.

Sainis, J.K., Mayne, R.G., Wallsgrove, R.M., Lea, P.J. & Miflin, B.J. (1981) *Planta* **152**, 491–496.

Saito, K., Miura, N., Yamazaki, M., Hirano, H. & Murakoshi, I. (1992) *Proc. Natl. Acad. Sci. U.S.A.* **89**, 8078–8082.

Saito, K., Tatsuguchi, K., Murakoshi, I. & Hirano, H. (1993) *FEBS Lett.* **324**, 247–252.

Sakakibara, H., Watanabe, M., Hase, T. & Sugiyama, T. (1991) *J. Biol. Chem.* **266**, 2028–2035.

Samach, A., Hareven, D., Gutfinger, T., Ken-Dror, S. & Lifschitz, E. (1991) *Proc. Natl. Acad. Sci. U.S.A.* **88**, 2678–2682.

Sangwan, I. & O'Brian, M.R. (1991) *Science*, **251**, 1220–1222.

Schloss, J.V. (1992) In *Chemistry and Biochemistry of Flavoenzymes* (F. Muller, ed.), Vol. 3, pp. 531–542. CRC Press, Boca Raton.

Schloss, J.V., Clskanik, L.M. & Van Dyke, D.E. (1988) *Nature* **331**, 360–362.

312   PETER J. LEA

Schultze, M., Quiclet-Sire, B., Kondorosi, E., Virelizier, H., Glushka, J.N., Endre, G., Gero, S.D. & Kondorosi, A. (1992) *Proc. Natl. Acad. Sci. U.S.A.* **89**, 192–196.

Schupp, R., Schatten, T., Willenbrink, J. & Rennenberg, H. (1992) *J. Exp. Bot.* **43**, 1243–1250.

Schuster, C. & Mohr, H. (1990) *Planta* **181**, 327–334.

Scott, D.J., May, H.D., Newton, W.E., Brigle, K.E. & Dean, D.R. (1990) *Nature* **343**, 188–190.

Sebastien, S.A. & Chaleff, R.S. (1987) *Crop Sci.* **27**, 948–952.

Sellstedt, A. & Atkins, C.A. (1991) *J. Exp. Bot.* **42**, 1493–1497.

Shah, S.P.J. & Cossins, E.A. (1970) *FEBS Lett.* **7**, 267–270.

Shargool, P.D. (1973) *FEBS Lett.* **33**, 348–351.

Shargool, P.D. & Jain, J.C. (1985) *Plant Physiol.* **78**, 795–798.

Shargool, P.D., Steeves, T., Weaver, M. & Russell, M. (1978a) *Can. J. Biochem.* **56**, 273–279.

Shargool, P.D., Steeves, T., Weaver, M. & Russel, M. (1978b) *Can. J. Biochem.* **56**, 926–931.

Shargool, P.D., Jain, J.C. & McKay, G. (1988) *Phytochemistry* **27**, 1571–1574.

Shaul, O. & Galili, G. (1992a) *Plant J.* **2**, 203–209.

Shaul, O. & Galili, G. (1992b) *Plant Physiol.* **100**, 1157–1161.

Shimamoto, K. & King. P.J. (1983) *Mol. Gen. Genet.* **189**, 69–72.

Siddiqi, M.Y., Glass, A.D.M., Ruth, T.J., & Fernando, M. (1989) *Plant Physiol.* **90**, 806–813.

Siddiqi, M.Y., Glass, A.D.M., Ruth, T.J. & Rufty, T.W. (1990) *Plant Physiol.* **93**, 1426–1432.

Sieceichowicz, K.A., Joy, K.W. & Ireland, R.J. (1988) *Phytochemistry* **27**, 663–671.

Siegel, L.M. & Wilkerson, J.O. (1989) In *Molecular and Genetic Aspects of Nitrate Assimilation* (J.L. Wray & J.R. Kinghorn, eds), pp. 263–283. Oxford University Press, Oxford.

Smith, B.E. (1990) In *Nitrogen Fixation: Achievements and Objectives* (P.M. Gresshoff, E.L. Roth, G. Stacey & W.E. Newton, eds), pp. 3–13. Chapman & Hall, New York.

Smith, B.E. & Eady, R.R. (1992) *Eur. J. Biochem.* **205**, 1–15.

Smith, I.K., Kendall, A.C., Keys, A.J., Turner, J.C. & Lea, P.J. (1985) *Plant Sci.* **41**, 11–17.

Smith, I.K., Vierheller, T.L. & Thorne, C.A. (1989) *Physiol. Plant.* **77**, 449–456.

Smith, R.J. & Gallon, J.R. (1993) In *Plant Biochemistry and Molecular Biology* (P.J. Lea & R.C. Leegood, eds), pp. 129–153.

Sodek, L. (1978) *Rev. Brazil Bot.* **1**, 65–69.

Sodek, L., Lea, P.J. & Miflin, B.J. (1980) *Plant Physiol.* **65**, 22–26.

Sodek, L. & Lea, P.J. (1993) *Phytochemistry* **34**, 51–56.

Solomonson, L.P. & Barber, M.J. (1990) *Annu. Rev. Plant Physiol. Mol. Biol.* **41**, 225–253.

Somerville, C.R. (1986) *Annu. Rev. Plant Physiol.* **37**, 467–507.

Somerville, C.R. & Ogren, W.L. (1979) *Nature* **280**, 833–836.

Somerville, C.R. & Ogren, W.L. (1980) *Nature* **286**, 257–259.

Sprent, J.I. & Sprent, P. (1990) In *Nitrogen Fixing Organisms: Pure and Applied Aspects*. Chapman & Hall, London.

Staton, A.L. & Mazelis, M. (1991) *Arch. Biochem. Biophys.* **290**, 46–50.

Steffens, J.C. (1990) *Annu. Rev. Plant Physiol. Mol. Biol.* **41**, 553–575.

Stewart, C.R. (1981) In *Physiology and Biochemistry of Drought Resistance in Plants* (L.G. Paley & D. Aspinall, eds), pp. 243–259. Academic Press, Sydney.

Stewart, G.R., Joly, C.A. & Smirnoff, N. (1992) *Oecologia* **91**, 511–517.

Streeter, J.G. (1973) *Arch. Biochem. Biophys.* **157**, 613–624.

Streit, L., Martin, B.A. & Harper, J.E. (1987) *Plant Physiol.* **84**, 654–657.

Susuki, H. & Verma, D.P.S. (1991) *Plant Physiol.* **95**, 384–389.

Suzuki, A., Audet, C. & Oaks, A. (1987) *Plant Physiol.* **84**, 578–591.

Swanson, E.B., Herrgesell, M.J., Arnoldo, M., Sippell, D.W. & Wong, R.S.C. (1989) *Theor. Appl. Genet*, **78**, 525–530.

Szamosi, I., Shaner, D. & Singh, B.K. (1992) In *Biosynthesis and Molecular Regulation of Amino Acids in Plants* (B.K. Singh, H.E. Flores & J.C. Shannon, eds), pp. 333–335. American Society of Plant Physiologists, Maryland.

Szoke, A., Miao, G.H., Hong, Z. & Verma, D.P.S. (1992) *Plant Physiol.* **99**, 1642–1649.

Ta, T.C. & Joy, K.W. (1985) *Can. J. Bot.* **63**, 881–884.

Ta, T.C. & Joy, K.W. (1986) *Planta* **169**, 117–122.

Ta, T.C., Joy, K.W. & Ireland, R.J. (1985) *Plant Physiol.* **78**, 334–337.

Taniguchi, M. & Sugiyama, T. (1990) *Arch. Biochem. Biophys.* **282**, 427–432.

Taniguchi, M., Sawaki, H., Sasakawa, H., Hase, T. & Sugiyama, T. (1992) *Eur. J. Biochem.* **204**, 611–620.

Taylor, A.A. & Stewart, G.R. (1981) *Biochem. Biophys. Res. Commun.* **101**, 1281–1289.

Thoen, A., Rognes, S.E. & Aarnes, H. (1978a) *Plant Sci. Lett.* **13**, 103–112.

Thoen, A., Rognes, S.E. & Aarnes, H. (1978b) *Plant Sci. Lett.* **13**, 113–119.

Thompson, G.A., Datko, A.H. & Mudd, S.H. (1982a) *Plant Physiol.* **70**, 1347–1352.

Thompson, G.A., Datko, A.H., Mudd, S.H. & Giovanelli, J. (1982b) *Plant Physiol.* **69**, 1077–1083.

Thompson, J.F. (1980) In *The Biochemistry of Plants* (B.J. Miflin, ed.), Vol. 5, pp. 375–402. Academic Press, New York.

Thornley, R.N. & Lowe, D.J. (1985) In *Molybdenum Enzymes* (T.G. Spiro, ed.), pp. 221–284. Wiley, New York.

Thummler, F. & Verma, D.P.S. (1987) *J. Biol. Chem.* **262**, 14730–14736.

Tiburcio, A.F., Kaur-Sawhney, R. & Galston, A.W. (1990) In *The Biochemistry of Plants* (B.J. Miflin & P.J. Lea, eds), Vol. 16, pp. 283–325. Academic Press, San Diego.

Tingey, S.V., Tsai, F.Y., Edwards, J.W., Walker, E.L. & Coruzzi, G.M. (1988) *J. Biol. Chem.* **263**, 9651–9657.

Tingey, S.V., Walker, E.L. & Coruzzi, G.M. (1987) *EMBO J.* **6**, 1–9.

Tonin, G.S. & Sodek, L. (1990) *Phytochemistry* **29**, 2829.

Tourneur, C., Jouanin, L. & Vaucheret, H. (1993) *Plant Sci.* **88**, 159–168.

Tsai, F-Y. & Coruzzi, G.M. (1990) *EMBO J.* **9**, 323–332.

Turano, F.J., Jordan, R.L. & Matthews, B.F. (1990) *Plant Physiol.* **92**, 395–400.

Turano, F.J., Weisemann, J.M. & Matthews, B.F. (1992) *Plant Physiol.* **100**, 374–381.

Tyagi, V.V.S., Henke, R.R. & Farkas, W.R. (1982) *Biochim. Biophys. Acta* **719**, 363–369.

Tyagi, V.V.S., Henke, R.R. & Farkas, W.R. (1983) *Plant Physiol.* **73**, 687–691.

Ullrick, W.R. (1992) In *Nitrogen Metabolism in Plants* (K. Mengel & D.J. Pilbeam, eds), pp. 120–138. Clarendon Press, Oxford.

Vance, C.P. & Gantt, J.S. (1992) *Physiol. Plant.* **85**, 266–276.

Van de Weil, C., Scheres, B., Franssen, H., Van Lierop, M-J., Van Lammeren, A., Van Kammen, A. & Bisseling, T. (1990) *EMBO J.* **9**, 1–7.

Vaucheret, H., Kronenberger, J., Lepingle. A., Vilaine, F., Boutin, J-P. & Caboche, M. (1992) *Plant J.* **2**, 559–562.

Verma, D.P.S. & Miao, G-H. (1992) In *Inducible Plant Proteins* (J.L. Wray, ed.), pp. 175–204. Cambridge University Press, Cambridge.

Vincentz, M., Moureaux, T., Leydecker, M-T., Vaucheret, H. & Caboche, M. (1993) *Plant J.* **3**, 315–324.

von Berg, K-H, L., Wosswinkel, R. & Bothe, H. (1992) In *Nitrogen Metabolism in Plants* (K. Mengel & D.J. Pilbeam, eds), pp. 39–54. Clarendon Press, Oxford.

Walker, E.L. & Coruzzi, G.M. (1989) *Plant Physiol.* **91**, 702–708.

Wallsgrove, R.M. (1990) In *Methods in Plant Biochemistry* (P.J. Lea, ed.), Vol. 3, pp. 325–333. Academic Press, London.

Wallsgrove, R.M. & Mazelis, M. (1980) *FEBS Lett.* **116**, 189–192.

Wallsgrove, R.M. & Mazelis, M. (1981) *Phytochemistry* **20**, 2651–2655.

Wallsgrove, R.M., Lea, P.J. & Miflin, B.J. (1982) *Planta* **154**, 473–476.

Wallsgrove, R.M., Lea, P.J. & Miflin, B.J. (1983) *Plant Physiol.* **71**, 780–784.

Wallsgrove, R.M., Risiott, R., King, J. & Bright, S.W.J. (1986) *Plant Sci.* **43**, 109–114.

Wallsgrove, R.M., Turner, J.C., Hall, N.P., Kendall, A.C. & Bright, S.W.J. (1987) *Plant Physiol.* **83**, 155–158.

Walsby, A.E. (1985) *Proc. R. Soc. Lond. B.* **226**, 345–366.

Walter, T.J., Connelly, J.A., Gengenbach, B.J. & Wold, F. (1979) *J. Biol. Chem.* **254**, 1349–1355.

Walton, N.J. & Woolhouse, H.W. (1983). *Planta* **158**, 469–471.

Weisemann, J.M. & Matthews, B.F. (1993) *Plant Mol. Biol.* **22**, 301–312.

Wenko, L.K., Treick, R.W. & Wilson, K.G. (1985) *Plant Mol. Biol.* **4**, 197–204.

White, P.J. (1983) *J. Gen. Microbiol.* **129**, 739–749.

Williamson, C.L. & Slocum, R.D. (1992) *Plant Physiol.* **100**, 1464–1470.

Wilson, B.J., Gray, A.C. & Matthews, B.F. (1991) *Plant Physiol.* **97**, 1323–1328.

Wittenbach, V.A., Rayner, D.R. & Schloss, J.V. (1992) In *Biosynthesis and Molecular Regulation of Amino Acids in Plants* (B.K. Singh, H.E. Flores & J.C. Shannon, eds), pp. 69–88. American Society of Plant Physiology, Maryland.

Wolk, C.P. (1982) In *The Biology of Cyanobacteria* (N.G. Carr & B.A. Whitton, eds), pp. 359–386. Blackwell, Oxford.

Won, T. & Mazelis, M. (1989) *Physiol. Plant.* **77**, 87–92.

Wong, Y.S. & Mazelis, M. (1981) *Phytochemistry* **20**, 1831–1834.

Wray, J.L. & Kinghorn, J.R. (1989) *Molecular and Genetic Aspects of Nitrate Assimilation.* Oxford University Press, Oxford.

Zehnacker, C., Becker, T.W., Suzuki, A., Carrayol, E., Caboche, M. & Hiel, B. (1992) *Planta* **187**, 266–274.

Zielke, H.R. & Filner, P. (1971) *J. Biol. Chem.* **246**, 1772–1779.

8 Nucleic Acids and Proteins

Eric Lam

8.1 INTRODUCTION

Nucleic acids and proteins are two major constituents of all living cells with critical roles. Nucleic acids contain the information necessary for the synthesis of proteins with specific structures and functions. On the other hand, proteins are needed for the metabolism of nucleic acids in addition to performing practically all the enzymic reactions in a living cell. With the advent of molecular techniques in every discipline of biology, our understanding of the chemical make-up and reactions within a living cell has undergone a major revolution in the last ten years. Although most of the fundamental work in the study of nucleic acids and proteins has been carried out in bacterial and animal model systems, many of the general principles appear to function in plants as well. In this chapter, our current state of knowledge on the properties of nucleic acids and proteins will be summarized, with special emphasis on some of the plant-specific aspects. For more detailed descriptions of some of the general properties of nucleic acids and proteins, the reader should consult Stumpf & Conn (1981, 1989), Grierson & Covey (1988), and Alberts *et al.* (1989).

8.1.1 Nucleic acids

There are two types of nucleic acid in living organisms: deoxyribonucleic acid (DNA) and ribonucleic acid (RNA). DNA and RNA are long polymers of subunits called nucleotides that are composed of a 5-carbon sugar backbone with one or more phosphate groups attached at the C-5 position. In addition, nitrogen-containing bases called purines and pyrimidines are also attached to the C-1 position of the sugar. Nucleotides for DNA contain a hydrogen atom at the C-2 position whereas those for RNA contain a hydroxyl group instead. Polymers of deoxyribose or ribose nucleotides are linked by a phosphodiester bond between the C-3 hydroxyl group of one nucleotide and the phosphate group attached to the C-5 position of the adjacent nucleotide. Due to the stereospecificity of this arrangement, nucleic acids are directional and are commonly diagrammatically displayed left to right from the 5'-phosphate containing end to the 3'-hydroxyl containing end (Fig. 8.1). For DNA, two different purines (adenine (A), guanine (G)) and two different pyrimidines (cytosine (C), thymine (T)) are found naturally. For RNA, the same bases are used with the exception that thymine is replaced by another pyrimidine uracil (U). The specific order of these bases on the sugar-phosphate backbone of DNA and RNA is the basis for their role as information storage devices. From these specific sequences, proteins with distinct structure and function are encoded and synthesized. In addition, regulatory information such as the precise expression pattern of the encoded protein is contained within the specific arrangement of these different nucleotides of DNA and RNA. For simplicity, a **gene** can be defined as a region of the DNA that contains the necessary information for the synthesis of one or more RNA transcripts.

An important property of nucleic acids is their ability to interact with each other through hydrogen bonding between the nitrogen and

A

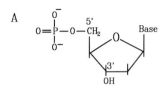

B

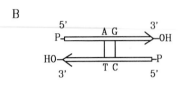

Figure 8.1 General structure of nucleic acids. (a) Chemical structure of the sugar-phosphate backbone. For simplicity, hydrogen atoms of the sugar moiety are not shown. The positions of the 5′ and 3′ carbon atoms are indicated. Four possible nitrogen bases can be attached to the C_1 carbon atom. (b) Some essential features of a double-stranded nucleic acid are depicted. The anti-parallel arrangement is emphasized. Base-pairing between the two strands is suggested by the hydrogen bonds (vertical lines) shown between two pairs of nucleotides.

oxygen atoms that are present in the keto and amino groups of the bases. Due to the structure of the different bases, interaction between them is highly specific. Thus, guanine can only interact with cytosine whereas adenine can interact with either thymine or uracil. These interactions, called **base-pairing**, are largely responsible for the secondary structures of DNA and RNA. Since three hydrogen bonds are formed between G and C as compared to two hydrogen bonds between A and T or U, G : C pairing is more stable. In nature, DNA is synthesized and maintained in a double-stranded form called the **double-helix**. This consists of two strands of DNA held together in an antiparallel fashion through base-pairing (see Fig. 8.1). In addition to hydrogen bonding between the bases, the double-helix structure is also stabilized by hydrophobic interactions between the planar aromatic surfaces of the bases that became stacked on each other in the center of the helix. Since interaction between bases is highly specific, the sequence of nucleotides from one strand of DNA will allow us to deduce the sequence of its complementary partner with great certainty. Due to the interaction between the bases, double-stranded nucleic acids have lower absorbance in ultraviolet light than the single-stranded forms. This behavior, called **hyperchromicity**, allows quantitative and kinetic measurement of the equilibrium between the two forms of nucleic acid. When a double-stranded nucleic acid in

aqueous solution is heated to 100°C or when the solution pH is increased to 13 or above, hydrogen bonds between the bases are disrupted and the two strands separate from each other. This process is called **denaturation**. Once denatured, the single-stranded nucleic acids can re-establish hydrogen bonds with other complementary sequences provided that the right conditions, such as temperature and pH, are available. This process, which essentially involves the formation of a double-stranded nucleic acid from the single stranded forms (e.g. DNA : DNA, DNA : RNA, or RNA : RNA) is called **hybridization** and is a simple but powerful tool in molecular biology. The specificity of base-pairing allows the identification and quantitation of a small amount of specific RNA or DNA sequences within highly heterogeneous mixtures. Techniques based on sequence-specific hybridization of nucleic acids, such as genome mapping and polymerase chain reaction (PCR), offer exciting possibilities of understanding and manipulating complex genetic materials. In the first case, the relative locations of different regions of the chromosomes in an organism may be determined (Little, 1992). In the latter, minute quantities of nucleic acids can be amplified millions of times through the action of an enzyme from a thermophilic bacteria called Taq DNA polymerase (Saiki *et al.*, 1988).

8.1.2 Proteins

Enzymatic reactions within living cells are typically mediated by proteins. The building blocks for proteins are amino acids that contain an identical backbone structure (Fig. 8.2) with one out of 20 possible side chains (R group) attached to it. Since their backbone structure is identical, the R group of amino acids distinguishes one amino acid from another. Each amino acid in a protein is connected to its neighbors via an amide linkage called a peptide bond, which involves the amino group of one residue and the carboxyl group of another. The primary structure of a protein is usually displayed left to right as the sequence of is constituent amino acids from the amino to the carboxyl termini. Like the purine and pyrimidine bases of nucleic acids, which are involved in specificity of base-pairing and other essential properties of DNA and RNA, the nature of the R-group in each of the 20 amino acids contributes to the functional and structural properties of proteins. However, instead of only four different building blocks as in the case of nucleic acids, there are 20 different possible side

A.

$$^-OOC - \overset{\overset{\displaystyle H}{|}}{\underset{\underset{\displaystyle NH_3^+}{|}}{C}} - R$$

B.

$$H_2N - \overset{\overset{\displaystyle H}{|}}{\underset{\underset{\displaystyle R1}{|}}{C}} - \overset{\overset{\displaystyle O}{||}}{C} - \underset{\underset{\displaystyle H}{|}}{N} - \overset{\overset{\displaystyle H}{|}}{\underset{\underset{\displaystyle R2}{|}}{C}} - COOH$$

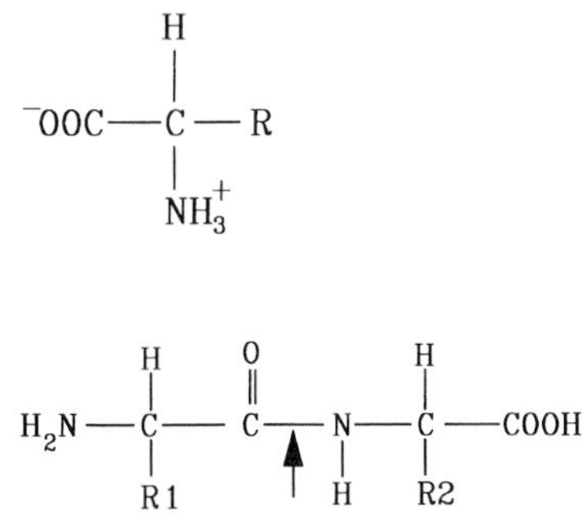

Figure 8.2 General structure of amino acids. (a) Backbone structure of amino acids. The identity of the R group varies between different amino acids. (b) The amide linkage (i.e. peptide bond) between two amino acids is indicated by an arrow.

chains for each amino acid residue in a protein. This fact gives rise to the tremendous variety of proteins that are found in nature. The R-groups of amino acids can be classified into three different categories under physiological conditions (Table 8.1). The charged amino acid side chains can be acidic, as in aspartic acid and glutamic acid, or basic, as in lysine and arginine. The other amino acids are distributed between the polar/uncharged and the hydrophobic groups. In general, hydrophobic amino acids of a polypeptide are sequestered within the core of the folded protein whereas the charged side chains are distributed on the surface of the protein. This serves to maximize the contact of charged groups with water molecules and avoid exposure of hydrophobic side chains to an aqueous environment. In addition, hydrogen bonds are also formed between polar side chains and with the peptide bonds in the backbone of the protein as well. In combination, these forces thus restrict the stable conformation of a particular amino acid sequence to a specific structure where

Table 8.1 Amino acids and their classification

Hydrophobic (Non-polar)	Polar: uncharged
Alanine (Ala)	Asparagine (Asn)
Isoleucine (Ile)	Cysteine (Cys)
Leucine (Leu)	Glutamine (Gln)
Methionine (Met)	Glycine (Gly)
Phenylalanine (Phe)	Serine (Ser)
Proline (Pro)	Threonine (Thr)
Tryptophan (Trp)	Tyrosine (Tyr)
Valine (Val)	
Polar: positively charged	Polar: negatively charged
Arginine (Arg)	Aspartic Acid (Asp)
Histidine (His)	Glutamic Acid (Glu)
Lysine (Lys)	

these interactions can be maximized. By interfering with the hydrogen bonding among the polar side chains and the hydrophobic interactions between non-polar amino acid residues, solvents such as phenol and detergents can unfold, or **denature**, proteins. Chemicals such as urea and guanidine hydrochloride are also mild denaturants that can be used to unfold proteins effectively. In addition, by removal of these highly water-soluble agents through dialysis, one can often renature the unfolded protein to its functional state.

In addition to the non-covalent hydrogen bonds and hydrophobic interactions among side chains, the secondary structure of a protein can also be stabilized by the formation of disulfide bonds between -SH groups of cysteines either in the same polypeptide chain or among different polypeptides of a multi-subunit protein. The formation of disulfide bonds is favored under oxidizing conditions and is disrupted by reducing agents, such as dithiothreitol or β-mercaptoethanol. Since the reducing conditions of the cytosol in eukaryotes are unfavorable to disulfide bond formation, these bonds are usually found in secreted or cell-surface proteins.

Two well-characterized folding patterns that contribute to the secondary structures of proteins are the α-helix and the β-pleated sheet. α-helices are rod-like structures in which the peptide bonds are regularly hydrogen-bonded to those of other neighboring amino acids. There are about 3.6 amino acids per turn in an α-helix. In this conformation, the R-groups are pointing outward from the axis of the helix and their interaction with those of other polypeptides or lipids helps to stabilize the helical structure. Amino acids such as proline, which cannot participate in hydrogen bonding because of its chemical properties, are rarely found in the middle of an α-helix. Aside from proline, amino acids with bulky side chains, such as tryptophan, and highly charged side chains, such as aspartate, are also unfavorable to α-helix formation.

The most common form of β-pleated sheet is the antiparallel form in which the polypeptide chain is folded back and forth upon itself. In this configuration, each section of the polypeptide will be flanked by two neighboring sections that run in the opposite direction. Hydrogen bonding between the peptide bonds of amino acids in neighboring sections holds the structure together. Proline, with its imino group, is an amino acid that is often found in the turns between adjacent peptide chains in β sheets. Parallel β sheets are formed when sections of the polypeptide are arranged so that they run in the same direction.

Together, α-helices and β-pleated sheets make up the fundamental structural units of most proteins. The determination of these structures has traditionally been achieved through X-ray crystallography. Recently, however, newer techniques of protein structure determination, such as multi-dimensional nuclear magnetic resonance (NMR) spectroscopy, are increasing the rate that complex structures are resolved.

8.1.3 Nucleic acid and protein relationships

A 'Central Dogma' in molecular biology is that RNA is **transcribed** from DNA and proteins in turn are **translated** from RNA. Although this is true in most instances, several important exceptions to this rule are known. The identification of reverse transcriptase is an intellectual as well as a technical milestone in molecular biology. This class of enzymes, found in both animal retroviruses and plant DNA viruses, can carry out the synthesis of DNA by using RNA as a template. Typically, this enzyme is encoded in the genome of the virus. The discovery of reverse transcriptases enabled the construction of complementary DNA (cDNA) libraries from RNA sources and is crucial to the rapid development of techniques for the cloning and analysis of DNA sequences that encode expressed genes. This is an important approach since a large percentage of the eukaryotic genomes contains DNA that is apparently not transcribed.

In addition to the synthesis of DNA from an RNA template, the linear sequence of information transfer between DNA, RNA and protein that is implicit in the Central Dogma has additional exceptions. A well-characterized phenomenon is the existence of sequences in the primary RNA transcript of many genes that are 'spliced' out in the mature form. The excised parts of the primary transcript are called **introns** whereas the parts that constitute the mature RNA product are called **exons**. The precise positions where splicing occurs in a transcript can be variable which may lead to different RNA species from a single gene. Another dramatic example of non-linear flow of information between DNA and RNA is the recent discovery of **RNA-editing**. In this process, specific nucleotides are deleted, inserted or substituted in the primary transcript. In many cases, these nucleotide changes alter substantially the amino acid sequence in the translated protein. This process was first observed in trypanosome kinetoplast mRNAs. Subsequently, evidence for its occurrence has been reported in mammalian cells, plant mitochondria and maize chloroplasts. Thus, it appears to be a conserved mechanism by which some primary mRNAs in eukaryotes are modified in its sequence specifically to produce a different protein when translated.

8.2 DNA

8.2.1 Content and diversity of DNA

The genome size for different organisms varies over a wide range (Table 8.2). Qualitatively, more complex organisms such as higher plants and mammals contain more DNA than simpler eukaryotes such as yeast and other fungi. In turn, the size of the yeast genome is about five times that of the bacteria *Escherichia coli*. Thus, difference in genome size roughly correlates with the apparent complexity of these divergent organisms. Among the higher eukaryotes, however, the reason for the large genomes in some species such as lily and wheat is unclear at present. In these plants, as well as in newts, the amount of nuclear DNA can be as much as ten times of that in humans. Since it is unlikely that there is less diversity in genes expressed in humans than in these other organisms, the quantity of nuclear DNA thus does not necessarily reflect the phenotypic complexity of an organism *per se*. One property of eukaryotic genomes is the presence of a large proportion of non-coding DNA. When denatured nuclear DNA is allowed to reassociate, at least two kinetically distinct populations are evident. The

Table 8.2   Genome size comparison between organisms

Species	Haploid genome size (kbp)
Escherichia coli	4.2×10^3
Yeast (*Saccharomyces cerevisiae*)	1.5×10^4
Nematode worm (*C. elegans*)	8.0×10^4
Fruit fly (*Drosophila melanogaster*)	1.5×10^5
Human (*Homo sapiens*)	2.0×10^6
Newts	5.0×10^7
Plants	
Arabidopsis thaliana	8.0×10^4
Cotton	7.2×10^5
Soybean	1.3×10^6
Tobacco	2.4×10^6
Pea	4.8×10^6
Wheat	5.2×10^6
Maize	5.7×10^6

fast-renaturing portion is composed of highly repetitive sequences of 300–1000 base pair (bp) in size that are represented many times in the genome. Unique sequences that represent functional genes are in the slow-renaturing portion of the DNA. One type of repeat sequences that has been well-characterized in different plants is called **satellite DNA**. These DNA sequences are isolated based on their different apparent density from the rest of the nuclear DNA during centrifugation. They are usually found as a minor band of DNA of slightly lower density than the bulk of the genomic DNA. Their amount in different plants is variable and their repeat unit sizes range from a few to several hundred base pairs. The origin of satellite DNA is still unclear. In general, plants that have larger genomes tend to have also a larger portion of repetitive sequences in their DNA. For example, in pea, which has a large genome with about 4.8×10^9 bp per haploid nucleus, about 85% of its DNA is repeated sequences (Murray et al., 1978). In contrast, there is only about 10% fast-renaturing DNA in the genome of Arabidopsis thaliana. In this plant, a 180 bp sequence has been identified to repeat between 4000 and 6000 times within the genome (Martinez-Zapater et al., 1986; Simoens et al., 1988). In addition, another 500 bp sequence is also found to be represented about 500 times per haploid nucleus. Like all eukaryotes, the telomeres at the ends of each of the five chromosomes in Arabidopsis are composed of a 7 bp sequence (CCCTAAA) repeated about 350 times. Together, these three types of characterized repeats account for about 20% of the repetitive sequences, or 2% of the total genomic DNA, that are present in this small plant. Aside from the telomeric repeats, which are likely to be involved in the maintenance of chromosomes, the function of the majority of the repetitive sequences present in the nuclear genome is largely unknown. In plants, however, species with a smaller size and a shorter life cycle tends to have smaller genome size and less repetitive sequences in their genome.

In addition to nuclear DNA, plants have at least two other centers of DNA metabolism which reside in the mitochondria and the plastids. In each of these compartments of the plant cell, DNA replication and protein synthesis are carried out by organelle-encoded as well as nuclear-encoded gene products. Although many of the basic properties of organellar DNAs and their metabolism are likely to be similar, if not identical to their counterpart in the nucleus, important differences also exist. The percentage of the plastid genome in the total DNA of a given cell is variable among different organs of a plant since the number of plastids is dependent on developmental as well as environmental cues. It can range from about 10 to 20% of the total DNA in a leaf cell whereas in roots, it constitutes only about 1% of the DNA. In each plastid, 10–200 copies of a double-stranded DNA exist as circular molecules with sizes ranging from 70 000 to about 180 000 base-pairs in different species. Plastid DNAs of several plant species have been completely sequenced. Potential and known protein-coding regions in these plastid genomes have been identified (Shimada & Sugiura, 1991). The plastid genome has thus been shown to encode 120–150 genes. These include many of the proteins involved in photosynthesis and housekeeping functions such as protein synthesis. These and other studies have also shown that plastid DNA of most plant species studied to date contains an inverted repeat region in which a significant portion of the genome is duplicated on the opposite sides of the circular DNA. An interesting exception is that of the pea plastid DNA where no evidence for an inverted repeat region is found. The size of this inverted repeat region constitutes about 10–20% of the total plastid genome and the number and identity of the genes located within it is somewhat variable between different species. Many of the ribosomal RNA and tRNA genes are located within this inverted repeat region. An interesting mechanism, called **copy correction**, apparently exists in the plastids to insure that the sequence on the inverted repeats are identical. Thus, if advantageous mutations and insertions are introduced to one of the repeats during plastid transformation procedures, the alterations will be duplicated onto the other repeat unit (Svab et al., 1990). This process may provide a buffer against mutations in critical genes for plastid biogenesis. At present, the mechanism for this interesting phenomenon is unknown. Mitochondrial DNA of higher plants has a rather complex mode of maintenance. The number as well as the size of its genome is variable in a given species and is in striking contrast to that found in yeast and mammalian cells where a well-defined circular DNA molecule is found. In the case of human mitochondria, the nucleotide sequence of all 16 569 bp has been determined (Anderson et al., 1981). Typically, mitochondrial DNA of different plant species contains about 200 000 to 2 million base pairs. In maize, seven different circular forms from 570 000 to 47 000 bp in size have been described. An important genetic trait, called cytoplasmic male sterility, has been shown to be caused by mutations in genes encoded in the mitochondrial genome of some plant species

(Levings, 1983). Unlike higher plants, the mito-chondrial genome of the liverwort *Marchantia polymorpha* is contained within a single circular molecule of 186 608 bp. The complete DNA sequence of this genome has been determined and genes encoding ribosomal RNAs, tRNAs and proteins involved in oxidative phosphorylation were identified (Oda *et al.*, 1992).

8.2.2 Cellular dynamics of DNA structure and organization

The nucleus of a eukaryotic cell is a dense struc-ture in which the genomic DNA is packaged in a complex nucleoprotein structure called **chroma-tin**. The structure of chromatin is not static and can be dramatically different in the life of a cell. In recent years, biochemical and molecular stud-ies have clearly shown that the organization of chromatin as well as DNA structure can have profound effects on gene expression (Grunstein, 1992). Thus, the understanding of DNA archi-tecture and chromatin organization is important from a functional as well as a structural point of view.

The double-helix of DNA can be envisioned as two tightly linked, antiparallel ribbons that are wound around an imaginary rod. One important consequence of this structure is that it is stereo-specific. Thus, one can wind these linked ribbons in two different directions and produce molecules that are mirror images of each other. Due to the geometry of the sugar-phosphate backbone, most double helical DNA in nature exists in the 'right-handed' form called **B-DNA**. There are about 10.5 base-pairs in each turn of the double helix in this conformation. Short stretches of DNA that con-tain alternating purine/pyrimidine sequences have been found to adopt the 'left-handed' form of the double helix called **Z-DNA**. Evidence for the presence of Z-DNA *in vivo* have been obtained in animal and plant systems (Wittig *et al.*, 1991; Ferl & Paul, 1992). However, the function of these Z-DNA structures remains obscure. In addition to double helical structures, new structures of DNA such as DNA triplex and quadruplex have been shown to be possible *in vitro* (Griffin & Dervan, 1989; Kang *et al.*, 1992). Recently, the structure of the telomeric repeats from the ciliated protozoan *Oxytricha* have been determined by X-ray crystal-lography (Kang *et al.*, 1992) and NMR spectro-scopy (Smith & Feigon, 1992). Both studies show that these G-rich repeats exist as a DNA quadru-plex. These examples serve to illustrate the diverse possibilities that DNA structure can assume.

The eukaryotic nucleus is a complex organelle that has a dynamic structure. Typically, it contains an envelope of two lipid bilayers with numerous pore structures involved in macromolecular trans-port in and out of the nucleus. Adjacent to the inside of this nuclear envelope is a matrix of proteins called **lamins**. Nuclear DNA packaged with basic proteins called **histones** are thought to be attached to this nuclear matrix. Evidence has been obtained that shows that at least one kind of lamin proteins, called lamin B, can interact specifically with certain DNA sequences that are thought to be involved in chromatin organization (Luderus *et al.*, 1992). There are also structural proteins such as tubulins and intermediate fila-ments within the nucleus. Their precise architec-ture and dynamics are complex and variable during the cell's life cycle. Another structure within the nucleus is the nucleolus. This is the site of most of the ribosomal RNA synthesis within the cell.

Nuclear DNA is normally packaged in protein complexes with highly basic proteins called histones. There are five distinct groups of histone proteins called H1, H2A, H2B, H3 and H4. The structure of these proteins is very well conserved among eukaryotes, especially in the case of histones H3 and H4 where their primary sequences differ very little between cows and peas. The sequence of the other three histones are more variable. Nucleosomes consist of a core of two molecules each of histones H2A, H2B, H3 and H4. About 150 base-pairs of double-stranded DNA are wound twice around this octameric protein and one molecule of histone H1 is bound to this nucleoprotein complex. This stage of DNA packaging produces fibers of about 10 nm in diameter and shortens the apparent length of the DNA about seven times. The next step of chromatin organization is the formation of a solenoid structure from the nucleosome arrays. This step is favored under conditions of higher ionic strengths *in vitro*, typically with Na^+ and Mg^{2+}, and involves the formation of a helical structure with six nucleosomes per turn. This process further condenses the DNA into fibers of about 30 nm in width. These 30 nm fibers of DNA can be organized into higher order structures that are less well characterized. Histone H1 phos-phorylation is likely to modulate at least part of this condensation process. The Packaged DNA are also tightly associated with an ill-defined network of proteinaceous material called the **nuclear scaffold**. Specific sequences in nuclear DNA, called scaffold attachment regions (SARs), are found to be tightly associated with this structure even after exhaustive extractions with salts to remove histones. SARs have been characterized from various eukaryotes including fruit flies and tobacco (Gasser & Laemmli, 1987; Hall *et al.*, 1991) and are rather A/T rich in their sequences. These DNA elements of about 300–1000 bp in length are thought to anchor different regions of the chromatin onto the nuclear scaffold with the intervening DNA organized into nucleosomes by interaction with histones. These intervening DNAs can be visualized by electron microscopy as loops of DNA that emanate from the nuclear scaffold and are commonly described as a **halo**. It is believed that the association of the SARs with the nuclear scaffold may have important functional consequences on gene expression (Gasser & Laemmli, 1987).

The fact that DNA molecules in nature are predominantly double-stranded and are either circular, as in the case of organellar DNAs, or anchored to nuclear proteins create topological problems for the process of replication and transcription. During these critical processes, multimeric protein complexes are thought to unravel the two strands while sliding along the double helix structure. This action compresses the DNA in front of the enzyme complexes and introduces torsional stress known as positive **supercoiling** while negative supercoiling is introduced to the DNA behind. Another topological problem arises during DNA replication when the double-stranded daughter molecules are required to separate from each other. An important class of proteins called **topoisomerases** are known to be involved in the relief of torsional stress in nuclear and organellar DNA. These enzymes catalyze transient breakage in the DNA to allow the two strands to rotate freely with respect to each other before rejoining them. They are critical to the survival of eukaryotic and prokaryotic cells, as demonstrated by mutants with conditional defects in genes that encode these enzymes (Goto & Wang, 1985; Wang, 1991). Topoisomerases have been characterized in plants and they share similarities with their counterparts in other systems (Lam & Chua, 1987; Kieber *et al.*, 1992). Since alterations in DNA topology have been shown to affect critical processes such as transcription and replication *in vitro* (Wang, 1991), topoisomerases may play a regulatory role in modulating these functions. The phosphorylation of yeast topoisomerase II has recently been shown to increase its enzyme activity *in vitro* and is correlated with the onset of mitosis in the cell cycle (Cardenas *et al.*, 1992).

Nuclear DNA of various organisms have also been shown to be methylated by enzymes called **methylases** present in eukaryotes. As much as 30% of the nuclear DNA in pea is methylated. The most common form of methylation is the addition of a methyl group at the 5-carbon position of the cytosine base. Usually, this takes place in the cytosine of a CpG site where p stands for the location of the phosphodiester bond. Because of the symmetry of this dinucleotide pair, the cytosine base of the opposite strand is also modified. It is important to note that not all CpG sites of the genome are methylated. In fact, methylated DNA is usually thought to be associated with regions of the genome that are inactive in gene expression. Many proteins involved in binding to DNA are known to be sensitive to methylation of their recognition sequences. Thus, differential modification of DNA can have profound effects on the expression of the information carried by the particular DNA sequence. In fact, treatment of cells with the cytosine analogue 5-azacytosine, which cannot be methylated, can cause dramatic phenotypic changes. At present, the regulation of DNA methylase activities remains to be elucidated.

Although the exact DNA sequence and the relative position of genes in a particular organism remains fairly stable, an important mechanism called **recombination** is used to maintain certain levels of genetic variation. This process insures that there is sufficient diversity in the genome to increase the chance of species survival in a changing environment. Two types of recombination events are commonly known: general genetic recombination and site-specific recombination. In general recombination, homologous DNA strands from two copies of the same chromosome can be exchanged. This process can be initiated by a broken phosphodiester bond on one of the strands of the two sister chromosomes and usually involves base-pairing between the strands of the two neighboring DNA. An important example of this process is the exchange of sections of chromosomes during the course of meiosis. This event, called **crossing-over**, involves the exchange of genetic information between different parts of the closely apposed chromosomes during the early development of gametes and allows for greater diversity of the resultant progenies. In contrast to general recombination, site-specific recombination involves the movement of genetic material in the genome without the requirement of extensive base-pairing. This event is catalyzed by a class of enzyme called **recombinases** and requires only a relatively small, but specific DNA sequence.

Recombination will occur between these specific sites in the presence of the particular recombinase and regions of the chromosome can either be deleted or inserted. By this mechanism, genetic information can be shuffled between and within chromosomes.

Genomic DNA can also be modified by an interesting class of DNA sequences called **transposable** elements. Originally predicted in the elegant maize cytogenetic work carried out by Barbara McClintock in the 1940s, these are DNA pieces that have the capacity to insert into various locations in the genome, either on their own or when activated by other insertion elements that are present. Since these DNA pieces can be excised from the site of insertion at a later time and then re-insert elsewhere in the genome, they are frequently referred to as 'jumping' genes, or **transposons**. Although they were originally discovered in plants, elements that behaved similarly have been characterized in both eukaryotes and prokaryotes. Many transposable elements have now been isolated from these organisms and they serve as valuable tools for the genetic analyses of complex organisms. When these elements insert into the control regions of a functional gene, they often inactivate its expression. Thus, transposable elements can be used as tags for insertional mutagenesis and identify genes that are responsible for complex functions. Recent studies with the maize transposable element *activator (Ac)* suggest that it can function in very different plant species and as such, it will be a valuable tool in the analysis of complex genetic traits in higher plants (Dooner *et al.*, 1991). Transposable elements frequently create short direct repeats at the site of insertion and their movement in the genome may be responsible for the evolution of repetitive DNA. In addition, since transposable elements frequently alter the insertion site sequence even after its excision, they may be a natural generator of random mutations in the genome.

DNAs of organelles are thought to be anchored to the internal sites of their membrane systems, in a way that is analogous to that of prokaryotes. Although the size of the organelle genomes are much smaller than that of their nuclear counterpart, they are also maintained in their own complex manner. In plastids, the DNA apparently exists as a mixture of monomeric and multimeric forms. The original observation (Kolodner & Tewari, 1975) of these forms of the plastid genome by electron microscopy was recently extended by the work of Deng *et al.* (1989) using pulse-field electrophoretic analysis of plastid DNA. In this work, spinach plastid DNA was shown to exist as monomers, dimers, trimers and tetramers in the ratio of $1:0.33:0.11:0.04$, respectively. The multimers appear to be true oligomers of the monomeric form and their ratio does not seem to differ significantly during leaf development.

The organization of the plant mitochondria genome has proved to be very complex. In higher plants studied to date, it appears to exist as a complex mixture of circular DNA molecules that result from recombination events with a larger genomic form called the **master circle**. In maize, for example, this master circle is a DNA molecule of about 570 000 base pairs and contains six sets of repeated sequences. Recombination events between these repeated sequences in the mitochondria are thought to generate circular DNA molecules of smaller sizes (Lonsdale *et al.*, 1984). The reason for this complex mode of DNA maintenance in plant mitochondria remains obscure.

Movement of DNA from the plastid and nuclear genome into that of plant mitochondria has also been suggested by DNA hybridization studies. For example, sequences of the plastid encoded 16S ribosomal RNA and large subunit of ribulose-1,5-bisphosphate carboxylase have been shown to be present in the maize mitochondrial genome (Stern & Lonsdale, 1982). Mitochondria encoded sequences, such as those for NADH dehydrogenase, have also been found in the plastid DNA of tobacco and liverwort (Shimada & Sugiura, 1991). Thus, mechanisms apparently exist by which genetic information between different compartments of a plant cell can be exchanged.

8.2.3 Biosynthesis of nucleotides: building blocks of nucleic acids

Nucleoside triphosphates, the fundamental building blocks for DNA and RNA, are also involved in essential processes such as sucrose synthesis and the formation of phospholipids. The pathway of uridine triphosphate (UTP) and cytidine triphosphate (CTP) biosynthesis is discussed in detail by Ross (1981). The two amino acids, glutamine and aspartate, play essential roles in the early steps of this scheme. Uridine monophosphate (UMP) is the first nucleotide product of this pathway and it is phosphorylated sequentially by the enzymes UMP-kinase and UDP-kinase to yield UTP. The enzyme

CTP synthetase converts UTP irreversibly into CTP and in the process, also converts glutamine into glutamate. The amino acids glutamine and glycine are involved in the early steps of purine nucleotide biosynthesis. The first true purine nucleotide product of this pathway is inosine 5'-monophosphate (IMP). From IMP, adenosine monophosphate (AMP) and guanosine monophosphate (GMP) can be formed. For the formation of AMP, aspartate and GTP are converted into fumarate and GDP as by-products of the pathway. In the pathway of GMP synthesis, glutamine and ATP are converted into glutamate and AMP simultaneously with the production of GMP. IMP can be synthesized from AMP or GMP via the AMP-deaminase or GMP-reductase enzymes, respectively. Many steps of these biosynthetic pathways are exothermic reactions that require the conversion of ATP into ADP as the driving force.

The reactions described above produce ribonucleotides, which are the building blocks for RNA. Deoxyribonucleotides, the subunits of DNA, are produced in cells through the reduction of ribonucleotides. Plants, as well as animal systems, carry out this reaction with the enzyme nucleotide diphosphate (NDP) reductase. Two additional enzymes are also needed for this reaction: thioredoxin and a flavoprotein that can reduce the thioredoxin with NADPH as the electron donor. The substrates for the conversion are the nucleoside diphosphates: UDP, ADP, GDP and CDP. Thioredoxin is a small protein of about 12 000 daltons in molecular weight, and is found in the cytoplasm as well as the stroma of the chloroplast. The deoxy-UDP that is formed in a cell is targeted for conversion into thymidylate monophosphate (TMP) by the enzyme TMP synthetase. This process is likely to be very efficiently coupled to the reduction of UDP since deoxy-UDP, if it is present, will be readily incorporated into DNA upon its phosphorylation. In addition to deoxy-UDP, TMP can also be formed from deoxy-CDP by the removal of ammonia from deoxy-CMP through the action of a deaminase. With the exception of IMP, nucleoside monophosphates (NMP) and NDP are converted into their respective triphosphate forms through the action of kinases. The absence of an IMP kinase is important because the presence of ITP will allow this nucleotide to be incorporated into nucleic acids. Since inosine can base pair with any of the other four bases, the incorporation of this nucleotide into nucleic acids will result in the loss of information present in the original sequence.

8.2.4 DNA replication

The replication of DNA is one of the most fundamental functions that a biological organism must carry out. The basic mechanism involved in this process appears to be conserved between eukaryotes and prokaryotes. In plant cells that are active in cell division, DNA replication occurs during the S phase of the cell cycle and typically last for about 10 hours. The rate of DNA synthesis in eukaryotes is approximately 50 nucleotides per second and is co-ordinated with the synthesis of histones.

Both strands of the DNA are used as templates for its duplication. Since each of the resultant daughter DNA molecules contains one strand of the original DNA and a newly synthesized strand, this process is called **semi-conservative replication**. DNA replication occurs in a structure called the **replication fork** and the synthesis of the new DNA strands are carried out by a family of enzymes called **DNA polymerases**. The replication fork is typically initiated at specific regions of the genome called **origins of replication** and the double-helix in this region is unwound by the action of an enzyme called **DNA helicase**. Following the unwinding of the double helix, an enzyme, **DNA primase**, catalyzes the polymerization of a short RNA primer on the exposed single strands. In addition, single-stranded DNA-binding proteins are then bound to the DNA in order to prevent formation of hairpin helices within each of the exposed single strands. The primers synthesized by DNA primase are about 10 nucleotides in length and they serve as the initiation sites for DNA polymerization. In nature, DNA polymerases as well as DNA primase will only catalyze the formation of new phosphodiester bonds in a 5' to 3' direction. However, unlike DNA primase, DNA polymerases cannot carry out this activity between two free nucleotides and require the 3'-hydroxyl group of a base-paired primer. Thus, the activity of DNA primase is indispensible for the initiation of DNA polymerization. Since both strands of DNA are used simultaneously during DNA replication, this fact dictates that one of the two strands must be replicated discontinuously. A current model of this process is shown in Fig. 8.3. The leading strand in the replication fork is thought to be synthesized continuously from the 5' to 3' direction (left to right in the figure). The lagging strand, however, is synthesized as short stretches of DNA from the RNA primers polymerized by DNA primase located at the replication fork. These fragments of about 200 nucleotides in length are called **Okazaki fragments**. Linkage of

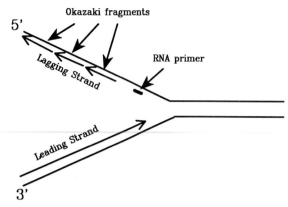

Figure 8.3 DNA synthesis at a replication fork. The discontinuous synthesis of DNA on the lagging strand is contrasted with that of the leading strand. The Okazaki fragments and the RNA primer synthesized by DNA primase are indicated with arrows.

the Okazaki fragments requires the removal of the RNA primers by a 5′ to 3′ exonuclease activity that is present in DNA polymerase α, followed with the repair of the resultant gap using a specialized DNA polymerase, and finally ligation of the fragments by the enzyme **DNA ligase**. The polymerization of DNA on RNA primers is likely to involve the soluble forms of DNA polymerases α and γ whereas DNA gap repair is likely to involve the chromatin-bound DNA polymerase β. One important activity of DNA polymerases is their ability to work as a 3′ to 5′ exonuclease. When an error is made during DNA polymerization and a wrong base is incorporated, DNA synthesis will be inhibited. This inhibition is due to the fact that DNA polymerase absolutely requires the 3′ hydroxyl end of a base-paired primer in order to add any new nucleotides and a mismatch at the 3′ end of the primer will not be tolerated. In this case, the 3′ to 5′ exonuclease activity intrinsic to DNA polymerase will remove the unpaired base and thus allow the polymerase activity to proceed. This 'proof-reading' capability of DNA polymerases is critical to their faithful duplication of large amounts of genetic information.

In plants, DNA replication has been studied by radiolabeling plants with radioactive thymidine followed by autoradiography of the isolated DNA (Van't Hof *et al.*, 1978). Each molecule of chromosomal DNA is found to contain many points of replication and they all proceed in both directions from each origin of replication. In total, there are tens of thousands of active replication forks in a plant cell nucleus during S phase. Apparently, different sets of these replication origins in the chromosome can be active at

different times. The control of DNA replication and its site of initiation in plants, and in eukaryotes in general, remain to be elucidated. The origins of DNA replication have been identified in *E. coli* and in yeast. It is believed that they act as binding sites for sequence-specific DNA binding proteins. The characterization of these origin-binding proteins is at present under active study. In a number of plants, the activity of DNA polymerase α has been found to increase significantly during seed germination, in meristematic cells at the shoot and root apices, and upon treatment with auxins and cytokinins (reviewed in Dunham & Bryant, 1985). These conditions are all known to be associated with high rates of cell division in plants and are in keeping with their requirement for higher levels of DNA polymerase activity. However, the regulatory mechanisms underlying these observations remain to be elucidated.

8.3 RNA

8.3.1 Property and diversity of plant RNAs

RNA is a molecule that is at once simpler and more complex than DNA. It is simpler than DNA because it consists of only a single-stranded polymer of nucleic acids rather than being double-stranded. This fact makes its synthesis and metabolism a more straightforward problem in comparison to that of DNA. However, the secondary structure of RNA is more difficult to predict since the possible locations of intramolecular base-pairing within an RNA molecule can be numerous. RNA is also more labile than DNA. *In vitro*, RNA is very sensitive to high pH since low levels of hydroxide ion will result in the formation of 2′,3′-cyclic phosphates from RNA. Nucleic acids without the 2′-hydroxyl group, as in the case of DNA, are much more stable under alkaline conditions. **RNases**, enzymes that degrade RNA avidly, are usually also very stable proteins. This fact makes it harder to isolate intact RNA compared to DNA. Cells contain three types of RNAs that can be distinguished by their structure, mode of synthesis and function. These are called **messenger RNA** (mRNA), **ribosomal RNA** (rRNA) and **transfer RNA** (tRNA). In a typical plant cell, rRNA accounts for about 85% of the total RNA. Of this pool of rRNA, the contribution from the plastids is highly variable. In photosynthetic tissues, where the copy number of plastids per cell is high, chloroplast rRNA can make up

25–40% of the total RNA. All nuclear-encoded genes are first transcribed within the nucleus and subsequently transported to the cytosol, where they are usually involved in protein synthesis. Thus, the process of translation is spatially and temporally uncoupled from that of transcription for nuclear genes. In contrast, organellar and bacterial transcription is thought to be closely linked to the process of translation.

(a) Messenger RNAs

mRNAs provide the templates on which proteins of specific amino acid sequences are synthesized. Most nuclear mRNAs carry information, via a triplet code called **codons**, for the production of a specific polypeptide sequence. This is referred to as **monocistronic**. In contrast, bacterial and plastid genes are often transcribed as **polycistronic** mRNAs that encode multiple polypeptides. In a typical cell, mRNA usually represents about 5–10% of the total RNA by weight. The number of distinct mRNA species is likely to be in the range of about 10 000 in a particular cell and the relative amounts of these different RNAs can differ by more than 100-fold at steady state. Thus, regulatory signals must be present to affect the differential accumulation of mRNA transcripts in the cytosol. A mature mRNA is typically between 400 and 4000 nucleotides in length. In higher eukaryotes such as plants, nuclear mRNA are usually synthesized in a precursor form that contains introns ranging in size from about 100 to a few thousand nucleotides. The introns are usually spliced out before the mature RNA is transported to the cytosol. Transcription of mRNA is carried out by the enzyme **RNA polymerase II** (Pol II) and proceeds from the 5′ end of the encoded transcript. After polymerization of about 20–30 nucleotides, the 5′ triphosphate group of the first ribonucleotide is covalently linked to a guanosine by a nuclear enzyme. In plants as well as in animals, this guanosine is then methylated at the N-7 nitrogen of the guanine base. This process is called **capping** and appears to be universal for nuclear-encoded mRNAs. The presence of a cap structure at the 5′ end of the mRNA is thought to promote translational initiation as well as to protect the RNA from 5′-exoribonucleases. However, it is not an essential for the translation reaction since uncapped mRNA can be translated *in vitro*. The 3′ end of most nuclear mRNA is also modified by the addition of multiple adenylate residues through the action of the enzyme **poly(A) polymerase**. A well-known exception is the transcripts encoding histones,

which do not contain polyadenylated 3′-ends. Transcription of mRNA normally proceeds about 50 to several thousand bases beyond the 3′-end residue that will eventually be polyadenylated. A conserved sequence AAUAAA that is about 30 nucleotides upstream from the poly(A) site is required to generate the proper 3′ end of the RNA with the co-operation of other sequence elements that are downstream from the adenylation site. Since RNAs with or without polyadenylation can be translated efficiently *in vitro*, it is unlikely that the poly(A) tails are required for translation *per se*. However, studies in animal systems have suggested that the length of the poly(A) tails is directly correlated with the stability of the mRNA in the cytosol (Bernstein & Ross, 1989). In plant protoplasts, it has been shown that the presence of 25 adenylate residues at the 3′-end of the mRNA can enhance the half-life of a transcript by about 3-fold (Gallie *et al.*, 1989). In addition, the presence of the poly(A) tail can also stimulate translation of the mRNA.

Another covalent modification of mRNA that takes place in the nucleus is the methylation of some adenylate residues. Up to two methylated adenine have been found per mRNA of 1600 nucleotides. At present, the significance of adenine methylation in mRNA is not clear since yeast mRNA apparently lacks this modification. Thus, it may not be an essential process that is conserved among eukaryotes. The mRNAs of mitochondria and plastids differ from those of the cytosol in several respects. (1) they do not contain any apparent 5′ methyl-guanosine cap; (2) they do not contain poly(A) tails at the 3′ end and (3) in plastids, polycistronic mRNAs are observed for many of the genes found in this organelle. Processing of these mRNA species in turn generates a complex pattern of transcripts (Gruissem, 1989).

(b) Ribosomal RNAs

The cytosol of eukaryotes contains 80S ribosomes as the protein complex responsible for the translation of mRNA into proteins. These ribosomes are either free in the cytoplasm or are membrane associated. Under appropriate conditions, the 80S ribosomes can be dissociated into two subunits one of which is twice the size of the other. The larger ribosomal subunit contains three RNA species of 25, 5.8 and 5S whereas the smaller subunit has one RNA component of 18S. Together, these RNAs make up about 50% in the total mass of the cytosolic ribosomes. The 70S ribosomes found in the plastids of higher plants have very similar characteristics to those of prokaryotes and they consist of 50S and 30S subunits. The 50S subunit of the plastid ribosome contains rRNAs with the size of 23, 5 and 4.5S. A single species of 16S rRNA is present in the smaller subunit. Plant mitochondrial ribosomes from various species sediment at 77–78S and contain three rRNAs. In the larger subunit of the ribosome, a 24–26S rRNA with sequences similar to those from the mitochondria of fungi and animals is found. In addition, a 5S rRNA apparently unique to this organelle in higher plants is also present. A single 18S rRNA is found in the smaller subunit of the plant mitochondria ribosomes. Although the sedimentation coefficients of the subunits of plant mitochondrial ribosomes are more similar to their counterparts in the cytosol, the sequence of their rRNA resembles more of those found in 70S ribosomes.

The size of the three classes of rRNAs (i.e. 23–26S; 16–18S; 4.5–5.8S) found in ribosomes of various compartments of the plant cell are approximately 3200, 1800 and 120 kilobases (kb) in length. The organization of rRNA genes in the nucleus is remarkably conserved among eukaryotes. They are typically arranged in large tandem arrays of a repeated transcription unit that consists of the 18S, 5.8S and 25S rRNAs with transcribed 'spacer' regions between these sequences. This organization is illustrated in Fig. 8.4. Intergenic regions between the repeats have also been found to contain internal repeating structures and may function to regulate expression of the rRNA genes. The rRNA repeat unit is present in a few hundred to several thousand copies in the nucleus and is organized into fibrillar centers in the nucleolus and in a condensed form in heterochromatin. In addition to the 18S–5.8S–25S rRNA gene clusters, the 5S rRNA genes also exist as long tandem arrays in various regions of the plant chromosomes. The sequence and predicted secondary structure of 5S rRNAs from diverse

sources in nature are highly conserved. Interestingly, the rate of 5S rRNA synthesis is apparently not correlated with that of the other rRNAs. Although the sequence of rRNAs are fairly well-conserved, the number of rRNA genes can be variable even within the same species. When active during interphase of the cell cycle, the condensed units in the nucleolus are uncoiled and transcription of many of the repeat units is initiated simultaneously. This can be visualized by electron microscopy as well as quantitated by radioisotope labeling of the newly synthesized rRNAs with [3H]uridine. From such studies, it has been shown that not only the transcription of multiple adjacent rRNA repeat units can be carried out simultaneously, each repeat unit can also initiate new RNA synthesis before termination of the earlier ones. These properties, in addition to the numerous numbers of these genes, accounts for the high proportion of rRNA within the total RNA pool of the cell. Since there are several million ribosomes per cell, the maintenance of high levels of rRNA is critical for the cell survival and proliferation.

The 18S–5.8S–25S repeat units are transcribed by *RNA polymerase I* to produce initial transcripts of about 45S. These 45S transcripts are then processed in the nucleolus to produce the three mature rRNAs before they are transported to the cytosol. However, methylation of highly conserved regions is known to take place even before the completion of 45S pre-rRNA synthesis. Although the exact role of rRNA methylation is not known for certain, it appears to be necessary for pre-rRNA processing. Ribosomal proteins are also known to be associated with the pre-rRNA during its synthesis, before its exit to the cytosol. After the transcription of pre-rRNA is complete, the 18S rRNA is cleaved from the precursor transcript. Subsequently, the 5.8S rRNA is hydrogen-bonded to complementary sequences at the 5' end of the 25S rRNA sequence. Cleavages at the spacer region between the 5.8S and 25S rRNAs then

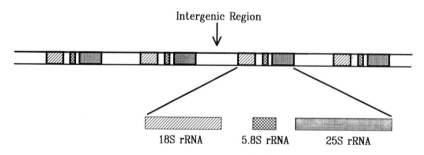

Figure 8.4   Structure of a rRNA gene cluster. The arrangement of the 18S, 5.8S and 25S rRNA-encoding genes are shown in an idealized form. An intergenic region between two rRNA repeats is indicated with an arrow.

complete the processing of the pre-rRNA. The 5.8S and 25S rRNAs interact strongly and will remain associated even in the absence of ribosomal proteins. In contrast, the 5S rRNA found in the same large ribosome subunit is released upon dissociation of the protein components.

The rRNA genes in the plastid genome are also organized in a similar fashion as their counterparts in the nucleus, with the exception that there are usually several tRNA genes that are co-transcribed. Since these genes are usually located in the inverted repeats of the plastid genome of most plants studied to date, there are thus two identical copies of these transcription units in these cases. The exact arrangement of the rRNA genes as well as the particular tRNA genes that are transcribed along with the rRNAs are variable among different species. In tobacco, the transcription unit contains three tRNA genes as well as the 16S, 23S, 4.5S and 5S rRNA genes. In spinach, the 5S rRNA gene is transcribed separately from the other rRNA genes. Plant mitochondria 18S and 26S rRNAs are larger than those of the plastids, fungi and animal mitochondria. They also have more sequence homologies to those of bacteria. In addition, a 5S rRNA is present in plant mitochondria that is not found in fungi and animal counterparts. The transcription units for the 18S and 5S rRNAs are closely linked in maize and are about 12 kb from the 26S rRNA gene. Comparison of the organization of these genes in the mitochondria genome of different species indicates that it is quite variable. Occurrences of introns in rRNAs, although quite rare, have also been reported for the 21S rRNA of the yeast mitochondria and the 23S rRNA of the *Chlamydomonas* chloroplast. As in the case of the nuclear-encoded rRNAs, the organellar transcripts are also known to be methylated and bound with ribosomal proteins during their synthesis and subsequent processing steps.

The 3′ end of the chloroplast 16S rRNA contains a CCUCC sequence that is thought to interact with a complementary sequence in the 5′ end of plastid mRNAs. A similar sequence in bacterial mRNAs is the well-known ribosome binding site called Shine–Dalgarno sequence (5′-AGGAGG-3′). This sequence is not found in the 18S rRNA of plant mitochondria or the 5′ untranslated regions of nuclear encoded mRNAs. In the case of plant mitochondria 18S rRNA, a UGAAU sequence has been suggested to function in an analogous fashion. At present, the exact structure and functional roles of each of the rRNAs in plants are still unclear. However, because of their high degrees of conservation

among all species, it is likely that rRNAs are intimately involved in the translation process.

(c) *Transfer RNAs and small nuclear RNAs*

The third major class of RNA present in the cell is tRNA. The structure and function of this group is perhaps the most well understood among RNAs. tRNAs make up 10–15% of the total cellular RNA and serve an essential function in the decoding process of translating mRNA sequences into proteins. Each cell and its organelles such as mitochondria and plastids contain a set of these tRNAs for their protein synthesis machinery. Typically, a plant cell contains about 110 distinct species of tRNAs, half of which are encoded and synthesized in the organelles. The size of tRNAs range from 70 to 90 nucleotides long and they always terminate at the 3′ end with the nucleotides CCA. Due to the small size and stability of tRNAs, it has been possible to obtain the crystal structure of several representative ones. It has been found that although the exact nucleotides are different among tRNA species, their folded three-dimensional structures are remarkably similar. As shown in Fig. 8.5, the single chain tRNA molecules forms four regions of double-stranded structures, commonly referred to as stems, that are held together by complementary base-pairing. In a planar drawing, this produces three loops and gives rise to the so-called 'cloverleaf' structure. As shown by the crystallographic studies, however, unusual base-pairing between loop 1, loop 3 and

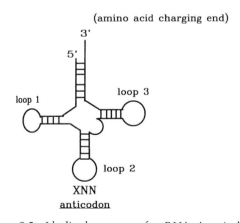

Figure 8.5 Idealized structure of a tRNA. A typical tRNA is shown diagrammatically in a clover-leaf-like structure. The position of the anticodon is shown as XNN, where X is the nucleotide that base-pairs with the third nucleotide of the codon triplet and N represents bases that can base-pair with the other two bases of the appropriate codon. The positions of loops 1, 2 and 3 are shown.

elsewhere in the molecule further compact the structure into an 'L' shaped form. These unusual interactions involve the formation of 'base triples' where the bases of two different nucleotides form hydrogen bonds with a third one simultaneously. Another important factor that contributes to the stability of the tRNA structure is the hydrophobic interactions among the planar surfaces of the bases that have been extensively stacked on each other. Together, these forces all combine to maintain the tRNA molecules in a highly ordered structure. One end of this L-shaped molecule contains the **anticodon** that will base-pair with the complementary codons in mRNAs whereas the other end is **charged** by covalently attaching a specific amino acid to the 3' end of the tRNA molecule through the action of enzymes called **aminoacyl-tRNA synthetases**. In this way, tRNAs serve as adaptors to translate information embedded within the mRNA sequences into specific amino acid sequences.

Of the 64 possible combinations of triplet codons using four different bases, three so-called **stop codons** serve as signals for the termination of translation. The remaining 61 nucleotide triplets thus out-number the 20 different amino acids that they represent. For methionine and tryptophan, two of the rare amino acids present in proteins, a single codon is used for each. For the other amino acids, however, multiple codons with corresponding tRNAs that have different anticodons exist. In these cases, the same aminoacyl-tRNA synthetase can charge the different tRNAs with the same amino acid. An interesting modification of tRNAs is the observation that adenine is not found in the first position in the anticodon. In this position, adenine is deaminated to form inosine which allows base-pairing with A, C or U. In addition, G and U may also undergo unusual base-pairing in this position. Modifications of these bases at this position, such as methylation of guanine or reduction of uracil, are also some of the unusual modifications that can affect the codon recognition specificities of tRNAs. These facts combine to cause the third base of each of the codons that specify certain amino acids to be variable and form the basis for the '**wobble hypothesis**' proposed by Crick. If one assumes that only the first two nucleotides in the codon are critical in tRNA recognition, a total of 24 tRNAs will be sufficient to correctly decode 61 codons. In fact, the tobacco plastid genome contains 30 tRNA genes and that of the human mitochondrial genome contains 22 tRNA genes. Covalent modifications of at least some of these tRNA species are thus necessary for them to have expanded codon recognition

properties. One interesting aspect about plant mitochondria- and plastid-encoded tRNAs is the fact that they do not contain the terminal CCA nucleotides that are found in tRNAs from other sources. This triplet is added to the ends of the mature transcripts by a **tRNA nucleotidyltransferase** in the organelles. Some of the mitochondrial tRNAs are highly homologous to their plastid counterparts (90–100%) whereas others have less than 80% of sequence similarities. Some evidence suggests that plastid tRNA genes may have been integrated into the mitochondria in plants. Another interesting characteristic of organellar tRNAs is the exclusive use of formylated methionyl tRNA as the initiator of translation, in contrast to cytoplasmic ribosomes where methionyl tRNA is used.

tRNAs are synthesized by a specific class of RNA polymerases, **RNA polymerase III**. Some of the tRNA genes contain introns and need correct splicing of the primary transcript in order to generate the functional form. Others, like some of the plastid tRNA genes, are transcribed as a polycistronic transcript and subsequent processing by RNases results in individual tRNA molecules. One particularly well-studied class of tRNA processing enzymes is RNase P. This enzyme requires an RNA component for its endonucleolytic activity and it appears to be present in plastids, mitochondria as well as in the nucleus.

Aside from the three main classes of RNAs, there are also small RNAs that are found in ribonucleoproteins (**snRNPs**). These RNAs (**snRNAs**) are found in all eukaryotic cells and are typically between 200 to 100 nucleotides long. Some of these snRNAs have been shown to contain trimethylguanosine 5' caps. The role of several snRNAs have been fairly well defined in yeast through mutagenesis approaches. Some of these RNAs, U1 and U5 for example, are associated with hnRNAs in the spliceosomes and are likely to perform critical functions in the splicing of introns from pre-mRNAs. U3 snRNA, on the other hand, is localized in the nucleolus and may be associated with pre-rRNAs. Although the mechanisms of RNA splicing are not well-characterized in plants, it is likely that they will be similar to those of the other eukaryotic systems.

8.3.2 Transcription mechanisms and regulation

Serving as the bridge between genes and proteins, the production of RNA is a critical point of control for cellular growth and differentiation. For example, some of the plant genes involved in flower and leaf morphogenesis have been found to encode putative transcription factors (Yanofsky et al., 1990; Vollbrecht et al., 1991). Thus, as in the well-documented case of segmentation genes of Drosophila, a cascade of transcriptional regulators is likely to underlie many of the pathways involved in developmental control. The central enzyme in transcription is called RNA polymerase. Mechanistically, the best-characterized RNA polymerase is that of the eubacteria which served as the model system in which the process of RNA synthesis has been studied in detail. The first step of this reaction involves the recognition and binding of RNA polymerase to a specialized region of the gene called a **promoter**. This is usually an asymmetric sequence located in the $5'$ end of the transcribed portion of the gene. Subsequent to the formation of a polymerase/promoter complex, local 'melting' of the transcription start site ensues to produce single strands of DNA within a short region. Base-pairing then occurs between free ribonucleoside triphosphate monomers and one of the two strands in this **open complex** before the adjacent nucleotides are joined by the formation of phosphodiester bonds. This creates a short region of DNA:RNA hybrid in the active site of the polymerase. The polymerase then moves along the DNA toward the $3'$-end of the growing RNA molecule while unwinding small stretches of the transcribed region one at a time. As the transcription complex passed through, the DNA double helix is rewound behind the polymerase and the RNA strand is displaced. When the polymerase encounters a special sequence in the gene called **termination signal**, the enzyme is released from the template DNA along with the completed RNA chain. To accomplish these tightly orchestrated reactions, some bacterial viruses encode a single subunit of RNA polymerase that can function without the aid of other proteins. The bacterial and eukaryotic enzymes, however, are large multi-subunit complexes of over 500 kilodaltons (kDa) in apparent molecular weight. It is probable that this difference reflects the need for tight regulation of transcription in most cells.

One of the most well-characterized RNA polymerases is that of E. coli. The holoenzyme consists of five distinct subunits called α, β, β', σ and ω. Two copies of the α subunit are present in each holoenzyme complex whereas the other subunits are present in single copies. The σ subunit acts as an initiation factor and provides the sequence specificity to the polymerase holoenzyme to preferentially transcribe from promoter elements of the bacterial genome. There are multiple σ subunits present in bacteria and they are involved in gene activation under specific conditions. One of the best known of these are the σ^{37} factors of E. coli, that are responsible for the activation of heat-shock responsive genes, and the multiple σ factors of B. subtilis that regulate genes involved in sporulation (Doi & Wang, 1986). Thus, there are likely to be distinct holoenzymes that have RNA polymerase activity in any given bacteria at a particular time. The predominant form of σ factor in E. coli is called σ^{70} and this protein has been shown to contain two different DNA-binding domains that recognize different sequence motifs in the prokaryotic promoter (Dombroski et al., 1991). These sequences are the so-called '-10' and '-35' elements located at the specified regions upstream of the transcription start-site and they have the consensus sequence of TATAAT and TTGACA, respectively. Genes that are tightly regulated in bacteria usually have very different -10 and -35 elements from this consensus and it is likely that a different σ factor will provide distinct sequence preference for the resultant RNA polymerase holoenzyme in order to transcribe them under a particular set of conditions. After transcription has been initiated in the open complex, σ factor is released from the polymerase. To promote efficient elongation and correct termination, **elongation factors** are then bound to the polymerase. Although sequence elements that are responsible for the regulation of bacterial gene expression is commonly thought to reside around the transcription initiation site, more recent studies have demonstrated that repressors and activators can also exert their effects over kilobases away to influence the action of RNA polymerase (Gralla, 1989). These modulators of transcription are typically sequence-specific DNA-binding proteins and at least in the case of the NR_1 regulator of the genes involved in nitrogen assimilation, protein phosphorylation has been shown to mediate the activation process.

In plant cells, at least three distinct compartments carry out RNA synthesis and each has their own special properties. These are the nucleus, plastids and mitochondria. In the nucleus, three different classes of RNA polymerase activities can be differentiated by their sensitivity toward the fungal toxin α-amanitin and the types of RNA that they produce. RNA polymerase I (Pol I) is localized in the nucleolus and is responsible for the synthesis of large rRNAs. This enzyme activity is insensitive to α-amanitin. RNA polymerase II (Pol II) is localized in the nucleoplasm, very sensitive to α-amanitin and is responsible for the synthesis of most, if not all mRNAs as well as some snRNAs in the nucleus. Lastly, RNA polymerase III (Pol III) is responsible for the synthesis of tRNAs, the small rRNAs, and some of the snRNAs. This enzyme is inhibited by α-amanitin at relatively high concentrations of the toxin. Organellar RNA polymerases, like those of prokaryotes, are not sensitive to α-amanitin and do not appear to differentiate between different types of RNA. Frequently, polycistronic transcripts are made and multiple processing steps are required to generate the correct RNA products. Biochemically, there is evidence that two distinct types of RNA polymerase may exist in chloroplasts. The soluble form of this enzyme behaves much like that of eubacteria whereas a chromosome-bound RNA polymerase activity, called transcriptionally-active chromosome (TAC), can be distinguished by its activity under high salt stress (100–200 mM NaCl) and insensitivity to heparin, an inhibitor of transcription initiation. The TAC form of chloroplast RNA polymerase appears to preferentially transcribe the rRNA operon in the inverted repeat region. The basis for the differences observed between the two RNA polymerase activities is still unclear.

The nuclear RNA polymerases are structurally more complex than their counterpart in prokaryotes. They contain large numbers of subunits and accessory factors that are required for correct initiation of transcription. In addition to having two large subunits with molecular weights of 190–150 kDa, purified nuclear RNA polymerases contain at least six or more additional subunits. Interestingly, the sequence of the two large subunits for Pol II and Pol III have been found to contain regions homologous with that of prokaryotic RNA polymerase (Allison et al., 1985). Thus, it is very likely that these enzymes have evolved from a common ancestor and mechanistically, the reactions catalyzed by these different enzymes are likely to be very similar. In the plastid genome, three genes with good homologies to bacterial RNA polymerase subunits α, β, and β' are found in the genome of tobacco but are absent in a nonphotosynthetic parasitic plant Epifagus virginiana (Wolfe et al., 1992). Thus, like the case of mitochondrial RNA polymerase, the plastids of E. virginiana must import the subunits of its RNA polymerase from the cytoplasm.

(a) Organellar transcription

Transcription of human and yeast mitochondria DNA have been studied in detail. The genome of human mitochondria is apparently transcribed only from two neighboring points. Two polycistronic mRNAs are generated with one of them covering the entire circular genome. These pre-mRNAs are then spliced to generate the various tRNAs and rRNAs. In addition, the few mRNAs are also polyadenylated. In yeast mitochondria, which are about five times the size of human mitochondria, individual promoters are used to transcribe each gene. In addition, multiple introns are found within some of the genes. An interesting observation in yeast mitochondria is that intron 2 of the cytochrome b gene encodes a protein called **maturase**. Mutations in intron 2 that disrupt the maturase also inhibited proper splicing of the cytochrome b transcript. In plant mitochondria, the situation appears more complicated than the other systems since multiple forms of the genome are present at any one time. In addition, it is difficult to determine whether the 5' termini of the multiple transcripts seen with some genes are the results of transcription initiation or RNA processing. However, most plant mitochondrial transcripts appear to begin within a consensus motif: TAAG(T/A)GA. Whether this apparent motif for transcription initiation has any functional significance has not been rigorously tested directly.

In contrast to the case of plant mitochondria, transcription in the plastids has been well-characterized due to the availability of in vitro transcription systems that can accurately produce plastid RNAs (Hanley-Bowdoin & Chua, 1988; Gruissem, 1989). Structurally, the plastid RNA polymerase consists of two large subunits with the apparent molecular weights of 180 to 140 kDa, which correspond to the β' and β subunits of prokaryotic RNA polymerase, in addition to about 10 other smaller subunits. Aside from the α, β, β' subunits, which are likely to be encoded in the plastid genome, the other subunits must all be imported. Although fractions from chloroplast extracts have been reported to have σ-like activity in vitro (Jolly & Bogorad, 1980; Lerbs et al., 1988; Tiller et al., 1991), the protein(s) involved has not been purified and its relationship to the well-known σ factors of E. coli remains elusive. Many mRNA and tRNA genes are transcribed as polycistronic transcripts in the plastid and their promoters are strikingly similar to that in bacteria. The consensus promoter element for the gene encoding the large subunit of ribulose-1,5-bisphosphate carboxylase (rbcL) from different species consisted of a TTGCGC and a TACAAT motif at the −35 and −10 region, respectively. These elements are strikingly similar to the −35/−10 elements found in many prokaryotic promoters. In fact, E. coli RNA polymerase can readily transcribe these plastid genes accurately *in vitro*. Deletion analyses have shown that a spacing of 18 bp between the −35 and −10 elements is crucial for optimal expression of a number of plastid promoters *in vitro*. Although there are many similarities between the transcription system of the plastid and bacteria, there are also a number of important differences. The gene encoding *trnS* has bee reported to be transcribed in the absence of its 5′ promoter region (Gruissem *et al.*, 1986). Also, a blue-light regulated promoter for the *psbD* gene does not contain obvious −35 or −10 motifs (Sexton *et al.*, 1990). Plastid RNA polymerase has been found to prefer supercoiled, circular DNA templates *in vitro* whereas *E. coli* enzyme can transcribe equally well from linearized DNA (Lam & Chua, 1987). These observations suggest that there may be plastid-specific factors or functional properties that distinguish the RNA polymerase from its bacterial counterpart. One additional difference in this regard is the insensitivity of the plastid enzyme toward rifampicin, which is a potent inhibitor of bacterial RNA polymerase. It should be pointed out that most of the detailed promoter analyses for plastid-encoded genes have so far been carried out *in vitro* only. Thus, if labile upstream factors are indeed required for their regulated expression, one may not be able to detect requirements of sequences in addition to the −35/−10 regions. Indeed, site-specific DNA-binding activities in chloroplast extracts have been reported in the last few years (Lam *et al.*, 1988; Baeza *et al.*, 1991). The function of the binding sites for these factors, however, has not been determined due to the lack of an *in vivo* assay. With the recent advent of a plastid transformation system in higher plants as well as in the alga *Chlamydomonas* (Svab *et al.*, 1990; Boynton *et al.*, 1988), these problems may be resolved in the near future.

Transcriptional rate measurements with isolated spinach plastids have suggested that the relative promoter strength of plastid-encoded genes does not vary significantly (Deng & Gruissem, 1987) in different tissues and environmental conditions. Regulation at the level of RNA stability and mRNA translation is concluded to be the predominant mechanisms that affect plastid gene expression. However, in addition to *psbD*, preferential transcription of *psbA* and *petE* genes in illuminated barley plants have been documented (Klein & Mullet, 1990; Haley & Bogorad, 1990). Methylation has also been reported as a mechanism for differential gene repression during the transition from chloroplasts to chromoplasts in tomato fruits (Kobayashi *et al.*, 1990). These results suggest that plastid gene expression may be regulated in a species-specific fashion.

(b) *Nuclear transcription control*

The initiation site of nuclear gene transcription is determined by promoter sequences that are distinct for each class of RNA polymerases and are very different from the −10/−35 element of prokaryote promoters. In spite of their differences in the sequence motifs used, these promoters are all organized in a roughly bipartite manner. In each case, a region of the promoter close to the start site of initiation is dedicated to the binding of crucial proteins that will recruit RNA polymerase to the correct start site. As one will predict from this function, the activities of these sequence motifs are distance- and orientation-dependent.

The promoter element for RNA polymerase I has been shown to reside in the region from −45 to +20 of the human rRNA promoter (Jantzen *et al.*, 1990) and it does not appear to be conserved among species. A factor SL1 from human nuclear extracts can apparently override this species-specificity when added to a mouse extract. Thus, species-specific factors may have evolved to selectively activate the promoters of rRNA genes. In the case of the wheat rRNA operons, the intergenic regions have been found to contain multiple repeats of a 135 bp sequence upstream of a putative promoter that resembles a sequence of the *Xenopus* rRNA promoter (Flavell *et al.*, 1986). The number of this A-repeat is variable among different rRNA encoding genes and may account for their variation in expression *in vivo*. In addition, methylation of the intergenic region may also modulate the activity of this class of promoters. Protein factors that bind to the upstream regions of rRNA genes have been detected in maize, radish and cucumber (Schmitz *et al.*, 1989; Echeverria *et al.*, 1992; Zentgraf & Hemleben, 1992). Their precise function and structure, however, is still not well understood.

In the case of Pol II promoters, two different non-exclusive elements are found in mRNA-encoding genes in diverse species. These are the TATA motif that is usually found at about 30 nucleotides upstream from the start site of transcription, and an initiator (Inr) motif that is absolutely required in Pol II-transcribed promoters that lack a TATA box. The Inr has a consensus sequence of $^{-3}CTCANTCT^{+5}$, where the central A residue is the start site of transcription and N is either T or C (Smale *et al.*, 1990). The basal factors that interact with Pol II-specific promoters have been especially well-characterized in the last few years (Zawel & Reinberg, 1992). One factor, TBP (TATA-binding protein) has received the most attention in the past few years and cDNA clones for similar proteins have been isolated from many organisms. Interestingly, although TBP in many organisms appear to be encoded by a single copy gene per haploid genome, several plants have been found to contain two different genes encoding TBP. The significance of this difference is at present unclear. The three dimensional structure of a TBP from *Arabidopsis thaliana* has recently been reported and found to be a 'saddle-like' protein that is likely to straddle the promoter region on one face of the helix (Nikolov *et al.*, 1992). The concave surface of this TBP protein is hypothesized to act as a landing-pad for other proteins that may serve to mediate many of the activities associated with the TATA region. From *in vitro* studies with TBP produced in bacteria, it has been proposed that TBP must interact with many different proteins in order to carry out its function as a basal factor for transcription. Perhaps the most extraordinary finding is the recent discovery that TBP may be involved in Pol III and Pol I transcription as well (Gill, 1992). For Pol III transcription, two different promoter types are known to exist that utilize different types of sequence motifs. For many of the tRNA genes that are transcribed by Pol III, mutagenesis studies have shown that two separate conserved elements within the coding region close to the transcription start site are involved in polymerase positioning and activation. These intragenic sequence elements, called ICR (intragenic control regions), are known to interact with at least three different protein factors. Interestingly, some of the snRNA genes are also transcribed by Pol III and their promoters are distinctively different from that of the tRNA genes. In the human U6 snRNA gene, for example, a functional TATA box and several upstream elements are used. In *Arabidopsis thaliana*, Waibel & Filipowicz (1990) have shown that the distance between the TATA element and a snRNA gene-specific upstream element (USE) are critical for the use of either Pol II or Pol III for U6 transcription. Thus, when the normal distance between these two elements within the U6 promoter is 23–24 bp, this gene is transcribed by Pol III. Insertion of 10 bases between the TATA and USE sequences converts the resultant promoter into one that is preferentially transcribed by Pol II. Interestingly, in both monocots and dicots, the U3 snRNA genes are also transcribed by Pol III (Marshallsay *et al.*, 1992). In all the other organisms studied so far, the U3 snRNA genes are transcribed by Pol II instead.

Although the promoter-proximal regions are sufficient to prime accurate transcription *in vitro*, usually under relatively high concentration of factors, more distant elements that interact with nuclear factors are required for high level gene expression *in vivo*. Many of these sites are located in so-called **enhancer** regions and function in a distance- and orientation-independent manner. In the last ten years, many of these upstream elements have been characterized and factors that interact with them isolated. Structure–function studies with numerous upstream factors have shown that to the first approximation, they are all bipartite in nature (Mitchell & Tjian, 1989). Thus, discrete domains of protein factors are involved in DNA-binding and transactivation. To a large extent, these domains appear to be independent and have evolved separately. Many plant DNA-binding proteins have now been characterized and many of these contain highly conserved motifs (Katagiri & Chua, 1992). Usually, the DNA-binding regions are rich in basic amino acids and distinct sequence motifs can often be found in factors that are involved in the same developmental program. A good example of this in plant systems is the genes that apparently control flower pigmentation and cellular identity in *Antirrhinum majus* and *Arabidopsis thaliana* (Coen, 1991). In these cases, a family of so-called MADS box-containing transcription factors has been found to regulate the developmental events that lead to proper formation of floral organs. The MADS box is a 56 amino acid residue region that is also found in certain transcription factors in yeast and mammals (Yanofsky *et al.*, 1990). Other well-conserved motifs that have been found in plant DNA-binding proteins and regulatory factors are: basic-leucine-zippers, zinc-fingers, homeoboxes, helix–loop–helix and Myb-like domains (reviewed in Katagiri & Chua, 1992). Among these different motifs, 3-dimensional structures of the zinc-finger and homeobox motifs are available from several representative members (Harrison, 1991). In addition to the presence of conserved DNA-binding domains, studies of plant DNA-binding proteins have also revealed novel juxtaposition of known motifs. One striking example is the cloning of multiple *Arabidopsis* cDNAs that contain a leucine-zipper adjacent to a homeobox (Ruberti *et al.*, 1991; Schena & Davis, 1992). This may suggest that plants have evolved different ways of combining ancient protein domains for performing similar or novel functions. In addition to well-known motifs, studies with plant trans-acting factors are also uncovering DNA-binding domains with new structures. Recent comparative studies of two new classes of plant factors involved in light-responsive gene expression, GT-1 and GT-2, suggest that a new tri-helical motif may be involved in DNA-recognition (Perisic & Lam, 1992; Dehesh et al., 1992; Gilmartin et al., 1992). It remains to be seen whether this structure will be found in DNA-binding proteins in other systems. Undoubtedly, crystallographic and functional studies with these cloned genes will enrich our understanding of transcription factors in the near future.

In contrast to the DNA-binding domain, the primary structure of activation domains are not well-defined. Some of the known activation domains are found to be rich in acidic amino acid residues while others are rich in glutamine or proline residues (Mitchell & Tjian, 1989). In order to affect transcription, it is likely that these regions are involved in protein–protein interaction with some of the basal factors or RNA polymerase, much in the same way that the elegant work on the lac repressor of lambda phage has demonstrated (Ptashne & Gann, 1990). Recent in vitro studies with purified upstream factors have suggested that the activation domain may either contact some of the basal factors involved in transcription initiation (Lin et al., 1990) or their action may be mediated by another class of factors called **coactivators** (Pugh & Tjian, 1992). This other class of transcription factor appears to be required for transcription initiation but does not bind DNA itself. It most likely interacts with the basal factor TBP and in this way is then anchored onto the TATA box region of the promoter. Other than a direct activation mechanism, some of the upstream factors may also activate transcription by a de-repression mechanism in which the normally chromatin-bound gene may be released from its quiescent state (Croston et al., 1991). In this case, the displacement of histone H1 from chromatin by an upstream factor can lead to gene activation in vitro. In the case of the tobacco transcription factor, TGA1a, it has been shown in vitro that it activates transcription by increasing the number of pre-initiation complex formation in the adjacent promoter (Katagiri et al., 1990). With the availability of many new tools and techniques, such as an in vitro transcription system that can respond to upstream factor (Yamazaki et al., 1990), we should learn much more in the next few years about how plant transcription factors may function.

From promoter analyses in which the function of different segments of the upstream region have been defined, it has been shown that different protein factors can synergistically interact with each other. In this way, the functional co-operation between transcription factors on the same promoter leads to a greater activity than the sum of each factor alone. Moreover, many DNA-binding proteins have been shown to bind to multiple promoters that are regulated differently. Thus, it appears that the diversity of gene expression patterns that one observes in vivo is likely the consequence of a combinatorial juxtaposition of different cis-acting elements. In the case of a limiting transcription factor, the in vivo occupancy of a binding site on a particular promoter by a DNA-binding protein will be dependent on its relative affinity to the specific sequence as well as its interaction with other adjacent factors. In this way, it is more realistic to consider the transcription profile of all genes in a particular cell as the result of the interaction among a network of factors that either directly or indirectly influence the initiation of transcription. In order to understand how this network functions in detail, it will be useful to characterize the properties and mode of action of many classes of trans-acting factors. From a combination of elegant genetics and molecular approaches, our understanding of how transcription factors can be modulated in their action has seen an explosion of information in the last few years. Some regulatory mechanisms that are known to modulate transcription factor activity are as follows.

1. Differential gene expression: this includes tissue-specific expression of the gene(s) that encode the transcription factor and differential splicing in which different forms of the factor can be generated from a single coding region.

2. Nuclear localization: in order to activate transcription through interaction with the promoter, a DNA-binding protein must be targeted into the nucleus. This step has been shown to be a regulatory switch for several transcription factors in Drosophila and mammalian cells.

3. Heterodimer formation: many DNA-binding proteins are known to form oligomers with other proteins with similar structural motifs, such as leucine-zippers that can form coiled-coils. Heterodimers between different proteins have been shown to either increase or decrease their DNA-binding activity. In addition, the sequence-specificity of the heterodimer can also be significantly different from that of the homodimers. Thus different promoters may be preferentially affected by the heterodimer and homodimers.

4. Covalent modification: protein phosphorylation has been the most well-studied mechanism
of transcription factor modulation by post-
translational mechanisms (Hunter & Karin,
1992). In some cases, the sites of phosphorylation have been mapped and the consequences
on transcription varies from inhibitory to
enhancement. Several plant DNA-binding proteins have also been shown to be phosphorylated *in vitro* by plant kinases. Their *in vivo*
functional significance, however, remains
unclear.

(c)   *Transcription elongation and termination*

In addition to transcription initiation, the processes of transcript elongation and termination site
selection are also possible points of gene control.
In bacteria, where these processes have been best
studied, two major classes of termination activities
are known. These are called the *rho*-dependent
and *rho*-independent types. In the first type, a
cellular factor *rho* is involved in sequence-specific
transcription termination. For the *rho*-independent
process, transcription usually terminates at regions
that have the following characteristics: a GC-rich
region that can form a stable hair-pin structure in
the RNA product followed by a stretch of A
residues on the template strand. Studies with
bacteriophages have shown that the action of *rho*
can be counteracted by **antitermination** factors
(Yager & von Hippel, 1987). For example, the Q
protein of lambda phage can bind to the elongating RNA polymerase complex and by doing so,
allows transcription to proceed beyond either *rho*-
dependent or -independent termination sites. The
binding of Q to the elongation complex is
enhanced by an elongation factor, NusA. Another
lambda protein, N, can also act as antiterminator
for *rho*-dependent termination sites. In this case,
co-operation with at least six other bacterial
proteins and binding of the N protein to a specific
site on lambda DNA are required. In eukaryotes,
the characterization of factors that influence the
elongation and termination of transcription have
made significant advances in recent years (Spencer
& Groudine, 1990). Two different types of
elongation factor activities have been identified
for mammalian Pol II transcription. The basal
transcription factor, TFIIF, has been found to
increase the rate of transcript elongation by the
ternary complex between RNA polymerase, template DNA and nascent RNA. The other type of
elongation factors, typified by the elongation
factor SII, enhances he progression of the ternary
complex through intrinsic termination sites. This

process has been especially well-studied with the
adenovirus system where transcription by Pol II
in vitro are known to be blocked at specific sites
along the major late gene. Addition of factor SII
apparently releases the halted ternary complex and
stimulates read-through of the natural termination
sites (Izban & Luse, 1992). Interestingly, PPR2, a
yeast gene identified as a transcription regulator
has been found to be homologous to factor SII.
This fact makes it very likely that elongation
factors for transcription are involved in the
regulated expression of certain endogenous
genes. Characterization of SII and TFIIF homologs
in plants will enhance our understanding of this
process in the future.

8.3.3   RNA processing

Nuclear and some organellar RNAs are synthesized as larger precursors and are usually associated with protein particles. These preRNAs are
polyadenylated at their 3' ends, for mRNAs, as
well as capped in their 5' end. Before their exit into
the cytoplasm, however, these transcripts have to
be properly processed into mature RNAs and then
transported through the nuclear membranes. From
pulse-labeling studies, it has been found that about
95% of the nuclear pre-RNA, called **heteronuclear
RNAs** (hnRNAs), are degraded within about an
hour after their synthesis and thus never make it to
the cytoplasm. Although the process of nucleo-
cytoplasmic RNA transport is poorly understood
in plants, some of the RNA processing reactions
are beginning to be characterized at the molecular
level. One of the most important of these is the
splicing of introns from pre-RNAs. For example,
regulation of intron splicing has been demonstrated in animal systems to be important for the
pathway of sex determination in *Drosophila*
(Baker, 1989). It is reasonable to expect that the
generation of different mRNAs from a primary
transcript will be an important point of regulation
for some pathways in plants as well.

There are two major types of intron splicing
mechanisms. The self-splicing introns of *Tetra-
hymena* rRNA and some fungal mitochondrial
genes are typical 'group I' introns. These introns
were among the first RNA molecules shown to
have enzymatic activities. Aside from the RNA
molecule itself, a guanosine is necessary for this
splicing reaction. As shown in Fig. 8.6, the first
step involves the attack of the phosphodiester
bond of a conserved uridine residue at the **5'-splice
site** by the guanosine nucleotide cofactor. This is
followed by another transesterification step in

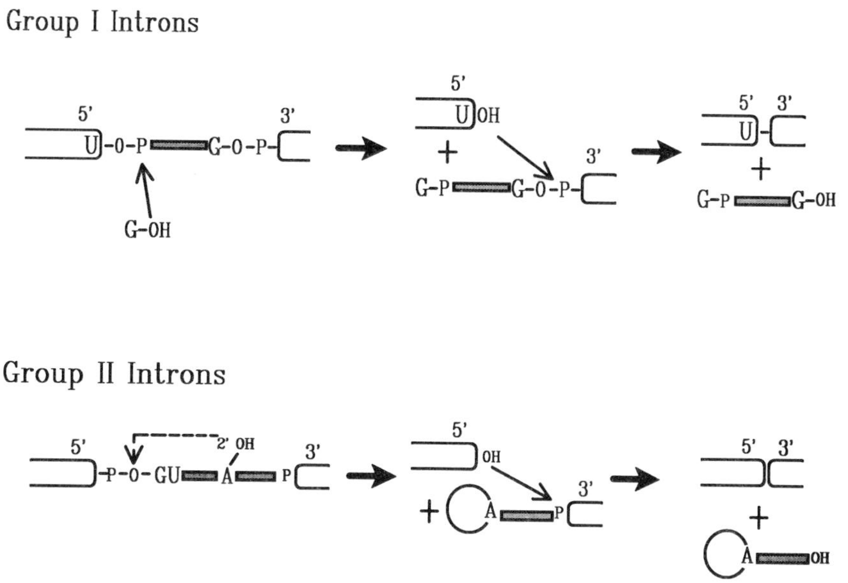

Group I Introns

Group II Introns

Figure 8.6 Mechanisms of group I and group II self-splicing introns. In group I introns, a guanosine cofactor attacks a phosphate group located in the 5′ splice-site of the pre-RNA. This results in a transient break in the RNA and the linkage of the guanosine cofactor to the 5′ end of the intron. Subsequent transesterification between the uridine residue at the 5′ splice site and the linkage at the 3′ splice site then results in the excision of the intron. In group II introns, the 2′ hydroxyl group of an internal adenosine attacks the phosphodiester linkage between the 5′ splice site and a conserved-GU-dinucleotide. This results in an unusual lariat structure at the 5′ end of the intron. Subsequent transesterification then results in the joining of the two exons and the excision of the intron in the lariat form. The region of the RNA that contains the intron is shaded.

which the 3′-hydroxyl group of the upstream exon displaces the 3′-hydroxyl group of a conserved guanosine residue of the intron at the **3′-splice site**. A consequence of this reaction pathway is that a guanosine is added to the 5′ end of the excised intron sequence.

'Group II' self-splicing introns are found in some structural genes of chloroplasts, plant mitochondria and yeasts. In contrast to group I introns, cofactors are not required for the self-catalyzed splicing reactions of this group. Instead, a conserved AU dinucleotide sequence within the intron near the 3′-splice site is used to catalyze the first step of the reaction. This so-called **branch point** involves the formation of a 2′,5,-phospho-diester bond between the adenosine residue and a guanosine residue at the 5′ end of the intron. The product of this reaction is the formation of a lariat-like RNA molecule. Subsequent attack of the 3′-splice site by the 3′-hydroxyl group of the upstream exon links the exons together and also releases the intron in the lariat form.

In addition to the self-splicing type, group II introns also include the introns of mRNA in the nucleus. However, the splicing process of these intervening sequences is dependent on a complex structure called the **spliceosome**. This is a ribonucleoprotein (RNP) complex that contains snRNAs as essential components in addition to proteinaceous components. Although the structure and function of the different components are beginning to be elucidated in yeast and animal model systems, the components of this important process in plants are less defined. Since introns of mRNAs from animals are not spliced, or only very poorly, in yeast and plants, it is likely that they may have very different requirements. The work of Goodall & Fillipowicz (1989, 1990) has established some important characteristics of the splicing system in higher plants. Their work using an artificial intron shows that although like all other introns, plant introns contain a GU and an AG dinucleotide at the 5′- and 3′-splice sites, respectively, the critical sequence elements are different. In vertebrates, a polypyrimidine tract is required between the branch point and the 3′-splice site whereas in yeast, a conserved UACUAAC sequence is found in the branch site of the intron. In plants, introns are found to be AU-rich (between 70 and 80%) and the branch site selection is not absolutely fixed. Introduction of GC-rich sequences within plant introns reduces splicing efficiency dramatically. A minimum length of 70–73 nucleotides (nt) for efficient splicing has also been defined for monocots and dicots. The fact that some monocot introns are not spliced efficiently in dicots indicates that there may also be species-specific mechanisms for intron recognition and processing (Keith & Chua, 1986).

Studies of the mechanism of replication for certain plant RNA viruses have uncovered a third type of RNA self-splicing reaction. In the replication cycle of the avocado sunblotch viroid and the satellite RNAs of tobacco ringspot virus, for example, RNA transcripts are initially synthesized as concatamers. Subsequent autocatalytic cleavage then occurs without the requirement of any cofactors to generate unit-length RNA molecules. From deletion and mutational analyses, it has been shown that only a small region of these self-splicing viral RNAs is required for the catalytic activity. This region can potentially form a secondary structure that has been called a **hammerhead**. On the basis of this information, a 19-nucleotide RNA molecule has been demonstrated to contain the ability to catalyze sequence-specific RNA cleavage (Uhlenbeck, 1987). This type of catalytic RNA molecule has been dubbed **ribozymes** and will likely find broad applications in basic and applied biology.

Two relatively new discoveries in RNA processing are **trans-splicing** and **RNA editing**. Trans-splicing involves the joining of two separately transcribed RNA molecules. This phenomenon has been found in trypanosomes, *Euglena*, nematodes, plant mitochondria and the chloroplast of *Chlamydomonas reinhardtii* and plants. Trans-splicing of nematode RNAs has also been demonstrated in mammalian cells (Bruzik & Maniatis, 1992), which indicates that the components of this reaction may be widely conserved. Based on systems that have been well characterized, such as that of the nematodes and trypanosomes, it is thought that *trans*-splicing involves reactions that are similar to those of group II introns of nuclear pre-mRNAs (Agabian, 1990). Similarities include: conserved GU and AG dinucleotides at the 5'- and 3'-splice sites, requirement of snRNAs, and formation of a Y-shaped intermediate that is structurally analogous to the lariat RNA. One of the most well-characterized *trans*-splicing systems is the assembly of the transcript for the gene encoding one of the reaction center subunits of photosystem I (*psaA*) in *C. reinhardtii*. Unlike in plant chloroplasts that have been studied to date, the coding region of *psaA* is transcribed in three separate units that are widely dispersed in the chloroplast genome of *C. reinhardtii*. Mutagenesis analysis has demonstrated that at least 14 nuclear genes and one chloroplast-encoded gene, *tscA*, are

required for the proper *trans*-splicing of *psaA* transcripts (Goldschmidt-Clermont *et al.*, 1990, 1991). *tscA* appears to be a small chloroplast RNA that is required for the first step of the *trans*-splicing reaction that joins introns 1 and 2 of the *psaA* transcript. Future characterization of this system is likely to uncover new insights into this exciting new area of biology. From the complete sequence of the plastid genome of tobacco, rice and liverwort, the *rps12* gene encoding a ribosomal protein in the chloroplasts also appears to require *trans*-splicing for its assembly. The *nad1* gene, encoding subunit 1 of the NADH dehydrogenase complex, of *Petunia*, wheat and *Oenothera* mitochondria, has been found to require at least three *trans*-splicing and one *cis*-splicing events to complete the assembly of five different exons (Chapdelaine & Bonen, 1991; Sutton *et al.*, 1991; Wissinger *et al.*, 1991). Thus, *trans*-splicing events are likely important reactions that take place in different compartments of the plant cell.

It has long been assumed that the sequence information located in the DNA level will be directly transferred to the RNA during transcription. Although RNA is known to be modified subsequent to their synthesis, such as methylation, these reactions usually do not alter the information inherent in the nucleotide sequence. It is thus a surprise when discrepancies between the genomic DNA sequences and mRNA sequences were discovered in parasitic protozoans such as trypanosomes and later in other systems such as plant mitochondria and chloroplasts. RNA editing in the kinetoplastid mitochondria of trypanosomes has been found to involve the insertion or deletion of U residues at multiple sites along the coding region of mRNAs (Simpson, 1990). In plant mitochondria, the *nad1* gene product of petunia and wheat mitochondria as well as the *coxII* transcript of petunia, wheat, pea and maize mitochondria has been found to be edited (Hiesel *et al.*, 1989; Sutton *et al.*, 1991). In the study with petunia mitochondria, evidence was obtained that indicates editing can proceed before the completion of *cis*- or *trans*-splicing. A total of 14 different C-to-U changes were found along the *coxII* coding region of petunia. Each of these changes causes an alteration in the amino acid encoded at that position. In the wheat mitochondrial *nad1* transcript, multiple editing sites are also found and most strikingly, the initiation codon for protein synthesis is created by RNA editing. Thus, a C-to-U change is required to generate the proper start site of translation. Evidence for RNA editing in higher plant chloroplasts has been reported recently. In maize, a single C-to-U conversion is required to generate the initiation codon for the *rpl2* transcript (Hoch *et al.*, 1991) whereas in tobacco chloroplasts, a similar editing process is required for a transcript encoded by the *psbL* gene (Kudla *et al.*, 1992). Furthermore, four editing sites within the coding region of the maize chloroplast *ndhA* transcript have been found (Maier *et al.*, 1992), thus demonstrating that non-AUG editing can take place in the plastids. These observations demonstrate that RNA editing is a common process in the two major organelles of higher plants.

The mechanism of RNA editing in higher plants is unknown at present. In trypanosomes, a small **guide RNA** that can partially base-pair with the target transcript may be involved in providing the sequence-specificity of this phenomenon. Models have been proposed which require concerted actions of a terminal uridylyl transferase, an RNA ligase and a site-specific endonuclease (Cech, 1991). Since the RNA editing that has been observed in plants is quite different from that of trypanosomes, it may not be surprising that a guide RNA equivalent has not been detected in plants. Thus, the mechanism for sequence-specific C-to-U substitutions in plants remains to be elucidated.

It should be pointed out that sequences within introns are not necessarily without function. For example, an **RNA maturase** required for splicing is encoded within the intron of a yeast mitochondrial gene (Carignani *et al.*, 1984). A DNA endonuclease has also been found to be encoded by intron sequences of the chloroplast 23S rRNA gene in the alga *Chlamydomonas reinhardtii* (Durrenberger & Rochaix, 1991). Along with *trans*-splicing and RNA editing, these observations illustrate the complexities of how genetic information may be organized in plants and other organisms.

8.3.4 RNA turnover

In order for transcription to serve as a versatile regulatory switch that can respond repeatedly to stimuli and developmental cues, degradation of the transcript is an important pathway. In addition, the steady-state concentration of a particular transcript is related to the transcription as well as RNA turnover rates. Thus, alterations in RNA levels can result from changes in the rate by which the particular transcript is being degraded. The average half-life of mRNAs in different organisms can vary dramatically. In *E. coli*, the average half-life is 3–5 min whereas it is about 20 min in the budding yeast *Saccharomyces cerevisiae*. In higher eukaryotes, such as vertebrates, the average is

about 10 h. In addition to 3'- to 5'-exoribonucleases, endoribonuclease activities are also likely to play important roles in the turnover processes. However, the specificity and regulatory properties of these enzymes are not defined in most systems.

Two well-characterized genes that are known to be regulated in their expression at the level of mRNA stability are casein and histones in mammalian cells. For example, the addition of the hormone prolactin can increase the stability of casein mRNA by a factor of 30 to 50. In plants, the involvement of mRNA stability in gene regulation has been implicated by a study which showed that the turnover of the pea ferredoxin transcript may be light-regulated (Dickey *et al.*, 1992). In these studies, ferredoxin transcripts produced by a viral promoter were found to accumulate in a light-dependent fashion in transgenic tobacco. RNA stability has also been proposed to be involved in the regulation of chloroplast gene expression (Deng & Gruissem, 1987). This conclusion is based on the dramatic differences observed for certain plastid mRNAs between light and dark conditions whereas their apparent transcription rates do not differ significantly. Recently, hairpin structures at the 3' end of the plastid transcripts for the *petD* and *rbcL* genes have been shown to be recognized by protein factors present in the chloroplasts of spinach (Chen & Stern, 1991; Schuster & Gruissem, 1991). These RNA binding proteins may be involved in the proper 3' processing and stability of these transcripts. Future biochemical and functional characterization of these proteins may shed light on mechanisms of plastid RNA processing.

Regulatory mechanisms that determine RNA half-lives are still poorly defined in plants at present. In other eukaryotes, an AUUUA motif near the 3' end of the transcript has been found to correlate with RNA instability (Shaw & Kamen, 1986) whereas the presence of a poly-A tail may help to prevent the rapid degradation of RNAs (Huez *et al.*, 1978). Consistent with this notion, non-polyadenylated mRNAs such as those encoding histones have been found to be extremely labile with half-lives in the range of minutes.

RNases, the enzymes that degrade RNAs, are beginning to be characterized systematically in higher plants. One class of plant RNases that have been studied quite well in recent years is that of the S-alleles involved in self-incompatibility in *Nicotiana alata*. These genes were initially identified as loci that determine gametophytic self-incompatibility. Subsequently, sequence analysis of the encoded proteins revealed homology with known RNases and their RNase activities were confirmed experimentally. Since homologs of these S-alleles have been found in self-compatible species such as *Arabidopsis thaliana*, these proteins are likely to have other conserved functions in addition to their role in determining successful pollination events of certain plants. In *Arabidopsis*, as many as 16 distinct RNases can be identified by gel electrophoresis (Yen & Green, 1991) and in wheat, distinct RNases were found to accumulate during senescence (Blank & McKeon, 1991). Biochemical and molecular characterization of these important enzymes should shed light on this area of plant biochemistry in the future.

8.4 PROTEINS

8.4.1 Genetic code: codon usage

Proteins are synthesized from the information encoded within mRNAs and the order of amino acids in the nascent polypeptide chains are determined by a triplet code of ribonucleotides called the codon. Thus, for each nucleotide sequence, there are three different possible ways of **reading** the encoded protein sequence. However, usually only one of these three **reading frames** is actually used for protein synthesis. This is due to the fact that the optimal binding site for ribosomes is often found near or includes an AUG codon that encodes the amino acid methionine. The correspondence of each codon to individual amino acids forms the **genetic code** and is shown in Fig. 8.7. Three codons are used to specify sites for termination of translation and aside from the codon for methionine, the other amino acid with a specific codon is tryptophan. The fact that multiple codons exist for most amino acids is the reason that the genetic code is called a **degenerate** code. Remarkably, aside from a few exceptions found in yeast and mammalian mitochondria, this code is essentially invariant in nature. This fact argues strongly that all organisms have evolved from a common ancestor.

Statistical analyses of the frequency that each codon is used in a particular organism have shown that they are not used in a random fashion. The preference of a certain codon for an amino acid is called **codon bias**. In most cases, this occurs mainly in the choice of bases for the third position of the codon. In higher plants, some of the known plant sequences in databases have been compiled and there is a striking difference in codon preference between monocots and dicots (Campbell & Gowri, 1990). In dicots, 44 codons are found to be used more frequently than others with the codons ending in U or A being more preferred than others.

5' end	second position				3' end
	U	**C**	**A**	**G**	
G	V (Val)	A (Ala)	D (Asp)	G (Gly)	U
	V	A	D	G	C
	V	A	E (Glu)	G	A
	V	A	E	G	G
C	L (Leu)	P (Pro)	H (His)	R (Arg)	U
	L	P	H	R	C
	L	P	Q (Gln)	R	A
	L	P	Q	R	G
A	I (Ile)	T (Thr)	N (Asn)	S (Ser)	U
	I	T	N	S	C
	I	T	K (Lys)	R	A
	M (Met)	T	K	R	G
U	F (Phe)	S	Y (Tyr)	C (Cys)	U
	F	S	Y	C	C
	L	S	Stop	Stop	A
	L	S	Stop	W (Trp)	G

Figure 8.7 The genetic code. A table illustrating the correspondence between the 64 possible codon triplets and the amino acids that they encode. Both the one letter amino acid representation and their three letter equivalents (in parentheses) are shown. The three 'Stop' codons that terminate translation are also indicated.

In monocots, a set of 38 frequently used codons are found and two classes of genes are revealed from the analysis. One class has the same U or A preference in the third position as that of dicots whereas the other class shows a clear bias toward codons with C and G in this position. Another observation is the avoidance of XCG and XUA codons, where X is any nucleotide. This may be the result of selective pressure during evolution of the genome. For example, the -CG- dinucleotide is a possible substrate for DNA- or RNA-methylases the action of which may lead to gene inactivation. In the organellar genomes of plants, codons ending in A or U are apparently preferred. Since optimal translation of the mRNA has been correlated wih the use of the proper codon bias of the particular system, this aspect of the gene structure is an important consideration for the transfer of genetic information from one source to another.

An important reaction in the process of decoding mRNA sequences is the attachment, or **charging**, of the correct amino acid to each tRNA species with different anticodons. The fidelity of this process is vital to the proper translation of nucleotide sequences into proteins. Since tRNAs all have similar 3-dimensional structures, the specificity of the aminoacyl-tRNA synthetases that carry out this reaction must depend at least in part on subtle sequence differences among each tRNA. Interestingly, it has been found that multiple tRNAs, called **isoacceptors**, that have the same amino acid specificity but have different anticodons, are aminoacylated by the same aminoacyl-tRNA synthetase. This fact implies that this class of enzymes must be able to recognize common features in the isoacceptors. In recent years, mutagenesis experiments with the *E. coli* tRNA^{ala} have demonstrated that a single base-pair (G-U) in the acceptor helix of the tRNA determines the specificity of the aminoacyl-tRNA^{ala} synthetase (Hou *et al.*, 1989). In fact, a synthetic RNA 'microhelix' with only seven base-pairs and without the anticodon can be properly aminoacylated if this major specificity determinant is included. In other tRNAs, notably the *E. coli* tRNA^{val} and tRNA^{met}, the specificity determinants apparently reside within their anticodons since the simple exchange of these three bases between the two tRNAs causes a switch in their amino acid identities. Thus, distinct structural features in the different tRNAs may be involved in their recognition by the proper aminoacyl-tRNA transferase. Lastly, overexpression of the Gln-tRNA synthetase in *E. coli* has been reported to alter its tRNA specificity and glutamine is found to be incorrectly

acylated onto another tRNA (Swanson *et al.*, 1988). This observation suggests that the relative level of tRNAs and the aminoacyl-tRNA transferases are also important parameters in the specificity of their interactions.

8.4.2 Translation

The process of translation takes place in a large, multimeric protein complex that is known as the ribosome. Ribosomes are found in both soluble and membrane-bound forms. They have been found to be associated with the endoplasmic reticulum (ER), thus giving rise to the so-called 'rough ER', and the thylakoid membranes of the chloroplasts. These membrane-associated ribosomes are thought to facilitate in part the synthesis and assembly of proteins that are destined either for secretion or membrane-localized.

The exact structural components and inhibitor sensitivity are different between ribosomes of eukaryotic and prokaryotic origins. In general, ribosomes consist of two subunits each comprising more than 50 polypeptides. In tobacco, for example, more than 70 distinct polypeptides associated with the 40S subunit and 47–50 with the 60S subunit of the cytoplasmic ribosome can be identified (Capel & Bourque, 1982). In addition to the enormous number of distinct proteins, about half the weight of the ribosomes is contributed by the rRNA components. The eukaryotic type of ribosomes are typically characterized by a sedimentation coefficient of about 80S and is sensitive to cycloheximide. The prokaryotic ribosomes, including those of the chloroplasts, have sedimentation coefficients of about 70S, are composed of fewer proteins, and are inhibited by chloramphenicol. In spite of these differences, the gross structural features of these ribosomes and many of the basic reactions involved in protein synthesis are remarkably conserved among different organisms.

Ribosomes consist of two multimeric subunits that can dissociate and reassociate *in vivo*. Each of these two non-identical subunits contains a different set of rRNAs that are likely to be involved in the organization of the associated polypeptides. mRNAs and tRNAs are associated with the small subunit whereas the formation of the peptide bonds between adjacent amino acids is catalyzed by the large subunit. There are three RNA binding sites in the small subunit that play important roles in translation. One of these sites binds to the mRNA that is to be translated while the other two, called **P** and **A** sites, interact with tRNAs that contain the proper anticodons for the particular mRNA.

The process of translation involves three distinct steps: initiation, elongation and termination. In the first step, multiple **initiation factors** (**IFs**) are required. These IFs are associated with the small ribosomal subunit in addition to an initiator tRNA^{met} bound at the P site. To initiate translation, this ribonucleoprotein complex has first to locate the proper AUG on the particular mRNA. In eukaryotic ribosomes, the cap structure at the 5′ end of mRNAs is known to promote initial ribosome binding. Two essential eukaryotic IFs, eIF-4A and eIF-4B, act together as an ATP-dependent RNA helicase that will unwind secondary structures at the 5′ end of the mRNA and are thought to facilitate ribosome docking. An important factor eIF-2 is known to bind to and position the initiator tRNA on the ribosome. In some cases, modification of eIF-2 by phosphorylation has been found to control the rate of protein synthesis. After docking onto the 5′ end of the mRNA, the small subunit of the ribosome is then thought to **scan** down the mRNA until an AUG codon is reached and translation can commence. The sequence requirement for an optimal initiation codon has been defined by mutational analyses. The consensus sequence found for animal translation systems was ACC<u>AUG</u>G (Kozak, 1986) whereas that in plant systems was AC<u>AA</u>UGG (Lutcke *et al.*, 1987). Interestingly, efficient initiation in plant cells has been observed with codons that differ from AUG (Gordon *et al.*, 1992). This phenomenon has also been found in animal systems and thus implies that there may exist another undefined mechanism by which peptide chain initiation can occur in the cytosol. In contrast to the initiation process of eukaryotes, only three initiation factors, IF-1, -2 and -3, are required in *E. coli*. Moreover, instead of the 5′ cap structure and ribosome scanning, the initiation site for translation is specified by a specific sequence about four to seven nucleotides upstream from the start site AUG. This initiator sequence, 5′-AGGAGG-3′, is called the **Shine–Dalgarno** sequence and is known to base-pair with a conserved 5′-CCUCCU-3′ sequence at the 3′ end of the 16S rRNA located in the small subunit of prokaryotic ribosomes. In this way, multiple sites for translation initiation can be specified in polycistronic mRNAs, which are frequently found in prokaryotes, and protein synthesis for different gene products can be carried out simultaneously on the same transcript.

Although the 'ribosome scanning' model of translational initiation appears to hold true for most eukaryotic genes, a noteworthy exception has been reported by Pelletier & Sonenberg (1988). Working with the poliovirus RNA, they discovered that translation on this naturally uncapped, dicistronic transcript is initiated by ribosome binding to specific sequences within the transcript. Sequences required for this internal initiation process may be as many as 500 bp. Similar observations were also made with encephalomyocarditis RNA and thus this phenomenon may be more widespread and a mechanism appears to exist in the cytoplasm to initiate translation within a transcript. In the future, polycistronic mRNAs may yet be found in the cytosol.

Once the initiation codon has been located by the smaller subunit of ribosomes, the large subunit is recruited with the help of eIF-5, a single polypeptide of 100–160 kDa. Concomitantly, a GTP that is bound to eIF-2 is hydrolyzed and the eIF-2GDP complex, P_i and eIF-3 are released from the small subunit. The catalysis of GTP hydrolysis by eIF-5 is a crucial step in the release of eIF-2 and eIF-3 as well as the joining of the two ribosomal subunits. If the GTPase reaction is inhibited by addition of non-hydrolyzable GTP analogs, these critical steps in translation are inhibited. The binding of the large ribosomal subunit and the release of the initiation factors thus complete the process of translation initiation. The final complex contains the two subunits of the ribosome, the Met-tRNA initiator and mRNA. The anticodon of the Met-tRNA is base-paired with the initiation codon AUG at the P-site whereas the A-site is empty. This ternary complex between mRNA, tRNA and ribosome is now ready for the elongation phase of translation.

The process of translation elongation can be considered as a three-step cycle. In the first step, an appropriate aminoacyl-tRNA molecule interacts with the codon exposed at the A-site of the ribosome/mRNA complex. This reaction is mediated by the elongation factor EF-1$_\alpha$. The function of this factor is similar to that of eIF-2 and its prokaryote counterpart, EF-Tu. EF-1$_\alpha$ first complexes with GTP before its association with aminoacyl-tRNAs. This ternary complex then interacts with the ribosome and the aminoacyl-tRNA is delivered to the A-site. Hydrolysis of the EF-1$_\alpha$-associated GTP then leads to the formation of a EF-1$_\alpha$GTP binary complex that dissociates from the ribosome. If this GTP hydrolysis reaction is inhibited by non-hydrolyzable GTP analogs, the subsequent reactions for peptide bond formation are inhibited. In this step of the cycle, the carboxyl group of the amino acid at the P-site is uncoupled from the tRNA and then joined by a newly formed peptide bond to the amino acid at the A-site. This reaction is catalyzed by a peptidyltransferase activity that is thought to be associated with the rRNA molecule in the large ribosomal subunit. No soluble protein or cofactor is apparently required. The last step of this ribosome cycle involves the translocation of the tRNA from the A-site to the P-site. Simultaneously, the free tRNA molecule that used to occupy the P-site is released and the ribosome migrates on the mRNA by exactly three bases. This movement of the ribosome is driven by protein conformational changes that accompany hydrolysis of a molecule of GTP to GDP and P_i. An important factor involved in this process is EF-2 which is a target for modification by an inhibitor of eukaryotic protein synthesis called diphtheria toxin. A post-translationally modified histidine residue, called diphthamide, in EF-2 is specifically ADP-ribosylated by this toxin. The resultant EF-2 is unable to catalyze GTP hydrolysis or ribosome translocation. The translocation of the ribosome to a new codon thus completes one cycle of the elongation reaction. The whole process requires about one twentieth of a second and for an average protein of 300 amino acid residues, about 15 s will be needed for its synthesis. Once the elongation phase of translation has begun and the ribosome starts to 'travel' down the mRNA, a new round of translational initiation with new ribosomes can take place. In this way, multiple ribosomes can be found to associate simultaneously with actively translated transcripts. These are called **polysomes** and can be isolated from free ribosomes based on the differences in their sedimentation coefficients.

The final step in protein translation involves the termination of the ribosome cycle at the proper termination codon. As shown in Fig. 8.7, tRNAs with anticodons complementary to UAA, UAG and UGA are normally not present in eukaryotes. When the ribosome reaches one of these termination codons, a protein called **release factor** (**RF**) binds to the A-site along with a molecule of GTP. The occupancy of the A-site stimulates the peptidyltransferase activity of the ribosome and the peptidyl-tRNA ester at the P-site is effectively hydrolyzed. This results in the release of the translated peptide and subsequently, the GTP bound to the RF protein is also hydrolyzed. Finally, RF dissociates from the ribosome and the free tRNA and ribosomal proteins are separated from the mRNA template. These components are then ready to participate in the synthesis of yet another protein. In rabbit reticulocytes, RF has an apparent size of 110 000 daltons and contains subunits of 55 000 daltons.

Little is known about the *in vivo* regulatory mechanisms of translation in plants, although wheat germ extracts have been used for *in vitro* translation assays for many years. However, since the pattern of proteins in plant cytoplasmic ribosomes is quite similar to that of mammals and the green alga *C. reinhardtii* (Capel & Bourque, 1982), it seems likely that the basic function and properties of cytoplasmic ribosomes are well conserved through evolution. In recent years, the study of replication mechanisms in several plant RNA viruses have revealed translational activation sequences. One of the best studied is a 68-base pair sequence called Ω in the untranslated leader of tobacco mosaic virus. This sequence when fused upstream of heterologous transcription units can activate translation of the resultant mRNA in both plants and bacteria (Gallie *et al.*, 1987). Interestingly, the effects of the Ω sequence in bacteria is much more dramatic in the absence of a Shine–Dalgarno sequence in the transcript and its effects appear to be independent of its position relative to the initiation site (Gallie & Kado, 1989). The mechanism by which this *cis*-acting element enhances translation remains to be elucidated.

An example of the more-characterized systems for study of translational regulation in eukaryotes is the transcript encoding ferritin, a protein conserved in eukaryotes (Klausner *et al.*, 1993). This protein is involved in the sequestration of iron within cells and its synthesis is regulated by the concentration of free iron in the cytosol. A stem–loop structure, called IRE (Iron Responsive Element), at the 5′-untranslated region of the ferritin transcript has been shown to be involved in sequence-specific interaction with a cellular protein, IRE-BP. Association of IRE-BP with the 5′ region prevents translational initiation of the ferritin mRNA. The binding of IRE-BP to the ferritin transcript is inhibited when iron is bound to this protein, thus IRE-BP is displaced from the ferritin transcripts under high concentrations of iron and translation of ferritin is activated. Interestingly, a similar IRE and the same IRE-BP are used for the regulation of mRNA stability of another gene encoding transferrin receptor (TfR), a protein that is expressed under low intracellular concentrations of iron and is involved directly in iron transport. In this case, the IRE is situated at the 3′-untranslated region of the TfR transcript and the association of IRE-BP with this site prevents mRNA degradation. When the iron concentration within the cell rises, IRE-BP will dissociate from the IRE and TfR mRNA turnover will be accelerated.

In contrast to that of cytoplasmic protein synthesis in plants, the regulation of translation in chloroplasts has received a lot more attention since it appears to play an important role in plastid biogenesis. An especially powerful approach is the identification of nuclear mutants in *C. reinhardtii* that apparently are defective in the translation of specific plastid-encoded genes (Rochaix *et al.*, 1989; Mayfield, 1990). Interestingly, each of these nuclear mutations appears to affect only one or a very small subset of plastid transcripts. Multiple nuclear genes may also be required for the translation of some plastid genes. These nuclear genes have been suggested to encode factors that interact with stem–loop structures in the 5'-untranslated regions of plastid mRNAs. Consistent with this hypothesis is the fact that plastid mutations that block translation of some chloroplast transcripts indeed map to these putative stem–loops. More recently, chloroplast proteins that interact with these stem–loop structures of the *psbA* gene of *C. reinhardtii* have been characterized (Danon & Mayfield, 1991). The results from this work show remarkable similarity to those observed for the regulation of ferritin production and argue for a similar principle by which mRNA translation may be regulated in general.

Although most polypeptides are synthesized in a continuous manner from an open reading frame, there is recent evidence that indicates the presence of protein splicing. The first case of this phenomenon was reported in 1985 by Carrington *et al.* in a study of the gene structure of concanavalin A (Con A), a legume seed protein. The nascent polypeptide that is translated from the Con A mRNA is apparently processed in such a way that results in the loss of the first 29 amino acid residues. In addition, the N-terminal half of the predicted protein sequence based on the cDNA information is found in the C-terminal half of Con A *in vivo*. The analysis suggested that this rearrangement occurs at the junction between amino acids 118 and 119 of the Con A protein. The process of protein splicing has also been implicated in the synthesis of the RecA protein from *Mycobacterium tuberculosis*, the DNA polymerase from a thermophilic archaebacterium, and a subunit for the yeast vacuolar H^+-ATPase (Shub & Goodrich-Blair, 1992). In the latter case, two portions of the vacuolar H^+-ATPase subunit TFP1, a 69 kDa protein, are separated by sequences that encode a protein with homology to a yeast endonuclease, a 50 kDa protein. However, a single un-rearranged mRNA species that contains these coding regions are produced, and translation of this full-length transcript results in a 69 kDa and

a 50 kDa protein. This processing, which involves the release of the internal 50 kDa protein and the proper splicing of the two halves of TFP1, can take place in either rabbit reticulocyte lysates or in *E. coli*, thus demonstrating that no eukaryote-specific factors are needed. It is also possible that this protein splicing process may be self-catalyzed, analogous to the self-splicing group I mRNA introns. The elucidation of the underlying mechanisms by which separate regions of a polypeptide may be rearranged will provide new information on how genetic information may be manipulated in nature.

8.4.3   Post-translational modification

The function of proteins can be regulated in many ways inside a cell. Many proteins are localized to specific sites of action within a cell after their synthesis. Some of these are even secreted out of the cell and are either transported to specific regions within the organism or are sent into the surrounding medium as signal molecules that mediate communication between organisms. Important examples of these types of proteins are the extracellular matrix proteins and peptide hormones of animal cells. In plants, the cell wall contains a number of interesting classes of proteins that consist of highly repetitive sequences. These are the hydroxyproline-rich glycoproteins, also called extensins, which can contain 20 or more repeats of short proline-rich peptide repeats such as Ser-(Pro)$_4$ (Showalter, 1993). The proline residues in these proteins are hydroxylated by an enzyme called prolyl hydroxylase after their synthesis. They are then transported to the cell wall where they are believed to play an important role in this dynamic structure that is of vital importance to the plant. Aside from extensins, there are also glycine-rich proteins and proline-rich proteins in the cell wall of plants. Each of these different types of proteins appears to have a distinct cell-specificity of expression, suggesting that they may have different functional roles in the plant (Showalter, 1993).

Recently, the first candidate for a plant peptide hormone was discovered in the laboratory of C. Ryan (Pearce *et al.*, 1991; McGurl *et al.*, 1992). This short polypeptide of 18 amino acid residues in length was found to be secreted by wounded plant cells of tomato. It is then transported to other parts of the plant where its presence appears to induce the synthesis of a number of plant defense genes such as proteinase inhibitors. The ability of this peptide to systemically induce defense-related genes thus gives rise to the name **systemin**. Interestingly, systemin is apparently synthesized first as a much larger precursor and subsequent proteolytic processing is assumed to be necessary to produce the active peptide. This mode of synthesis is strikingly similar to the well-characterized biosynthetic pathways for animal peptide hormones, suggesting a common origin from which this type of signalling system has evolved. In the near future, the elucidation of the mechanisms of secretion, transport and signalling for peptides such as systemin will likely uncover novel aspects of plant biochemistry.

In addition to their proper localization, protein functions are in many cases modulated at the level of post-translational modification such as glycosylation, methylation, phosphorylation, acetylation and acylation. Also, the rate by which particular proteins are degraded can play an important regulatory role. Combining with the various points by which RNA synthesis and processing can be controlled, the cell thus has an enormous repertoire of pathways by which genetic information can be expressed in a dynamic fashion.

(a) *Transport and secretion*

Aside from the relatively small number of proteins that are synthesized in the mitochondria and plastids, most of the proteins in a plant cell are translated in the cytosol and then transported to the various organelles. In addition to the mitochondria and plastids, the other membrane-enclosed compartments within a plant cell include the nucleus, the Golgi complex, secretory vesicles, vacuoles, tonoplasts and the endoplasmic reticulum. In order for the numerous proteins, each with a distinct destination, to be routed correctly to these subcellular compartments, there must be 'signals' that allow for their proper identification. These signals are likely specific amino acid sequences or structural features that can be recognized by regulatory components, such as membrane-bound receptors, in each of these different compartments. Indeed, analyses in various eukaryotic systems have demonstrated that this is the case. This section discusses two plant systems that have been well-characterized in terms of the signal components that are involved in protein traffic control.

(i) Translocation of proteins into organelles

Mitochondria and plastids are two of the important organelles within plant cells. Although each of these compartments contains its own genetic material, most of the proteins within these semi-autonomous structures are synthesized in the cytosol. Proteins that are destined for these organelles are synthesized as a larger precursor with an amino terminal extension that is not present in the mature protein. This extra sequence, called a **signal peptide** or **transit peptide**, can vary in length from about 30 to 100 amino acid residues. The attachment of these signal peptides to heterologous proteins has demonstrated that they can indeed direct foreign proteins to the proper organelles. Thus, there must be information inherent in these sequences that allow for their proper localization within the cell once they are synthesized. Sequence comparison between many such signal peptides reveals that there is no obvious sequence motif which is absolutely conserved among proteins that are destined for the same compartment. Instead, there are common features which are found in various signal peptides that have been studied to date.

1. There is usually an abundance of serine and threonine in the signal peptide;
2. Small hydrophobic amino acids, such as valine and alanine, are found quite often;
3. The overall charge of the transit peptide is usually basic with very few acidic amino acid residues;
4. Regions are present in the transit peptide that can form amphiphilic α-helices or β-sheets.

These are thought to facilitate the interaction between the transit peptide and the membranes that the protein will have to traverse. In fact, synthetic peptides, the structures of which are based on mitochondrial transit peptides, have been shown to partition between aqueous and membrane phases (Allison & Schatz, 1986). Thus, signal peptides that can target proteins to the mitochondria may interact spontaneously with the membrane. It is possible, however, that membrane-bound proteins that interact with these targeting signals may also be required for the proper discrimination between proteins destined for different subcellular locations. Signal peptide-binding proteins have been detected in chloroplasts, although their functional role in protein targeting still remains uncertain (reviewed in Keegstra *et al.*, 1989).

At present, some differences in the initial process of signal peptide binding to mitochondria and chloroplasts are known. Although ATP is known to be required for transport into both organelles, mitochondrial transport has an additional requirement for a membrane potential.

Moreover, the hydrolysis of ATP in the case of mitochondria occurs in the cytoplasmic side whereas that for chloroplast transport appears to take place inside the organelle. The binding of precursors to the chloroplasts also requires ATP whereas the interaction between mitochondria and the signal peptides appears to be spontaneous. Although these differences suggest that the process of protein recognition and transport may be different between these two organelles, a report by Hurt *et al.* (1986) suggests that there must also be common mechanisms involved. In this work, the transit peptide of *C. reinhardtii* RbcS (small subunit of ribulose-1,5-bisphosphate carboxylase), a chloroplast-localized protein, was fused to mouse dihydrofolate reductase (a cytoplasmic protein) and a truncated yeast mitochondrial protein, subunit IV of cytochrome oxidase. The RbcS signal peptide was able to target both proteins into yeast mitochondria, although with less efficiency than an authentic mitochondrial signal peptide. These results argue that there must also be common functional properties between targeting signals for these organelles.

The transport of proteins into mitochondria and chloroplasts may be cotranslational *in vivo*. Thus, as these proteins are being synthesized on cytoplasmic ribosomes, their translocation into organelles may be initiated and both processes are carried out simultaneously. However, it is clear from *in vitro* experiments with isolated organelles that a cotranslation mechanism is not obligatory for the proper recognition and processing of these proteins. Upon uptake of the preproteins into the organelles, the signal peptide is cleaved by a peptidase activity. Several chloroplast proteins have been reported to be associated with this signal peptidase activity (Oblong & Lamppa, 1992). Their characterization may provide more information on the events and mechanisms involved in the processing of signal peptides.

(ii)   Translocation of proteins into the ER

The endoplasmic reticulum (ER) is involved in the routing of many types of proteins within a cell to different destinations. Mechanistically, the two most well-characterized target compartments are the lumen of the ER and the vacuole. Like the case for mitochondria and plastids, the uptake of the preproteins into the ER involves a signal peptide at the N-terminal end. These are also short N-terminal extensions of about 13–30 amino acids long and are removed upon uptake into the ER by an endopeptidase. The composition of these ER-targeted signal peptides are similar to those for

mitochondria and plastids as well. They contain basic regions near the amino terminus and also an abundance of hydrophobic residues. Once synthesized on cytoplasmic ribosomes, these signal peptides interact with a ribonucleoprotein called **signal recognition particle (SRP)** and an ER-bound receptor to initiate their transfer into the lumen of the ER. SRP has now been well-characterized in animals and yeast systems and is a protein complex of six different polypeptides. It contains a 7S RNA that appears to be highly conserved in eukaryotes, including plants. A 54 kDa subunit of SRP is found to bind both the 7S RNA as well as the signal peptides destined for the ER. Upon binding to these signal peptides, SRP prevents any further translation of the mRNA by the associated ribosomes. This inhibition, however, is relieved once this complex is associated with a 'docking-protein' in the ER membrane. This protein is also known as the **SRP receptor**. Upon positioning the signal peptide at the SRP receptor site, SRP is released and translation of the mRNA is resumed with the concomitant translocation of the protein into the ER. This provides an elegant mechanism by which transfer into the ER is coupled with translation. The tight co-ordination between these two processes may be crucial for intrinsic membrane proteins that otherwise will not be soluble in the cytosol and thus will never reach the receptor sites.

Once taken up in the ER, a protein is either secreted or is delivered to a specific subcellular compartment. For proteins known to be localized in the vacuoles, it has been shown that multiple signals are likely required for their proper targeting. Thus, although their signal peptide is sufficient to affect ER-uptake when fused to a heterologous protein, the final product is secreted rather than targeted to the vacuoles. The specific structural properties of vacuolar proteins that mediate their localization remain to be determined. For proteins that are retained on the ER, a common carboxy-terminal motif is found in eukaryotes. This motif consists of either KDEL or HDEL. Although the mode of action for this motif is not known, its introduction to the C-terminus of a vacuolar protein results in substantial retention of the chimeric gene product in the ER (Chrispeels, 1991). This result implies that this motif may play an important role in the sorting process of proteins within the ER.

In addition to acting as a 'point-of-departure' for many proteins that are either secreted or targeted to specific compartments within a cell, the ER is also a place where many of these proteins are glycosylated. Typically, an asparagine residue of the protein that is being translated is modified by glycans, such as mannose, through the action of oligosaccharyl transferases. A preferred site for this process is NH_3-Asn-X-Thr/Ser-COOH. If the translocation process for a protein involves passing through the Golgi apparatus, the N-linked glycans can be further modified by glycosidases and glycosyltransferases. This results in the linkage of different types of sugar moieties, such as fucose and xylose, to the simple glycans that were added to the polypeptide in the ER. Glycosylation of some proteins has been implicated in facilitating their transport through the ER to their destinations. These protein-bound glycan residues may also be involved in protein folding or the turnover of the protein.

(iii) Molecular chaperones and protein folding

The synthesis of a functional protein requires that the polypeptide produced by translation can be folded in the proper conformation. In order for the active structure to be stable under normal conditions, it is thought that this should represent a thermodynamically favored state, an energy minimum. This idea then predicts that given the proper conditions, a newly synthesized or denatured polypeptide should be able to self-assemble into a functional form of unique structure. Indeed, this is the case for some small proteins, such as ribonuclease A. However, for most proteins, the probability of incorrect interactions among different regions of the polypeptide can occur. This is even more problematic for multimeric proteins with different subunits that require precise assembly. In the last 10 years, a new class of proteins has emerged from studies on requirements for protein assembly *in vivo* (reviewed in Ellis & van der Vies, 1991). These are called **molecular chaperones** and are defined as proteins that prevent incorrect interactions between parts of other protein(s), but are not involved in the final assembled protein or protein complex. The most well-characterized chaperone in plants is the chaperonin 60 (chap60) located in the plastids. This protein was first discovered in 1980 as a protein that is associated with the large subunit of RbcL after its synthesis in the chloroplasts. It was subsequently found to bind to a variety of newly synthesized proteins, including the small subunit of ribulose-1,5-bisphosphate carboxylase (rbcS) and the β-subunit of the chloroplast ATPase. Interestingly, chap60 does not appear to bind to the mature processed form of rbcS, thus indicating that the signal peptide and/or an unfolded structure is required for chaperone interaction. Plastid chap60 consists of

two kinds of subunit, α and β, that are related to each other in protein sequence. In addition, they are homologous to the bacterial chap60 protein (also known as groEL). The functional characteristics of the bacterial chap60 protein are much better defined than its plastid homolog. In *E. coli*, another chaperonin, chap10, is known to interact with chap60 to facilitate assembly of various proteins. However, chap60 alone has been shown to be capable of facilitating transport of proteins across membranes. One likely mode of action for chap10 is that by binding to chap60, it affects the release of the properly folded polypeptide. Both bacterial chap60 and chap10 are oligomeric proteins with 14 and 7 identical subunits each, respectively. They are known to be arranged in a seven-membered ring configuration. In plants, chap60 is encoded by a small gene family and the expression of these genes is enhanced by light as well as heat shock. The bacterial and plastid chap60s have been shown to mediate the folding of denatured rbcL and rbcS subunits from *Rhodospirillum rubrum*. Elucidation of the mechanisms and components involved in chaperone function will provide new information on how protein assembly is carried out *in vivo*.

(b) Regulation of proteins by covalent modification

Post-translational modification of proteins plays an important role in regulatory pathways of all cells. Covalent modifications such as phosphorylation, acylation, acetylation and carboxymethylation are some of the well-known examples. Proteins can also be regulated by processing events that are carried out by highly specific proteases. This section discusses two of the more widely studied protein modifications that are known to be involved in the modulation of protein functions.

(i) Phosphorylation

This is probably the best characterized and most widely applied mode of protein modification. A particularly versatile feature of protein phosphorylation is its reversible nature, which is achieved by the competitive action of two classes of enzymes, kinases and phosphatases. *In vivo*, the sensitivity of these enzymes to external and internal signals allows for rapid as well as long-term cellular responses. The demonstration that the activity of some protein kinases and phosphatases are themselves regulated by phosphorylation suggests that a cascade of phosphorylation events is involved in the transduction of extracellular signals or developmental decisions. Elegant examples of these phosphorylation cascades are those involved in the onset of mitosis in eukaryotes (Nurse, 1990) and the activation of oncogenesis in animal cells by mitogens (Roberts, 1992). Two main types of protein kinases are known in eukaryotes: those that phosphorylate serine and threonine residues, and those that modify tyrosine residues. A phosphate group is attached by a phosphodiester bond to the hydroxyl oxygen of the R-group in these amino acids. The specificity of each kinase is governed by the particular amino acid sequence surrounding the target residue. The actions of these kinases are counteracted by two classes of phosphatase that act on phosphoserine/phospho-threonine and phosphotyrosine, respectively. In bacteria, phosphorylation of histidine and aspartic acid residues is used for the regulation of such diverse functions as chemotaxis and osmoregula-tion of transcription (Stock et al., 1989).

In higher plants, the study of protein phosphory-lation has been well documented (Ranjeva & Boudet, 1987; Roberts & Harmon, 1992). Protein phosphorylation responsive to stimuli such as light and phytohormones has been observed, although in most cases their functional significance remains unclear. However, there are a number of examples where the role of protein phosphorylation has been well characterized in plants. The multi-subunit enzyme pyruvate dehydrogenase (PDC), involved in the oxidative decarboxylation of pyruvate to acetyl-CoA, is a key enzyme for metabolite control. This enzyme complex has been purified from plant mitochondria and plas-tids. Like its counterpart in animal systems, phosphorylation of PDC leads to deactivation of this enzyme and a decrease in acetyl-CoA.

The major light-harvesting chlorophyll binding protein in the chloroplasts, LHCPII, is also known to be reversibly phosphorylated at a threonine residue located in its N-terminus. In this case, phosphorylation is catalyzed by a thylakoid-bound kinase that is regulated by the redox potential of the plastoquinone (PQ) pool of the chloroplast membrane. This kinase is activated when the PQ pool is reduced, such as the case under high light intensities. The phosphorylation status of LHCPII apparently modulates the distribution of excita-tion energy between photosystems (PS) I and II complexes involved in photosynthesis. More light energy is diverted to PSI when LHCPII is phosphorylated, thus enhancing the rate at which the reduced PQ pool can be oxidized and electrons transferred to ferredoxin. A thylakoid-bound phosphatase is involved in the de-excitation pro-cess. This enzyme, however, appears to be

constitutive and is not affected by the redox state of PQ. Several protein kinases have been identified and purified from thylakoid membranes. It remains to be demonstrated unequivocally which enzyme corresponds to that which actually phos-phorylates LHCPII in vivo.

In the last few years, many plant genes that encode protein kinases or putative kinases have been reported. These are typically obtained by one of two methods. The first is a biochemical approach in which the enzyme activity of interest is purified to homogeneity. The corresponding gene is then cloned by using either degenerate oligonucleotides synthesized according to the partial protein sequence, or antibodies raised against the protein (Roberts & Harmon, 1992). The second is a direct screening procedure using a set of degenerate oligonucleotides that is based on the conserved sequences of protein kinases studied so far (Lawton et al., 1989). The study of calcium-dependent protein kinases (CDPKs) is especially interesting since many cellular pro-cesses in plants appear to involve calcium as a signal transduction intermediate. These include phytochrome-dependent seed germination and abscisic acid-induced stomatal closure. CDPK-like enzyme activities have been found in a variety of plants and algal systems. In general, they appear to be monomers of 40–90 kDa. The enzyme binds calcium and is activated by 50–100-fold. In contrast, calmodulin or phospho-lipids have no obvious effects on CDPK activities, unlike protein kinase C or the Ca^{2+}/calmodulin-dependent kinases of animal systems. Like most kinases that have been studied, CDPK can also catalyze at a slow rate autophosphorylation on certain of its own serine and threonine resi-dues. The genes for CDPK of soybean, carrot and Arabidopsis have been cloned recently. They appear to be members of a multigene family in these plants and their sequences show significant similarities to the catalytic domain of Ca^{2+}/calmodulin-dependent protein kinases of animal systems. Moreover, the deduced amino acid sequence of the cDNAs shows that the car-boxyl-terminus of CDPK is homologous to the calcium binding domain of calmodulin. This unique feature is consistent with the Ca^{2+}-dependent, but calmodulin-independent behavior of CDPKs. Although CDPKs are now well-characterized at the molecular level, their in vivo role and mechanism of regulation remains unknown. Future studies using a combination of biochemical and molecular approaches will hopefully elucidate the function of this class of proteins.

(ii) Protein acylation

The covalent attachment of lipid moieties to proteins represents another widespread mechanism by which protein functions can be regulated. In some cases, the attachment of fatty acids is critical for the targeting of the protein to the membranes whereas in other cases, the role of acylation appears to be involved in proper folding or protein–protein interaction. In addition, to glycophospholipid, residues with a chain length of 12 (laurate), 14 (myristate), 15 (farnesyl), 16 (palmitate) and 20 (geranylgeranyl) have been found to be attached to proteins. In eukaryotes, three different modes of acylation have been characterized.

1. N-myristoylation of a glycine residue at the amino terminus during translation. This involves the formation of an amide linkage between the carboxyl group of a fatty acid and the amino group of glycine.
2. Formation of a thioester or ester bond between the fatty acid and an internal cysteine residue or serine and threonine residues. This is usually attached after translation has been completed.
3. Covalent attachment of a phosphatidyl-inositol group to the carboxyl-terminus after translation has completed.

Of these different mechanisms, N-myristoylation is one of the best characterized. The enzyme, myristoyl CoA : protein N-myristoyl transferase (NMT), has been purified from yeast and has also been detected in wheat germ and rat liver (Towler et al., 1988). A number of important enzymes in animal and yeast systems have been demonstrated to be critically dependent on N-myristoylation. These include cAMP-dependent protein kinase and the family of α-subunit in trimeric G-proteins involved in the signal transduction pathway of many membrane-bound receptors. For these enzymes to be acylated, the methionine residue at the N-terminus has to be removed before the linkage of fatty acids by NMT can commence. Interestingly, although N-myristoylation is required for the transforming function of a dominant mutant of $G\alpha_{12}$ in rat fibroblasts, it is not essential for membrane association (Gallego et al., 1992).

Another well-studied acylation system is that which modifies the cysteine residue at the C-terminus. These modification sites are usually of the type C-A-A-X where A is any aliphatic residue and X is any residue. Farnesyltransferase and geranylgeranyltransferase are enzymes that catalyze some of these acylation events with the latter preferring substrates with the sequence of C-A-A-L. In yeast and rat, these enzymes are found each to contain two different subunits called α and β. The β subunit is thought to interact with the C-terminal end of the protein substrate whereas the α subunit binds the fatty acid. The last three amino acids are removed prior to the linkage of the fatty acid moiety to the terminal cysteine via a thioester bond. The C-A-A-X motif has been found in the C-terminus of many Gαi and Gγ subunits of heterotrimeric G-proteins, the ras-oncogene and related small GTP-binding proteins in animal and fungal systems. These proteins are all known to play important roles in signal transduction and protein secretion in these systems. The covalent linkage of a geranylgeranyl isoprenoid group to a Gγ-dependent kinase was found to induce membrane association and activation of kinase activities in mammalian cells (Inglese et al., 1992). In this case a direct correlation is obtained between protein acylation, membrane localization and enzyme function. Since plant Gα subunit also has been found to contain the N-terminal glycine residue that is N-myristoylated (Ma et al., 1990), as well as the C-A-A-X motif in the C-terminus of some small GTP-binding proteins (Palme et al., 1992), it is very likely that these types of protein acylation will also be conserved in higher plants. One documented case of protein acylation in plants was reported by Mattoo & Edelman (1987), who showed that in the aquatic angiosperm Spirodela, a 32 kDa protein, which most likely is the reaction center protein for photosystem II, is palmitoylated. The acylation of this membrane protein is correlated with its movement from the stromal lamellae of the thylakoid into the stacked granal regions. However, the actual site(s) of protein acylation and the functional consequence of the palmitoylation remains to be established. The study of how protein acylation can be regulated in plants and its effects on protein functions will be an important area of study.

(c) Protein turnover

Protein degradation, or turnover, plays an important role in the life of plants from seed germination to senescence (reviewed in Vierstra, 1989). Catabolism of storage proteins in a germinating seed provides a readily available source of essential substrates for the growth of a seedling. Proteins that became damaged due to free-radical attacks or heat stress need to be removed efficiently. This housekeeping function also extends to specific regulators of development whose expression needs to be tightly controlled. In addition, proteolysis is a way to desensitize activated components of signal transduction pathways in order to respond rapidly to changes in the environment. In plants, a number of proteins are known to be relatively unstable with measured half-lives in the range of hours or less. The ability to rapidly degrade these proteins allows for their regulation at the level of protein concentration in response to changes in the environment. Nitrate reductase, an enzyme that is involved in nitrogen assimilation from nitrate, has a half-life of about 4 h in tobacco. Its synthesis is responsive to the level of intracellular nitrate and ammonia. Thus, expression of this enzyme is repressed under low concentration of nitrate and high levels of ammonia. In turn, the rapid degradation process quickly lowers the enzyme activity within the cell. The photoreceptor phytochrome is also under proteolytic control in etiolated seedlings. Thus, the inactivated phytochrome is in a stable state after its synthesis and the protein accumulates in dark-grown tissues. However, upon activation by red light, this photoreceptor is rapidly degraded. This is a good example in which the stability of an important signal sensor is regulated. Protein turnover also appears to play an important role in the biogenesis and maintenance of the plastids. Many of the enzymes in the plastids consist of multiple subunits. It is well-documented that defects which prevent accumulation of one of the subunits will also decrease the level of the other subunits in the complex, even though their rate of RNA and protein synthesis appears to be unchanged (Bruce & Malkin, 1991). Thus, the completed complexes in these cases must be degraded much more rapidly than those that do have all the proper subunits. One striking example of such a phenomenon is the inhibition of rbcS expression by antisense RNA (Rodermel et al., 1988). In this case, a decrease in the level of rbcS, a nuclear gene product, is reflected in a decrease in the accumulation of rbcL, a plastid-encoded protein. A specific membrane-associated protease has also been proposed to carry out the rapid turnover of the 32 kDa photosystem II reaction center protein (Ohad et al., 1985). In this case, the trigger for the turnover appears to be protein damage under high light intensities. An important point to be stressed is that the above examples are highly regulated processes. Thus, there must be proteolytic pathways that have high specificity in substrate recognition. In the following, we will discuss some of our knowledge on the signals that direct protein turnover in eukaryotes.

(i) PEST sequences and the N-end rule: signals for rapid protein degradation

Since both short-lived and stable proteins co-exist in the same cell, it is obvious that there must be special signals that can allow the cell to turnover one class of proteins more rapidly than another. An economical way of achieving regulated protein turnover within a cell is to identify or mark proteins that will become the preferred targets of intracellular proteases. By comparing the amino acid sequence of 12 rapidly degraded proteins with half-lives of 2 h or shorter, Rogers et al. (1986) found that they all have one or multiple regions rich in the amino acids proline, glutamic acid, serine and threonine. They named these types of regions as PEST sequences, after the one letter amino acid code for these residues. Usually, these PEST sequences are found to be flanked by basic amino acids. Rogers et al. (1986) propose that exposure of PEST-containing regions on a protein will render it much more susceptible to intracellular proteolysis. This hypothesis takes into account that PEST sequences are also present in a variety of proteins that are known to be regulated in their stability. In these cases, association with other proteins or a specific conformation of the protein itself may sequester the PEST sequence from the recognition component of the proteolysis pathway. At present, not much is known about the mechanism involved in PEST recognition and the protease(s) that may be responsible for the subsequent proteolysis. However, it is noteworthy that a 19S protein particle has been reported to be a ubiquitous 'proteasome' in many eukaryotic cells (Arrigo et al., 1988). This enzyme complex has an apparent molecular weight of about 700 kDa and prefers alkaline pH for its function. In addition, it contains three endoproteolytic sites that can catalyze hydrolysis of proteins and peptides. It will be interesting in the future to see whether there is any functional link between the PEST sequence and these proteasomes.

Although most proteins are synthesized with a methionine residue at their N-terminus, the identity of the amino acid at the N-terminus can be altered. A class of enzymes called **aminoacyl-tRNA-protein transferases** can catalyze the addition of specific amino acid residues to the N-terminus of the acceptor protein and thus causes a post-translational modification of the protein sequence. Working with the yeast S. cerevisiae, Bachmair et al. (1986) demonstrated that the identity of the N-terminal amino acid can profoundly affect the stability of a protein that is otherwise identical. This work introduced the so-called **N-end rule** for protein stability. The amino acids M, S, A, T, V and G (in the one letter amino acid code), when placed at the N-terminus of the bacterial protein β-galactosidase, are found to result in stable proteins with half-lives of greater than 20 h. When substituted with the other amino acids at this position, half-lives of 30 min to less than 2 min were found. These results demonstrate that the identity of the N-terminal amino acid is a critical signal for the destabilization of proteins. More recent work has shown that a second determinant of the N-end rule is a critical lysine residue within the protein itself. No specific sequence context for this critical lysine is apparently required. The optimal location of this lysine residue is within a relatively disordered region that is close to the N-terminal domain. The lack of a specific amino acid sequence that is required for the recognition of this lysine residue and the PEST sequence by the targeting components is reminiscent of the situation with transit peptides for protein translocation. In both cases, the structural characteristics of the peptide rather than its specific sequence act as the determinant of specificity. For the N-end rule, a family of related proteins called **E3** is known to recognize and bind to the destabilizing amino acids at the N-terminus. Subsequently, the interaction with the internal lysine residue then targets the protein for degradation by a proteolysis pathway that involves a protein component called ubiquitin (Bachmair & Varshavsky, 1989). This pathway of targeted proteolysis is discussed in more detail in the following section.

(ii) Ubiquitin: a universal targeting system of protein turnover

Ubiquitin is a 76 amino acid-long polypeptide that is found with few changes in all eukaryotes examined to date. Its amino acid sequence is one of the best conserved in nature and differs by only two to three residues between yeast, mammals and plants. This protein can be detected in both the nucleus and the cytoplasm. However, it does not appear to be in the mitochondria or plastids. The role of ubiquitin in protein turnover was first elucidated in animal systems and subsequently verified in other eukaryotic systems, including plants (reviewed in Vierstra, 1989). Essentially, the main function of ubiquitin is to 'tag' the proteins which are destined to be proteolyzed quickly in the cytosol. It achieves this goal by covalently attaching its C-terminal glycine carboxyl group to the ϵ-NH$_2$ group of the N-terminal or internal lysine

residues of the target protein. Once attached, the protein–ubiquitin conjugate is then degraded rapidly by an ATP-dependent protease complex and free ubiquitin is also released in the process.

In order to participate in the conjugation reaction, free ubiquitin needs to be activated by an enzyme, **E1**. E1 catalyzes the ATP-dependent adenylation of the C-terminus of ubiquitin. Ubiquitin is first linked to E1 via a thiol ester bond and is then transferred to a carrier protein called **E2** by a transesterification reaction. The E2–ubiquitin conjugate is then used as a substrate for the ligation of ubiquitin to lysine residues in proteins complexed with E3. This is catalyzed by a **ubiquitin–protein lyase**. Interestingly, this enzyme can also carry out the opposite reaction of specifically cleaving the amide linkage between ubiquitin and the target protein. This reverse reaction may be important in the regulation and proof-reading of the pathway. In any case, this lyase does not require ATP, unlike the ubiquitin–conjugate protease complex.

The ubiquitin-dependent pathway of protein turnover appears to be the primary avenue of targeted proteolysis in cells. Greater than 90% of short-lived proteins have been shown to be degraded via the ubiquitin conjugate pathway in a mouse cell line. In plants, the photoreceptor phytochrome has been found to be proteolyzed in its active form via an ubiquitinated intermediate. However, the mechanisms involved in the differential targeting of plant cell proteins for this and other pathways of proteolysis remain to be elucidated. Little is known about the mechanism for protein catabolism in plastids and mitochondria. A nuclear mutant of maize shows an accelerated rate of turnover for two polypeptides associated with photosystem II (Leto *et al.*, 1985). This suggests that there are likely to be nuclear-encoded components which are involved in the regulation of protein stability in the plastids. Although we are beginning to understand more about protein turnover in plants, it is quite clear that much more fundamental questions will need to be addressed in the near future in order for us to really appreciate the intricacies of the pathways involved.

8.5 SUMMARY

In writing this chapter, I have tried to cover as much as possible of the general aspects of nucleic acids and proteins, with emphasis on regulatory mechanisms. Data obtained in various plant systems are compared with those of other eukaryotes to illustrate the striking conservation of many pathways and modes of regulation. Thus, the beauty of the apparent complexity of an organism lies within the simple principles that are reiterated in many different forms and used in a combinatorial fashion. This is illustrated in the common architecture of transcription factors and promoters, the role of RNA in translation and splicing, and the structural properties of peptides as determinants for protein transport and degradation. With a combination of molecular genetics and biochemical approaches, and applying the tools and knowledge gained from other systems, advances in plant biochemistry will more than likely increase its pace in the near future.

REFERENCES

Agabian, N. (1990) *Cell* **61**, 1157–1160.

Alberts, B., Bray, D., Lewis, J., Raff, M., Roberts, K. & Watson, J.D. (1989) *Molecular Biology of the Cell*, 2nd edn. Garland Publishing, New York.

Allison, D.S. & Schatz, G. (1986) *Proc. Natl. Acad. Sci. U.S.A.* **83**, 9011–9015.

Allison, L.A., Moyle, M., Shales, M. & Ingles, J.C. (1985) *Cell* **42**, 599–610.

Anderson, S., Bankier, A.T., Barrell, B.G., de Bruijn, M.H.L., Coulson, A.R., Drouin, J., Eperon, I.C., Nierlich, D.P., Roe, B.A., Sanger, F., Schreier, P.H., Smith, A.J.H., Staden, R. & Young, I.G. (1981) *Nature* **290**, 457–465.

Arrigo, A.-P., Tanaka, K., Goldberg, A.L. & Welch, W.J. (1988) *Nature* **331**, 192–195.

Bachmair, A. & Varshavsky, A. (1989) *Cell* **56**, 1019–1032.

Bachmair, A., Finley, D. & Varshavsky, A. (1986) *Science* **234**, 179–186.

Baeza, L., Bertrand, A., Mache, R. & Lerbs-Mache, S. (1991) *Nucleic Acids Res.* **19**, 3577–3581.

Baker, B.S. (1989) *Nature* **340**, 521–524.

Bernstein, P. & Ross, J. (1989) *Trends Biochem. Sci.* **14**, 373–377.

Blank, A. & McKeon, T.A. (1991) *Plant Physiol.* **97**, 1409–1413.

Boynton, J.E., Gillham, N.W., Harris, E.H., Hosler, J.P., Johnson, A.M., Jones, A.R., Randolph-Anderson, B.L., Robertson, D., Klein, T.M., Shark, K.B. & Sanford, J.C. (1988) *Science* **240**, 1534–1538.

Bruce, B.D. & Malkin, R. (1991) *Plant Cell* **3**, 203–212.

Bruzik, J.P. & Maniatis, T. (1992) *Nature* **360**, 692–695.

Campbell, W.H. & Gowri, G. (1990) *Plant Physiol.* **92**, 1–11.

Capel, M.S. & Bourque, D.P. (1982) *J. Biol. Chem.* **257**, 7746–7755.

Cardenas, M.E., Dang, Q., Glover, C.V.C. & Gasser, S.M. (1992) *EMBO J.* **11**, 1785–1796.

Carignani, G., Groudinsky, O., Frezza, D., Schiavon, E., Bergantino, E. & Slonimski, P.P. (1984) *Cell* **35**, 733–742.

Carrington, D.M., Auffret, A. & Hanke, D.E (1985) *Nature* **313**, 64–67.

Cech, T.R. (1991) *Cell* **64**, 667–669.

Chapdelaine, Y. & Bonen, L. (1991) *Cell* **65**, 465–472.

Chen, H-C. & Stern, D.B. (1991) *Mol. Cell. Biol.* **11**, 4380–4388.

Chrispeels, M.J. (1991) *Annu. Rev. Plant Physiol. Mol. Biol.* **42**, 21–53.

Coen, E.S. (1991) *Annu. Rev. Plant Physiol. Plant Mol. Biol.* **42**, 241–279.

Croston, G.E., Kerrigan, L.A., Lira, L.M., Marshak, D.R. & Kadonaga, J.T. (1991) *Science* **251**, 643–649.

Danon, A. & Mayfield, S.P. (1991) *EMBO J.* **10**, 3993–4001.

Dehesh, K., Hung, H., Tepperman, J.M. & Quail, P.H. (1992) *EMBO J.* **11**, 4131–4144.

Deng, X-W. & Gruissem, W. (1987) *Cell* **49**, 379–387.

Deng, X-W., Wing, R.A. & Gruissem, W. (1989) *Proc. Natl. Acad. Sci. U.S.A.* **86**, 4156–4160.

Dickey, L.F., Gallo-Meagher, M. & Thompson, W.F. (1992) *EMBO J.* 2311–2317.

Doi, R.H. & Wang, L.-F. (1986) *Microbiol. Rev.* **50**, 227–243.

Dombroski, A.J., Walter, W.A., Record, Jr, T.M., Siegele, D.A. & Gross, C.A. (1992) *Cell* **70**, 501–512.

Dooner, H.K., Keller, J., Harper, E. & Ralston, E. (1991) *Plant Cell* **3**, 473–482.

Dunham, V.L. & Bryant, J.A. (1985) In *The Cell Division Cycle in Plants* (J.A. Bryant & D. Francis, eds), pp. 37–59. Cambridge University Press, London.

Durrenberger, F. & Rochaix, J-D. (1991) *EMBO J.* **10**, 3495–3501.

Echeverria, M., Delcasso-Tremousaygue, D. & Delseny, M. (1992) *Plant J.* **2**, 211–219.

Ellis, J.R. & van der Vies, S.M. (1991) *Annu. Rev. Biochem.* **60**, 321–347.

Ferl, R.J. & Paul, A-L. (1992) *Plant Mol. Biol.* **18**, 1181–1184.

Flavell, R.B., O'Dell, M., Thompson, W.F., Vincent, M., Sardana, R. & Baker, R.F. (1986) *Phil. Trans. R. Soc. London, B* **314**, 386–397.

Gallego, C., Gupta, S.K., Winitz, S., Eisfelder, B.J. & Johnson, G.L. (1992) *Proc. Natl. Acad. Sci. U.S.A.* **89**, 9695–9699.

Gallie, D.R. & Kado, C.I. (1989) *Proc. Natl. Acad. Sci. U.S.A.* **86**, 129–132.

Gallie, D.R., Sleat, D.E., Watts, J.W., Turner, P.C. & Wilson, T.M.A. (1987) *Science* **236**, 1122–1124.

Gallie, D.R., Lucas, W.J. & Walbot, V. (1989) *Plant Cell* **1**, 301–311.

Gasser, S.M., & Laemmli, U.K. (1987) *Trends Genet.* **3**, 16–22.

Gill, G. (1992) *Curr. Biol.* **2**, 565–567.

Gilmartin, P.M., Memelink, J., Hiratsuka, K., Kay, S.A. & Chua, N-H. (1992) *Plant Cell* **4**, 839–849.

Goldschmidt-Clermont, M., Girard-Bascou, J., Choquet, Y. & Rochaix, J-D. (1990) *Mol. Gen. Genet.* **223**, 417–425.

Goldschmidt-Clermont, M., Choquet, Y., Girard-Bascou, J., Michel, F., Schirmer-Rahire, M. & Rochaix, J-D. (1991) *Cell* **65**, 135–143.

Goodall, G.J. & Filipowicz, W. (1989) *Cell* **58**, 473–483.

Goodall, G.J. & Filipowicz, W. (1990) *Plant Mol. Biol.* **14**, 727–733.

Gordon, K., Futterer, J. & Hohn, T. (1992) *Plant J.* **2**, 809–813.

Goto, T. & Wang, J.C. (1985) *Proc. Natl. Acad. Sci. U.S.A.* **82**, 7178–7182.

Gralla, J.D. (1989) *Cell* **57**, 193–195.

Grierson, D. & Covey, S.N. (1988) *Plant Molecular Biology*, 2nd edn. Blackie, Glasgow.

Griffin, L.C. & Dervan, P.B. (1989) *Scince* **245**, 967–971.

Gruissem, W. (1989) In *The Biochemistry of Plants*, vol. 15 (P.K. Stumpf & E.E. Conn, eds), pp. 151–191. Academic Press, New York.

Gruissem, W., Elsner-Menzel, C., Latshaw, S., Narita, J.O., Schaffer, M.A. & Zurawski, G. (1986) *Nucleic Acids Res.* **14**, 7541–7556.

Grunstein, M. (1992) *Sci. Am.* **267**, 68–74B.

Haley, J. & Bogorad, L. (1990) *Plant Cell* **2**, 323–333.

Hall, Jr, G., Allen, G.C., Loer, D.S., Thompson, W.F. & Spiker, S. (1991) *Proc. Natl. Acad. Sci. U.S.A.* **88**, 9320–9324.

Hanley-Bowdoin, L. & Chua, N-H. (1988) *Trends Biochem. Sci.* **12**, 67–70.

Harrison, S.C. (1991) *Nature* **353**, 715–719.

Hiesel, R., Wissinger, B., Schuster, W. & Brennicke, A. (1989) *Science* **246**, 1632–1634.

Hoch, B., Maier, R.M., Appel, K., Igloi, G.L. & Kossel, H. (1991) *Nature* **353**, 178–180.

Hou, Y-M., Francklyn, C. & Schimmel, P. (1989) *Trends Biochem. Sci.* **14**, 233–237.

Huez, G., Marbaix, G., Gallwitz, D., Weinberg, E., Devos, R., Hubert, E. & Cleuter, Y. (1978) *Nature* **271**, 572–573.

Hunter, T. & Karin, M. (1992) *Cell* **70**, 375–387.

Hurt, E.C., Slotanifar, N., Goldschmidt-Clermont, M., Rochaix, J-D. & Schatz, G. (1986) *EMBO J.* **5**, 1343–1350.

Inglese, J., Koch, W.J., Caron, M.G. & Lefkowitz, R.J. (1992) *Nature* **359**, 147–150.

Izban, M.G. & Luse, D.S. (1992) *Genes Dev.* **6**, 1342–1356.

Jantzen, H-M., Admon, A., Bell, S.P. & Tjian, R. (1990) *Nature* **344**, 830–836.

Jolly, S.O. & Bogorad, L. (1980) *Proc. Natl. Acad. Sci. U.S.A.* **77**, 822–826.

Kang, C., Zhang, X., Ratliff, R., Moyzis, R. & Rich, A. (1992) *Nature* **356**, 126–131.

Katagiri, F. & Chua, N-H. (1992) *Trends Genet.* **8**, 22–27.

Katagiri, F., Yamazaki, K., Horikoshi, M., Roeder, R.G. & Chua, N-H. (1990) *Genes Dev.* **4**, 1899–1909.

Keegstra, K., Olsen, L. & Theg, S.M. (1989) *Annu. Rev. Plant Physiol. Plant Mol. Biol.* **40**, 471–501.

Keith, B. & Chua, N-H. (1986) *EMBO J.* **5**, 2419–2425.

Kieber, J.J., Tissier, A.F. & Signer, E.R. (1992) *Plant Physiol.* **99**, 1493–1501.

Klausner, R.D., Rouault, T.A. & Harford, J.B. (1993) *Cell* **72**, 19–28.

Klein, R.R. & Mullet, J.E. (1990) *J. Biol. Chem.* **265**, 1895–1902.

Kobayashi, H., Ngernprasirtsiri, J. & Akazawa, T. (1990) *EMBO J.* **9**, 307–313.

Kolodner, R. & Tewari, K.K. (1975) *Biochim. Biophys. Acta* **402**, 372–390.

Kozak, M. (1986) *Cell* **44**, 283–292.

Kudla, J., Igloi, G.L., Metzlaff, M., Hagemann, R. & Kossel, H. (1992) *EMBO J.* **11**, 1099–1103.

Lam, E. & Chua, N-H. (1987) *Plant Mol. Biol.* **8**, 415–424.

Lam, E., Hanley-Bowdoin, L. & Chua, N-H. (1988) *J. Biol. Chem.* **263**, 8288–8293.

Lawton, M.A., Yamamoto, R.T., Hanks, S.K. & Lamb, C.J. (1989) *Proc. Natl. Acad. Sci. U.S.A.* **86**, 3140–3144.

Lerbs, S., Brautigam, E. & Mache, R. (1988) *Mol. Gen. Genet.* **211**, 459–464.

Leto, K.J., Bell, E. & McIntosh, L. (1985) *EMBO J.* **4**, 1645–1653.

Levings, C.S. (1983) *Cell* **32**, 659–661.

Lin, Y-S., Carey, M., Ptashne, M. & Green, M.R. (1990) *Nature* **345**, 971–981.

Little, P. (1992) *Nature* **359**, 367–368.

Lonsdale, D.M., Hodge, T.P. & Fauron, C.M-R. (1984) *Nucleic Acids Res.* **12**, 9249–9261.

Luderus, E.M.E., de Graaf, A., Mattia, E., den Blaauwen, J.L., Grande, M.A., de Jong, L. & van Driel, R. (1992) *Cell* **70**, 949–959.

Lutcke, H.A., Chow, K.C., Mickel, F.S., Moss, K.A., Kern, H.F. & Scheele, G.A. (1987) *EMBO J.* **6**, 43–48.

Ma, H., Yanofsky, M.F. & Meyerowitz, E.M. (1990) *Proc. Natl. Acad. Sci. U.S.A.* **87**, 3821–3825.

Maier, R.M., Hoch, B., Zeltz, P. & Kossel, H. (1992) *Plant Cell* **4**, 609–616.

Marshallsay, C., Connelly, S. & Filipowicz, W. (1992) *Plant Mol. Biol.* **19**, 973–983.

Martinez-Zapater, J.M., Estelle, M.A. & Somerville, C.R. (1986) *Mol. Gen. Genet.* **204**, 417–423.

Mattoo, A.K. & Edelman, M. (1987) *Proc. Natl. Acad. Sci. U.S.A.* **84**, 1497–1501.

Mayfield, S.P. (1990) *Curr. Opin. Cell Biol.* **2**, 509–513.

McGurl, B., Pearce, G, Orozco-Cardenas, M. & Ryan, C.A. (1992) *Science* **255**, 1570–1573.

Mitchell, P.J. & Tjian, R. (1989) *Science* **245**, 371–378.

Murray, M.G., Cuellar, R.E. & Thompson, W.F. (1978) *Biochemistry* **17**, 5781–5790.

Nikolov, D.B., Hu, S-H., Lin, J., Gasch, A., Hoffmann, A., Horikoshi, M., Chua, N-H., Roeder, R.G. & Burley, S.K. (1992) *Nature* **360**, 40–46.

Nurse, P. (1990) *Nature* **334**, 503–508.

Oblong, J.E. & Lamppa, G.K. (1992) *EMBO J.* **11**, 4401–4409.

Oda, K., Yamato, K., Ohta, E., Nakamura, Y., Takemura, M., Nozato, N., Akashi, K., Kanegae, T., Ogura, Y., Kohchi, T. & Ohyama, K. (1992) *Plant Mol. Biol. Rep.* **10**, 105–111.

Ohad, I., Kyle, D.J. & Hirschberg, J. (1985) *EMBO J.* **4**, 1655–1659.

Palme, K., Diefenthal, T., Vingron, M., Sander, C. & Schell, J. (1992) *Proc. Natl. Acad. Sci. U.S.A.* **89**, 787–791.

Pearce, G., Strydom, D., Johnson, S. & Ryan, C.A. (1991) *Science* **253**, 895–898.

Pelleter, J. & Sonenberg, N. (1988) *Nature* **334**, 320–325.

Perisic, O. & Lam, E. (1992) *Plant Cell* **4**, 831–838.

Ptashne, M. & Gann, A.A.F. (1990) *Nature* **346**, 329–331.

Pugh, F.B. & Tjian, R. (1992) *J. Biol. Chem.* **267**, 679–682.

Ranjeva, R. & Boudet, A.M. (1987) *Annu. Rev. Plant Physiol.* **38**, 73–93.

Roberts, T.M. (1992) *Nature* **360**, 534–535.

Roberts, D.M. & Harmon, A.C. (1992) *Annu. Rev. Plant Physiol. Plant Mol. Biol.* **43**, 375–414.

Rochaix, J-D., Kuchka, M., Mayfield, S., Schirmer-Rahire, M., Girard-Bascou, J. & Bennoun, P. (1989) *EMBO J.* **8**, 1013–1021.

Rodermel, S.R., Abbott, M.S. & Bogorad, L. (1988) *Cell* **55**, 673–681.

Rogers, S., Wells, R. & Rechsteiner, M. (1986) *Science* **234**, 364–369.

Ross, C.W. (1981) In *The Biochemistry of Plants* vol. 6 (P.K. Stumpf & E.E. Conn, eds), pp. 169–205. Academic Press, New York.

Ruberti, I., Sessa, G., Lucchetti, S. & Morelli, G. (1991) *EMBO J.* **10**, 1787–1791.

Saiki, R.K., Gelfand, D.H., Stoffel, S., Scharf, S.J., Higuchi, R., Horn, G.T., Mullis, K.B. & Erlich, H.A. (1988) *Science* **239**, 487–491.

Schena, M. & Davis, R.W. (1992) *Proc. Natl. Acad. Sci. U.S.A.* **89**, 3894–3898.

Schmitz, M.L., Maier, U., Brown, J.W. & Feix, G. (1989) *J. Biol. Chem.* **264**, 1467–1472.

Schuster, G. & Gruissem, W. (1991) *EMBO J.* **10**, 1493–1502.

Sexton, T.B., Christopher, D.A. & Mullet, J.E. (1990) *EMBO J.* **9**, 4485–4494.

Shaw, G. & Kamen, R. (1986) *Cell* **46**, 659–667.

Shimada, H. & Sugiura, M. (1991) *Nucleic Acids Res.* **19**, 983–995.

Showalter, A.M. (1993) *Plant Cell* **5**, 9–23.

Shub, D.A. & Goodrich-Blair, H. (1992) *Cell* **71**, 183–186.

Simoens, C.R., Gielen, J., Van Montagu, M. & Inze, D. (1988) *Nucleic Acids Res.* **16**, 6753–6766.

Simpson, L. (1990) *Science* **250**, 512–513.

Smale, S.T., Schmidt, M.C., Berk, A. & Baltimore, D. (1990) *Proc. Natl. Acad. Sci. U.S.A.* **87**, 4509–4513.

Smith, F.W. & Feigon, J. (1992) *Nature* **356**, 164–168.

Spencer, C.A. & Groudine, M. (1990) *Oncogene* **5**, 777–786.

Stern, D.B. & Lonsdale, D.M. (1982) *Nature* **299**, 698–702.

Stock, J.B., Ninfa, A.J. & Stock, A.M. (1989) *Microbiol. Rev.* **53**, 450–490.

Stumpf, P.K. & Conn, E.E. (1981) *The Biochemistry of Plants* vol.6. Academic Press, New York.

Stumpf, P.K. & Conn, E.E. (1989) *The Biochemistry of Plants*, vol. 15. Academic Press, New York.

Sutton, C.A., Conklin, P.L., Pruitt, K.D. & Hanson, M.R. (1991) *Mol. Cell. Biol.* **11**, 4274–4277.

Svab, Z., Hajdukiewicz, P. & Maliga, P. (1990) *Proc. Natl. Acad. Sci. U.S.A.* **87**, 8526–8530.

Swanson, R., Hoben, P., Sumner-Smith, M., Uemura, H., Watson, L. & Soll, D. (1988) *Science* **242**, 1548–1551.

Tiller, K., Eisermann, A. & Link, G. (1991) *Eur.J. Biochem.* **198**, 93–99.

Towler, D.A., Gordon, J.I., Adams, S.P. & Glaser, L. (1988) *Annu. Rev. Biochem.* **57**, 69–99.

Uhlenbeck, O.C. (1987) *Nature* **328**, 596–598.

Van't Hof, J., Kiniyuki, A. & Bjerknes, C.A. (1978) *Chromosoma* **68**, 269–285.

Vierstra, R. (1989) In *The Biochemistry of Plants*, vol. 15 (P.K. Stumpf & E.E. Conn, eds), pp. 521–536. Academic Press, New York.

Vollbrecht, E., Veit, B., Sinha, N. & Hake, S. (1991) *Nature* **350**, 241–243.

Waibel, F. & Filipowicz, W. (1990) *Nature* **346**, 199–202.

Wang, J.C. (1991) *J. Biol. Chem.* **266**, 6659–6662.

Wissinger, B., Schuster, W. & Brennicke, A. (1991) *Cell* **65**, 473–482.

Wittig, B., Dorbic, T. & Rick, A. (1991) *Proc. Natl. Acad. Sci. U.S.A.* **88**, 2259–2263.

Wolfe, K.H., Morden, C.W. & Palmer, J.D. (1992) *Proc. Natl. Acad. Sci. U.S.A.* **89**, 10648–10652.

Yager, T.D. & von Hippel, P.H. (1987) In *Escherichia coli and Salmonella typhimurium: Cellular and Molecular Biology* (F.C. Neidhardt & H.E. Umbarger, eds), pp. 1241–1275. American Society for Microbiology, Washington DC.

Yamazaki, K., Chua, N.-H. & Imaseki, H. (1990) *Plant Mol. Biol. Rep.* **8**, 114–123.

Yanofsky, M.F., Ma, H., Bowman, J.L., Drews, G.N., Feldmann, K.A. & Meyerowitz, E.M. (1990) *Nature* **346**, 35–39.

Yen, Y. & Green, P.J. (1991) *Plant Physiol.* **97**, 1487–1493.

Zawel, L. & Reinberg, D. (1992) *Curr. Opin. Cell Biol.* **4**, 488–495.

Zentgraf, U. & Hemleben, V. (1992) *Nucleic Acids Res.* **20**, 3685–3691.

9 Regulation of Gene Expression in Plants

J.A. Gatehouse

9.1 INTRODUCTION

Although the regulation of gene expression in plants has been studied at the phenotypic level for many years, by observations of the responses of plants to external factors such as stress, pests and pathogens, or applied chemicals, until recently the underlying bases of these responses at the molecular level were essentially unknown. Similarly, although classical genetics, from its origin in Mendel's studies on garden peas, had identified genes which affected the development and other phenotypic characteristics of plants, the mechanism by which such genes functioned was also unknown. The revolution in biology, brought about by the techniques of biochemistry and molecular genetics, has opened up the possibility of being able to understand the connections between genetics and physiology, or perhaps more properly between genotype and phenotype, by investigating the molecular mechanisms that link the two.

An earlier chapter in this volume has described the basic structure of the genetic material in plants, and the flow of information from DNA to RNA to proteins. This chapter concentrates on the processes which determine whether particular genes are to be expressed (i.e. the 'switches' that turn genes on and off, or qualitative regulation), and how much of their products will accumulate (i.e. the quantitative regulation of gene expression). As

will be seen, this division between 'switches' and 'regulators' is often arbitrary, but it is a useful starting point for more sophisticated models.

9.2 NUCLEAR GENES

Plant cells contain three distinct genomes; those of the nucleus, the plastid (i.e. the chloroplast in green tissue), and the mitochondrion. The nucleus contains almost all the functional genes of the plant, and its genome is organized on the general model for eukaryotes. However, the plastid and mitochondrion contain genomes which have many features in common with prokaryotes, and are thus considered separately.

9.2.1 Structure of a 'typical' nuclear gene

The definition of a gene at the molecular level is not quite as simple as in classical genetics. For present purposes, we shall consider a gene to contain a transcribed sequence of DNA, which is bounded by bases at which transcription is initiated and terminated. By convention, the initiation point is termed the $5'$ end of the gene, and the termination point the $3'$ end. This reflects the formation of the transcript (from the $5'$ end to the $3'$ end) rather than the direction in which the

PLANT BIOCHEMISTRY
ISBN 0-12-214674-3

DNA strand is transcribed (from 3' to 5'). Similarly, all DNA sequences are given in the same direction as the transcript (5' to 3') so that the sequences reported are in fact those of the non-transcribed strand of DNA. This convention is adopted so that the sequences reported for genes can be directly compared to those of transcripts (i.e. RNA molecules) or gene products (polypeptides), without the necessity for manipulation. Since it is possible for a transcribed unit to contain more than one RNA molecule (as in the ribosomal RNA genes, where the initial transcript is cleaved into different ribosomal RNAs), or to encode more than one polypeptide (as is common in prokaryotes), more than one 'classical' gene can correspond to a single transcribed unit. The reader is warned that usage of the term 'gene' is very far from uniform throughout the scientific community!

The transcribed unit of a nuclear gene gives rise to a primary transcript (sometimes collectively referred to as 'heteronuclear RNA'), which is modified before attaining its mature form in the cytoplasm. As mentioned above, the structural RNAs of ribosomes are modified by cleavage from a precursor. tRNA molecules are subject to extensive base modification. Most polypeptide-encoding mRNAs are modified by a common route, which involves 'capping' at their 5' ends through a 5' to 5' triphosphate linkage with 7-methylguanosine, removal of intron sequences by splicing, and addition of poly(A) 'tails' (possibly involving removal of some of the original transcript) at the 3' end. These modifications have been described earlier, and are determined by sequences within the transcribed unit.

Associated with the transcribed unit are sequences of DNA that play other functional roles, concerned with transcription initiation, termination and regulation. The transcription initiating and terminating sequences are normally considered to be part of the gene, as are those functional and regulatory sequences found in the region of DNA up to approximately 1000 bp 5' to the transcription start, which can be postulated to be involved in the transcription initiation process. These sequences constitute the gene promoter. Genes only constitute a small fraction of the total genomic DNA in most plants; even in *Arabidopsis thaliana*, which has one of the smallest nuclear DNA contents per cell in plants (approx. 1.5×10^8 bp per haploid genome), repetitive non-transcribed DNA has been estimated to occupy 20% of the genome. Individual genes are thus dispersed in a background of non-transcribed DNA, and although clusters of related sequences

have been observed, as a general rule each gene maintains its own promoter sequences and behaves independently. The regulation of expression of genes in eukaryotes thus differs fundamentally from that in the typical prokaryote, where sets of genes are organized into operons which are co-ordinately regulated from a single promoter. An idealized gene structure is shown in Fig. 9.1, where the relationship of the different functional elements of the gene is represented.

9.2.2 Transcription

Transcription of nuclear genes is carried out by one of three different types of RNA polymerase, according to the type of gene product. The ribosomal RNA genes, which give rise to the two large ribosomal RNA molecules, 18S and 25S, and a small 5.8S ribosomal RNA, are transcribed by RNA polymerase I. RNA polymerase III is responsible for transcription of the genes for transfer RNAs and other small, stable RNAs, such as those involved in intron processing. All messenger RNAs (i.e. those that encode polypeptides) are transcribed by RNA polymerase II. Each of the RNA polymerases is a multi-subunit enzyme, of complex structure; there are at least 10 subunits, and the whole molecule has an M_r of over 500 000. There are similarities between the enzymes in both structure and polypeptide sequences, since the large core subunits show some sequence homology between different polymerases, and some small subunits are common. Each polymerase catalyzes the addition of nucleotide triphosphates (NTPs) to the 3' OH of a nucleotide ribose sugar ring, forming a 3'–5' phosphodiester linkage, the NTP being determined by complementarity to the transcribed template strand. The RNA is thus produced in a 5'–3' direction.

Experiments carried out *in vitro* with purified enzyme have shown that RNA polymerase by itself will transcribe DNA starting from any free 3' ends, but it is not capable of binding to specific DNA sequences, and is thus not sufficient for accurate, efficient transcription initiation. For this to occur *in vivo*, a specific multiprotein complex (the transcriptional activation complex) which includes RNA polymerase, must assemble in such a way that the polymerase is aligned with its active site fitting onto the transcription start point. This is done by binding of proteins, called transcription factors, to specific DNA sequences in the gene. The transcription factors then form a complex with RNA polymerase which aligns the enzyme

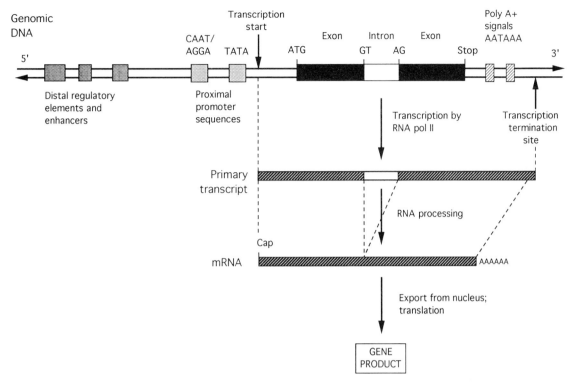

Figure 9.1   Idealized representation of a polypeptide-encoding nuclear gene. The double line represents double-stranded genomic DNA; various functional parts of the gene are represented by boxes superimposed on this line. The hatched line represents (single-stranded) RNA produced from the gene as a primary transcript, and as the mature messenger RNA.

with the correct point on the DNA, and stimulates transcription. The complex then dissociates as transcription proceeds, allowing the enzyme to move along the gene in the elongation complex, which may involve other factors *in vivo*, although, as stated above, RNA polymerase alone will carry out this reaction *in vitro*. The remainder of the transcriptional activation complex, without RNA polymerase, does not immediately dissociate, but remains available to bind a further molecule of polymerase, and thus initiate another round of transcription. This process is illustrated diagrammatically in Fig. 9.2.

Correct initiation of transcription therefore depends on both protein–DNA and protein–protein interactions. Transcription factors initially recognize DNA sequences, and bind to them, whereas correct assembly of the initiation complex, and the positioning of RNA polymerase, requires precise interactions between proteins. The components of the transcriptional activation complexes, and the order in which they assemble, differ slightly between genes transcribed by RNA polymerases I, II and III. Since genes encoding polypeptides make up the vast majority of genes,

complexes containing RNA polymerase II have been subjected to the most study, and will be described in more detail.

The core of the RNA polymerase II (pol II) initiation complex is a protein termed the TATA-binding protein, or TBP. This protein binds strongly and specifically to a conserved DNA sequence, called the TATA box, present in most genes transcribed by pol II, and found between 19 and 27 bp 5′ to the transcription start (see Fig. 9.1), with the consensus sequence TATA(A/T)A(A/T). TBP forms part of a multiprotein transcription factor called TFIID, which can contain up to 10 other polypeptides besides TBP. These proteins are necessary for formation of the transcription initiation complex. A second transcription factor, TFIIA, is necessary for the assembly of TFIID. The TFIIA–TFIID complex, bound to the TATA box, is then recognized by a further transcription factor, TFIIB; RNA polymerase is now bound, in association with a further factor, TFIIF. Transcription is then initiated by the entry of factors TFIIE, -H and -J into the complex.

TBP also plays a key role in the transcription initiation complexes involving RNA polymerases I

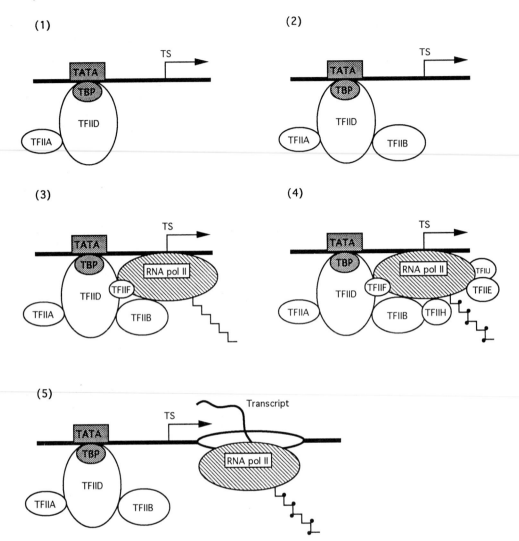

Figure 9.2   Transcription initiation process for genes transcribed by RNA polymerase II. Stage 1, binding of transcription factors TFIID and TFIIA to the TATA box; stage 2, addition of TFIIB to form the 'DAB complex'; stage 3, entry of RNA polymerase in association with TFIIF; stage 4, entry of TFIIE, -H and -J, accompanied by phosphorylation of RNA pol II in its C-terminal domain (represented by black dots), mediated by TFIIH; stage 5, transcription under way; the open area adjacent to RNA polymerase represents separation of the DNA strands.

and II. Although there is no TATA box 5′ to ribosomal RNA genes (the specific protein–DNA interaction necessary for transcription involves a species-specific DNA sequence 5′ to the transcription initiation site) the factor essential for transcription of rRNA genes by pol I contains TBP. Possibly one of the other polypeptides in this complex is responsible for DNA sequence recognition; this could also be the case in those (rare) genes transcribed by pol II which lack a TATA box. The specific DNA recognition sequence in 5S and tRNA genes transcribed by pol III is 3′ to the transcription start, i.e. actually within the gene,

and is recognized by a transcription factor called TFIIIA. However, two further factors, TFIIIB and -C, are necessary to form an active transcription initiation complex, and seem to be common to all genes transcribed by pol III. TFIIIB also contains TBP, and in some pol III genes, e.g. the U6 small nuclear RNA, a 5′ TATA box to which TBP binds is present. TBP from *Arabidopsis thaliana* (land cress) has been characterized, and its structure has been determined at low resolution. It is a saddle-shaped protein, where the underside of the saddle is structurally complementary to duplex DNA, so that the antiparallel β strands making up this region can interact with the grooves in the double-stranded DNA, thus gaining access to the bases 'inside' the helix. Other types of DNA-binding protein are discussed later in this chapter.

Some plant genes contain another sequence recognized as being conserved throughout the eukaryotes, the 'CAAT' box, with the consensus sequence GG(C/T)CAATCT, at approximately 75 bp 5' to the transcription site (see Fig. 9.1). This may be another target sequence for a common component of transcription factors, as yet unidentified. However, this sequence seems to be much more variable, and less strongly conserved in plants than in other higher eukaryotes, and differing consensus sequences in this region have been proposed for plant genes.

(a) Regulatory elements

The structure of a gene promoter can formally be divided into two parts; the minimum promoter, which constitutes the minimum amount of DNA sequence, additional to the transcribed unit, to allow the gene to express, and the regulatory elements, which increase and control the expression of the minimum promoter. For genes transcribed by pol II, the minimum promoter consists of approximately 100 bp of DNA 5' to the transcription start (the 5' flanking sequence), including the TATA box, and the CAAT box if present. On the model described above, these are the DNA sequences which are necessary for assembly of the transcription initiation complex. In addition to the minimum promoter, further regulatory elements are present. DNA-binding proteins are specifically bound at these regulatory elements, and control the expression of specific genes by interacting with components of the transcription initiation complex (see Fig. 9.3). The regulatory elements in the DNA sequence are termed cis-acting sequences (since they affect the activity of transcription units they are adjacent to), and the DNA-binding proteins are referred to as trans-acting factors (since they are the products of genes removed from their site of action). The action of the specifically bound trans-acting factors can be to increase the rate of formation of functional transcription initiation complexes, and thus initiation, or to decrease it, thus giving rise to positive or negative regulation of gene expression. However, most of the interactions so far detected have a positive effect on gene expression, and it seems that in eukaryotes, activation of transcription initiation is a much more common mechanism of gene control than repression (in contrast to prokaryotes, where

repression is widely used, e.g. in the lac operon of Escherichia coli). The cis-acting sequences are required to be at a (more or less) fixed distance from the transcription initiation point, and in the correct orientation relative to it, in order for the protein–protein interactions between the trans-acting factors and components of the transcription initiation complex to form correctly. The helical nature of the DNA double strand can limit the possibility of interactions, since the spacing of binding sites will determine whether proteins are bound in line with each other, or on opposing faces of the cylinder described by the helix. Distances between binding sites can also be important in determining the correct spacing of proteins for interactions to take place. Although oversimplification based on conventional structural models must be avoided, since DNA-binding proteins are known to induce local alterations in DNA structure at and around the binding site, spatial relationships between DNA-binding proteins, determined by the linear spacing of cis-acting sequences along the DNA strand, will be important in determining their interactions.

Although terminology in this area is not fixed, the cis-acting sequences that bind factors, which interact directly with the transcription activation complex, are considered as part of the gene promoter in this chapter. In the literature they are also referred to as 'enhancers', but the requirement for a (more or less) fixed distance from, and orientation to, the transcription start point distinguishes these sequences from classical enhancer sequences, considered in a subsequent section.

(b) Cis-acting sequences

The cis-acting sequences so far identified in plant genes conform to two general types. The first type is a relatively short sequence, often with a core which contains internal symmetry, which is associated with particular groups of genes. Two examples of this type are cis-acting sequences with the core CACGTG, found in genes responsive to light or abscisic acid, and the 'heat shock module', with the consensus sequence GAAnnTTCnnGAAnnTTC (core sequence in bold type) found in plant genes whose expression is induced by heat shock. Both these cis-elements are located in the region 70–200 bp 5' to the transcription start. The second type has a less well-defined sequence, extending over a greater span than the first type, but is A–T-rich. These A–T-rich regions are found associated with many different plant genes, and are active both over relatively short distances (up to 500 bp 5' to the transcription start), and at greater distances,

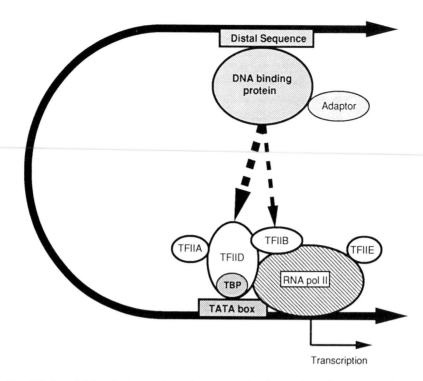

Transcription

Figure 9.3 Simplified model for the interaction of a *trans*-acting factor, bound to its specific recognition sequence (the *cis*-acting sequence), with the transcription initiation complex. The double-stranded DNA helix is represented by a heavy black line; the curve in this line is for diagrammatic purposes only. Interactions are represented by dashed arrows. The factor can act to stimulate the activity of a pre-formed complex, or (possibly more likely) to assist the formation of an active complex. The interaction of the *trans*-acting factor with the transcription initiation complex can in turn be modulated (to give positive or negative regulation) by a further protein–protein interaction; either with an adaptor protein, as shown here, or with another *trans*-acting protein molecule. In the complex formed including RNA polymerase II, as shown here, current evidence favors TFIID (which includes the TATA box binding protein; TBP) as the target for interactions with *trans*-acting factors, although TFIIB is also a possible target. The structure of the transcription initiation complex has been simplified, and only one *trans*-acting DNA-binding protein has been shown; in real examples, multiple *cis*-acting sequences are present in the 5' flanking sequences of plant genes, leading to the binding of multiple *trans*-acting factors and the possibility of interactions between the different factors.

although these far upstream binding sites are likely to act more like enhancers (section 9.2.3c) than as components of the promoter.

All genes are examined to date contain multiple *cis*-acting sequences, which interact with multiple *trans*-acting factors. Some of these interactions have been readily detectable, others less so, possibly corresponding to the relative abundances and binding strengths of the different DNA-binding proteins. The accepted model is to assign particular functions to specific *cis*-acting sequences, so that a gene will contain in its promoter different *cis*-acting sequences conferring, for example, tissue specificity, response to environmental conditions, response to developmental cues (acting as 'switches' for expression), with other sequences, such as the A–T rich regions described above, acting to up-regulate the activity of the gene.

This model is supported by evidence showing that deletion of regions of the promoter containing putative *cis*-acting sequences removes the function assigned to that element. For example, in the wheat Em gene promoter, the effects of removal of different regions of the promoter was assayed by constructing promoters with various deletions, and fusing them to a 'reporter' gene, which gave an easily detectable product (in this case, the bacterial enzyme β-glucuronidase, detectable by hydrolysis of chromogenic or fluorescent substrates). The chimaeric promoter/reporter constructs were tested by transient expression assay in transformed protoplasts. Removal of A–T rich regions between 554 and 168 bp 5' to the transcription start decreases activity of the promoter, but does not affect its response to abscisic acid. The gene product is a small polypeptide induced by desiccation. These regions thus function as 'regulators'. A further region, from 168 to 102 bp 5' to the transcription start functions as a switch, in that if it is removed, response to abscisic acid and expression are essentially abolished. This region contains a sequence conserved in all gene promoters regulated by abscisic acid that have been examined, which became an obvious candidate as a *cis*-acting sequence. Its identity was confirmed by constructing an oligonucleotide containing the putative sequence, and fusing this to a constitutively expressed promoter (i.e. a promoter expressed in all cells, and not responsive to environmental stimuli). The oligonucleotide confers responsiveness to abscisic acid on this constitutive promoter, proving that it contains the *cis*-acting sequence responsible. This is the sequence containing the core CACGTC described above.

(c) Trans-*acting factors*

The proteins that interact with specific DNA sequences and regulate gene expression are now beginning to be characterized in higher eukaryotes in general. A number of classes of DNA-binding proteins have been identified on the basis of common sequence (or structural) motifs, and the interaction between polypeptide and DNA is reasonably well understood for several types of these proteins.

The majority of plant *trans*-acting factors characterized so far belong to the so-called 'basic leucine zipper' (bZIP) type of DNA-binding protein. These proteins contain two functional regions; a conserved 'structural' domain comprising the carboxy-terminal of the protein, and a variable N-terminal region, which is thought to be responsible for interaction with other proteins in transcriptional activation. The function of the conserved domain in these proteins is reasonably well understood. A basic region of approximately 30 amino acids, strongly conserved between different plant bZIP proteins, is responsible for DNA binding. Adjacent to this region is the 'leucine zipper' which is characterized by a sequence where every seventh amino acid is leucine. The 'leucine zipper' forms an amphipathic twisted α-helix, where, because of the pitch of the helix (one turn per 3.5 residues), the leucine residues are all aligned with each other down one side of the cylinder described by the helix. (The α-helix thus has a hydrophobic 'side' and a hydrophilic 'side'.) The protein functions by subunit dimerization, where two polypeptides are brought together by hydrophobic interaction

between the leucine zipper regions. In doing so the basic DNA-binding regions are made to fit around a DNA helix, with their α-helices fitting into the major groove (Fig. 9.4). Since the two basic helices bind in opposite orientations relative to the DNA double strand, this accounts for the symmetric nature of many *cis*-acting sequences, which are the targets for binding this type of protein. A complication is that dimerization of the leucine zipper protein may not involve two identical subunits, and in some animal transcriptional activators (e.g. *fos-jun*), heterodimerization is required for correct function. Basic leucine zipper proteins in plants will also undergo heterodimerization, although its functional role is not yet clear.

Plants also contain homologs of other types of eukaryotic transcription factors previously identified in animals or yeast, such as the products of genes that regulate plant development (the homeotic genes), which contain the MADS DNA-binding domain (a helix–turn–helix motif), and the products of genes regulating anthocyanin biosynthesis in maize, which are homologous to the Myc basic helix–loop–helix proteins and the Myb proteins. Interestingly, the maize genes, like some animal bZIP proteins, need to assemble into heterodimers to function, with a Myb type protein being paired with an Myc type. It is perhaps less

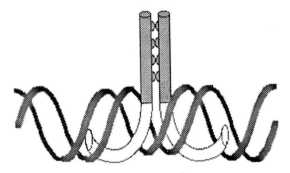

Figure 9.4 Diagrammatic representation of a bZIP (basic leucine zipper) type DNA binding protein (shaded) interacting with its target binding site in a double-stranded DNA helix (black). The protein dimerizes through the leucine zipper region (represented by 'bobbles'), and interacts with the DNA through the basic DNA binding domains (curved), which fit into the major groove of the helix. Both the protein domains are made up of α-helix (represented as a cylinder), the leucine zipper region being distorted slightly so that the helices coil about each other; the axis of this region is perpendicular to the axis of the DNA helix. The DNA binding domains determine specificity of binding to the symmetrical binding site, each helix binding to half the site.

expected that a type of eukaryotic transcriptional regulator protein common in animals, the 'zinc finger' proteins (so called because they contain DNA-binding loops in the polypeptide chain which are stabilized by a zinc atom co-ordinated to cysteine and histidine residues at the base of the loop), appear to be much less frequently found in plants.

The presence of *trans*-acting factors which interact with specific *cis*-acting DNA sequences in gene promoters is readily demonstrated by carrying out electrophoretic mobility shift analysis (EMSA), where a labeled double-stranded DNA fragment containing a *cis*-acting sequence is incubated with a protein extract from nuclei (usually) of cells where the gene is being expressed. The DNA fragment can be cloned from the gene of interest, or synthesized as complementary oligo-nucleotides. Binding between a component of the extract and the DNA sequence results in the formation of a complex, which has lower mobility (due to its larger size) on electrophoresis through an acrylamide or agarose gel. Comparison with an adjacent gel track containing free probe is used to show the band (or bands) due to complex formation. Because the DNA probe can be end-labeled with radioactive phosphorus (^{32}P) to high specific activity, very small amounts of complex can be detected by autoradiography, and the assay is very sensitive. This technique must be used with care to distinguish specific interactions from non-specific protein–DNA interactions, and it is necessary to show that a large excess of competitor non-specific DNA does not abolish the observed binding. Such assays have also been used to define more closely the *cis*-acting elements themselves, by producing probe sequences which contain dele-tions, or sequence alterations. The exact site of interaction between the DNA-binding protein and the probe sequence can be determined (in favor-able cases) by a 'footprinting' assay (Fig. 9.5). In this technique, a DNA probe, including a *cis*-acting sequence, is labeled at one end only, and then allowed to form a complex with a DNA-binding protein. The DNA–protein complex is then treated with a reagent which cleaves DNA at random, such as DNase I, or the *o*-phenanthroline-copper reagent. The partially cleaved DNA is dissociated from the complex, and run on a sequencing gel. The region of the DNA probe bound to the protein is protected from the cleaving reagent, and DNA fragments corresponding to the protected region are absent from the resulting 'ladder' of fragments on the sequencing gel.

The demonstration of the presence of DNA-binding proteins which interact specifically with *cis*-acting sequences is only a first step in characterizing the proteins and their functional roles. The observed binding activity can be assayed in extracts from different tissues and conditions, to test whether transcriptional activity of the gene correlates with the presence of the putative *trans*-acting factor. For example, binding to *cis*-acting sequences in seed storage protein genes is observed in extracts from seed nuclei (i.e. from tissues where the genes are being expressed), but not in extracts from leaf nuclei (i.e. from tissues where the genes are inactive). Once it is established that the interaction is of importance in controlling gene expression, then the techniques of molecular genetics can be employed to isolate cDNAs or genes encoding the *trans*-acting factor, thus allowing the factor to be characterized. Perhaps the most obvious method to isolate these encoding genes is to screen an expression library of cDNA clones with a labeled DNA probe containing the *cis*-acting sequence. However, although this method has proved successful in isolating a number of genes encoding *trans*-acting factors in plants (e.g. the gene encoding the factor that binds to the *cis*-element controlling responsiveness to abscisic acid in the wheat *Em* gene; see above), it suffers from many technical problems. Other methods used include 'tagging' of mutants with transposon sequences, and using probes encoding homologous proteins in other species.

(d)   Control systems

Although the functions of individual components of the mechanisms used to control transcription initiation of genes are becoming clear, it is apparent that an integrated description of how any particular gene is controlled still poses severe problems to the scientist. The interaction of several *trans*-acting factors with the transcription initiation complex and with each other, and with potential modifiers, is of considerable complexity. In addition, in identifying the genes that encode *trans*-acting factors, it may appear that the problem of explaining the control of gene expression has merely been postponed, since something must control these genes in turn. A related problem concerns fine control; when gene expression is examined in detail, even very closely related genes show differences in their expression levels, and responses of expression to developmental or environmental stimuli. For example, in a subfam-ily of pea seed storage protein genes, one member of the family shows an increase in expression level during seed desiccation whereas another shows a decrease. However, it is not possible for each gene

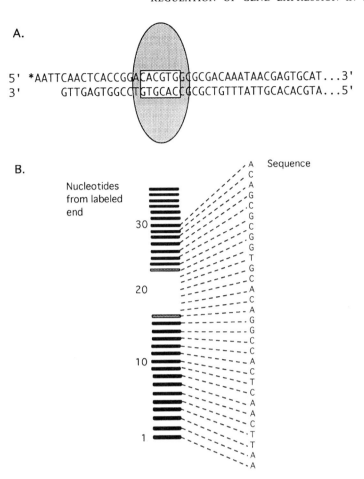

A.

5' *AATTCAACTCACCGGACACGTGGCGCGACAAATAACGAGTGCAT...3'
3'   GTTGAGTGGCCTGTGCACCGCGCTGTTTATTGCACACGTA...5'

B.

Nucleotides
from labeled
end

Sequence

A
C
A
G
C
G
C
G
G
T
G
C
A
C
A
G
G
C
C
A
C
T
C
A
A
C
T
T
A
A

30

20

10

1

Figure 9.5 Diagrammatic representation of a 'footprinting' experiment to define the region of a *cis*-acting sequence which interacts with a DNA-binding protein. A double-stranded DNA fragment with a radioactive label (*) at one end is prepared, and allowed to interact with the DNA-binding protein, as a component of a nuclear extract (a). The complex is treated with a non-specific nuclease, such as DNase I, and the labeled fragments from partial digestion with the nuclease are separated by denaturing gel electrophoresis (b). The resulting 'ladder' of labeled fragments detected by autoradiography (i.e. single-stranded fragments from the 'top' strand of DNA, extending from the labeled 5' end to the site of DNase cleavage) shows the absence of bands corresponding to nucleotides in the binding region, which are protected from the nuclease by the DNA-binding protein (a). The core binding sequence is boxed in (a).

to have its own individual controlling *trans*-acting factor(s), since this would lead to an infinitely large genome.

These problems are far from being solved, but several approaches to their solution are gaining general acceptance. First, the number of *trans*-acting factors is thought to be small compared to the total number of gene products (i.e. genes encoding these factors must be a small proportion of the total genes). From this it follows that a single *trans*-acting factor can be involved in the control of a number of genes. There is some experimental evidence that this is the case, since *cis*-acting sequences from different genes will compete with each other in binding to a single

factor. Second, control of a single gene involves more than one *trans*-acting factor, and precise control can depend on interactions between the factors. In this way, a large number of combinations, corresponding to different patterns of gene expression, can be built up from a limited number of factors. This view is also supported by experimental evidence of multiple binding sites for *trans*-acting factors in the promoters of many genes. The combination of different *trans*-acting factors in functional heterodimers (see above) is another method of increasing the range of effects possible from a limited number of DNA-binding proteins. Finally, rather than a rigid hierarchy of genes which are controlled by the products of other genes, which are in turn controlled by the products of more genes, control of gene expression seems likely to be a much more dynamic process. This may involve the full range of feedback and feedforward controls already explored in enzyme regulation, competition of different genes for limited amounts of factors (so that the strength of binding at the *cis*-acting sequences becomes important in controlling expression), and of different factors for limiting amounts of modifiers. As with many other biological problems, it is apparent that a reductionist approach can only go a certain distance towards explaining the phenomena studied, before the system becomes too complex to understand. Nevertheless, reductionism remains the most useful tool for constructing models which, to a greater or lesser extent, explain and predict natural phenomena.

A major problem in studying transcription in plants has been the absence of a functional *in vitro* transcription system, which would allow the effects of specific factors on individual genes to be assayed. To some extent, this difficulty has been addressed by the use of transgenic plant technology (q.v.), but the recent description of a cell- (and organelle-)free transcription system from wheat germ in which correct initiation occurs, suggests that detailed analyses of the interactions of *trans*-acting factors with plant genes will be possible in the near future.

9.2.3 Physical and chemical modifications of DNA which regulate transcription

The previous discussion has focused on the gene promoter and the way in which transcription itself is directly regulated. A further important method of regulating transcription of nuclear genes is the availability of the DNA in the gene promoter for binding to the various functional and regulatory proteins. If the *cis*-acting sequences in the gene promoter are not accessible to, or modified so that they will not bind to the transcription and *trans*-acting factors, transcription is effectively prevented. Similarly, the presence of sequences that make the DNA more available, by disrupting normal nucleosome formation, will enhance transcription. These phenomena are considered in this section.

(a) *Chromatin structure*

The structural organization of nuclear DNA in plants has been discussed in Chapter 8. The degree

of condensation of DNA has a determining effect on availability for transcription. Most markedly, the packing of the entire genome into the condensed state of a metaphase chromosome abolishes transcription altogether, and no RNA synthesis takes place during mitosis. Transcriptional activity is maximal during the interphase stage of the cell cycle, when the chromosomes have extended and unwound structures, and are dispersed within the nucleus. Control of individual genes is therefore dependent on the lower orders of DNA structure, rather than the higher orders, from the extended chromosome down to the level of individual nucleosomes.

The extended chromosome is essentially composed of a chromatin fiber, a structure in which the DNA is still highly condensed. In this structure, the DNA is first packed by being wound around an octameric complex of histone molecules (two molecules each of H2A, H2B, H3 and H4). The nucleosome thus formed contains 146 bp of DNA wrapped approximately 1.75 turns around the surface of the histone octamer. Formation of these nucleosomes is not dependent on DNA sequence. Nucleosomes are normally spaced by approximately 100–200 bases of 'free' DNA (this DNA is in fact complexed with histone H1 and other non-histone proteins) to give the familiar 'beads on a string' model. Histone H1 is thought to be responsible for assembly of the chromatin fiber, a structure in which the 'beads on a string' structure is itself coiled up into a helix approximately 30 nm in diameter, with about six nucleosomes per turn. The current view is that the chromatin fiber must be unfolded, or opened, to allow transcription of genes, and that closed chromatin is transcriptionally inactive.

The opening of chromatin causes both local and long-range structural alterations to the chromatin fiber, which can be assayed biochemically by a sensitivity of 'open' chromatin, containing actively expressing genes, to digestion by a non-specific nuclease such as DNase I. This general sensitivity to digestion extends over regions beyond actual genes, and suggests an organization of the chromatin into functional domains. Chromatin opening can thus act to facilitate expression of genes within an entire domain, and as such forms a necessary early step in controlling gene expression. These functional chromatin domains are thought to correspond to structural loops of chromatin fiber. The loops themselves are formed by the interaction of DNA sequences (scaffold-associated regions, or SARs) with structural proteins in the nucleus. Dynamic changes in chromatin structure in these loops could then be brought about by the proteins anchoring the loops, and by co-operative interactions between the anchored regions and proteins, like histone H1, which control fiber assembly. Chromatin loops up to 100 kb have been observed in animals, which is certainly large enough to accommodate a gene cluster, and allowing a degree of co-ordinated regulation of the genes in the cluster. A well-studied example is the 5 kbp histone gene repeat in *Drosophila*, which contains a complete set of histone genes (H1, 2a, 2b, 3 and 4) delimited by scaffold-associated DNA regions. Many examples of members of the same gene family being found in clusters are known in plants, although plant chromatin has been comparatively little studied.

(b) Nucleosome spacing

Even after the chromatin has been opened, DNA sequences may still be inaccessible to proteins which have to be bound for transcription to occur. It is clear from early studies in which chromatin was extensively digested with non-specific nucleases that the DNA in contact with the octamer of histones in the nucleosome is relatively inaccessible, since it is protected from digestion. The position of nucleosomes relative to *cis*-acting sequences in the gene promoter thus becomes of some significance in controlling gene expression.

A general observation has been that although most DNA is covered by nucleosome arrays, the promoters of active genes often have nucleosome-free regions (detected by hypersensitivity to digestion with DNase I) associated with *cis*-acting sequences, to allow full access by *trans*-acting factors. It has been suggested that in some genes the normal spacing of nucleosomes allows free access to the TATA box and the *cis*-acting sequences. However, in others a change in nucleosome spacing or structure, brought about by interactions between *trans*-acting factors and *cis*-acting sequences, is necessary to allow the promoter to function. Such interactions would not be subject to the same strict spatial constraints as those involved in the formation of the transcription initiation complex, and could involve sequences within, or 3' to, active genes.

(c) Enhancers

Enhancer sequences have been described in a number of eukaryotic systems, including the plant viral promoter on the Cauliflower Mosaic Virus (CaMV) 35S RNA gene. An enhancer sequence shares many of the properties of the previously described *cis*-acting DNA sequences. It can be shown to up-regulate transcription, with the important difference that enhancers can function at variable distances from the transcription start point, can be in either orientation relative to the direction of transcription, and do not necessarily need to be located in the near 5' flanking region of the gene. They can act at a distance, and in some cases have been described as being within the transcribed sequence of the gene, or 3' to the transcribed sequence. DNA-binding proteins which interact specifically with these sequences have been described and characterized. One mechanism by which these regulatory elements could function is by alteration of chromatin conformation, as described above. This would account for action at a distance, and a relative independence of position relative to the gene.

(d) DNA methylation

A further mechanism of repression of nuclear gene expression is based on chemical modification of bases in DNA, by methylation. In eukaryotes, this methylation takes place on the 5-carbon of the cytosine base (Fig. 9.6); modification of C residues in the dinucleotide CG is universal throughout eukaryotes, and in plants modification of C residues in the trinucleotides CNG also occurs. The alteration in the chemical structure of the base can interfere with the binding of proteins which recognize the C residue as part of their target site for binding. The most easily demonstrated example of prevention of binding is with restriction endonucleases which contain CG (or CNG) as part of their recognition sequence at which cleavage of double-stranded DNA takes place. For example, the enzyme Msp I, which cuts at the recognition sequence CCGG, will not cleave DNA that is methylated on either C residue. However, a second enzyme, Hpa II (from a different bacterial species), which also cuts at the same recognition sequence,

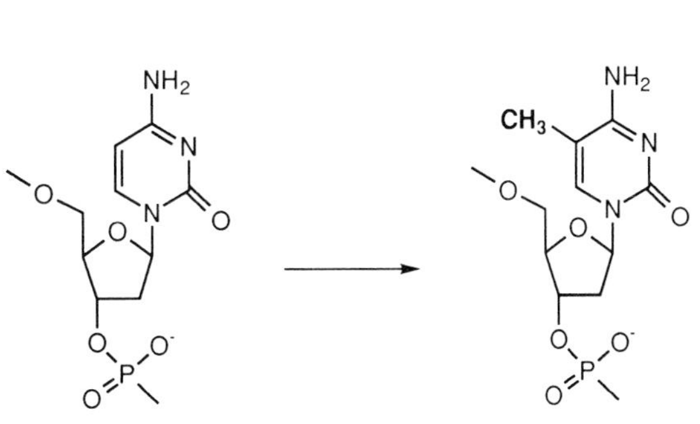

Figure 9.6 Methylation of cytidine residues in DNA of eukaryotes. The methyl group added at the 5-position of the base is shown in bold type.

will cleave DNA methylated on the second C in CCGG. This difference in behavior forms the basis of an assay for DNA methylation at CCGG sequences, and also points to variability in the effects of DNA methylation on different DNA-binding proteins. The methylation of DNA can take place on one strand (hemimethylation) or both strands, and, by conversion of hemimethylated CG dinucleotides into fully methylated, methylation patterns can be inherited from mother to daughter cells.

The exact role of DNA methylation remains obscure. It is not required for cell viability, but is essential for correct execution of the fetal developmental program in mice. Patterns of methylation have been observed to fluctuate during development, which clearly points to a dynamic role in gene regulation. However, the clearest demonstrations of the effects of DNA methylation, all relate to a role in the repression of transcription. Where the *cis*-acting sequences involved in control of expression contain CG (or CNG) sequences, methylation of the C residue may prevent binding of a *trans*-acting factor, and thus repress transcription. In agreement with this hypothesis, many transcription factors have target sites containing C residues that can be methylated, and show inhibition of binding if the C is methylated. However, other *cis*-acting sequences do not contain methylatable C-residues, and some *trans*-acting factors are insensitive to C-methylation. A further complication is that promoter regions of genes are often observed to be relatively free of C methylation (such regions are sometimes referred to as 'CpG islands'), although there is no reason why DNA methylation should be a regulatory mechanism for all genes. Transcription could also be repressed by DNA methylation if proteins were present which specifically bound to methylated C residues, and blocked access to *cis*-acting sequences. Such proteins have been found in both animals and plants.

Two examples of the effects of DNA methylation can be provided from plant systems. In the ribosomal RNA genes of hexaploid wheat, differential expression of different sets of RNA genes can be observed due to the presence of distinguishable ribosomal gene clusters on different chromosomes. Assay of methylation levels on the different sets of RNA genes by restriction endonucleases (see above) has shown that the level of expression is inversely correlated with methylation level; in other words, the sets of genes where methylation levels are higher have repressed transcription relative to less methylated genes. A second example of transcription repression concerns the fate of genes introduced into transgenic plants. In a few cases, genes that in initially transformed plants were actively expressed, have become 'silent' in progeny plants. The repression of transcription has been shown in these cases to be due to methylation of the introduced DNA. In an analogous situation in animal cell lines, where introduced genes in transgenic cell cultures are silenced as the cells pass through successive generations, methylation has been shown to 'spread' into the introduced DNA from surrounding sequences.

Methylation of nuclear DNA appears to be a fairly random process, both in animals and plants, and it is apparent that many methylation events have no effect on gene transcription at all; nevertheless, the role of methylation in gene control is now well established.

(e)   *Insertion and excision of transposable elements*

Transposable elements are DNA sequences that can move to different locations in the genome. Two major classes of these mobile elements are known in plants (and other higher eukaryotes); transposons, which move in the form of DNA, and retrotransposons, which, as the name suggests, move via an RNA intermediate (and require the activity of reverse transcriptase to form the double-stranded DNA that integrates back into the genome). The size of these mobile elements varies from approximately 600 bp to over 10 kbp; the smaller elements do not contain enough sequence to encode all the functions necessary for transposition, and thus rely on fully functional elements to provide the required enzyme activities (as in the *Ac-Ds* elements in maize, where the *Ac* elements are fully functional, and can provide the enzymes to move the non-functional *Ds* elements). Some of the more spectacular phenotypes of plants are due to the activities of transposable elements. The transposons were originally identified in a plant species, maize, as a result of the spontaneous and unstable mutations they caused. A typical example would be the 'streaking' of pigmentation seen in some varieties of snapdragon (*Antirrhinum majus*) flowers, caused by the transposable elements *Tam1* and *Tam3*. A full discussion of mobile elements, even restricted to plants, is beyond the scope of the present chapter; however, the effects of insertion of a mobile element on gene expression do need to be considered.

The insertion of a mobile element is a more-or-less random process, as the integrating enzymes have little or no sequence specificity; however, physical accessibility of DNA, as described in the preceding sections, will have an important influence, and there will thus be a strong tendency for insertions to occur in genes that are being actively expressed, since the DNA in these areas will be exposed. (Similarly, excision of an inserted element will also occur when the region of DNA is in 'open' chromatin, i.e. when the gene(s) in that region are actively expressed.) The effect on gene expression, and thus phenotype, of an insertion, is usually to abolish expression if the insertion occurs in the transcribed sequence of the gene, or its promoter. If the inserted sequence is in the transcribed region, it is likely that either the coding sequence of the gene product will be disrupted, or intron processing will be interfered with. If the insertion is in the promoter region, then it is likely to interfere with the protein–DNA and protein–protein interactions necessary for expression. However, gene expression is not always abolished; a retrotransposon in an intron of the potato starch phosphorylase gene has no apparent effect on expression, and enhancer sequences in transposons may actually increase expression in some cases, although examples have yet to be described in plants. The effects of an insertion are also affected by the cell in which the integration event takes place. Insertion in a vegetative cell leads to no lasting change, since it will only affect part of an individual plant, but insertion into germ-line cells results in a heritable phenotype which can affect the whole of a progeny plant.

The effects of retrotransposons on gene expression are generally irreversible, since the originating element is still present even after an RNA copy is produced for transposition. It might therefore be predicted that the plant genome would be littered with retrotransposon sequences, and this has been found to be the case, even if most of the sequences are no longer functional. Transposons are more interesting, since the inserted sequence does actually excise when it transposes, and thus if the element moves, activity may be restored to a gene inactivated by the original insertion. This is the cause of the streaks of pigmentation seen in snapdragon flowers, mentioned above. Here the insertion has occured in one of the structural genes encoding an enzyme on the biosynthetic pathway to the anthocyanin floral pigment (in the best characterized case, in chalcone synthase, but insertions in enzymes further down the flavonoid biosynthetic pathway can have the same effect on phenotype). When the gene would normally be activated, i.e. in floral tissue, the element becomes active due to chromatin opening, and excision can

occur. Activity is now restored to the gene, so that in the cell in which the excision has occurred, and its vegetative progeny, pigment production takes place, leading to a streak of pigmented tissue. Insertion of a transposon causes a duplication of several (3–8) bp of host sequence at the site of insertion, which is not removed on excision, so that transposons which insert into coding sequences of genes usually cause permanent inactivation. However, the effect of the small extra sequence in promoter regions (or in transcribed, non-translated sequence) is usually not so important, and restoration of activity takes place.

In contrast to the mechanisms of gene control discussed above, the effects of inserted sequences are highly specific to particular lines of particular plant species. Only a few varieties of snapdragon have 'streaky' flowers; in most varieties the floral pigmentation is stable. Selection for the 'unstable' phenotype has specifically selected the rare events where a transposon has inserted into a specific gene, in germ-line cells, in such a way that excision restores activity. However, mobile elements are present in the genomes of all plants (although in many cases they are inactive), and represent one of the ways in which genetic diversity can be generated. Other types of chromosomal rearrangements, such as deletions caused by unequal crossing-over, can also affect gene expression by physically damaging the DNA sequences of genes, but their effects are less-well characterized. The special case of DNA insertions in the genomes of transgenic plants is considered later in this chapter.

9.2.4 Post-transcriptional regulation of gene expression

Control of the transcriptional activity of genes may be considered as the primary means of controlling expression. However, further control over the accumulation of gene products can be exercised at subsequent stages of the synthetic process, i.e. the formation of the mature RNA molecule, and, where relevant, the translation of mRNA species. In addition, the rate of turnover of RNA molecules can play an important role in regulating gene expression, since the shorter-lived an RNA species, the less of it accumulates (at a given rate of synthesis). Even protein turnover has an effect on gene expression, as measured by accumulation of the final gene product. All these effects have been loosely grouped together under the term 'post-transcriptional regulation'.

For convenience, a rather arbitrary distinction may be drawn between regulation of gene expression, and metabolic regulation through control of enzyme activities. Although this distinction is certainly not drawn in a living organism, where the whole biochemical system is interactive (that is, each component part is influenced to a greater or lesser extent by all the other parts), it follows the reductionist principles which enable scientific analysis to be carried out. This section will therefore consider the factors which affect levels of RNA species as relevant to 'post-transcriptional control', and ignore any event subsequent to the involvement of mRNA in protein translation.

(a)   *Effects on levels of specific RNA species*

The level of a particular RNA species as a proportion of total RNA (and thus, by estimating total RNA per cell, on a per cell basis) can be readily determined by quantitative hybridization techniques. Excess of a specific labeled probe DNA sequence, complementary to the RNA species under analysis, is allowed to form double strands with the RNA. After removal of non-hybridized probe, the amount of double-stranded hybrid formed is estimated. If a radioactive label is used on the probe, this can be done by scintillation counting or autoradiography. The normal method is to separate the RNA species in a sample of total RNA by size, using electrophoresis on an agarose (or acrylamide) gel after denaturation of the RNA, and then immobilize the resulting separated nucleic acids on a nitrocellulose or nylon filter (a 'Northern' blot). The amount of hybridization is then estimated by autoradiography, using known amounts of synthetic RNA to get an absolute quantitation. In practice, better quantitation after densitometry is given if the RNA species are not separated by electrophoresis before immobilization ('dot' or 'slot' blots), but it is necessary to check such an assay carefully for binding to other RNA species, which are usually separated by the electrophoresis.

The determination gives the steady-state level of the RNA species, which is a result of both RNA synthesis and RNA degradation. It may be expressed as,

$$[\text{RNA species}] = \int (k_s - k_d [\text{RNA species}]) \, dt$$

where k_s is the rate of synthesis of the RNA species, as determined by transcriptional activity of the corresponding gene, and k_d its rate of degradation, assumed to follow first order kinetics.

Since both k_s and k_d can be affected by a variety of controlling factors, care must be excercised in assuming that, for example, a decrease in RNA level necessarily means a decrease in transcriptional activity of the corresponding gene. Unfortunately, it is not straightforward to separate the different contributions to the observed RNA level. An estimate of the relative rates at which RNA species are synthesized may be obtained from 'run-on' transcription assays, using isolated nuclei. In this technique, transcripts produced by nuclei incubated in the presence of radioactively labeled NTPs (nucleotide triphosphates) are hybridized to unlabeled specific DNA probe sequences (in excess). The amount of specific hybridization, estimated by the radioactivity of the transcript, then gives a measure of the amount of specific transcript produced. Since little or no transcription initiation has been shown to take place in isolated nuclei, this type of assay gives a 'snapshot' of the transcripts that were being synthesized when the nuclei were isolated. Attempts to estimate the rate of degradation of specific mRNA species in plant cells have been less successful, and very little is known about the degradative processes involved.

Despite the foregoing discussion, in most cases that have been examined in plants, there is a clear qualitative correlation between estimates of transcriptional activity of genes, and the measured levels of the corresponding RNA species. For example, in the case of seed storage protein genes examined in many species, when different stages of seed development are examined, transcription of the storage protein genes is very low until significant amounts of storage protein mRNAs are detected. Transcription is at maximal activity while the amount of mRNAs is increasing rapidly, but declines as the amount of mRNA peaks and subsequently declines. It is when quantitative comparisons are made between different genes, or different developmental stages, that the effects of post-transcriptional control become apparent. For example, for two subfamilies of pea legumin genes (*legA* and *legJ*), the relative rates of gene transcription were 1.1 : 1 during the cell expansion phase of seed development, but the relative levels of the corresponding mRNA species were 5.0 : 1. Control of this type is also apparent in the effect of sulfur deprivation on storage protein gene expression, where relatively small effects on the transcriptional activity of genes encoding sulfur-rich proteins are amplified by post-transcriptional effects to result in large decreases in the levels of the corresponding mRNAs. A further example, showing how activity of one gene can affect the expression of another by post-transcriptional effects, is the suppression of legumin synthesis in peas with a defective gene at the *r* locus (which encodes starch branching enzyme). No effect on legumin gene transcription due to the defective *r* gene could be observed, but legumin mRNA levels were reduced. Post-transcriptional effects can also be important in regulating gene expression in response to external stimuli, as in the case of the pea gene *Fed-1*, which encodes ferredoxin. In contrast to all other light-regulated genes examined, activation of expression of this gene in response to light is not dependent on the 5' flanking region of the gene (where interactions between *trans*-acting factors and *cis*-acting sequences in the promoter would take place), but is determined by the transcribed unit itself. Since effects of light on transcription of this gene are small, this cannot be due to an 'enhancer' sequence in the DNA, but must reflect a mechanism acting on the transcript.

The mechanisms by which post-transcriptional control takes place are not fully understood, since many different processes could be contributing during processing of the primary transcripts, export of RNA to the cytoplasm, and survival of the cytoplasmic RNA. However, mRNA stability does seem to play a major role, and in some examples, this has been shown to be clearly related to translation of the encoded polypeptide. A mutant form of the soybean Kunitz trypsin inhibitor gene containing three base changes in the protein coding region, leading to nonsense codons, was transcribed at the same rate as the normal gene, but protein accumulation decreased by at least 100-fold due to suppression of mRNA levels, presumably due to an increased rate of mRNA degradation. Several examples of mutant genes which have the start codon altered, but are otherwise identical to functional genes, are known; no protein products are produced by these genes, and the corresponding mRNAs are undetectable, or present at very low levels. The most clear demonstration of the importance of 'translatability' for mRNA stability has been shown by a mutant form of the *Phaseolus vulgaris* lectin gene. This mutant contained an in-frame stop codon in the protein coding sequence due to a frame shift mutation, and was not expressed to give detectable levels of mRNA. When the frame-shift error was repaired by *in vitro* mutagenesis, and the repaired gene was introduced into a transgenic plant, it proved to be as active as a wild-type gene. Repair of the coding sequence error was estimated to have increased the mRNA level by at least 40-fold, again presumably by decreasing the rate of mRNA degradation. In these examples,

post-transcriptional control is the major factor in determining gene expression.

Results from animal systems have shown that polyadenylation plays an important role in determining mRNA stability, since the half-life of an mRNA is related to the length of the poly(A) tail. A similar relationship has been assumed in plants, although little experimental evidence is available to support the assumption. An example may be provided by seed storage protein mRNAs, where the poly(A) tails have been shown to shorten as seed development proceeds into the desiccation phase, and mRNA levels decline.

(b)   *Effects at protein translation*

Some limited evidence for further regulation of gene expression at the translational level in plants has been put forward. In this type of regulation, the rate of synthesis of a polypeptide from its encoding mRNA is altered so that although the levels of mRNAs may not change, production of protein does. An example is the expression of heat-shock genes during carrot somatic embryogenesis, where at the mid-globular stage of embryo development, the level of an mRNA species encoding a small heat-shock protein was found to be lower than at other stages, and to show no increase in level, due to gene activation, on heat shock. Nevertheless, a full complement of heat-shock proteins was synthesized by these embryos, presumably by more active translation of pre-existing mRNA. A different mechanism of translation-level control is shown by rice seed storage protein genes, where a more efficient partition of the glutelin mRNA species into membrane-bound polysomes results in much higher levels of synthesis of glutelin than prolamine in 15-day-old seeds. This is despite the fact that the two types of mRNAs are present in similar amounts, and transcriptional levels from the two classes of genes were also similar.

9.3   ORGANELLAR GENES

The multiple levels of control of gene expression outlined for nuclear genes in plants will apply also to the genes in the two organellar genomes of plants, those of the plastid and the mitochondrion. Consequently, in this section the differences between the organellar and nuclear genes will be emphasized, particularly those which affect control of expression.

9.3.1 Prokaryotic features

The prokaryotic nature of the organellar genomes, and the transcription and translation machinery in organelles, described in Chapter 8, result in some major differences in the regulation of gene expression, compared to nuclear genes, and result in some major simplifications. The comparatively small size of these genomes (208–2400 kbp for mitochondria, and 120–270 kbp for plastids, depending on plant species) mean that a far more limited number of transcripts can be produced, and the absence of chromatin limits higher order DNA structures as a control mechanism (although plastid RNA polymerase requires supercoiled DNA as a template to initiate transcription). The complex array of transcription initiation factors in the nuclear genome have no counterparts in the organellar genomes, and RNA polymerase itself is simpler in structure. The gene promoters in organellar genes therefore consist only of recognition sequences for the polymerase, analogous to the '−35' and '−10' regions in E. coli promoters, and the transcription start point. The polymerase forms a non-specific complex with DNA, moves along until it locates the recognition sites, and is then correctly positioned to start transcription. The transcriptional activities of promoters in E. coli depend on how well the '−10' and '−35' regions fit an optimum consensus sequence, and similar controls appear to operate in the organellar genomes. However, the repression or activation of promoter activity by trans-acting DNA binding proteins, which is a feature of operons in E. coli, is not observed in organellar genomes. The necessary proteins are not encoded in organelles, and the overall control of gene activity in plastids and mitochondria has been taken over by the nucleus (section 9.3.4). A common feature of prokaryotic genomes is polycistronic mRNAs, which are the product of transcription of linked coding sequences from a common promoter; these are observed in organellar genomes also, as in the production of a transcript containing coding sequences of two of the subunits of the ATP synthase complex in wheat mitochondria.

9.3.2 Control of gene expression in mitochondria

The mitochondrial genomes of plants, like those of other eukaryotes, encode a very limited number of polypeptides, together with the rRNA and tRNA components of the translation machinery necessary for their expression. A feature unique to plants is that some of the mitochondrial tRNAs are produced from the nuclear genome and imported into the mitochondrion. The polypeptides produced in the mitochondrion include several of the ribosomal proteins, and components of cytochrome oxidase, cytochrome b, ATP synthetase and NADH : CoQ reductase. Mitochondria do not have genes for RNA polymerase polypeptides, and thus are inherently transcriptionally inactive. Notable features of the mitochondrial genome are the large amount of non-coding DNA it contains, and its extreme variability between plant species.

Control of the activity of RNA polymerase in mitochondria is not well characterized. Although transcription of regions containing genes occurs at a much higher level than non-coding regions, virtually the whole genome appears to be transcribed at a low level. It is not known whether this is the result of false initiation events by RNA polymerase, or failure to terminate transcripts from gene regions. Analysis of the system is hampered by a failure to determine consensus cis-acting promoter sequences in the mitochondrial genome, which suggests that transcription initiation is not under strict control. Many genes appear to have multiple sites for transcription initiation. Extensive modification of transcripts occurs at the post-transcriptional level, including processing and selective degradation of unwanted sequences. Unlike their prokaryotic equivalents, some mitochondrial genes contain introns, and RNA splicing is thus another postranscriptional modification. A complication of the organellar system is that some transcripts have to be assembled from several different RNA transcripts, first joining the RNAs together by base-pairing in the introns, before splicing the introns out (referred to as trans-splicing). For example, the complete transcript of the nad1 gene (encoding a polypeptide of NADH dehydrogenase) of Oenothera must be assembled from three different RNA species, joined by base pairing in introns 1 and 3, before all 4 introns are spliced out to give the mature RNA. Finally, a feature again unique to plant mitochondria is a form of 'editing' of transcripts in which some C residues on RNA are converted to U, predominantly in coding sequences so that amino acid codons are altered. It had originally been thought that plant mitochondria might not use the universal genetic code, but the discovery of RNA editing has proved this idea to have been mistaken. This process is necessary and specific, but its role in controlling gene expression is unclear.

The large amounts of non-coding DNA in the mitochondrial genome, along with the presence of extrachromosomal elements, make rearrangements of the genome a fairly frequent occurrence. These rearrangements can lead to changes in gene expression due to physical damage to the genes (see above), or can alter promoter regions. The phenomenon of cytoplasmic male sterility, which is carried by the mitochondria and causes a failure of pollen development, has been shown to be associated with changes in the structure of the mitochondrial genome which lead to the formation of aberrant or non-functional proteins.

9.3.3 Control of gene expression in plastids

Although the plastid genome is smaller than the mitochondrion, it contains more gene sequences, and can produce a more extensive range of products. It contains genes for all the ribosomal and tRNAs, and most of the proteins necessary for the machinery of protein synthesis, as well as the α- and β-subunits of its own RNA polymerase, and components of the enzyme systems necessary for fixing CO_2, light harvesting, and energy generation. The organization of genes shows strong prokaryotic features, with clustering of related genes to allow expression as polycistronic messages from single promoters. The complete sequencing of the chloroplast genomes from a liverwort, and tobacco has allowed the genome to be studied in considerable detail. Perhaps the major non-prokaryotic feature of the chloroplast genome is the presence of introns in some genes, which necessitates RNA splicing to occur as a post-transcriptional modification mechanism. In a few genes, the mature RNA is assembled by *trans*-splicing, as described above for mitochondria. Post-transcriptional processing of RNAs also occurs; for example, the gene for the large subunit of ribulose bisphosphate carboxylase (*rbcL*) gives rise to a 1.8 kb mRNA which is processed to 1.6 kb by removal of the non-coding sequence at the 5' end. As in mitochondria, it is likely that much of the fine control of chloroplast gene expression occurs post-transcriptionally.

Many plastid genes have regions 5' to the transcription initiation site in which prokaryotic-like *cis*-acting promoter sequences are present, and, like mitochondria, these will function by direct interaction with RNA polymerase. Although many gene promoters contain sequences similar to the '−10' and '−35' elements found in *E. coli*, not all do, suggesting that another factor may be required for initiation in these cases. Compared to the nuclear genome, genes are very close together on the plastid genome (as is typical for prokaryotes). For example, the transcription start points for *rbcL* and the gene encoding the β-subunit of the ATP synthetase complex, *atpB*, are only 159 bp apart (the opposite strands of DNA in the two genes are transcribed, so the two genes are arranged with 5' ends facing each other). The two promoters thus directly influence the activity of each other, since binding of RNA polymerase at one promoter interferes with binding at the other.

9.3.4 Interaction of nuclear and organellar genes

The nuclear genome in plant cells maintains an absolute control over expression of genes in the organelles by two means. Firstly, it has taken over many of the genes encoding vital organellar functions (such as DNA replication), which are thus controlled by nuclear factors, and secondly, it controls expression of genes within the organelles by the protein products of nuclear encoded genes. This is perhaps most dramatically shown by the changes in plastid morphology in different plant tissues, where in leaves the plastids are in the form of chloroplasts, and are filled with the membrane systems of the light harvesting complexes, whereas in storage tissues the plastids contain no pigments or membrane systems, but instead are full of starch grains. These changes are determined by the nucleus, not the organellar genome. There is now considerable evidence that exchange of DNA sequences is possible between the nuclear and organellar genomes, since chloroplast DNA sequences have been found in the nucleus and mitochondrion. This process seems to have led to a steady removal of functional genes from the organelles, which are added to the nuclear genome. For example, the *coxII* gene normally present in mitochondria, which encodes a subunit of cytochrome oxidase, is absent in cowpea, but a copy of the gene is present in the nucleus, modified by the addition of a transit peptide for import into mitochondria, and given a eukaryotic promoter. The organellar genomes have thus been left with little or no autonomy.

At a coarse level, control of organellar gene expression can be achieved by the nucleus by controlling expression of the nuclear gene products necessary for the overall processes of DNA replication and expression (DNA and RNA polymerases, ribosomal proteins, translation factors). However, fine control of individual genes is achieved through further nuclear genes, many of which are necessary for expression of specific organellar genes. This fine control seems to be exercised predominantly at the post-transcriptional level, rather than by direct effects on transcription initiation, and, in mitochondria at least, often involves processing of primary transcripts (as in the case of nuclear restorer genes for cytoplasmic male sterility caused by defects in mitochondrial genes). Proteins which are assembled from subunits encoded by separate genomes, such as ribulose bisphosphate carboxylase, where the large subunits are encoded by plastids (*rbcL* gene), and the small subunits by the nucleus (*rbcS* genes), would seem to require that nuclear and organellar genes are controlled together, so that similar amounts of the two subunits (which are present in equimolar amounts in the complete enzyme) are produced. Although both *rbcL* and *rbcS* genes are regulated primarily by light at the transcriptional level, quantitative regulation of gene expression occurs post-transcriptionally, at translation, since the mRNAs produced by *rbcL* and *rbcS* accumulate in maize leaves in a ratio of 10:1, whereas the corresponding polypeptides are present in a ratio of 1:1.

9.4   SIGNALING MECHANISMS IN GENE REGULATION

In contrast to the relatively advanced state of knowledge of the control of plant gene expression from gene transcription onwards, and the vast body of work describing the responses of plants to external stimuli such as biochemicals of various types ('plant hormones'), environmental conditions, etc., knowledge of what links the two phenomena in plants is very limited. The biochemistry of what happens between a stimulus arriving at the outside of the plant cell, and the change in gene expression it brings about, is only just beginning to be explored, and other eukaryotic signal transduction systems are being used as models for what takes place in plants. Difficulties in working with signaling mechanisms in plants include the complexity of the responses evoked by a single stimulus, the number of different chemicals that plants respond to, the differences in response to the same chemical in different plant tissues, and the difficulty of trying to determine whether a particular chemical is the primary cause of a response, or has a pleiotrophic effect which in turn induces the response studied. The highly specific responses by specific tissues in higher animals to circulatory hormones have not been

observed in plants, and in general this model has not proved helpful in understanding signaling in the plant kingdom.

9.4.1   Stimuli and receptors

In order to respond to a stimulus, the plant cell must first be able to perceive it. Characterizing the receptors for physicochemical stimuli in plants has proved to be a major obstacle, and to date only the mechanism by which light is perceived is at all well understood.

Although at least three types of photoreceptors are present in plants (for blue/uv, red, and far-red light), the only well-characterized receptor is phytochrome, which responds to red and far-red light. Phytochrome is a protein of molecular weight 124 000, which contains a tetrapyrrole chromophore. As synthesized, this chromophore has an absorption maximum at 660 nm; irradiation with red light (i.e. with a wavelength near 660 nm; this is the predominant wavelength reaching lower leaves of the plant, as a result of absorption by chloroplast pigments in the leaf canopy) alters the structure of the chromophore, which causes a conformational change in the protein, altering a number of its properties including accessibility to proteases and other enzymes. The two forms of phytochrome are designated P_r (as synthesized) and P_{fr} (after irradiation). Irradiation of P_{fr} with far-red light (at the new absorption maximum of the chromophore, 730 nm) will cause it to revert to P_r; probably more important physiologically is that P_{fr} will slowly revert to P_r in the dark. P_{fr} is also subject to degradation *in vivo*. P_{fr} is responsible for gene activation (not directly; q.v.), and has been shown to control the activities of many light-regulated genes, such as those encoding ribulose bisphosphate carboxylase, chlorophyll *a/b* binding protein, etc. The system is summarized in Fig. 9.7.

One reason for the progress that was made with the phytochrome system is that a receptor for light does not need to be in the cell wall, and that phytochrome is a cytoplasmic protein, and thus could be purified (not necessarily the case for other light receptors). A soluble receptor for salicylic acid, which stimulates the activity of genes involved in plant resistance to pathogens, has also been identified. Perception of heavy metal ions (brought in through the uptake systems for micronutrients), other small inorganic ions, and metabolic intermediates such as simple sugars or amino acids could also involve cytoplasmic receptors, but most other stimuli do not enter

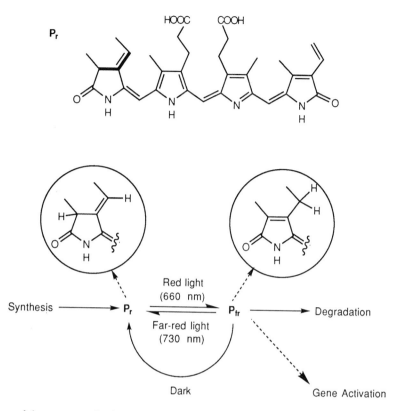

Figure 9.7 Structure of the tetrapyrrole chromophore in phytochrome (P_r form); irradiation with red light causes a double bond shift at the bonds highlighted in bold. A simplified scheme for the light-dependent changes in phytochrome, resulting in gene activation, is also shown.

the plant cell, and thus the receptors must be located in the cell wall. Such proteins are very difficult to purify from plants where the outer cellulose and xyloglucan cell wall hinders any attempt at solubilization. However, use of the photoaffinity label, 5-azido-[7-³H]indole-3-acetic acid, an active homolog of the plant hormone indole acetic acid (IAA), on membrane preparations from plants has shown that a number of membrane-associated polypeptides (in both the plasma membrane and the endoplasmic reticulum) can specifically bind auxins such as IAA. Unfortunately, such polypeptides are merely identified as bands on gel electrophoresis, and their physiological function is not clear; correlation of binding activity with a physiological response, or a mutant phenotype, is necessary for characterization. The amazing variety of signals which can affect gene expression (more or less specifically) in plants demonstrate that elaborate receptor systems must be present; for example, expression of proteinase inhibitor genes in tomato is specifically activated by certain oligosaccharides from plant or fungal cell walls, an 18 amino acid peptide systemin, and

jasmonic acid (a terpenoid); it is also regulated by abscisic acid, IAA, sucrose, carbohydrates in general, and tissue type and development.

9.4.2 Signal transduction

Perception of a stimulus is only a first step towards altering gene expression as a response. In order for the perception to have an effect on the expression of specific genes, transduction of the signal is necessary; that is, a molecular mechanism which changes the initial perception into a regulatory effect on transcription, or post-transcriptional processes. In higher animals, signal transduction in genes regulated by steroid and thyroid hormones has proved relatively simple to understand (at least, in outline). The small size of these hormones, and their ability to pass through the outer membranes of animal cells, allows the use of an intracellular receptor. The receptors belong to a family of proteins that contain a conserved DNA-binding domain, and which function directly as transcriptional activators, i.e. the receptor itself is the transducing molecule. Binding of the hormone to the receptor causes a change in its properties, which results in the complex being localized in the nucleus and showing high-affinity binding to specific sites in DNA, in the promoter regions of genes that are activated by the hormone. The actual mechanism of gene activation by the hormone–receptor complex *trans*-acting factor is still the subject of investigation. In the case of the chicken, ovalbumin gene seems to involve the release of the gene promoter from a repressed state. Unfortunately, in plants, examples of such a simple signal transduction mechanism have yet to be demonstrated. In the case of genes regulated by phytochrome, protein synthesis is normally necessary in order for the regulatory effect to be produced. This requires the involvement of proteins synthesized *de novo* in the transduction mechanism. However, the effect of phytochrome on transcription of its own encoding gene suggests a more direct signal transduction mechanism. Red-light activation of the P_r form of phytochrome to the P_{fr} form (see Fig. 9.7) represses the active transcription of the phytochrome-encoding *phyA* gene in dark-grown tissue of oats, without requiring protein synthesis. Repression is dependent on the amount of P_{fr} present, and can thus be relieved by returning the tissue to the dark. This process has a physiological role in preventing the receptor for light from accumulating when it is not wanted, but making sure it is present when needed. However, the sequence of phytochrome does not suggest that it interacts directly with DNA, as do the animal hormone receptors, nor has specific binding of P_{fr} to DNA been shown to occur. Presumably other parts of this signal transduction mechanism are present in the cell prior to the P_r to P_{fr} conversion. One suggestion has been that P_{fr} is involved in phosphorylation of nuclear proteins.

Plant signal transduction mechanisms must thus, in general, involve the participation of second messengers, or intermediary signal molecules. These molecules are used to control the activity of protein kinases, which phosphorylate target proteins and (possibly through a number of stages) bring about a change in metabolic activity or gene expression. The process is outlined in Fig. 9.8.

A number of reasonably well-characterized second messenger systems have been established in animal systems, but work on plants has made comparatively little progress. The best-characterized second messenger system in animals, involving cyclic AMP as the intracellular molecule produced as a response to extracellular binding of a hormone at a receptor in the cell membrane, has not been found in plants; although most of the components of this system are present, plants do not contain protein kinase A, the target for activation by cyclic AMP. There is also some evidence to suggest that activation of protein kinase C by diacylglycerol, a second messenger in animal cells derived from phophatidylinositol, is not a functional signaling route in plants. What transducing systems are present in plants? There is abundant evidence of intracellular changes in Ca^{2+} concentration being a common response to many stimuli in plants, such as phytochrome activation, guard cell stimulation and even wind pressure on leaves. The response is rapid and transient, as would be required. Plant cells contain the necessary components of the Ca^{2+}-involving signal transduction system, including the calcium-binding protein calmodulin, phosphatidlyinositol-4,5-bisphosphate, inositol-1,4,5-trisphosphate phopholipase C, membrane-associated heterotrimeric GTP-binding proteins which link the perception of the stimulus by a cell wall receptor to activation of enzymes acting within the cell (although most of the plant G_α-polypeptides identified to date are more similar to GTP-binding proteins involved in vesicular transport than the components of true G-proteins) and Ca^{2+}-phospholipid activated protein kinase. In addition, the vacuole is used as a Ca^{2+} store, whose membrane permeability can be specifically changed. Evidence has been presented that in the phytochrome-mediated activation of light-regulated genes, such as that encoding the chlorophyll *a/b*-binding protein (see above), both a G-protein and changes in intracellular calcium are necessary for signal transduction.

Whether or not intracellular Ca^{2+} levels do prove to be the major signal transducing mechanism in plants, the targets of the cytoplasmic protein kinases remain to be investigated before the complete chain of events from perception to gene expression can be established.

9.5 GENE REGULATION IN PLANT DEVELOPMENT

Although some genes, such as those encoding enzymes of the glycolytic pathway, or ribulose bisphosphate carboxylase, are expressed in all, or many tissues of the plant, the expression of most plant genes is restricted to certain organs, tissues or cell types, and certain developmental stages. This type of gene regulation is thus a consequence of the tissue differentiation, and the tissue-specific

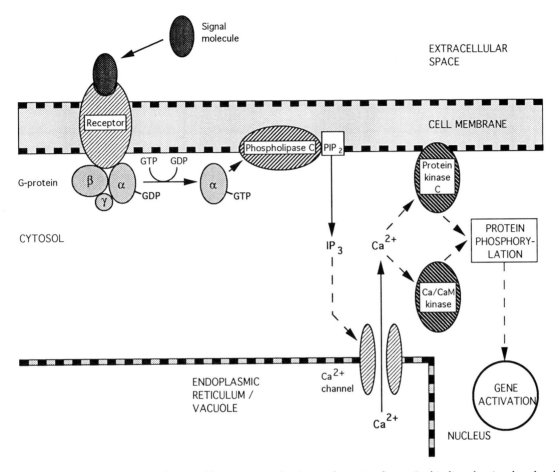

Figure 9.8 Simplified version of a possible signal transduction pathway in plants. On binding the signal molecule at a trans-membrane receptor located in the cell wall, GDP bound to the α-subunit of a specific heterotrimeric G-protein associated with the receptor is converted into GTP by phosphorylation with the γ-phosphoryl group of cytoplasmic GTP. The α-subunit of the G-protein dissociates, and interacts with phospholipase C. Phospholipase produces inositol trisphosphate, which acts on the cellular calcium stores to increase the cytoplasmic calcium level. The calcium level then triggers the protein kinases (Ca/CaM kinase = calcium/calmodulin regulated protein kinase) that result in conversion of *trans*-acting factors from inactive into active forms, and thus cause gene activation. Many steps have been omitted for clarity, including all the processes regenerating the original forms of the signal molecules.

genes have their expression controlled by development. On the other hand, study of various naturally occurring mutations in plants, by classical genetics, has identified a number of genetic loci that control developmental processes. Mutations at these loci alter the form of organs of the plant (for example, converting leaflets into tendrils), and thus these genes themselves control development. Both these types of regulation, representing early and late steps in a progression of causes and results, are the subject of intense study at present. A further factor is the interplay between cell lineage and environmental factors in determining the fate of a meristematic cell, and thus the genes that are expressed in it. While it is clear that communication between cells is an important factor in regulating gene expression, the processes by which this is achieved are not understood at present.

9.5.1 Tissue-specific expression

The ability to study the expression of single characterized genes and their promoters by the use of promoter–reporter gene constructs in transgenic plants (section 9.6), taken in conjunction with techniques for immunolocalization of gene products, and localization of mRNA species by *in situ* hybridization, has greatly enhanced the level of knowledge available on the precise control of the expression of genes in different tissues of the plant. Whereas previously tissue-specific expression was a phenomenon studied mainly at the macroscopic level, it has now become refined to the point at which a particular gene may have its expression restricted to just a few cells, for example in floral meristems, or in the tapetal layer. At the same time, detailed study of genes expressed abundantly in a 'tissue-specific' manner has shown that control is not always absolute. For example, although the major legumin seed proteins in legumes do seem to have their expression restricted to tissues of the developing embryo (and endosperm), many other proteins accumulated in seeds, such as trypsin inhibitors, are also expressed at low levels in other parts of the plant. In pea, the active lectin gene is expressed at high levels in the seed (where lectin protein accumulates), and at much lower levels in the root, where the gene product is required for establishing the symbiotic relationship with *Rhizobia*. Although attempts can be made to derive some general principles for tissue-specific expression, or even to look for common factors that control the expression of multiple genes with similar expression patterns, the (almost) infinite variability of gene expression means that control of each gene must be considered individually, in relation to the function of the gene product. The reader is referred to more specialized texts for descriptions of the expression patterns of individual genes.

The *cis*-acting sequences which determine tissue-specific expression can be shown to be present in the 5' flanking regions of suitable genes, by deleting these regions and demonstrating that expression from the truncated promoter is abolished. However, it has proved difficult to specify which sequences are actually responsible for tissue-specificity. One technique has been to compare the sequences of promoter regions of genes sharing a common expression pattern, to attempt to find common regions of sequence, which were presumed to be the *cis*-acting sequences. These comparisons have been carried out with both cereal and legume seed storage protein genes, and a number of conserved sequences (described as various 'boxes', with sequences ranging in length from 8 to 30 bp) have been identified. However, the functional analysis of these sequences has not yielded clear-cut results. If the sequences really are *cis*-acting elements that determine tissue-specific expression, then it should be possible to demonstrate specific interaction with *trans*-acting factors, which should correlate with expression of the genes. Furthermore, the sequences should be shown to be functional when tested as promoter–reporter constructs in transgenic plants, i.e. deletion or mutation of the sequences should abolish tissue specific expression, and addition of the sequences to a constitutively expressed promoter should give tissue-specific enhancement of expression. A number of cereal prolamin genes contain a conserved sequence approximately 300 bp 5' to the transcription start, which has the consensus core sequence TGTAAAG. A nuclear factor has been shown to bind to this sequence in the maize zein genes, but removal of the sequence did not abolish tissue specific expression of the promoter in a transgenic tobacco plant. Mutation of a conserved sequence in legume storage protein genes, the so-called legumin box, did affect expression of the promoter in transgenic plants, but interaction of this region of the promoter with *trans*-acting factors could not be demonstrated. It has been suggested that 'tissue-specificity' may be determined by a complex interaction between regulatory elements, and metabolic and environmental signals (which affect gene expression through a series of *cis*-acting sequences and *trans*-acting factors). Complex control mechanisms of this type have been considered above.

Figure 9.9 Determination of the phenotypes of floral organs by homeotic genes in *Arabidopsis thaliana*. The floral organs are each derived from a single whorl of cells in the floral meristem. Expression of homeotic genes in these four whorls results in the production of transcription factors which control the phenotype of the organs derived from them. The observed phenotypes in normal flowers, and in various floral mutants, can be explained by the action of three gene classes. The genes involved can be defined by the phenotypes of different homeotic floral mutations; they are *ag* (agamous), *ap2* (apetala2) and *ap3/pi* (apetala3/pistilata; both gene products must be present to give the normal phenotype, since a mutation to either gene causes the mutant phenotype). A whorl of cells in which *ag* alone is expressed gives rise to sepals; expression of *ag* and *ap3/pi* gives rise to petals; expression of *ap2* and *ap3/pi* gives rise to stamens; and expression of *ap2* alone gives rise to carpels. Expression of *ap3/pi* is confined to whorls 2 and 3; *ag* is normally expressed in whorls 1 and 2, and *ap2* is normally expressed in whorls 3 and 4. The expression patterns of *ag* and *ap2* are mutually exclusive, and control each other; if the expression of either gene is prevented by a mutation, the product of the other gene is present not only in the whorls where its encoding gene is normally expressed, but also in the whorls where the expression is normally prevented by the product of the inactivated gene. The diagrams show the phenotypes of floral structures formed by both the wild-type plant, and by mutants where expression of homeotic genes of different classes is inactivated.

Origin of the organs in a flower of *Arabidopsis thaliana*
from four whorls of cells in the flowering meristem

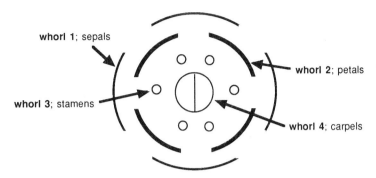

Expression of homeotic genes in whorls in wild type and mutant *Arabidopsis*,
with resulting phenotypes of cells

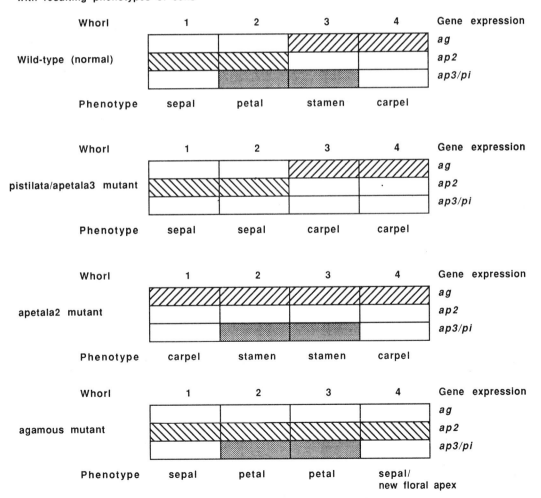

Figure 9.9   (Caption opposite)

9.5.2 Genes which control development

The presence of genes which control developmental processes in higher eukaryotes has been inferred from homeotic mutations, which are defined as mutations at a single locus which cause a change in the morphology, or developmental pattern, of an organ. The action of these homeotic genes must thus be to control the action of all the genes whose expression determines the phenotype of the organ. The best characterized genes in plants are those that affect floral morphology, and parallel homeotic mutations have been studied in *Arabidopsis* and *Antirrhinum*.

The flowers of *Arabidopsis* and *Antirrhinum* consist of four concentric whorls of organs, designated whorls 1–4, with whorl 1 corresponding to the cell type on the outside of the flower (Fig. 9.9). The four whorls thus produce the stamens, petals, stamens (pollen-containing) and the carpels (ovule-containing), respectively, the carpels being located in the center of the flower. Homeotic mutants in floral patterning alter the organ type in the different whorls of the floral structures, as detailed in Fig. 9.9. A typical example would be the *agamous* mutant of *Arabidopsis*, where the stamens and carpels are replaced by petals and sepals, respectively. A model for the way in which organ type is determined is now generally accepted, and is based on the action of complementary gene products encoded by the different homeotic genes. Three classes of homeotic gene, A, B and C, determine organ identity by their expression in different whorls of cells. In the wild-type flower, whorl 1 is determined to be sepals by the product of gene A alone; whorl 2 is determined as petals by the combined action of genes A and B; whorl 3 is determined as stamens by the combined actions of genes B and C, and whorl 4 is determined as carpels by the action of gene C alone. Gene A is thus expressed in whorls 1 and 2, B in whorls 2 and 3, and C in whorls 3 and 4. The expression of genes A and C is postulated to be mutually antagonistic, so that expression of C in whorls 3 and 4 prevents expression of A. The model is supported by the isolation of homeotic genes of classes A, B and C from both *Arabidopsis* and *Antirrhinum*, and demonstrations that mutations in genes corresponding to A and C, leading to aberrant expression (since inactivation of one gene allows the other to be expressed in all whorls), do indeed produce the expected phenotypes (Fig. 9.9). For example, in the *agamous* mutant in *Arabidopsis*, gene C is defective, and thus the four whorls contain the products of genes A, AB, AB and A, respectively (since C no longer prevents A being expressed), leading to the observed phenotype of sepals, petals, petals, sepals.

Sequence analysis of the gene products encoded by floral homeotic genes shows them to contain a domain very similar to that in several known DNA-binding proteins, which function as specific transcriptional activators, the MADS domain. These proteins thus are likely to function as transcriptional regulators of the subordinate genes which determine the organ phenotypes. In turn the floral homeotic genes may themselves be controlled by other transcriptional regulators, corresponding to homeotic genes which affect the floral meristems, and can prevent flower formation altogether. One such gene, *floricula* in *Antirrhinum*, has been cloned, and also encodes a protein with some similarity to transcriptional activators.

9.6 TRANSGENIC PLANTS

Transgenic plants can be used both to study the regulation of gene expression, and to manipulate it. Although anything more than a brief outline of plant transformation techniques is beyond the scope of this chapter, many results obtained using transgenic plant material are directly relevant.

Foreign DNA can be introduced into the plant genome by two methods; direct uptake of DNA and vector-mediated transfer. Direct uptake is conceptually simpler, since it relies on the inherent ability of the plant genome (through endogenous recombination systems) to incorporate DNA that is introduced into the plant cell, albeit at low efficiency. This incorporation may exploit the mechanisms that are responsible for transfer of genetic information between organellar and nuclear genomes. Many different techniques of DNA introduction have been tried in the search for viable transformation systems, where transformed plants can be produced at a reasonable frequency. Most methods have failed, either because the efficiency of DNA incorporation is too low, or because of technical problems, such as regenerating whole plants from tissue cultures. Methods such as soaking pollen grains in DNA solutions, or injecting DNA into the vascular tissue leading to the anthers, are too inefficient, whereas preparing protoplasts, and causing them to take up DNA after permeabilization or by electroporation (electrophoresis across the cell membrane), is efficient, but for many species regeneration of plants from the protoplasts cannot be achieved. A compromise method, in which tissues that can be regenerated, such as meristems, or embryogenic cultures, are treated with DNA either by electroporation, or by particle bombardment (where the DNA is coated onto microprojectiles, typically gold or tungsten particles 1–2 μm in diameter), has achieved some more general success, although it still remains technically difficult.

Vector-mediated transfer of DNA seeks to exploit naturally occurring organisms that are capable of transferring DNA to the plant genome. Unfortunately, the obvious candidates for this task, plant viruses, are of little or no value, since most are RNA viruses, do not transfer their genetic material into the host genome (unlike many animal and bacterial viruses), and cannot be manipulated to incorporate foreign DNA. Even those plant viruses that have a DNA genome, such as cauliflower mosaic virus (CaMV), replicate via an RNA intermediate, and do not transfer DNA. Fortunately, a soil-living bacterium, *Agrobacterium* (several species are known; the organism is very closely related to the symbiotic nitrogen-fixing bacterium, *Rhizobium*, found in association with legumes), which causes galling on stems and other tissues, was found to be capable of DNA transfer. The genes which caused uncontrolled proliferation of plant tissues in the gall, and production of metabolites for the bacterium to use, were found to be bacterial in origin. Further investigation showed that in *Agrobacterium tumefaciens*, the organism causing crown gall, the bacterium was capable of transferring a part of its genetic material to the plant nuclear genome, if it had access to a site of wounding (where the physical barriers of lignin and cellulose layers were breached). The gall-forming bacterial genes which were transferred formed a contiguous region on a plasmid carried by the bacterium, called a Ti-plasmid, and were separate from the genes which directed transfer, so that any foreign DNA put into the transferred region would also be transferred. From this point to the development of a fully functional transfer system for foreign DNA required considerable technical ingenuity. Unwanted *Agrobacterium* genes (which caused gall formation) were removed, leaving only the sequences necessary to specify the region to be transferred (the so-called left- and right-border sequences). Genes that would allow transformed tissue to be selected by growth on antibiotic-containing media were introduced, and methods for introducing the foreign DNA of choice were developed. The resulting transfer system allows foreign DNA to be engineered into the transferred region of a 'disarmed' *Agrobacterium* plasmid (i.e. that contains no gall-forming genes) reasonably easily; incubation of plant tissue pieces, or tissue cultures, or protoplasts with *Agrobacteria* containing the foreign DNA results in the transfer taking place. Transformed tissue can then be selected by the appropriate antibiotic, and plants regenerated from these transformed tissues.

Agrobacterium-mediated gene transfer to plants is technically easier than direct transfer, but suffers from several limitations. First, *Agrobacterium* does not infect all host plants with equal efficiency, and will not infect monocots at all. Genetic engineering of cereal crops is thus not possible by this route at present, although it is possible to force the bacterium to infect some monocot tissues by chemical treatment. Secondly, plants still have to be regenerated from callus tissue formed on tissue pieces, or tissue cultures, or protoplasts, and this is not possible for many species. Nevertheless, well-established procedures for obtaining transgenic plants of solanaceous species, particularly tobacco, and brassicas, are now in use in laboratories throughout the world. The ease with which tobacco may be transformed has made it the preferred host for many experiments involving expression of foreign genes, and although there has been considerable discussion of the validity of the assumption that this (or any other) species is a good model for gene control in plants in general, conclusions reached on the basis of experiments in a single species do seem to have general validity. This suggests that gene control mechanisms are strongly conserved throughout the plant kingdom.

9.6.1 Expression of gene constructs in transgenic plants

A frequent aim of plant transformation experiments is to achieve expression of a foreign gene in the transformed plant. The experimenter may wish to determine the effect of the gene product on the phenotype of the plant, in which case it will often be desirable to express the gene in all tissues, at a high level. This will require the replacement of the existing gene promoter with a more suitable one. Alternatively, the tissue-, developmentally-, or environmentally-regulated expression of the gene may be studied, in which case it is advantageous to replace the coding sequence of the gene with a sequence encoding a product that can be readily and specifically detected (a 'reporter' gene). Only relatively rarely is an intact foreign gene used in plant transformation experiments. With genes from other kingdoms, expression will often not be achieved at all, due to differing requirements for promoter sequences. With other plant genes, expression of the foreign gene can be masked by endogenous genes, unless the host is carefully chosen, or sensitive and specific assay techniques are used. The production of gene constructs, containing sequences from more than one gene, has become routine in molecular biology, and the reader is referred to appropriate texts for the methodologies employed.

Constitutive expression of a gene product in all tissues of the plant is readily achieved by replacing the normal gene promoter with a highly expressed constitutive promoter such as that from the 35S gene in cauliflower mosaic virus (referred to as the CaMV 35S promoter). This promoter is up-regulated by cell division, and thus is expressed at some time in the life of all cells. Replacement of the normal gene promoter is usually achieved by cutting the DNA in the transcribed region, 5' to the start of the coding sequence (Fig. 9.10). Abolition of the normal expression pattern of genes, whose promoters have been replaced by the CaMV 35S promoter in transgenic plants, demonstrates that the promoter region is responsible for the original tissue- or stimulus-determined pattern of expression. Where expression of a foreign gene is required to show tissue- or stimulus-specificity, promoters from highly expressed plant genes with the desired expression pattern can be used. For example, the promoter from the ribulose bisphosphate carboxylase *rbcS* gene will give expression in green tissues, seed storage protein gene promoters will give expression in seed storage tissue, and suitable protease inhibitor gene promoters can be used to give expression that is inducible on wounding. These possibilities allow the expression pattern of a foreign gene to be manipulated to the experimenter's requirements, provided suitable promoters can be found or engineered.

The investigation of plant gene promoters has been greatly advanced by the use of reporter gene constructs in transgenic plants (Fig. 9.10). The coding sequence 'driven' by a promoter sequence under investigation can be chosen for ease of assay, to allow the activity of the promoter to be determined in both a qualitative (i.e. in which cells, in response to which stimuli, the promoter is active) and quantitative (i.e. the level of expression of the gene product) sense. The coding sequence from the *E. coli* bacterial β-glucuronidase (GUS) gene is the most common reporter gene used

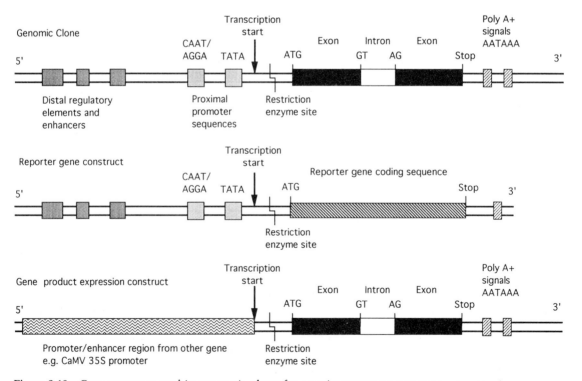

Figure 9.10 Gene constructs used in transgenic plants for assaying promoter sequences, or expressing gene products. The coding sequence from a cDNA clone may be used in the expression construct as an alternative to the genomic sequences.

in plants. It is readily assayed both qualitatively (by histochemical staining using the substrate 5-bromo-4-chloro-3-indolyl glucuronide; X-gluc) and quantitatively (using the fluorescent substrate methylumbelliferyl-β-D-glucuronide). β-Glucuronidase is not present as an endogenous activity in plant tissues. A number of studies have shown that the normal activities of gene promoters are retained in promoter–reporter fusions where the fusion is made between the transcriptional and translational start points of each gene, although a few exceptions are known where sequences within the transcribed sequence of the gene appear to have non-specific enhancer-like activity (such as the maize sucrose synthase gene, where the first intron, located in the 5′ transcribed non-translated region of the gene, contains an enhancer sequence). Assays of this type can be used to study the functions of different regions of the promoter, by mutating the promoter sequences *in vitro* and determining the effects of the mutation by assay of a promoter–reporter fusion in transgenic plants. This is a lengthy procedure, due to the time taken to produce the transgenic plants, and some investigators have preferred to carry out promoter assays by transient expression of directly introduced DNA in protoplasts. However, there is a drawback that the responses observed in protoplasts may not represent the responses *in planta*. The work necessary to analyze fully even a single promoter in detail by functional assays has deterred all but the most determined researchers, but in one or two genes (notably the *rbcS*, histone genes, and leghemoglobin) a reasonably full description of all the *cis*-acting sequences has been built up.

9.6.2 Suppression of endogenous gene expression

Besides wanting to express foreign genes in transgenic plants, a further tool of use both to researchers and biotechnologists would be to be able to suppress selectively the expression of specific genes in a transgenic plant. For the researcher, this type of 'reverse genetics' enables the function of genes to be precisely determined. In biotechnology, this technique could be used to prevent the accumulation of unwanted gene products, such as toxins, unpleasant flavors, or proteins that interfered with processing of material for the food industry. Although removal of some gene products which are encoded by gene families with a number of active members, such as storage proteins, is a difficult goal, inactivation

of single genes is achievable with present methods.

A simple method of inactivating a gene is to damage it by insertion of foreign DNA, either into the transcribed sequence, or the promoter region; this is analogous to inactivation by transposable elements discussed earlier. Insertion of foreign DNA by *Agrobacterium*, like insertion of elements, is essentially a random process, with selectivity being conferred by the chromatin conformation. Even though insertion of foreign DNA by *Agrobacterium* shows a very strong bias towards insertion at or near active genes, inactivation of specific genes is not feasible by these methods, unless a very efficient method for generating transgenic plants, and for selection of the desired inactivation, is available. A better scheme would be to introduce a functional transposing element via *Agrobacterium*, and screen progeny plants of the transformed individuals for the desired inactivation. This strategy is feasible, as shown by the isolation of the *Bz* gene in maize, which determines kernel pigmentation, by the method of transposon tagging, although in this case an endogenous transposable element was used. Several groups are currently using the transgenic transposon tagging strategy to isolate genes in *Arabidopsis*. Although this technique is useful for isolating genes that control a defined phenotype, it cannot be used to inactivate specific genes. Some research has been done to investigate the possibilities of a directed recombination strategy to inactivate specific genes, using an introduced DNA sequence with homology to a target in the host genome, but results have not been very encouraging.

The most promising method of gene inactivation in present use involves post-transcriptional control of gene expression, by antisense RNA. In this technique, an RNA complementary to the normal transcript of the target gene is generated by an introduced gene construct which contains all, or part of the transcribed sequence of the gene, driven by a promoter of choice (usually CaMV 35S), with the difference to a normal expression construct that the gene sequence is reversed relative to the promoter, so that it is transcribed on the opposite strand to that normally used (Fig. 9.11). This is equivalent to saying that the gene is transcribed in a 3′ to 5′ direction, rather than 5′ to 3′; the reader should refer back to the discussion of gene transcription. The mechanism of the subsequent suppression of gene expression is not yet clear, but it is likely that the antisense RNA molecule interferes with the post-transcriptional processing of the normal RNA, by hybridizing to it, and that the RNA–RNA hybrid is degraded in the nucleus so that no mature mRNA, and thus no gene product, is produced.

This technique has been used in petunia to alter floral pigmentation, by suppression of chalcone synthase, the first enzyme on the flavonoid biosynthetic pathway leading to anthocyanin production, and in tomato to alter the fruit quality phenotype by suppressing polygalacturonase (cell wall softening enzyme) production. In the tomato example, the gene chosen was expressed during fruit ripening in normal fruit, and the effect of the antisense introduction gene was to reduce dramatically the level of normal mRNA. Progeny of the initially transformed plants which were homozygous for the introduced antisense gene had levels of polygalacturonase in ripe fruit approximately 1% of normal wild-type fruit. The technique has also been used to elucidate the function of unidentified genes. In tomato, a cDNA expressed in ripening fruit was used to make an antisense gene construct, and the effect of this antisense gene on the phenotype of transgenic fruit was determined. These plants (if kept isolated) failed to ripen the fruit at all, and this was shown to be due to a failure to produce ethylene, the environmental signal (or 'plant hormone') used to activate genes involved in the ripening process. The unknown cDNA thus encoded an enzyme of ethylene biosynthesis, probably the enzyme which converts 1-aminocyclopropane-1-carboxylic acid into ethylene, which was confirmed by subsequent expression of the cDNA in bacteria. This technique is clearly a very powerful tool for elucidating gene function, but is only usable in those plant species which are amenable to transformation technology. A curious observation has been that transforming plants with a 'sense' construct containing a partial transcript sequence can also lead to inactivation of an endogenous gene; this phenomenon is yet to be fully explained, but may be an artefact involving spurious reading of the inserted DNA in an antisense direction, either from other promoters in the chunk of foreign DNA that is inserted, or from endogenous promoters in the host genome.

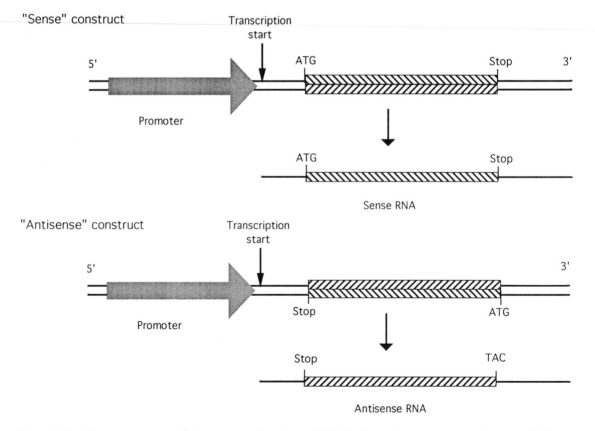

Figure 9.11 Gene constructs producing sense and antisense RNA. In the antisense construct, the transcribed sequence is reversed with respect to the promoter, so the other strand of DNA is transcribed. The resulting RNA is thus complementary to the normal 'sense' RNA.

9.6.3 Applications in biotechnology

Some of the practical applications of transgenic plant technology have been outlined above. Broadly, proposed applications for transgenic plants can be divided into different types, depending on the level of complexity of the engineering that is proposed. First, applications that involve the addition of, or suppression of a single gene have already been shown to be feasible. The simplest cases are those where a single foreign gene is added to the target plant, and over-expressed, usually from a constitutive promoter. Transgenic plants that show resistance to virus attack, by expression of the viral coat protein, to herbicides, by expression of engineered bacterial detoxification genes, and to insects, by expression of genes encoding insecticidal bacterial proteins, or insecticidal plant proteins, have all been shown to work in the laboratory, and the herbicide- and insect-resistant transgenic plants have been successfully trialled in the field. Suppression of single genes, as in the tomato example above, has also been achieved, but is more difficult to put into practice as the target genes for suppression, in order to produce a desired phenotype, are often not characterized, or even not known. A more complex type of application involves manipulating a biosynthetic pathway, to achieve goals such as altering the quality of oil produced by crops such as oilseed rape, or altering protein quality in seeds, or investigating and manipulating carbohydrate or nitrogen metabolism. Where a control step in the pathway can be clearly identified, this has been successful in altering phenotype, as in the case of removal of floral pigmentation by suppressing chalcone synthase expression as described above. However, in most cases control points in the pathways remain to be identified, and attempts to manipulate metabolite partitioning can produce unexpected results; for instance, overexpression of invertase in transgenic tobacco plants in order to inhibit export of sucrose from source leaves, results in plants of very much reduced size, with reduced root formation and the appearance of chlorotic lesions on the leaves. Explaining results such as these can give new insights into metabolism. There is a need for much detailed biochemistry to be studied in concert with molecular genetics before pathways can be manipulated in a predictable manner; for many of the pathways of secondary metabolism, not even all the steps have been characterized. Alteration of the levels of expression of several genes, in a controlled manner, is beyond present technology. Finally, applications can be envisaged in which developmental processes in plants are themselves altered, to produce new organs (such as nitrogen-fixing nodules on cereals, as has been proposed), or altered physiology. These applications have yet to be addressed, as in almost all cases the genes that control developmental processes as unknown.

9.7 CONCLUSION

The regulation of gene expression in plants, as in other higher eukaryotes, is a subject of daunting complexity. Nevertheless, even a partial understanding of how plant genes work, in conjunction with the methods of molecular biology and plant tissue cultures, opens the door to a dazzling array of techniques for manipulating various aspects of the phenotypes of plants. Man uses plants as food, as a source of building materials, for the extraction of scents, dyes, drugs and many other biochemicals, and for decoration of his environment. At present we are at a very early stage of realizing many of the goals that have been ambitiously (and, in some cases, unrealistically) proposed for manipulating gene expression in plants, such as nitrogen-fixing cereals, because the systems are not sufficiently understood at the molecular and genetic levels. Nevertheless, herbicide- and insect-resistant plants produced by foreign gene expression in transgenic plants are already a reality, and on the verge of commercial use. For plant biochemists, study of plant genes, their functions and regulation, is resulting in a quantum leap in the understanding of familiar biochemical pathways, and the elucidation of less familiar ones. It is easy both to overestimate the progress made since molecular genetics was applied to plant systems, and to underestimate how much is yet to be done; but there is no doubt that molecular biology is here to stay, and will form one of the best tools available for investigating the molecular bases of plant phenotypes.

FURTHER READING

An exhaustive bibliography for this chapter would occupy more space than the chapter itself. The author has therefore given some suggestions for further reading, and selected references (to mid-1993) below.

Background

Lewin, B. (1987) *Genes*, 3rd edn. John Wiley, New York.

Mathews, C.K. & van Holde, K.E. (1990) *Biochemistry*. Benjamin/Cummings, Redwood City, California. Ch. 28 is especially relevant.

Old, R.W. & Primrose, S.B. (1989) *Principles of Genetic Manipulation* 4th edn. Blackwell Scientific Publications, Oxford.

Watson, J.D., Hopkins, N.H., Roberts, J.W., Steitz, J.A. & Weiner, A.M. (1987) *Molecular Biology of the Gene*, 4th edn., Vols. I and II. Benjamin/Cummings, Redwood City, California.

Collections of review chapters

Grierson, D. (ed.) (1991) *Developmental Regulation of Plant Gene Expression*. Blackie, Glasgow and London.

Leaver, C.J., Boulter, D. & Flavell, R.B. (eds) (1986) *Differential Gene Expression and Plant Development*. Royal Society, London.

Verma, D.P.S. & Goldberg, R.B. (eds) (1989) *Temporal and Spatial Regulation of Plant Genes*. Springer-Verlag, Vienna.

Verma, D.P.S. (ed.) (1993) *Control of Plant Gene Expression*. CRC Press, Boca Raton, FL.

Review articles and primary literature (see also collections above)

Nuclear genes

RNA polymerase and transcription initiation

Nikolov, D.B., Hu, S-H., Lin, J., Gasch, A., Hoffmann, A., Horikoshi, M., Chua, N-H., Roeder, R.G. & Burley, S.K. (1992) Crystal structure of TFIID TATA box binding protein. *Nature* 360, 40–46.

Grasser, K.D., Maier, U., Haass, M.M. & Feix, G. (1990) Maize high mobility group proteins bind to CCAAT and TATA boxes of a zein gene promoter. *J. Biol. Chem.* 265, 4185–4189.

Cis-Acting sequences and enhancers

Barros, M.D.C., Czarnecka, E. & Gurley, W.B. (1992) Mutational analysis of a plant heat shock element. *Plant Mol. Biol.* 19, 665–674.

Bustos, M.M., Guiltinan, M.J., Jordano, J., Begum, D., Kalkan, F.A. & Hall, T.C. (1989) Regulation of β-glucuronidase expression in transgenic tobacco plants by an A/T-rich, *cis*-acting sequence found upstream of a French bean β-phaseolin gene. *Plant Cell* 1, 839–846.

Donald, R.G.K., Schindler, A.B. & Cashmore, A.R. (1990) The plant G box promoter sequence activates transcription in *Saccharomyces cerevisiae* and is bound *in vitro* by a yeast activity similar to GBF, the plant G box binding factor. *EMBO J.* 9, 1727–1734.

Litts, J.C., Colwell, G.W., Chakerian, R.L. & Quatrano, R.S. (1991) Sequence analysis of a functional member of the Em gene family of wheat. *DNA Sequence* 1, 263–272.

Marcotte, W.R., Russell, S.H. & Quatrano, R.S. (1989) Abscisic acid-responsive sequences from the Em gene of wheat. *Plant Cell* 1, 969–976.

Pederson, T.J., Arwood, L.J., Spiker, S., Guiltinan, M.J. & Thompson, W.F. (1991) High mobility group chromosomal proteins bind to AT-rich tracts flanking plant genes. *Plant Mol. Biol.* 16, 94–103.

Staiger, D., Kaulen, H. & Schell, J. (1989) A CACGTG motif of the *Antirrhinum majus* chalcone synthase promoter is recognised by an evolutionarily conserved nuclear protein. *Proc. Natl. Acad. Sci. U.S.A.* 86, 6930–6934.

Trans-Acting factors

Alber, T. (1992) Structure of the leucine zipper. *Curr. Opin. Genet. Dev.* 2, 205–210.

Branden, C. & Tooze, J. (1991) *Introduction to Protein Structure*. Garland Publishing, New York, Ch. 8.

Brunelle, A.N. & Chua, N-H. (1993) Transcription regulatory proteins in higher plants. *Curr. Opin. Genet. Dev.* 3, 254–258.

Gill, G. & Tjian, R. (1992) Eukaryotic coactivators associated with the TATA box binding protein. *Curr. Opin. Genet. Dev.* 2, 236–242.

Kuwabara, M.D. & Sigman, D.S. (1987) Footprinting DNA-protein complexes *in situ* following gel retardation assays using 1,10-phenanthroline-copper ion; *E. coli* RNA-polymerase-lac promoter complexes. *Biochemistry* 26, 7234–7239.

Ludwig, S.R. & Wessler, S.R. (1990) Maize *R* gene family: tissue-specific helix–loop–helix proteins. *Cell* 62, 849–854.

Schmidt, R.J., Burr, F.A., Aukermann, M.J. & Burr, B. (1990) Maize regulatory gene opaque-2 encodes a protein with a 'leucine zipper' motif that binds to DNA. *Proc. Natl. Acad. Sci. U.S.A.* 87, 46–48.

Control systems

Yamazaki, K., Chua, N. & Imaseki, H. (1990) Accurate transcription of plant genes *in vitro* using a wheat germ-chromatin extract. *Plant Mol. Biol. Rep.* 8, 114.

Physical and chemical modifications of *DNA* which regulate transcription

Greider, C.W. (1992) Telomere chromatin and gene expression. *Curr. Biol.* 2, 62–64.

Inamdar, N.M., Ehrlich, K.C. & Ehrlich, M. (1991) CpG methylation inhibits binding of several sequence-specific DNA binding proteins from pea, wheat, soybean and cauliflower. *Plant Mol. Biol.* 17, 111–123.

Laemmli, U.K., Käs, E., Poljak, L. & Adachi, Y. (1992) Scaffold-associated regions: *cis*-acting determinants of chromatin structural loops and functional domains. *Curr. Opin. Genet. Dev.* 2, 275–285.

Svaren, J. & Hörz, W. (1993) Histones, nucleosomes and transcription. *Curr. Opin. Genet. Dev.* 3, 219–225.

Tate, P.H. & Bird, A.P. (1993) Effects of DNA methylation on DNA-binding proteins and gene expression. *Curr. Opin. Genet. Dev.* 3, 226–231.

Transposable elements and retrotransposons

Coen, E.S., Robbins, T.P., Almeida, J., Hudson, A. & Carpenter, R. (1989) In *Mobile DNA* (D.E. Berg & M.M. Howe, eds), pp. 413–436. American Society of Microbiology, Washington, DC.

Gierl, A., Saedler, H. & Peterson, P.A. (1989) Maize transposable elements. *Annu. Rev. Genet.* 23, 71–85.

Smyth, D.R. (1993) Plant retrotransposons. In *Control of Plant Gene Expression* (D.P.S. Verma, ed.), pp. 1–15. CRC Press, Boca Raton, FL.

Weil, C.F. & Wessler, C.R. (1990), The effects of plant transposable element insertion on transcription initiation and RNA processing. *Annu. Rev. Plant Physiol. Mol. Biol.* 41, 527–552.

Post-transcriptional regulation of gene expression

Elliot, R.C., Dickey, L.F., White, M.J. & Thompson, W.F. (1989) *cis*-Acting elements for light regulation of pea ferredoxin 1 gene expression are located within transcribed sequences. *Plant Cell* 1, 691–698.

Jofuku, K.D., Schipper, R.D. & Goldberg, R.B. (1989) A frame shift mutation prevents Kunitz trypsin inhibitor mRNA accumulation in soybean embryos. *Plant Cell* 1, 427–435.

Thompson, A.J., Evans, I.M., Boulter, D., Croy, R.R.D. & Gatehouse, J.A. (1989) Transcriptional and post-transcriptional regulation of storage protein gene expression in pea (*Pisum sativum* L.). *Planta* 179, 279–286.

Vellanoweth, R.L. & Okita, T.W. (1993) Regulation of expression of wheat and rice seed storage protein genes. In *Control of Plant Gene Expression* (D.P.S. Verma, ed.), pp. 377–392. CRC Press, Boca Raton, FL.

Voelker, T.A., Moreno, J. & Chrispeels, M.J. (1990) Expression analysis of a pseudogene in transgenic tobacco – a frameshift mutation prevents mRNA accumulation. *Plant Cell* 2, 255–262.

Organellar genes

Chloroplast genes

Briat, F.J., Lescure, A.M. & Mache, R. (1986) Transcription of chloroplast DNA: a review. *Biochimie* 66, 981–990.

Bryant, D. (1992) Puzzles of chloroplast ancestry. *Curr. Biol.* 2, 240–242.

Mullet, J.E. (1988) Chloroplast development and gene expression. *Annu. Rev. Plant Physiol.* 39, 475–502.

Koller, B., Fromm, H., Galun, E. & Edelman, M. (1987) Evidence for *in vivo trans* splicing of pre-mRNAs in tobacco chloroplasts. *Cell* 48, 111–119.

Suguira, M. (1989) The chloroplast genome. In *The Biochemistry of Plants*, vol. 15 (A. Marcus, ed.), pp. 133–150. Academic Press, San Diego, CA.

Westhoff, P. & Herrmann, R.G. (1988) Complex RNA maturation in chloroplasts. The *psbB* operon from spinach. *Eur. J. Biochem.* 171, 551–564.

Mitochondrial genes

Brown, G.G. (1993) Regulation of mitochondrial gene expression. In *Control of Plant Gene Expression* (D.P.S. Verma, ed.), pp. 175–198. CRC Press, Boca Raton, FL.

Conklin, P.L., Wilson, R.K. & Hanson, M.R. (1991) Multiple *trans*-splicing events are required to produce mature *nad1* transcript in a plant mitochondrion. *Genes Dev.* 5, 1407.

Covello, P.S. & Gray, M.W. (1989) RNA editing in plant mitochondria. *Nature* 341, 662–666.

Finnegan, P.M. & Brown, G.G. (1990) Transcriptional and post-transcriptional regulation of RNA levels in maize mitochondria. *Plant Cell* 2, 71–76.

Grienenberger, J-M. (1993) RNA editing in plant mitochondria. In *Control of Plant Gene Expression* (D.P.S. Verma, ed.), pp. 199–210. CRC Press, Boca Raton, FL.

Hanson, M.R. (1991) Plant mitochondrial mutations and male sterility. *Annu. Rev. Genet.* 25, 461.

Lonsdale, D.M. (1989) The plant mitochondrial genome. In *The Biochemistry of Plants* (P.K. Stumpf & E.E. Conn, eds), Vol. 15, pp. 229–254. Academic Press, New York.

Walbot, V. (1991) RNA editing fixes problems in plant mitochondrial transcripts. *Trends Genet.* 7, 37–39.

Interaction with nuclear genes

Cheung, A.Y., Bogorad, L., van Montagu, M. & Schell, J. (1988) Relocating a gene for herbicide tolerance: a chloroplast gene is converted into a nuclear gene. *Proc. Natl. Acad. Sci. U.S.A.* 85, 391–395.

Nugent, D. & Palmer, J.D. (1991) RNA mediated transfer of the gene *coxII* from the mitochondrion to the nucleus during flowering plant evolution. *Cell* 66, 473–481.

Schuster, W. & Brennicke, A. (1987) Plastid, nuclear and reverse transcriptase sequences in the mitochondrial genome of *Oenothera*; is genetic information transferred between organelles via RNA? *EMBO J.* 6, 2857–2861.

Sheen, J-Y. & Bogorad, L. (1986) Expression of the ribulose-1,5-bisphosphate carboxylase large subunit gene and three small subunit genes in two cell types of maize leaves. *EMBO J.* 5, 3417–3422.

Signaling mechanisms in gene regulation

Barbier-Brygoo, H., Ephritikhine, G., Klambt, D., Maurel, C., Palme, K., Schell, J. & Guern, J. (1991) Perception of the auxin signal at the plasma membrane of tobacco mesophyll protoplasts. *Plant J.* 1, 83–93.

Farmer, E.E. & Ryan, C.A. (1992) Octadecanoid precursors of jasmonic acid activate the synthesis of wound-inducible proteinase inhibitors. *Plant Cell* **4**, 129–134.

Gilman, A.G. (1987) G-proteins: transducers of receptor generated signals. *Annu. Rev. Biochem.* **56**, 615–649.

Hesse, T., Feldwisch, J., Balshüsemann, D., Bauw, G., Puype, M., Vandekerckhove, J., Löbler, M., Klämbt, D., Schell, J. & Palme, K. (1989) Molecular cloning and structural analysis of a gene from *Zea mays* (L.) coding for a putative receptor of the plant hormone auxin. *EMBO J.* **8**, 2453–2461.

Lissemore, J. & Quail, P.H. (1988) Rapid transcriptional regulation by phytochrome of the genes for phytochrome and chlorophyll a/b-binding protein in *Avena sativum*. *Mol. Cell. Biol.* **8**, 4840–4850.

Lumsden, P.J. (1993) Mechanisms of signal transduction. In *The Molecular Biology of Flowering* (B.R. Jordan, ed.), pp. 21–45. CAB International Wallingford, Oxon.

Ma, H., Yanofsky, M.F. & Meyerowitz, E.M. (1990) Molecular cloning and characterisation of GPA1, a G protein α-subunit gene from *Arabidopsis thaliana*. *Proc. Natl. Acad. Sci. U.S.A.* **87**, 3821–3825.

Nagy, F., Kay, S.A. & Chua, N-H. (1988) Gene regulation by phytochrome. *Trends Genet.* **4**, 37–42.

Pearce, G., Strydom, D., Johnson, S. & Ryan, C.A. (1991) A polypeptide from tomato leaves induces wound-inducible proteinase inhibitor proteins. *Science* **253**, 895–898.

Ryan, C.A. & Farmer, E.E. (1991) Oligosaccharide signals in plants: a current assessment. *Annu. Rev. Plant Physiol. Mol. Biol.* **42**, 651–674.

Stein, J.C., Howlett, B., Boyes, D.C., Nasrallah, M.E. & Nasrallah, J.B. (1991) Molecular cloning of a putative receptor protein kinase gene encoded at the self-incompatibility locus of *Brassica oleracea*. *Proc. Natl. Acad. Sci. U.S.A.* **88**, 8816–8820.

Gene regulation in plant development

Tissue-specific expression

Bevan, M., Colot, V., Hammond-Kossack, M., Holdsworth, M., Torres de Zabala, M., Smith, C., Grierson, C. & Beggs, K. (1993) Transcriptional control of plant storage protein genes. In *The Production and Uses of Genetically Transformed Plants* (M.W. Bevan, B.D. Harrison & C.J. Leaver, eds), pp. 21–27. Chapman & Hall/Royal Society, London.

Chamberland, S., Daigle, N. & Bernier, F. (1992) The legumin boxes and the 3' part of a soybean β-conglycinin promoter are involved in seed gene expression in transgenic tobacco plants. *Plant Mol. Biol.* **19**, 937–949.

Mason, H.S., DeWald, D.B., Creelman, R.A. & Mullet, J.E. (1992) Coregulation of soybean vegetative storage protein gene expression by methyl jasmonate and soluble sugars. *Plant Physiol.* **98**, 859–867.

Okamuro, J.K., Jofuku, K.D. & Goldberg, R.B. (1986) Soybean seed lectin gene and flanking non-seed protein genes are developmentally regulated in transformed tobacco plants *Proc. Natl. Acad. Sci. U.S.A.* **83**, 8240–8243.

Shirsat, A.H., Meakin, P.J. & Gatehouse, J.A. (1990) Sequences 5' to the conserved 28bp leg box element regulate the expression of pea seed storage protein gene *legA*. *Plant Mol. Biol.* **15**, 685–689.

Genes controlling development

Coen, E.S. (1991) The role of homeotic genes in floral development and evolution. *Annu. Rev. Plant Physiol. Mol. Biol.* **42**, 241–279.

Coen, E.S. & Meyerowitz, E.M. (1991) The war of the whorls: genetic interactions controlling flower development. *Nature* **353**, 31–37.

Mayer, U., Torres Ruiz, R.A., Berleth, T., Miséra, S. & Jürgens, G. (1991) Mutations affecting body organisation in the *Arabidopsis* embryo. *Nature* **353**, 402–407.

Oppenheimer, D.G., Herman, P.L., Sivakumaran, S., Esch, J. & Marks, M.D. (1991) A *myb* gene required for leaf trichome differentiation is expressed in stipules. *Cell* **67**, 483–493.

Vollbrecht, E., Veit, B., Sinha, N. & Hake, S. (1991) The developmental gene *knotted-1* is a member of a maize homeobox gene family. *Nature* **350**, 241–243.

Weigel, D. (1993) Patterning the *Arabidopsis* embryo. *Curr. Biol.* **3**, 443–445.

Transgenic plants

Bevan, M.W., Harrison, B.D. & Leaver, C.J. (1993) *The Production and Uses of Genetically Transformed Plants*. Chapman & Hall/Royal Society, London.

Bowen, B.A. (1993) Markers for plant gene transfer. In *Transgenic Plants* Vol. 1 *Engineering and Utilisation* (S. Kung & R. Wu, eds), pp. 89–123. Academic Press, San Diego, CA.

Coomber, S.A. & Feldmann, K.A. (1993) Gene tagging in transgenic plants. In *Transgenic Plants* Vol. 1 *Engineering and Utilisation* (S. Kung & R. Wu, eds), pp. 225–240. Academic Press, San Diego, CA.

Hamilton, A., Lycett, G.W. & Grierson, D. (1990) Antisense gene that inhibits synthesis of the hormone ethylene in transgenic plants. *Nature* **346**, 284–287.

Jefferson, R.E., Kavanagh, T. & Bevan, M.W. (1987) GUS fusions: β-glucuronidase as a sensitive and versatile gene fusion marker in higher plants. *EMBO J.* **6**, 3901–3906.

Miki, B.L.A. & Lyer, V.N. (1990) Fundamentals of gene transfer in plants. In *Plant Physiology, Biochemistry and Molecular Biology* (D.T. Dennis & D.H. Turpin, eds), pp. 473–489. Longman Scientific and Technical, Singapore.

Schuch, W. (1991) Using antisense RNA to study gene function. In *Molecular Biology of Plant Development* (G.I. Jenkins & W. Schuch, eds), pp. 117–128. Society for Experimental Biology.

Smith, C.J.S., Watson, C.F., Ray, J., Bird, C.R., Morris, P.C., Schuch, W. & Grierson, D. (1988) Antisense RNA inhibition of polygalacturonase gene expression in transgenic tomatoes. *Nature* **334**, 724–726.

Walden, R. (1993) Cell culture, transformation and gene technology. In *Plant Biochemistry and Molecular Biology* (P.J. Lea & R.C. Leegood, eds), pp. 275–295. John Wiley, Chichester.

10 Phenolic Metabolism

D. Strack

10.1 INTRODUCTION

10.1.1 Structural variety of phenolics

A phenolic is a chemical compound characterized by at least one aromatic ring (C_6) bearing one or more hydroxyl groups. Many phenolics occur as derivatives formed by condensation or addition reactions. Most of the thousands of phenolics known to date are of plant origin.

Three different biogenetic routes lead to plant phenolics.

1. The **shikimate/arogenate pathway** leads, through phenylalanine, to the majority of plant phenolics, the phenylpropane (C_6-C_3) derivatives (phenylpropanoids). Some phenolics are formed from intermediates of the shikimate/arogenate pathway, e.g. the hydroxybenzoate gallate from dehydroshikimate and some quinones from chorismate through succinylbenzoate (**succinylbenzoate pathway**).
2. The **acetate/malonate pathway** (**polyketide pathway**) leads to some plant quinones but also to various side-chain-elongated phenylpropanoids, e.g. the large group of the flavonoids (C_6-C_3-C_6).

3. The **acetate/mevalonate pathway** leads by dehydrogenation reactions to some aromatic terpenoids, mostly monoterpenes (see Chapter 11).

The shikimate/arogenate and the polyketide pathways are the most important ones in biosyntheses of the plant phenolics. Table 10.1 gives a survey of the structural classes of phenolics, from the very simple to the most complex.

The present chapter serves as an introduction to the most commonly found phenolics. Of these, the predominant structures in higher plants are the flavonol glycosides, hydroxycinnamate conjugates and condensed tannins in the leaves, and the anthocyanins in petals and fruits.

10.1.2 Functions of phenolics

There is increasing evidence that the functional aspects of phenolics must be considered in making any distinction between primary and secondary compounds. For instance, it is known that many secondary compounds are valuable storage vehicles and indispensable elements in anatomical and morphological structures. But there is increasing knowledge that they are also of prime ecological

PLANT BIOCHEMISTRY
ISBN 0-12-214674-3

Table 10.1. The major classes of phenolics in plants. For structures of example compounds see p. 389. The polymeric structures are shown in the respective chapters

C No.	C Skeleton	Compound class	Compound example	Example of compound occurrence
6	C_6	Simple phenols	Hydroquinone	Arbutin in Rosaceae and Ericaceae
			Catechol	Urushiol in *Toxicodendron radicans*
7	C_6-C_1	Hydroxybenzoates	4-Hydroxybenzoate	Mainly wall-bound in most plant families
8	C_6-C_2	Acetophenones	4-Hydroxyacetophenone	Picein in *Picea abies*
		Phenylacetates	4-Hydroxyphenylacetate	In *Taraxacum officinale*
9	C_6-C_3	Hydroxycinnamates	Caffeate	Chlorogenate in Solanaceae
		Phenylpropenes	Eugenol	In essential oils of various families
		Coumarins	Esculetin	Cichoriin in *Cichorium intybus*
		Chromones	2-Methyl-5-hydroxy-7-methoxychromone	Eugenin in *Eugenia aromatica*
10	C_6-C_4	Naphthoquinones	Juglone	As 5-O-glucoside in Juglandaceae
13	C_6-C_1-C_6	Xanthones	1,3,6,7-Hydroxyxanthone	Mangiferin in *Mangifera indica*
14	C_6-C_2-C_6	Stilbenes	Resveratrol	In *Eucalyptus* trees
		Anthraquinones	Emodin	Emodin 6-O-glucoside in *Rheum palmatum*
15	C_6-C_3-C_6	Flavonoids	Quercetin	Rutin in various families
18	$(C_6$-$C_3)_2$	Lignans	Pinoresinol	In *Picea* and *Pinus* trees
30	$(C_6$-C_3-$C_6)_2$	Biflavonoids	Amentoflavone	In most gymnosperms
n	$(C_6)_n$	Catechol melanins	Naphthalene polymer	In the ascomycete *Daldinia concentrica*
	$(C_6$-$C_1)_n$: Glc	Hydrolyzable tannins	Gallotannins	Chinese Tannin in galls of *Rhus semialata*
	$(C_6$-$C_3)_n$	Lignins	Guaiacyl lignins	Lignins in gymnosperms
			Guaiacyl-syringyl lignins	Lignins in angiosperms
	$(C_6$-C_3-$C_6)_n$	Condensed tannins	Catechin polymers	Tannin in the bark of *Quercus robur*

$(C_6$-$C_1)_n$: Glc = polyester with glucose (Glc) as the common central polyol moiety; arbutin = hydroquinone-O-β-glucoside; urushiol = catechol with a C_{15} hydrocarbon side chain [-(CH$_2$)$_7$CH=CH(CH$_2$)$_5$CH$_3$]; picein = 4-glucosyloxyacetophenone; chlorogenate = 5-O-caffeoylquinate; cichoriin = esculetin 7-O-β-glucoside; eugenin = 2-methyl-5-hydroxy-7-methoxychromone; mangiferin = 1,3,6,7-tetrahydroxyxanthone 2-C-β-glucoside; rutin = quercetin 3-O-rutinoside.

importance for the improvement in plants' survival. Without the 'discovery' of secondary compounds plants would never have evolved to such a wide range of different species and they would never be able to maintain coexistence in their habitats. This paragraph will briefly summarize some of the important functions of phenolics for plants. In addition, some examples of their importance for human welfare will be given.

Phenolics are of great importance as cellular support materials. They form an integral part of cell wall structures, mainly in polymeric materials such as lignins and suberins for mechanical support and barriers against microbial invasion. Lignins are, after cellulose, the second most abundant organic structures on earth. With their formation, plants became adapted to terrestrial life by building rigid organs such as woody stems and conducting cell elements for water transport.

Phenolics are of great ecological importance along with various toxic nitrogen-containing compounds as well as attractant or repellent terpenoids. The most significant function of the phenolic flavonoids, especially the anthocyanins, together with flavones and flavonols as co-pigments, is their contribution to flower and fruit colors. This is important for attracting animals to the plant for pollination and seed dispersal. Phenolics may accumulate as a result of stress (see Chapter 14). There is strong evidence that DNA-damaging ultraviolet (UV) light induces the accumulation of UV light-absorbing flavonoids and other phenolics, predominantly in dermal tissues of the plant body. Flavonoids, especially anthocyanins, may appear transiently during plant ontogeny. This can be observed in seedlings and young leaves, and suggests, yet speculative, physiological functions in light perception.

Phenolics may protect plants against predators (see Chapters 13 and 14). Herbivores react sensitively to the phenolic content in plants. Simple phenolic acids, as well as complex tannins and phenolic resins at the plant surface, are effective deterrents, e.g. in plant–bird interactions where they interfere with digestion through interaction with the microbial flora of the cecum. The phenolic-dependent feeding behavior of the Canada goose (*Branta canadensis*) is also well known. Phenolics may accumulate as inducible low-molecular-weight compounds, called 'phytoalexins', as a result of

Hydroquinone   Catechol   4-Hydroxybenzoate   4-Hydroxyacetophenone   4-Hydroxyphenylacetate   Caffeate

Eugenol   Esculetin   Eugenin   Juglone

1,3,6,7-Tetrahydroxyxanthone   Resveratrol   Emodin

Quercetin   Pinoresinol   Amentoflavone

microbial attack. The plant recognizes this attack by detection of parasite-derived molecules, called elicitors. Thus phytoalexins are postinfectional, i.e. although they might already be present at low concentrations in a plant, they rapidly accumulate upon attack as induced compounds. In contrast, the preinfectional toxins are constitutive compounds. They are present in healthy tissues in concentrations high enough to ward off attack, either already as free toxins or in conjugated forms from which they are released after attack. Among the phenolic phytoalexins and toxins, hydroxycoumarins and hydroxycinnamate conjugates are of major importance. They contribute to disease resistance mechanisms in plants.

Phenolics may influence competition among plants, a phenomenon called 'allelopathy'. Besides the well-known effects of volatile terpenoids acting as allelopathic compounds, there are toxic water-soluble phenolics, such as simple phenols (e.g. hydroquinone), hydroxybenzoates and hydroxycinnamates. A known toxin from *Juglans* species (Juglandaceae) is juglone (5-hydroxynaphthoquinone), which is highly toxic for a wide range of plants. It occurs in the plant as a non-toxic

glucoside and is made active by deglucosylation and oxidation after leaching from the leaves into the soil.

There are also phenolics in root exudates which may act as toxins. It is known that the rubber plant guayule (*Parthenium argentatum*; Asteraceae) exudes cinnamate, which is toxic to the plant itself (autotoxicity). Cinnamate might reduce competition between members of the same and other species.

The most recent findings concerning the function of phenolics, especially the flavonoids, is that they can act as signal molecules (host-recognition substances) in the interaction between nitrogen-fixing bacteria in certain members of the Fabaceae (leguminous plants). These plants exude flavonoids which act selectively in *Rhizobia* as inducers of nodulation (*nod*) gene transcription by activating regulatory proteins. *Nod* gene products, in turn, cause the host to form root nodules, which enable the plant to use atmospheric nitrogen through the action of the bacterial nitrogen-activating enzyme nitrogenase (see Chapter 7).

Along with many other secondary compounds of plant origin, phenolics are of great importance to humans. Biological activities of secondary compounds were found empirically by our ancestors to be not only unsavoury, poisonous or hallucinogenic, but also of possible pharmacological value. They learned to use plants in folk medicine, and today we have a rich source of phenolic-bearing medicinal herbs that are used as drugs in medicine. Polyphenols are important in foodstuffs and nutrition, in wines and herbal teas, because of their astringent taste. Plants rich in polyphenols were used for centuries in leather-making because of the ability of tannins to complex with proteins. Phenolic pigments of flowers and fruits are also of aesthetic value. Thus plant breeders have great interest in the production of a wide range of differently colored ornamental plants. These pigments are probably the most widespread food colors occurring as anthocyanin-based red colors in diverse edible plant material and in fruit juices, wines and jams. Anthocyanins have considerable potential in the food industry as safe and effective food additives.

10.2 SHIKIMATE/AROGENATE PATHWAY

The shikimate/arogenate pathway (Japanese *shikimino-ki*, for *Illicium anisatum*, Illiciaceae, the plant from which shikimate was first described by Eykmann in 1885) leads to the three aromatic amino acids L-phenylalanine, L-tyrosine and L-tryptophan. They are important precursors for auxin-type plant hormones (see Chapter 12) and various secondary compounds including phenylpropanoids. Figure 10.1 gives an outline of the shikimate/arogenate pathway to the aromatic amino acids. The sequence of reactions, leading

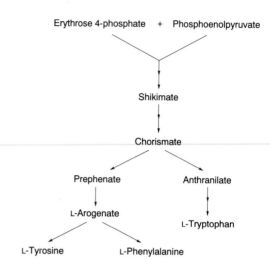

Figure 10.1   Outline of the shikimate/arogenate pathway leading to the aromatic amino acids in higher plants. Note the two branch points occur at chorismate and arogenate.

to phenylalanine and tyrosine, needs in all 11 enzymes (see E1-1 to E1-11 in Enzyme Box 1).

The pathway begins with two condensation reactions leading to the basic cyclohexane skeleton. The first reaction (intermolecular aldol condensation), catalyzed by the enzyme E1-1, is the condensation of erythrose 4-phosphate (a pentose phosphate pathway intermediate) with phosphoenolpyruvate (PEP, a glycolytic intermediate). The product is the open-chain C_7 sugar 2-dehydro-3-deoxyarabinoheptulosonate-7-phosphate (DAHP). The second step in the shikimate/arogenate pathway is the conversion of DAHP into 3-dehydroquinate (E1-2). This is a complex sequence of reactions involving an oxidation, a β-elimination, a reduction

---

ENZYME BOX 1
SHIKIMATE/AROGENATE PATHWAY

E1-1:   2-Dehydro-3-deoxyphosphoheptonate aldolase (= DAHP synthase); i.e. 2-dehydro-3-deoxy-D-*arabino*-heptonate-7-phosphate D-erythrose-4-phosphate-lyase

E1-2:   3-Dehydroquinate synthase; i.e. 2-dehydro-3-deoxyarabino-heptulosonate-7-phosphate-lyase

E1-3:   3-Dehydroquinate dehydratase; i.e. 3-dehydroquinate hydro-lyase

E1-4:   Shikimate 3-dehydrogenase; i.e. shikimate : NADP⁺ 3-oxidoreductase

E1-5:   Shikimate kinase; i.e. ATP : shikimate 3-phosphotransferase

E1-6:   3-Phosphoshikimate 1-carboxyvinyltransferase (= EPSP synthase); i.e. phosphoenolpyruvate : 3-phosphoshikimate 5-O-(1-carboxyvinyl)-transferase

E1-7:   Chorismate synthase; i.e. 5-O-(1-carboxyvinyl)-3-phosphoshikimate phosphate-lyase

E1-8:   Chorismate mutase; i.e. chorismate pyruvate mutase

E1-9:   Prephenate aminotransferase; i.e. glutamate : prephenate aminotransferase

E1-10:   Arogenate dehydrogenase; i.e. arogenate : NAD⁺ oxidoreductase (decarboxylating)

E1-11:   Arogenate dehydratase; i.e. arogenate hydro-lyase (decarboxylating)

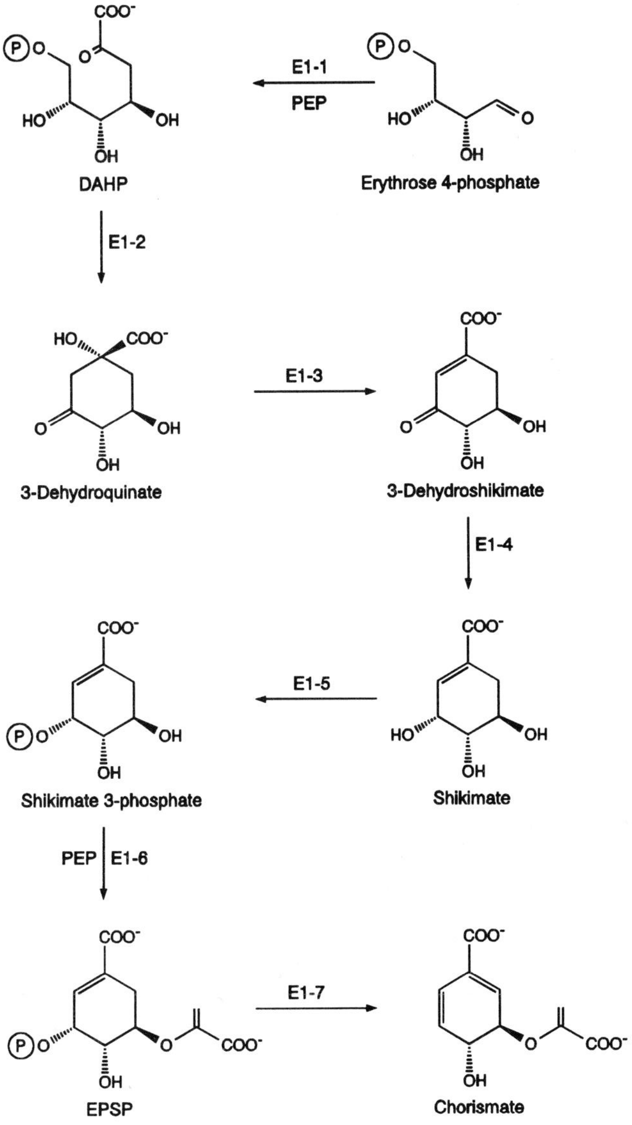

DAHP

Erythrose 4-phosphate

3-Dehydroquinate

3-Dehydroshikimate

Shikimate 3-phosphate

Shikimate

EPSP

Chorismate

and an intramolecular aldol condensation to give the cyclic structure. A final NADH-dependent reduction leads to 3-dehydroquinate which is then *cis* dehydrated by the enzyme E1-3 to 3-dehydroshikimate. (The stereospecific *cis* elimination of water is in contrast to many other *trans* dehydrations in plant metabolism.) 3-Dehydroshikimate is reduced by a NADPH-dependent dehydrogenase (E1-4) to shikimate. After C-3 phosphorylation (E1-5), shikimate 3-phosphate reacts with PEP (EPSP synthase; E1-6) to produce the enolether 5-enolpyruvylshikimate 3-phosphate (EPSP).

EPSP synthase is inhibited by the herbicide glyphosate [N-(phosphonomethyl)glycine]. This is a phosphoenolpyruvate analog, which is used extensively as a non-selective, broad spectrum,

postemergence herbicide in many countries in the world.

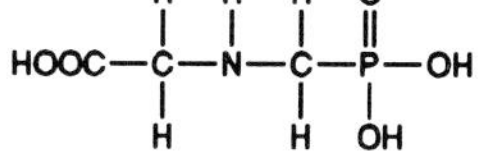

Glyphosate

Application of glyphosate causes depletion of the aromatic amino acid pools, which results in reduction of protein synthesis, cessation of growth and finally leads to death. Glyphosate has been successfully used in biochemical and physiological experiments from which considerable information on the regulation of phenolic formation has been obtained. Some plants are known to become resistant to this compound. This can be explained by an overproduction of EPSP synthase, which enables the shikimate/arogenate pathway to function normally.

The next step in the shikimate/arogenate pathway is the elimination of phosphate from EPSP, catalyzed by chorismate synthase (E1-7), which gives chorismate (Greek, fork). The two subsequent enzyme activities create the first major bifurcation (Figure 10.1): the anthranilate synthase reaction leads to the branch of tryptophan synthesis whereas the chorismate mutase (E1-8) reaction leads to the branch of phenylalanine and tyrosine syntheses.

Chorismate mutase catalyzed a pericyclic Claisen-rearrangement to the quinonoid prephenate. In this reaction the pyruvate side chain is transferred through a bicyclic chair-like transition state from C-5 to C-1, giving the basic phenylpropanoid skeleton. (Claisen-rearrangement is so-called after the German chemist L. Claisen (1851–1930) and is a sigmatropic reaction in which an phenylallyl ether is converted into a allylphenolic. The reaction may occur thermally, but the rate is dramatically increased (about 10^6-fold) in the presence of chorismate mutase.

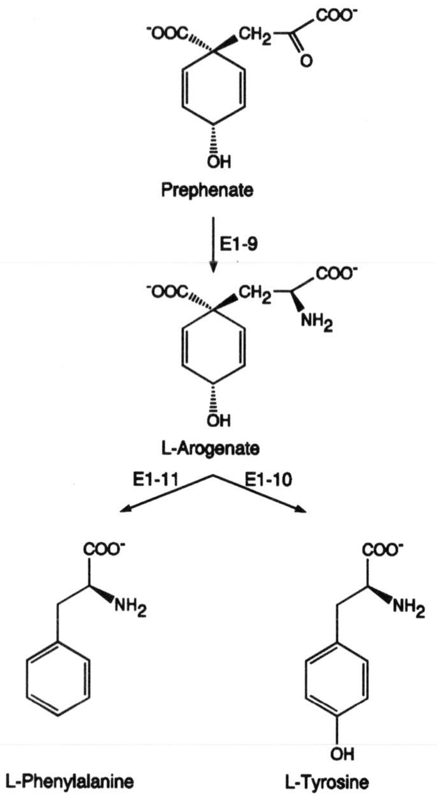

Prephenate

E1-9

L-Arogenate

E1-11 / E1-10

L-Phenylalanine   L-Tyrosine

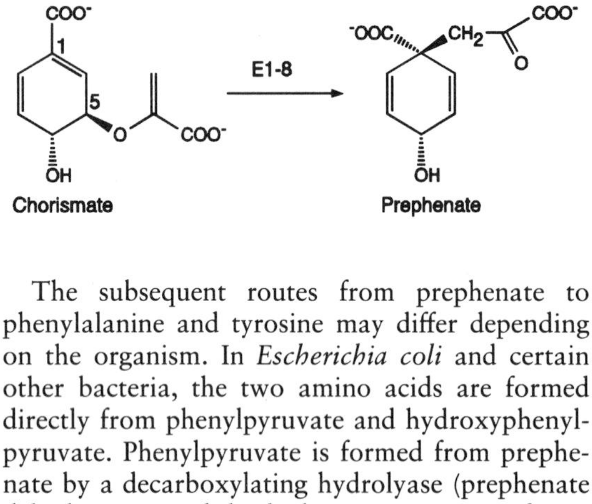

Chorismate   Prephenate

The subsequent routes from prephenate to phenylalanine and tyrosine may differ depending on the organism. In *Escherichia coli* and certain other bacteria, the two amino acids are formed directly from phenylpyruvate and hydroxyphenylpyruvate. Phenylpyruvate is formed from prephenate by a decarboxylating hydrolyase (prephenate dehydratase) and leads by an aminotransferase reaction to phenylalanine, whereas hydroxyphenylpyruvate is formed by a decarboxylating NAD^+-dependent oxidoreductase and leads by an aminotransferase reaction to tyrosine. It became apparent, however, that eukaryotes and many prokaryotes use arogenate as the immediate precursor for these amino acids. The arogenate pathway is the characteristic route to phenylalanine and tyrosine in higher plants.

The conversion of prephenate into arogenate is catalyzed, in the presence of glutamate, by a pyridoxal 5′-phosphate-dependent prephenate aminotransferase (E1-9). The subsequent two branches from arogenate are controlled by the enzymes arogenate dehydrogenase (E1-10) and arogenate dehydratase (E1-11), which catalyze the formation of tyrosine and phenylalanine, respectively.

This bifurcation is an important point of regulation (feedback inhibition). Here the partitioning of the carbon flow between the two amino acids is controlled. The dehydratase is inhibited by phenylalanine and the dehydrogenase by tyrosine. Both amino acids also inhibit chorismate mutase (E1-8) activity. Further control possibly results from the fact that phenylalanine fluxes into protein and phenylpropanoid biosyntheses may be coordinated by parallel pathways that are located in different compartments and include several isoenzymes. One shikimate/arogenate pathway is thought to be located in chloroplasts (most likely also in proplastids), in which the aromatic amino acids are produced mainly for protein biosynthesis, whereas the second is probably membrane associated (channeling of intermediates!) in the cytosol, in which phenylalanine is also produced for the formation of the phenylpropanoids. However, there is urgent need to clarify this hypothesis.

10.3 PHENYLALANINE/ HYDROXYCINNAMATE PATHWAY

The phenylalanine/hydroxycinnamate pathway is defined as 'general phenylpropanoid metabolism'. It includes reactions leading from L-phenylalanine to the hydroxycinnamates and their activated forms, the coenzyme A (CoA) thioesters and the 1-O-acylglucosides. The latter accumulate in many plants as typical hydroxycinnamate conjugates. Enzyme Box 2 lists seven enzymes (E2-1 to E2-7) which catalyze the important reactions of the phenylalanine/hydroxycinnamate pathway.

10.3.1 Phenylalanine ammonia-lyase

The interface between phenylalanine and the secondary phenylpropanoid metabolism is controlled by the enzyme phenylalanine ammonia-lyase (PAL; E2-1). This enzyme catalyzes a non-oxidative deamination of phenylalanine to form the first secondary phenylpropane structure. In a concerted process, ammonia is eliminated, including the *pro-3S* hydrogen atom of C-3, to give *E*-cinnamate.

PAL belongs to the class of carbon-nitrogen lyases (C-N cleavage) that form a double bond, in contrast to dehydrogenation and hydrolysis. The active side of the enzyme contains a dehydroalanine residue whose methylene group binds to the amino group of phenylalanine (β-addition). The product elimination process of PAL activity generates, after a prototropic shift, *E*-cinnamate and

the 'amino enzyme'. The enzyme is finally regenerated by the release of ammonia. This ammonia that is generated in large amounts, e.g. in lignifying tissues, might be reassimilated through the action of glutamine synthetase. This possible nitrogen recycling needs to be studied.

PAL exists as a tetrameric structure. Typical properties are (i) M_r 270 000 to 330 000 with probably four identical subunits, (ii) negative cooperativity with respect to phenylalanine, i.e. two K_m values, one at low and the other at high phenylalanine concentrations, and (iii) a pH optimum in the range 8–9. Natural inhibitors are known which affect the catalytic activity. A macromolecular inhibitor has been identified as a hydrophobic protein with M_r of 19 000.

Several studies on the physiology of phenolic metabolism have blocked PAL activity *in vivo* by using synthetic PAL-specific inhibitors such as phenylalanine analogs, e.g. the widely used L-2-amino-oxy-3-phenylpropionic acid (L-AOPP) or the recently introduced 2-amino-indan-2-phosphonic acid (AIP). They are effective at the nanomole level. Such inhibitors proved to be useful tools for studying phenylpropanoid metabolism. From the perspective of the biochemical study of PAL action, the application of AOPP gave insight into the transition state binding of phenylalanine, as it is a transition state analog. It has been assumed that conformational changes of the enzyme may contribute to lowering the free energy of activation for the product elimination process.

In certain grasses and fungi, PAL may also act on tyrosine, which leads directly to 4-coumarate. The earlier notion that this reaction might be due

---

ENZYME BOX 2
PHENYLALANINE/HYDROXYCINNAMATE PATHWAY

E2-1: Phenylalanine ammonia-lyase (PAL)

E2-2: Cinnamate 4-hydroxylase; i.e. *E*-cinnamate, NADPH : oxygen oxidoreductase

E2-3: 4-Coumarate 3-hydroxylase (= monophenol monooxygenase; phenolase); i.e. 4-coumarate, ascorbate (or NADPH) : oxygen oxidoreductase

E2-4: Ferulate 5-hydroxylase; i.e. ferulate, NADPH : oxygen oxidoreductase

E2-5: Caffeate/5-hydroxyferulate methyltransferase; i.e. SAM : caffeate/5-hydroxyferulate 3-O-methyltransferase

E2-6: Hydroxycinnamoyl-CoA ligases; i.e. HCA : CoA ligase (AMP-forming)

E2-7: Hydroxycinnamate O-glucosyltransferases; i.e. UDP-glucose : HCA 9-O-acylglucosyltransferase to a separate enzyme, the tyrosine ammonia-lyase (TAL), had to be abandoned. There is no report of a plant exhibiting higher ammonia-lyase activity with tyrosine than with phenylalanine. Enzyme preparations are always more active with phenylalanine and it has never been shown that only tyrosine is accepted.

In some plants PAL exists as a single enzyme. In others, however, multiple forms are found. For example, in *Arabidopsis thaliana* (Brassicaceae), PAL isoenzymes are synthesized *de novo* in response to UV irradiation, microbial attack and wounding. However, possible specific metabolic roles of the individual isoenzymes have not definitely been proven. In *Medicago sativa* (alfalfa; Fabaceae), a multigene family of up to six PAL genes, or possible allelic variants, has been detected. It was shown with alfalfa cell cultures, that the PAL transcripts are rapidly induced on treatment with a fungal elicitor derived from the cell walls of the plant pathogen *Colletotrichum lindemuthianum*.

10.3.2   Biosynthesis of hydroxycinnamates

A series of hydroxylation and methylation reactions, catalyzed by the enzymes E2-2 to E2-5, leads to the sequential formation of the common hydroxycinnamates (Fig. 10.2). These are 4-coumarate, caffeate, ferulate and sinapate. The rare 5-hydroxyferulate is the substrate for sinapate formation. The hydroxycinnamates usually occur

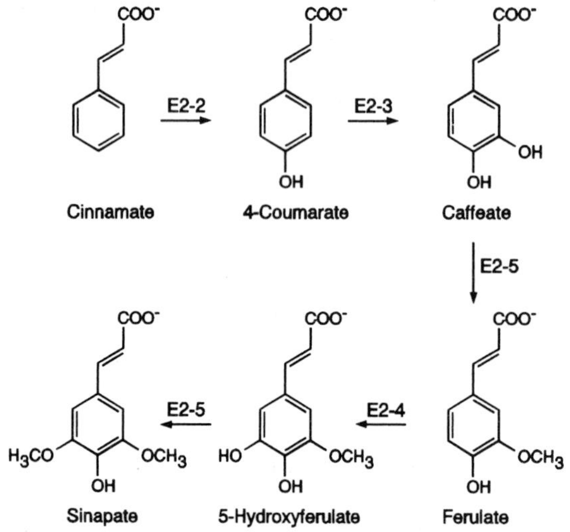

Figure 10.2   Biosynthesis of hydroxycinnamates by sequential hydroxylation and methylation reactions.

as the *trans (E)*-isomers, derived from *E*-cinnamate (see PAL action). However, *cis (Z)*-isomers may appear by either photochemically or enzymatically mediated isomerization. This applies also for other phenolics, such as the monolignols.

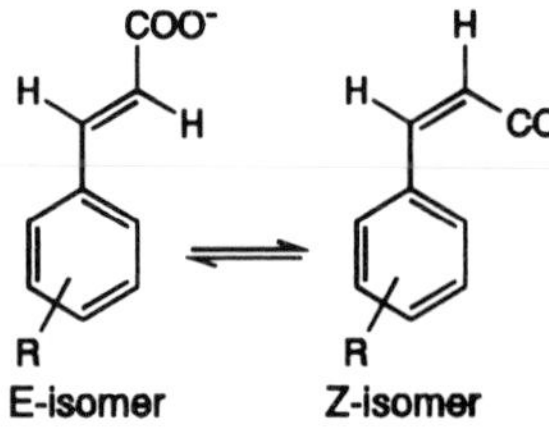

The hydroxylases E2-2, E2-3 and E2-5 belong to the monooxygenases, also referred to as mixed-function oxidases, which introduce a single atom of molecular oxygen into the substrate. The oxygen is split during this reaction and the second oxygen atom is reduced to H_2O. A second oxidizable substrate is required, e.g. NADPH.

There are two types of these hydroxylases; (i) the membrane-bound (microsomal) cytochrome *P*-450-dependent oxygenases (E2-2 and E2-4) and (ii) the soluble phenolase which catalyzes the introduction of a second hydroxyl group into a monophenol, as in caffeate formation (E2-3). Cinnamate 4-hydroxylase (E2-2) introduces a hydroxyl group specifically in the 4-position of *E*-cinnamate. The enzyme is inactive with the *Z*-isomer. The 4-hydroxylation occurs with a concomitant 'NIH shift'. This is an intramolecular migration of the proton that is displaced to an adjacent *ortho* position on the aromatic ring. There is strong evidence that the reaction proceeds through arene oxide (epoxide) intermediates. The NIH shift is named after the National Institute of Health (NIH), Bethesda, USA, where enzymatic hydroxylations of aromatic compounds have been extensively studied.

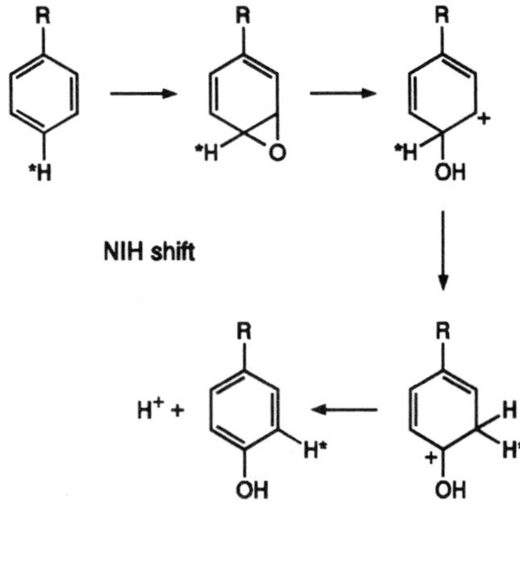

The phenolase 4-coumarate hydroxylase (E2-3) introduces a second hydroxyl group *ortho* (3-position) to the hydroxyl group of 4-coumarate. This leads to the loss of the *ortho* proton. The third hydroxylase activity is the ferulate hydroxylase (E2-4), which is again a membrane bound cytochrome *P*-450-dependent oxygenase.

There are various reports about low substrate specificities and the observation of a second reaction leading to quinonoid structures. It is a matter of dispute to what extent many of the described phenolase activities are involved in the pathway for the formation of the free hydroxycinnamates. Further research will be needed to show whether there are different and highly specific hydroxylase activities leading to the hydroxycinnamates.

Regarding the problem discussed above, there are interesting results showing that hydroxylases exist that specifically accept hydroxycinnamate conjugates. This is realized in the formation of caffeate at the 4-coumaroyl moiety of 4-coumaroyl-CoA or of 5-O-(4-coumaroyl)-shikimate:

$$\text{4-Coumaroyl-CoA} \xrightarrow[-\text{H}_2\text{O} \ -\text{NADP}^+]{+\text{O}_2+\text{NADPH}+\text{H}^+} \text{Caffeoyl-CoA}$$

5-O-(4-Coumaroyl)-shikimate

$$\xrightarrow[-\text{H}_2\text{O} \ -\text{NADP}^+]{+\text{O}_2+\text{NADPH}+\text{H}^+} \text{5-}O\text{-Caffeoylshikimate}$$

O-Methylation of caffeate and hydroxyferulate yields ferulate and sinapate, respectively. The enzyme (E2-5) involved uses *S*-adenosyl-L-methionine (SAM) as methyl donor. (It is not certain whether one or two enzymes are involved in these reactions.) During the methyl transfer SAM is converted into SAH (*S*-adenosyl-L-homocysteine). Interestingly, SAH may be the substrate for a specific hydrolase leading to adenosine and L-homocysteine. This favors the overall formation of the methoxylated products.

As shown for the hydroxylase reactions, methylations can also occur with hydroxycinnamate conjugates. Thus there are examples that the ferulate substitution pattern can be established at the level of the CoA derivatives:

$$\text{Caffeoyl-CoA} \xrightarrow[-\text{SAH}]{+\text{SAM}} \text{Feruloyl-CoA}$$

10.3.3 Activation of hydroxycinnamates

The final step of the 'general phenylpropanoid metabolism' is the carboxyl activation of hydroxycinnamates, catalyzed by hydroxycinnamate : CoA ligases (E2-6) or by the action of *O*-glucosyl-

transferases (E2-7). The hydroxycinnamate-CoAs formed enter various specific phenylpropanoid reactions, such as condensations with malonyl-CoA leading to the flavonoids, NADPH-dependent reductions leading to lignins or conjugation reactions in ester and amide formation. The hydroxycinnamate 1-*O*-acylglucosides serve as acyl donor molecules in various transferase reactions leading to *O*-esters. These are alternative reactions to the CoA thioester-dependent transferases.

The formation of hydroxycinnamoyl-CoAs is analogous to the activation of fatty acids with ATP and CoA:

$$\text{Hydroxycinnamate} \xrightarrow[-\text{PP}_i]{+\text{ATP}}$$

$$\text{Hydroxycinnamoyl-AMP} \xrightarrow[-\text{AMP}]{+\text{CoA·SH}}$$

$$\text{Hydroxycinnamoyl-CoA}$$

It is likely that there are specific ligases (isoforms) for specific pathways, e.g. in flavonoid, lignin, and ester formation, but this has not yet been unambiguously proven. The ligase that has been most thoroughly investigated is the 4-coumarate : CoA ligase (E2-6) involved in flavonoid biosynthesis. The often mentioned discrepancy between substrate specificities of some ligases and the phenolic substitution patterns of the products may partly be explained by the fact that secondary reactions with hydroxycinnamoyl-CoAs, such as hydroxylations or *O*-methylations, can generate the final phenolic substitution patterns.

The formation of 1-*O*-hydroxycinnamoylglucose (hydroxycinnamate 1-*O*-acylglucosides) is catalyzed by uridine 5'-diphosphate (UDP)-glucose-dependent glucosyltransferases:

$$\text{Hydroxycinnamate} \xrightarrow[-\text{UDP}]{+\text{UDP-glucose}}$$

$$\text{1-}O\text{-Hydroxycinnamoylglucose}$$

The presence of other specific glycosyltransferases that utilize UDP-bound sugars as glycosyl donors in reactions leading to *O*-glycosides is well established. These reactions yield a wide range of phenolic glycosides, such as the flavonoid glycosides.

10.4 PHENYLPROPANOID PATHWAYS

The hydroxycinnamates are utilized in various pathways that use four predominant types of side-chain reactions (Fig. 10.3).

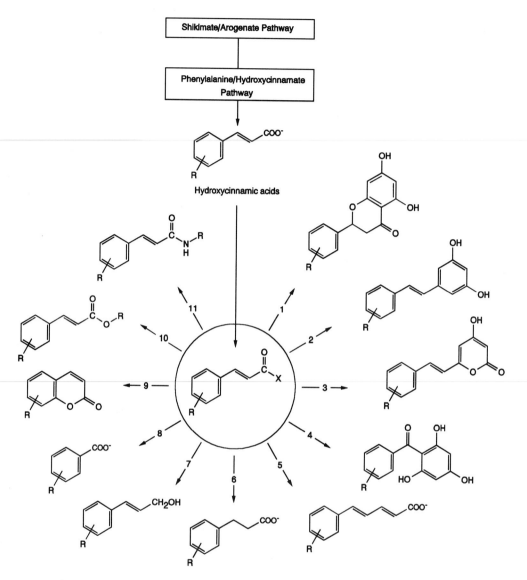

Figure 10.3 Scheme of the central role of hydroxycinnamates in the formation of various phenylpropanoids: side-chain elongation (reactions 1–5; 1 leads to flavonoids, 2 to stilbenes, 3 to styrylpyrones, 4 to benzophenones which undergo cyclization to xanthones, 5 to chain-elongated hydroxycinnamates); side-chain reduction (reaction 6 leads to dihydrocinnamates, 7 to hydroxycinnamyl alcohols); side-chain degradation (reaction 8 leads to hydroxybenzoates); 2-hydroxylation and lactonization (reaction 9 leads to hydroxycoumarins); conjugation (reaction 10 leads to hydroxycinnamate esters, 11 to hydroxycinnamate amides). X = OH, SCoA (thioester) or glucose (1-O-acylglucoside).

1. **Condensation** (side-chain elongation) with malonyl-CoAs (addition of acetate units with liberation of CO_2), e.g. sequential reactions with three malonyl-CoAs, leads to the flavonoids.
2. **Degradation** (side-chain shortening) by removal of an acetate unit leads to the hydroxybenzoates.
3. **Reduction** (NADPH-dependent) leads to the lignin precursors, hydroxycinnamyl alcohols.
4. **Conjugation** through attachment to a hydroxyl- or amino-group-bearing molecule with liberation of H_2O leads to esters or amides in rare cases, also attachment to the phenolic hydroxyl groups is known, which leads e.g. to glycosides.

10.5 HYDROXYCINNAMATE CONJUGATES

Hydroxycinnamates often accumulate as conjugates, preserving the C_6-C_3 structure. Esters and amides are the most common types of conjugates, whereas glycosides occur rarely. The term 'conjugation', originally used in plant hormone metabolism, is defined as attachment, mostly condensation, of various monomeric or oligomeric compounds, which markedly alter the chemical properties of the phenolic compound. Conjugating moieties can be carbohydrates, proteins, lipids, amino acids, amines, carboxylic acids, terpenoids, alkaloids or flavonoids. In addition, insoluble forms of hydroxycinnamates occur bound to polymers such as cutins and suberins, lignins or polysaccharides in cell wall fractions.

The systematic distribution of hydroxycinnamate conjugates may show correlation with the systematic arrangements of plant families. Thus, the most widespread caffeate esters occur as caffeoyl-disaccharides in Scrophulariaceae and Oleaceae, 5-O-caffeoylquinate (chlorogenate) in Asteraceae, Solanaceae and Rubiaceae, and caffeoyldihydroxyphenyllactate (rosmarinate) in Lamiaceae and Boraginaceae. Another example is the characteristic occurrence of the sinapate ester sinapoylcholine (sinapine) in Brassicaceae.

10.5.1 Conjugation reactions

Conjugation reactions may determine whether a compound is converted into a metabolically inactive final end product (permanent storage) or into a transiently accumulating intermediate that may be subject to further metabolic pathways, e.g. further conjugation, interconversion or degradation. Besides UDP-glucose-dependent transferases involved in the formation of hydroxycinnamate glucose esters (1-O-acylglucosides) or glucosides, there are three types of transferases which have been described to date that catalyze the formation of hydroxycinnamate conjugates: (i) hydroxycinnamoyl-CoA thioester-, (ii) hydroxycinnamate 1-O-acylglucoside- and (iii) hydroxycinnamate O-ester-dependent transferases. Currently it is known that the pathway utilized is not dependent on the nature of the conjugating moiety, but rather on the source of enzyme used, i.e. the plant species investigated. A good example of converging lines for the formation of hydroxycinnamate O-esters is O-caffeoylglucarate which accumulates in the leaves of different plants (Fig. 10.4). In *Secale cereale* (Poaceae) the biosynthesis proceeds via the

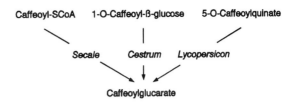

Figure 10.4 Alternative biosyntheses of caffeoylglucarate.

caffeoyl-CoA thioester, in *Cestrum elegans* (Solanaceae) via 1-O-caffeoylglucose, and in *Lycopersicon esculentum* (Solanaceae) via 5-O-caffeoylquinate, i.e. chlorogenate.

Chlorogenate is another example of such alternative pathways. Usually it is formed through caffeoyl-CoA, as found in various members of the Solanaceae, but it may also be formed through 1-O-caffeoylglucose, e.g. in *Ipomoea batatas* (Convolvulaceae).

10.5.2 Conjugation with glucose

Conjugations of phenolics with sugars are catalyzed by transferases that utilize nucleotide sugars, e.g. the UDP-glucose-dependent glucosyltransferases. A general reaction scheme of a glucosyltransfer reaction can be formulated as follows:

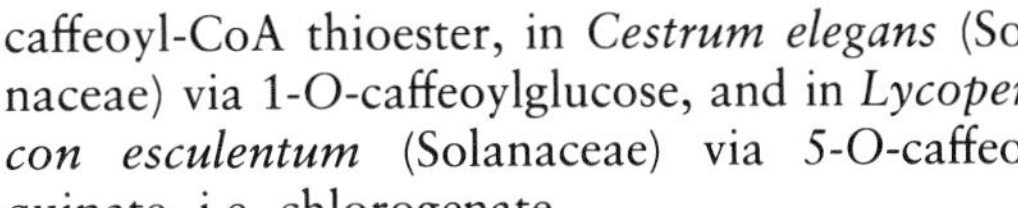

$$R—OH \quad \xrightarrow[- UDP]{+ UDP\text{-}glucose} \quad R—O—glucose$$

$$R—COOH \qquad\qquad R—\underset{\underset{O}{\parallel}}{C}—O—glucose$$

One would expect that the free energy of hydrolysis of nucleotide sugars would exceed that of the phenolic glycoside formed. It has

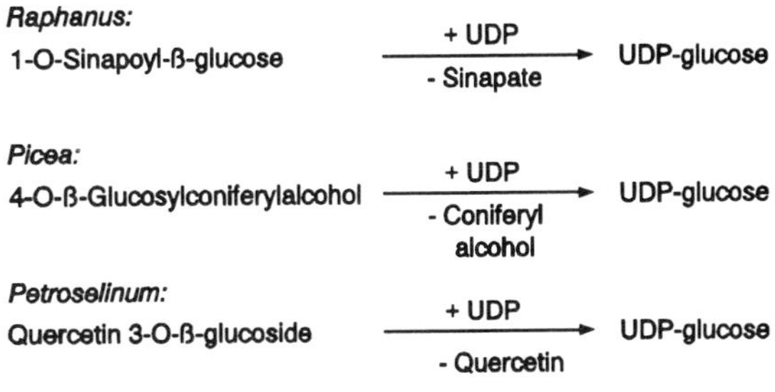

Raphanus:

$$1\text{-}O\text{-}Sinapoyl\text{-}\beta\text{-}glucose \quad \xrightarrow[- Sinapate]{+ UDP} \quad UDP\text{-}glucose$$

Picea:

$$4\text{-}O\text{-}\beta\text{-}Glucosylconiferylalcohol \quad \xrightarrow[\substack{- Coniferyl \\ alcohol}]{+ UDP} \quad UDP\text{-}glucose$$

Petroselinum:

$$Quercetin\ 3\text{-}O\text{-}\beta\text{-}glucoside \quad \xrightarrow[- Quercetin]{+ UDP} \quad UDP\text{-}glucose$$

Figure 10.5 *In vitro* formation of UDP-glucose from an acylglucoside (ester) with a protein preparation from *Raphanus sativus* (Brassicaceae), as well as from phenolic glucosides with protein preparations from *Picea abies* (Pinaceae) and *Petroselinum crispum* (Apiaceae).

been found, however, that the reactions leading to phenolic glucosides as well as 1-O-acylglucosides (esters) of phenolic acids can be freely reversible. This indicates that the glycosidic bond of glucose (C-1) carries a high group transfer potential. *In vitro* studies prove that UDP-glucose can be formed in appreciable amounts from phenolic glucosides and free UDP (Fig. 10.5). This free reversibility might be of physiological importance in possible catabolic pathways (turnover, degradation) by preserving the hydrolytic energy in the form of UDP-glucose.

10.6 HYDROXYCOUMARINS

Some members of the plant families Rutaceae, Solanaceae and Apiaceae accumulate large amounts of hydroxycoumarins. These are lactones derived formally from 2-hydroxycinnamate to give coumarin (unsubstituted aromatic ring) or from 2-hydroxylated hydroxycinnamates to give hydroxycoumarins. Lactone formation takes place between the 2-hydroxyl group and the carboxyl group of the C_3 side chain. A prerequisite for this reaction is the isomerization of *E*-hydroxycinnamate to *Z*-hydroxycinnamate.

The point of divergence from the phenylalanine/ hydroxycinnamate pathway to coumarin and hydroxycoumarin formations is the 2-hydroxylation. The presence of a specific 2-hydroxylase has not yet been unambiguously proven, although there are reports of such an activity that exhibits similar properties to those of enzyme E2-2, possibly including an NIH shift. The pathway to coumarin itself (Fig. 10.6), a well-known constituent in wood ruff (*Galium odoratum*; Rubiaceae) and *Melilotus alba* (Fabaceae), proceeds through a 2-O-glucoside, probably of both the *E*- and *Z*-cinnamate. It is assumed that the *E*-form is transported into the vacuole where it is transformed to a great extent into the *Z*-form. This isomerization may be induced by UV light, which can easily be demonstrated by *in vitro* experiments. However, there is evidence for a light-independent isomerization in the vacuole.

The final formation of coumarin by lactonization occurs spontaneously after deglucosylation, which is catalyzed by a β-glucosidase when the tissues are disrupted, e.g. by mechanical injury. It causes the characteristic smell of new mown hay. In the undamaged tissue, the β-glucosidase and the 2-O-glucoside are separated from each other at the cell level. The glucoside is located in the vacuoles and the β-glucosidase is either associated with the cell walls or located within the intercellular spaces.

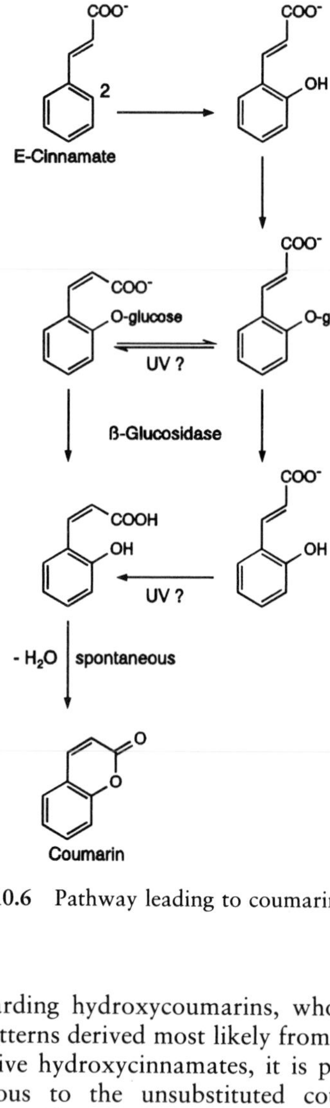

Figure 10.6 Pathway leading to coumarin in *Melilotus alba*.

Regarding hydroxycoumarins, whose substitution patterns derived most likely from those of the respective hydroxycinnamates, it is possible that, analogous to the unsubstituted coumarin, the 2-O-glucosides also occur in living cells. Most hydroxycoumarins known, however, exist as aglycones or as conjugates, e.g. cichoriin (esculetin 7-O-glucoside) in *Cichorium intybus* (Asteraceae). Coumarate leads to umbelliferone, caffeate to esculetin and ferulate to scopoletin. However, there is also evidence that the substitution pattern of hydroxycoumarins may be determined at the coumarin level, analogous to the sequential substitution reactions in the phenylalanine/hydroxycinnamate pathway. One example is the biosynthesis of cichoriin (Fig. 10.7) from umbelliferone, derived from *Z*-4-coumarate. Hydroxylation of umbelliferone at C-6 leads to esculetin. The final step is the UDP-glucose-dependent glucosylation at C-7.

There are more complex hydroxycoumarins that are mainly derived from umbelliferone. These are

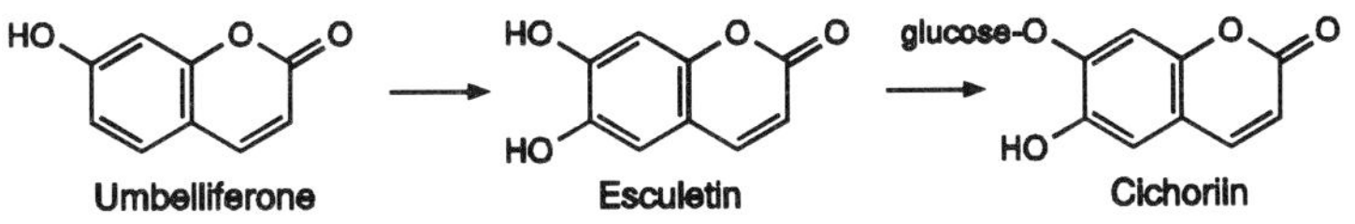

Figure 10.7 Formation of cichoriin from umbelliferone.

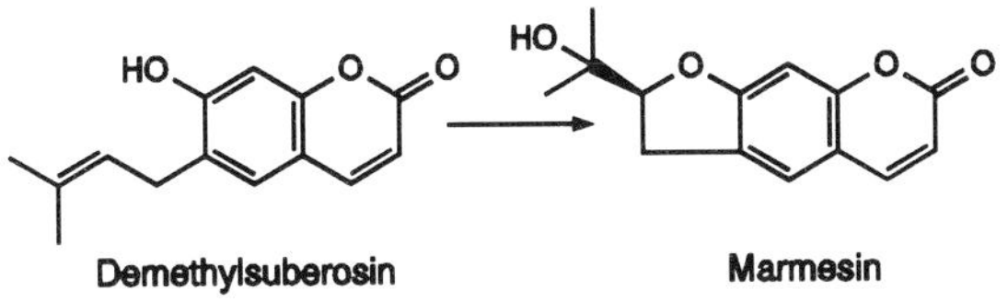

Figure 10.8 Biosynthesis of marmesin catalyzed by a NADPH- and oxygen-dependent marmesin synthase.

various prenylated hydroxycoumarins that lead to several hundred lipophilic furanocoumarins and dihydropyranocoumarins. They are frequently excreted into schizolysigenous channels or onto waxy surfaces. Many furanocoumarins are toxic and some of them are true phytoalexins which inhibit the germination of spores from fungal pathogens. Cell cultures from members of the Apiaceae (e.g. *Petroselinum crispum* or *Ammi majus*), treated with fungal elicitors from *Phytophthora megasperma* f. sp. *glycinea* or *Alternaria carthami*, proved to be excellent model systems for the investigation of the enzymology and regulation of the biosynthesis of hydroxycoumarin derivatives.

As in the formation of flavonoidal pterocarpans, dimethylallyltransferase and cyclase activities (cytochrome *P*-450-monooxygenases) are involved in their biosyntheses. Although the crucial steps in these pathways have not yet been unambiguously elucidated, it is likely that oxidative cyclizations through epoxides lead to the different heterocycles attached to the hydroxycoumarins. A key enzyme

in these biosyntheses is marmesin synthase (a cytochrome *P*-450-monooxygenase) which catalyzes the conversion of demethylsuberosin into marmesin (Fig. 10.8). This product is thought to be the pivotal precursor for both the furano- and the dihydropyranocoumarins.

Examples of the different classes of compounds formed are bergapten, a linear furanocoumarin ('psoralens'), in bergamot oil (essential oil from fruits of *Citrus aurantium bergamia*; Rutaceae), pimpinellin, an angular furanocoumarin ('isopsoralens'), in *Pimpinella* species (Apiaceae), and visnadin, a biologically active complex dihydropyranocoumarin derivative, from fruits of *Ammi visnaga* (Apiaceae).

10.7 HYDROXYBENZOATES

As with the hydroxycinnamates, the hydroxybenzoates (C_6-C_1) seem to be universally distributed in plants. These are the common 4-hydroxybenzoate, protocatechuate, vanillate, gallate and syringate. They may be present in soluble conjugated forms as well as bound to cell wall fractions, e.g. linked to lignin. The trihydroxyl derivative gallate is frequently present in high molecular structures such as gallotannins. Another common hydroxybenzoate is salicylate (2-hydroxybenzoate). It is widespread in Ericaceae and may be present as its methyl ester in essential oils, e.g. in the pungent-smelling oil of *Gaultheria procumbens* (wintergreen).

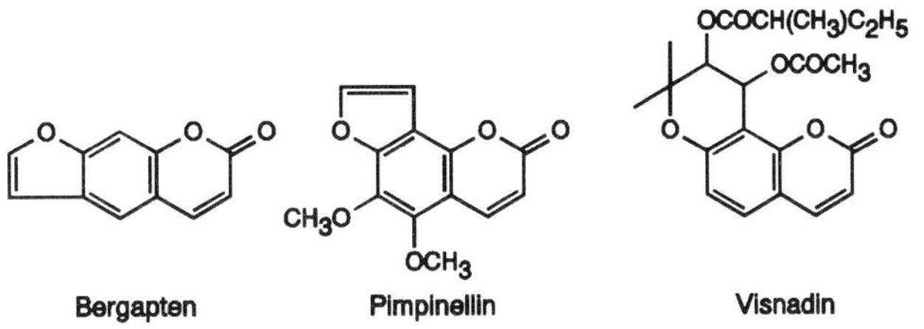

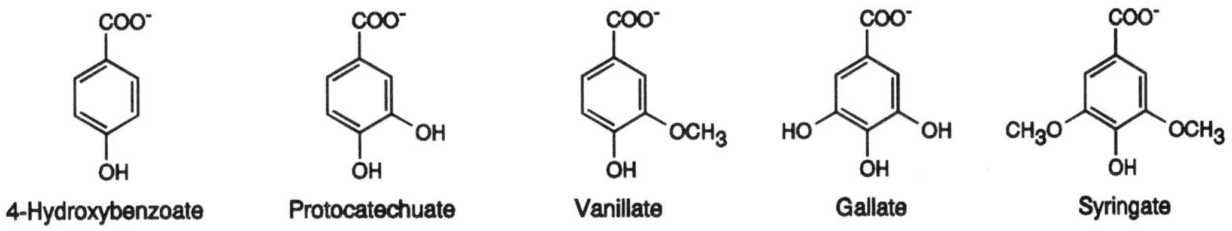

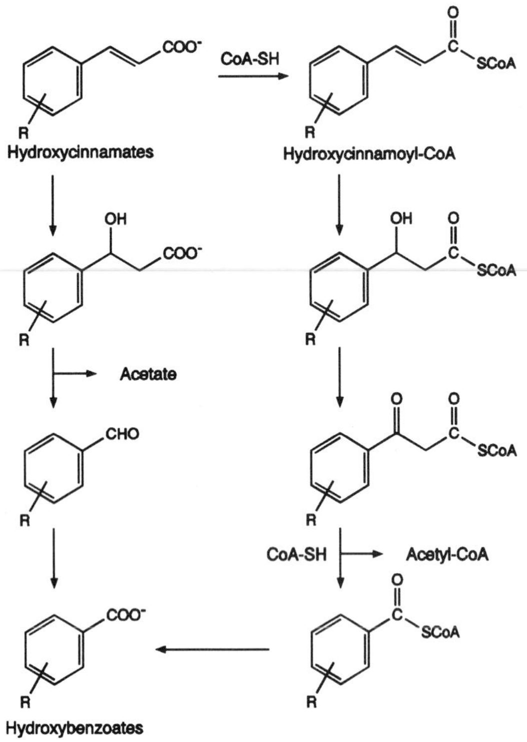

Figure 10.9 Postulated biosynthetic pathways to hydroxybenzoates.

There are conflicting results concerning the biosynthesis of hydroxybenzoates. It is likely that there are several pathways leading to the individual hydroxybenzoates, depending on the plant. One major route is the side-chain degradation of hydroxycinnamates by removal of acetate. The reaction sequence has been proposed to proceed through the CoA-ester (Fig. 10.9), analogous to the β-oxidation of fatty acids (see Chapter 6). However, a second possible pathway, that does not require the CoA-activated hydroxycinnamates, cannot be excluded. (Hydroxybenzoates are also produced occasionally by degradation of flavonoids.)

The substitution patterns of hydroxybenzoates may be determined by the hydroxycinnamate precursor. However, hydroxylations and methylations may occur with the hydroxybenzoates, analogous to the phenylalanine/hydroxycinnamate pathway.

Hydroxybenzoate with trihydroxyl substitution (gallate) can also be formed by a pathway which branches from the shikimate/arogenate pathway at 3-dehydroshikimate. Direct aromatization of the enol form of 3-dehydroshikimate is likely to be the predominant route leading to gallate in plants. This is supported by results from feeding experiments with [14C]shikimate in the presence of the PAL inhibitor AOPP and the EPSP synthase (E1-6) inhibitor, glyphosate. AOPP had no impact on the incorporation of radioactive shikimate into gallate. Glyphosate, however, enhanced its incorporation in *Quercus robur* (Fagaceae) cell suspension cultures.

Hydroxybenzoates may be utilized as precursors for other products. This may occur, as with the hydroxycinnamates, through the CoA esters, e.g. reduction to benzaldehydes and benzylalcohols or conjugations, such as the ester formation in the biosynthesis of cocain (unsubstituted benzoate as ester moiety). Hydroxybenzoates are also found as acyl moieties in acylated flavonoids.

10.8 FLAVONOIDS

More than 5000 different flavonoids are known. The C_{15} aglycone skeleton appears in various structural classes according to the oxidation state of the central pyran ring. Structures within these classes are modified by hydroxylation and methoxylation. They are usually glycosylated, and furthermore many of them are acylated with aliphatic and aromatic acids. The enzymes that are necessary for the biosynthesis of the flavonoid aglycone structures (flavonoid aglycone pathway) are listed in Enzyme Box 3.

10.8.1 Chalcone synthase and chalcone isomerase

Chalcone synthase (E3-1), considered to be the rate-limiting enzyme in flavonoid synthesis, catalyzes the formation of the basic C_{15} skeleton and thus channels hydroxycinnamates into flavonoid biosynthesis. The enzyme condenses 4-coumaroyl-CoA with three molecules of malonyl-CoA to form 'naringenin chalcone' (2',4,4',6'-tetrahydroxychalcone; note the different numbering system compared to that of the other flavonoids). Chalcone synthase is a dimeric protein with M_r 78 000–88 000 and probably two identical subunits. The pH optimum of the reaction is in the range 7.5–8.5 for 4-coumaroyl-CoA. The enzyme from most sources is highly specific for 4-coumaroyl-CoA, although there are examples where other hydroxycinnamoyl-CoAs are accepted.

ENZYME BOX 3
FLAVONOID AGLYCONE PATHWAY

E3-1:   Chalcone synthase; i.e. malonyl-CoA : 4-coumaroyl-CoA malonyltransferase
E3-2:   Acetyl-CoA carboxylase; i.e. acetyl-CoA : CO_2 ligase
E3-3:   Chalcone isomerase
E3-4:   Flavone synthase I (= 2-hydroxyflavanone synthase); i.e. flavanone, 2-oxo-glutarate : oxygen oxidoreductase (dioxygenase); includes a dehydratase reaction
E3-5:   Flavone synthase II (= 2-hydroxyflavanone synthase); i.e. flavanone, NADPH : oxygen oxidoreductase; includes a dehydratase reaction
E3-6:   Isoflavone synthase (= 2-hydroxyisoflavanone synthase); i.e. flavanone, NADPH : oxygen oxidoreductase; includes a dehydratase reaction
E3-7:   Flavanone 3-hydroxylase (dioxygenase); i.e. naringenin, 2-oxoglutarate : oxygen oxidoreductase
E3-8:   Dihydroflavonol/dihydroflavone 4-reductase; i.e. dihydroflavonol/dihydroflavone : NADPH oxidoreductase
E3-9:   Flavonol synthase (dioxygenase); i.e. dihydroflavonol, 2-oxoglutarate : oxygen oxidoreductase
E3-10:   Flavan-3,4-diol 4-reductase; i.e. flavan-3,4-diol : NADPH oxidoreductase

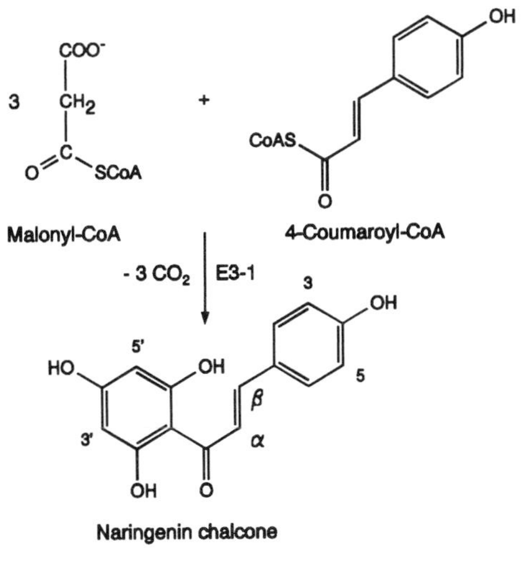

Naringenin chalcone

Naringenin chalcone is the first intermediate possessing a monohydroxyl B-ring, typical of virtually all flavonoids. The proposed mechanism of the malonyl transfers is a stepwise addition of acetate from malonyl-CoA to 4-coumaroyl-CoA. The stepwise addition of acetate units becomes obvious under suboptimal chalcone reaction conditions, where side products appear that contain one or two acetate units less that naringenin chalcone.

A 'random orientation' of the acetate units forms the flavonoid A-ring. This leads to the 5-hydroxyflavonoid pathway forming the typical 5,7-hydroxylation pattern of the phloroglucinol type (pattern of polyketide ring hydroxylation) of the flavonoid A-rings. The chalcone synthase uses the CoA esters as immediate substrates, in contrast to the use of enzyme-bound esters in the fatty acid–synthase complex with its acyl carrier protein (see Chapter 6).

Malonyl-CoA is supplied for the chalcone synthase by the ATP-dependent action of acetyl-CoA carboxylase (E3-2). This is a typical enzymatic reaction in fatty acid synthesis. It is also, as seen here and in the next section on flavonoid conjugation, an integral part of flavonoid synthesis. It supplies malonyl-CoA as an important building block for the flavonoid aglycones and as the acyl donor for malonylation of the glycosides.

There is a further enzyme that has a very similar action to that of chalcone synthase. It also uses 4-coumaroyl-CoA and three molecules of malonyl-CoA to catalyze the formation of a C_{15} phenolic. This is a synthase whose catalysis leads through a different folding mechanism with the enzyme-bound intermediate to stilbenes, as in the formation of resveratrol (3 malonyl-CoA + 4-coumaroyl-CoA) or pinosylvin (3 malonyl-CoA + cinnamoyl-CoA). Based on comparisons of cloned gene sequences and analysis of the reaction mechanism it is clear that chalcone and stilbene synthase are closely related enzymes.

There are some flavonoids that lack a 5-hydroxyl group, such as the flavanone liquiritigenin or the isoflavonoid glyceollin (resorcinol-type pattern). These structures are formed by the action of a NADPH-dependent reductase prior to ring-A closure at the chalcone synthase-bound intermediate, and this formation leads to the 5-deoxyflavonoid pathway. The A-ring of the obligatory 2'-deoxy product trihydroxychalcone (isoliquiritigenin) is considered to be formed by a 'nonrandom orientation' of the acetate units.

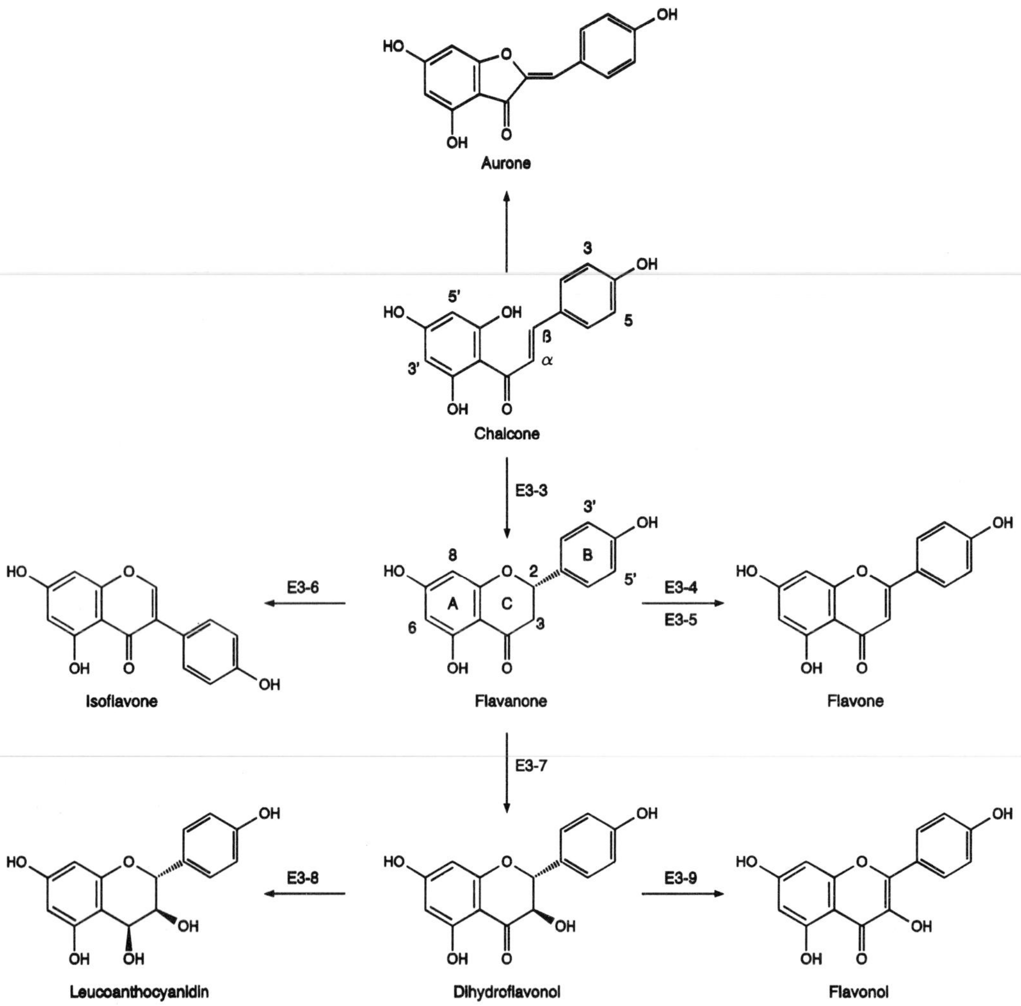

Figure 10.10   Pathway leading to major classes of flavonoid aglycones.

Going forward in the 5-hydroxyflavonoid pathway (Fig. 10.10), the cyclization of chalcone to the laevorotatory (−)-(2S)-flavanone naringenin is catalyzed by chalcone isomerase (E3-3) which is found in a tight complex with the chalcone synthase during flavanone synthesis. Such a complex prevents the possible nonenzymatic cyclization of the chalcone to the R configuration at C-2. Chalcone isomerase can occur in multiple forms (true isoenzymes) which differ considerably concerning pH optima and K_m values.

10.8.2   Biosynthesis of the flavonoid classes

The flavonoid aglycones are classified according to the oxidation state of the heterocyclic ring C (pyran ring) which connects the two benzene rings

A and B. Fig. 10.10 summarizes the enzymatic reactions involved in production of the various flavonoid classes (Enzyme Box 3). Aurones are derived directly from the chalcone intermediate. The enzymology of this reaction, probably containing a peroxidase-mediated step, is still unknown. The flavanone naringenin is the precursor of three different flavonoid classes, i.e. flavones, isoflavones and dihydroflavonols.

(a)   Isoflavones

Oxidative rearrangement of naringenin with a 2,3-aryl shift yields the isoflavone genistein. The reaction mechanism in Fig. 10.11 has been proposed. The initiating step in isoflavone formation may be an epoxidation catalyzed by a cytochrome P-450-dependent monooxygenase (E3-6). After structural rearrangement, aryl shift

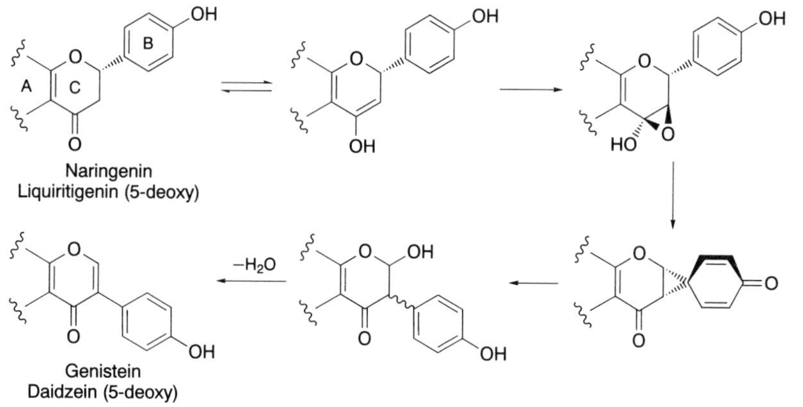

Figure 10.11 Proposed mechanism of isoflavone synthase activity (E3-6) in the formation of genistein (5-hydroxyflavonoid pathway) and daidzein (5-deoxyflavonoid pathway).

and addition of a hydroxyl ion to C-2, elimination of water by a dehydratase gives the isoflavone structure. The dehydration is likely to be catalyzed by a separate soluble enzyme. Analogous reactions in flavone and flavonol syntheses might also be catalyzed by dehydratases; however, this has not yet been unequivocally proven. Elimination of water in these reactions could occur spontaneously.

(b) Pterocarpans

An important subgroup of the isoflavones are the pterocarpans that are infection-induced phytoalexins. They are, together with the isofla-

vones, characteristic constituents in members of the Fabaceae possessing fungicidal and bactericidal activities. Among the plants thoroughly investigated with respect to production of pterocarpans are the leguminous plants *Cicer arietinum* (chickpea), *Glycine max* (soybean), *Medicago sativa* (alfalfa) and *Pisum sativum* (pea). As an example of pterocarpan biosynthesis, the pathway to the glyceollins in *Glycine max* is shown in Fig. 10.12. The important enzymes are listed in Enzyme Box 4.

The pathway starts with daidzein, a product of the 5-deoxyflavonoid pathway. The oxygenases involved are membrane-bound cytochrome

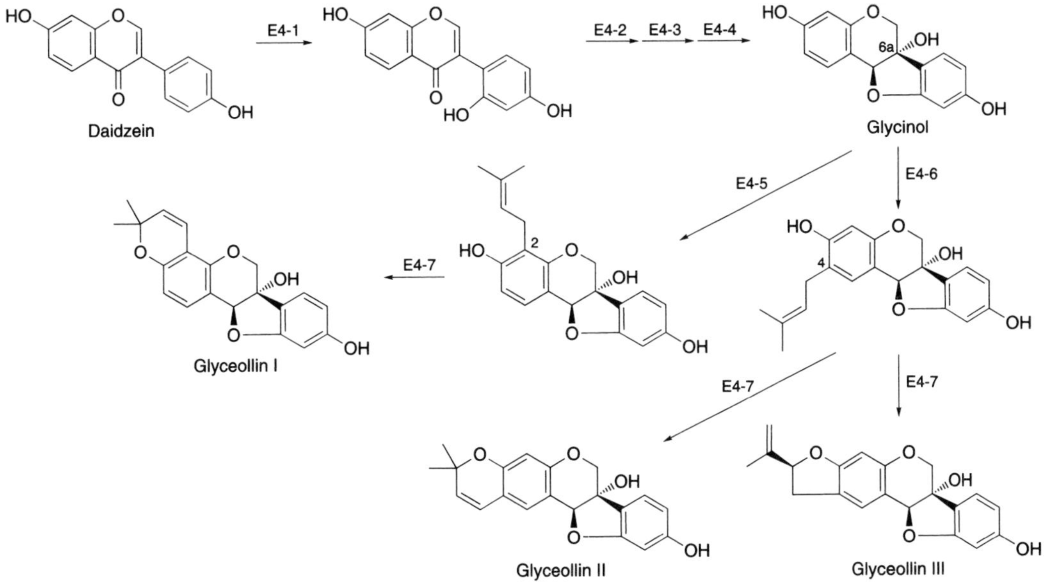

Figure 10.12 Biosynthesis of glyceollins in *Glycine max*.

ENZYME BOX 4
PTEROCARPAN/GLYCEOLLIN PATHWAY

E4-1:   Isoflavone 2′-hydroxylase; i.e. isoflavone, NADPH : oxygen oxidoreductase
E4-2:   2′-Hydroxyisoflavone reductase; i.e. 2′-hydroxyisoflavone : NADPH oxidoreductase
E4-3:   Pterocarpan synthase; i.e. 2′-hydroxyisoflavanone : NADPH oxidoreductase
E4-4:   Pterocarpan 6a-hydroxylase; i.e. pterocarpan, NADPH : oxygen oxidoreductase
E4-5:   Prenyltransferase I; i.e. DMAPP : glycinol (= 3,6a,9-trihydroxypterocarpan) 2-dimethylallyl-
        transferase
E4-6:   Prenyltransferase II; i.e. DMAPP : glycinol 4-dimethylallyltransferase
E4-7:   Pterocarpan cyclase; i.e. dimethylallylglycinol, NADPH : oxygen oxidoreductase

P-450-monooxygenases. The terminal steps are the prenylations with or without cyclization. Two distinct Mn^{2+}-dependent prenyltransferases (E4-5 and E4-6) have been described. They catalyze the transfer of a dimethylallyl moiety from dimethyl-allylpyrophosphate (DMAPP) to C-2 or C-4 of the pterocarpan nucleus (Dimethylallyltransferases also catalyze the formation of prenylated coumarins and prenylated dihydroxynaphthoate in anthraquinone biosynthesis. They are well known in isoprenoid biosynthesis as described in chapter 11). Cyclization of the pterocarpan isoprenyl side chains is catalyzed by a pterocarpan cyclase (E4-7) which also appears to be a cytochrome P-450-monooxygenase. It catalyzes the cyclization of both 2- and 4-dimethylallylglycinols.

For most of the enzymes of the pterocarpan/glyceollin pathway, including those of the general phenylpropanoid metabolism, induction by elicitor preparations has been observed. The usual preparations are mycelia or zoospores from the pathogen *Phytophthora megasperma* f. sp. *glycinea*, which causes stem and root rot of soybean.

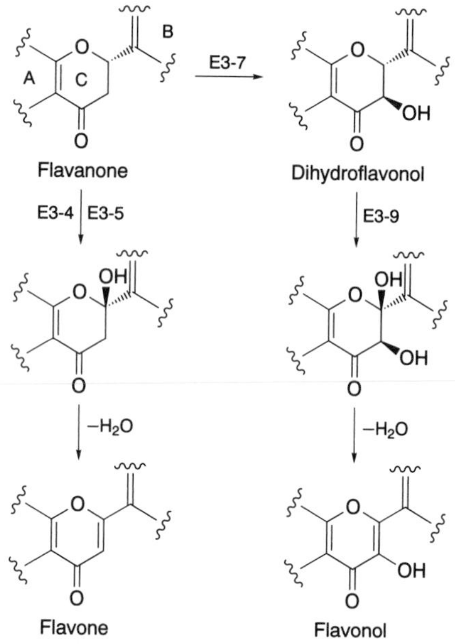

(c)   *Flavones and flavonols*

In the pathways leading to the more common flavonoid classes (Fig. 10.13), a second branch from naringenin leads to the flavones by the introduction of a double bond between C-2 and C-3. Two reaction steps are necessary for this conversion. Although this has not yet been definitely proven, it has been suggested that in the first step 2-hydroxyflavanone is formed (E3-4, E3-5). In the second step, water is eliminated, possibly catalyzed by a dehydratase. Two different flavone synthases have been described; flavone synthase I (e.g. in *Petroselinum crispum*; Apiaceae) is a typical dioxygenase and flavone synthase II (e.g. in *Antirrhinum majus*; Scrophulariaceae) is a monooxygenase.

An analogous reaction sequence leads to the most widespread flavonols. A flavanone

Figure 10.13   Two-step reactions leading from flavanone to flavones and from dihydroflavonol to flavonols, catalyzed by flavone synthases (E3-4, E3-5) and flavonol synthase (E3-9).

3-hydroxylase (E3-7, a soluble dioxygenase) leads to 3-hydroxyflavanone (dihydroflavonol). Then, as already described for flavone formation, a 2-hydroxylase activity (E3-9) is followed by water elimination. The two products are the flavone apigenin and the flavonol kaempferol.

(d)   *Anthocyanins and flavanols*

Dihydroflavonol may enter another pathway leading to the anthocyanins. An NADPH-dependent dihydroflavonol 4-reductase (E3-8) catalyzes the formation of a flavan-3,4-*cis*-diol, a leucoanthocyanidin structure. The steps to the anthocyanidins have not yet been demonstrated *in vitro*. A hypothetical three-step pathway includes a hydroxylation and two dehydrations

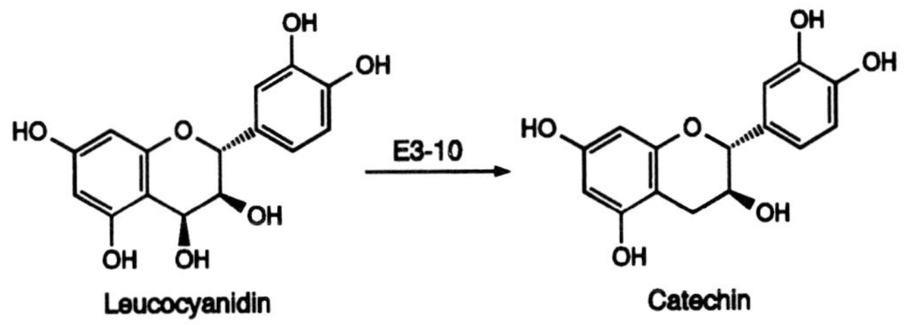

Leucocyanidin → (E3-10) → Catechin

(Fig. 10.14). The produced labile flavylium structure is stabilized by a glucosylation of the 3-OH group giving pelargonidin 3-O-β-glucoside. This is catalyzed by a 3-O-glucosyltransferase (Fig. 10.15). The 4-reductase (E3-8) also catalyzes the reduction of flavanones (dihydroflavones) to flavan-4-ols. They enter reactions analogous to those leading to the anthocyanidins. The products are the rare yellow-colored 3-deoxyanthocyanidins that occur in mosses, ferns and in some grasses and members of the Gesneriaceae and Bignoniaceae.

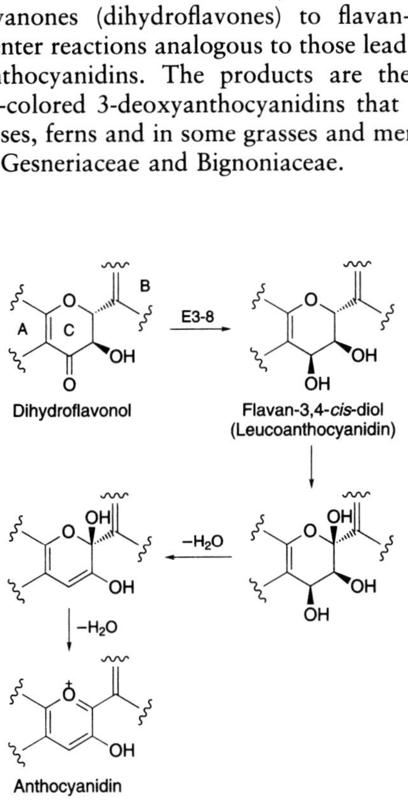

Flavan-3,4-cis-diol is the substrate for another biosynthetic step, leading to flavan-3-ols. This is catalyzed by a NADPH-dependent flavan-3,4-diol 4-reductase (E3-10), which is involved in the biosynthesis of catechin and its relatives, important structural elements of condensed tannins. The reaction leading to catechin itself uses the flavan-3,4-cis-diol leucocyanidin as substrate.

The dihydroflavonol 4-reductase (E3-8) involved in anthocyanin formation, has been the subject of genetic transformation of flower color. The gene from maize (*Zea mays*; Poaceae) encoding dihydroquercetin 4-reductase, an enzyme which also accepts dihydrokaempferol as substrate, has been introduced into a white mutant of petunia (*Petunia hybrida*; Solanaceae). It resulted in the formation of pelargonidin derivatives which gave flowers with brick-pink coloration, a hue not observed in petunia before. This is an outstanding example of the introduction of a new enzymatic reaction into a plant by interspecific gene transfer. However, there are numerous limitations, such as variegation and unstable expression of newly introduced genes, and considerable effort is still required to establish effective modulation of heterologous gene expression in plants.

Figure 10.14 Postulated sequence of reactions leading to the anthocyanidin nucleus.

10.8.3 Substitution of flavonoids

The basic substitution patterns of flavonoids are partly determined by the mode of action and substrate preference of chalcone synthase (5,7-hydroxyl groups of the A-ring and the 4'-hydroxyl group of the B-ring) and on enzymes

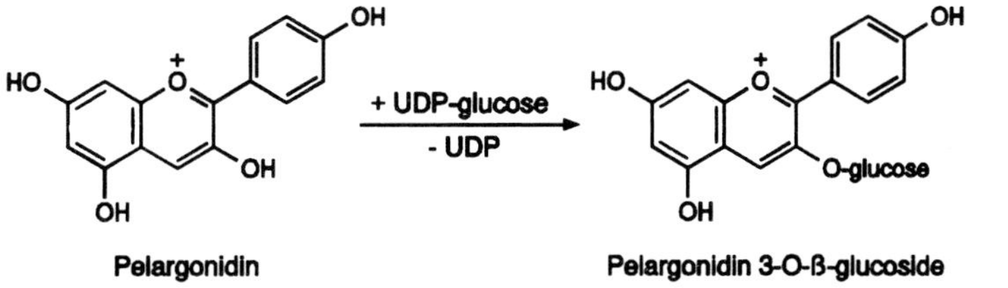

Pelargonidin → (+ UDP-glucose, - UDP) → Pelargonidin 3-O-β-glucoside

Figure 10.15 Glycosylation of an anthocyanidin to the first stable anthocyanin, exemplified with the formation of pelargonidin 3-O-glucoside.

of the flavonoid aglycone pathway (hydroxyl groups of the C-ring). Further hydroxylations as well as methylations of the hydroxyl groups, particularly at the B-ring, are catalyzed by flavonoid-specific hydroxylases and O-methyl-transferases. The 3'- and 5'-hydroxylations are catalyzed by cytochrome P-450-enzymes. Distinct position-specific SAM-dependent O-methyltransferases lead to some common B-ring methoxylated flavonoids. The common substitution patterns of flavonols and anthocyanidins are shown in Table 10.2.

Methyltransferase activities also lead to some rare polymethoxylated flavonoids, such as the flavonol glucosides in *Chrysosplenium americanum* (Saxifragaceae), e.g. a 5-hydroxy-3,6,7,2',4'-penta-methoxyflavone-5'-O-β-glucoside. It has been shown that there are five flavonol-position-specific O-methyltransferases which catalyze the sequential methylation leading to this pentamethyl ether.

Table 10.2 Naturally occurring flavonols and anthocyanidins

		Substitution	
		3'	5'
Kaempferol	Pelargonidin	H	H
Quercetin	Cyanidin	OH	H
Myricetin	Delphinidin	OH	OH
Isorhamnetin	Peonidin	OCH$_3$	H
Larycitrin[a]	Petunidin	OCH$_3$	OH
Syringetin[a]	Malvidin	OCH$_3$	OCH$_3$

[a]Rare flavonols.

10.8.4 Conjugation of flavonoids

Except for some polymethoxylated derivatives, which may occur in lipophilic secretions such as bud exudates, flavonoids usually occur as water-soluble glycosides in vacuoles. There is evidence that these glycosides are synthesized at the endoplasmic reticulum, and then transported in vesicles to the vacuole where the vesicles coalesce with the tonoplast. An earlier convenient hypothesis of a tonoplast-associated flavonoid glycosylation during flavonoid transfer into the vacuole has not been verified.

Sugar conjugations extend from the common simple 3-O-glycosides (monosides, e.g. the common monoglucosides) through the widespread 3-O-diglycosides (biosides) to glycosides, bearing sugars at C-3 of the C-ring as well as at C-5 and C-7 of the A-ring. Flavonoids with sugars attached to the B-ring are also known, but they are rare. There are several hundred different glycosides known, with glucose, galactose, rhamnose, xylose and arabinose as the most frequently found moieties. The two major types of linkages are O- and C-glycosides. For example, rutin (quercetin 3-O-rutinoside) and vitexin (apigenin 8-C-glucoside) both occur in *Fagopyrum esculentum* (Polygonaceae). All known flavonoid glycosyltransferases use UDP-sugars as glycosyl donors.

Flavonoid conjugation is not restricted to glycosylation. Many flavonoids bear acylated sugars. The acyl groups are either hydroxycinnamates or aliphatic acids such as malonate. The respective transferases use CoA-activated acids as acyl donors. Regarding hydroxycinnamate-acylated flavonoids, hydroxycinnamate 1-O-acylglucosides may also serve as donors.

Aromatic acylations of anthocyanins have been shown to stabilize the anthocyanidin structure in vacuoles. Acylation of flavonoids with malonate, that is the most common aliphatic acid moiety, may be responsible for vacuolar trapping of flavonoids, due to changes in the molecular symmetry caused by the acidic environment within the vacuole.

An increasing number of publications report the occurrence of flavonoid sulfate esters. To date more than 60 structures of flavone and flavonol sulfates have been recorded. The enzymes involved are position-specific sulfotransferases which catalyze the transfer of sulfate groups from 3'-phosphoadenosine 5'-phosphosulfate (PAPS) to different hydroxyl groups of flavonoids.

10.8.5 Conjugation of anthocyanidins

The anthocyanin structures require special attention since they show the most complex conjugated structures among the flavonoids. The basic structure is the flavylium cation (oxonium ion), but, depending on the pH, anthocyanins can exist in various structural forms (Fig. 10.16). At pH 4–6, which is the pH range of most plant vacuoles, hydration of the anthocyanidin nucleus leads to the colorless carbinol pseudobase. Thus there must be a mechanism for protection of anthocyanins against hydration *in vivo*. The most efficient protection mechanism is copigmentation: (i) intermolecular copigmentation with other non-covalently bound colorless substances (flavonol or flavone glycosides as well as hydroxycinnamate esters, e.g. chlorogenate); (ii) intramolecular copigmentation with aromatic acyl groups (namely hydroxycinnamates) which are part of the anthocyanin structure (e.g. sandwich-type stacking) (Fig. 10.17).

The intramolecular copigmentation of complex acylated anthocyanin structures shows an *in vivo* conformation with the acyl groups on either side of the anthocyanidin nucleus, thus protecting the flavylium cation from nucleophilic water addition. Such structures exemplify the potential complexity of conjugation (polyglycosylation and polyacylation) in anthocyanins. The 'heavenly blue anthocyanin' from blue petals of *Ipomoea tricolor* is composed of peonidin with six molecules of glucose and three molecules of caffeic acid (Fig. 10.17). The caffeates are esterified with one glucose molecule and glycosylated at one of their aromatic hydroxyl groups with a second glucose molecule. Besides the common acylation with hydroxycinnamates, aliphatic dicarboxylates are also commonly found as acyl moieties, such as malonate, succinate, malate or oxalate in the so-called 'zwitterionic anthocyanins'.

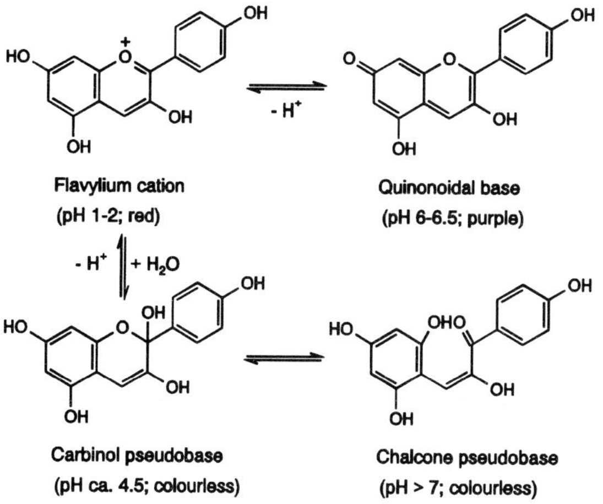

10.8.6 Turnover and degradation of flavonoids

Most flavonoids are true end products and may remain unaltered throughout the lifetime of a plant. However, there is evidence that turnover and degradation of flavonoids may also occur. Among various degradation reactions are glycoside hydrolysis (e.g. action of β-glucosidases), demethylation, hydration and oxidation. Oxidation reactions can also lead to polymerization, catalyzed by peroxidases. One manifestation of this process is the appearance of polymeric 'browning substances' in injured tissues upon pathogen attack or senescence.

Peroxidases are present in a large number of isoforms in different tissue and cell compartments. The substrate specificity of peroxidases is

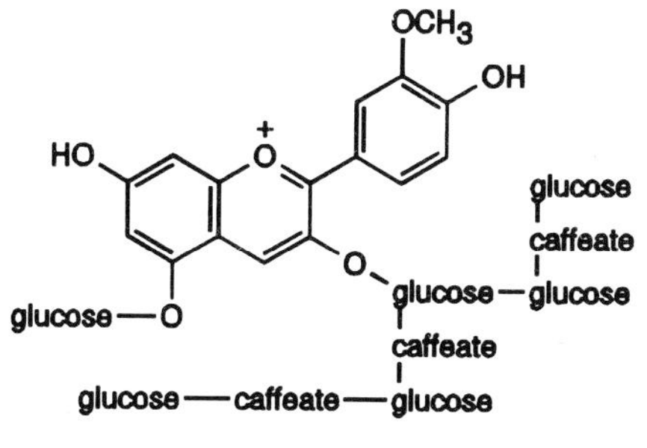

Figure 10.16 Structural transformation of anthocyanins in aqueous solution at various pH values (shown for the anthocyanidin pelargonidin).

Flavylium cation
(pH 1-2; red)

Quinonoidal base
(pH 6-6.5; purple)

− H⁺

+ H₂O

Carbinol pseudobase
(pH ca. 4.5; colourless)

Chalcone pseudobase
(pH > 7; colourless)

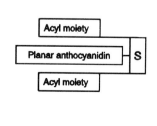

'Heavenly blue anthocyanin'

Intramolecular copigmentation

Figure 10.17 Structure of 'heavenly blue anthocyanin' from blue petals of *Ipomoea tricolor*. The scheme of the intramolecular copigmentation is a generalized structure of such conformations; S = sugar moiety.

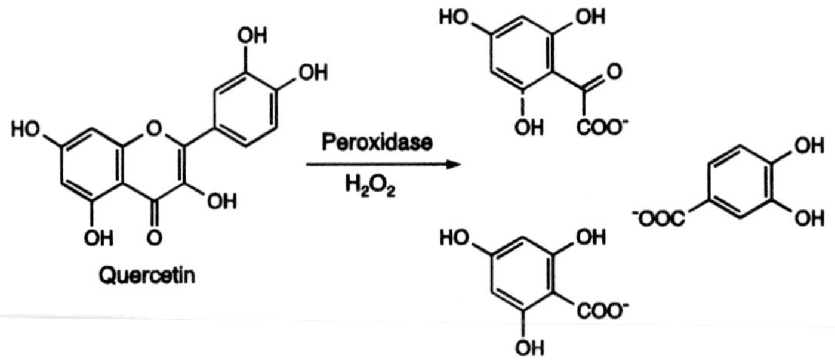

Figure 10.18 Peroxidase-mediated flavonol degradation exemplified with quercetin leading to hydroxybenzoates as initial catabolites (non-stoichiometrical scheme).

relatively low, whereas their specificity for H_2O_2 is high. Flavonoid destruction has also been attributed to peroxidase activity, a process that leads to hydroxybenzoate formation formed from rings A and B of the flavonoid (Fig. 10.18). However, the overall importance of flavonoid degradation, and the means by which it is regulated, are unknown.

An obvious example of flavonoid disappearance is seen with the anthocyanins. Anthocyanins may accumulate transiently during plant ontogeny, for instance in seedlings or young leaves, suggesting some, yet, speculative physiological functions (e.g. light perception or light filtration). Besides true degradation, anthocyanin disappearance could involve conversion to colorless forms. In general, one should be aware of the disappearance of soluble flavonoids through conversion into insoluble forms. It is known that in needles of members of the Pinaceae water-soluble flavonol glycosides can become attached to cell wall components. The transport mechanism and the exact location are unknown.

10.9 LIGNINS

Lignins constitute a group of heterogeneous phenylpropane polymers in plants, where they form important constituents of cell walls within supporting and conducting tissues (e.g. xylem cells). These tissues enable vascular plants to develop into large upright structures, such as trees. Lignin is always associated with wall polysaccharides (cellulose, hemicellulose). Cell walls from sapwood and heartwood of woody species have similar compositions of the major components of cellulose–hemicellulose–lignin–pectic substances in the ratio of about 4 : 3 : 1 : 0.1.

The following scheme shows a hypothetical section of a growing lignin polymer with its possible attachment to wall polysaccharides.

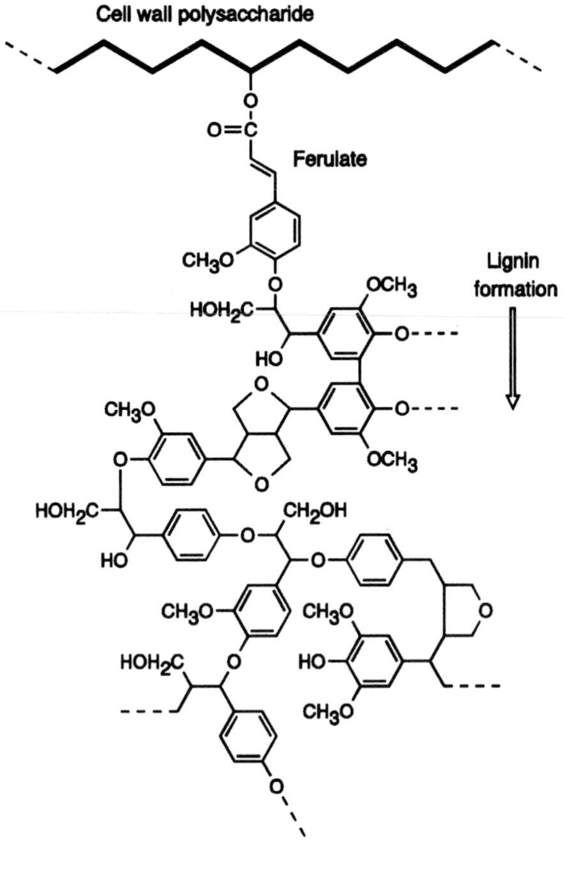

The monomeric constituents of lignins are three hydroxycinnamyl alcohols (monolignols). These are 4-coumaryl, coniferyl and sinapyl alcohols. Caffeyl alcohol has not been found as a lignin constituent. Whereas gymnosperm lignins (guaiacyl lignins) are typically derived mainly from coniferyl alcohol together with small amounts of 4-coumaryl alcohol, angiosperm lignins (guaiacyl-syringyl lignins) generally contain all three alcohols. In early stages of lignification, 4-coumaryl alcohol is deposited initially, followed by coniferyl alcohol and, in angiosperms, finally by sinapyl alcohol.

Lignin is a characteristic component of secondary cell walls. Its deposition starts towards the end of primary cell growth at the cell corners and the middle lamella. It is postulated that the initial steps might be attachments (ester bonds) of hydroxycinnamates to specific sites in the wall polysaccharides. Then peroxidase-mediated couplings of monolignols to these hydroxycinnamates may form the links between the growing lignin polymers and the polysaccharides. There are five important enzymes known that catalyze the formation of monolignol radicals which then polymerize to yield lignin structures (Enzyme Box 5).

Fig. 10.19 illustrates the formation of coniferyl alcohol and its dimer, the lignan pinoresinol, including initiation of lignin formation. The key reactions leading to the lignin monomers, the monolignols, are reductions catalyzed by two NADPH-dependent oxidoreductases, the hydroxycinnamoyl-CoA reductase (E5-1) and monolignol dehydrogenase (E5-2). The monolignols produced are thought to be glucosylated (4-O-β-glucosides; E5-3) to give transportable forms which are secreted into the lignifying cell wall. Here the glucose is removed by a β-glucosidase (E5–4) and the liberated monolignol enters the lignin pathway. Formation of lignin structures from monolignols is initiated by peroxidase activity (E5-5) followed by a chemically driven dehydrogenative polymerization. Interconversion of the mesomeric free radicals leads to random coupling, followed by intramolecular reaction of a quinone methide-type dimer. Further intermolecular reactions with other phenolics and polysaccharides in the wall form stable cross-links.

An interesting phenomenon is the fact that in some lignifying tissues, such as the monolignol-rich bark tissue of the Fabaceae, only the Z-isomers of the monolignols were detected (compare E/Z-isomerization of hydroxycinnamates), e.g. Z-coniferyl and Z-sinapyl alcohols in *Fagus grandifolia* (Fagaceae). Their role in lignification and the

Z-Coniferyl alcohol   Z-Sinapyl alcohol

Futoquinol

possible involvement of an E/Z-monolignol isomerase has yet to be documented.

10.10 LIGNANS AND NEOLIGNANS

Lignans are generally defined as phenylpropane dimers. The phenylpropane units are C-C linked, mostly through their C_3-side chains (tail-to-tail such as pinoresinol). Neolignans, such as futoquinol, are phenylpropane dimers that are linked head-to-tail.

Although lignans and neolignans occur as lignin substructures, there is no conclusive evidence that they serve as direct precursors of lignins. Instead, they represent a distinct class of phenolics, rather than being merely lignin precursors. They are widely distributed in plants accumulating as soluble components, some of them as glycosides, and there is increasing knowledge about their biological activities. There are examples about their antimicrobial, antifungal and antifeedant properties. For example, the neolignan futoquinol from *Piper chamaejasme* (Piperaceae) serves as an insect antifeedant. Furthermore, since many known lignans are optically active, at least their biosynthesis is stereoselective and cannot be the products of a typical H_2O_2-requiring peroxidase activity. The latter leads to racemic products. This point awaits further investigation. We also need to know more

---

ENZYME BOX 5
MONOLIGNOL RADICAL FORMATION

E5-1:   HCA-CoA reductase; i.e. HCA : NADP$^+$ oxidoreductase
E5-2:   Monolignol dehydrogenase; i.e. monolignol : NADP$^+$ oxidoreductase
E5-3:   Monolignol glucosyltransferase; i.e. UDP-glucose : monolignol 4-O-glucosyltransferase
E5-4:   Monolignol glucoside ß-glucosidase; i.e. monolignol glucoside ß-glucohydrolase
E5-5:   Peroxidase; i.e. monolignol : H$_2$O$_2$ oxidoreductase

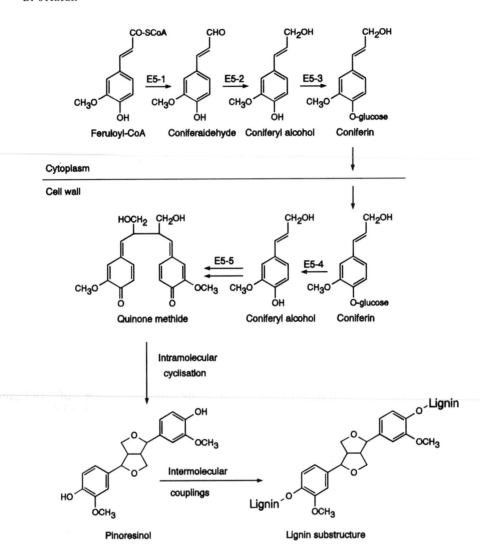

Figure 10.19 An example of early stages of lignin formation including biosynthesis of coniferyl alcohol, transport of its 4-O-glucoside (coniferin) into the cell wall, and initiation of lignin formation.

about the catalytic mechanisms of the lignan- and neolignan-specific enzymes and their locations.

10.11 TANNINS

Water-soluble plant polyphenols which cause protein precipitation from aqueous solutions are called tannins. They are oligo- and polymeric phenolics that are divided into two classes: hydrolyzable and condensed (nonhydrolyzable) tannins. The latter are also named proanthocyanidins due to their behavior to give anthocyanidins in the presence of acids. In contrast to the polyphenolic polymer lignin, both tannin groups are located in vacuoles.

10.11.1 Hydrolyzable tannins

The typical structure of the hydrolyzable tannins, the gallotannins and the ellagitannins, is characterized by a central polyhydroxyl moiety that is usually β-D-glucopyranose. Hamamelose, shikimate, quinate or cyclitols have also been found. The hydroxyl groups are most often esterified with gallate, but hydroxycinnamates have also been observed. The gallotannins are gallate polyesters. An important gallotannin component is pentagalloylglucose, which reacts further with gallate molecules to give *meta*-depside bonds. A galloyltannin that is a heptagalloyltannin with *meta*-depside bonds has been isolated, among other gallate esters, from the tannins of *Rhus semialata*

Heptagalloylglucose (2,3-di-O-digalloyl-1,4,6-tri-O-galloylglucose)

(Anacardiaceae), and indeed, in this plant total substitution can give 10–12 galloyl residues per glucose molecule. These structures have also been reported for Aleppo galls (*Quercus infectoria*; Fagaceae) or for *Paeonia albiflora* (Paeoniaceae).

Gallate

Glc-T | UDP-Glc

1-Galloylglucose

Gall-T1 | 1-Galloyl-Glc

1,6-Digalloylglucose

Gall-T2 | 1-Galloyl-Glc

1,2,6,-Trigalloylglucose

Gall-T3 | 1-Galloyl-Glc

1,2,3,6-Tetragalloylglucose

Gall-T4 | 1-Galloyl-Glc

1,2,3,4,6-Pentagalloylglucose

– n [H] / n x 1-Galloyl-Glc

Ellagitannins   **Gallotannins**

Figure 10.20 Outline of the gallotannin pathway leading to the structure subclasses gallotannins and ellagitannins. Gallotannins are formed by galloyltransfer from 1-O-galloylglucose to the galloyl residues of the central glucose leading to *meta*-depside bonds (Glc = glucose). Ellagitannins are formed by oxidative C-C coupling of adjacent galloyl residues.

The biosynthesis of polygalloylglucoses, including the formation of intergalloyl depside bonds or oxidation (dehydrogenation) to ellagitannin structures, is outlined in Fig. 10.20. This pathway starts by the formation of 1-O-galloyl-β-D-glucose from gallate and UDP-glucose, catalyzed by UDP-glucose:gallate O-glucosyltransferase (Glc-T). The following four reaction steps leading to 1,2,3,4,6-pentagalloylglucose are catalyzed by galloyltransferases that are dependent on 1-O-galloylglucoses as acyl donors (enzymes Gall-T1 to Gall-T4). Further galloylations (depside formation) leading to the gallotannins are also catalyzed by such 1-galloylglucose-dependent acyltransferases. Ellagitannins arise from secondary C-C linkages between adjacent galloyl groups. Ellagitannins may combine further to yield oligomeric derivatives coupled through C-O-C linkages.

10.11.2 Condensed tannins

Flavonoids may bind to each other to form dimers (biflavonoids), as exemplified by amentoflavone, found in most gymnosperms. The interflavonoid linkage is usually between C-5′ of the B-ring and C-8 of the A-ring. The carbonyl group at C-4 of the pyran ring is always free. These biflavonoids are rather lipophilic and are located outside the living cell (i.e. on the cell wall surface).

The interflavonoid linkage of the oligo- and polymeric flavonoids such as the condensed tannins (proanthocyanidins), on the other hand, always includes bonds to C-4. These products

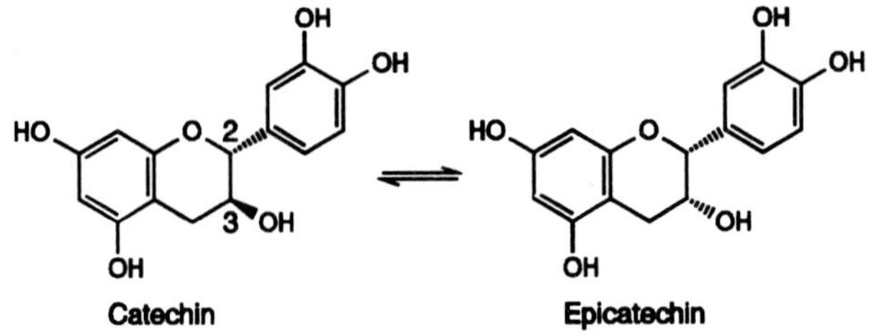

Catechin   Epicatechin

accumulate in the vacuoles and have a much higher molecular weight than the gallotannins (M_r between 2000 and 7000). The monomeric flavonoid subgroups involved in proanthocyanidin

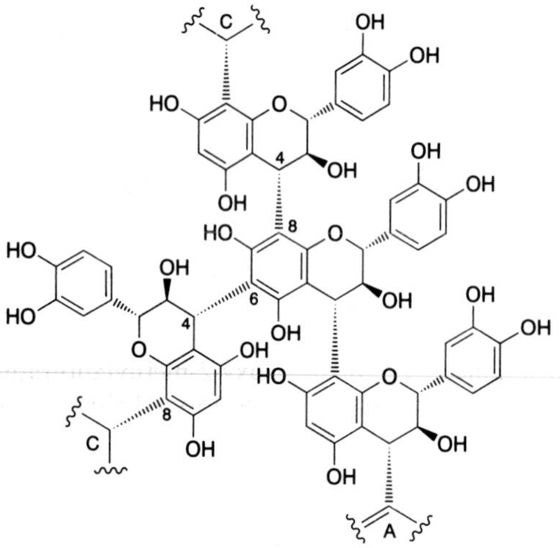

Figure 10.21 Substructure of procyanidin with flavan-3-ol 2,3-*trans* units showing the typical inter-flavan C-4–C-8 bond of the linear backbone with a C-4–C-6 branch.

biosyntheses are 3,4-diols, although 4-ols are also frequently found. They form polymerized flavan-3-ols and flavans, respectively. The flavan-3-ol units may occur in four isomeric structures, although only two of them, the 2,3-*trans* (2R, 3S)- and the 2,3-*cis* (2R, 3R)-isomers, are commonly found in nature. The structures of these isomers are exemplified with catechin (*trans*) and epicatechin (*cis*) (see above).

The common C-C bond in dimeric proanthocyanidins is between C-4 and C-8, which gives linear structures. Bonds between C-4 and C-6 are also possible, and these result in branched globular structures. Figure 10.21 shows a substructure of a typical proanthocyanidin with the C-4–C-8 inter-flavan bond and a C-4–C-6 branch. Such a compound is called procyanidin if it yields cyanidin when heated in the presence of acid.

The key enzymatic steps in proanthocyanidin biosynthesis are unknown. Only the reaction leading from the flavan-3,4-*cis*-diols (leucoanthocyanidins) to the flavan-3-ols, which is catalyzed by a stereospecific reductase, has been thoroughly described. Thus catechin itself, which is known to accumulate among other related structures, is formed through the 2,3-*trans* pathway. The

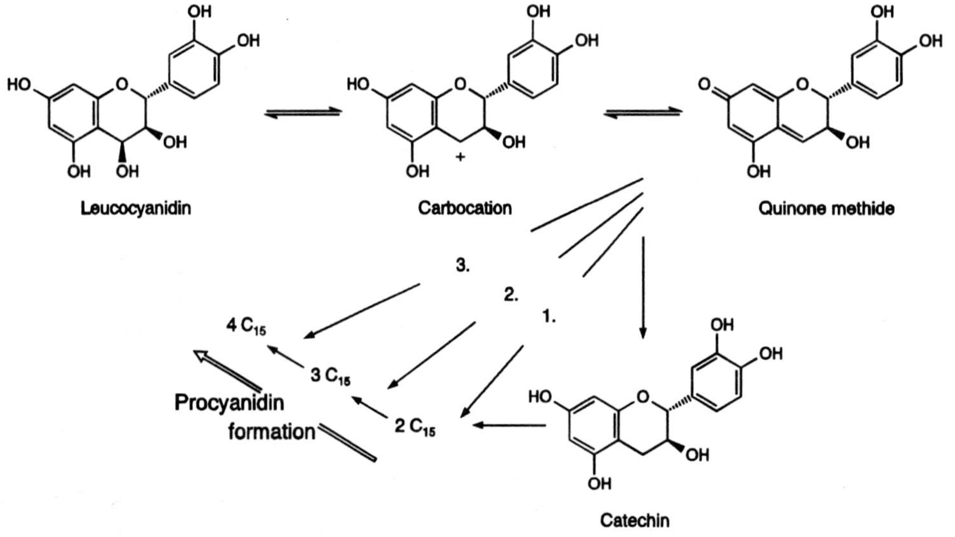

Figure 10.22 Postulated pathway leading to proanthocyanidins, exemplified with procyanidin.

analogous enzyme, using the 2,3-*cis* pathway to give epicatechin, is still unknown.

The proanthocyanidins may be formed as by-products of the reactions leading to the flavan-3-ols. It is thought that there is a stereospecific capture of the intermediate carbocations or their quinone methides by the end products flavan-3-ols (Fig. 10.22). The resulting oligomeric structure grows by further additions of these intermediates. This is most likely a highly co-ordinated process, but virtually nothing is known about the enzymes involved.

10.12 QUINONES

The quinones commonly occurring in nature are benzo-, naphtho- and anthraquinones. Although they derive from different biosynthetic routes their structures contain phenolic moieties and they will therefore be dealt with here. Most of the biosynthetic sequences in higher plants have been established from isotope incorporation studies and still await enzymatic proof. However, some of the key enzymes leading to the naphthoate intermediate are well described from bacterial sources. Recent advances of plant cell culture techniques have had an important impact on current studies about quinone biosynthesis.

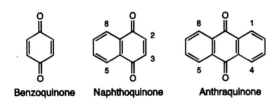

Benzoquinone   Naphthoquinone   Anthraquinone

Benzoquinone isoprenoid derivatives such as plastoquinone and ubiquinone (lipoquinones) play important roles in primary metabolism (see Chapter 2). Secondary benzoquinones are typical constituents in fungi, mainly in Hyphomycetes and Basidiomycetes, but they are rare in higher plants. An example of their occurrence in higher plants is the reduced benzoquinone conjugate, hydroquinone glucoside (arbutin), that regularly occurs in Rosaceae and Ericaceae. Substituted hydroquinones, such as methoxylated structures (2,6-dimethoxyquinone) occur in wheat grains and seedlings. Hydroquinones are formed by oxidative decarboxylation of 4-hydroxybenzoate derivatives that are derived most likely from hydroxycinnamates.

Many naphtho- and anthraquinones occur in higher plants. Naphthoquinones are often responsible for the pigmentation of colored heartwood

and bark, e.g. shikonin in the root bark of *Lithospermum erythrorhizon* (Boraginaceae). An example of their presence in living cells is juglone glucoside, an allelopathic agent in leaves and roots of members of the Juglandaceae. The sporadic occurrence of naphthoquinones is known from about 20 families of higher plants, including besides the two mentioned the Ebenaceae, Droseraceae, Balsaminaceae, Plumbaginaceae and Bignoniaceae.

Anthraquinones, of which more than 200 are known, occur almost as frequently in bacteria, fungi and lichen as in members of higher plant families. Important anthraquinone-bearing plant families are the Caesalpiniaceae, Polygonaceae, Rhamnaceae and Rubiaceae. There are at least two structural types of anthraquinones, one bearing only hydroxyl groups in the third ring C, whereas the other is hydroxylated in both aromatic rings A and C. The former pattern normally reflects biosynthesis by the succinylbenzoate pathway, while the latter pattern is typical of biosynthesis by the acetate/malonate pathway (polyketide pathway). However, one should be aware of exceptions to this rule, e.g. morindone (1,5,6-trihydroxy-2-methylanthraquinone) has been proven to be formed through the succinylbenzoate pathway. The hydroxyl groups of ring A of morindone are in fact introduced at a late stage of the biosynthetic pathway.

Of the different pathways leading to quinonoid structures, the polyketide pathway is found mostly in micro-organisms but also exists in some higher plants; for example it is widespread in members of the Rhamnaceae and Polygonaceae. The second pathway, proceeding through succinylbenzoate, is widely distributed in higher plants. The occurrence of both pathways in plants is exemplified by the following: the naphthoquinone plumbagin from *Plumbago europaeus* (Plumbaginaceae) and the anthraquinone emodin from *Rhamnus frangula* (Rhamnaceae) are formed via the polyketide (acetate/malonate) pathway, but the naphthoquinone lawsone from *Impatiens balsamina* (Balsaminaceae) and the anthraquinone alizarin from *Rubia tinctoria* (Rubiaceae) are derived from the succinylbenzoate pathway.

Acetate/malonate derived quinones:

Plumbagin   Emodin

Succinylbenzoate-derived quinones:

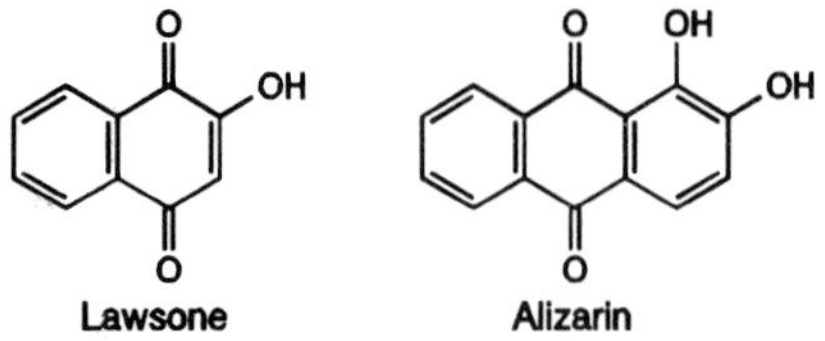

Lawsone   Alizarin

Like the benzoquinones, some naphthoquinones may also be formed from 4-hydroxybenzoate. This has been demonstrated in shikonin biosynthesis (Fig. 10.23) where the second ring of the shikonin structure derives from isoprenoid metabolism. The key enzyme is a specific geranylpyrophosphate (GPP)-dependent geranyltransferase that catalyzes the formation of geranylhydroquinone. The reaction steps leading through cyclization to the naphthoquinone nucleus await investigation.

Shikonin is widely used in countries of Eastern Asia because of its wound-healing activities. Because of its intense red color it is also used as an important dyestuff for fabrics and cosmetics. Cell cultures of *Lithospermum erythrorhizon* are capable of producing large amounts of shikonin, and in Japan, a biotechnological process with high-yielding culture strains has been developed for the industrial production of this pigment. This is a prominent example of a successful biotechnological process using plant cell cultures.

10.12.1 Polyketide pathway

Acetyl-CoA is the starting molecule for most polyketides. Linear Claisen condensation with several acetyl residues derived from malonyl-CoA leads, with concomitant loss of CO_2 to the polyketide (acetogenin) structures $[-(CH_2\text{-}CO)_n\text{-}]$. Direct condensation (without reduction) and cyclization give various aromatic structures (compare biosynthesis of flavonoid A-ring). The aromatic polyketide biosynthesis has to be distinguished from the fatty acid biosynthetic pathway, where the polyketides undergo reduction and dehydration to form aliphatic hydrocarbons (see Chapter 6). The

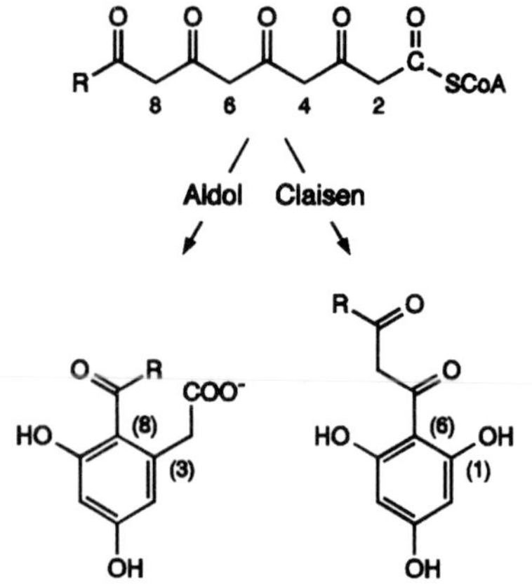

Figure 10.24 Two examples of polyketide cyclization by aldol condensation (e.g. C-3 to C-8) or Claisen condensation (e.g. C-1 to C-6).

enzyme-bound postulated polyketide structures are very reactive, and must be temporarily stabilized by hydrogen bonding or chelation on the enzyme surface. Different cyclization mechanisms are probably guided by the topology of the different enzymes that catalyze the formation of different phenolics (Fig. 10.24). The aromatic ring systems are either formed by aldol condensation, i.e. reaction of a carbonyl group with an CH_2-group, or Claisen condensation, i.e. reaction of the thioester group with an CH_2-group.

A typical folding mechanism involving aldol condensations of a hypothetical polyketomethylene compound can lead to the anthraquinone chrysophanol in *Rhamnus frangula* (Rhamnaceae).

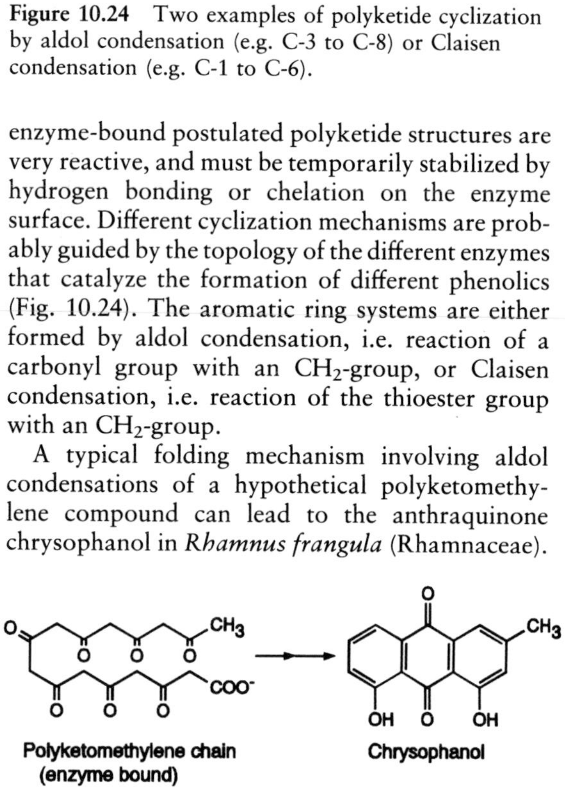

Polyketomethylene chain   Chrysophanol
(enzyme bound)

The variation among polyketide-derived phenolic structures derives from (i) the chain initiating unit, e.g. use of propionyl-CoA, butyryl-CoA,

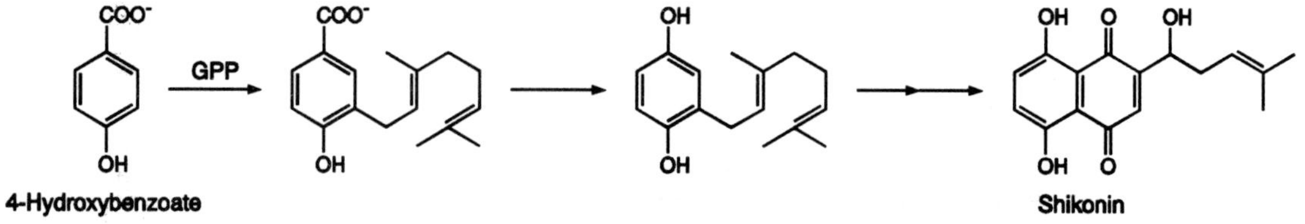

4-Hydroxybenzoate   Shikonin

Figure 10.23 Formation of shikonin.

malonyl-CoA or hydroxycinnamoyl-CoA (e.g. 4-coumaroyl-CoA in flavonoid biosynthesis) instead of acetyl-CoA, (ii) the number of malonyl-CoAs involved, (iii) further modification by oxidation or reduction or (iv) introduction of substituents and conjugations (e.g. O- and C-glycosylations).

10.12.2 Succinylbenzoate pathway

There is another widely occurring pathway leading to quinones. This is a pathway of mixed origin with 2-succinylbenzoate as the key precursor. Through this pathway, several naphthoquinones (e.g. juglone) and anthraquinones (e.g. alizarin which is confined to plants of the family Rubiaceae), are formed (Fig. 10.25). The biosynthetic pathways leading to these products are extensions of the shikimate/arogenate pathway, branching from chorismate, which is the precursor of isochorismate. The formation of this isomer is catalyzed by the enzyme isochorismate hydroxymutase. Isochorismate in turn is converted into 2-succinylbenzoate in the presence of 2-oxoglutarate and thiamine pyrophosphate in a Michaelis-type reaction. This reaction sequence constitutes a new and hitherto unprecedented aromatization process.

Succinylbenzoate is subsequently activated at the succinyl residue to give a mono-CoA ester. This activation requires ATP and yields AMP in bacteria, but ADP in higher plants. The initial steps in the succinylbenzoate pathway have been partly characterized at the enzyme level.

Ring closure of the CoA-ester to give 1,4-dihydroxy-2-naphthoate is the next step in bacteria but may not be operative in higher plants. Most of the metabolic steps beyond the CoA-ester remain unknown in higher plants, except, in juglone biosynthesis, 1,4-naphthoquinone is known to be an intermediate metabolite. It is also likely that the third ring in alizarin biosynthesis is generated from dimethylallylpyrophosphate (DMAPP).

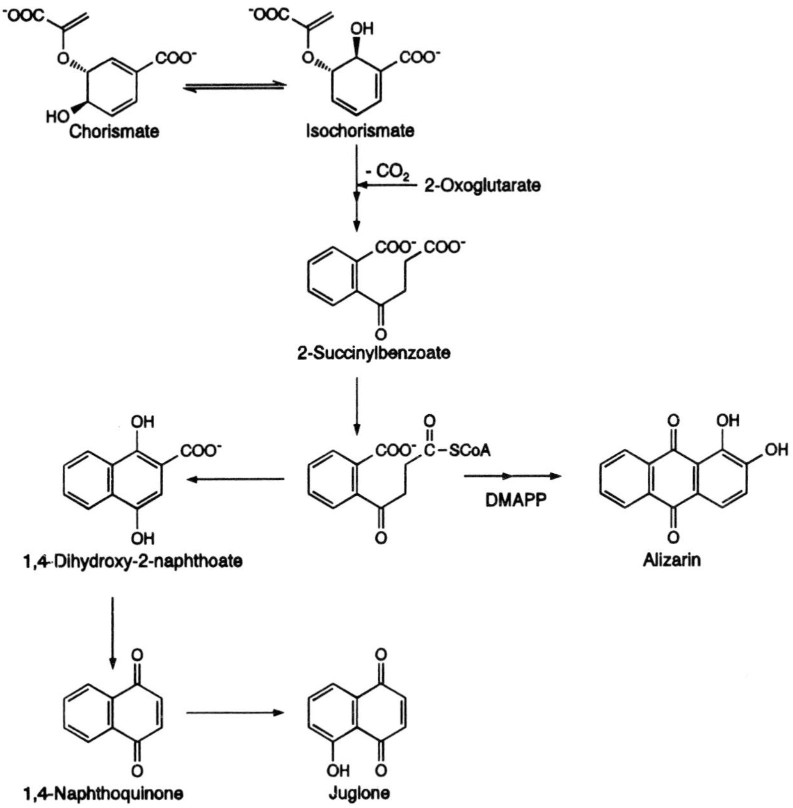

Figure 10.25 Postulated biosynthesis of juglone and alizarin.

FURTHER READING

Anderson, J.W. & Beardall, J. (1991) *Molecular Activities of Plant Cells. An Introduction to Plant Biochemistry*. Blackwell Scientific Publications, Oxford.

Bell, E.A. & Charlwood, B.V. (eds) (1980) *Encyclopedia of Plant Physiology*. New Series, Vol. 8, Secondary Plant Products. Springer-Verlag, Berlin, Heidelberg.

Conn, E.E. (ed.) (1981) In *The Biochemistry of Plants. A Comprehensive Treatise* (P.K. Stumpf & E.E. Conn, eds), Vol. 7, Secondary Plant Products. Academic Press, London, New York.

Constabel, F. & Vasil, I.K. (eds) (1988) *Cell Culture and Somatic Cell Genetics of Plants* (I.K. Vasil, ed.), Vol. 5, Phytochemicals in Plant Cell Cultures. Academic Press, London.

Dewick, P.M. (1984–1990) The shikimate pathway. Series of reviews in *Natural Product Reports*.

Duke, S.O. & Hoagland, R.E. (1985) Effects of glyphosfate on metabolism of phenolic compounds. In *The Herbicide Glyphosate* (E. Grossbard & D. Atkinson, eds), pp. 75–91. Butterworths, London.

Forkmann, G. (1991) Flavonoids as flower pigments: the formation of the natural spectrum and its extension by genetic engineering. *Plant Breeding* 106, 1–26.

Goodwin, T.W. (ed.) (1988) *Plant Pigments*. Academic Press, London, San Diego.

Goodwin, T.W. & Mercer, E.I. (1990) *Introduction to Plant Biochemistry*, 2nd edn. Pergamon Press, Oxford.

Harborne, J.B. (1988) *Introduction to Ecological Biochemistry*, 3rd edn. Academic Press, London, San Diego.

Harborne, J.B. (1989) Recent advances in chemical ecology. *Nat. Prod. Rep.* 6, 85–109.

Harborne, J.B. ed. (1989) *Methods in Plant Biochemistry* (P.M. Dey & J.B. Harborne, eds), Vol. 1, Plant Phenolics. Academic Press, London.

Haslam, E. (1974) *The Shikimate Pathway*. Butterworth, London.

Hiroyuki, I. & Leistner, E. (1988) Biochemistry of quinones. In *The Chemistry of Quinonoid Compounds*, Vol. II (S. Patai & Z. Rappoport, eds), pp. 1293–1349. Wiley, Chichester.

Hrazdina, G. & Jensen, R.A. (1990) Multiple parallel pathways in plant aromatic metabolism? In *Structural and Organizational Aspects of Metabolic Regulation*

(P.A. Srere, M.E. Jones & C.K. Mathews, eds), pp. 27–42. Wiley-Liss, New York.

International Union of Biochemistry and Molecular Biology (IU BMB), (1992) *Enzyme Nomenclature*. Academic Press, London.

Jensen, R.A. (1985) The shikimate/arogenate pathway: Link between carbohydrate metabolism and secondary metabolism. *Physiol. Plant.* 66, 164–168.

Lea, P.J. (ed.) (1992) *Methods in Plant Biochemistry* (P.M. Dey & J.B. Harborne, eds), Vol. 9, Enzymes of Secondary Metabolism. Academic Press, London.

Lewis, N.G. & Paice, M.G. (eds) (1989) *Plant Cell Wall Polymers. Biogenesis and Biodegradation*. ACS Symposium Series 399. American Chemical Society, Washington, DC.

Luckner, M. (1990) *Secondary Metabolism in Microorganisms, Plants, and Animals*, 3rd edn. Springer-Verlag, Berlin, Heidelberg, New York.

Matern, U., Strasser, H., Wendorff, H. & Hamerski, D. (1988) Coumarins and furanocoumarins. In *Cell Culture and Somatic Cell Genetics of Plants*, Vol. 5, Phytochemicals in Plant Cell Cultures (F. Constabel & I.K. Vasil, eds), pp. 3–21. Academic Press, New York.

Molgaard, P. & Ravn, H. (1988) Evolutionary aspects of caffeoyl ester distribution in dicotyledons. *Phytochemistry* 27, 2411–2421.

Stafford, H.A. (1990) *Flavonoid Metabolism*. CRC Press, Boca Raton.

Stafford, H.A. & Ibrahim, R.K. (eds) (1992) *Recent Advances in Phytochemistry*, Vol. 26, Phenolic Metabolism in Plants. Plenum Press, New York.

Swain, T., Harborne, J.B. & Van Sumere, C.F. (eds) (1979) *Recent Advances in Phytochemistry*, Vol. 12, Biochemistry of Plant Phenolics. Plenum Press, New York.

Torssell, K.B.G. (1983) *Natural Product Chemistry. A Mechanistic and Biosynthetic Approach to Secondary Metabolism*. Wiley, Chichester.

Van Sumere, C.F. & Lea, P.J. (eds) (1985) *Ann. Proc. Phytochem. Soc. Europe*, Vol. 25, The Biochemistry of Plant Phenolics. Clarendon Press, Oxford.

Williams, D.H., Stone, M.J., Hauck, P.R. & Rahman, S.K. (1989) Why are secondary metabolites (natural products) biosynthesized? *J. Nat. Prod.* 52, 1189–1208.

11 Isoprenoid Metabolism

Peter M. Bramley

11.1 INTRODUCTION

Plants contain an enormous range of isoprenoid compounds with a wide variety of structures and functions. Some are primary metabolites, e.g. the steroids, but the majority are synthesized as secondary metabolites that are uniquely plant products. The high levels of some of these in turpentine oil gave rise to the alternate generic name 'terpenoid' and the two terms are used interchangeably in the literature. The term isopentenoid has also been used to describe the group (Nes & MacLean, 1977).

Isoprenoids have been known since antiquity as ingredients of perfumes, soaps, flavorings and as food colorants. Consequently, they have been studied since the beginning of modern chemistry in the 1800s. It was not until after 1945, however, that it was appreciated that the individual isoprenoid classes were related biosynthetically. The development of chromatographic and spectroscopic techniques has led to a general understanding of the structure, biosynthesis and properties of terpenoids. New terpenoids are being found continually, as testified by reports in periodicals such as *Phytochemistry* and in regular series published by the British Royal Institute of Chemistry, entitled Specialist Periodical Reports on Terpenoids and Sterols.

Valiant attempts to publish the structures and properties of all the terpenoids have proved to be impractical due to the sheer number of molecules involved. The latest collation, a three-volume Dictionary of Terpenoids (Connolly & Hill, 1991) contains some 22 000 structures, but excludes the sterols that can be found in the *Dictionary of Steroids* (Hill *et al.*, 1991).

It is beyond the scope of this chapter to cover all aspects of the chemistry and biochemistry of plant terpenoids as they are widely distributed in all 94 orders of flowering plants. Consequently, only representative structures and the salient features of the biosynthesis of each class of isoprenoid will be described. Similarly, details of techniques for the isolation and characterization of the terpenoids are not included; they can be found, for example, in a comprehensive text edited by Charlwood & Banthorpe (1991).

11.2 NOMENCLATURE, CLASSIFICATION AND OCCURRENCE

The isoprenoids are built up of C_5, isoprene units and the nomenclature of the main classes reflects the number of isoprenoid units present (Table 11.1). In 1987 Wallach proposed the 'isoprene rule',

Table 11.1   Main classes of isoprenoids found in plants

Carbon atoms	Name	Parent isoprenoid	Subclasses
10	Monoterpenoids	GPP	Iridoids
15	Sesquiterpenoids	FPP	Abscisic acid, sesquiterpenoid lactones
20	Diterpenoids	GGPP	Gibberellins
25	Sesterterpenoids	GFPP	None
30	Triterpenoids	Squalene	Phytosterols, saponins, cardenolides
40	Tetraterpenoids	Phytoene	None
>40	Polyprenols, rubbers	GGPP + $(C_5)_n$	None

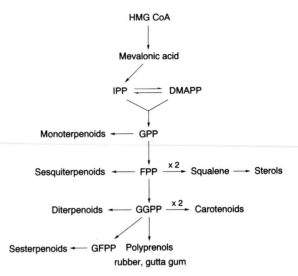

Figure 11.1   General pathway of isoprenoid biosynthesis. Abbreviations are given in the text.

which was later extended to the 'biogenetic isoprene rule' by Ruzicka, who suggested that all isoprenoids are synthesized from a biological 'active isoprene' (Ruzicka *et al.*, 1953; Ruzicka, 1959). This rule was based on only fragmentary knowledge of the pathways, but has turned out to be largely correct in its conception. It stated that each class of isoprenoid was derived from a single parent compound unique to that class (listed in Table 11.1). Terpenoids are thus seen as being formed by various cyclizations, rearrangements and even the loss or addition of carbon atoms from linear arrangements of isoprene units. It can, however, be confusing since some classes have more than one name for the carbon skeleton and several numbering systems are in use. In this chapter, the most accepted numbering system for the skeletal types will be used.

11.3   GENERAL PATHWAY OF TERPENOID BIOSYNTHESIS

The pathway of isoprenoid biosynthesis (Fig. 11.1) was originally elucidated from studies on sterols in animals and yeast. It is virtually only in the Plant Kingdom that the side branches occur to yield the terpenoids characteristic of a particular species. The key to the pathway was the discovery of the C_6 compound mevalonic acid (MVA) and the later elucidation of the metabolic pathway from acetyl-CoA, through MVA (Fig. 11.2), to the biological 'active isoprene' isopentenyl pyrophosphate (IPP), which in turn is isomerized to dimethylallyl pyrophosphate (DMAPP; Fig. 11.3).

The formation of MVA is catalyzed by hydroxymethylglutaryl-CoA (HMG-CoA) reductase

(HMGR; EC 1.1.1.34), an enzyme which has been extensively studied in animals due to its role in the regulation of cholesterol formation (Preiss, 1985;

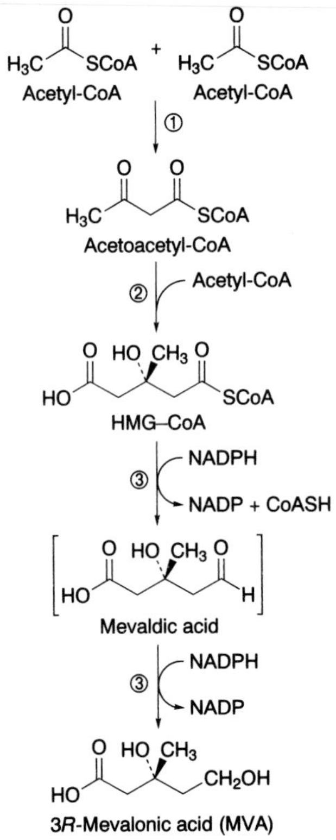

Figure 11.2   Conversion of acetyl CoA into mevalonic acid. Enzymes: 1, acetoacetyl CoA thiolase; 2, HMG-CoA synthase; 3, HMG-CoA reductase. Abbreviations are given in the text.

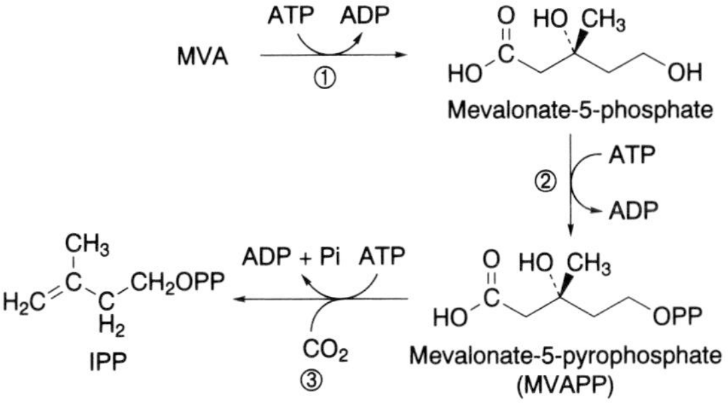

Figure 11.3 Metabolism of mevalonic acid into isopentenyl pyrophosphate. Enzymes: 1, MVA kinase; 2, mevalonate-5-phosphate kinase; 3, mevalonate-5-diphosphate decarboxylase. Abbreviations are given in the text.

Brown & Goldstein, 1986). The *in vitro* conversion of HMG-CoA into MVA has been achieved by various membrane fractions from higher and lower plants (listed by Bach *et al.*, 1991). The data available on plant HMGR suggest that it is present in the plastids and in mitochondria, but this is still a matter of debate (see section 11.11). The reaction (Fig. 11.2) involves two molecules of NADPH per MVA produced, via mevaldic acid (Sabine, 1983). The plant enzyme has two membrane-spanning domains, as predicted from sequences of cloned HMGR cDNAs (Narita & Gruissem, 1989;

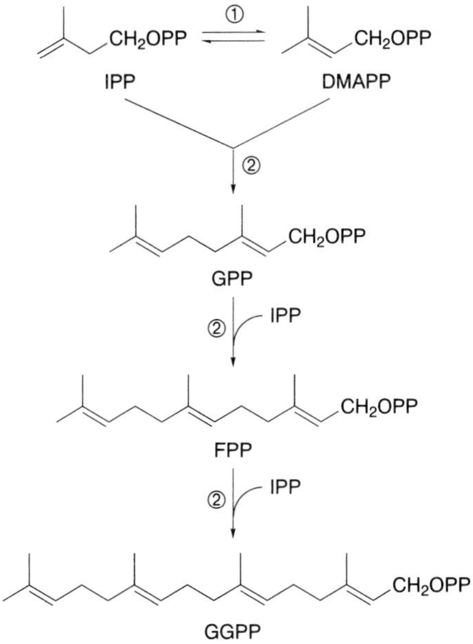

Figure 11.4 Isomerization of isopentenyl pyrophosphate and prenyl transferase reactions. Enzyme: 1, IPP isomerase; 2, prenyl transferase. Abbreviations are given in the text.

Learned & Fink, 1989; Caelles *et al.*, 1989; Wettstein *et al.*, 1989), with considerable conservation at the C-terminal region. Details of the protein sequences are given in the review of Bach *et al.* (1991).

The conversion of MVA into IPP occurs in three steps, each requiring one mole of ATP per mole of substrate (Fig. 11.3). These reactions have been demonstrated in many plant species (reviewed by Gray, 1987).

DMAPP acts as the prenyl donor to a molecule of IPP, producing geranyl pyrophosphate (GPP) by a 'head to tail' condensation reaction. GPP in turn acts as a prenyl donor to another IPP molecule to form farnesyl pyrophosphate (FPP). This condensation of allylic pyrophosphates with IPP produces the higher prenyl pyrophosphates (Fig. 11.4). The prenyl transferases (EC 2.5.1.1) catalyze these reactions to produce the all-*trans* prenyl pyrophosphates. The isolation, assay, purification and characterization of prenyl transferases from higher plants have been comprehensively reviewed by Alonso & Croteau (1993). Protein sequence comparisons of these enzymes (also termed isoprenyl diphosphate synthases) have been made in order to construct a phylogenetic tree (Chen *et al.*, 1994). Within the central pathway, chain lengthening is by head-to-tail condensations, but the formation of tri- and tetraterpenoids involves final head-to-head condensations of FPP and geranylgeranyl pyrophosphate (GGPP), respectively, to yield squalene and phytoene, the first tri- and tetraterpenoids (Fig. 11.1).

11.4 MONOTERPENOIDS

Apart from the relatively rare hemiterpenoids (e.g. isoprene), the monoterpenoids are the simplest class of isoprenoids with a C_{10} structure constructed of two isoprene units. They are components of essential oils and particularly accumulate in certain Umbelliferae and Pinaceae, but are probably ubiquitous in higher plants and algae. Interest in them originated due to their use in perfumes and as food flavors, but they may also have important pharmacological properties. Details of their extraction and purification from plants have been reviewed by Charlwood & Charlwood (1991a).

Monoterpenoids have been classified according to some 38 different types (Dev *et al.*, 1982). They can be divided into four broad structural categories: acyclic, cyclopentanoid, cyclohexanoid and irregular monoterpenes (Fig. 11.5). The first three categories arise from 'head-to-tail' linkage of two

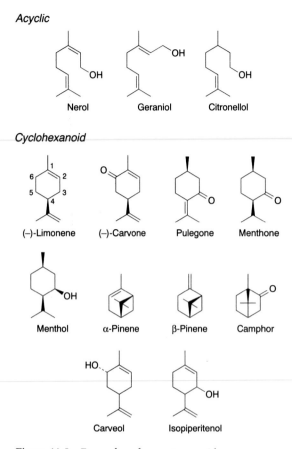

Acyclic

Nerol   Geraniol   Citronellol

Cyclohexanoid

(−)-Limonene   (−)-Carvone   Pulegone   Menthone

Menthol   α-Pinene   β-Pinene   Camphor

Carveol   Isopiperitenol

Figure 11.5 Examples of monoterpenoids.

C_5 units, but the irregular monoterpenoids do not fit this pattern and are apparently formed by a 'head-to-middle' joining of isoprene units. General reviews on their chemistry and biosynthesis are numerous (e.g. Banthorpe & Branch, 1983; Grayson, 1984, 1987, 1988; Charlwood & Charlwood, 1991b).

11.4.1 Biosynthesis

The role of GPP as the precursor of monoterpenoids was established in the 1960s and a chain length-specific GPP synthase associated with monoterpene synthesis in plants has been reported (Croteau & Purkett, 1989; Alonso & Croteau, 1993). Considerable efforts have been made to show *in vitro* that GPP is efficiently converted into cyclic monoterpenoids (see Croteau, 1987). GPP synthase has been purified and characterized (Clastre *et al.*, 1993) and found to be located in the glandular trichomes of sage (Croteau & Purkett, 1989). The monoterpene cyclases (synthases) have a similar reaction mechanism to

the prenyl transferases, in that they involve an initial ionization of the allylic pyrophosphate with electrophilic attack of the resulting allylic carbocation on a double bond (Alonso & Croteau, 1993).

The cyclases (synthases) are usually operationally soluble proteins, M_r 50–100 kDa (Croteau & Cane, 1985), but may be associated with the endoplasmic reticulum (Belingheri *et al.*, 1988) or plastids (Kreuz & Kleinig, 1984) *in situ*. More than 20 monoterpenoid cyclases have been identified (Charlwood & Charlwood, 1991a). A number have now been purified (see Alonso & Croteau, 1993). Most catalyze the formation of alkenes, but a few yield oxygenated products. The cyclic, oxygenated monoterpenoids are formed from cyclization products by oxidations and reductions, which can involve cytochrome *P*-450-dependent oxidases (Karp *et al.*, 1987). Such reactions are typified by the monoterpenoids found in peppermint. The genetic and biochemical basis of monoterpene composition in the mints (*Mentha*) have recently been reviewed by Croteau & Gershenzon (1994). Comparisons between the monoterpenes of *M. spicata* (spearmint) and *M. piperita* (peppermint) show that they are distinguished by the position of oxidation of the *p*-menthane ring (Lawrence, 1981). Peppermint produces almost exclusively monoterpenoids with oxygen at C-3 (e.g. pulegone, menthone and menthol; Fig. 11.5), whereas spearmint species produce those with oxygen at C-6 (e.g. carvone; Fig. 11.5). It has been shown that (−)-(4S) limonene, the first cyclic terpene formed from GGPP is hydroxylated at C-3 to form *trans*-isopiperitenol in peppermint, or at C-6 to form (−) *trans*-carveol in spearmint. The remaining enzymes in the two types of mint are similar. The catalytic activity of these cytochrome *P*-450-dependent hydroxylases determines the type of monoterpenoid found in most species of *Mentha* (Karp *et al.*, 1990; Croteau *et al.*, 1991). Limonene synthase (GPP : limonene cyclase) is a soluble protein, purified to homogeneity from both spearmint and peppermint (Alonso *et al.*, 1992). Its basic properties are the same in both species, with a mechanism of action similar to other terpenoid cyclases of higher plants (Croteau, 1987) and histidine and cysteine residues at the active site (Rajaonarivony *et al.*, 1992a,b). This step is rate-limiting in monoterpene biosynthesis in peppermint (Gershenzon & Croteau, 1990).

Pinene biosynthesis has been studied extensively in soluble extracts of sage (*Salvia officinalis*). Three monoterpene synthase (cyclase) activities that catalyze the conversion of GPP into monoterpene olefins have been detected. Pinene cyclase I converts GPP into the bicyclic (+)-α-pinene and (+)-camphene; cyclase II forms (−)-α-pinene and (−)-β-camphene, whereas cyclase III transforms GPP into (+)-α-pinene (+)-β-pinene and to monocyclic and acyclic olefins (Wagschal *et al.*, 1994). The stereochemistry of these conversions has been reported (Pyun *et al.*, 1994).

11.4.2 Biological activities of monoterpenoids

Monoterpenoids are cytotoxic to plants and play an important role in plant–plant interactions. For example, α- and β-pinene, limonene and citronellol (Fig. 11.5) inhibit the growth of *Amaranthus retroflexus* near to the orange tree, *Citrus aurantium* (Al Saadawi *et al.*, 1985). A number of monoterpenoids possess antimicrobial activity (Deans & Ritchie, 1987). The therapeutic properties of essential oils and their individual components have been reviewed by Schilcher (1985).

11.4.3 Iridoids

Iridoids are monoterpenoids with a methylcyclopentane skeleton that can be converted into secoiridoids by cleavage of the cyclopentane ring (Fig. 11.6). They are found virtually exclusively in dicotyledonous plants, many of them as glycosides. Details of their formation from geraniol and their properties are detailed by Inouye (1991).

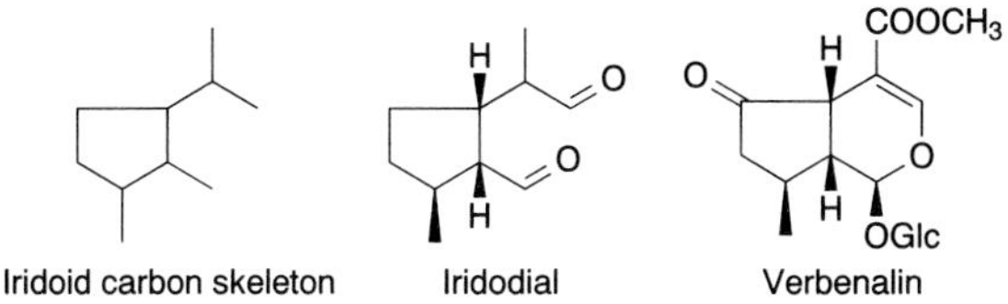

Figure 11.6 Structures of some iridoids.

11.5 SESQUITERPENOIDS

This group form the largest class of terpenoids and are found in plants, liverworts, mosses, fungi and algae. More than 100 sesquiterpenoid skeletons are known and thousands of compounds identified (see Connolly & Hill, 1991; Fraga, 1991). They commonly occur with the monoterpenoids in essential oils, with the sesquiterpenoids in the lower amounts. Their accumulation in higher plants is often associated with the presence of specialized secretory structures, such as oil glands (Loomis & Croteau, 1973).

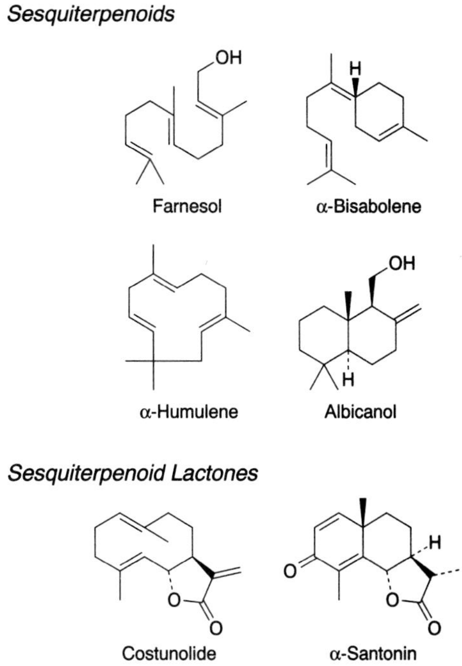

Sesquiterpenoids

Sesquiterpenoid Lactones

Figure 11.7 Sesquiterpenoids and sesquiterpenoid lactones.

The first sesquiterpenoids to be studied, over a century ago, were from essential oils of plants such as clove and cedar oils. Structural elucidations showed them to be derived from three isoprene units (C_{15}). Many further structures were soon determined and it is now clear that their structural diversity is remarkable, as shown in the representative structures in Fig. 11.7. Methods for the isolation and determination of sesquiterpenoids are comprehensively covered by Fraga (1991).

11.5.1 Biosynthesis

The sesquiterpenoids are formed by the condensation of IPP with GPP to yield all-*trans*-farnesyl pyrophosphate (FPP) (Fig. 11.1). The biosynthesis of cyclic sesquiterpenoids has been explained using hypothetical routes with cationic intermediates, with all-*trans*-FPP, 2-*cis*, 6-*trans*-FPP and nerolidyl pyrophosphate (NPP) as precursors. Cyclizations, hydride shifts, rearrangements etc., governed by steric and electronic considerations, lead to the formation of many sesquiterpenoids. Schemes of this type can be found in several reviews (e.g. Banthorpe & Charlwood, 1980; Torsell, 1983; Fraga, 1991). Comprehensive details of these pathways are contained in reviews by Cordell (1976) and Cane (1981, 1984, 1990).

11.5.2 Functions

A very broad range of biological properties has been reported for the sesquiterpenoids, including insect antifeedant substances, insect juvenile hormones and pheromones, phytoalexins, mycotoxins, antibiotics and plant growth regulators. Details can be found in Fraga (1991).

11.5.3 Sesquiterpenoid lactones

Over 300 sesquiterpenoid lactones have been characterized (Connolly & Hill, 1991). Early interest in this group of terpenoids derived from the bitter principles of herbal remedies. These bitter principles were found to contain sesquiterpenoid lactones. Most contain the α-methylene-γ-lactone group, as shown in the examples in Fig. 11.7. Their chemistry, physical properties and isolation have been reviewed by Fischer (1991). The presence of epoxy and carbonyl functions enables the lactones to cause the irreversible alkylation of enzymes and this property probably plays an important role in the plant's interactions with other organisms. Very few biosynthetic data have been obtained, but it is generally thought that they arise from the cyclization of FPP or NPP and a biogenetic relationship of the major types has been suggested (Fischer, 1991).

11.5.4 Abscisic acid

Abscisic acid (ABA, Fig. 11.8) is found in all monocots and dicots and in gymnosperms, but not in liverworts and mosses. Its concentration in tissues varies widely, with high amounts in the plastid. It is the only biologically active member of its group, which includes xanthoxin and phaseic

acid (Fig. 11.8). Details of its isolation and purification are described by Milborrow & Netting (1991).

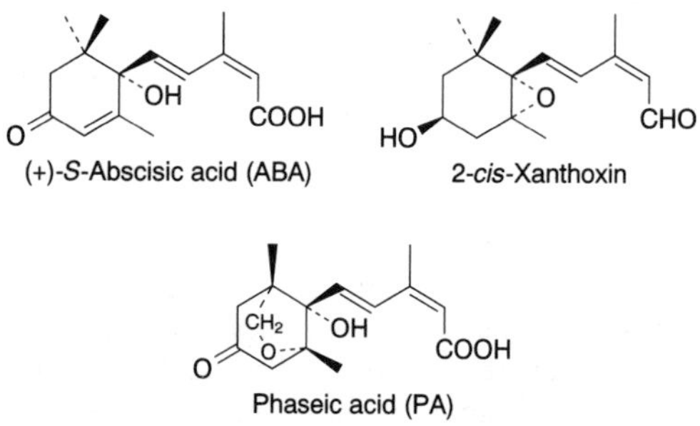

(a) Biosynthesis and metabolism

ABA has the structure of a sesquiterpenoid and was expected to be derived from the fusion of three C_5 units. Theoretically, there are two routes for the conversion of MVA into ABA: a direct route via FPP, or an indirect one involving the cleavage of a C_{40} carotenoid. There has been considerable controversy over the years as to which pathway operates in plants, with proponents for each pathway (e.g. Milborrow, 1983; Zeevart & Creelman, 1988). Since the late 1980s, however, the experimental evidence, as reviewed by Parry (1993), has been increasingly in favor of the indirect pathway. The generally accepted route, from all-trans violaxanthin, via all-trans neoxanthin and 9'-cis neoxanthin is outlined in Fig. 11.9.

ABA is inactivated by its catabolism through two routes: oxidation and conjugation. The primary pathway, at least in tomatoes, is via 8'-hydroxy ABA and phaseic acid (PA). The latter is reduced to dihydrophaseic acid (DPA) which in turn is glycosylated (Milborrow & Vaughan, 1982). ABA itself also conjugates to glucose (Koshimizu et al., 1968) leading to sequestration into the vacuole (Bray & Zeevart, 1985). The increase in ABA levels, typically introduced by water stress, is also linked to an increase in PA and DPA (Parry et al., 1990), with oxidative catabolism apparently the route for rapid inactivation (Zeevart & Creelman, 1988). This process occurs in the cytosol (Hartung et al., 1982), although most ABA is found in the chloroplast and hence protected from catabolism. Stress and the dark decrease the stromal pH, causing an increase in ABA in the cytosol and hence its catabolism.

(b) Biological role of ABA

ABA is a plant growth regulator, best known to function in response to stress such as drought and waterlogging (Addicott, 1983). Externally applied ABA initiates stomatal closure, thus acting as an endogenous 'antitranspirant' (Mittelhauser & Van Steveninck, 1969). ABA-less mutants have confirmed this role (Neill & Horgan, 1985). ABA also has a key role in developmental processes as it up- and down-regulates specific genes (Zeevart & Creelman, 1988; Skriver & Mundy, 1990; Cohen & Bray, 1990). A review of recent advances in ABA action and signaling has been published (Giraudat et al., 1994).

(+)-S-Abscisic acid (ABA)

2-cis-Xanthoxin

Phaseic acid (PA)

Figure 11.8 Abscisic acid and its metabolites.

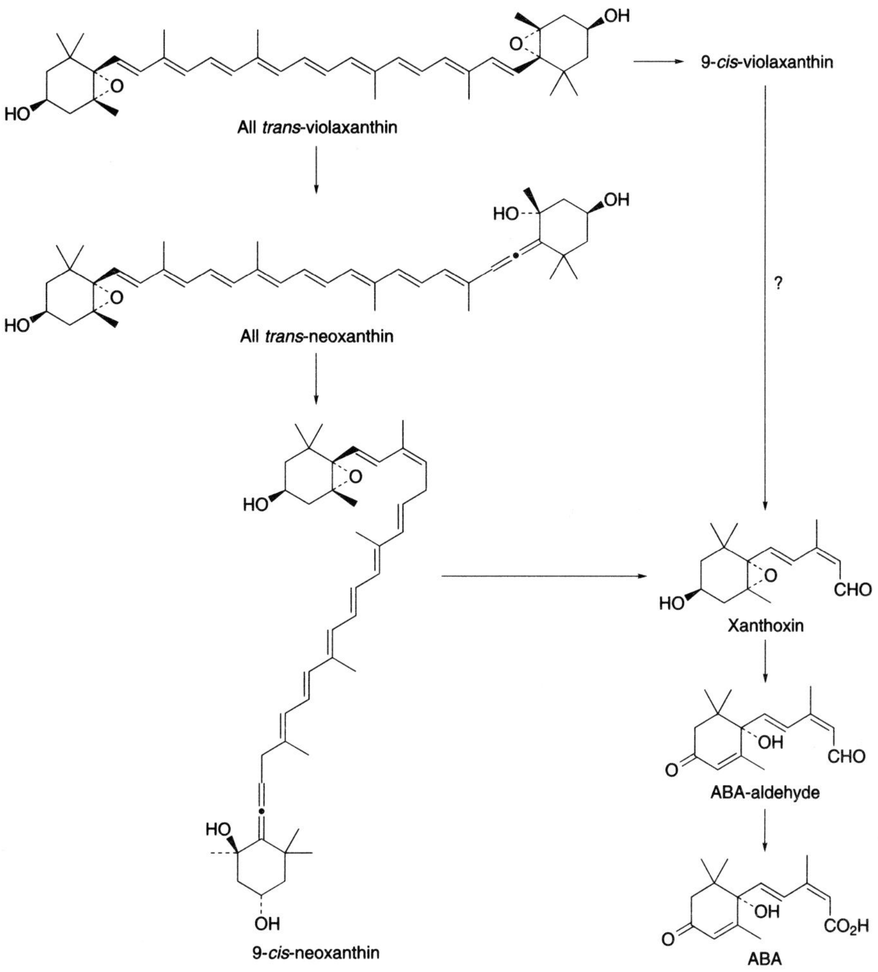

Figure 11.9 Biosynthesis of abscisic acid.

11.6 DITERPENOIDS

Diterpenoids are C_{20} compounds, derived from GGPP, often with skeletal rearrangements. They are principally found in higher plants and fungi and include the gibberellins (GAs) and the gymnosperm resin acids. Most are cyclic compounds, with a few exceptions such as phytol. Typical examples are shown in Fig. 11.10.

11.6.1 Gibberellins

The GAs are tetracyclic diterpenoids, essential for the normal growth and development of plants. To date some 90 GAs have been identified; some 79 in higher plants and 27 in fungi (Beale & Willis,

1991). They can be divided into two groups: (a) those that retain the full C_{20} atoms of $(-)$-*ent*-kaurene (the C_{20} GAs), e.g. GA_{13} (Fig. 11.10); and (b) those that have lost one carbon atom (C-20), with the formation of a 19,10-γ-lactone structure (the C_{19} GAs), e.g. GA_3 (Fig. 11.10). A trivial nomenclature has evolved, based upon 'A numbers', e.g. GA_1. The prefix *ent*- is used for enantiomeric structures and it reverses the stereochemical designation at each chiral center (e.g. *ent*-kaurene, Fig. 11.10). GAs are produced in minute quantities ($\mu g/kg$ fresh weight) with the highest concentrations in embryos or endosperm, compared with low levels in vegetative shoots. Protocols for their extraction and analysis have been reviewed in detail (Beale & Willis, 1991; Hedden, 1993).

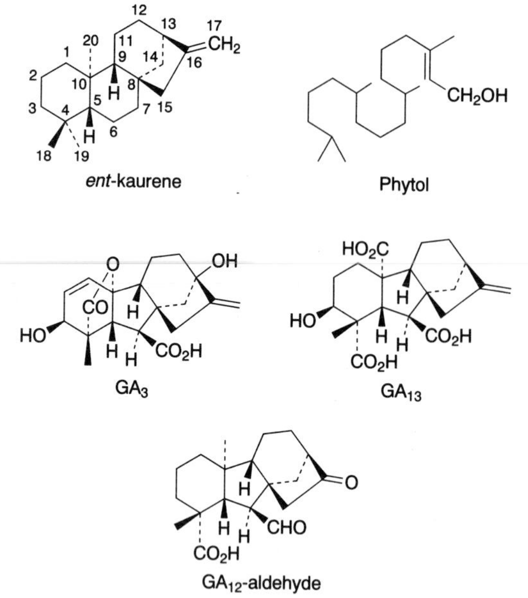

Figure 11.10 Examples of diterpenoids.

(a)   *Biological function of GAs*

GAs act as regulators (hormones) of developmental processes such as stem elongation, fruit and seed development, seed germination and dormancy. The most studied role of GAs is the control of internode elongation in stems. Current thinking is that a diverse range of cellular events are modulated by GAs including ion channel activity to gene expression, as reviewed by Hooley (1994). Not all GAs are biologically active, and among the active compounds certain structural features seem to increase activity. The C_{20}-GAs are more active than the C_{19}-GAs. The presence of a 2β-hydroxyl group removes biological activity, whereas the 3β-hydroxyl in the C_{19}-GAs gives the highest activity (e.g. GA_1, GA_3, GA_4, GA_7).

(b)   *Biosynthesis*

GA formation has been studied extensively in whole cells and cell extracts and comprehensive reviews of these studies have been published (Hedden *et al.*, 1978; Graebe, 1987; Spray & Phinney, 1987; Lange & Graebe, 1993). The early steps from MVA to GA_{12}-aldehyde are identical in higher plants and fungi, but the pathways differ beyond GA_{12} in the number and position of the hydroxyl groups introduced before oxidation of the C-20 to give the GA_{19}-GAs (Beale & Willis, 1991).

The first committed step in the pathway is the conversion of geranylgeranyl pyrophosphate

(GGPP) into *ent*-kaurene, first shown by Graebe (1969) in a cell-free preparation of *Marah macrocarpus* seeds. Subsequently, this reaction has been demonstrated in many cell extracts, e.g. *Zea mays* (Hedden & Phinney, 1976). The reaction is a 2-step cyclization catalyzed by *ent*-kaurene synthetase (Fig. 11.11). The enzyme has two activities, named A and B, which catalyze GGPP to copalyl pyrophosphate (CPP) and CPP to *ent*-kaurene, respectively, probably at different sites on the protein (Fall & West, 1971; Duncan & West, 1981). Kaurene synthetase is thought to be the key regulatory enzyme in the GA pathway (Hedden & Phinney, 1976, 1979; Duncan & West, 1981). The *An1* gene, coding for kaurene synthesis, has recently been cloned (Bensen *et al.*, 1995).

Ent-kaurene is then converted into *ent*-kaurenoic acid by three sequential oxidations at the C-19 of *ent*-kaurene (Fig. 11.11). This is a membrane-bound enzyme, known as kaurene oxidase. Each step requires NADPH and is a cyt *P*-450-dependent reaction (Murphy & West, 1969; Hedden & Graebe, 1981). It is still uncertain whether one protein with three active sites or separate enzymes are involved in the oxidations (reviewed by Archer, 1994). The enzyme(s) has still to be purified to homogeneity in a catalytically active form.

Ent-kaurenoic acid is hydroxylated to 7-hydroxy-kaurenoic acid, catalyzed by kaurenoic acid hydroxylase and requiring oxygen, NADPH and cyt *P*-450. GA_{12}-aldehyde (Fig. 11.10) is formed via contraction of the β-ring from 6 to 5 carbon atoms, with C-7 being extruded (Hedden *et al.*, 1978). The pathways beyond GA_{12}-aldehyde have been elucidated in several higher plants and shown to yield the vast array of GAs (see Lange &

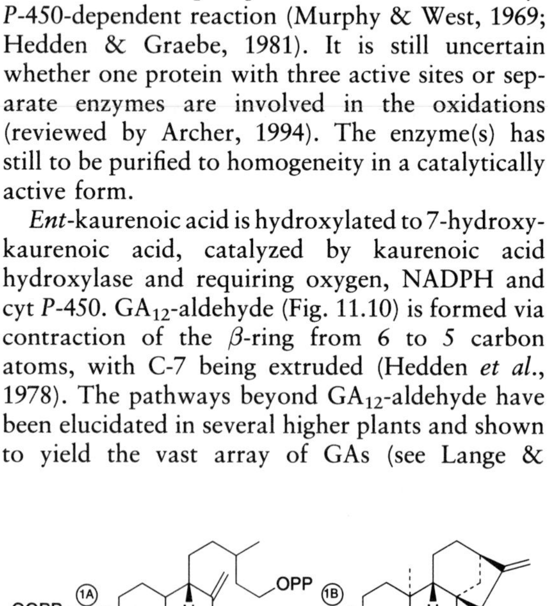

Figure 11.11 Reactions involved in the conversion of GGPP into *ent*-kaurenoic acid. Enzymes: 1A and 1B, *ent*-kaurene synthetase; 2, kaurene oxidase.

Graebe, 1993). The reactions are catalyzed by both microsomal and soluble enzymes, with distinct cofactor requirements. The former are cyt P-450-dependent, whereas the soluble enzymes require 2-oxoglutarate and Fe^{2+} (see Lange & Graebe, 1993).

(c) Control of GA biosynthesis

Photoinduction of GA biosynthesis has been shown to occur in higher plants and fungi. Dark-grown plants contain less GAs than those in the light (Lange, 1970). Moore & Ecklund (1974) showed, in pea seedlings, an increase in kaurene synthesis during de-etiolation and phytochrome has been implicated in the control of levels of extractable GAs (see Hedden et al., 1978).

11.7 TRITERPENOIDS

The C_{30} terpenoids are formed by the head-to-head condensation of two FPP molecules to yield squalene, the precursor of all triterpenoids (Fig. 11.12). More than 4000 different molecules have been isolated, incorporating 40 skeletal types, the most common being the pentacyclic ring, as in β-amyrin (Fig. 11.12).

11.7.1 Phytosterols

Sterols, a member of the triterpenoid isoprenoids, are characterized by a 3β-monohydroxy perhydro-1,2-cyclopentanophenanthrene ring system (Fig. 11.12). The numbering system shown is based upon the IUPAC-IUB rules of 1967, although revised rules have been published (IUPAC-IUB, 1989). The older nomenclature will be used in this section, as will trivial names of the phytosterols. Systematic names can be found in the review by Goad (1991).

Over 4000 triterpenoids have been characterized (Connolly & Hill, 1991), but only around 300 sterols and closely related compounds occur in plants. These are listed by Akihisa et al. (1991). Sterols are often esterified by a fatty acid at the C-3 hydroxyl group. Steryl esters have been reported to be in membranes (Dyas & Goad, 1993) and also in the soluble cellular fraction (Duperon & Duperon, 1973). The esterification process may allow regulation of the amount of free sterols in membranes by subcellular compartmentation (Gondet et al., 1994).

In general, most plants produce sterols that are alkylated at C-24, as typified by sitosterol and stigmasterol (Fig. 11.12). Cholesterol rarely exceeds a few per cent of phytosterol mixtures. A combination of chromatography (GC, HPLC),

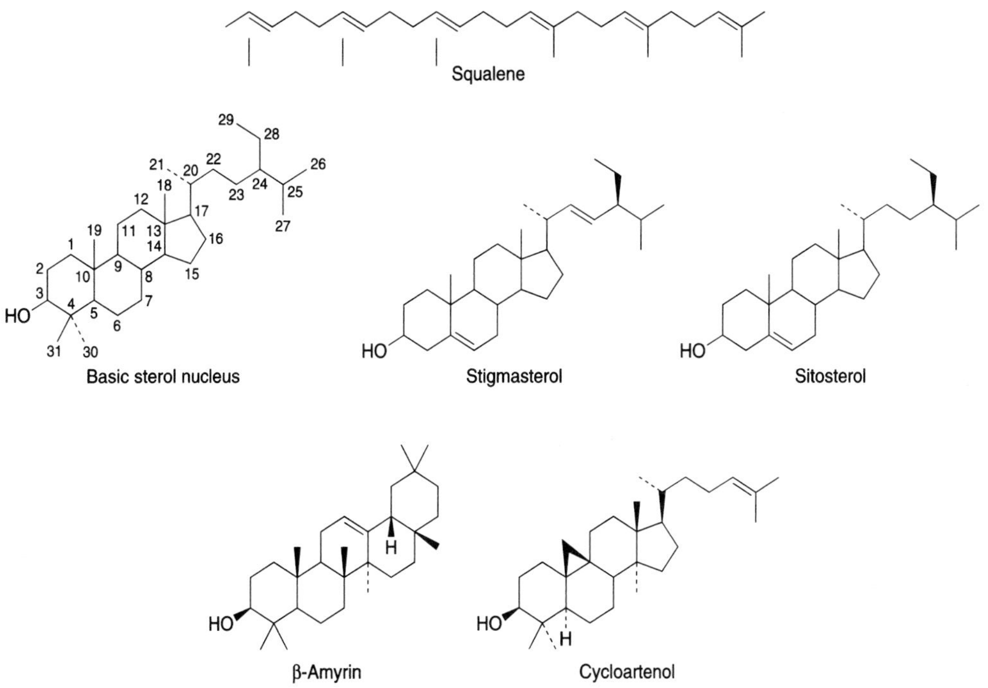

Figure 11.12 Structures of phytosterols originating from squalene. The numbering of the sterol nucleus is according to the IUPAC-IUB (1976) recommendations.

mass spectrometry and NMR is used for the purification and identification of sterols (Akihisa *et al.*, 1991; Goad, 1991).

(a)   *Biosynthesis*

It is generally accepted that the steps leading to squalene from HMG-CoA (Fig. 11.1) are the same in animals, plants and fungi. Squalene synthetase has been partially purified from tobacco cell cultures (Hanley & Chappell, 1992). Squalene epoxidase, which catalyzes the conversion of squalene into its 2,3-epoxide, has been well characterized in animals, but less so in plants. The rat enzyme is microsomal, requires O_2, NADPH and a cytosolic fraction for activity (Scallen *et al.*, 1968; Yamamoto & Bloch, 1970). The purified enzyme has a dissociable FAD and requires a supernatant protein factor (SPF). The latter may promote transfer of squalene from one membrane to another. The cyclization of squalene 2,3-epoxide into polycyclic triterpenoids varies between species. The enzymes involved are called squalene epoxide cyclases (EC 5.4.99.7) and are numerous. They are named according to the product of the reaction, e.g. cycloartenol cyclase in higher plants and algae, involved in the formation of phytosterols (Rees *et al.*, 1968). The product of squalene epoxide cyclization depends on the conformation that squalene epoxide assumes in binding to the cyclase and the nature and position of the nucleophile or base in the enzyme. Two types of cyclase can be envisaged: the one responsible for cyclization of the chair–boat–chair–boat folded squalene epoxide yielding tetracyclic triterpenoids such as lanosterol and cycloartenol and the cyclase producing β-amyrin through squalene epoxide having a chair–chair–chair–boat conformation (Cattel *et al.*, 1986).

The lengthy pathway from cycloartenol (Fig. 11.12) to the other phytosterols (see Goad, 1991), has been validated by labeling studies and assays of the enzymes involved (reviewed by Goad, 1977; Benveniste, 1986). The pathway involves the removal of three methyl groups (two at C-4, one at C-14) from the pentacyclic cycloartenol. The enzymology of the sequence is not well understood, mainly because of the difficulty in obtaining cell extracts able to catalyze the demethylations.

Results from *in vivo* labeling show that the 14α-methyl group is lost after elimination of one 4-methyl group and cleavage of the 9,19-cyclopropane ring (Benveniste, 1986). More recently, microsomal preparations from maize have been used (Pascal *et al.*, 1990, 1993, 1994; Taton *et al.*, 1994). These studies have shown that, in higher plants, both demethylations share a common property: the loss of the methyl groups is a result of successive monooxygenase reactions, performed by a single terminal oxidase. This is a cyt *P-450* for the C14 demethylation (Rahier & Taton, 1986) and an unidentified, cyanide-sensitive protein capable of activating oxygen for the C4 demethylation (Pascal *et al.*, 1993). The demethylations occur in the sequence 4, 14, 4 and proceed via a α-face attack. Cytochrome b_5 has been implicated as an intermediate electron carrier in both oxidations (Pascal *et al.*, 1990, 1994). Additional carbon atoms of the side chains are derived from *S*-adenosylmethionine (Janssen *et al.*, 1991).

The factors that regulate differences in the sterol content of plants from species to species are not well characterized, nor are the mechanisms controlling changes in sterols during plant development (Goad, 1983; Gray, 1987).

(b)   *Biological functions of phytosterols*

Sterols play a structural role in plant membranes analogous to cholesterol in animal cells. The plasma membrane has the greatest sterol content and the highest sterol:phospholipid ratio (e.g. Brown & Dupont, 1989). Chloroplast membranes contain only a small amount of free sterol.

Sterols are important for the growth of the plant, as shown from studies with synthetic plant growth regulators. For example, the 2R,3S-enantiomer of paclobutrazol reduced the shoot height of barley seedlings and showed a concomitant reduction in 4-demethyl sterols, but an increase in 14α-methyl sterols (Burden *et al.*, 1987). The retardation in growth was partially overcome by cholesterol, but full restoration of growth required a trace of 24-ethylsterol, e.g. stigmasterol (Haughan *et al.*, 1989). It has been concluded that the cell has two sterol requirements, one for 'bulk' sterol for new membrane production by dividing cells and the other for a 24-ethylsterol for a specific, stimulatory function (Haughan *et al.*, 1988; Goad *et al.*, 1988). Cycloartenol has been shown to be the plant equivalent of cholesterol in terms of biological function in the membranes (reviewed by Ourisson, 1994). Ourisson suggests that cycloartenol is a primitive cholesterol surrogate and a precursor, historically, of lanosterol.

11.7.2   Saponins

Saponins are glycosidic triterpenoids widely found in the plant kingdom. They are soluble in water, giving stable foams. Three major classes are found: (a) steroid glycosides; (b) steroid alkaloid glycosides and (c) triterpene glycosides, the largest group.

All three classes have one or more linear or branched carbohydrate chains attached to an aglycone (sapogenin). The sugar residues are added to the aglycone later as ether or ester linkages and acylation of the sugars is possible. Their isolation and purification has been reviewed by Hostettmann *et al.*, (1991).

(a) *Biological activities of saponins*

Their function in plants is not totally understood, although they do protect against fungal attack (Defago, 1977). They are able to lyse erythrocytes and this is the standard method for their detection (Hostettmann *et al.*, 1991), but they have very low oral toxicity to humans, probably due to the low rates of absorption from the intestine.

11.7.3 Cardenolides

Cardenolides are C_{23} steroid derivatives which have cardiac activity (May, 1990) and hence are also called cardiac glycosides. Digitoxin, a component

of *Digitalis*, is used clinically as a heart stimulant. They occur in several plant families and also in some insects, e.g. Monarch butterflies, which makes them unpalatable to predators. Their isolation and identification using chromatography and NMR has been reviewed in detail by Connolly & Hill (1991).

11.8 CAROTENOIDS

The carotenoids are an abundant group of naturally occurring pigments, present in all green tissues, where they are constituents of the chloroplast, as well as being responsible for most of the yellow to red colors of flowers and fruits. Over 600 structures have been established, of which some 150 are found in photosynthetic organisms (Straub, 1987; Britton, 1991). All carotenoids have both trivial and semisystematic names. The latter convey structural information and are listed by Britton (1991). Carotenoid hydrocarbons are called carotenes, whereas derivatives containing oxygen functions are the xanthophylls. They may be acyclic (e.g. lycopene, Fig. 11.13), or contain 5- or 6-membered rings at one or both ends of the molecule, e.g. β-carotene, lutein (Fig. 11.13). Because of the extensive double bond system in

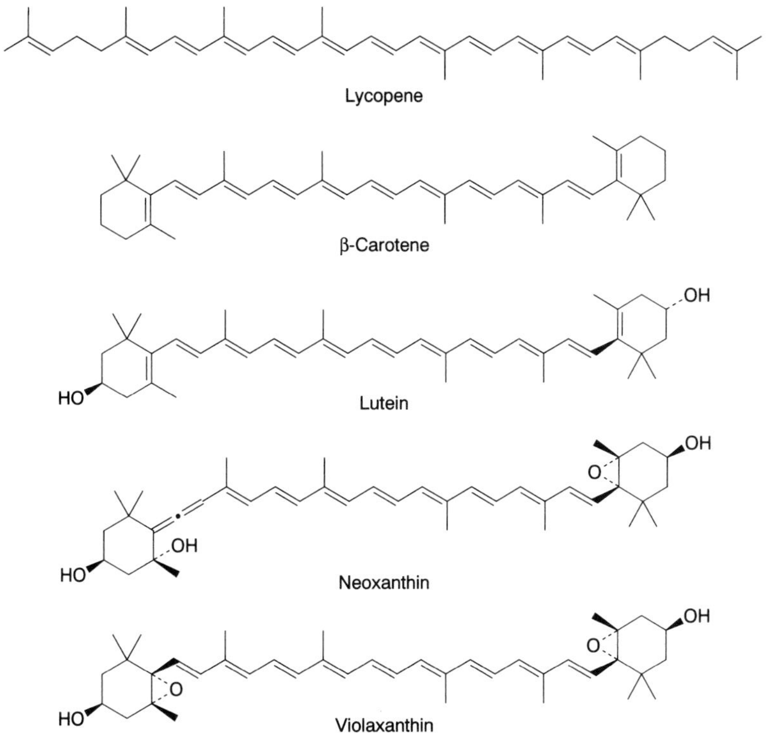

Figure 11.13 Typical carotenoids of higher plants.

the carotenoid molecule, a carotenoid can, in theory, exist in a large number of geometric isomers (*cis/trans* isomers). Most carotenoids are, in fact, found to be in the all-*trans* form, but *cis* isomers do exist (see Young, 1993a).

The most obvious structural feature of a carotenoid molecule is the chromophore of conjugated double bonds which, in carotenoids of plant tissues, varies from three in the colorless phytoene to 13 in canthaxanthin, which is red. This double-bond system also renders them susceptible to isomerization and oxidative degradation, especially during extraction and purification procedures. Detailed descriptions of isolation and purification procedures and the identification of pure carotenoids are comprehensively covered elsewhere (Davies, 1976; Britton, 1985, 1991). Regular reviews of advances in carotenoid chemistry are found in publications of The Royal Society of Chemistry (e.g. Britton, 1989) and also in the proceedings of the International Symposia on Carotenoids, which are held every three years.

11.8.1   Distribution

This section does not provide a comprehensive coverage of the distribution of carotenoids in Nature, and only describes photosynthetic and non-photosynthetic tissues of higher plants. More detailed reviews have been published elsewhere (Goodwin, 1980; Britton, 1991; Young, 1993a).

(a)   *Photosynthetic tissues*

Carotenoids accumulate in the photosynthetic tissues of all higher plants. Both carotenes and xanthophylls are found in leaves, with the same four major carotenoids, β-carotene, lutein, violaxanthin and neoxanthin (Fig. 11.13). Minor components include α-carotene, β-cryptoxanthin, zeaxanthin, antheraxanthin and lutein 5,6-epoxide. In general, the carotenes comprise some 25% of the total and lutein is 45% of the total carotenoids. During leaf senescence the chloroplast disintegrates and esterification of the xanthophylls occurs. Unusual carotenoids found in leaves are given by Goodwin & Britton (1988) and Britton (1991).

In photosynthetically-active chloroplasts, the carotenoids are part of the pigment–protein complexes (PPCs) in the thylakoid membranes. The core complex, CC1, contains one β-carotene per 40 chlorophyll *a* molecules, whereas LHC1 is associated with lutein, violaxanthin and neoxanthin. β-Carotene is also in CCII, while LHCII

contains xanthophylls. Further details of the carotenoids in PPCs is given by Young (1993b), based upon the nomenclature of Peter & Thornber (1991), and see also Chapter 2.

There are also small quantities of carotenoids in the chloroplast envelope (Joyard *et al.*, 1991) and the envelope of amyloplasts (Fishwick & Wright, 1980). In all plastid envelope membranes, violaxanthin is the major carotenoid (Jeffrey *et al.*, 1974). Dark-grown plants accumulate predominantly xanthophylls in their etioplasts (Goodwin, 1958).

(b)   *Non-photosynthetic tissues*

The carotenoids of flower petals can be divided into three main groups (Goodwin, 1980): (a) highly oxygenated carotenoids such as auroxanthin and flavoxanthin; (b) carotenes, sometimes in high concentrations, e.g. β-carotene in *Narcissus* and (c) species-specific, e.g. crocetin in the *Crocus*. Flower carotenoids are frequently esterified.

The distribution of carotenoids in fruits is extremely complex and subject to considerable variation. Patterns, characteristic of each species and variety, often occur at each ripening stage and can vary from a few to over 50 in citrus fruits. Their biosynthesis is autonomous in most fruit and continues after the fruit has been removed from the parent plant. Unripe (green) fruit contains the same pigments as other photosynthetic tissues but upon ripening the chloroplasts differentiate into chromoplasts and there is often, but not always, *de novo* synthesis of carotenoids.

According to Goodwin (1980) eight main groups of fruits can be distinguished according to carotenoid content: (i) insignificant amounts, e.g. strawberry; (ii) large amounts of chloroplast carotenoids, e.g. blueberry; (iii) large quantities of lycopene, its hydroxy derivatives and more saturated carotenoids, e.g. tomato; (iv) large amounts of β-carotene and its hydroxy derivatives, e.g. peach; (v) large amounts of carotenoid epoxides, e.g. carambola; (vi) unusual carotenoids, e.g. capsanthin in red pepper; (vii) poly-*cis* carotenoids, e.g. prolycopene in tangerine tomatoes; and (viii) apocarotenoids, e.g. persicaxanthin in *Citrus* species.

Anthers and pollen contain carotenoids (Goodwin, 1980; Goodwin & Britton, 1988). Most seeds have only traces of carotenoids, with the exception of maize which contains large quantities of β-carotene, β-cryptoxanthin and zeaxanthin. The best known example of a root carotenoid is β-carotene in the carrot and sweet potato.

11.8.2 Biosynthetic pathway and enzymology

In this section, only the salient features of the pathway and the enzymes involved in higher plants will be described. Comprehensive reviews can be consulted elsewhere (Jones & Porter, 1986; Britton, 1990, 1993; Bramley, 1993a; Sandmann, 1991, 1994a,b).

(a) Formation of phytoene from GGPP

The first dedicated step in the formation of carotenoids is the head-to-head condensation of two molecules of all-*trans* geranylgeranyl pyrophosphate (GGPP) via the cyclopropylcarbinyl pyrophosphate, prephytoene pyrophosphate (PPPP), to form phytoene (Fig. 11.14). The configuration of phytoene in eukaryotic plants is predominantly 15-*cis* (see Jones & Porter, 1986). Numerous plant extracts are able to form phytoene from precursors such as MVA, IPP and

GGPP and possible cofactor requirements have been reported (Bramley 1993a). The general consensus is that Mn^{2+} is required. The purified phytoene synthase from *Capsicum annuum* fruit requires no other cofactor (Dogbo et al., 1988) but the tomato fruit enzyme has an absolute requirement for ATP (Fraser and Bramley, unpublished observations).

Phytoene synthase is located in the stroma of chloroplasts, chromoplasts and amyloplasts of *Capsicum* (Dogbo et al., 1987), but is reported to be a peripheral membrane protein of the inner membrane of *Narcissus* chromoplasts (Kreuz et al., 1982). It is synthesized on the 80S ribosomes prior to post-translational processing and import into the plastid.

(b) Desaturation and isomerization reactions

The sequence from phytoene to lycopene involves four stepwise dehydrogenations, occurring alternatively to either side of the chromophore to form

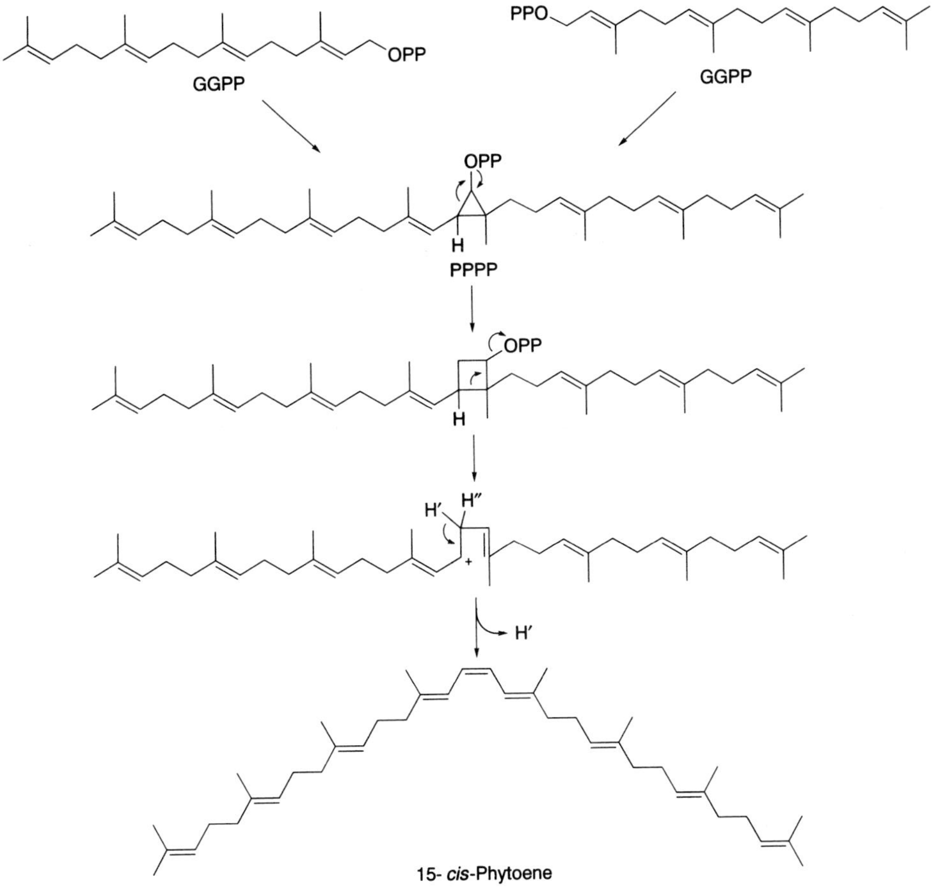

Figure 11.14 The formation of 15-*cis* phytoene from GGPP. Abbreviations are given in the text.

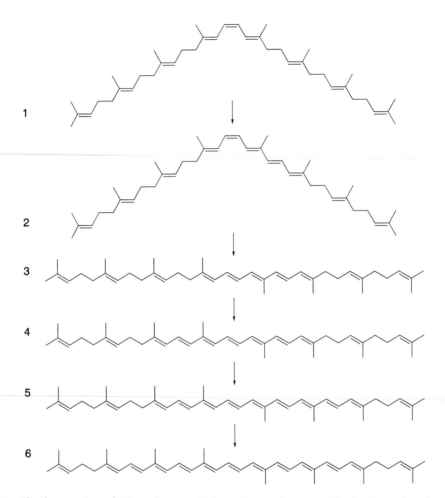

Figure 11.15   The desaturation of 15-*cis* phytoene (1) into all-*trans* lycopene (6), via 15-*cis*-phytofluene (2), all-*trans*-phytofluene (3), ζ-carotene (4) and neurosporene (5).

phytofluene, ζ-carotene, neurosporene and lycopene (Fig. 11.15). In addition, in higher plants, the 15-15′ double bond is isomerized from the *cis* to *trans* configuration. This isomerization is probably not enzymically catalyzed, but may result from a strained, saturated carotene at the active site of the desaturase enzyme (Goodwin, 1983) and probably occurs at the stage of phytofluene (Camara & Moneger, 1982; Kushwaha *et al.*, 1970).

In higher plants there are two desaturases: first, phytoene desaturase, which converts phytoene into phytofluene and ζ-carotene and then ζ-carotene desaturase which yields neurosporene and lycopene (see Bramley, 1993a). The loss of two hydrogen atoms at each step in the sequence is most likely to be a dehydrogenation reaction, perhaps requiring an oxidized pyridine nucleotide and molecular oxygen as the final electron acceptor, with an oxidoreductase as a redox mediator (Beyer *et al.*, 1989). The H atoms are lost

stereospecifically in a *trans* fashion (McDermott *et al.*, 1973). Phytoene desaturase is a common target site for bleaching herbicides (Bramley, 1993b).

The desaturase enzymes are membrane bound within the plastid and have been reported to be associated with the chloroplast envelope of spinach leaves (Lutke-Brinkhaus *et al.*, 1982) and chromoplast membranes of *Narcissus* flowers (Kreuz *et al.*, 1982).

(c)   *Cyclization reactions*

Alicyclic end groups are common features of many carotenoids, with two rings, the β- and ε- found in higher plant carotenoids. The two rings are formed independently, although probably from the same precursor. The stereochemistry of cyclization has been reviewed by Britton (1990). A range of cyclization products is probably formed

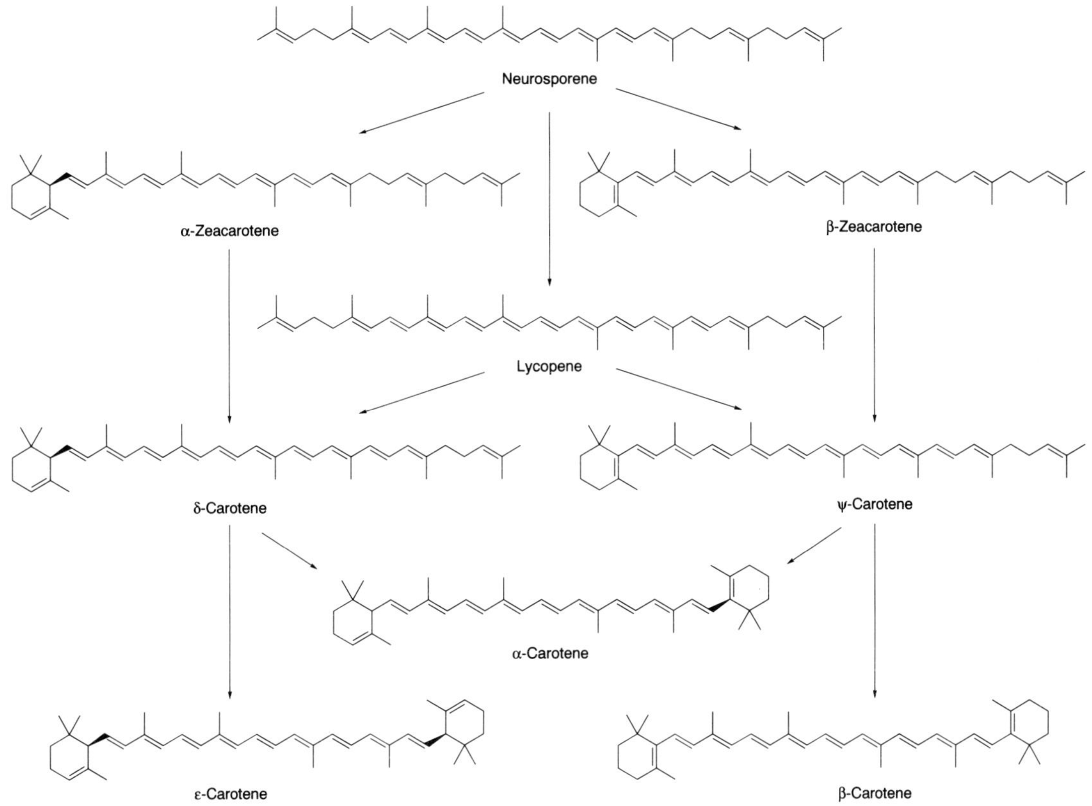

Figure 11.16 Cyclization reactions in carotenogenesis.

as shown in Fig. 11.16. The number of enzymes required for these reactions is unknown, but some evidence suggests that separate enzymes for the formation of each ring of a bicyclic carotene (see Jones & Porter, 1986).

(d) *Xanthophyll formation*

Alicyclic xanthophylls with oxygen substituents on the rings are common in higher plants, e.g. hydroxy groups at C3 and C3′ (lutein and zeaxanthin, Fig. 11.13) or epoxy groups at 5,6 or 5′,6′ (violaxanthin and neoxanthin, Fig. 11.13). The oxygen moieties originate from molecular oxygen, and hydroxylation may be catalyzed by a cyt *P*-450-dependent mixed-function oxidase (Britton, 1990). Little else is known about the enzymes of xanthophyll biosynthesis. The conversion of zeaxanthin into antheraxanthin and violaxanthin (the 'violaxanthin cycle') requires O_2 and NADPH. Details of this cycle are in the review by Demmig-Adams & Adams (1993), and see also Chapter 2.

11.8.3 Genetics of carotenoid biosynthesis

The first studies on the genetics of carotenoid biosynthesis used tomato mutants with a range of color phenotypes. These formed the basis of the Porter–Lincoln series of carotene desaturation (Fig. 11.15). Details of these pioneering studies can be found in Goodwin (1980). All of the carotenoid genes are nuclear encoded (Kirk & Tilney-Bassett, 1978) and therefore include a transit peptide sequence enabling import of the protein into the plastid. Over the last five years very rapid progress has been made on the molecular genetics of carotenoid formation in higher plants. The early gene cloning studies focused on photosynthetic and non-photosynthetic bacteria such as *Rhodobacter* and *Erwinia* spp. This work has resulted in the characterization of all the genes encoding the enzymes of the carotenoid pathway in these organisms. Reviews on these investigations have been published (Sandmann, 1991, 1994a,b; Britton, 1993; Armstrong, 1994).

More recently, the corresponding genes have been cloned from higher plants. The lack of homology between the bacterial and plant genes precluded the use of the former as DNA probes for the latter, so alternative approaches have been used. In particular, the genes from cyanobacteria such as *Synechococcus* have sufficient homology to the higher plant homolog to be used as probes, and this has enabled several phytoene desaturase (Pds) genes to be cloned. Similarly, the sequences of lycopene cyclase and phytoene synthase genes have been obtained. This expanding area of carotenoid research has been reviewed by several of the key workers in the field (Sandmann, 1991, 1994a,b; Bartley & Scolnik, 1994; Armstrong, 1994).

11.8.4 Regulation in higher plants

Since carotenoids are an essential part of the pigment–protein complexes of chloroplast thylakoids (see section 11.8.1a), the regulation of carotenogenesis is linked to the formation of chlorophylls, proteins, lipids and chloroplast development. Clearly, there must be regulatory mechanisms that control the supply of carotenoids to the PPCs and also at the level of the individual pathway, as different carotenoids are located at different locations in the chloroplast. Britton (1993) has described several phases or variations of carotenoid biosynthesis in chloroplasts and points out the danger of oversimplification in assuming that all plant species have the same mechanism of regulation. Environmental factors such as light intensity, including sun/shade adaptation and photobleaching also have profound effects on carotenoid levels (see Young, 1993c), but the molecular mechanisms underlying such changes are not well understood. The availability of gene probes (section 8.3) and specific *in vitro* enzyme assays (e.g. Fraser *et al.*, 1994) will facilitate future studies in this area.

11.8.5 Functions of carotenoids in higher plants

A vital role of carotenoids in photosynthetic tissues is photoprotection by quenching the triplet state of chlorophyll and scavenging for singlet oxygen. This function is associated with the ability of the carotenoid molecule to participate in photochemical reactions such as singlet–singlet energy, triplet–triplet energy, oxidation, reduction and isomerization. These properties have been reviewed in detail by Frank & Cogdell (1993),

whilst general reviews on photoprotection can be found elsewhere (Cogdell & Frank, 1987; Koyama, 1991). A second, essential function is that of acting as accessory light-harvesting pigments, as their presence in PPCs indicates (Siefermann-Harms, 1985; Young, 1993b; Frank & Cogdell, 1993). Finally, it should be remembered that carotenoids are also precursors of abscisic acid (section 11.5.4), an essential plant hormone.

11.9 POLYTERPENOIDS

Polyterpenes of high molecular weight and formula $(C_5H_8)_n$ are widely distributed in the Plant Kingdom. In natural rubber the isoprene units are in the *cis* configuration, whereas they are all-*trans* in gutta percha. The biosynthesis of natural rubber, studied mainly in the 1950s and 1960s, using latex from *Hevea brasiliensis*, showed that the chain-extension was by the successive addition of IPP to pre-existing rubber (Lynen & Hemming, 1960; Archer *et al.*, 1961). The transferase is a soluble enzyme, located on the surface of the rubber particles (Archer & Audley, 1973). Although rubber was originally thought to be all-*cis*, it may be that other linkages such as *trans*-1,4 may be present (Campos-Lopez & Palacios, 1976). Details of structural studies on rubber and gutta by NMR have been reviewed by Tanaka (1991).

Open-chain terpenoid alcohols are often referred to as 'prenols' or as 'prenyl' compounds. Higher plants, algae and some fungi contain linear unsaturated polyprenols which have 45–115 carbon atoms and varying *cis* and *trans* bonds. The first such long-chain terpenoid alcohol to be discovered was the C_{45} solanesol (now renamed all-*trans* nonaprenol), and plastoquinones and ubiquinone have prenyl side chains. The function of all these prenyl alcohols is not certain, but the monophosphate esters are involved in glycan biosynthesis. General reviews have been published by Threlfall (1980), Hemming (1983), Langworthy & Pond (1986) and Charlwood & Banthorpe (1991).

11.10 MINOR CLASSES OF TERPENOIDS

11.10.1 Degraded terpenoids

Some terpenoids have been found to be degraded *in vivo* to yield nor- or apo-products. Nor refers to

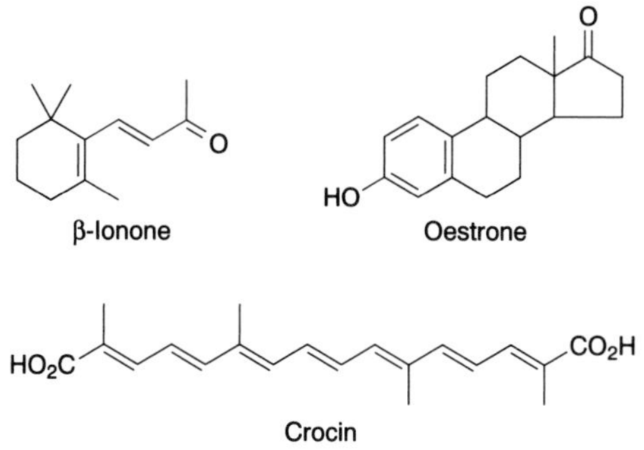

Figure 11.17 Degraded terpenoids.

loss of methyl group(s), whereas apo- indicates the cleavage of larger skeletal fragments. Some examples are shown in Fig. 11.17. Apocarotenoids are common. For example, the C_{20} crocetin is thought to be formed by degradation of zeaxanthin.

11.10.2 Sesterterpenoids

Since 1965, the C_{25} group of isoprenoids, the sesterterpenoids, has been recognized. They are believed to be formed by the extension of GGPP by condensation with IPP, to give geranylfarnesyl pyrophosphate (GFPP), the parent compound of the class. They are usually found with diterpenoids, particularly in fungi, lichens, seaweeds and higher flowering plants. Structural types have been reviewed by Hanson (1986) and their occurrence by Crews & Naylor (1985).

11.11 CONTROL AND COMPARTMENTATION OF ISOPRENOID BIOSYNTHESIS

The previous sections of this chapter have exemplified the bewildering array of isoprenoids found in plants. These compounds must be produced in different amounts, in several parts of the plant and at various stages of plant growth and development. Since all isoprenoids are derived from the same central pathway (Fig. 11.1), the plant must have control mechanisms that regulate how much, where and when isoprenoids are biosynthesized.

The subcellular locations of the enzymes of isoprenoid formation are clearly of fundamental importance in understanding the regulation of synthesis of these molecules. Extensive studies have been carried out on the location of the enzymes catalyzing the reactions from acetyl CoA. Within the plant cell there are three compartments

that form isoprenoids: the cytosol, the plastid and the mitochondrion. Within these compartments some enzymes are membrane-bound, whereas others are soluble or loosely attached to membranes. Largely based upon investigations in the 1960s and 1970s, two opposing hypotheses have been proposed concerning the subcellular location of the individual enzymes of the pathway.

The pioneering studies of Goodwin and co-workers in the 1960s used labeling experiments with $^{14}CO_2$ and [2-^{14}C]MVA to investigate the sites of isoprenoid biosynthesis in maize. They concluded that the three different compartments are separated by membranes that are impermeable to the intermediates of the pathway (Goodwin & Mercer, 1963; Goodwin, 1965). In contrast, the later, alternative suggestion is that the formation of IPP occurs solely in the cytosol and then it enters the other subcellular compartments as a precursor of specific isoprenoids (Kreuz & Kleinig, 1984). The experimental evidence which formed the basis of the two hypotheses has been critically reviewed by Gray (1987) and Kleinig (1989). Two difficulties in any interpretations of the experimental data are that it is extremely difficult to isolate pure, undamaged organelles and, secondly, different plant tissues may have alternative patterns of compartmentation, e.g. those containing chloroplasts or chromoplasts.

Whatever model for the compartmentation of intermediates is correct, the mechanisms that regulate the flux through the pathway must also be understood. The many branches in the pathway may suggest that control is exerted at each branch point, perhaps by feedback inhibition. This may be so, but insufficient experimental evidence is available on the properties of the enzymes at each branch point of the pathway, especially in a single species. In addition, measurements such as flux through the pathway using for example the technique of Croteau & Gershenzon (1994), maximum catalytic activities *in vivo* and substrate concentrations in the cell are required.

More recently, considerable progress has been made on cloning genes of the pathway including those encoding enzymes of putative rate-limiting steps. These can be used to measure expression patterns of genes during cell development and differentiation. They include genes for HMG-CoA reductase (HMGR: e.g. Narita & Gruissem, 1989; Choi *et al.*, 1992), phytoene synthase (*Psy*, Ray *et al.*, 1992; Bartley *et al.*, 1992; Bartley & Scolnik, 1993) and the cyclases for the mono-, sesqui-, and diterpenes (Colby *et al.*, 1993; Facchini & Chappel, 1992; Mau & West, 1994). Gene cloning has also led to the discovery of isoprenoid gene families. For example, a family of HMGR genes has been isolated from individual higher plants (Yang *et al.*, 1991; Choi *et al.*, 1992; Enjuto *et al.*, 1994) which have separate roles in the plant cell and unlike animals which contain a single HMGR gene (Chin *et al.*, 1984). These patterns of gene expression suggest that isoenzymes of HMGR, dedicated to the formation of specific isoprenoid classes, are regulated independently from each other. This possibility has been discussed recently by Chappell (1995). A similar level of sophistication of the control of isoprenoid gene expression is shown by the formation of carotenoids in the tomato. There are at least two phytoene synthase (*Psy*) genes, one of which is expressed in green tissues, whereas the other is up-regulated during fruit ripening (Bartley *et al.*, 1992; Bartley & Scolnik, 1993; Fray & Grierson, 1993). The products of the reactions catalyzed by the two gene products are identical (Fraser and Bramley, unpublished observations).

It is very likely that our understanding of the control of isoprenoid biosynthesis will advance significantly in the next few years, as more gene families are characterized and accurate, specific assays of enzyme levels and metabolic fluxes are developed. The increasing availability of transgenic plants, with altered levels of isoprenoids, will also aid in the detection of rate-limiting steps.

REFERENCES

Addicott, F.T. (1983) *Abscisic Acid*. Praeger, New York.

Akihisa, T., Kokke, W.C.M.C. & Tamura, T. (1991) In *Physiology and Biochemistry of Sterols* (G.W. Patterson & W.D. Nes, eds), pp. 172–228. American Oil Chemists' Society, Champaign, Illinois.

Alonso, W.R. & Croteau, R. (1993) In *Methods in Plant Biochemistry* (P.J. Lea, ed.), vol. 9, pp. 239–260. Academic Press, London.

Alonso, W.R., Rajaonarivory, J.M., Gershenzon, J. & Croteau, R. (1992) *J. Biol. Chem.* **267**, 7582–7587.

Al-Saadawi, I.S., Arif, M.B. & Al Rubeaa, A.J. (1985) *J. Chem. Ecol.* **11**, 1527–1534.

Archer, C.J. (1994) PhD thesis, University of London.

Archer, B.L. & Audley, B.G. (1973) In *Phytochemistry* (G.P. Miller, ed.), vol. 2, pp. 310–343. Van Nostrand Reinhold, New York.

Archer, B.L., Ayrey, G., Cockbain, E.G. & McSweeney, G.P. (1961) *Nature* **189**, 663–664.

Armstrong, G.A. (1994) *J. Bacteriol.* **176**, 4795–4802.

Bach, T.J., Wettstein, A., Boronat, A., Ferrer, A., Enjuto, M., Gruissem, W. & Narita, J.O. (1991) In *Physiology and Biochemistry of Sterols* (G.W. Patterson & W.D. Nes, eds), pp. 29–49. American Oil Chemists' Society, Champaign, Illinois.

Banthorpe, D.V. (1991) In *Methods in Plant Biochemistry* (B.V. Charlwood & D.V. Banthorpe, eds), vol. 7, pp. 1–41. Academic Press, London.

Banthorpe, D.V. & Branch, S.A. (1983) *Terpenoids and Steroids*. Specialist Periodical Report of the Chemical Society, Vol. 12, pp. 1–74. Royal Institute of Chemistry, London.

Banthorpe, D.V. & Charlwood, B.V. (1980) In *Encyclopaedia of Plant Physiology*, New Series, Vol. 8 (E.A. Bell & B.V. Charlwood, eds), pp. 185–220. Springer-Verlag, Berlin.

Bartley, G.E. & Scolnik, P.A. (1993) *J. Biol. Chem.* **268**, 25718–25721.

Bartley, G.E. & Scolnik, P.A. (1994) *Annu. Rev. Plant Physiol. Plant Mol. Biol.* **45**, 287–301.

Bartley, G.E., Viitanen, P.V., Bacot, K.D. & Scolnik, P.A. (1992) *J. Biol. Chem.* **267**, 5036–5039.

Beale, M.H. & Willis, C.L. (1991) In *Methods in Plant Biochemistry* (B.V. Charlwood & D.V. Banthorpe, eds), Vol. 7, pp. 289–330. Academic Press, London.

Belingheri, L., Pauly, G., Gleizes, M. & Marpeau, A. (1988) *J. Plant Physiol.* **132**, 80–85.

Bensen, R.J., Johal, G.S., Crane, V.C., Tossberg, J.J., Schnable, P.S., Meeley, R.B. & Briggs, S.P. (1995) *Plant Cell* **7**, 75–84.

Benveniste, P. (1986) *Annu. Rev. Plant Physiol.* **37**, 275–308.

Beyer, P., Mayer, M. & Kleinig, H. (1989) *Eur. J. Biochem.* **184**, 141–149.

Bramley, P.M. (1993a) In *Methods in Plant Biochemistry*, Vol. 9 (P.J. Lea, ed.), pp 281–297. Academic Press, London.

Bramley, P.M. (1993b) In *Carotenoids in Photosynthesis* (A.J. Young & G. Britton, eds), pp. 127–159. Chapman & Hall, London.

Bray, E.A. & Zeevart, J.A.D. (1985) *Plant Physiol.* **79**, 719–722.

Britton, G. (1985) *Methods Enzymol.* **111**, 113–149.

Britton, G. (1989) *Nat. Prod. Rep.* **6**, 359–404.

Britton, G. (1990) In *Carotenoids: Chemistry and Biology* (N.I. Krinsky, M.M. Mathews-Roth & R.F. Taylor, eds), pp. 167–184. Plenum, New York.

Britton, G. (1991) In *Methods in Plant Biochemistry*, Vol. 7 (B.V. Charlwood & D.V. Banthorpe, eds), pp. 473–518. Academic Press, London.

Britton, G. (1993) In *Carotenoids in Photosynthesis* (A.J. Young & G. Britton, eds), pp. 96–126. Chapman & Hall, London.

Brown, D.J. & Dupont, F.M. (1989) *Plant Physiol.* **90**, 955–961.

Brown, M.S. & Goldstein, J.L. (1986) *Science* **232**, 34–47.

Burden, R.S., Carter, G.A., Clark, T., Cooke, D.T., Croker, S.J., Deas, A.H.B., Hedden, P., James, C.S. & Lenton, J.R. (1987) *Pestic. Sci.* **21**, 253–267.

Caelles, C., Ferrer, A., Balcells, L., Hegardt, F.G. & Boronat, A. (1989) *Plant Mol. Biol.* **13**, 627–638.

Camara, B. & Moneger, R. (1982) *Physiol. Veg.* **20**, 757–773.

Campos-Lopez, E. & Palacios, J. (1976) *J. Polymer Sci. Polymer Chem. Ed.* **14**, 1561–1563.

Cane, D.E. (1981) In *Biosynthesis of Isoprenoid Compounds*, Vol. 1 (J.W. Porter & S.L. Spurgeon, eds), pp. 283–374. Wiley-Interscience, New York.

Cane, D.E. (1984) In *Enzyme Chemistry: Impact and Applications* (C.J. Suckling, ed.), pp. 210–231. Chapman & Hall, London.

Cane, D.E. (1990) *Chem. Rev.* 90, 1089–1103.

Cattel, L., Ceruti, M., Delprino, L., Balliano, G., Duriatti, A. & Bouvier-Nave, P. (1986) *Lipids* 21, 31–38.

Chappell, J. (1995) *Plant Physiol.* 107, 1–6.

Charlwood, B.V. & Banthorpe, D.V. (1991) In *Methods in Plant Biochemistry* (B.V. Charlwood & D.V. Banthorpe, eds), Vol. 7, pp. 537–542. Academic Press, London.

Charlwood, B.V. & Charlwood, K.A. (1991a) In *Methods in Plant Biochemistry* (B.V. Charlwood & D.V. Banthorpe, eds), Vol. 7, pp. 43–98. Academic Press, London.

Charlwood, B.V. & Charlwood, K.A. (1991b) *Proc. Phytochem. Soc. Eur.* 31, 95–132.

Chen, A., Kroon, P.A. & Poulter, C.D. (1994) *Protein Sequence* 3, 600–607.

Chin, D.J., Gil, G., Russell, D.W., Liscum, L., Laskey, K.L., Basu, S.K., Okayama, H., Berg, P., Goldstein, J.L. & Brown, M.S. (1984) *Nature* 308, 613–617.

Choi, D., Ward, B.L. & Bostock, R.M. (1992) *Plant Cell* 4, 1333–1344.

Clastre, M., Bantignies, B., Feron, G., Soler, E. & Ambid, C. (1993) *Plant Physiol.* 102, 205–211.

Cogdell, R.J. & Frank, H.A. (1987) *Biochim. Biophys. Acta* 895, 63–79.

Cohen, A. & Bray, E.A. (1990) *Planta* 182, 27–33.

Colby, S.M., Alonso, W.R., Katahira, E.J. McGarvey, D.J. & Croteau, R. (1993) *J. Biol. Chem.* 268, 23016–23024.

Connolly, J.D. & Hill, R.A. (1991) In *Methods in Plant Biochemistry* (B.V. Banthorpe & D.V. Banthorpe, eds), Vol. 7, pp. 361–368. Academic Press, London.

Cordell, G.A. (1976) *Chem. Rev.* 76, 425–460.

Crews, P. & Naylor, S. (1985) *Fortschr. Chem. Org. Naturst.* 48, 203–273.

Croteau, R. (1987) *Chem. Rev.* 87, 929–954.

Croteau, R. & Cane, D.E. (1985) *Methods Enzymol.* 110, 383–405.

Croteau, R. & Gershenzon, J. (1994) In *Genetic Engineering of Plant Secondary Metabolism* (B.E. Ellis, ed.), pp. 193–229. Plenum Press, New York.

Croteau, R. & Purkett, P.J. (1989) *Archiv. Biochem. Biophys.* 271, 525–535.

Croteau, R., Karp, F., Wagschal, K.C., Satterwhite, D.M., Hyatt, D.C. & Skotland, C.B. (1991) *Plant Physiol.* 96, 744–753.

Davies, B.H. (1976) In *Chemistry and Biochemistry of Plant Pigments*, Vol. 2 (T.W. Goodwin, ed.), pp. 38–165. Academic Press, London.

Deans, S.G. & Ritchie, G. (1987) *Int. J. Food Microbiol.* 5, 165–180.

Defago, G. (1977) *Ber. Schweiz Bot. Gaz.* 87, 79–132.

Demmig-Adams, B. & Adams, W.W. (1993) In *Carotenoids in Photosynthesis* (A.J. Young & G. Britton, eds), pp. 206–251. Chapman & Hall, London.

Dev, S., Narula, A.P.S. & Yadav, J.S. (1982) *CRC Handbook of Terpenoids*, vols 1 and 2. CRC Press, Boca Raton, Florida.

Dogbo, O., Bardat, F., Laferriere, A., Quennemet, J., Brangeon, J. & Camara B. (1987) *Plant Sci.* 49, 89–101.

Dogbo, O., Laferriere, A., d'Harlingue, A. & Camara, B. (1988) *Proc. Natl. Acad. Sci. U.S.A.* 85, 7054–7058.

Duncan, J.D. & West, C.A. (1981) *Plant Physiol.* 68, 1128–1134.

Duperon, P. & Duperon, R. (1973) *Physiol. Veg.* 11, 487–505.

Dyas, L. & Goad, L.J. (1993) *Phytochemistry* 34, 17–29.

Enjuto, M., Balcells, L., Campos, N., Caelles, A.M. & Boronat, A. (1994) *Proc. Natl. Acad. Sci. U.S.A.* 91, 927–931.

Facchini, P.J. & Chappell, J. (1992) *Proc. Natl. Acad. Sci. U.S.A.* 89, 11088–11092.

Fall, P.R. & West, C.A. (1971) *J. Biol. Chem.* 246, 6913–6928.

Fischer, N.H. (1991) In *Methods in Plant Biochemistry* (B.V. Charlwood & D.V. Banthorpe, eds), Vol. 7, pp. 187–211. Academic Press, London.

Fishwick, M.J. & Wright, A.J. (1980) *Phytochemistry* 19, 5–59.

Fraga, B.M. (1991) In *Methods in Plant Biochemistry* (B.V. Charlwood & D.V. Banthorpe, eds), Vol. 7, pp. 145–185. Academic Press, London.

Frank, H.A. & Cogdell, R.J. (1993) In *Carotenoids in Photosynthesis* (A. Young & G. Britton, eds), pp. 252–326. Chapman & Hall, London.

Fraser, P.D., Truesdale, M.R., Bird, C.R., Schuch, W. & Bramley, P.M. (1994) *Plant Physiol.* 105, 405–413.

Fray, R. & Grierson, D. (1993) *Plant Mol. Biol.* 22, 584–602.

Gershenzon, J. & Croteau, R. (1990) *Rec. Adv. Phytochem.* 24, 99–160.

Giraudat, J., Parcy, F., Bertauche, N., Gosti, F., Leung, J., Morris, P-C. Bouvier-Durand, M. & Vertanian, N. (1994) *Plant Mol. Biol.* 26, 1557–1577.

Goad, L.J. (1977) In *Lipids and Lipid Polymers in Higher Plants* (M. Tevini & H.K. Lichtenthaler, eds), pp. 146–168. Springer-Verlag, Berlin.

Goad, L.J. (1983) *Biochem. Soc. Trans.* 11, 548–552.

Goad, L.J. (1991) In *Methods in Plant Biochemistry* (B.V. Charlwood & D.V. Banthorpe, eds), Vol. 7, pp. 369–434. Academic Press, London.

Goad, L.J., Haughan, P.A. & Lenton, J.R. (1988) *British Plant Growth Regulator Group*, Monograph 17, pp. 91–105.

Gondet, L., Bronner, R. & Benveniste, P. (1994) *Plant Physiol.* 105, 509–518.

Goodwin, T.W. (1958) *Biochem. J.* 70, 612.

Goodwin, T.W. (1965) In *Biosynthetic Pathways in Higher Plants* (J.B. Pridham & T. Swain, eds), pp. 55–71. Academic Press, London.

Goodwin, T.W. (1980) *Biochemistry of Carotenoids*, Vol. 1. Chapman & Hall, London.

Goodwin, T.W. (1983) *Biochem. Soc. Trans.* 11, 473–483.

Goodwin, T.W. & Britton, G. (1988) In *Plant Pigments* (T.W. Goodwin, ed.), pp. 61–132. Academic Press, London.

Goodwin, T.W. & Mercer, E.I. (1963) In *The Control of Lipid Metabolism* (J.K. Grant, ed.), pp. 37–40. Academic Press, London.

Graebe, J.E. (1969) *Planta* 85, 171–174.

Graebe, J.E. (1987) *Annu. Rev. Plant Physiol.* 38, 419–465.

Gray, J.C. (1987) *Adv. Bot. Res.* 14, 25–91.

Grayson, D.H. (1984) *Nat. Prod. Reports* 1, 319–335.

Grayson, D.H. (1987) *Nat. Prod. Reports* 4, 377–397.

Grayson, D.H. (1988) *Nat. Prod. Reports* 5, 419–464.

Hanley, K. & Chappell, J. (1992) *Plant Physiol.* 98, 215–220.

Hanson, J.R. (1986) *Nat. Prod. Reports* 3, 123–132.

Hartung, W., Gimmler, H., Heilman, B. & Kaiser, G. (1982) In *Plant Growth Substances* (P.F. Wareing, ed.), pp. 325–333. Academic Press, London.

Haughan, P.A., Lenton, J.R. & Goad, L.J. (1988) *Phytochemistry* 27, 2941–2500.

Haughan, P.A., Burden, R.S., Lenton, J.R. & Goad, L.J. (1989) *Phytochemistry* 28, 781–787.

Hedden, P. (1993) *Annu. Rev. Plant Physiol. Plant Mol. Biol.* 44, 107–129.

Hedden, P. & Phinney, B.O. (1976) *Plant Physiol.* 57, suppl. 86, 1–107.

Hedden, P. & Phinney, B.O. (1979) *Phytochemistry* 18, 1475–1479.

Hedden, P. & Graebe, J.E. (1981) *Phytochemistry* 20, 1011–1015.

Hedden, P., MacMillan, J. & Phinney, B.O. (1978) *Annu. Rev. Plant Physiol.* 29, 149–192.

Hemming, F.W. (1983) *Biochem. Soc. Trans.* 11, 497–504.

Hill, R.A., Kirk, D.N., Makin, H.L.J. & Murphy, G.M. (1991) *Dictionary of Steroids.* Chapman & Hall, London.

Hohn, T.M. & Beremand, P.D. (1989) *Gene* 79, 131–138.

Hooley, R. (1994) *Plant Mol. Biol.* 26, 1529–1555.

Hostettmann, K., Hostettmann, M. & Marston, A. (1991) In *Methods in Plant Biochemistry* (B.V. Charlwood & D.V. Banthorpe, eds), Vol. 7, pp. 435–471. Academic Press, London.

Inouye, H. (1991) In *Methods in Plant Biochemistry* (B.V. Charlwood & D.V. Banthorpe, eds), Vol. 7, pp. 99–143. Academic Press, London.

IUPAC-IUB (1967) *Eur. J. Biochem.* 10, 1–9; *Eur. J. Biochem.* 25, 1–3; *Pure Appl. Chem.* 31, 285–322.

IUPAC-IUB (1975) *Pure Appl. Chem.* 41, 407–413.

IUPAC-IUB (1989) *Eur. J. Biochem.* 186, 13–30.

Janssen, G.G., Kalinowska, M., Noeton, R.A. & Nes, W.D. (1991) In *Physiology and Biochemistry of Sterols* (G.W. Patterson & W.D. Nes, eds), pp. 83–117. American Oil Chemists' Society, Champaign, Illinois.

Jeffrey, S.W., Douce, R. & Benson, A.A. (1974) *Proc. Natl. Acad. Sci.* 71, 807–810.

Jones, B.L. & Porter, J.W. (1986) *CRC Crit. Rev. Plant Sci.* 3, 295–324.

Joyard, J., Bloch, M.A. & Douce, R. (1991) *Eur. J. Biochem.* 199, 489–509.

Karp, F., Harris, J.L. & Croteau, R. (1987) *Archiv. Biochem. Biophys.* 256, 179–193.

Karp, F., Mihaliak, C.A., Harris, J.L. & Croteau, R. (1990) *Arch. Biochem. Biophys.* 276, 219–226.

Kirk, J.T.O. & Tilney-Bassett, R. (1978) *The Plastids*, 2nd edn. Freeman, San Francisco.

Kleinig, H. (1989) *Annu. Rev. Plant Physiol.* 40, 39–59.

Koshimizu, K., Inui, M., Fukui, H. & Mitsui, T. (1968) *Agric. Biol. Chem.* 32, 789–791.

Koyama, Y. (1991) *J. Photochem. Photobiol.* 9B, 265–280.

Kreuz, K. & Kleinig, H. (1984) *Eur. J. Biochem.* 141, 531–535.

Kreuz, K., Beyer, P. & Kleinig, H. (1982) *Planta* 154, 66–69.

Kushwaha, S.C., Subbarayan, C., Beeler, D.A. & Porter, J.W. (1970) *J. Biol. Chem.* 245, 4708–4717.

Lange, T. (1970) *Annu. Rev. Plant Physiol.* 21, 531–570.

Lange, T. & Graebe, J.E. (1993) In *Methods in Plant Biochemistry* (P.J. Lea, ed.), Vol. 19, pp. 403–430. Academic Press, London.

Langworthy, T.A. & Pond, J.L. (1986) *Appl. Microbiol.* 7, 253–258.

Lawrence, B.M. (1981) In *Essential Oils* (B.D. Mookherjee & C.J. Mussiham, eds), pp. 1–81. Allwed, Wheaton, USA.

Learned, R.M. & Fink, G.R. (1989) *Proc. Natl. Acad. Sci. U.S.A.* 86, 2779–2783.

Loomis, W.D. & Croteau, R. (1973) *Recent Adv. Phytochem.* 6, 147–185.

Lutke-Brinkaus, F., Liedvogel, B., Kreuz, B. & Kleinig, H. (1982) *Planta* 156, 176–180.

Lynen, F. & Hemming, U. (1960) *Angew. Chem.* 72, 820–829.

Macaskill, D. & Croteau, R. (1993) *Anal. Biochem.* 215, 142–149.

Mau, C.J. & West, C.A. (1994) *Proc. Natl. Acad. Sci. U.S.A.* 91, 8479–8501.

May, P.M. (1990) In *Comprehensive Medicinal Chemistry* (C. Hansch, P.G. Summes & J.B. Taylor, eds), pp. 206–209. Pergamon Press, Oxford.

McDermott, J.C.B., Britton, G. & Goodwin, T.W. (1973) *Biochem. J.* 134, 1115–1117.

Milborrow, B.V. (1983) In *Abscisic Acid* (F.J. Addicott, ed.), pp. 79–111. Praeger, New York.

Milborrow, B.V. & Netting, A.G. (1991) In *Methods in Plant Biochemistry*, Vol. 7 (B.V. Charlwood & D.V. Banthorpe, eds), pp. 231–261. Academic Press, London.

Milborrow, B.V. & Vaughan, G. (1982) *Aust. J. Plant Physiol.* 9, 361–372.

Mittelhauser, C.J. & Van Steveninck, R.F.M. (1969) *Nature* 221, 281–282.

Moore, T.C. & Ecklund, P.R. (1974) In *Plant Growth Substances* pp. 252–259. Hirokawa, Tokyo.

Murphy, P.J. & West, C.A. (1969) *Archiv. Biochem. Biophys.* 133, 395–407.

Narita, J.O. & Gruissem, W. (1989) *Plant Cell* 1, 181–190.

Neill, S.J. & Horgan, R. (1985) *J. Exp. Bot.* 36, 1222–1231.

Nes, W.R. & MacLean, M.L. (1977) *Biochemistry of Steroids and Other Isoprenoids.* University Park Press, Baltimore, Maryland.

Ourisson, G. (1994) *J. Plant Physiol.* 143, 434–439.

Parry, A.D. (1993) In *Methods in Plant Biochemistry*, Vol. 9 (P.J. Lea, ed.), pp. 381–402. Academic Press, London.

Parry, A.D., Babiano, M.J. & Horgan, R. (1990) *Planta* **182**, 118–128.

Pascal, S., Taton, M. & Rahier, A. (1990) *Biochem. Biophys. Res. Commun.* **172**, 98–106.

Pascal, S., Taton, M. & Rahier, A. (1993) *J. Biol. Chem.* **268**, 11639–11654.

Pascal, S., Taton, M. & Rahier, A. (1994) *Archiv. Biochem. Biophys.* **312**, 260–271.

Peter, G.F. & Thornber, J.P. (1991) In *Methods in Plant Biochemistry*, Vol. 9 (L.J. Rogers, ed.), pp. 195–210. Academic Press, London.

Preiss, B. (1985) *Regulation of HMG-CoA Reductase*. Academic Press, New York.

Pyun, H-J., Wagschal, K.C., Jung, D-I., Coates, R.M. & Croteau, R. (1994) *Arch. Biochem. Biophys.* **308**, 488–496.

Rahier, A. & Taton, M. (1986) *Biochem. Biophys. Acta* **1125**, 215–222.

Rajaonarivony, J.I.M., Gershenzon, J. & Croteau, R. (1992a) *Arch. Biochem. Biophys.* **296**, 49–57.

Rajaonarivony, J.I.M., Gershenzon, J., Miyazaki, J. & Croteau, R. (1992b) *Arch. Biochem. Biophys.* **299**, 77–82.

Ray, J., Moureau, P., Bird, C.R., Bird, A., Grierson, D., Maunders, M., Trausdale, M., Bramley, P.M. & Schuch, W. (1992) *Plant Mol. Biol.* **19**, 401–404.

Rees, H.H., Goad, L.J. & Goodwin, T.W. (1968) *Tetrahedron Lett.* 723–725.

Ruzicka, L. (1959) *Proc. Chem. Soc. Lond.* 341–360.

Ruzicka, L., Eschenmoser, A. & Heusser, H. (1953) *Experientia* **9**, 357–367.

Sabine, J.R. (ed.) (1983) *Monographs on Enzyme Biology: HMG-CoA Reductase*. CRC Press, Boca Raton, Florida.

Sandmann, G. (1991) *Physiol. Plant.* **83**, 186–193.

Sandmann, G. (1994a) *J. Plant Physiol.* **143**, 444–447.

Sandmann, G. (1994b) *Eur. J. Biochem.* **223**, 1–7.

Scallen, T.J., Dean, W.J. & Schuster, M.W. (1968) *J. Biol. Chem.* **243**, 5202–5206.

Schilcher, H. (1985) In *Essential Oils and Aromatic Plants* (A. Baerheim Svendsen & J.J.C. Scheffer, eds), pp. 217–231. Martinus Nijhoff/Dr W. Junk, Dordrecht.

Siefermann-Harms, D. (1985) *Biochim. Biophys. Acta*.

Skriver, K. & Mundy, J. (1990) *Plant Cell* **2**, 505–512.

Spray, C.R. & Phinney, B.O. (1987) In *Ecology and Metabolism of Plant Lipids* (G. Fuller & W.D. Nes, eds), pp. 25–43. American Chemistry Society.

Straub, O. (1987) *Key to Carotenoids*. Birkhauser Verlag, Basel.

Tanaka, Y. (1991) In *Methods in Plant Biochemistry* (B.V. Charlwood & D.V. Banthorpe, eds), Vol. 7, pp. 519–536. Academic Press, London.

Taton, M., Salmon, F., Pascal, S. & Rahier, A. (1994) *Plant Physiol. Biochem.* **32**, 751–760.

Threlfall, D. (1980) In *Secondary Plant Products. Encyclopedia of Plant Physiology* (E.A. Bell & B.V. Charlwood, eds), Vol. 8, pp. 288–398. Springer Verlag, Berlin.

Torsell, K.B.G. (1983) *Natural Product Chemistry. A Mechanistic and Biosynthetic Approach to Secondary Metabolism*. Wiley, Chichester.

Wagschal, K.L., Pyun, H-J., Coates, R.M. & Croteau, R. (1994) *Arch. Biochem. Biophys.* **308**, 477–487.

Wettstein, A., Caelles, C., Boronat, A., Jenke, H-S. & Bach, T.J. (1989) *Biol. Chem. Hoppe-Seyler* **370**, 806–807.

Yamamoto, S. & Bloch, K. (1970) *J. Biol. Chem.* **245**, 1670–1674.

Yang, Z., Park, H., Lacy, G.H. & Cramer, C.L. (1991) *Plant Cell* **3**, 397–405.

Young, A.J. (1993a) In *Carotenoids in Photosynthesis* (A.J. Young & G. Britton, eds), pp. 16–71. Chapman & Hall, London.

Young, A.J. (1993b) In *Carotenoids in Photosynthesis* (A.J. Young & G. Britton, eds), pp. 72–95. Chapman & Hall, London.

Young, A.J. (1993c) In *Carotenoids in Photosynthesis* (A.J. Young & G. Britton, eds), pp. 161–205. Chapman & Hall, London.

Zeevart, J.A.D. & Creelman, R.A. (1988) *Annu. Rev. Plant Physiol. Plant Mol. Biol.* **39**, 439–473.

12   Special Nitrogen Metabolism

Michael Wink

12.1   INTRODUCTION

Nitrogen is the major limiting nutrient for most plants. The main exogenous sources are atmospheric N_2 or soil-derived nitrate (from fertilizers, manure, degradation of organic matter) which enter plant metabolism by nitrogen fixation and nitrate reduction (see Chapter 7). The resulting NH_4^+ is used to build up the various amino acids. Twenty amino acids constitute the basic building units for all cellular proteins and also provide the nitrogen and/or carbon skeletons for various metabolites of low molecular weight.

Nitrogen-containing compounds which are important for the functioning of every plant include purine and pyrimidine bases and derived nucleosides and nucleotides (used in DNA and RNA synthesis), cofactors (NAD^+, FAD, CoA, TPP etc.) and growth factors ('hormones') such as auxins, cytokinins and ethylene. Their biosynthesis and function is covered in this chapter (sections 12.6 and 12.7).

In addition, plants produce a wide variety of secondary metabolites. Many of them function as allelochemicals and serve as chemical defense compounds against herbivores, micro-organisms or competing plants. On the other hand, colors or scents attract pollinating insects and seed- or fruit-dispersing animals (Harborne, 1993; Schlee, 1992). It had been assumed earlier in this century that secondary metabolites have no specific function or were considered to be nitrogenous waste products, similar to the function of urea or uric acid in animals. Since nitrogen is a limiting nutrient, the latter assumption is highly unlikely and not supported by experimental data. In addition, alkaloids are often mobilized in senescing plant parts. This would be a surprising fate for any metabolic waste product. Furthermore, experimental or circumstantial evidence indicates that many secondary metabolites are important for the fitness of plants (Wink, 1988, 1993a). Examples are given later in this chapter.

Over 13 000 nitrogen-containing secondary metabolites have been described so far (Southon & Buckingham, 1989). Because only 10% of all plants have been analyzed with modern phytochemical techniques (HPLC, capillary GLC, mass spectrometry, NMR) it can safely be assumed that the real number is significantly larger. Alkaloids, amines, non-protein amino acids, cyanogenic glycosides and glucosinolates are the main compounds in this category which are discussed in this chapter (sections 12.2–12.5) in terms of biosynthesis, compartmentation and function. Because of the large number of compounds concerned, only a few representative examples have been selected here. More detailed overviews can be found in Bell & Charlwood (1980), Conn (1981), Robinson (1981), Teuscher & Lindequist (1994), Luckner (1990), Mothes *et al.* (1985), Rosenthal & Janzen (1979), Rosenthal & Berenbaum (1991).

Most nitrogen-containing secondary compounds are derived from amino acids which donate the carbon skeleton and/or the nitrogen. Besides the 20 protein-building amino acids, others such as ornithine are also important precursors. Very often, the amino acid is decarboxylated by a pyridoxal phosphate containing decarboxylase in the first step. Other enzymes which have been found to be associated with the biosynthesis of nitrogen-containing secondary metabolites include transaminases, amine oxidases, oxidoreductases, peroxidases, hydrolases, N- or O-methyltransferases, acyl-CoA transferases, mono- and dioxygenases, phenolases and several 'synthases' which, in general, are specific and stereo-selective enzymes showing a high affinity for their particular substrate (Waller & Dermer, 1981). Many N-containing natural compounds have evolved in Nature as defense compounds, and they affect molecular targets in animal cells, such as receptors of neurotransmitters (e.g. nicotine, ephedrine, ergot alkaloids, β-carboline alkaloids) (see Section 12.5.8). To fulfil this function the molecules need to be synthesized in a stereochemically 'correct' configuration (Floss, 1981). Therefore, the involvement of highly specific enzymes in the biosynthesis of alkaloids is not surprising. Even the β-glucosidases of plants show substantial specificity. This is plausible since they must liberate preformed defense chemicals, such as glucosinolates or cyanogens, in case of emergency (Fig. 12.15). Less specific are degrading enzymes only, such as some esterases, other hydrolases and peroxidases, which may serve to break down microbial phytotoxins and thus must cope with a broader substrate spectrum. Earlier views, which were not supported by enzymic studies, had assumed that secondary metabolites are either formed without the aid of enzymes or are made by some general and unspecific enzymes. Compared to the large number of secondary metabolites, the enzymatic details leading to their biosynthesis have been worked out for a limited number of instances only. Here, a wide field for research lies wide open.

Because alkaloids and other nitrogen-containing allelochemicals mainly serve as defense compounds, they need to be present in sufficiently high concentrations and at the right place and time. This demands that secondary metabolism must be highly co-ordinated which can be seen from the observation that alkaloid formation often depends on the developmental stage of a plant. Their contents and composition may vary in an annual or even a diurnal cycle. Allelochemical effects are dose-dependent. To be effective, plants need to store sufficient amounts of their defense compounds, usually at a strategic site (often epidermal tissues or plant parts important for reproduction and/or survival, such as bark, flowers, seeds or fruits) and to prevent any intoxication of its own metabolism by storing the allelochemicals in vacuoles, latex, or resin ducts, or dead tissues (Wink, 1993b).

As a common theme, we can observe that plants which produce seeds rich in energy supplies (carbohydrates, lipids, proteins) concomitantly accumulate potent chemical defense compounds, often alkaloids, non-protein amino acids, cyanogenic glycosides, glucosinolates, protease inhibitors, lectins or other toxalbumins. Their presence in seeds can be mutually exclusive, i.e. legume seeds either store alkaloids (e.g. quinolizidines, pyrrolizidines) or non-protein amino acids but not both at the same time. During germination, the breakdown of nutrient reserves is a general procedure and usually includes the nitrogenous defence compounds. They serve a double purpose, i.e. that of N-storage and of protection. They are thus both degradable and toxic N-storage compounds.

In conclusion, besides studying the biosynthesis of these metabolites, knowledge on intra- and inter-cellular transport, cell- or tissue-specific sequestration/accumulation and compartmentation is of prime importance in order to understand and appreciate the functional role of such secondary metabolites in the biology of a plant. Ecological aspects of nitrogen-containing secondary constituents are also considered in Chapter 14.

12.2 NON-PROTEIN AMINO ACIDS

Whereas proteins of all organisms are based on the 20 common L-amino acids (which can be modified post-translationally), more than 900 secondary compounds have been identified in plants, which are classified as amino acids, because they carry a carboxyl and an amino or imino group (Fowden, 1981; Bell, 1980; Rosenthal, 1982, 1991). These amino acids are not building blocks of proteins in plants (therefore called 'non-protein amino acids') but are often antinutrients or antimetabolites, i.e. they may interfere with the metabolism of micro-organisms or herbivores.

12.2.1 Structures in relation to protein amino acids

Many non-protein amino acids resemble protein amino acids (Fig. 12.1) and quite often can be considered to be their structural analogs. They often derive biosynthetically from protein amino acids. Others are made by biosynthetic routes which show little similarity to those concerned in the synthesis of protein amino acids (Fowden 1981;

Protein amino acid · Nonprotein amino acid

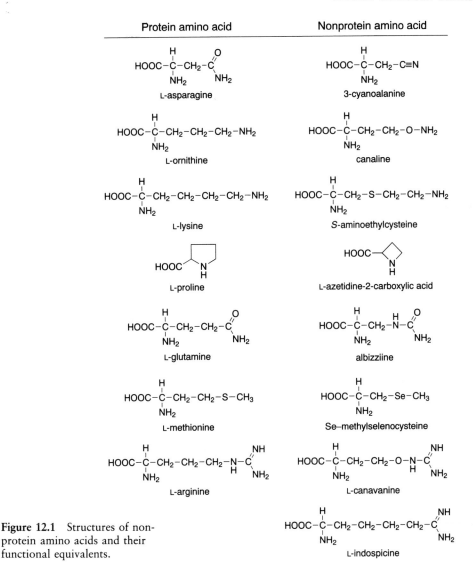

Figure 12.1 Structures of non-protein amino acids and their functional equivalents.

Luckner, 1990). Figure 12.2 presents examples of biogenetic pathways of a few non-protein amino acids which derive from L-cysteine and L-asparagine.

12.2.2 Occurrence and role in plant metabolism

Non-protein amino acids are especially abundant in Leguminosae, Liliaceae and in several higher fungi (e.g. *Amanita muscaria*, *A. pantherina*, *Coprinus* spp.) and marine algae. It has been postulated that they occur more widely but that they are usually not noticed because their concentrations in other plants are extremely low. For example, L-azetidine-2-carboxylic acid, a typical non-protein amino acid of Liliaceae, could be recovered from sugar beet when it was extracted on a sufficiently large scale (Fowden, 1981). Nevertheless, the distribution of non-protein amino acids may be a valuable tool to establish systematic relationships between plant taxa (Fowden, 1981). Organs rich in these metabolites are seeds (Leguminosae) or rhizomes (Liliaceae). Concentrations in seeds can exceed 10% of dry weight and up to 50% of the nitrogen present could be attributed to them. Since non-protein amino acids are often (at least partly) remobilized during germination they certainly function as N-storage compounds in addition to their role as defense chemicals (Rosenthal, 1982).

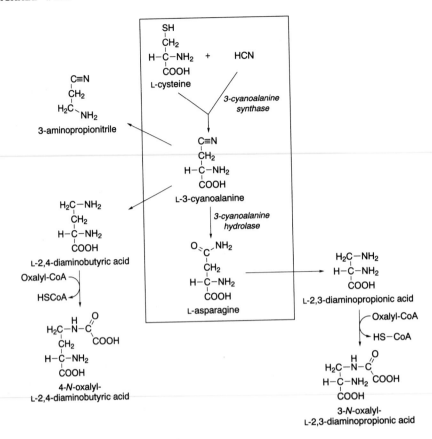

Figure 12.2 Biosynthesis of L-3-cyanoalanine, 3-aminopropionitrile, 2,4-diaminobutyric acid, N^4-oxalyl-L-2,4-diaminobutyric acid, L-2,3 diaminopropionic acid and N^3-oxalyl-L-2,3-diaminopropionic acid in taxa of the genus *Lathyrus*.

12.2.3 Allelochemical effects and ecological roles of non-protein amino acids

If non-protein amino acids are taken up by herbivores, micro-organisms or other plants, they may interfere with their metabolism (Rosenthal, 1982, 1991) in the following ways.

1. They can be accepted in ribosomal protein biosynthesis in place of the normal amino acid leading to defective proteins (e.g. canavanine, azetidine-2-carboxylic acid, 2-amino-4-methylhexenoic acid, etc.).
2. They may inhibit the activation of aminoacyl-tRNA synthetases or other steps of protein biosynthesis.
3. They may competitively inhibit uptake systems for amino acids (e.g. azetidine-2-carboxylic acid, 3,4-dehydroproline).
4. There may be inhibition of amino acid biosynthesis by substrate competition or by

mimicking end-product mediated feedback inhibition of earlier key enzymes in the pathway (e.g. azaserine, albizzine, S-aminoethylcysteine).
5. They may affect other targets, such as DNA-, RNA-related processes (canavanine, mimosine), receptors of neurotransmitters, inhibit collagen biosynthesis (mimosine), or β-oxidation of lipids (L-hypoglycine) (Fowden, 1981; Teuscher & Lindequist, 1994).

In vertebrates, effects may be among others: fetal malformations (L-azetidine-2-carboxylic acid, L-indospicine, mimosine), neurotoxic disturbances (β-cyanoalanine, N-oxalyl-diaminopropionic acid, quisqualic acid), hallucinogenic effects (muscimol, ibotenic acid), hair loss (mimosine), diarrhoea (Se-methyl-L-selenocysteine), paralysis (β-cyano-alanine, cirrhosis of the liver (indospicine), hypoglycemia (L-hypoglycin), and arrhythmia (coprine). Severe intoxications are known from man and domestic animals. For some compounds administered intravenously LD_{50} values are available

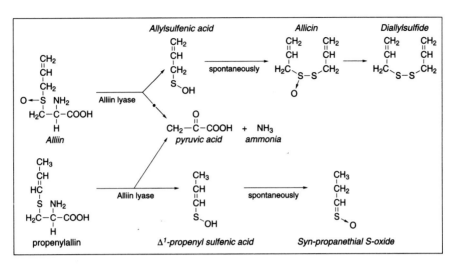

Figure 12.3 Formation of allicin and syn-propanethial-S-oxide after tissue decompartmention.

(Teuscher & Lindequist, 1994): e.g. β-cyano-alanine: rat, 13.4 mg/kg; N-oxalyl-L-α,β-diamino-propionic acid: mouse, 300 mg/kg; muscimol: rat, 4.5 mg/kg, (intraperitoneally, 45 mg/kg). In plants, micro-organisms, and insects non-protein amino acids can reduce growth or even lead to death.

The non-protein amino acids of onions and garlic (*Allium cepa, A. sativum, A. ursinum*) which are derivatives of cysteine are a special case. Alliin and derivatives are not incorporated into protein, but hydrolyzed by alliin lyase to allicin, when tissues are damaged (Fig. 12.3). Another product is propenylsulfenic acid which spontaneously rearranges to *syn*-propanethial S-oxide (Fig. 12.3). Allicin and diallylsulfide display strong antimicrobial activity. *syn*-Propanethial S-oxide functions as the tear-promoting substance (well known to every cook) which is further hydrolyzed to propionaldehyde, sulfuric acid and H_2S. Allicin and *syn*-propanethial S-oxide are strong feeding repellents for herbivores. Thus alliin can be considered as a preformed defense chemical which is activated upon wounding or infection.

It is remarkable that plants which produce non-protein amino acids seem to tolerate these antimetabolites. Their amino acyl-tRNA synthetases discriminate between the protein amino acid and its analogs whereas enzymes from non-adapted organisms do not (Fowden, 1981; Rosenthal, 1991). A similar case has been reported for an adapted bruchid beetle (*Caryedes brasiliensis*) and a weevil (*Sternechus tuberculatus*) which feed on seeds rich in canavanine, which is toxic to non-specialized insects (Rosenthal, 1982, 1991).

From the point of view of plant metabolism and ecology, non-protein amino acids can be considered therefore as 'multipurpose' defense and N-storage compounds. Since they affect a basic target, protein biosynthesis and amino acid metabolism, present in all organisms, they can be ecologically important in plant–plant ('allelopathy'), plant–microbe (bacteria, fungi, plant–virus) and plant–herbivore (insects, vertebrates) interactions. Because of their antimetabolite activities, some non-protein amino acids (e.g. canavanine) are of pharmaceutical interest, for example as antitumor compounds. More details are available in Bell (1980), Rosenthal (1982, 1991) and Teuscher & Lindequist (1994).

12.3 AMINES

Depending on the degree of substitution, it is possible to distinguish primary, secondary, tertiary or quaternary amines and as chemical classes: aliphatic mono- and diamines, polyamines, aromatic amines and amine conjugates. Some tertiary amines may be oxidized to the corresponding N-oxides, which are charged molecules that cannot diffuse easily across biomembranes. Amines behave as cations at physiological hydrogen ion concentrations and show pK_a values between 9 and 11. Amines are widely distributed secondary metabolites in plants and occur as independent products in their own right or as intermediates in the biosynthesis of alkaloids.

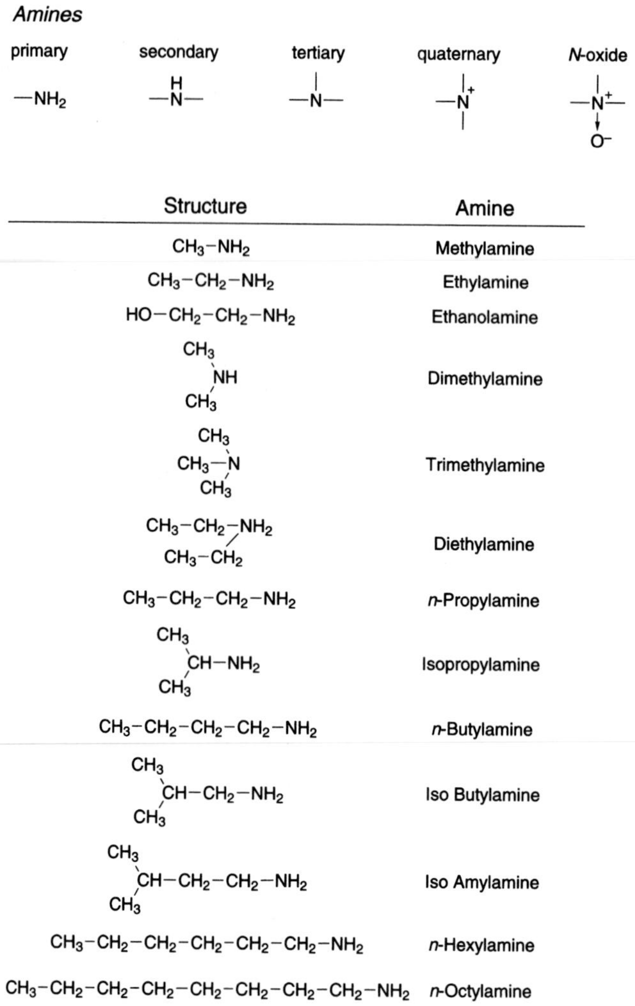

Amines

primary	secondary	tertiary	quaternary	N-oxide
—NH₂	$\overset{\text{H}}{-\text{N}-}$	$-\overset{\mid}{\text{N}}-$	$-\overset{\mid}{\underset{\mid}{\text{N}^+}}-$	$-\overset{\mid}{\underset{\downarrow}{\text{N}^+}}-$ O^-

Structure	Amine
CH_3-NH_2	Methylamine
$CH_3-CH_2-NH_2$	Ethylamine
$HO-CH_2-CH_2-NH_2$	Ethanolamine
$\overset{CH_3}{\underset{CH_3}{\diagdown}}NH$	Dimethylamine
$\overset{CH_3}{\underset{CH_3}{CH_3-N}}$	Trimethylamine
$\overset{CH_3-CH_2-NH_2}{\underset{CH_3-CH_2}{\diagup}}$	Diethylamine
$CH_3-CH_2-CH_2-NH_2$	n-Propylamine
$\overset{CH_3}{\underset{CH_3}{\diagdown}}CH-NH_2$	Isopropylamine
$CH_3-CH_2-CH_2-CH_2-NH_2$	n-Butylamine
$\overset{CH_3}{\underset{CH_3}{\diagdown}}CH-CH_2-NH_2$	Iso Butylamine
$\overset{CH_3}{\underset{CH_3}{\diagdown}}CH-CH_2-CH_2-NH_2$	Iso Amylamine
$CH_3-CH_2-CH_2-CH_2-CH_2-CH_2-NH_2$	n-Hexylamine
$CH_3-CH_2-CH_2-CH_2-CH_2-CH_2-CH_2-CH_2-NH_2$	n-Octylamine

Figure 12.4 Structures of some common aliphatic monoamines.

12.3.1 Aliphatic monoamines

(a) *Structures and biosynthesis*

Aliphatic monoamines (*ca* 30 compounds known from plants) are especially common in flowers of a number of plant families (Rosaceae, Araceae etc.) and in fruiting bodies of some fungi, e.g. *Phallus impudicus* (Fig. 12.4) (Smith, 1980, 1981). In higher plants, aliphatic aldehydes are predominantly aminated by an L-alanine-aldehyde aminotransferase which is dependent on pyridoxal phosphate. This enzyme is mainly localized in mitochondria and seems to be independent from other ubiquitous transaminases (Wink & Hartmann, 1981). The enzyme shows higher affinities for aldehydes with longer chain lengths (>4).

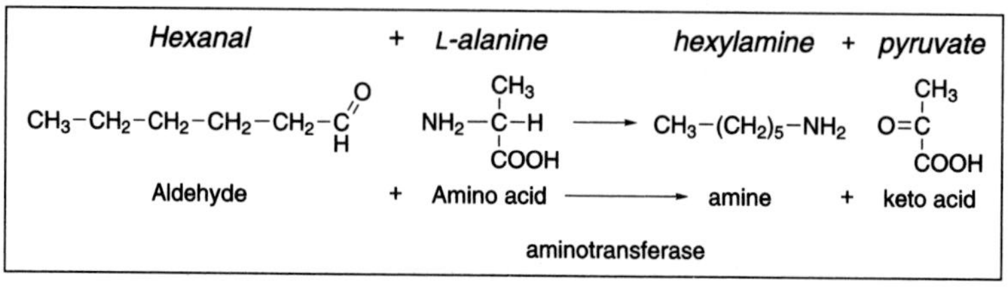

Hexanal	+	L-alanine		hexylamine	+	pyruvate
$CH_3-CH_2-CH_2-CH_2-CH_2-C\overset{\diagup O}{\diagdown_H}$		$NH_2-\overset{CH_3}{\underset{COOH}{C-H}}$	$\longrightarrow$	$CH_3-(CH_2)_5-NH_2$		$O=\overset{CH_3}{\underset{COOH}{C}}$
Aldehyde	+	Amino acid	$\longrightarrow$	amine	+	keto acid

aminotransferase

In many taxa of the Rhodophyceae (red algae) amines are produced from amino acids by aid of an insoluble and rather unspecific pyridoxal phosphate dependent amino acid decarboxylase for which leucine was the best substrate, followed by norleucine, isoleucine, valine, norvaline, 2-aminobutyric acid, phenylalanine, methionine, cysteine and homocysteine.

```
┌─────────────────────────────────────────────────────┐
│  L-Valine                    iso Butylamine          │
│                                                       │
│  CH₃    NH₂                 CH₃                        │
│   |      |                   \                         │
│    CH-C-COOH    ──────►       CH-CH₂-NH₂ + CO₂        │
│   |      |                   /                         │
│  CH₃     H                  CH₃                        │
│                                                       │
│  Amino acid     ──────────►   amine      + CO₂        │
└─────────────────────────────────────────────────────┘
```

(b) Function of aliphatic monoamines

Because these compounds mimic the smell of rotten meat, carrion-feeding insects are attracted which visit the flowers or fruiting bodies of fungi and carry away pollen or spores and thus contribute to pollination or spore distribution. A highly evolved system can be found in members of the Araceae. In *Arum maculatum* (Fig. 12.5), flower morphology, primary and secondary metabolism are highly co-ordinated to achieve a successful fertilization. The spadix is rich in starch. When the flower is mature, starch is broken down rapidly, resulting in pronounced thermogenesis. As a result, the club of the spadix becomes quite warm. This is important for the dissipation of aliphatic monoamines (especially dimethylamine, ethylamine, dibutylamine, isobutylamine and isoamylamine), which are synthesized just at this time by action of a mitochondrial alanine-aldehyde aminotransferase (Wink & Hartmann, 1981). The smell of the amines attracts pollinating insects, which follow the smell and end up on the spathe. From there they glide into the main vessel containing the male and female reproductive organs. The insects cannot escape easily from this cavity because the entrance is blocked by special hairs (Fig. 12.5). If insects already carry pollen, they come into contact with the stigma

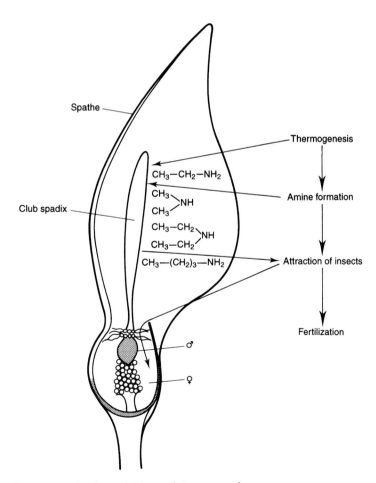

Figure 12.5 Role of amines in the flower biology of *Arum maculatum*.

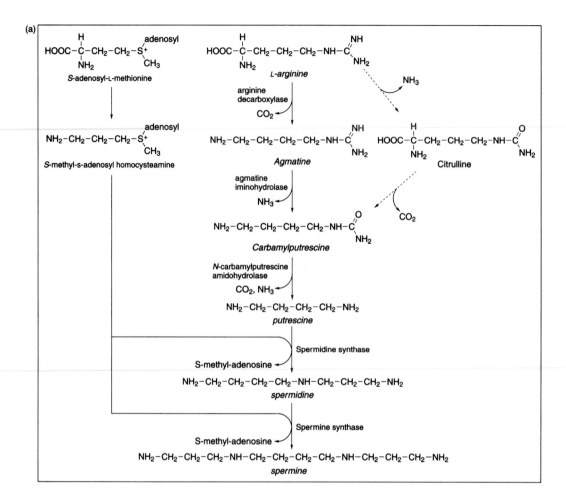

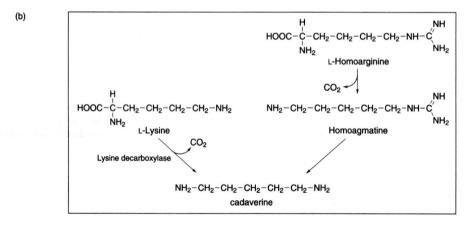

Figure 12.6 Biosynthetic pathways for putrescine, spermidine and spermine (a), cadaverine (b) and the reaction products of conversion by amine oxidases (c).

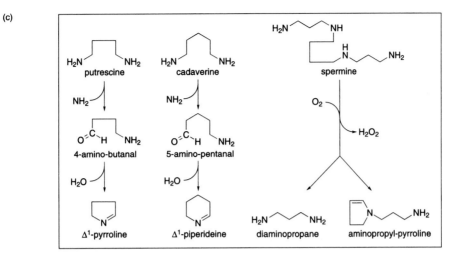

Figure 12.6   (Continued)

Putrescine can be formed via three pathways (Fig. 12.6a). These are: (1) from arginine via agmatine and N-carbamylputrescine; enzymes involved are arginine decarboxylase, agmatine iminohydrolase and N-carbamylputrescine amidohydrolase; (2) from arginine via citrulline, as found in *Sesamum indicum*; and (3) from ornithine by decarboxylation in the case of putrescine-derived alkaloids, such as nicotine or tropane alkaloids. Cadaverine is either generated from lysine with aid of a lysine decarboxylase or alternatively from homoarginine via homoagmatine in *Lathyrus sativus* (Fig. 12.6b).

The biosynthesis of the polyamines spermidine and spermine in plants probably follows a scheme found in bacteria and animals (Fig. 12.6). In the first step, S-adenosyl-L-methionine (SAM) is decarboxylated by SAM decarboxylase to S-methyl-S-adenosylcysteamine. S-methyl-S-adenosylcysteamine and putrescine are condensed by spermidine synthase to spermidine. Adding a further unit of S-methyl-S-adenosylcysteamine to spermidine by spermine synthase yields spermine. Di- and polyamines are oxidized by amine oxidases (Fig. 12.6c) which occur in several families of higher plants (Smith, 1980). Diamine

and thus fertilize the flower. The insects are dusted with pollen which matures only after fertilization of the female flowers. After fertilizations the entrance hairs become weak and release their 'victims' which then carry pollen to the next *Arum* flower.

Besides an important role in the flower biology of certain plants, amines may be feeding deterrents against herbivores. In the case of *Phallus impudicus*, it is likely that amines serve a double purpose: first, to attract spore-dispersing insects and secondly, to protect against predation by herbivores (fungivores). In addition, ammonia and amines can be considered as cellular poisons and are found to be antibacterial. A role in antimicrobial defence might be another function, as already postulated for these compounds in marine red algae (Hartmann & Aufermann, 1973).

12.3.2   Di- and polyamines

(a)   *Structures and biosynthesis*

The diamine putrescine and the polyamines spermidine and spermine are present in probably all plants, whereas the diamine cadaverine occurs within the Leguminosae.

putrescine	*spermidine*
NH$_2$–CH$_2$–CH$_2$–CH$_2$–CH$_2$–NH$_2$	NH$_2$–CH$_2$–CH$_2$–CH$_2$–NH–CH$_2$–CH$_2$–CH$_2$–CH$_2$–NH$_2$
cadaverine	*spermine*
NH$_2$–CH$_2$–CH$_2$–CH$_2$–CH$_2$–CH$_2$–NH$_2$	NH$_2$–CH$_2$–CH$_2$–CH$_2$–NH–CH$_2$–CH$_2$–CH$_2$–CH$_2$–NH–CH$_2$–CH$_2$–CH$_2$–NH$_2$

oxidase is a common enzyme in the Leguminosae, although its functional role is still unclear. The general equation of amine oxidation is:

$$R\text{-}CH_2\text{-}NH_2 + O_2 + H_2O$$
$$\rightarrow R\text{-}CHO + NH_3 + H_2O_2$$

Diamine oxidases of *Lathyrus sativus* and *Pisum sativum* have molecular weights (dimers) of 148 000 or 185 000, respectively, and putrescine and cadaverine are the best substrates, although spermidine and spermine are also accepted. They contain copper, which can be easily removed by chelating agents, such as diethyldithiocarbamate (DIECA) leading to complete enzyme inhibition. A polyamine oxidase from the Gramineae has an apparent molecular weight of 100 000 and is relatively resistant to carbonyl inhibitors and chelating reagents.

Reaction products are 4-aminobutanal for putrescine, 5-aminopentanal for cadaverine and 1,3-diaminopropylpyrroline and 1,3-diaminopropane for spermine. The latter two products occur in leaves of barley seedlings. In aqueous systems, 4-aminobutanal can cyclize to Δ^1-pyrroline and 5-aminopentanal to Δ^1-piperideine (Fig. 12.6c).

(b)   *Function of di- and polyamines*

Di- and polyamines are poly-cations and known to bind to poly-anions such as DNA. Whereas DNA is stabilized by histones in eukaryotic organisms, putrescine and polyamines take over the role of histones in bacteria. DNA in plant mitochondria and chloroplasts (which are of microbial origin according to the endosymbiosis theory, see Chapter 1) is not regulated and stabilized by histones. Putrescine and polyamines could be of similar importance in these organelles as in bacteria. *In vitro*, also many steps of protein biosynthesis are stimulated by polyamines, maybe through interaction with nucleic acids. In addition, putrescine and polyamines can stabilize biomembranes. In consequence of all these interactions, putrescine, spermine and spermidine are probably important for the regulation of growth in plants which would explain their ubiquitous distribution.

Furthermore, putrescine is known to respond to environmental stress conditions and accumulates especially under K^+ and Mg^{2+} deficiency. Under these conditions, the corresponding enzyme activities are substantially enhanced. Since polyamines occur as cations under physiological hydrogen ion concentrations, they cannot cross biomembranes

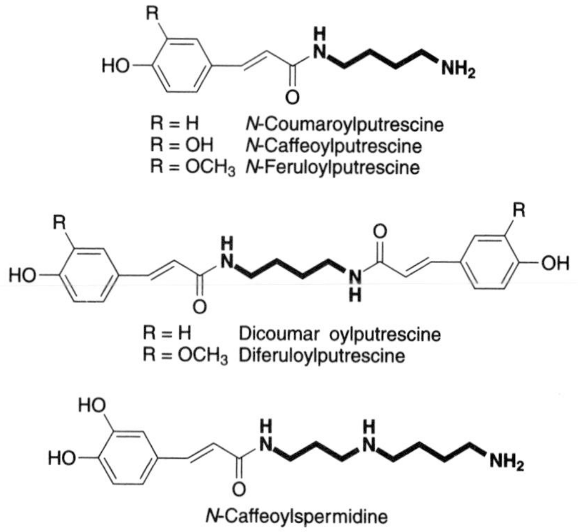

Figure 12.7   Aromatic amides of putrescine and spermidine.

by simple diffusion. From studies on the uptake of polyamines in cell cultures, a carrier-mediated process has been suggested instead (Pistocci *et al.*, 1988).

In addition to their overall function as 'growth modulators', putrescine and cadaverine serve as precursors for several groups of alkaloids, such as pyrrolizidine, nicotine, tropane, quinolizidine, *Lycopodium* and *Sedum* alkaloids (see section 12.5.4).

Di- and polyamines are also found as conjugates with cinnamic acid and its derivatives in several higher plants. Some examples are listed in Fig. 12.7. In *Nicotiana tabacum*, these compounds accumulate in the reproductive organs. In addition, the formation of putrescine conjugates is stimulated by viral or fungal infections (Smith, 1980). Furthermore, they accumulate in tobacco cell suspension cultures, maybe as a sort of stress response. Coumaroyl-, dicoumaroyl- and caffeoyl-putrescine have been shown to inhibit viral multiplication. And coumaroylagmatine dimers (called hordatines) have antifungal properties. Therefore, these amine conjugates seem to be inducible defense compounds.

12.3.3   Aromatic amines

In general, aromatic amines derive from amino acids via decarboxylation, a step which is catalyzed by a respective amino acid decarboxylase (Fig. 12.8). Some of the relevant enzymes have

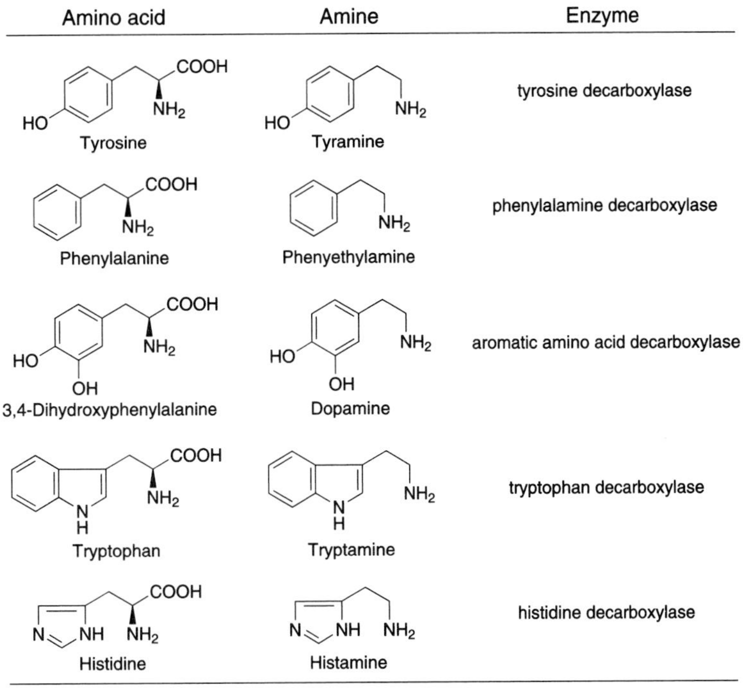

Amino acid	Amine	Enzyme
Tyrosine	Tyramine	tyrosine decarboxylase
Phenylalanine	Phenyethylamine	phenylalamine decarboxylase
3,4-Dihydroxyphenylalanine	Dopamine	aromatic amino acid decarboxylase
Tryptophan	Tryptamine	tryptophan decarboxylase
Histidine	Histamine	histidine decarboxylase

Figure 12.8 Biosynthesis of some important aromatic amines by decarboxylation of amino acids.

been purified and a few of the respective genes have been cloned (ornithine decarboxylase, tryptophan decarboxylase). Transformation of plants with recombinant decarboxylases resulted in an enhanced pool of the respective biogenic amines and in some cases in the stimulated production of secondary metabolites.

(a) Histamine and other imidazole-alkylamines

Histamine has been recorded from a number of plants across the whole range of the plant kingdom (Smith, 1980). It occurs as a free amine or as N-methylated, N-acetylated derivatives and as amides with organic acids (Luckner, 1990). Histamine is present in the stinging hairs of nettles (*Urtica dioica, U. urens, U. parviflora*) of *Jatropha urens* and *Laportea* spp. where it is accompanied by other neurotransmitters, serotonin and acetylcholine (Fig. 12.9). In this case, most vertebrates, including *Homo sapiens*, can easily experience their roles as chemical defense compounds.

In vertebrates, histamine is a neurotransmitter and stored in secretory vesicles of mast cells (basicytes). If these cells release their histamine, a painful and itching swelling of the skin immediately results. Injection of histamine into the skin by the stinging hairs of *Urtica* has the same effect. It should be recalled that many animals use

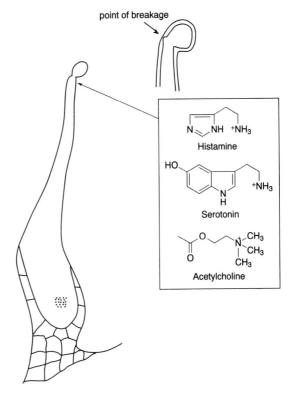

Figure 12.9 Illustration of a stinging hair of *Urtica dioica*.

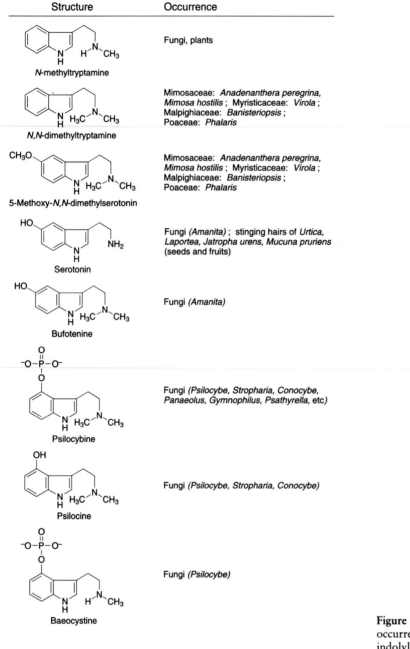

Structure	Occurrence
N-methyltryptamine	Fungi, plants
N,N-dimethyltryptamine	Mimosaceae: *Anadenanthera peregrina*, *Mimosa hostilis* ; Myristicaceae: *Virola* ; Malpighiaceae: *Banisteriopsis* ; Poaceae: *Phalaris*
5-Methoxy-N,N-dimethylserotonin	Mimosaceae: *Anadenanthera peregrina*, *Mimosa hostilis* ; Myristicaceae: *Virola* ; Malpighiaceae: *Banisteriopsis* ; Poaceae: *Phalaris*
Serotonin	Fungi (*Amanita*) ; stinging hairs of *Urtica*, *Laportea*, *Jatropha urens*, *Mucuna pruriens* (seeds and fruits)
Bufotenine	Fungi (*Amanita*)
Psilocybine	Fungi (*Psilocybe, Stropharia, Conocybe, Panaeolus, Gymnophilus, Psathyrella*, etc)
Psilocine	Fungi (*Psilocybe, Stropharia, Conocybe*)
Baeocystine	Fungi (*Psilocybe*)

Figure 12.10 Structures and occurrence of some indolylalkylamines.

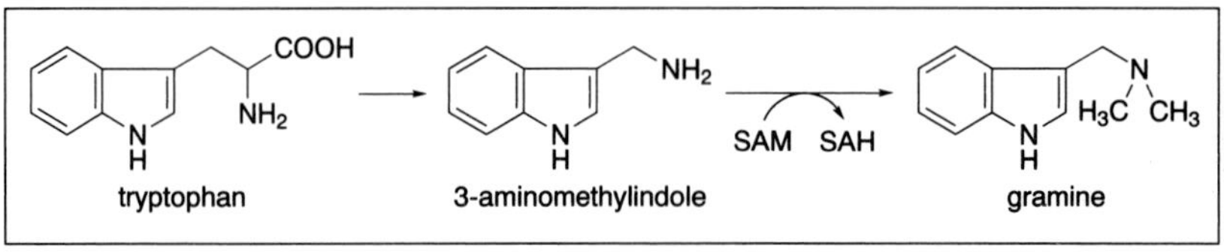

Figure 12.11 Biosynthesis of gramine.

histamine in their poison cocktails, such as Cnidaria, Mollusca, stinging insects (wasps), Porifera, Myriapoda (*Scolopendra*), or fish (*Trachinus draco*) (Teuscher & Lindequist, 1994). In case of the stinging hairs of *Urtica*, both morphology and chemistry evolved concomitantly to exercise their defensive function (Fig. 12.9). It is not clear whether the histamine found in other plant cells and tissues also has an allelochemical function, because histamine is hardly resorbed after oral administration.

(b) Tryptamine and other indole-alkylamines

Indole-alkylamines have been recorded from a number of plants and higher fungi (Smith, 1980, 1981; Luckner, 1990). They may serve as precursors for auxins (see section 12.6.1), alkaloids (see section 12.5.4) or are allelochemicals in their own right in that they have deterrent or hallucinogenic properties. Fig. 12.10 lists some of the relevant derivatives of tryptamine.

A role of indole-alkylamines in the primary metabolism of plants is not evident. Serotonin figures as an important neurotransmitter in the central nervous system (CNS) of animals. Since the compounds illustrated in Fig. 12.10 are structurally similar to that of serotonin, they may modulate serotonin receptors and be thus physiologically active in animals. Hallucinogenic effects, which have been reported for bufotenine, psilocybine, psilocine and *N,N*-dimethyltryptamine, can be a consequence of ingestion of plants or fungi which contain these metabolites. It is plausible that herbivores are highly vulnerable under these conditions: they may be attacked by predators or just fall down from trees or rocks or drown in water. Indolylalkylamines thus serve as chemical defense compounds against herbivores. *Homo sapiens* has used these hallucinogenic drugs (intoxicating snuffs and decoctions) for many years, especially in South America (more details are given in Schultes & Hoffmann, 1980, 1987). The negative effects of taking hallucinogenic drugs are well-known and do not need to be discussed further.

Serotonin, bufotenine and derivatives also occur in several animal poisons (Cnidaria, spiders, scorpions, wasps, bees), the latter substance especially in the skin of toads and frogs.

Gramine is derived from tryptophan (Fig. 12.11) and widely distributed among Poaceae, Leguminosae and Aceraceae. Gramine was shown to be active against fungi (mildew, *Erisyphe graminis*), other plants and insects (Wippich & Wink, 1985; Wink, 1993a) and can be classified as a wide-ranging chemical defense compound.

(c) Phenylethylamine, tyramine and other phenylalkylamines

A wide variety of phenylalkylamines occur in plants (Fig. 12.12). Some of them are precursors of certain alkaloids (section 12.5), others function as active allelochemicals. These phenylalkylamines are biogenetically derived from phenylalanine or tyrosine. The biosynthesis of mescaline and capsaicine (Luckner, 1990; Yeoman, 1989) is exemplified in Figs 12.13 and 12.14. N-methylations are

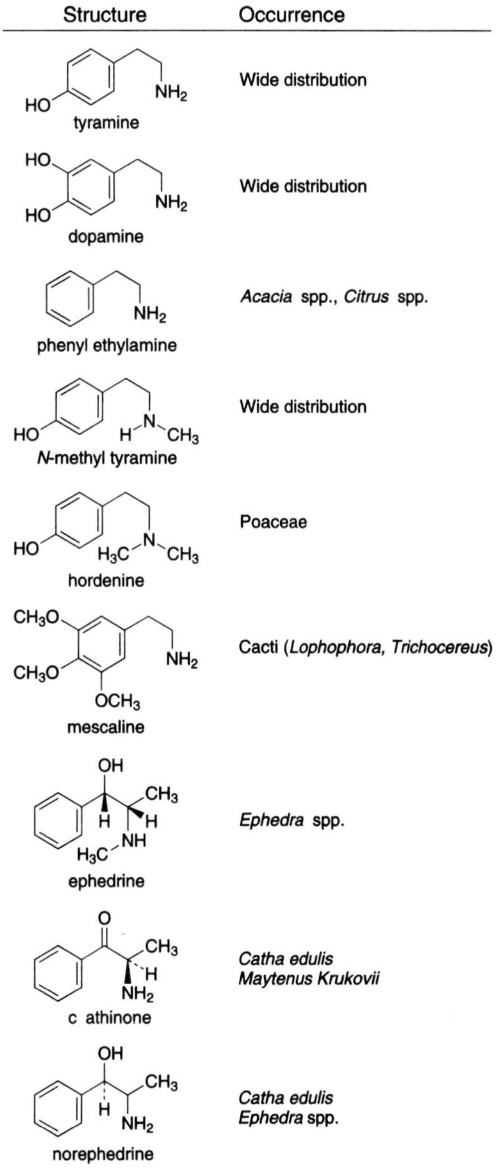

Figure 12.12 Some examples for phenylalkylamines.

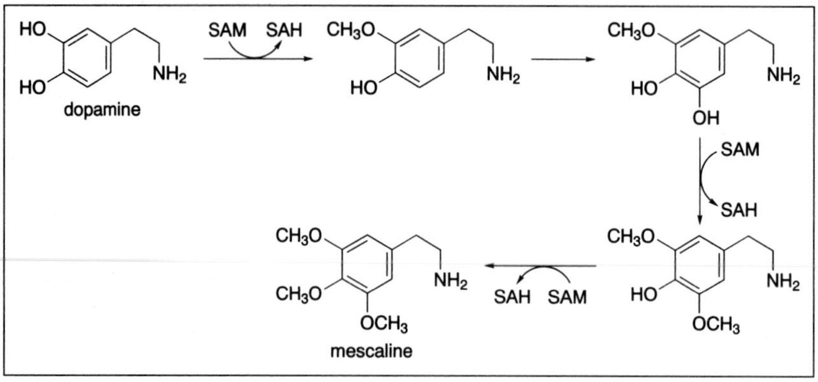

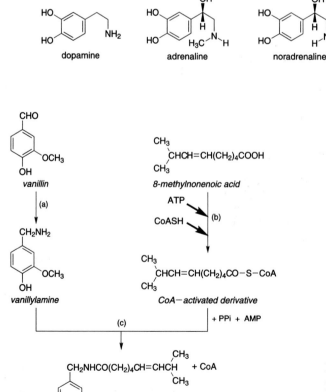

Figure 12.13   Formation of mescaline in *Lophophora williamsii*.

catalyzed by specific *S*-adenosylmethionine *N*-methyltransferases. The reaction can be reversed by demethylation which may occur during turnover of these compounds.

These aromatic amines mimic some important neurotransmitters, such as dopamine and norepi-

nephrine (noradrenaline) or hormones, such as epinephrine (adrenaline).

Mescaline is a known hallucinogen from several cacti of the genera *Lophophora* (peyotl) and *Trichocereus* which have been used by indians for ritual purposes (Schultes & Hofmann, 1980,

Figure 12.14   Biosynthesis of capsaicin in *Capsicum annuum*. Proposed terminal steps in the biosynthetic pathway are (a) transaminase; (b) acyl-CoA synthase; (c) acyltransferase.

1987). The psychomimetic phenylalkylamines interact with the receptors of the neurotransmitters serotonin, norepinephrine and dopamine and interfere with storage, release and metabolism of the natural biogenic amines. Ephedrine and derivatives (Fig. 12.12) activate the release of norepinephrine, dopamine and serotonin and inhibit their re-uptake, leading to euphoric effects in higher vertebrates. A number of phenylalkylamines occur in animals poisons (Hymenoptera, bees and wasps; skins of toads (*Bufo* spp.) and frogs). As discussed above these hallucinogenic and euphoric compounds can be classified as antiherbivore defense chemicals. Because of their biological activities, some of them are valuable medicinal compounds.

Some fruits, such as banana are especially rich in the neurotransmitters dopamine, norepinephrine, serotonin and histamine. Whether these aromatic amines function as allelochemicals in this instance remains to be elucidated, since these amines are easily degraded (amine oxidases) after oral intake.

12.4 CYANOGENIC GLYCOSIDES AND GLUCOSINOLATES

Both groups of nitrogen-containing secondary metabolites share a number of common features. They derive biogenetically from amino acids and occur as glycosides which are stored in the vacuole. They function as prefabricated defense compounds which are activated by action of a β-glucosidase in case of emergency, releasing the deterrent: toxic cyanide from cyanogens or isothiocyanates from glucosinolates (Fig. 12.15).

12.4.1 Cyanogenic glycosides

(a) *Structures and occurrence of cyanogenic glycosides*

Cyanogens are derivatives of 2-hydroxynitriles which often form glycosides with β-D-glucose. They are optically active because of chirality of the hydroxylated C-atom 2 which either occurs in the

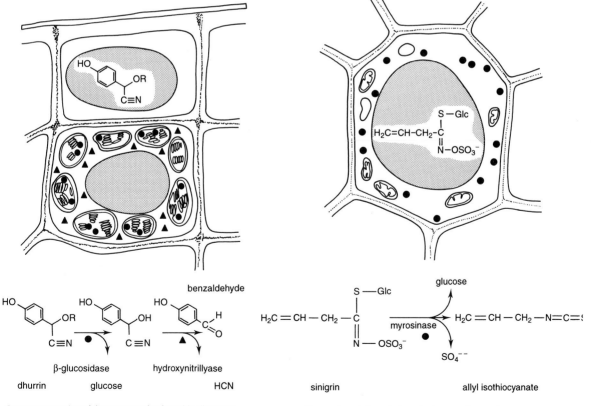

Compartmentation of the cyanogenic glycoside dhurrin in *Sorghum* and its degrading enzymes (after Kojima *et al.*, 1979).

Compartmentation of glucosinolates and myrosinase in horse radish (after Matile, 1984).

Figure 12.15 Compartmentation of cyanogenic glycosides (a) and glucosinolates (b) and the respective activating enzymes, β-glucosidase, hydroxynitril-lyase and myrosinase.

S- or *R*-configuration (Fig. 12.16) (Conn, 1980; Seigler, 1975, 1991; Nahrstedt, 1989). More than 60 cyanogenic glycosides are known which are widely distributed among plants (more than 2600 cyanogenic taxa have been reported) especially among members of the Rosaceae, Leguminosae, Gramineae, and Araceae (Conn, 1980; Seigler, 1991). A few insects, such as the Zygaenidae (Lepidoptera) feed on plants with cyanogens and sequester these compounds but are also capable of synthesizing the same cyanogens independently themselves (Nahrstedt, 1989). In members of the Sapindaceae cyanolipids have been described and certain bacteria and fungi produce HCN by a mechanism which does not involve cyanogenic glycosides or cyanolipids (Conn, 1980). Conn (1980) and Nahrstedt (1989) provide extensive information on the phytochemical analysis (spectroscopy, chromatography, distribution) of cyanogens. Cyanogenic glycosides are easily hydrolyzed by dilute acid to nitriles which further decompose to aldehydes or ketones and HCN.

(b) *Biosynthesis and function of cyanogenic glycosides*

Cyanogenic glycosides are derived from L-amino acids (Fig. 12.16) and biosynthesis seems to be catalyzed by a multienzyme complex (Conn, 1980) (Fig. 12.17). Dhurrin biosynthesis was achieved in a microsomal membrane preparation and seemed to be a channeled process (Conn, 1980, 1981). In the first step, the amino group of L-amino acids is hydroxylated by a L-amino acid *N*-monooxygenase. Upon oxidative decarboxylation, the *N*-hydroxy-L-amino acid is converted into an aldoxime. The step from aldoxime to nitrile is catalyzed by an aldoxime dehydratase. The nitrile is then hydroxylated at the C2-position by a nitrile monooxygenase to yield the key intermediate 2-hydroxynitrile (or cyanohydrin). Glucosyltransferase forms the β-glucoside using activated glucose, i.e. UDP-glucose.

In case of emergency, i.e. when plants are wounded by herbivores or other organisms, the cellular compartmentation breaks down and cyanogenic glycosides come into contact with an active β-glucosidase of broad specificity, which hydrolyzes them to yield 2-hydroxynitrile (cyanohydrin) (Figs 12.15 and 12.17). 2-Hydroxynitrile is further cleaved into the corresponding aldehyde or ketone and HCN by a hydroxynitrile lyase. In cyanolipids, the fatty acid is hydrolyzed by an esterase to 2-hydroxynitrile to yield the corresponding aldehyde and HCN (Fig. 12.17).

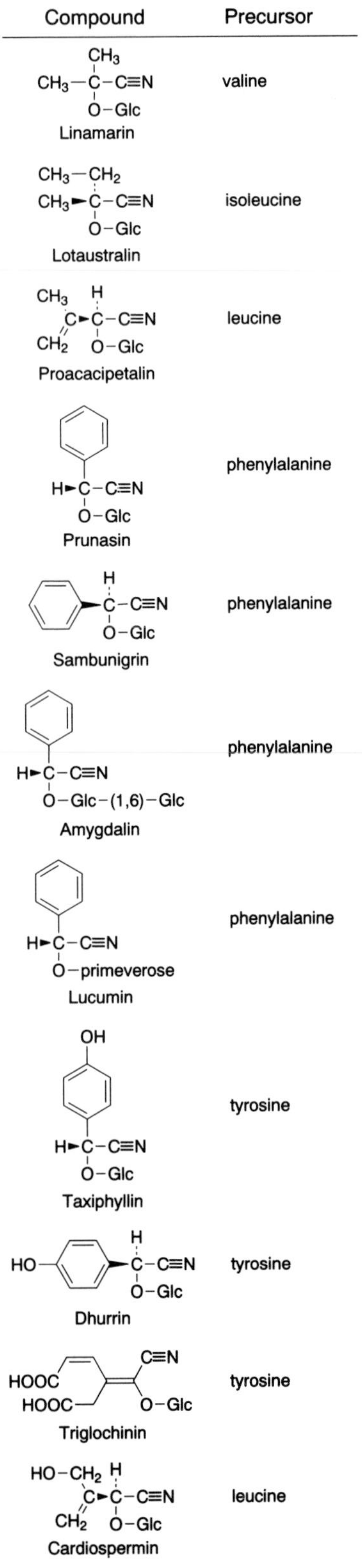

Figure 12.16 Structures of a few common cyanogenic glycosides.

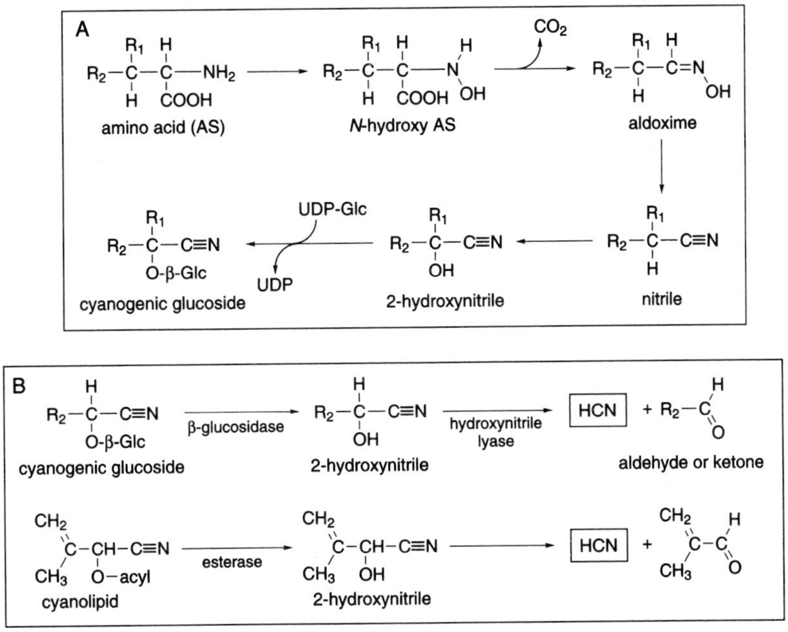

Figure 12.17 Biosynthesis and defensive activation (degradation) of cyanogens. (a) Biosynthesis of cyanogenic glucosides; (b) activation of cyanogens.

HCN is highly toxic for animals or microorganisms due to its inhibition of enzymes of the respiratory chain (i.e. cytochrome oxidases) and its binding to other enzymes containing heavy metal ions. The lethal dose of HCN in man is 0.5–3.5 mg/kg after oral application and death of man or animals have been reported after the consumption of plants with cyanogenic glycosides, whose concentrations can be up to 500 mg HCN/100 g seeds. Normally 50–100 mg HCN/100 g seeds and 30–200 mg HCN/100 g leaves have been recorded (Teuscher & Lindequist, 1994). Foods which contain cyanogens, such as manihot (*Cassava esculenta*), have repeatedly caused intoxications and even deaths in man and animals. Animals can rapidly detoxify small amounts of HCN by rhodanese (Fig. 2.18). Furthermore, a number of herbivores have been studied which can

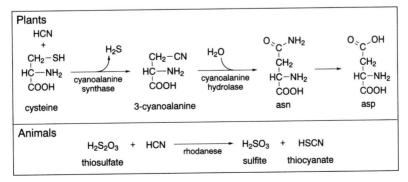

Figure 12.18 Detoxification of HCN by plants and animals.

tolerate HCN (Seigler, 1991), at least in lower concentrations. Nevertheless, cyanogens should be considered as active and potent chemical defense compounds.

Cyanogenic glycosides are formed in the cytoplasm but stored in the central vacuole. In *Sorghum* leaves, storage is tissue-specific and takes place in epidermal vacuoles (Fig. 12.15a). The corresponding β-glucosidase and hydroxynitrile lyase were demonstrated in the adjacent mesosphyll cells, safely away from the cyanogenic glycosides. By such a spatial arrangement, cyanogenic glycosides are present in a strategically favored position, since the epidermal tissues are attacked by small herbivores or micro-organisms in the first place and in addition, can be activated rapidly when needed. Since HCN cannot be stored by plants as the free hydrocyanic acid, its accumulation as a preformed defense chemical is an ingenious solution. Since cyanogenic glycosides are polar substances they do not diffuse across biomembranes, such as the tonoplast by simple diffusion. Although not yet demonstrated experimentally, a carrier-mediated transport system or vesicle fusion might exist to facilitate an exchange from cytoplasm into the vacuole.

Besides their role as respiratory toxins, cyanogenic glycosides as such or as the corresponding aldehyde or ketone are often feeding deterrents. Additionally they can serve as mobile nitrogen storage compounds in seeds of several plants (e.g. in *Prunus* and in Lima bean *Phaseolus lunatus*), a property already noted for non-protein amino acids (section 12.2) or some alkaloids (see section 12.5.7). Cyanogenic glycosides are not end products of metabolism but undergo a steady turnover. To achieve a certain rate of accumulation, biosynthesis and degradation are in a state of equilibrium. Apparently a β-glucosidase, not necessarily the one involved in cyanogenesis, has limited access to cyanogenic glycosides in these plants. A turnover rate of about 0.05 μmol/h in seedlings with 1.0 μmol dhurrin/shoot has been estimated (Conn, 1981).

HCN is also a toxin for the plants which synthesize them. To prevent autotoxicity a detoxification pathway exists, combining HCN with L-cysteine to yield 3-cyanoalanine by β-cyanoalanine synthase, a pyridoxal phosphate-containing enzyme (Fig. 12.18). Cyanoalanine is hydrolyzed by β-cyanoalanine hydrolase to L-asparagine. β-Cyanoalanine synthase occurs in all plants but

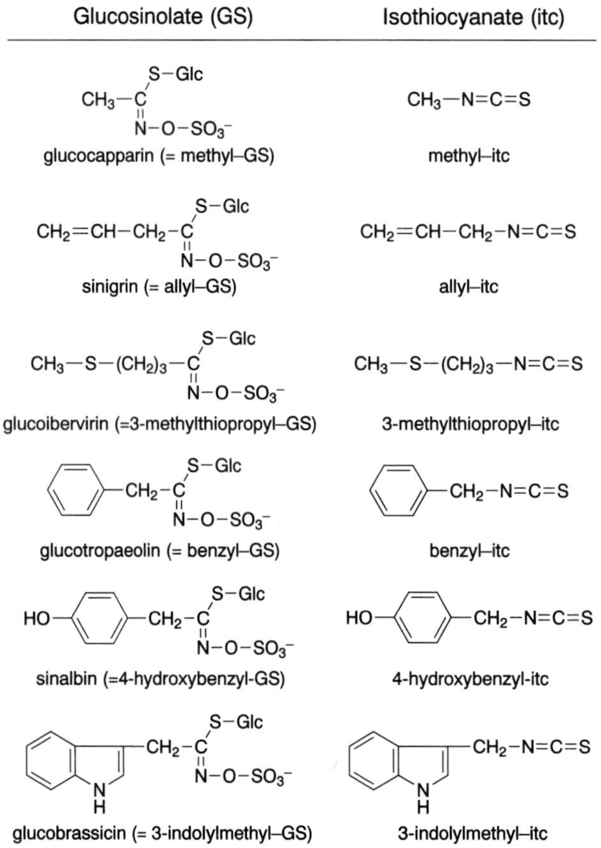

Figure 12.19 Structures of some glucosinolates (gs) and corresponding isothiocyanates (itc).

seems to be more pronounced in strongly cyano-genic species (Conn, 1981).

12.4.2 Glucosinolates

Glucosinolates resemble cyanogens in many respects, but contain sulfur as an additional atom. They can be classified as thioglucosides. When hydrolyzed, glucosinolates liberate D-glu-cose, sulfate and an unstable aglycone, which may form an isothiocyanate (common name 'mustard oil') as the main product under certain conditions, or a thiocyanate, a nitrile or cyano-epithioalkane. Isothiocyanates are responsible for the distinc-tive, pungent flavor and odor of mustard and horseradish.

(a)   Structures and occurrence of glucosinolates

More than 80 different glucosinolates have been found in higher dicotyledonous plants (Fig. 12.19) in the order Capparales, which include the families Capparidaceae, Cruciferae (syn. Brassicaceae), Resedaceae, Moringaceae, Tropaeolaceae and others (Larsen, 1981).

(b)   Biosynthesis, breakdown and function of glucosinolates

Both, protein or non-protein amino acids serve as precursors for the biosynthesis of glucosinolates (Fig. 12.19). The pathway leading to the corre-sponding aldoxime is analogous to that of cyanogenic glycosides (Fig. 12.20). Aldoxime is converted into a thiohydroximic acid, a key intermediate, using L-cysteine as a sulfur donor. A thioglucoside is formed in the next step with aid of a UDP-glucose:thiohydroximate glucosyltrans-ferase. The transfer of a sulfate group is cata-lyzed by sulfotransferase using phosphoadenosine-

phosphosulfate (PAPS) as active sulfate donor. Glucosinolates may be further modified by hydro-xylation, oxidation and other substitutions; e.g. sinigrin derives from glucoibervirin.

Glucosinolates are polar molecules which are formed in the cytoplasm and stored in vacuoles, as illustrated for horseradish (Fig. 12.15b). They are fairly stable at neutral pH values and occur as salts with cations such as K^+ or sinapine to balance the negative charges on the glucosinolate anions. All plant parts may accumulate glucosi-nolates, but seeds and roots are often especially rich in these allelochemicals. Concentrations are in the range 0.1–0.2% fresh weight in leaves, up to 0.8% in roots and up to 8% in seeds. The profile of glucosinolates which can comprise up to 15 different compounds, may differ between developmental stages, organs and populations. Glucosinolates are dynamic metabolites and a half-life of 2 days has been reported from *Isatis tinctoria*.

All plants which sequester glucosinolates also possess thioglucoside glucohydrolases (common name myrosinase; molecular weight 125 000–150 000) which can hydrolyze glucosinolates to D-glucose and an aglycone, spontaneously rearran-ging to isothiocyanate. These hydrolases, which can be activated by ascorbic acid, are stored away from the vacuole in the cell wall and in endoplasmic reticulum, Golgi vesicles and mito-chondria. Upon cell and tissue disintegration (wounding) or increase of membrane permeability, the enzyme and its substrate come together liberating the pungent, repellent and antibiotic isothiocyanate. Depending on the environmental conditions, enzymes and other compounds pres-ent, the aglycone can rearrange to isothiocyanate (in a Lossen-type arrangement) as the most common product, or to nitriles, thiocyanates or cyano-epithioalkanes, or oxazolidine-2-thiones (Fig. 12.21).

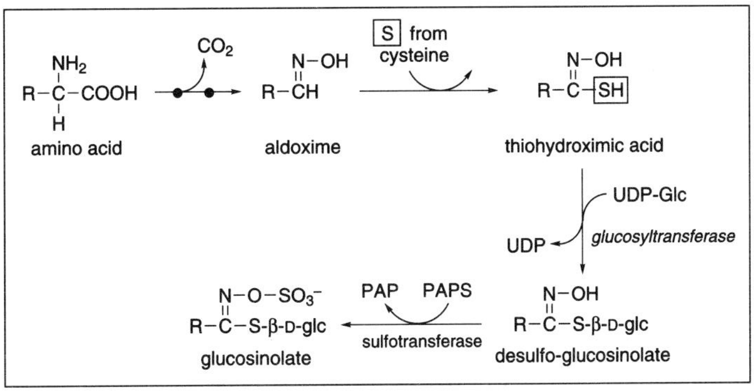

Figure 12.20 Pathway of glucosinolate biosynthesis.

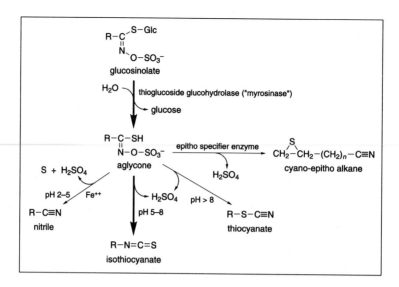

Figure 12.21 Degradation (defensive activation) of glucosinolates.

Alkyl-, alkenyl-, methylthioalkyl- and benzyl-isothiocyanates are lipophilic, volatile allelochemicals with a pungent smell and taste. 4-Hydroxybenzyl- and methylsulfinyl and methylsulfonylaklyl-isothiocyanates are not volatile and pungent smelling, but have otherwise similar properties. Because isothiocyanates can easily penetrate biomembranes, they can interact with epidermal and mucosal skin leading to painful irritations. In addition, they can lead to bronchitis, pneumonia, gastroenteritis, diarrhoea, heart and kidney disorder and even to abortions. Nitriles are liver and kidney toxins. Oxazolidine-2-thione derivatives (such as progoitrin, glucoconringin, glucobarbarin and glucosisymbrin) inhibit the oxidation of iodate to iodine which leads to goitre. Rhodanide ions interact with various enzymes and also affect thyroid functions (Teuscher & Lindequist, 1994). Isothiocyanates are antibiotic; besides making bacterial or fungal cells leaky, interactions with cellular targets are also possible.

Glucosinolates can thus be considered as pre-formed defense chemicals which are activated in case of emergency. They have a wide range of activities and seem to be important especially in plant–herbivore but also in plant–plant and plant–microbe interactions. As with other allelochemicals, a few specialists have evolved during evolution, which are attracted by glucosinolates or mustard oils, respectively. Some of them tolerate these compounds in their diets (e.g. *Pieris brassicae* (Lepidoptera) (Louda & Mole, 1991)).

Homo sapiens can be considered a glucosinolate 'addict', since plants containing them are used as spices and condiments (cabbage, radish, horseradish, mustard, capers, flowers and unripe seeds of *Tropaeolum*). But nevertheless, plant breeders have reduced the levels of glucosinolates in some agricultural species (e.g. *Brassica napus*) in order to improve palatability and reduce toxicity.

12.5 ALKALOIDS

12.5.1 Definition

The definition of alkaloids, of which more than 12 000 structures have been described already, has seen a number of changes throughout the years. Formerly, this class of secondary compounds was restricted to plant bases with a heterocyclic nitrogen atom. Exocyclic nitrogen bases were termed 'pseudoalkaloids'. Other definitions demanded that the skeleton of alkaloids should derive from amino acids or that these bases have pharmacological activities. At present, the definition is much more pragmatic and includes all nitrogen-containing natural products which are not otherwise classified as peptides, non-protein amino acids, amines, cyanogenic glycosides, glucosinolates, cofactors, phytohormones or primary metabolites (such as purine and pyrimidine bases). Even a number of antibiotics produced by bacteria or fungi are therefore included as alkaloids. This makes sense from the point of view of chemistry and biosynthesis, since 'antibiotic' is a functional nominator which includes substances of all biogenetic classes.

12.5.2 Occurrence

Alkaloids have been detected in about 15% of plants, bacteria, fungi, and even in animals. Within the plant kingdom, they occur in primitive groups such as *Lycopodium* or *Equisetum*, in gymnosperms and angiosperms. In higher plants (angiosperms) some families contain more alkaloidal taxa than others: alkaloid-rich are Papaveraceae, Berberidaceae, Leguminosae, Boraginaceae, Apocynaceae, Asclepiadaceae, Liliaceae, Gnetaceae, Ranunculaceae, Rubiaceae, Solanaceae and Rutaceae (Hegnauer, 1962–1990; Mothes *et al.*, 1985).

It has been suggested that alkaloids evolved early in evolution and were already present at the time (ca. 200 million years ago) when the angiosperms began to radiate (Swain, 1977). In general, specific alkaloid types are restricted to particular systematic units and are therefore of some importance for the systematics, taxonomy and phylogeny of plants. For example, benzylisoquinoline alkaloids are typical for Papaveraceae, Berberidaceae and Ranunculaceae which seem to be phylogenetically related. However, it is difficult to interpret the situation when specific alkaloids are synthesized in the correct stereochemical fashion in systematic units which are not related. Some examples are: ergot alkaloids occur in fungi (*Claviceps*) but also in members of the Convolvulaceae; quinolizidine alkaloids are typical for some Leguminosae, but have also been detected in Berberidaceae (*Caulophyllum, Leontice*). In addition, traces of these compounds were detected in cell cultures of unrelated families when challenged with elicitors. It has been speculated that the enzymes leading to the main quinolizidine alkaloid skeleton are much more widely distributed in the plant kingdom, but are normally repressed (Wink & Witte, 1983). Alternatively, convergent evolution and horizontal gene transfer have been suggested (Luckner, 1990). Since a number of genes in alkaloid formation have been cloned already or will be cloned in the future, this question should be answered in the near future.

12.5.3 Detection

Plant extracts are often analyzed by thin-layer chromatography as a first screening to determine whether alkaloids are present or not. A number of reagents give typical color reactions, such as Dragendorff's, Mayer's reagent etc. (Waterman, 1992). Because alkaloids are typically present in complex mixtures consisting of 2–5 major and 20–50 minor alkaloids, TLC is unable to separate them all. Better methods are HPLC and capillary-GLC. The latter method is extremely useful because it has a good resolving power, is very sensitive and selective if a nitrogen-specific detector is used. Furthermore, GLC can be directly coupled with a mass spectrometer (MS), to provide the mass spectra even for very minor components. Since many alkaloids have been analyzed by MS, a large collection of mass spectra is available which makes it possible to identify many of the known alkaloids unambiguously. Usually mass spectra are recorded in the electron impact (EI) mode, which promotes fragmentation. If information of the molecular ions are needed (which can be elusive in EI-MS), other MS-techniques, such as chemical ionization (CI), field desorption (FD) or Fast Atom Bombardment (FAB) are the methods of choice (Waterman, 1992).

HPLC tends to be less sensitive and of lower separation capacity than capillary GLC. Theoretically, also HPLC can be directly coupled to a mass spectrometer, but these instruments are not widely distributed. However, modern photo diode array detectors are very helpful to identify known metabolites by UV-VIS spectroscopy. In addition, HPLC has the major advantage that it is possible to isolate a compound in milligram quantities which allow a structure elucidation by NMR (^{1}H, ^{13}C).

If small quantities (fg or ng amounts) of known alkaloids need to be detected routinely, immunological procedures such as radioimmunoassay (RIA) or enzyme immunoassays (EIA, ELISA) should be appropriate. To obtain specific antibodies, the alkaloid in question has to be coupled chemically to a large protein, such as albumin, usually by some sort of spacer (e.g. succinic acid).

12.5.4 Biosynthesis

The skeleton of most alkaloids is derived from amino acids (Fig. 12.22) although moieties from other pathways, such as terpenoids are often combined. In addition, in a number of alkaloids (e.g. steroid alkaloids) the nitrogen (deriving from glutamine or other NH_2 sources) is added in the final steps of a biosynthetic pathway, i.e. the alkaloid skeleton does not stem from amino acids.

Precursor Amino acid	Alkaloid	Examples		Main occurrences
Ornithine	pyrrolizidine A.	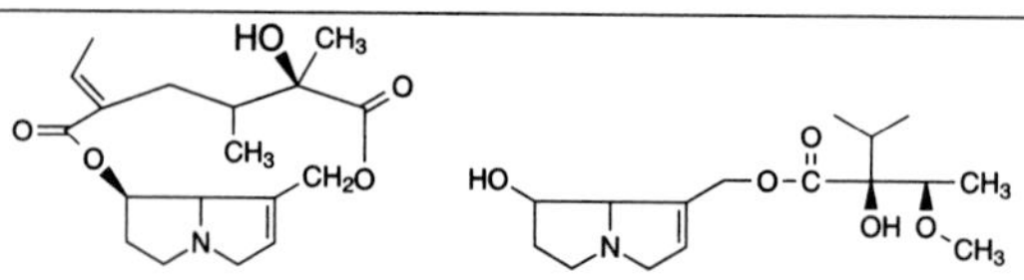 *senecionine*	*heliotrine*	*Senecio* spp., *Crotalaria* spp, *Heliotropium*, and other Boraginaceae
	tropane A.	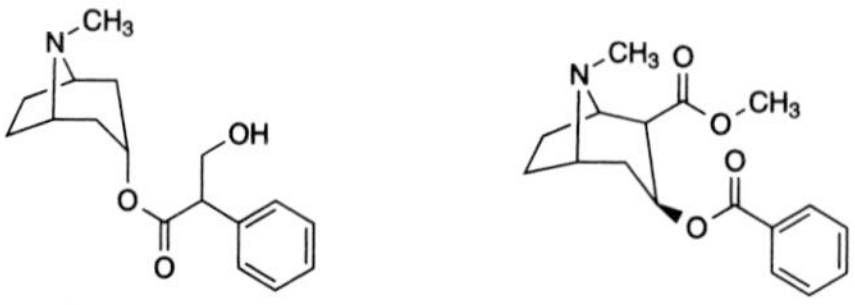 *hyoscyamine*	*cocaine*	Solanaceae, *Erythroxylum*, (*Hyoscyamus, Datura, Atropa, Duboisia*)
	nicotine	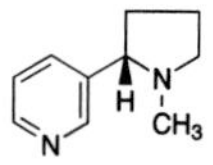 *nicotine*		*Nicotiana* spp. (traces in many other plants)
Lysine	punica A.	Pseudopelletierine	Pelletierine	*Punica granatum, Duboisia, Sedum.*
	sedum A.	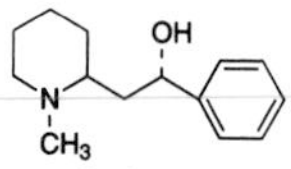 Sedamine		*Sedum* spp.
	lobelia A.	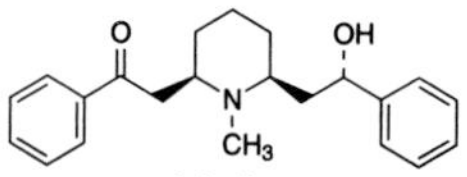 lobeline		*Lobelia* spp.
	quinolizidine A.	lupanine	cytisine	*Lupinus, Cytisus, Genista, Baptisia, Thermopsis, Laburnum* and other Genisteae, *Caulophyllum*
	lycopodium A.	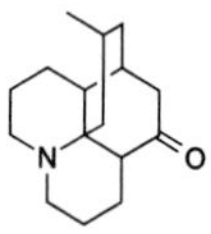 lycopodine		*Lycopodium* spp.
Aspartic acid	areca A.	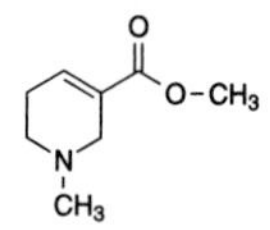 arecoline		*Areca catechu*
Histidine	pilocarpine	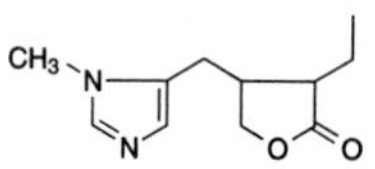 pilocarpine		*Pilocarpus* spp.

Figure 12.22 Biosynthesis and occurrence of alkaloids.

Precursor Amino acid	Alkaloid	Examples	*heliotrine*	Main occurrences
Anthranilic acid	benzoxazines	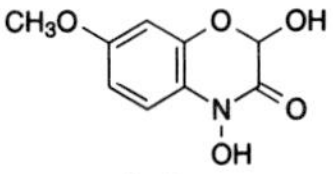 dimboa		Gramineae
	furoquinoline A.	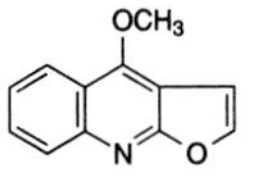 dictamnine		Rutaceae
	acridone A.	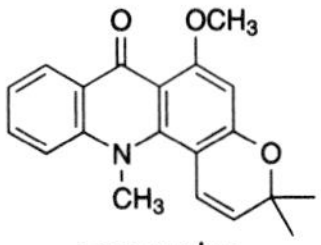 acronycine		*Acronychia*, *Melicope*, Rutaceae
Tryptophan	indole A.	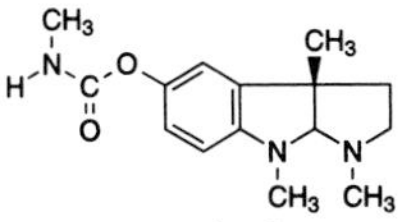 physostigmine		*Physostigma venenosum*
	ergoline A.	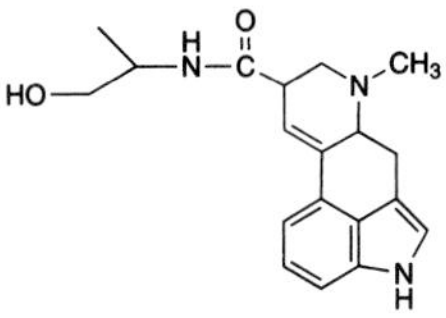 ergometrine		*Claviceps*, Convolvulaceae
	β-carboline A.	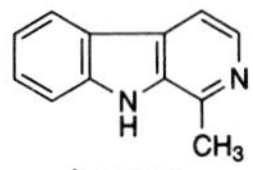 harman	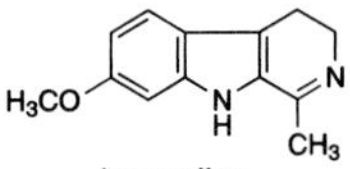 harmaline	Loganiaceae, Apocynaceae, Zygophyllaceae
	monoterpene indole A.	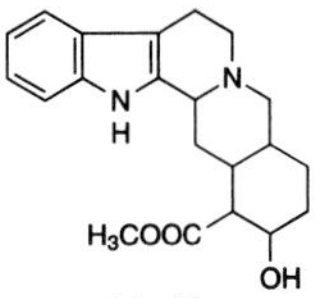 yohimbine	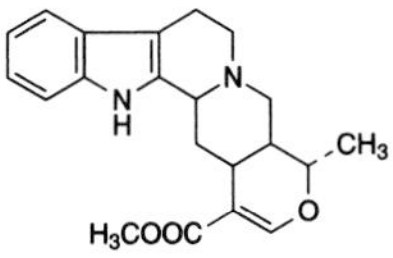 Ajmalicine	Apocynaceae, Loganiaceae, Rubiaceae
		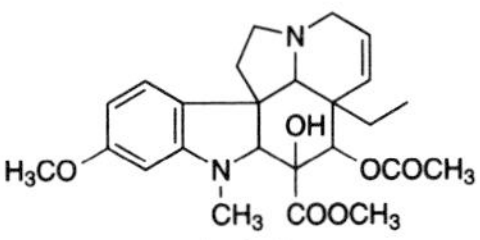 vindoline	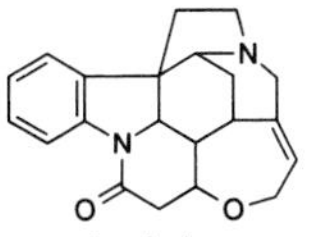 strychnine	
	quinoline	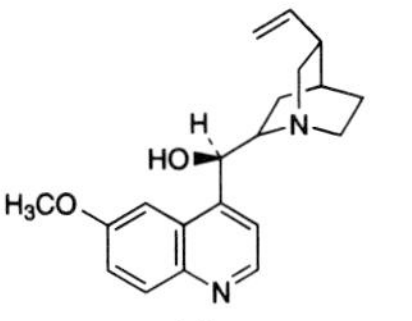 quinine	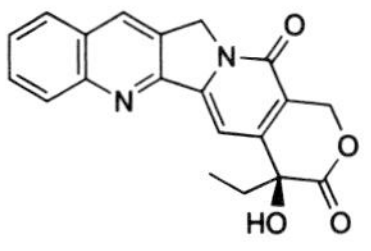 camptothecin	*Cinchona*, *Camptotheca acuminata*

Figure 12.22   (*Continued*)

Precursor Amino acid	Alkaloid	Examples		Main occurrences
Phenylalanine/ Tyrosine	Amaryllidaceae A.	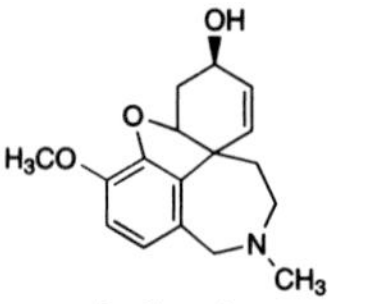 galanthamine 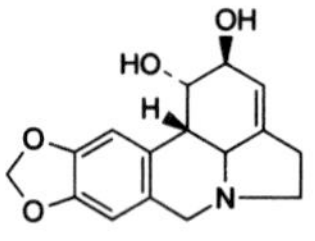	lycorine	Amaryllidaceae
	Tetrahydroisoquinoline A.	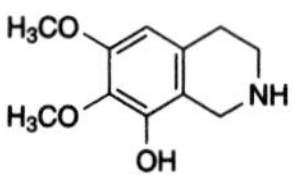 anhalamine 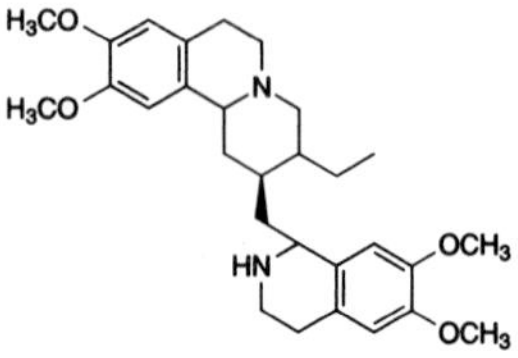	emetine	Lophophora , Cephaelis
	Benzylisoquinoline A.	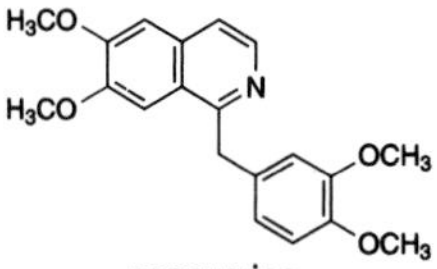 papaverine		Papaveraceae, Apocynaceae
	Protoberberine A.	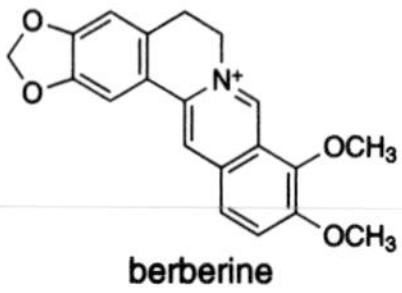 berberine 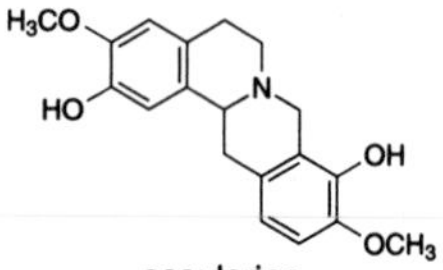	scoulerine	Berberidaceae Papaveraceae
	Benzophenanthridine A.	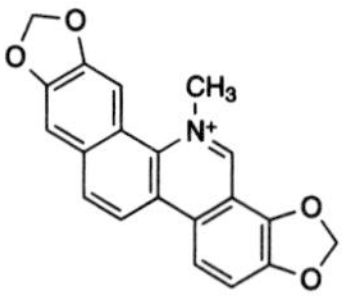 sanguinarine 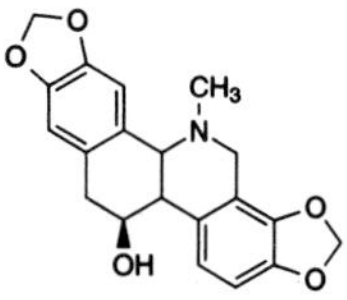	chelidonine	Papaveraceae
	Morphinane A.	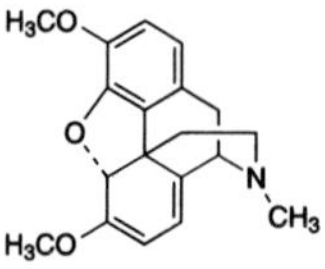 thebaine 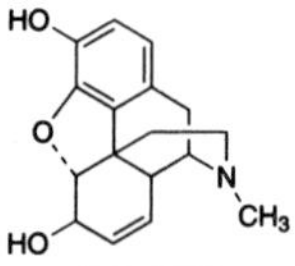	morphine	Papaveraceae
	Aporphine A.	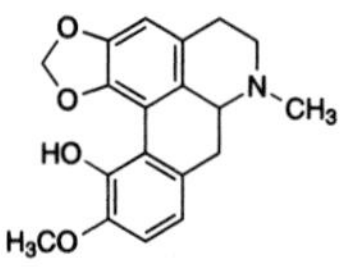 bulbocapnine 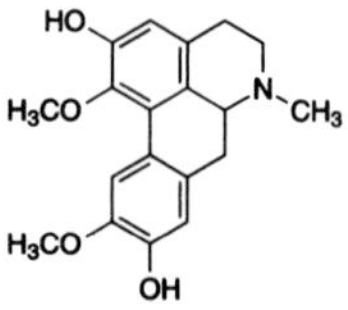	boldine	Papaveraceae, Monimiaceae Magnoliaceae Lauraceae
	Erythrina A.	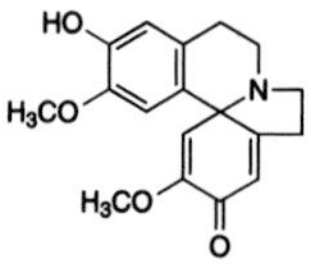 erysodienone		Erythrina

Figure 12.22   (*Continued*)

Precursor Amino acid	Alkaloid	Examples	Main occurrences
Phenylalanine/ Tyrosine *continued*	Phenylethyl-isoquinoline A.	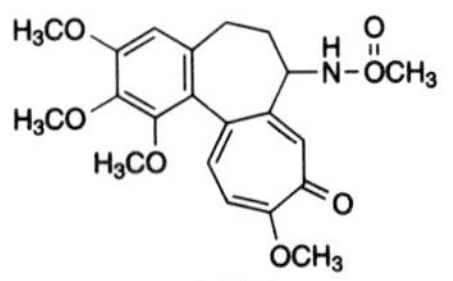 colchicine	*Colchicum Gloriosa superba*
	Aristolochia A.	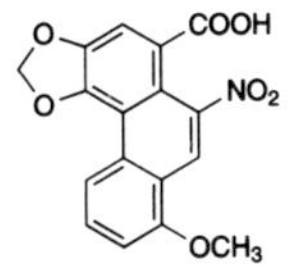 aristolochic acid	*Aristolochia , Asarum*
	Betacyanin/ betaxanthin	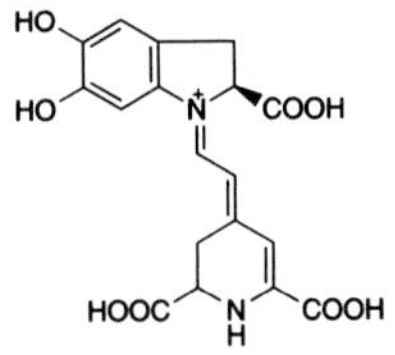 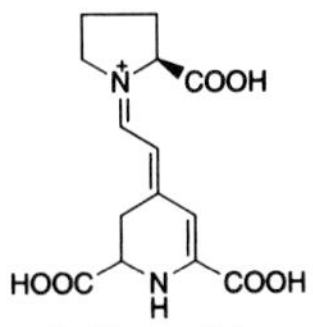 Betanidin Indicaxanthin	*Centrospermae fungi*
NH$_2$ (gly, ala, arg)	Steroid A.	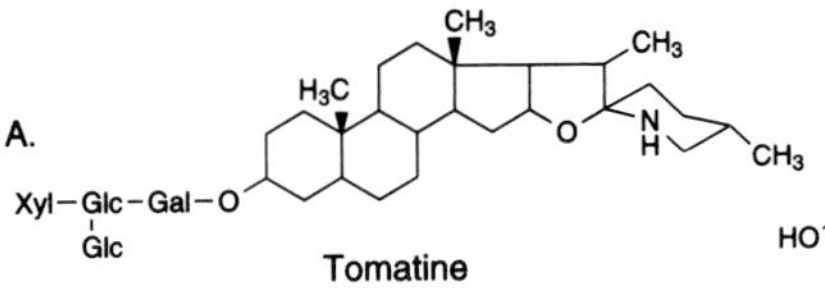 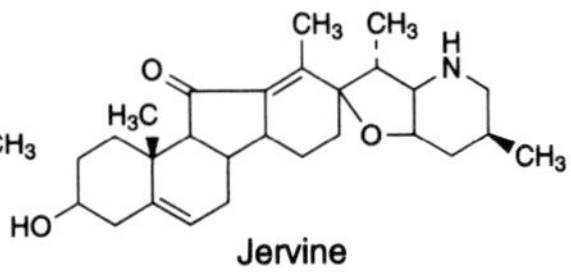 Tomatine Jervine	*Solanum , Veratrum*

Figure 12.22 *(Continued)*

Biosynthetic pathways have been worked out in detail for a few alkaloids so far (Mothes *et al.*, 1985; Luckner, 1990).

In many other instances pathways are based on 'intelligent guesses' and some preliminary precursor experiments. Precursor experiments often involve the feeding of radioactively labeled compounds (^{3}H or ^{14}C) to intact plants or more recently to cell and tissue cultures. After hours to days or weeks, the alkaloids are isolated and it is determined by autoradiography, scintillation counting, radio- TLC, -GLC or -HPLC whether and how much of the precursor has been incorporated (Mothes *et al.*, 1985; Luckner, 1990). More recently, non-radioactive ^{13}C-labeled precursors have been employed. Using ^{13}C-NMR

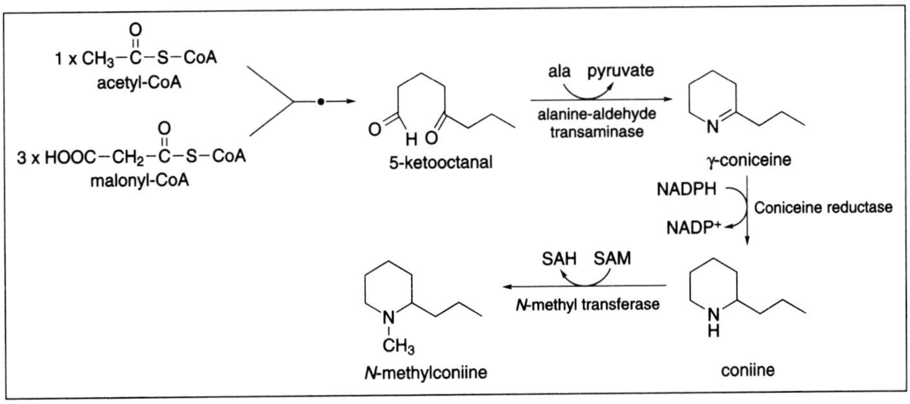

Figure 12.23 Biosynthesis of *Conium* alkaloids (Roberts, 1981).

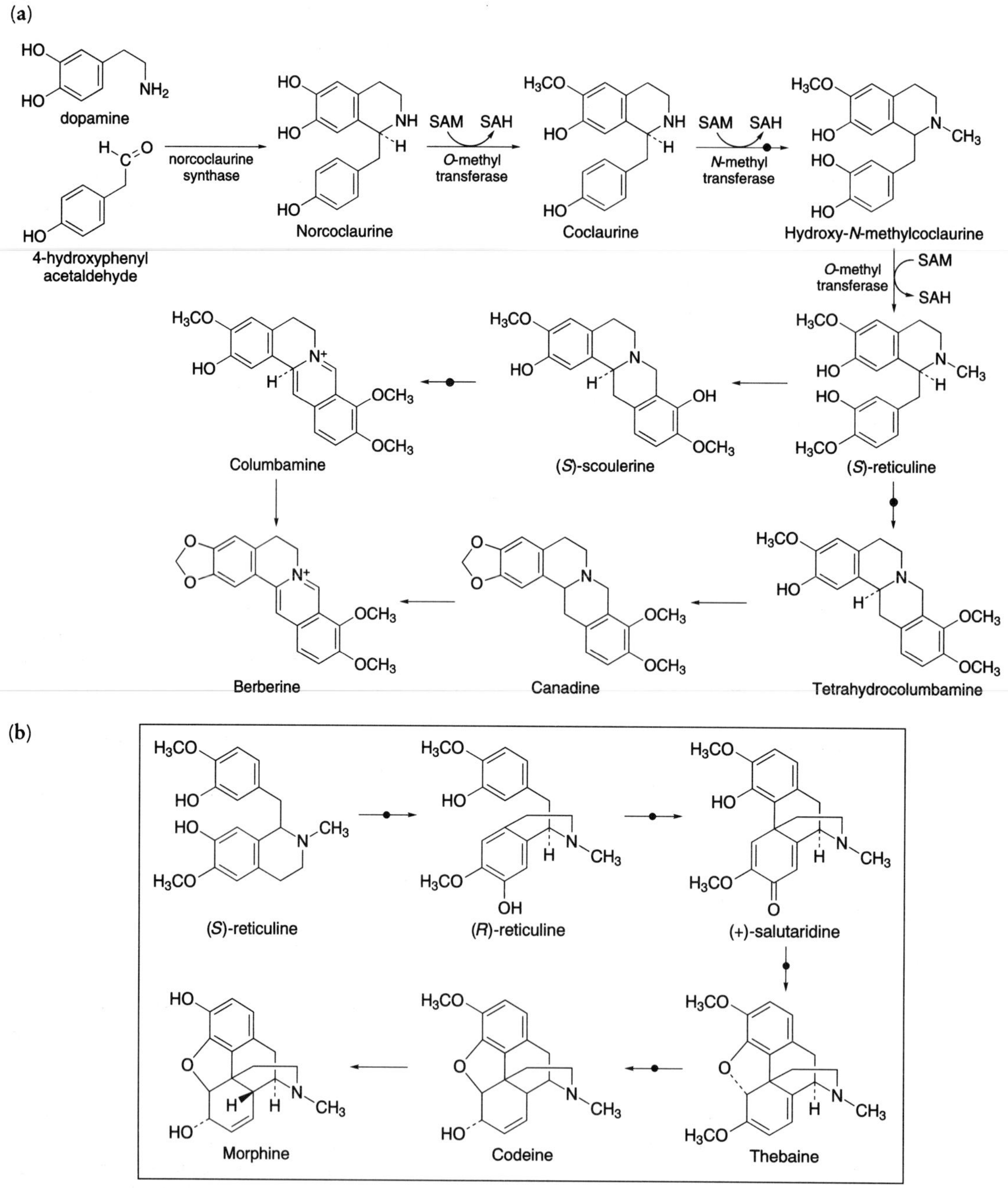

(a)

(b)

Figure 12.24 Biosynthesis of (a) protoberine and (b) morphinane alkaloids (Zenk, 1989; Okada *et al.*, 1988). Steps between scoulerine and berberine have been localized in cytoplasmic vesicles.

analysis, the incorporation of any carbon atom can be directly followed. The limitations of this method are the availability of precursors which are labeled at a particular carbon atom and the relative insensitivity of ^{13}C-NMR spectroscopy.

Another more laborious approach, which has definitively solved a number of biosynthetic

pathways, includes the isolation and characterization of the enzymes catalyzing the particular reactions of a biosynthetic sequence. Using enzyme preparations from plants but especially from cell suspension cultures a number of pathways leading to benzylisoquinoline, protoberber-ine, tropane and monoterpene-indole alkaloids have been elucidated at the enzymic level (in the laboratories of Zenk, Stöckigt, Yamada and co-workers).

More recently, the first genes encoding enzymes of alkaloid biosynthesis have been isolated and cloned, such as tryptophan decarboxylase, ornithine decarboxylase, strictosidine synthase, berberine-bridge enzyme, and hyoscyamine 6β-hydroxylase. Using these enzymes it will be possible to influence alkaloid metabolism of transgenic plants. A nice example has been presented by Yamada *et al.* (1990), who transformed *Atropa belladonna*, a plant producing atropine, with the gene of hyoscyamine 6β-hydroxylase. The transformants expressed the enzyme and in these plants hyoscyamine was completely converted into scopolamine, which is normally present in low amounts only (Hashimoto & Yamada, 1992). Those biosynthetic pathways (Luckner, 1990; Hartmann, 1991) which have been elucidated at the enzymic level leading to coniine, morphine, berberine, ajmalicine and hyoscyamine are exemplified in some detail in Figs 12.23–12.26.

12.5.5 Accumulation and storage

(a) Compartmentation

Although the exact site of alkaloid formation in a plant cell has been elucidated for a few species only, it is certainly correct to assume that most compounds are synthesized in the cytoplasm. Membrane-enclosed vesicles have been implied in the biosynthesis of berberine (Amann *et al.*, 1987). The chloroplast is not only the site of photosynthesis and related processes but also harbors a number of biosynthetic pathways, such as those for fatty acids and amino acids (see Chapter 7). In addition, quinolizidine alkaloids (QA) are synthesized in the chloroplast stroma (Wink & Hartmann, 1982); thus both the alkaloid and its amino acid precursor, L-lysine, share the same compartment. QA formation is regulated by light and displays a diurnal rhythm (Wink & Witte, 1984). Light regulation seems to be triggered by (1) lysine availability (it is also made during the day), (2) change of the stromal hydrogen ion concentration to pH 8 (enzymes of QA formation have a pH optimum at pH 8) and (3) by reduction of QA enzymes by thioredoxin (reduced thioredoxin is generated under illumination).

Alkaloids are not formed in the extracellular space nor in the vacuole. However, the vacuole seems to function as a major compartment of alkaloid storage (Table 12.1). A number of plants produce latex which in addition to its glueing properties often contain defense chemicals, such as alkaloids (morphine and related benzylisoquinoline alkaloids in Papaveraceae; protoberberine and benzophenanthridine alkaloids in *Chelidonium*; lobeline and other piperidine alkaloids in *Lobelia*) and terpenoids (e.g. phorbolesters). The alkaloids are selectively sequestered in small latex vesicles (diameter <1 μm) and reach local concentrations of up to 1 M (Hauser & Wink, 1990; Roberts, 1987; Roberts *et al.*, 1991).

(b) Sites of alkaloid formation and storage

Whereas a few alkaloids are formed ubiquitously in all plant organs, an organ- or even tissue-specific formation seems to be more common (Table 12.2). Since most eukaryotic genes are regulated in a cell-, tissue- and organ-specific manner, genes of alkaloid formation seem to be no exception. We should expect that the corresponding promotors are regulated by respective regulatory proteins (see Chapter 1). This conclusion is important for the interpretation of results obtained with cell suspension cultures. Whereas alkaloid formation is active in differentiated systems (root or shoot cultures), it is usually reduced or even absent in undifferentiated suspension cultured cells. It is likely that the corresponding alkaloid genes are just 'switched off' (Wink, 1987, 1989). The production of berberine and sanguinarine in cell cultures seems to be more the exception than the rule.

Alkaloids are stored predominantly in tissues which are important for survival and reproduction, which include actively growing young tissues, root and stem bark, flowers (especially seeds), seedlings and photosynthetically active tissues. Alkaloid contents in storage organs can be quite high, reaching up to 10% of dry weight in some instances which is important if the alkaloids function as effective defense compounds. In several herbaceous plants, alkaloids are stored in epidermal and subepidermal tissues (e.g. cocaine, colchicine, aconitine, steroidal alkaloids, nicotine, veratrine, buxine, coniine; Wink, 1987) which have to ward off small enemies (insects, micro-organisms) in the first place. In lupin and broom, quinolizidine alkaloid concentrations are up to 200 mM in epidermal tissue, whereas mesophyll tissue have values below 5 mM (Wink, 1986). To reach these sites of accumulation, short- and long-distance transport is often required.

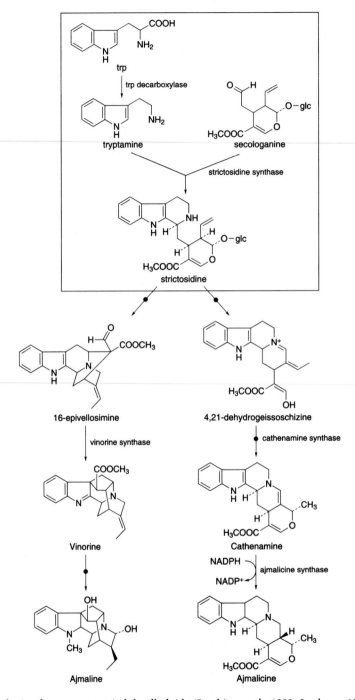

Figure 12.25 Biosynthesis of monoterpene indole alkaloids (Stöckigt *et al.*, 1992; Luckner, 1990).

Some plants possess typical alkaloid-storing cells, called 'idioblasts'. These idioblasts have been detected in *Corydalis* (for corydaline), *Sanguinaria* (for sanguinarine), *Ruta* (for rutacridones), *Catharanthus* (for indole alkaloids) and in *Macleaya* (for protopine) (Wink, 1987).

(c) *Factors influencing alkaloid patterns and alkaloid contents*

Alkaloid patterns usually vary between the site of synthesis and the sites of accumulation, since a number of secondary substitutions may take place

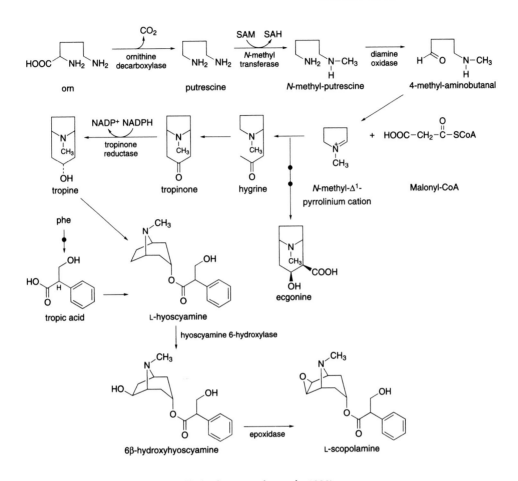

Figure 12.26 Biosynthesis of tropane alkaloids (Yamada *et al.*, 1990).

Table 12.1 Storage of alkaloids in the vacuole

Alkaloid	Plant genera
Leaf vacuoles:	
Lupanine	*Lupinus*
Sparteine	*Cytisus, Lupinus*
Hyoscyamine	*Atropa*
Nicotine	*Nicotiana*
S-scoulerine	*Fumaria*
S-reticuline	*Fumaria*
Ajmalicine	*Catharanthus*
Serpentine	*Catharanthus*
Catharanthine	*Catharanthus*
Betalaines	*Beta, Chenopodium*
Senecionine-*N*-oxide	*Senecio*
Capsaicin	*Capsicum*
Latex vesicles:	
Sanguinarine	*Chelidonium*
Berberine	*Chelidonium*
Morphine and other morphinanes	*Papaver*

Table 12.2 Examples for organ-specific biosynthesis of alkaloids

Alkaloid	Organ	Plant genera
Tropane alkaloids	roots	*Atropa, Datura, Hyoscyamus, Mandragora*
Nicotine	roots	*Nicotiana*
Senecionine	roots	*Senecio*
Emetine	roots	*Cephaelis*
Sanguinarine	roots	*Sanguinaria*
Betalaines	roots, shoots	*Beta*
Quinine	stem bark	*Cinchona*
Berberine	stem and root bark	*Berberis, Mahonia*
Caffeine	green tissue	*Coffea*
Quinolizidine alkaloids	leaves and other photosynthetic tissues	*Lupinus, Cytisus, Laburnum, Baptisia*

Table 12.3 Evidence for a diurnal cycle of alkaloid formation

Alkaloids	Maximum of alkaloid formation	Plant sources
Quinolizidines	noon–early evening	*Lupinus, Cytisus, Baptisia, Laburnum*
Tropanes	evening, midnight	*Atropa*
Nicotine	midnight	*Nicotiana*
Morphine	noon	*Papaver*

Table 12.4 Long-distance transport of alkaloids by phloem or xylem

Alkaloid	Xylem	Phloem	Taxa
Lupanine, sparteine	—	X	*Lupinus, Cytisus*
Cytisine	—	X	*Laburnum, Petteria, Genista, Spartium*
Matrine	—	X	*Sophora*
Senecionine (N-oxide)	—	X	*Senecio*
Seneciphylline (N-oxide)	—	X	
Aconitine	?	X	*Aconitum*
Nicotine	X	—	*Nicotiana*
Hyoscyamine	X	—	*Atropa*
Scopolamine	X	—	*Datura, Hyoscyamus*
Swainsonine	—	X	*Astragalus*

in the latter tissues. Alternatively, transport may be selective, distributing differing cocktails. In addition, alkaloid profiles of seeds and seedlings often differ from those of the mature plant. Both patterns and concentrations usually change during the development of plants and the annual cycle. In general, alkaloid levels are markedly reduced in senescing tissues, so that shed leaves are often nearly alkaloid-free (Waller & Nowacki, 1978). In some plants, alkaloid concentrations may even fluctuate in a diurnal cycle (Table 12.3). Alkaloid formation and storage may be influenced by environmental stress, such as wounding or infection (Waller & Nowacki, 1978) (see section 12.5.8).

12.5.6 Transport

A number of alkaloids are synthesized and stored in all parts of plants, whereas others are restricted to a particular organ. Alternatively, alkaloids are synthesized and then accumulate in many organs which usually demands a long-distance transport. Theoretically, this transport could proceed sym- or apoplastically. However, utilization of established transport routes, such as xylem or phloem seem to be more common. Because it is technically difficult to sample and to analyze xylem and phloem sap, only a limited number of appropriate data are available (Table 12.4).

Besides long-distance transport, short-distance and intracellular transport need to be reviewed. In general, alkaloids are synthesized in the cytosol or in membrane-enclosed vesicles (ER, mitochondria, chloroplasts), but are accumulated and sequestered in the vacuole (Table 12.1).

The storage of high concentrations of alkaloids is a prerequisite for their allelochemical roles as defense compounds. Since these concentrations would interfere with the normal metabolism, the allelochemicals are safely stored in the vacuole of often specialized cells or tissues (such as

the epidermis). The vacuole of these alkaloid-accumulating cells has been termed 'toxic' or 'defense' compartment (Matile, 1984; Wink, 1993b).

Storage in the vacuole or in vesicles requires that alkaloids have to pass the tonoplast and to accumulate within the vacuole against a concentration gradient. For the passage across the tonoplast, three mechanisms are possible: (1) simple diffusion (takes place in case of lipophilic alkaloids, e.g. nicotine, ajmalicine, vinblastine, colchicine); (2) carrier-mediated transport (in case of polar and charged alkaloids, which is the rule for most alkaloids under physiological conditions, e.g. hyoscyamine, lupanine, reticuline, scoulerine, senecionine); and (3) membrane fusion (in case of alkaloids that are formed in vesicle-enclosed compartments, e.g. berberine).

Because alkaloids are sequestered against a concentration gradient in the vacuole or in latex vesicles, the driving force of uphill accumulation needs to be determined. In some instances, vesicles or vacuoles contain alkaloid binding or complexing compounds. For example, latex vesicles of *Chelidonium majus* contain 500–1200 mM chelidonic acid which binds or complexes berberine and benzophenanthridine alkaloids. It can be shown experimentally that chelidonic acid acts as a trapping agent which causes the apparent uphill transport resulting in alkaloid concentrations in vesicles of 500–1200 mM (Hauser & Wink, 1990). Protonation, organic acids or specific peptides may constitute other trapping mechanisms.

There is some evidence that amino acids and some ions are transported into the vacuole by transporters which are fuelled by proton–substrate antiport mechanisms. Protons are enriched in the vacuole which generally have hydrogen ion concentrations of 0.001–1 mM (= pH 6–3), by proton-translocating ATPases and pyrophosphatases. By analogy, it has been assumed (supported by experimental evidence) that carrier-transport of alkaloids is achieved by a proton-alkaloid antiport mechanism (Wink, 1993b).

12.5.7 Recycling

In general, alkaloids are not end products of metabolism but can be degraded which seems to be plausible because nitrogen is a limiting nutrient for plants. As discussed in the introduction, alkaloids stored in seeds are partly degraded during germination and seedling development and their nitrogen is probably used for the synthesis of amino acids. Degradative pathways have not yet been worked out. In addition to this developmentally specific recycling, there is evidence that a number of alkaloids are turned over all the time with half-lives of between 2 and 48 h (Luckner, 1990). Examples are nicotine, quinolizidine alkaloids and tropane alkaloids. Alkaloid turnover is often quite substantial in cell cultures: *Lupinus* callus cultures are even able to live on the QA sparteine as a sole nitrogen source for more than 6 months (Wink & Witte, 1985). How can this phenomenon be explained? A number of alkaloids are allelochemicals and affect molecular targets such as receptors of neurotransmitters (tropanes, nicotine, etc.). For this interaction a correct stereochemical configuration is required. Because alkaloids may oxidize or racemize spontaneously (and thus lose their activity), a continuous turnover would ensure that there is always a sufficient concentration of active compound available, similar to the situation of protein turnover.

2.5.8 Functions

(a) *Role as defense compounds*

As described above, alkaloid biology is tightly connected with the basic physiology of plants. Many of the features described here would make no sense if these compounds lack a vital function for the producer. The main function is that of chemical defense against herbivores (both insects, other arthropods and vertebrates) which can be seen from the fact that many alkaloids have a high affinity for receptors of neurotransmitters which

are only present in animals. In some instances, alkaloids play a role (additionally) in the anti-microbial defense (against bacteria, fungi, viruses) and even in the interaction between one plant and another (allelopathy) (Harborne, 1993; Whitacker & Feeney, 1971; Hartmann, 1991; Verpoorte, 1987; Wink, 1993a).

Alkaloids are certainly multipurpose compounds which, depending on the situation, may be active in more than one environmental interaction. For example, quinolizidine alkaloids are the most important defense chemicals in Leguminosae against insects and other herbivores, but they also influence bacteria, fungi, viruses and even the germination of other plants. In addition, they are employed as degradable N-transport and N-storage compounds (Wink, 1987, 1988, 1992).

Alkaloids repel or deter the feeding of many animals (for humans and other vertebrates, many have a bitter or pungent taste) or if ingested they are toxic. In micro-organisms and competing plants, a reduction of growth and antibiosis are usually the visible effects of alkaloid intoxication. How are these diverse effects being achieved? Although most compounds have not been studied in all details, nevertheless an impressive list of cellular and molecular targets have been identified which are selectively inhibited or modulated by alkaloids (overview Wink, 1993a) (Table 12.5). As a consequence of such interactions, organ malfunctions (heart, lung, liver, kidney, CNS) result which may reduce reproduction, and fertility in animals and other organisms or just kill them (Robinson, 1981; Wink, 1993a).

Because many alkaloids appear to have been shaped during evolution by 'molecular modeling', many of them are used by *Homo sapiens* as medicinal compounds. Using them at non-toxic concentrations, the allelochemicals may have positive effects. It was formulated long ago by Theophrastus Bombastus von Hohenheim, called Paracelsus (1493–1541) 'Was ist das nit Gifft ist? Alle Ding sind Gifft und nichts ohn Gifft. Allein die Dosis macht, das ein Ding kein Gift ist' (Everything is a poison and nothing is without toxicity. Only the dose makes that a thing is no poison) (Luckner, 1990).

(b) *Presence in the right concentration at the right time and place*

To be effective, alkaloids need to be present at the right time, site and concentration. Alkaloid metabolism and biochemistry seem to have been optimized and co-ordinated in most systems to fulfill this prerequisite (see sections 12.5.1–12.5.7).

Table 12.5 Allelochemical effects of alkaloids

Target	Alkaloids which affect this target
DNA/RNA	
(intercalation)	berberine, quinine, β-carbolines, sanguinarine, chelerythrine, coptisine, fagaronine, dictamine, ellipticine
(alkylation)	pyrrolizidines, aristolochic acid
(mutations)	anagyrine, anabasine, coniine, ammodendrine, cyclopamine, jervine, veratrosine, theobromine
(inhibition of DNA/RNA polymerases)	vincristine, chelilutine, fagaronine, amanitine, hippeastrine, lycorine
Cytoskeleton/microtubules	vinblastine, colchicine, taxol, maytansine, maytansinine
Protein biosynthesis (inhibition)	emetine, quinolizidines, vinblastine, tubulosine, cryptopleurine, harringtonine, haemanthamine, lycorine, pretazettine, tylocrebine, tylophorine
Membrane stability	steroidal alkaloids, tetrandine, ellipticine, berbamine, cepharanthine
Electron transport chains (inhibition)	ellipticine, alpigenine, syanguinarine, tetrahydropalmatine, capsaicin, DIMBOA
Ion channels, carrier (inhibition)	aconitine, sparteine, acronycine, harmaline, quinine, reserpine, colchicine, salsolinol, sanguinarine, caffeine, monocrotaline, capsaicin, cassaine, palytoxin, saxitoxin, tetrodotoxin
Receptors of neurotransmitters	
Acetylcholine (ACh)	nicotine, hyoscyamine, scopolamine, coniine, cytisine, pilocarpine, arecoline, muscarine, anabasine, lobeline, tubocurarine, C-toxiferine, heliotrine, eupanine, sparkine, senecionine
Dopamine	ergot alkaloids, cocaine
Serotonin	β-carbolines, ergot alkaloids
Epinephrine/norepinephrine	ephedrine, ergot alkaloids, yohimbine
Glycine	strychnine
GABA	β-carbolines, muscimol, biculline
opiate	morphine
Transport or degradation of neurotransmitters	
	reserpine, cocaine, β-carbolines, physostigmine, galanthamine
Enzyme inhibitors	
adenylate cyclase	anonaine, isoboldine, tetrahydroberberine
phosphodiesterase	papaverine, caffeine, theobromine, theophylline, β-carbolines
hydrolases	castanospermine, swainsonine, other polyhydroxyalkaloids

For further examples, see Wink (1993a).

An interesting variation can be seen in some plants, where alkaloid formation is enhanced by wounding or microbial attack, i.e. in case of emergency the production of defense compounds is stimulated. A few examples are recorded in Table 12.6.

How effective are alkaloidal defenses? In lupins, which normally produce high amounts of quinolizidine alkaloids, mutants have been selected which accumulate only trace amounts of alkaloids. When both, alkaloid-rich and low-alkaloid lupins are planted in the field without fences or other protection, then a dramatic effect can be regularly seen (Table 12.7). Whereas the low-alkaloid lupins were selectively grazed or infested by herbivores, their alkaloid-rich counterparts remained almost unmolested. More experimental data are certainly needed but the defensive role of alkaloids seems to be beyond doubt (Wink, 1993a).

(c) *The exceptions of the rule: adaptations of specialists*

No defense is absolute. Whereas chemical defense works against the majority of potential enemies, usually a few specialists have emerged during evolution which have specialized on the toxin protected ecological niche. This phenomenon can be clearly seen in those insects which are highly host-plant specific. Some of these insects take up the dietary alkaloids, store and exploit them for their own chemical defense or that of their offspring. Others additionally transform the alkaloids into pheromones or utilize them as

Table 12.6 Stimulation of alkaloid production by wounding or microbial elicitors (Wink, 1993a)

Alkaloids	Effector	Plant material
Ajmalicine/catharanthine	fungal elicitor	*Catharanthus* cell cultures
Canthin-6-one	yeast/fungal elicitor	*Ailanthus* cell cultures
Harringtonia alkaloids	fungal elicitor	*Harringtonia* cell cultures
Hyoscyamine	wounding	*Atropa* plants
Liriodenine	wounding	*Liriodendron* plants
Methylxanthines	NaCl	*Coffea* cell cultures
Nicotine	wounding	*Nicotiana* plants
Quinolizidines	wounding	*Lupinus* plants
	elicitors (low MW)	*Lupinus* cell cultures
Rutacridones	fungal elicitors	*Ruta* cell cultures
Sanguinarine	fungal elicitors	*Papaver, Eschscholtzia* cell cultures

morphogens. Well-studied examples have been published for pyrrolizidine and quinolizidine alkaloids (Boppré, 1990; Wink, 1985, 1992, 1993a; Nickisch-Rosenegk *et al.*, 1990; Hartmann, 1991; Meinwald, 1990; Blum, 1981; Harborne, 1993; Schlee, 1992).

12.6 AUXINS, CYTOKININS AND ETHYLENE

Growth and development (germination, formation of roots and shoots, senescence) of plants is governed by special growth factors or phytohormones as in other complex eukaryotic organisms. Of those which contain nitrogen in their molecule, we distinguish auxins, cytokinins and

Table 12.7 Herbivory in bitter and sweet lupins

Lupin	Herbivore	Alkaloid content[a]	Herbivory
Lupinus albus	hares/rabbits	high	0–5%
		low	100%
	aphids	high	0%
		low	100%
	leaf miners	high	0%
		low	100%
	beetles (*Sitona*)	high	0%
		low	100%
	Macrosiphum albifrons[b]	high	80%
		low	5%

[a]High = 1–3 mg QA/g fresh wt; low = <0.1 mg/g fresh wt.
[b]*M. albifrons* is a specialist which stores QA and uses them for defence against predators.
See also Wink (1988, 1992).

ethylene. In contrast to the animal system, phytohormones are not formed in special organs and then transported through a transport system, but are more or less synthesized at the place where they are needed. Therefore, some authors prefer not to use the term hormones for these compounds but use growth substances instead.

12.6.1 Auxins: biosynthesis and function

The main active auxin is indole-3-acetic acid (IAA) which is synthesized in the shoot tips of growing plants, e.g. of a seedling. The biosynthesis starts with the amino acid tryptophan (Fig. 12.27) and can follow three different routes, i.e. via indole-3-acetamide, tryptamine or indolepyruvic acid.

Auxin induces the elongation of cells which lie underneath the tips. In addition, auxin mediates the various tropisms in plants (i.e. bending of climbers, movement of leaves). Auxin seems to regulate gene expression in various growth and developmental processes, i.e. some specific mRNAs are 500-fold enhanced after IAA treatment. Receptor proteins which bind auxin with high affinity have been described from the cytoplasmic membrane and characterized genetically.

For cell and tissue culture of plants, auxins are necessary in the growth medium: in general, the synthetic analogs 2,4-dichlorophenoxyacetic acid (2,4-D) and naphthylacetic acid (NAA) are employed besides the natural indole-3-acetic acid.

Some insects induce gall formation (and elicit the formation of specific allelochemicals, e.g. gallotannins) in a number of plants (e.g. galls of

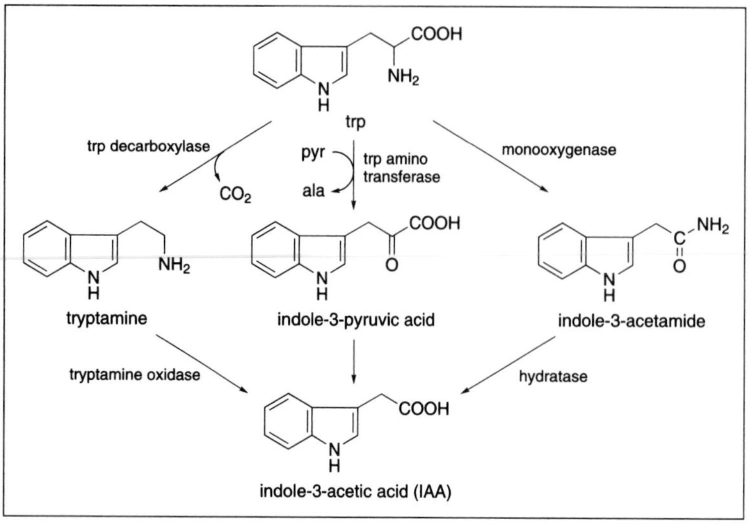

Figure 12.27   Pathway of auxin formation.

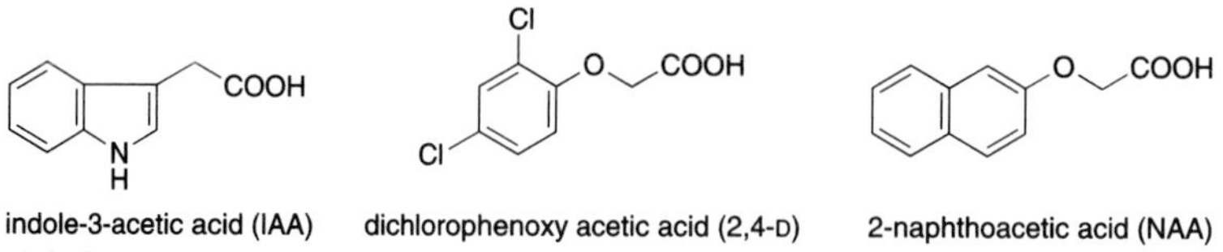

indole-3-acetic acid (IAA)   dichlorophenoxy acetic acid (2,4-D)   2-naphthoacetic acid (NAA)

oak leaves). It is assumed that these organisms produce auxin which in turn activates certain genes to build up the complex galls which safely harbor the growing insect larvae. The complex and interesting interactions still need to be elucidated in detail.

Agrobacterium rhizogenes contains the Ri-plasmid with genes coding for enzymes of auxin biosynthesis. Upon transformation of plants (shoots or leaves), indole-3-acetic acid is produced which leads to the formation of hairy roots. Hairy roots can easily be cultured *in vitro*. They are a useful system to study the biochemistry of those secondary compounds which are made by roots *in vivo* (see Chapter 9). *Agrobacterium tumefaciens* contains the Ti-plasmid with genes coding for auxin and a cytokinin (N^6-dimethylallyladenine) formation. Upon transformation, plant cells grow rapidly and form 'crown-gall' tumors. If cytokinin genes are deleted 'hairy roots' similar to the situation after *A. rhizogenes* infection appear, indicating the complex interactions of auxin and cytokinin. The Ti plasmid has become a useful tool to transfer foreign genes into new plants (see Chapter 9). Auxin is rapidly degraded by peroxidases and so-called 'IAA-oxidases' which have been detected in many plants (Barz & Köster, 1980).

12.6.2   Cytokinins: biosynthesis and function

The cytokinin zeatin which is characterized by a hydroxylated prenyl substituent has been isolated from maize (*Zea mays*). N^6-dimethylallyladenine is closely related and derives from adenosine monophosphate condensed with dimethylallyl diphosphate (Fig. 12.28). A DNA degradation product, 6-furfuryladenine (=kinetin) exerts a strong cytokinin activity and is used as a phytohormone in cell and tissue culture of several plants (see Chapter 15).

Cytokinins are phytohormones which stimulate cell division (activation of replication and transcription) and delay senescence. In buds they function as an antagonist of auxin; for example, the local application of cytokinins to soybean flowers prevents their premature abortion and increases seed yield. It has been assumed that cytokinins bind to membrane receptors and elicit a series of cellular signals.

12.6.3   Ethylene: biosynthesis and function

Ethylene is derived from the amino acid L-methionine (Fig. 12.29) via SAM, and 1-amino-

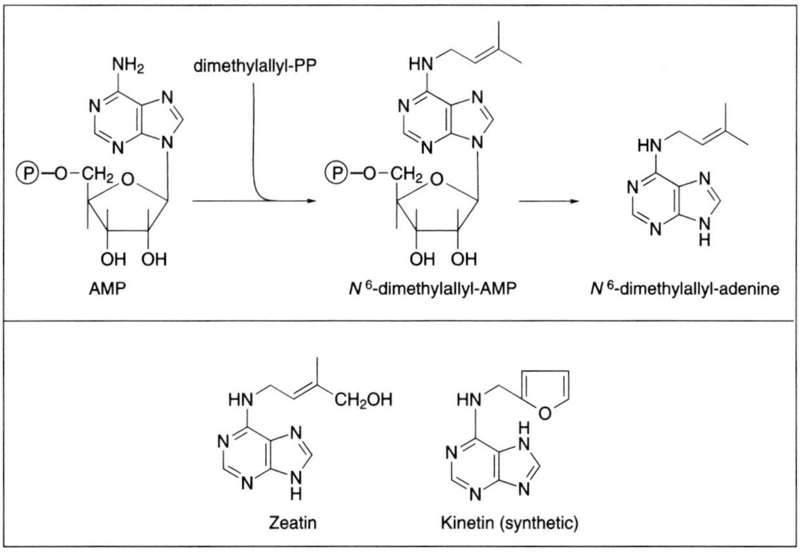

Figure 12.28 Possible biosynthetic pathway of cytokinin formation.

cyclopropane-1-carboxylic acid (ACC). The latter step is catalyzed by 1-aminocyclopropane-1-carboxylic acid (ACC)-synthase, an enzyme whose gene has already been isolated and cloned. 1-aminocyclopropane-1-carboxylic acid (ACC) is oxidized by ACC oxidase to ethylene and cyanoformic acid (which decarboxylates spontaneously to HCN and CO_2) (Luckner, 1990). Ethylene can be degraded by oxidation to ethylene oxide (ethylene monooxygenase) and further to ethylene-glycol. ACC-synthase is a key enzyme in that ethylene formation is controlled by the expression of ACC-synthase, which can be induced by certain stress conditions.

Ethylene is a small volatile phytohormone which can easily diffuse through biomembranes and may influence their integrity. Ethylene affects several developmental processes such as the ripening of fruits or the defoliation of leaves. Because its synthesis is generally enhanced in stressed tissues (i.e. by wounding, infection, water stress, etc.) ethylene may activate the biosynthesis of some defense chemicals. In *Hevea*, ethylene stimulates the flow of latex, which can be considered as a sort of defense reaction. It has been suggested that aerial communication between plants might be mediated by ethylene. In molds, ethylene is used to allow hyphae to pass obstacles without touching (Luckner, 1990; Schlee, 1992).

2-Chloroethylphosphonic acid (= ethephon, ethrel) decomposes to ethylene in neutral or alkaline solution and is used commercially to exercise ethylene effects (e.g. in fruit ripening).

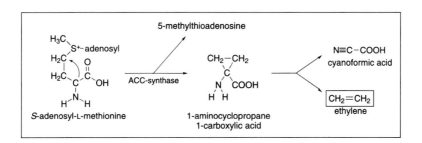

Figure 12.29 Formation of ethylene.

12.7 NITROGEN BASES AND COFACTORS

12.7.1 Biotin

Biotin contains two fused five-membered heterocyclic rings, with N and S as the heteroatoms. Its biosynthesis in micro-organisms and plants is complex, incorporating L-alanine, pimelyl-CoA, CO_2, ATP and the sulfur atom of L-methionine as precursors (Fig. 12.30). Biotin is a cofactor of enzymes which transfer carboxyl groups (carboxylases, transcarboxylases). It is covalently bound through a peptide bond between the carboxyl group of biotin and an amino group of lysine. Using ATP as an energy source and CO_2 as a substrate, protein-bound biotin is converted into N^1-carboxybiotin (Fig. 12.30). Carboxybiotin can transfer its carboxyl group on CH-acidic substrates, e.g. acetyl-CoA to yield malonyl-CoA, which is an important precursor for fatty acid, flavonoid and polyketide biosynthesis (see Chapter 10). Animals and *Homo sapiens* cannot synthesize biotin and need vitamin H in their plant-derived diet. Biotin deficiency may lead to hair loss and dermatitis. Biotin is frequently used in molecular genetics and immunology to label oligonucleotides and proteins, which can be identified by enzyme- or fluorescence-labeled avidin or streptavidin binding to biotin-residues with high affinity.

12.7.2 Tetrahydrofolic acid

Tetrahydrofolic acid (THF) is a pteridine derivative and synthesized in micro-organisms, plants and animals. Its biosynthesis starts with GTP, leading to 7,8-dihydroneopterin as a key intermediate. After the loss of two carbon atoms, 6-hydroxymethyl-7,8-dihydropterin diphosphate is condensed with 4-aminobenzoic acid and L-glutamine to yield dihydrofolic acid. Dihydrofolate reductase reduces this substrate to tetrahydrofolic acid (Fig. 12.31). Tetrahydrofolic acid is converted into an activated coenzyme by incorporation of a C1-unit. This new C-atom is present at differing oxidation levels. These coenzymes are involved in the transfer of one-carbon units, such as hydroxymethyl groups ('activated formaldehyde'), formyl groups ('activated formic acid') and methyl groups ('activated methanol') (Fig. 12.31): N^5,N^{10}-methenyl-THF is involved in the formation of serine, N^5,N^{10}-methylene-THF in that of thymidine 5-phosphate and purines, N^5-methyl-THF in that of methionine and methylcobalamine (in Archaebacteria) and N^{10}-formyl THF in that of purines.

Folic acid functions as a vitamin in Man, which is reduced in the human body to THF. Tetrahydrobiopterin is also derived from 7,8-dihydroneopterin and serves as cofactor in monooxygenases (Fig. 12.31), such as phenylalanine 4-monooxygenase. Pterins are colored compounds and some of them are exploited as pigments in insects, fishes, reptiles and birds.

12.7.3 S-Adenosyl-L-methionine

S-Adenosylmethionine (SAM) is another important donor of methyl groups and figures as coenzyme in N-, O-, C-, and S-methyltransferase

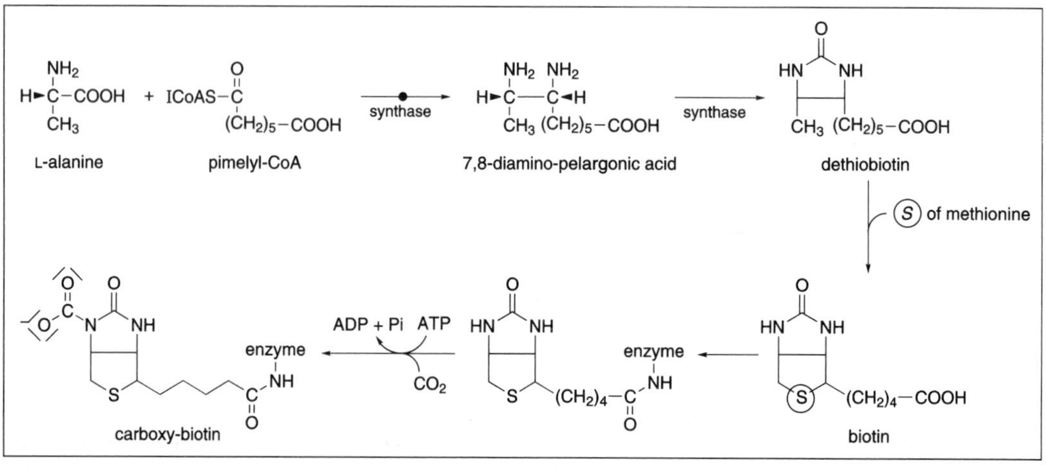

Figure 12.30 Biosynthesis of biotin and carboxy-biotin.

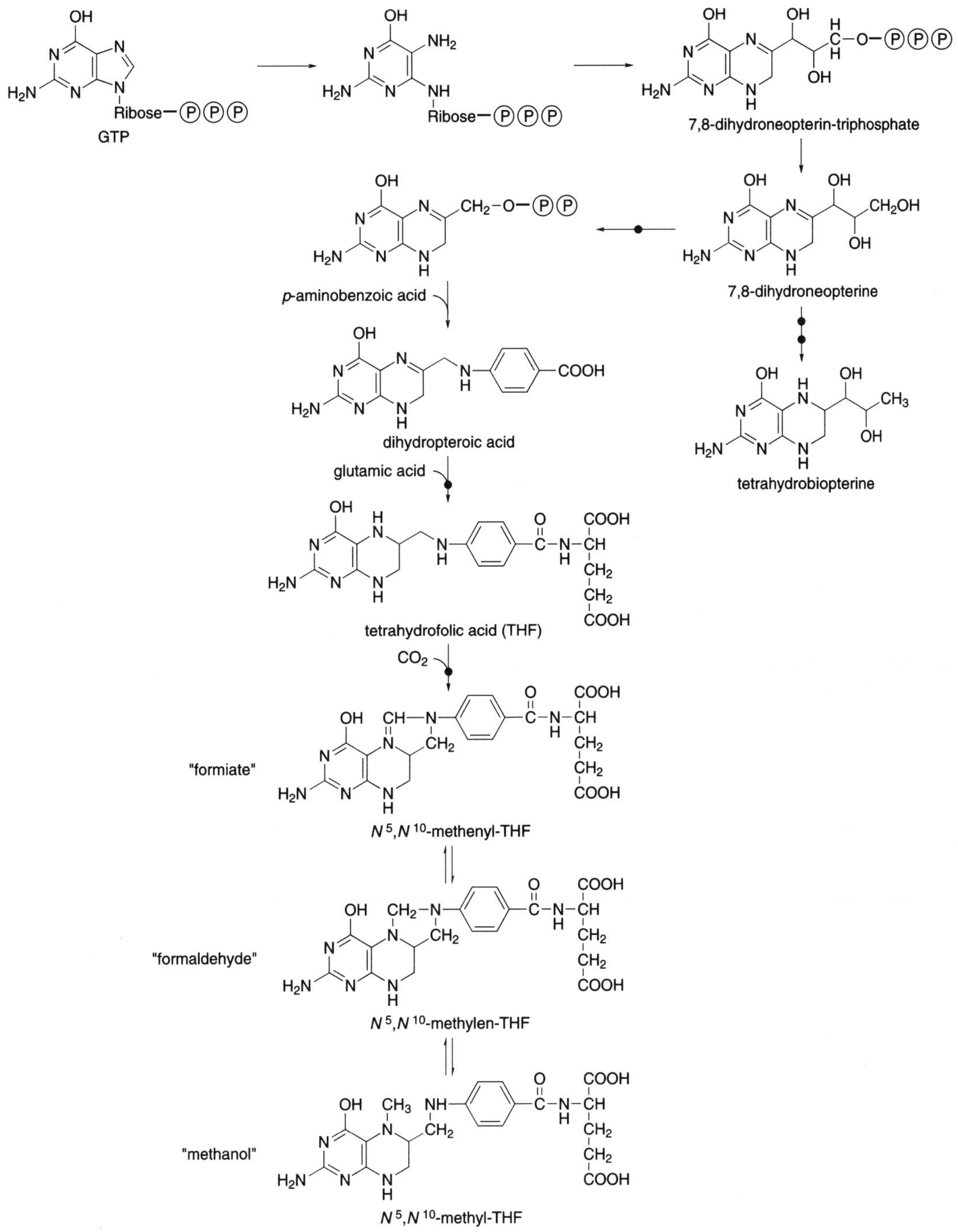

Figure 12.31 Biosynthetic pathway leading to tetrahydrofolate and derivatives.

reactions. ATP and L-methionine are condensed by methionine adenosyltransferase resulting in a positively charged and thus reactive sulfonium ion (Fig. 12.32). As a consequence, the $^+CH_3$ group is easily transferred to other metabolites forming O-methyl, N-methyl, C-methyl and S-methyl derivatives. The remaining S-adenosyl-L-homocysteine (SAH) is a strong competitive inhibitor of the transferase reaction.

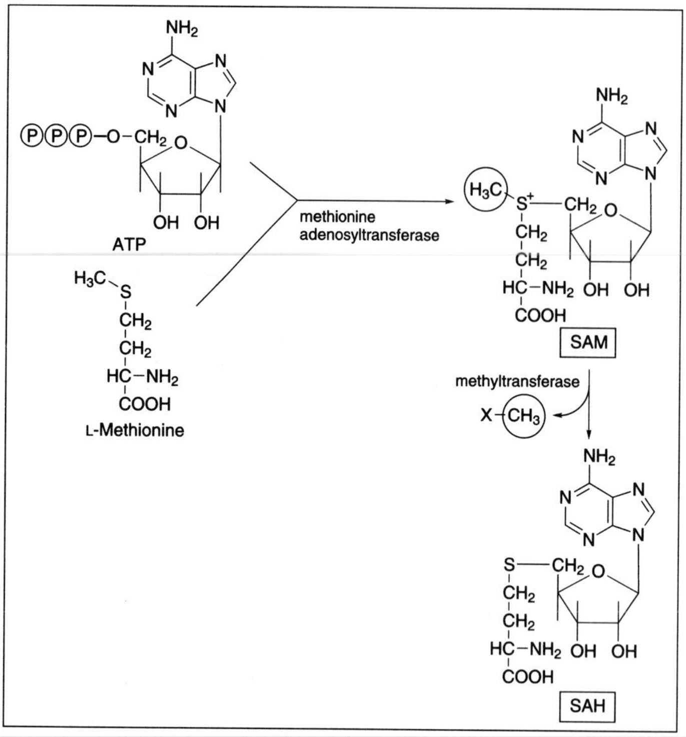

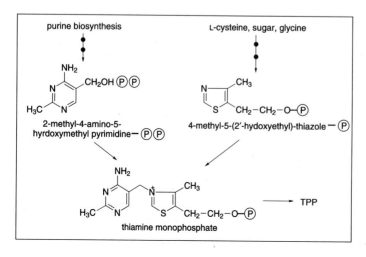

Figure 12.32 Formation of *S*-adenosyl-L-methionine (SAM).

12.7.4 Thiamine diphosphate

Thiamine diphosphate (TPP) consists of a pyrimidine and a thiazole ring. TPP is synthesized in micro-organisms and plants and figures as a vitamin (vitamin B_1) for vertebrates. The thiazole ring is derived from L-cysteine, glycine or tyrosine and a sugar moiety. The pyridine ring comes from 5-aminoimidazole ribonucleotide (Fig. 12.33). TPP is an important cofactor of 2-oxo acid decarboxylases, such as pyruvate and 2-oxoglutarate decarboxylase, and of transketolases. C-2 of the thiazole ring figures as the active center of TPP. Upon losing a proton, a carbanion is generated which reacts and binds to the carbonyl group of 2-oxo acids. The positively charged nitrogen attracts electrons thus weakening the bond of the carboxyl group and liberating CO_2. The remaining aldehyde can be transferred as a nucleophilic group. In case of the pyruvate dehydrogenase complex, the acceptor is lipoic acid (see Chapter 3). As mentioned above, vitamin B_1 is essential for *Homo sapiens*. Its deficiency leads to beriberi.

Figure 12.33 Pathway of TPP formation.

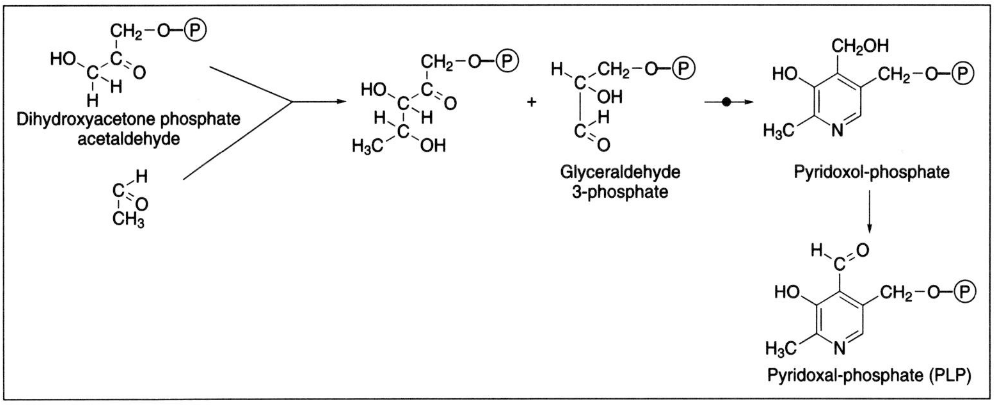

Figure 12.34 Biosynthesis of PLP.

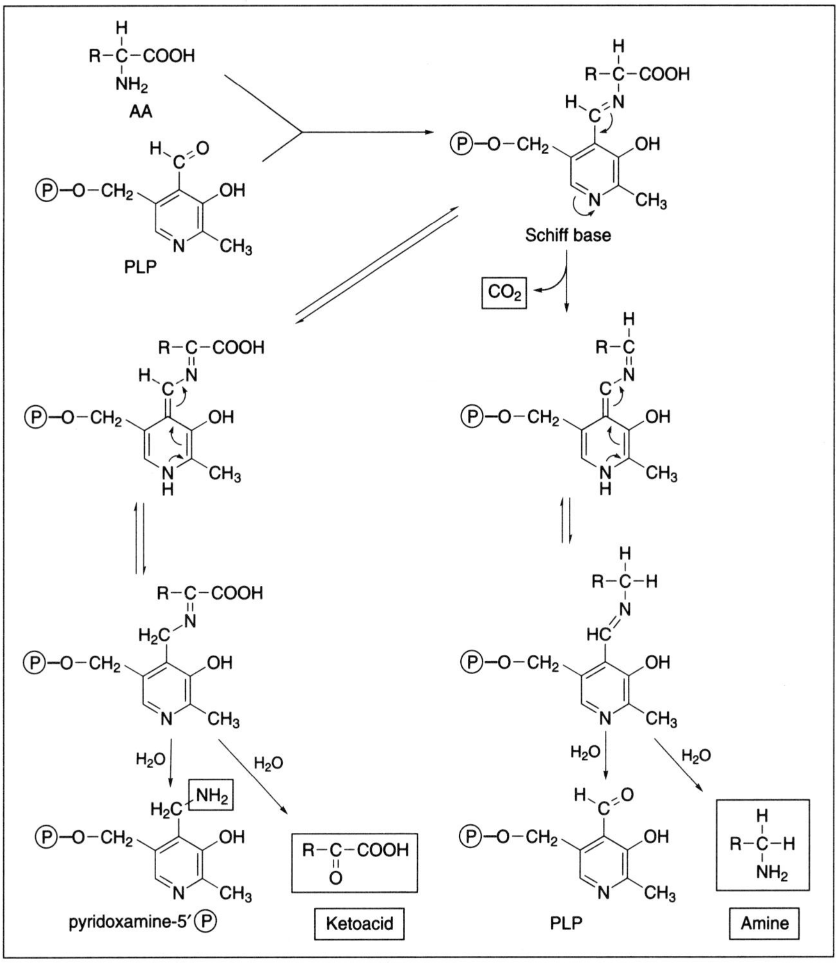

Figure 12.35 Transamination and decarboxylation of amino acids by PLP-containing enzymes.

12.7.5 Pyridoxal phosphate

Pyridoxal phosphate (PLP) is a pyridine derivative, carries a free aldehyde as a functional group and is synthesized in most micro-organisms and plants. It has been suggested that acetaldehyde, dihydroxy-acetone phosphate and D-glyceraldehyde-3-phosphate figure as precursors (Fig. 12.34). The resulting pyridoxol-phosphate is reduced to PLP. Pyridoxin, pyridoxal and pyridoxamine act as vitamins (vitamin B_6) and are transformed to PLP. PLP is the cofactor of aminotransferases and amino acid decarboxylases which play a central role in the metabolism of nitrogenous compounds. It is covalently, although reversibly, bound to a lysine of the active center. As the main reaction, a Schiff's base is formed between the aldehyde of PLP and the amino group of amino acids (Fig. 12.35). The pyridine nitrogen atom is a strong electrophile and induces the displacement of a pair of electrons adjacent to the α-carbon of the respective amino acid, resulting in the loss of a substituent at the α-C atom. Depending on the enzyme involved, the same cofactor can mediate quite different reactions, i.e. decarboxylation and transamination.

12.7.6 Coenzyme A

CoA can be considered as an adenosine-3-mono, 5-diphosphate which is esterified with pantetheine ($=$ panthotenic acid $+$ cysteamine). Its functional group is the thiol of the cysteamyl residue which can form energy-rich thiol esters and thus activate organic acids. Its biosynthesis in plants and micro-organisms proceeds from pantoic acid (Fig. 12.36) and β-alanine to yield after phosphorylation 4'-phosphopantothenic acid. This reaction is catalyzed by pantothenate synthetase.

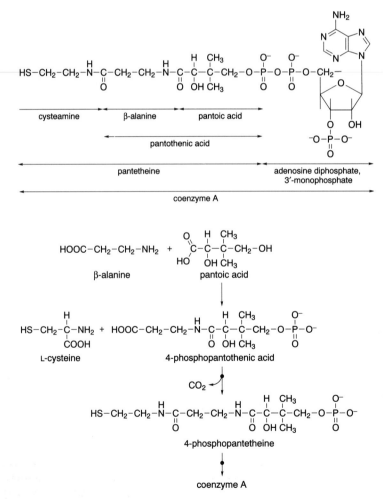

Figure 12.36 Biosynthesis of Coenzyme A.

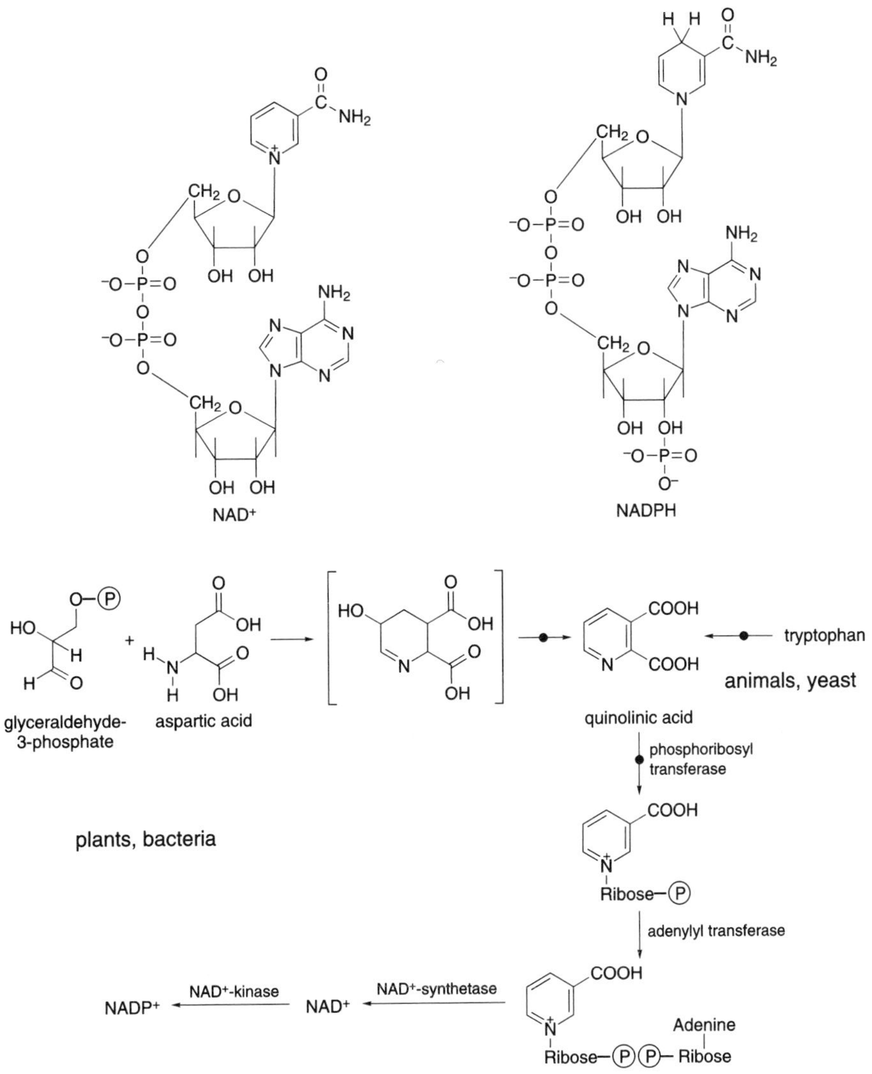

Figure 12.37 Biosynthesis of nicotinic acid and NAD$^+$.

4'-phosphopantothenic acid forms a peptide with L-cysteine. After decarboxylation of the cysteine residue to cysteamine, 4'-phosphopantetheine is generated. Then AMP is added and phosphorylated in the 3'-position.

Coenzyme A (A stands for 'acyl') is an important cofactor for CoA-transferases to yield esters or amides. Acetyl-CoA which is generated by oxidative decarboxylation or fatty acid degradation, is a most important C2-fragment in the primary metabolism of plants (citric acid cycle, fatty acid biosynthesis (see Chapter 6), polyketides, etc.). Many secondary metabolites are esters with a wide range of organic acids, such as propionic acid, butyric acid, angelic acid, tiglic acid, benzoic acid and cinnamic acid. They are activated by acid-thiol ligases, a reaction involving ATP and CoA. The acyl-CoA is then combined with a hydroxyl or amino group to yield esters or amides. Pantothenic acid is a vitamin for animals and can be channeled into CoA biosynthesis.

12.7.7 Nicotinamide adenine dinucleotide

NAD$^+$/NADH or NADP$^+$/NADPH$^+$ are the cofactors of many hydrogen-transferring oxidoreductases. These two nucleotides carry a nicotinamide as a reactive moiety, which is linked to ribose forming a N-glycosidic bond. NADP is characterized by a third phosphate group in the 2' position of ribose. Because of the positive charge of the pyridine ring, these coenzymes are abbreviated NAD$^+$ or NADP$^+$ (Fig. 12.37). NAD$^+$ and NADP$^+$ are coenzymes or better cosubstrates of many dehydrogenases which reduce primary and secondary alcoholic hydroxyl groups or oxidize aldehydes or ketones to hydroxyl groups. In most bacteria and higher plants nicotinic acid is derived from L-aspartic acid and probably D-glyceraldehyde-3-phosphate (Fig. 12.37). Quinolinic acid is decarboxylated and enters the 'nicotinic acid cycle' which generates, among others, NAD$^+$. The

enzymes involved have been characterized in suspension cultured cells of tobacco.

These coenzymes can bind hydrogen reversibly: the pyridine ring is being reduced and the nitrogen loses its positive charge. Mechanistically, a hydride ion is transferred from the substrate to the 4-position of the pyridine ring. Since this carbon atom is prochiral then, the hydride ion can be introduced either in the pro R or pro S configuration. Depending on the oxidoreductases involved, the reaction procedes in a stereochemically controlled and enzyme-specific fashion (Luckner 1990). The following reaction scheme can be formulated:

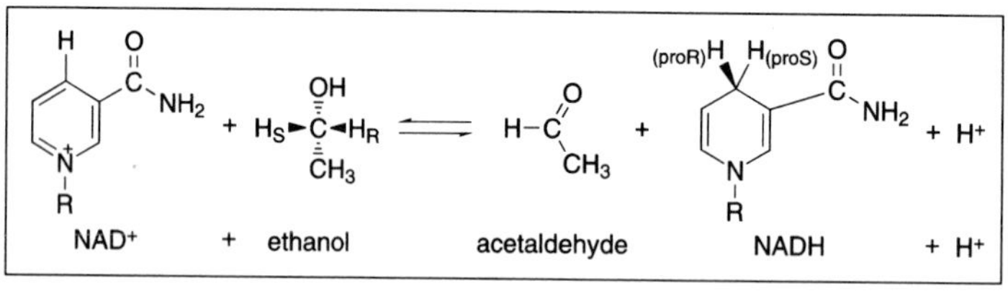

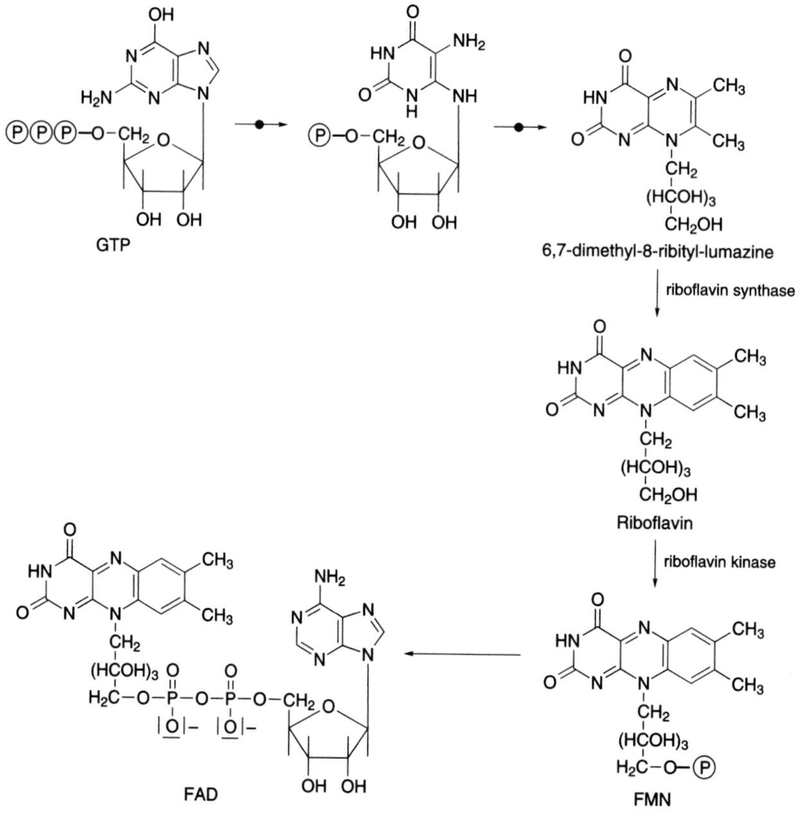

Figure 12.38 Biosynthesis of riboflavin, FMN and FAD.

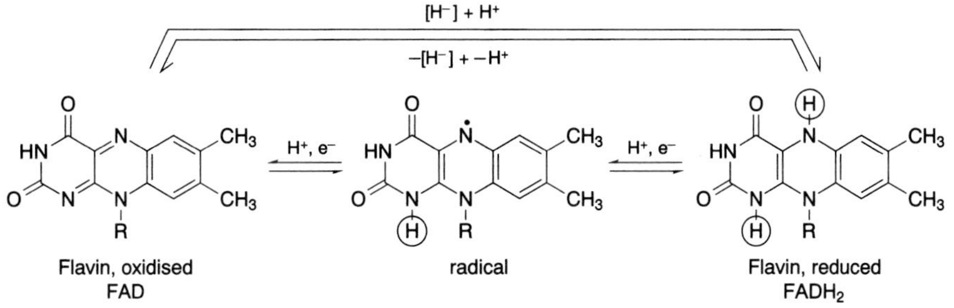

Figure 12.39   Oxidoreduction of FAD/FADH$_2$.

According to this scheme, we write NADH$^+$ + H$^+$ for the reduced form of NAD$^+$. NADH is mainly generated in mitochondria, NADPH during photosynthesis in chloroplasts (see Chapter 2). Whereas, as a rule of thumb, NAD$^+$/NADH is generally involved in reactions of energy metabolism ('reduction equivalents' see Chapter 3), NADP$^+$/NADPH is used in synthetic reactions. The hydrogen of NADPH can be transferred to NADH by transhydrogenases under certain metabolic conditions.

Nicotinic acid (syn. niacine) and nicotinamide (syn. niacinamide) function as vitamins for *Homo sapiens*, although they can be synthesized from tryptophan (Fig. 12.37) if this amino acid is available. Otherwise, deficiency syndromes, such as pellagra, diarrhoea and delirium can occur. Many biochemical reactions are oxidoreductions

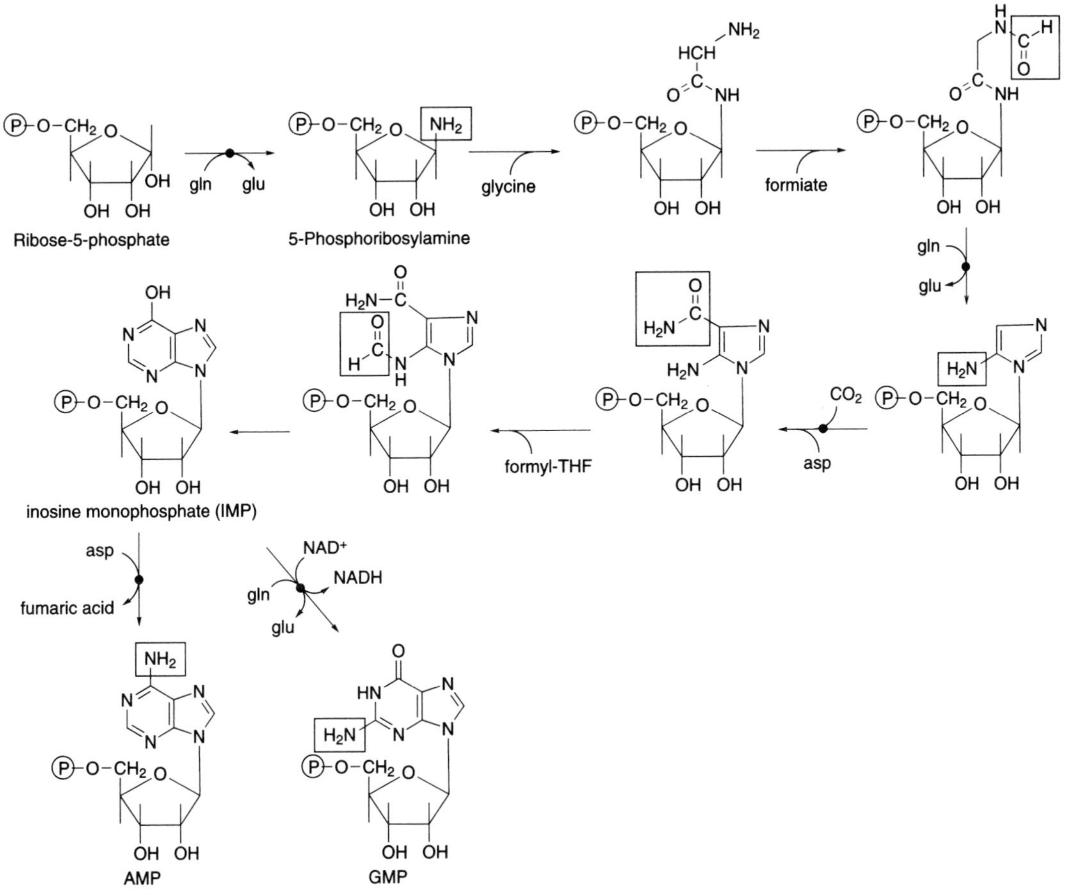

Figure 12.40   Biosynthesis of purines.

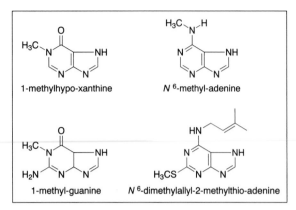

Figure 12.41   Structures of modified purines.

or can be coupled with these processes. Since the reduction of NAD^+ to NADH can easily be followed photospectroscopically at 340 nm, corresponding tests are widely used in biochemistry and clinical laboratories.

12.7.8   Flavin adenine nucleotides

Riboflavin, a yellow-colored vitamin of the vitamin B_2 complex provides the basic skeleton for the cofactors flavin mononucleotide (FMN) or flavin adenine dinucleotide (FAD) (Fig. 12.38). Flavins are either covalently bound to proteins (via one methyl group) or stick to their protein because they have a high affinity for it. These proteins are called flavoproteins since they have a yellow color when purified. Flavins can be classified as isoalloxazine derivatives having a pteridine nucleus. As with tetrahydrofolic acid (THF), the biosynthesis

starts with GTP. After elimination of C8, the ribosyl moiety is cleaved (Fig. 12.38). A four carbon unit is added (tetrolose which derives from pentose phosphate) to yield 6,7-dimethyl-8-ribityl-lumazine. Two units of 6,7-dimethyl-8-ribityl-lumazine are fused and riboflavin results. Upon phosphorylation, FMN is generated. Adding an AMP yields FAD. FMN and FAD function as cofactors of many oxidoreductases in that they carry hydrogen or electrons (for example, in electron transport chains; see Chapter 2). The isoalloxazine molecule functions as a reversible redox system. The oxidized form can be considered as a quinone whereas the reduced form is a hydroquinone. Uptake of one electron leads to a semiquinone (Fig. 12.39). In one step reactions, the addition and elimination of a hydride ion takes place (reaction 1), whereas radical formation is typical in two-step reactions (II) (Luckner, 1990).

12.7.9   Purines and pyrimidines

(a)   *Purines*

Purines occur in all organisms as the bases of DNA and RNA (e.g. adenine, guanine), of nucleotides (ATP and GTP), and last but not least in a few organisms as alkaloids (see section 12.5.1). Purine biosynthesis seems to be similar in all organisms (Fig. 12.40). Ribose-5-phosphate functions as a precursor and the purine skeleton is set together 'piece by piece'. Methenyl-THF and formyl-THF function as donors of carbon atoms.

Inosine-5-monophosphate is a key intermediate, from which guanosine-5-monophosphate (GMP) is

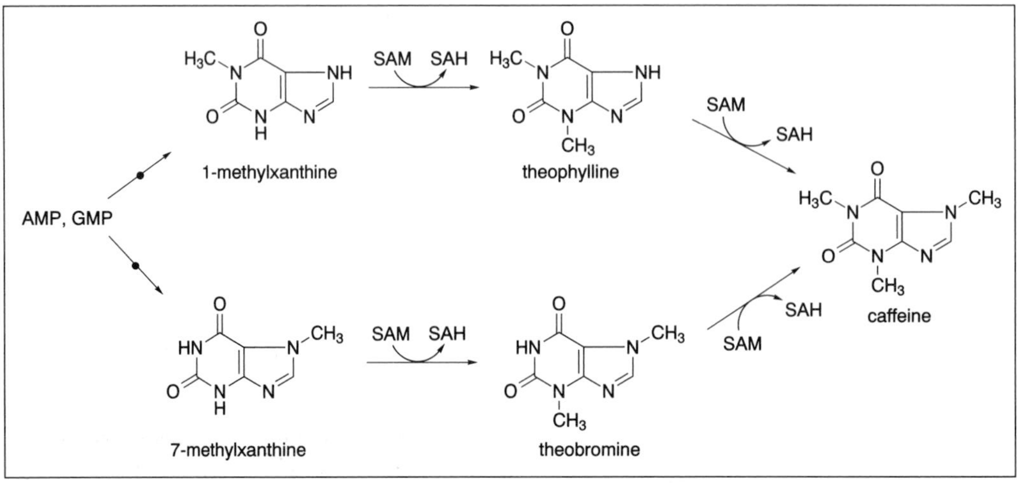

Figure 12.42   Biosynthesis of caffeine.

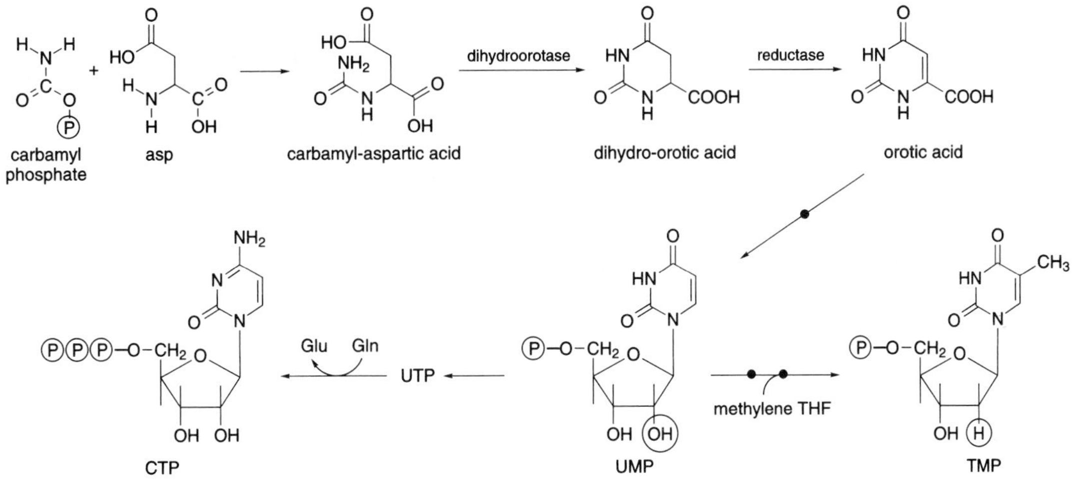

Figure 12.43 Biosynthesis of pyrimidine bases.

derived via xanthosine-5-monophosphate (XMP). IMP is condensed with aspartic acid with the aid of adenosylsuccinate synthase. Upon cleavage of fumaric acid, adenosine-5-monophosphate (AMP) is obtained. Purine biosynthesis is regulated at the level of amido-phosphoribosyl transferase, which is allosterically inhibited by the end products AMP, GMP and IMP. If the supply of monophosphates is sufficient, their biosynthesis comes to a stop. There is some evidence that part of the biosynthetic pathway is catalyzed by a multifunctional protein.

In nucleic acids, methylated and prenylated purine bases occur (Fig. 12.41) which are formed transcriptionally by methyl or prenyl transferases using SAM or dimethylallyl-PP as substrates. N^6-prenylated adenine derivatives have been discussed in section 12.6.2. Methylated xanthines (theophylline = 1,3-dimethylxanthine, caffeine = 1,3,7-trimethylxanthine) are an offshoot of purine

biosynthesis in a few plants (e.g. *Coffea*, *Theobroma*, *Camellia*, *Ilex*, *Cola*). They are synthesized by specific N-methyltransferases which use SAM as a cosubstrate (Fig. 12.42).

(b) Pyrimidines

Pyrimidines are present in all organisms as the DNA and RNA bases (uracil, thymine and cytosine). The biosynthesis (Fig. 12.43) starts by combining carbamyl phosphate and L-aspartic acid with aspartate carbamyltransferase, the resulting intermediate is cyclized to dihydroorotic acid by dihydroorotic acid cyclase. Orotic acid forms an N-glycoside with 5-phospho-ribosyl-1-diphosphate yielding orotidine monophosphate. Through decarboxylation, uridine monophosphate (UMP) is formed which functions as a precursor for cytidine triphosphate and thymidine monophosphate. The methyl group of TMP derives from methylenetetrahydrofolic acid (Fig. 12.32). Because some of the key enzymes (carbamoylphosphate synthetase, aspartate carbamoyltransferase) have been localized in the chloroplast, it has been suggested that the main site of UMP synthesis in plants might be the plastids.

About one third of all cytosine residues in plant nuclear DNA is present as the 5-methylcytosine derivative (Fig. 12.44), which is formed as a postreplicational modification by a SAM-dependent methyltransferase. Methylation of cytosine is thought to be related to gene expression in that hypermethylated genes are 'silent'. In tRNA molecules a number of substituted pyrimidine and purine bases are typically present which regulate the spatial structure tRNAs and their

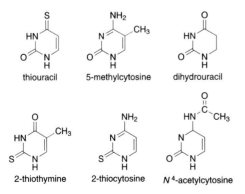

Figure 12.44 Substituted pyrimidines of DNA and RNA.

Table 12.8 Nucleosides and nucleotides

Base	+sugar = nucleoside	+phosphate = nucleotide monophosphate	+diphosphate	+triphosphate
Adenine	adenosine	AMP	ADP	ATP
Guanine	guanosine	GMP	GDP	GTP
Cytosine	cytidine	CMP	CDP	CTP
Thymine	thymidine	TMP	TDP	TTP
Uracil	uridine	UMP	UDP	UTP

binding to acyl tRNA synthetases (Fig. 12.44). The unusual nucleotides are formed transcriptionally. Some pyrimidines serve as precursors for non-protein amino acids (e.g. lathyrine, erythro-4-hydroxy-L-homoarginine, willardiine) in some plants (Luckner, 1990).

12.7.10 Nucleosides and nucleotides

Depending on the degree of derivatization, bases and corresponding derivatives have their own nomenclature which is given in Table 12.8. Adenine, guanine, cytosine and thymine are the pyrimidine bases which form DNA. In RNA, thymine is replaced by uracil (see Chapter 8). The biosynthesis of the bases, which in general yield ribonucleotides, has been dealt with in section 12.7.8. To generate di- and triphosphates, the monophosphates are phosphorylated by specific kinases using ATP as a phosphate donor. To generate deoxyribonucleotides for DNA, the $2'$OH group of ribose in ribonucleotide diphosphates is eliminated by ribonucleotide reductase, involving thioredoxin (which has 2 free HS-groups) as a reductant. The oxidized thioredoxin (in -S-S- form) can be recycled through reduction with NADPH.

ATP and GTP can be transformed to cyclic phosphate esters, such as adenosine-$3':5'$-phosphate (cAMP) and guanosine-$3':5'$-phosphate (cGMP) by adenylate or guanylate cyclase. Cyclic phosphate esters are degraded by phosphodiesterases (Fig. 12.45). Cyclic AMP has been detected in plants, but it certainly does not play a role as a second messenger as in animals. It has been suggested that cAMP and cGMP are involved in the regulation of gene expression.

ATP and also other ribonucleotide tri- and diphosphates are anhydrides of phosphoric acid. The phosphates are bound as 'high energy phosphates' which liberate about $30-40\,\text{kJ}\,\text{mol}^{-1}$ under standard conditions (see Chapter 2). Phosphotransferases (kinases) can exploit this energy to transfer phosphate groups on acceptor molecules which become more active or less diffusible in turn. This activation is common in many biosynthetic pathways leading to primary and secondary metabolites. ATP also fuels proton-, ion- and metabolite transport across biomembranes (ATPases), often against a concentration gradient. Proteins change their conformation when phosphorylated. The conformational change can be used to switch on or off enzymes, to perform movements and to transfer information (regulatory and

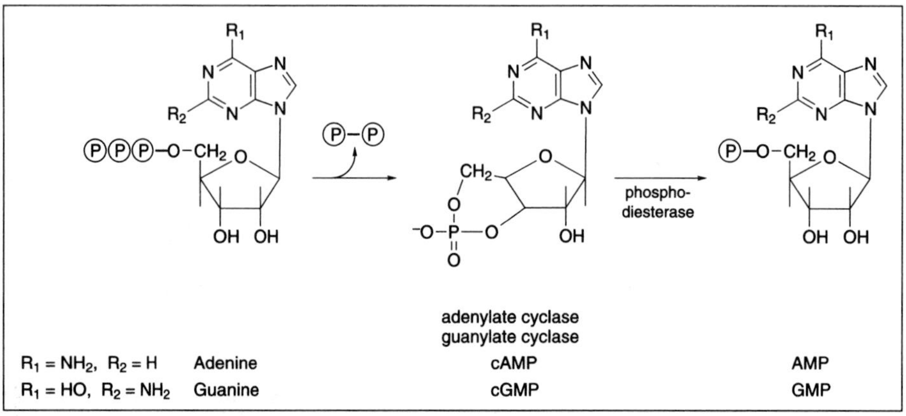

Figure 12.45 Biosynthesis and degradation of cAMP and cGMP.

gene activating proteins), for example between cytosol and nucleus (see Chapter 8).

ATP is generated from AMP and ADP predominantly in mitochondrial electron transport chains and during photosynthesis (see Chapter 2). ATP can also transfer its phosphate groups on other nucleotide mono- and diphosphate to yield GTP, TTP, CTP etc. To transfer sugar moieties onto acceptor molecules in order to form O, N- or C-glycosides, they need to be activated, usually in the form of UDP-sugars, but GDP-, ADP-, TDP- and dUDP derivatives are needed for some reactions (Luckner, 1990). Glucosyltransferases need UDP-glucose as a cofactor. In analogy, choline is activated as CDP-choline.

REFERENCES

Amann, M., Wanner, G. & Zenk, M.H. (1987) *Planta* **167**, 310–320.

Barz, W. & Köster, J. (1981) In *The Biochemistry of Plants* (P.K. Stumpf & E.E. Conn, eds), Vol. 7, *Secondary plant products*, pp. 35–85. Academic Press, New York.

Bell, E.A. (1980) In *Encyclopedia of Plant Physiology*, (E.A. Bell & B.V. Charlwood, eds), Vol. 8, pp. 403–432. Springer-Verlag, New York.

Bell, E.A. & Charlwood, B.V. (1980) *Secondary Plant Products. Encyclopedia of Plant Physiology*. Vol. 8. Springer, New York.

Blum, M.S. (1981) *Chemical Defenses of Arthropods*. Academic Press, New York.

Boppré, M. (1990) *Journal of Chemical Ecology*, **16**, 1–20.

Conn, E.E. (1980) In *Secondary Plant Products. Encyclopedia of Plant Physiology* (E.A. Bell & B.V. Charlwood, eds), Vol. 8, pp. 461–492. Springer, New York.

Conn, E.E. (1981) In *The Biochemistry of Plants* (P.K. Stumpf & E.E. Conn, eds), Vol. 7, *Secondary plant products*, pp. 279–501. Academic Press, New York.

Floss, H.G. (1981) In *The Biochemistry of Plants* (P.K. Stumpf & E.E. Conn, eds), Vol. 7, *Secondary Plant Products*, pp. 177–214. Academic Press, New York.

Fowden, L. (1981) In *The Biochemistry of Plants* (P.K. Stumpf & E.E. Conn, eds), Vol. 7, *Secondary Plant Products*, pp. 215–248. Academic Press, New York.

Harborne, J.B. (1993) *Introduction to Ecological Biochemistry*, 4th edn. Academic Press, London, New York.

Hartmann, T. (1991) In *The Chemical Participants*, Vol. 1 *Herbivores–Their Interactions with Secondary Plant Metabolites* (C.A. Rosenthal & M.R. Berenbaum, eds), pp. 79–121. Academic Press, New York.

Hartmann, T. & Aufermann, B. (1973) *Mar. Biol.* **21**, 70–74.

Hashimoto, T. & Yamada, Y. (1992) In *Plant Tissue Culture and Gene Manipulation for Breeding and Formation of Phytochemicals* (K. Oono et al., eds), pp. 255–259. NIAR, Tsukuba.

Hauser, M.T. & Wink, M. (1990) *Z. Naturforsch.* **45c**, 949–957.

Hegnauer, R. (1962–1990) *Chemotaxonomie der Pflanzen*. Vols. 1–10. Birkhäuser Verlag, Basel.

Kojima, M., Poulton, J.E., Thayer, S. & Conn, E.E. (1979) *Plant Physiol.* **63**, 1022–1028.

Larsen, P.O. (1981) In *The Biochemistry of Plants* (P.K. Stumpf & E.E. Conn, eds), Vol. 7, *Secondary Plant Products*, pp. 502–526. Academic Press, New York.

Levin, D.A. (1976) *Annu. Rev. Ecol. Syst.* **7**, 121–159.

Louda, S. & Mole, S. (1991) In *The Chemical Participants*. Vol. 1 *Herbivores–Their Interactions with Secondary Plant Metabolites*. (C.A. Rosenthal & M.R. Berenbaum, eds), pp. 123–164. Academic Press, New York.

Luckner, M. (1990) *Secondary Metabolism in Microorganisms, Plants, and Animals*, 3rd edn. Springer, Berlin.

Matile, P. (1984) *Naturwissenschaften* **71**, 18–24.

Meinwald, J. (1990) *Pure Appl. Chem.* **62**, 1325.

Mothes, K., Schütte, H.R. & Luckner, M. (1985) *Biochemistry of Alkaloids*. Verlag Chemie, Weinheim.

Nahrstedt, A. (1989) *Planta Med.* **55**, 333.

Nickisch-Rosenegk, E.V., Schneider, D. & Wink, M. (1990) *Z. Naturforsch.* **45c**, 881–894.

Okada, N., Shinmyo, A., Okada, H. & Yamada, Y. (1988) *Phytochemistry* **27**, 979–982.

Pham, T.D.T. & Roberts, M.F. (1991) *Phytochem. Anal.* **2**, 68–73.

Pistocci, R., Keller, F., Bagni, N. & Matile, P. (1988) *Plant Physiol.* **87**, 514–518.

Roberts, M.F. (1987) In *Plant Vacuoles* (B. Marin, ed.), NATO ASI series, vol. 134, pp. 513–528. Plenum, New York, London.

Roberts, M.F., Homeyer, B.C. & Pham, T.D.T. (1991) *Z. Naturforsch.* **46c**, 377–388.

Robinson, T. (1981) *The Biochemistry of Alkaloids*, 2nd edn. Springer, Heidelberg, New York.

Rosenthal, G.A. (1982) *Plant Nonprotein Amino Acids and Imino Acids*. Academic Press, London, New York.

Rosenthal, G.A. (1991) In *Herbivores. Their Interactions with Secondary Plant Metabolites* (G.A. Rosenthal & M.R. Berenbaum, eds). Academic Press, London, New York.

Rosenthal, G.A. & Berenbaum, M.R. (1991) *The Chemical Participants*. Vol. 1 *Herbivores – Their Interactions with Secondary Plant Metabolites*. Academic Press, New York.

Rosenthal, G.A. & Janzen, D.H. (1979) *Herbivores: Their Interaction with Secondary Plant Metabolites*, Academic Press, London, New York.

Schlee, S. (1992) *Ökologische Biochemie*, 2nd edn. Fischer, Jena.

Schultes, R.E. & Hofmann, A. (1980) *The Botany and Chemistry of Hallucinogens*. Charles Thomas, Springfield.

486　MICHAEL WINK

Schultes, R.E. & Hofmann, A. (1987) *Pflanzen der Götter*. Hallwag Verlag Bern.

Seigler, D.S. (1975) *Phytochemistry* **14**, 9–29.

Seigler, D.S. (1991) *The Chemical Participants*. Vol. 1 *Herbivores – Their Interactions with Secondary Plant Metabolites* (G.A. Rosenthal & M.R. Berenbaum, eds), pp. 35–77. Academic Press, New York.

Smith, T.A. (1980) In *Secondary Plant Products. Enyclopedia of Plant Physiology*. Vol. 8 (E.A. Bell & B.V. Charlwood, eds), pp. 433–460. Springer, Berlin.

Smith, T.A. (1981) In *Secondary Plant Products. The Biochemistry of Plants* (P.K. Stumpf & E.E. Conn, eds), Vol. 7, pp. 249–268. Academic Press, New York.

Southon, I.W. & Buckingham, J. (1989) *Dictionary of Alkaloids*. Chapman & Hall, London.

Stöckigt, J., Lansing, A., Falkenhagen, H., Endreß, S. & Ruyter, C.M. (1992) In *Plant Tissue Culture and Gene Manipulation for Breeding and Formation of Phytochemicals* (K. Oono *et al.*, ed.), pp. 277–292. NIAR, Tsukuba.

Swain, T. (1977) *Annu. Rev. Plant Physiol.* **28**, 479–501.

Teuscher, E. & Lindequist, U. (1994) *Biogene Gifte*. Fischer Verlag, Stuttgart, New York.

Verpoorte, R. (1987) In *Antiseptika* (A. Kramer *et al.*, eds). VEB Verlag Volk und Gesundheit, Berlin.

Waller, G.R. (1987) *Allelochemicals: Role in Agriculture and Forestry*, ACS Symp. Ser. 330.

Waller, G.R. & Dermer, O.C. (1981) In *Secondary Plant Products. The Biochemistry of Plants* (P.K. Stumpf & E.E. Conn, eds), Vol. 7. Academic Press, New York.

Waller, G.R. & Nowacki, E. (1978) *Alkaloid Biology and Metabolism in Plants*. Plenum Press, New York, London.

Waterman, P. (1992) *Methods in Plant Biochemistry*, Vol. 8. Academic Press, London.

Whitacker, R.W. & Feeney, R.P. (1971) *Science* **171**, 751–770.

Wink, M. (1985) *Plant Syst. Evol.* **150**, 65–81.

Wink, M. (1986) *Z. Naturforsch.* **41c**, 375–380.

Wink, M. (1987) In *Cell Culture and Somatic Cell Genetics of Plants* (F. Constabel & I.K. Vasil, eds), Vol. 4, pp. 17–42. Academic Press, London, New York.

Wink, M. (1988) *Theor. Appl. Genet.* **75**, 225–233.

Wink, M. (1989) In *Primary and Secondary Metabolism of Plant Cell Cultures* (W.G.W. Kurz, ed.), pp. 237–251. Springer-Verlag, Berlin.

Wink, M. (1992) In *Plant–Insect Interactions* Vol. IV (E.A. Bernays, ed.), pp. 133–169. CRC Press, Boca Raton.

Wink, M. (1993a) In *The Alkaloids* (G. Cordell, ed.), Vol. 43, pp. 1–118. Academic Press, London.

Wink, M. (1993b) *J. Exp. Bot.* **44**, 231–246.

Wink, C. & Hartmann, T. (1981) *Z. Naturforsch.* **36c**, 625–632.

Wink, M. & Hartmann, T. (1982) *Plant Physiol.* **70**, 74–77.

Wink, M. & Twardowski, T. (1992) In *Allelopathy: Basic and Applied Aspects* (S.J.H. Rizvi & V. Rizvi, eds), pp. 129–150. Chapman & Hall, London.

Wink, M. & Witte, L. (1983) *FEBS Lett.* **159**, 196–200.

Wink, M. & Witte, L. (1984) *Planta* **161**, 519–524.

Wink, M. & Witte, L. (1985) *Z. Naturforsch.* **40c**, 767–775.

Wippich, C. & Wink, M. (1985) *Experientia* **41**, 1477–1479.

Yamada, Y., Hashimoto, T., Endo, T., Yukimune, Y., Kohno, J., Hamaguchi, N. & Dräger, B. (1990) In *Secondary Products from Plant Tissue Culture* (B.V. Charlwood & M.J.C. Rhodes, eds), pp. 227–242. Clarendon Press, Oxford.

Yeoman, M.M., Holden, M.A., Hall, R.D., Holden, P.R. & Holland, S.S. (1989) In *Primary and Secondary Metabolism of Plant Cell Cultures* II (W.G.W. Kurz, ed.), pp. 162–174. Springer Verlag, Heidelberg.

Zenk, M.H. (1989) *Recent Adv. Phytochem.* **23**, 429–457.

13 Biochemical Plant Pathology

Jonathan D. Walton

13.1 INTRODUCTION

A major subspecialty of plant biology is the study of the interactions between plants and micro-organisms. Biological stress due to pathogenic micro-organisms has had a strong influence on the evolution of all organisms, including plants (Hamilton *et al.*, 1990). Plant diseases are often a serious limitation on agricultural productivity and have therefore influenced the history and development of agricultural practises. Plant pathologists and plant breeders have been remarkably successful in controlling many crop diseases through the development of appropriate cultural practises, pesticides, and genetically resistant varieties. However, much of this work has been and continues to be empirical and therefore has not required an understanding of plant diseases at the cellular and subcellular levels. As numerous and as valuable as they are, only very recently have Mendelian genes controlling disease resistance been isolated, and elucidation of their biochemical functions remains a major challenge.

The development of new strategies to control diseases is the primary purpose of research on plant/pathogen interactions. These might include, for example, the identification of essential pathogen virulence factors and the development of means to block them, or the transfer of resistance genes into crop plants from unrelated species. A secondary benefit is a better understanding of the physiology of the healthy plant through a study of the metabolic disturbances caused by plant pathogens, much as the study of human diseases such as diabetes and cancer has contributed to our understanding of normal human metabolism.

Research on plant/pathogen interactions is distinguished from other fields of plant biochemistry because there are two genomes involved, that of the host plant and that of the pathogen. The reciprocal selection pressure over evolutionary time of the pathogen and host genomes on each other, a process known as co-evolution, is reflected in the high degree of complexity of many plant/pathogen interactions. The series of actions and reactions by the pathogen and by the host that occurs during pathogenesis constitutes a form of dialog written in the language of chemistry and biochemistry. It is a rather formidable task to dissect the elements of this dialog and to distinguish those elements that are important from those that are simply secondary effects or symptoms.

13.1.1 The major groups of pathogens

Only a handful of plant species are cultivated, and the responses of these plants to pathogens have been, not surprisingly, the most intensively studied. On the other hand, our cultivated plants are susceptible to attack by a great many types of pathogens, which have very different life cycles and pathogenic strategies.

Viruses, of which over a thousand infect higher plants, require a living plant cell in which to replicate. Some viruses depend on animal vectors, typically insects, for movement between plants, and some have quite intricate relationships with their vectors.

About 1600 species of bacteria cause plant diseases. Genera include the Gram-negative *Erwinia, Pseudomonas, Xanthomonas, Xylella,* and

PLANT BIOCHEMISTRY
ISBN 0-12-214674-3

Agrobacterium, and the Gram-positive *Clavibacter (Corynebacterium)* and *Streptomyces*. *Agrobacterium tumefaciens*, the cause of crown gall, is unique among plant pathogens because it subverts host cells by transferring part of its DNA (the T-DNA residing on the Ti plasmid) to the host. All pathogenic bacteria can be cultured on simple nutrient media. Mycoplasma-like organisms (MLOs) and spiroplasmas are wall-less prokaryotes that cause about 200 diseases.

Fungi, as a group, are the most important plant pathogens. More than 8000 species of fungi parasitize plants. They demonstrate great diversity in their morphology, in their life cycles, and in the complexity of their relationships with plants. Some fungi are normally saprophytes (i.e. they live on already dead organic material), but can opportunistically pathogenize plants. Such fungi typically infect many different plant species. At the other extreme, some fungi are obligate pathogens (i.e. they will not grow except on the host plant) and are highly specialized.

Nematodes, although actually animals, cause a number of important plant diseases, especially in the tropics, and are traditionally studied by plant pathologists. Some genetic and biochemical aspects of diseases caused by nematodes are similar to those of diseases caused by fungi and bacteria. Nematodes are also important vectors of some pathogenic viruses.

13.1.2 Susceptibility and resistance

The concepts of susceptibility and resistance are central to plant pathology. If a particular pathogen can cause disease on a particular plant, that plant is susceptible; if not, the plant is resistant to that particular pathogen. Of course, a plant that is susceptible to one pathogen is not necessarily susceptible to another, although there are a few cases known in which a single plant resistance gene controls reaction to two unrelated pathogens.

Compatibility is used as a synonym for a susceptible disease interaction. The term recognizes the contribution of both the pathogen and the host to any disease interaction. An incompatible reaction is one in which no disease occurs, i.e. the plant is resistant and/or the pathogen is nonpathogenic. It is important to note that economically a plant is resistant if yields are unaffected in spite of pathogen growth and damage. Thus a crop plant can be susceptible at the biochemical level but resistant at the economic one.

Biochemical plant pathologists frequently work with simplified model systems in which the distinction between a compatible and an incompatible reaction is clear-cut, but in the real world this is not always the case. For example, the reactions of oats to the fungus *Puccinia coronata*, which causes crown rust, are by convention described using seven degrees, from 'immune' to 'completely susceptible'. For simplicity, but with the danger of being simplistic, four are lumped together as 'susceptible' and three as 'resistant'. Furthermore, the degree of disease that results from any particular interaction depends on many factors, including the age and health of the plant, the genetic background of the two organisms beyond any particular resistance and pathogenicity genes, the amount of inoculum and method of inoculation, and environmental conditions.

Pathogenicity is sometimes used to describe an all-or-none disease situation, and virulence to describe gradations of pathogenicity, e.g. one isolate of a pathogen is more or less virulent than another. However, it is also common to say that a particular isolate is more or less pathogenic than another.

Specificity is another central concept in plant/pathogen interactions. Specificity describes both the extent to which a pathogen is restricted to particular host plants, and, conversely, the extent to which a plant is or is not susceptible to a particular pathogen. No pathogen can infect all plants, and no plant is susceptible to all pathogens. Some pathogens, such as fungi in the oomycetous genus *Pythium* and the ascomycetous genus *Sclerotinia*, attack hundreds of different plant species. Many different micro-organisms, including saprophytes, can infect fleshy plant parts such as fruits, tubers, and roots.

Most pathogens are restricted to one or a few host species, and, conversely, most host plants are susceptible to only a few pathogens. For example, *Puccinia graminis* is a pathogen of cereals and never soybeans, and *Phytophthora megasperma* is a pathogen of soybeans and never wheat. Specificity is envisaged to have evolved through the co-evolution of pathogens and plants as they reciprocally developed specialized mechanisms of attack and defense. Those pathogens that are more specific presumably have a more specialized biochemistry devoted to penetrating, colonizing and overcoming the resistance mechanisms of their particular hosts.

Specificity extends far below the species level in both pathogens and hosts. Different isolates of a pathogen typically differ in their ability to infect different host species, varieties, or genotypes. For example, although *P. graminis* is a pathogen of wheat, any particular interaction between a single spore and a single wheat plant will be compatible or incompatible depending on the genotypes of that spore and that plant. When specificity of a pathogen to the level of host species cannot be correlated with any of the morphological and/or chemical tests used taxonomically to classify the pathogen, the different specialized forms are called pathovars (for bacteria), or formae speciales (for fungi). Both pathovars and formae speciales can be further divided into races. For example, the large bacterial species *Pseudomonas syringae* is divided into many pathovars, each infecting a different plant species. On the basis of host-range these pathovars are clearly different, but they cannot be reliably differentiated on any other basis. By definition, all isolates of *Ps. syringae* pv. *tomato* infect at least one variety of tomato, but different races of *Ps. s.* pv. *tomato* infect different varieties of tomato. Likewise, the rust pathogen *Puccinia graminis* attacks a number of species of grasses, whereas the formae speciales of *P. graminis* are restricted to certain species. *P. graminis* f. sp. *tritici* and *P. g.* f. sp. *avenae*, for example, are morphologically indistinguishable but attack only wheat and oats, respectively. *P. graminis tritici* is delimited into races on the basis of infectivity on particular varieties or genotypes of wheat. Using 12 standard wheat lines, some 350 races of *P. graminis tritici* have been described.

Species are difficult to define for many microorganisms, especially those that reproduce asexually. The classification of pathovars, formae speciales, and races of pathogens is even more biologically arbitrary and is subject to frequent (human) revision. Conceptual problems arise especially when pathogens are classified on the basis of their host range. For example, is a single-gene mutant of *Ps. syringae* pv. *tomato* that no longer parasitizes any variety of tomato still pv. *tomato*? Nonetheless, some effort, no matter how unsatisfactory, must be made. The goal of plant pathologists as microbial taxonomists is to define and monitor pathogen populations in a way that is as useful as possible to growers and breeders.

13.2 GENETICS OF HOST/PATHOGEN INTERACTIONS

As a consequence of the highly successful efforts of plant breeders to find and utilize genetic disease resistance (the most economical as well as environmentally benign way to control plant diseases), our knowledge of the genetics of host/pathogen interactions is extensive. Hundreds of specific resistance genes have been described and mapped in virtually all of the major crop plants. The genetics of a number of plant pathogens have also been analyzed. Until the advent of genetic engineering, this could only be done with pathogens with a sexual stage, namely, rusts, smuts, powdery mildews, and a few other fungi. It is now possible to do molecular genetics with viral, bacterial and fungal pathogens that lack a true sexual cycle.

13.2.1 Genetics of resistance

Genetic resistance that is effective at preventing successful attack only by certain races of a pathogen is called specific (or vertical) resistance, whereas resistance that is effective at preventing successful attack by all races of a pathogen is called general (or horizontal) resistance. Specific resistance is typically inherited monogenically and is not 'durable', i.e. pathogens are observed to evolve to overcome it. For example, specific resistance genes against *Puccinia graminis tritici* incorporated into new wheat varieties frequently become obsolete within a few years due to the emergence of new races of the pathogen. General resistance is assumed to be polygenic (and therefore is less well-understood than specific resistance) and is 'durable' (i.e. pathogens have not been observed to overcome it). Another term for general resistance is non-host resistance. Whether the biochemical mechanisms underlying general and specific resistance are similar or not is controversial (Heath, 1991).

The known Mendelian resistance genes of higher plants are genes for specific (vertical) resistance. Resistance genes found in tomato, for example, include the nine *Cf* genes, which give resistance to different races of the filamentous fungus *Cladosporium fulvum*; *Pto*, which gives resistance to pathovar *tomato* of the bacterium *Pseudomonas syringae*; *Mi*, which gives resistance to the nematode *Meloidogyne incognita*; and *Tm2a*, which gives resistance to tobacco mosaic virus (TMV). Note that since none of the above-mentioned pathogens are amenable to Mendelian genetic analysis, prior to molecular genetics the corresponding specificity in the pathogens could only be ascribed to the level of species or race, not to particular Mendelian genes. It has now been shown that *C. fulvum* has a single gene, *avr9*, that controls reaction on tomato plants containing *Cf 9* (see section 13.4.2c), and *P. syringae* pv. *tomato* has a single gene, *avrPto*, that controls reaction on tomato plants containing *Pto*.

Other types of plant gene involved in plant/ pathogen interactions are commonly called defense genes. In contrast to Mendelian resistance genes, defense genes are thought to be involved in general (horizontal), as opposed to specific, resistance. Defense genes have been isolated either via the protein products or by differential screening of cDNA libraries made from mRNA isolated from infected vs. uninfected plants. Examples of defense genes include the genes for enzymes in phytoalexin biosynthetic pathways (see section 13.4.2b), for pathogenesis-related (PR) proteins (see section 13.4.2d), and for some hydroxyproline-rich glyco-proteins (see section 13.4.2e). In contrast to resistance genes, defense genes can have functions in addition to their putative defense roles. For example, the early enzymes in the isoflavonoid phytoalexin pathway are also necessary for the synthesis of lignins and anthocyanins.

13.2.2   Genetics of pathogenicity

Among plant pathogenic organisms, only a few fungi can be studied by conventional Mendelian genetics, and even among these, field isolates frequently show a high degree of infertility and have to be 'tamed' in the laboratory. Extensive studies of the inheritance of pathogenicity have been done with the Ascomycete *Erysiphe graminis*, which causes powdery mildews on cereals, and with Basidiomycetes that cause rust and smut diseases on cereals, such as *Puccinia graminis* and *Ustilago hordei*. Study of the inheritance of pathogenicity in *Melampsora lini*, cause of flax rust, led to the development of the gene-for-gene hypothesis (see section 13.2.3).

With the development of the techniques of DNA-mediated transformation, it is now possible to do 'genetics' with many pathogenic bacteria and imperfect fungi. These molecular studies have found further examples in which specific pathogenicity is inherited monogenically.

13.2.3   The gene-for-gene hypothesis

Especially valuable genetic information has come from studies in which both the host and the pathogen can be manipulated genetically, which is, unfortunately, a small subset of all plant/pathogen interactions. From his work with flax rust, Flor proposed the gene-for-gene hypothesis, which states that (specificity) genes in the pathogen that determine pathogenicity have a one-to-one corresponding match to (specificity) genes in the host

that determine resistance. The phenotype of a particular host resistance gene depends on the genotype of the pathogen at the complementing gene, and vice versa. The particular complementing gene pairs are frequently said to 'interact', which must be interpreted genetically, not necessarily biochemically.

Flor further found that plant resistance is usually dominant to susceptibility, and pathogen avirulence is usually dominant to virulence. (Flax is diploid, and *M. lini* is dikaryotic). Therefore, when a plant has the dominant allele of a particular resistance gene, and the pathogen has the dominant avirulence allele of the corresponding gene, a state of no disease (i.e. incompatibility) will result. Incompatibility is epistatic to compatibility: regardless of the number of non-interacting gene pairs in an interaction, any one interacting gene pair will suffice to give incompatibility.

Many biochemical models of gene-for-gene interactions have been proposed and debated. One popular model is that the primary gene products of a complementary pair of resistance and avirulence genes do physically interact (or, to put it another way, the plant resistance gene product 'recognizes' the pathogen avirulence gene product), and that this leads to the induction of plant defense responses and hence to resistance (incompatibility). In the absence of recognition, plant defenses are not induced and susceptibility results (Ellingboe, 1976; Keen, 1982). This model is consistent with the dominant alleles in both the plant and the pathogen being functional (i.e. encoding proteins), and the recessive alleles being non-functional.

The gene-for-gene hypothesis has become a dominant paradigm in molecular plant pathology. It has been extended by analogy to other disease interactions in which genetic analysis of the pathogen, or the host, or of both, is not possible. For example, on the basis of the patterns of compatibility and incompatibility among potato resistance genes and races of *Phytophthora infestans*, cause of late blight, one can assign a putative monogenic pathogenicity genotype to each race. The validity of the gene-for-gene concept for systems not amenable to classical genetical analysis is testable using molecular genetics.

It has been proposed that all established or putative gene-for-gene interactions might have a common biochemical basis. Supporting this possibility is the fact that many plant genes giving resistance to a particular pathogen are allelic, each different allele controlling resistance to a particular genotype of the pathogen. This suggests that the proteins encoded by at least some resistance genes might be variations on a theme. Some models invoke analogies to mammalian immunoglobulin genes or the hypervariable surface antigens of animal parasites. The recent demonstration that the products of genes conferring resistance against diverse organisms including viruses, bacteria, and fungi share common structural motifs lends strong support to the idea of common mechanisms of disease resistance (see section 13.4.2g). On the other hand, the predicted products of cloned resistance genes also show significant structural differences. Furthermore, diverse types of incompatible reactions have been described cytologically and biochemically, which argues against common mechanisms of resistance. For example, specific resistance in wheat to different races of *P. graminis tritici* has been shown to be due to different factors, including a reduction in the number of successful fungal penetrations, a longer latent period between penetration and appearance of symptoms, and reduced sporulation.

13.3 MECHANISMS OF PATHOGENICITY: PENETRATION AND SPREAD

The degree of specificity of a pathogen is related to the biochemical specialization of the disease interaction. The most fundamental level of pathogenic specialization consists of those traits that distinguish a pathogenic organism from its saprophytic relatives. Pathogens must have an array of biochemical specializations that are required for the ecological niche of living at the expense of another living organism. Collectively, these traits are called basic compatibility factors. All levels of specificity above the saprophyte/pathogen dichotomy are superimposed on basic compatibility. Such traits might include the ability to colonize and survive inside a seed, to remain dormant during crop rotations, or to produce spores that can travel long distances in the wind. Basic compatibility traits with a simpler biochemical basis might include the ability to adhere to a plant leaf, to germinate in response to plant signals, to penetrate the plant cuticle, or to degrade widespread noxious plant compounds.

The cuticle and epidermal cell walls of plants constitute an effective environmental barrier. The ability to get inside the plant is therefore an important attribute of all pathogens. Viruses infect plants through wounds or through animal (insect and nematode) vectors. Bacteria enter through wounds, leaf and secondary root scars, or hydathodes. Nematodes penetrate roots by

mechanical force. Fungi are more diverse: some enter plants only through wounds or natural openings such as stomata, but others have specialized means of penetration. Germinating spores of *Magnaporthe grisea*, cause of rice blast, form a specialized penetration structure called an appressorium (Fig. 13.1a). Melanization of the appressorium restricts the flow of water and allows it to develop a tremendous osmotic pressure (in excess of 80 bars) that drives a penetration peg through the epidermis by mechanical force (Fig. 13.1b) (Howard *et al.*, 1991). Consistent with the importance of hydrostatic pressure in penetration by *M. grisea*, melanin-deficient mutants are not pathogenic. However, melanin-deficient mutants of other fungi are still pathogenic, indicating that fungal penetration does not always rely on mechanical force. These fungi probably use enzymes to disrupt the epidermis.

The development of appressoria of the bean rust fungus, *Uromyces appendiculatus*, is triggered by minute topological features of the stomata of bean leaves (Hoch *et al.*, 1987).

Whether an interaction will be compatible or incompatible is determined by the events that occur when a penetrating pathogen first encounters living cytoplasm. In both types of interaction, a few cells are invaded. In an incompatible interaction, growth of the pathogen ceases at this point, but in a compatible reaction the pathogen continues to grow and invade new tissue. Thus, ability to spread from the initial site of infection is a fundamental hallmark of a successful pathogen. The importance of spread has been directly demonstrated for viruses. Some viruses can replicate in non-host cells, but are 'nonpathogenic' because they cannot spread to other cells. The 30-kDa 'movement protein' of TMV permits spread by altering the size exclusion properties of the plasmodesmata (Ding *et al.*, 1992).

13.3.1 Enzymes

All cellular plant pathogens make one or more enzymes capable of degrading the polymers of plant cell walls (Walton, 1994). At least 20 different cell-wall-degrading enzymes have been described from plant pathogens. Pathogens that cause 'soft-rot' diseases, especially of storage organs such as carrots and potato tubers, are known for their high production of pectinases, including endopolygalacturonase, pectic lyases, and pectin methylesterase. There is no strong evidence that any cell-wall-degrading enzyme has

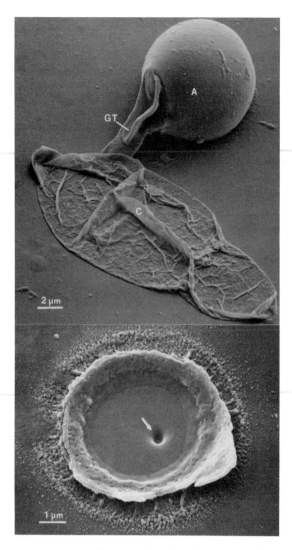

any role in disease specificity, and thus they are considered basic compatibility factors.

A number of genes encoding cell-wall-degrading enzymes have been cloned and mutated by targetted gene disruption in plant pathogenic organisms. Enzymes which have been analyzed for a role in disease from one or more plant pathogens include cutinase, α1,4-polygalacturonase, β1,4-xylanase, mixed-linked (β1,3-β1,4)-glucanase, β1,3-glucanase, pectic lyase, cellulase (β-glucosidase, cellobiohydrolase and β1,4-glucanase), pectin methylesterase, lipase and protease. In some disease interactions single genes encoding enzymes have been shown to be required for pathogenicity or to contribute to virulence (defined as the quantitative level of disease) and in other cases they appear to be superfluous. However, since most enzymes exist as multiple isoenzymes (encoded by different genes) in most pathogens, there have been few definitive conclusions about the importance of any class of enzyme as a whole, as opposed to individual isoenzymes. For example, *Erwinia chrysanthemi*, a cause of soft rot of storage crops, has more than five pectate lyase structural genes, and *Cochliobolus carbonum*, a pathogen of maize, has at least four xylanase genes. There is emerging evidence that pathogens make different isoenzymes in the plant or in response to plant extracts than they do when grown in culture.

Enzymes of the pathogen that can degrade plant cell walls have also received attention as triggers of plant defenses (i.e. elicitors, see section 13.4.2c). Fungal xylanase causes necrosis, ethylene synthesis, and electrolyte leakage in tobacco, symptoms associated with the induction of active resistance, and fungal endopolygalacturonase causes necrosis and induces phytoalexin biosynthesis (Walton, 1994). Therefore, cell-wall-degrading enzymes might have a dual role in the plant/pathogen interaction. A number of plants have proteins that inhibit microbial cell-wall-degrading enzymes. The role of a polygalacturonase-inhibiting protein (PGIP) from bean in modulating the balance between wall degradation and phytoalexin induction is under investigation (Toubart *et al.*, 1992).

Figure 13.1 Scanning electron microscopy of the germination and development of a conidium (spore) of the fungal rice pathogen *Magnaporthe grisea* on an artificial membrane of Mylar. (a) The conidium (C) has germinated to give rise to a germ tube (GT) and an appressorium (A). The conidium and the germ tube have collapsed. The appressorium, a specialized penetration structure, is highly turgid, single-celled and is completely separated from the germ tube by a septum (R.J. Howard and N.D. Read, unpublished micrograph). (b) The appressorium has been ruptured by sonication near the plane of attachment to the Mylar membrane. The upper and lower portions of the appressorium have separated revealing the underlying surface. Driven by an extremely high turgor pressure, the penetration peg, which arose from the appressorium, has created a hole (arrow) in the Mylar substratum. (Reprinted with permission from Howard *et al.*, 1991.)

13.3.2 Toxins

Micro-organisms and plants are prodigious producers of secondary metabolites, and some of these compounds are known or suspected to be important in various facets of plant/microbe interactions (Ballio & Graniti, 1991). Most pathogenic organisms make at least one compound that is

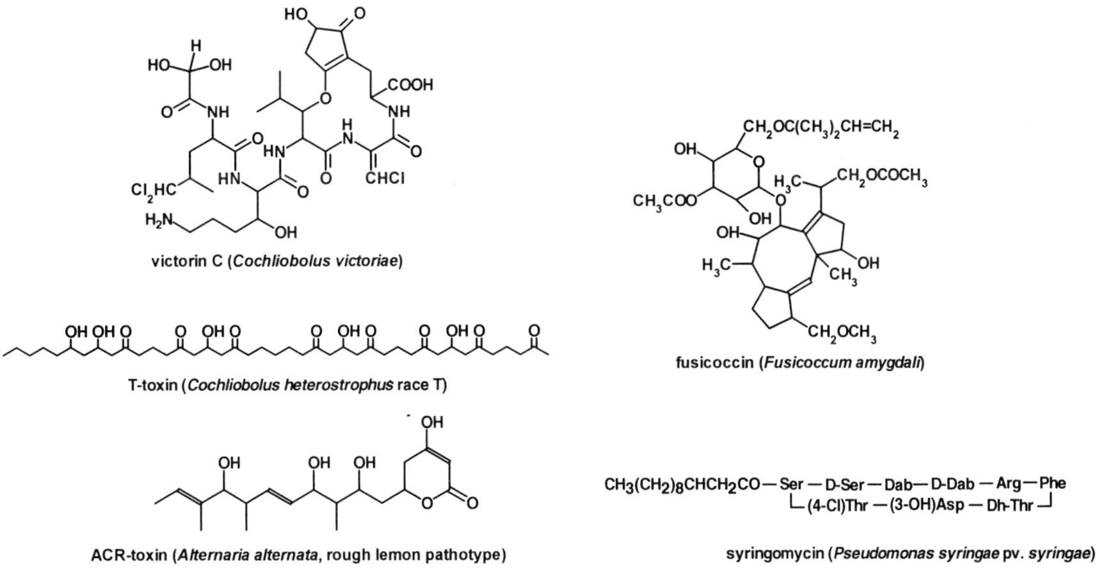

Figure 13.2 Structures of representative selective and non-selective toxins involved in plant diseases. Only the major forms of each toxin are shown. Victorin C, T-toxin and ACR-toxin are host-selectively active against oats with the *Vb* gene, against maize with Texas male-sterile cytoplasm (Tcms) and against rough lemon, respectively. Fusicoccin and syringomycin are non-selective. Abbreviations used for syringomycin: Ser, L-serine; D-Ser, D-serine; Dab, L-2,4-diaminobutyric acid; D-Dab, D-2,4-diaminobutyric acid; Arg, L-arginine; Phe, L-phenylalanine; (4-Cl)Thr, 4-chloro-L-threonine; (3-OH)Asp, 3-hydroxy-L-aspartic acid; Dh-Thr, dehydro-L-threonine.

deleterious to plants, and some pathogens make many. Although most phytotoxins are secondary metabolites, some are ribosomally synthesized toxic oligopeptides and proteins. New phytotoxic compounds from plant pathogens are described on a regular basis, but for most an involvement in pathogenesis has not been established. The compounds of diverse structures known as elicitors (see section 13.4.2c) are also often phytotoxic.

Some toxins are known to be required for pathogenicity, whereas others contribute to the particular symptoms of a disease but are not absolutely required. Some microbial secondary metabolites traditionally considered to be phytotoxins are not toxic in all bioassays, e.g. HC-toxin. Some pathogens, such as the bacteria *Pseudomonas savastanoi* and *Agrobacterium tumefaciens*, are notable for producing auxins and cytokinins, which cause characteristic growth abnormalities. Gibberellins were first discovered as secondary metabolites from the fungal pathogen of rice, *Gibberella fujikuroi*, which causes excessive stem elongation of infected plants.

The majority of phytotoxic compounds are non-specific, i.e. they are active against all plants and often other organisms as well. The structures of non-specific toxins are as diverse as secondary metabolites in general (Fig. 13.2). Non-specific toxins are made by both phytopathogenic bacteria

and fungi. In many cases non-selective toxins contribute to the symptoms or quantitative virulence of pathogens, but in other cases appear to have no role in disease. The sites of action of some non-specific toxins are known, and some are useful inhibitors for biochemical research. For example, fusicoccin, produced by the fungus *Fusicoccum amygdali*, mimics many of the physiological effects of auxin. Tagetitoxin from the bacterium *Pseudomonas syringae* pv. *tagetis* specifically inhibits the RNA polymerase of chloroplasts. The gene clusters involved in the biosynthesis of the non-selective toxins phaseolotoxin, coronatine, syringomycin, syringotoxin, and tabtoxin, all made by pathovars of *P. syringae*, have been isolated (Gross, 1991).

In contrast to the non-specific toxins, other toxins are active only against the same species, cultivar, or genotype of host plant that the producing pathogen attacks. Such 'host-selective' or 'host-specific' toxins are known only from fungi. The evidence that host-selective toxins are critical agents of specificity in the disease interactions in which they occur has come mainly from genetic analyses of sensitivity in the hosts and of toxin production in the pathogens. Approximately 20 host-selective toxins have been characterized. The ascomycetous fungal genera *Cochliobolus* (synonym *Helminthosporium*) and *Alternaria* in particular exploit this pathogenic strategy. Host-selective toxins, like non-selective toxins, are secondary metabolites of diverse structures (Fig. 13.2). T-toxin, made by *C. heterostrophus* (*H. maydis*), was the proximal cause of the devastating Southern corn leaf blight epidemic in 1970 in the United States. The molecular basis of sensitivity to T-toxin is well-understood (Levings & Siedow, 1992).

Production of and resistance to host-selective toxins is typically inherited monogenically in the pathogen and host, respectively. The *TOX2* locus of *C. carbonum* race 1, which controls production of the cyclic tetrapeptide HC-toxin, is large (greater than 100 kb) and complex, and includes a gene encoding a 570-kDa enzyme called HC-toxin synthetase (Walton & Panaccione, 1993). The complementary resistance gene in maize, called *Hm*, encodes an enzyme called HC-toxin reductase that gives resistance to *C. carbonum* race 1 by reductive metabolism of HC-toxin (Meeley *et al.*, 1992; Johal & Briggs, 1992; Fig. 13.3).

13.3.3 Suppressors

Suppressors are microbial compounds that are not themselves injurious to plant cells but are necessary for successful invasion because they prevent the induction of active plant defense responses. Suppressors are conceptually attractive in light of the finding that a large number of omnipresent microbially derived compounds, called elicitors, can trigger plant defense responses (see section 13.4.2c). Induction of plant defenses would appear to be almost inevitable during infection by any pathogen unless those defenses were somehow specifically suppressed. A peptidic suppressor from the fungal pea pathogen *Mycosphaerella pinodes* has been characterized, but the evidence for specific suppressors is still inconclusive.

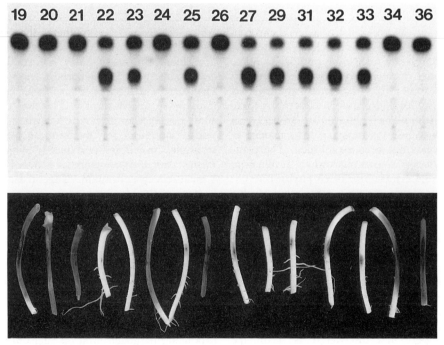

Figure 13.3   Gene-specific detoxification by extracts of etiolated maize of the host-selective toxin HC-toxin, and co-segregation with whole plant reaction to the producing fungal pathogen, *Cochliobolus carbonum* race 1. Top: tritiated HC-toxin and NADPH were incubated with cell-free extracts of single maize shoots from a population segregating one to one for resistance and susceptibility controlled by the *Hm* gene. Susceptible plants (19–21, 24, 26, 34, 36) have genotype *hm/hm*; resistant plants (22, 23, 25, 27–33) have genotype *Hm/hm*. Native HC-toxin (upper spot) and HC-toxin in which an essential carbonyl group has been enzymatically reduced (lower spot) were resolved by TLC and detected by autoradiography. Bottom: pathogenicity tests of the same plants shown in the upper figure. After the shoot had been removed for the enzyme assays, the remaining mesocotyl was inoculated with *C. carbonum* race 1 and photographed after 5 days. In a susceptible reaction almost the entire mesocotyl is necrotic; in a resistant reaction, discoloration is limited to the site of inoculation. (Reprinted with permission from Meeley *et al.*, 1992.)

13.3.4 Molecular biology of bacterial pathogenicity and avirulence

(a) avr *genes*

Bacterial *avr* genes convert a normally pathogenic (compatible) bacterium into an avirulent one, as determined by the conversion of the symptoms from water-soaking, indicative of susceptibility, to a hypersensitive response, indicative of resistance (see section 13.4.2a). For example, when the *avrA* gene of *Pseudomonas syringae* pv. *glycinea* race 6 is transformed into race 5, the transformant is no longer pathogenic on soybean cultivar Harasoy. It can, however, still infect cultivar Hardee. Therefore, *avrA* is a specific pathogenicity gene. The fact that the *avr* genes control avirulence and not virulence indicates that the gene-for-gene hypothesis, originally developed from Mendelian studies of dikaryotic fungi, is applicable to bacteria as well. More than 16 *avr* genes have been identified (Dangl, 1994).

avr genes in bacteria are sometimes found on plasmids and sometimes on the bacterial chromosomes. Some *avr* genes are physically as well as functionally absent in pathogenic strains, and in other cases exist as apparently non-functional 'alleles'. For example, *avrA*, *avrB* and *avrC* of *Ps. syringae* pv. *glycinea* have restricted distribution among isolates of *Ps. syringae*. On the other hand, DNA sequences cross-hybridizing to *avrD* of *Ps. syringae* pv. *tomato* are also present in other races. The *avrXa10* gene of *Xanthomonas oryzae* pv. *oryzae* contains 15.5 copies of a 102-bp repeat that is also found, but in 17.5 copies, in the *avrBs3* gene from *Xanthomonas campestris* pv. *vesicatoria*. The numbers of repeats in different *avr* genes of the latter pathogen are correlated with host-range.

The biochemical functions of *avr* genes, with the exception of *avrD* of *Ps. syringae* pv. *tomato* (Keen *et al.*, 1990), are not known, and this remains a major unsolved problem in molecular plant pathology. In light of recent work on the function of *hrp* genes (see section 13.3.4b) and the structure of resistance (R) genes that correspond to particular *avr* genes (see section 3.4.2g), it remains an attractive hypothesis that *avr* gene products interact directly with R gene products, but this has not yet been demonstrated. Such a model of *avr* gene action is complicated by the fact that *avr* gene products are apparently not secreted, at least from bacteria grown in culture.

(b) hrp *genes*

Bacterial pathogenicity genes other than *avr* genes have been found using random transposon mutagenesis. *hrp* (for *hypersensitive reaction and pathogenicity*) genes are gene clusters that are required both for pathogenicity on compatible hosts and, like the bacterial *avr* genes, for induction of the hypersensitive reaction (see section 13.4.2c) on incompatible hosts (Bonas, 1994). *hrp* genes have been found in all of the major groups of plant pathogenic bacteria. The *hrp* cluster of *Ps. syringae* pv. *phaseolicola* spans 22 kb and contains nine complementation groups. The *hrp* cluster of *Pseudomonas solanacearum* is 23 kb in size and contains at least 18 open reading frames. As with many *avr* genes, *hrp* gene transcription is nutritionally regulated.

Several *hrp* clusters have been sequenced, and *hrp* genes from one species of plant pathogenic bacterium are often similar in organization and sequence to *hrp* genes in other species. Strikingly, some of the predicted protein products also show strong sequence similarity to membrane-localized proteins of animal pathogenic bacteria, such as *Shigella*, *Yersinia*, and others, that control the secretion of proteins necessary for virulence (Fenselau & Bonas, 1995). Therefore, *hrp* proteins might, among other things, be necessary for the secretion of the *avr* gene products and/or other proteins necessary for pathogenicity.

The *hrpN* gene of *Erwinia amylovora* encodes a cell-surface associated protein, called harpin, that by itself causes necrosis and other symptoms of the hypersensitive reaction. Harpin is also required for pathogenicity and therefore has characteristics of both an elicitor and a toxin (Wei *et al.*, 1992). Analogs of harpin have been found in other bacterial pathogens (He *et al.*, 1993; Arlat *et al.*, 1994).

13.4 MECHANISMS OF PLANT RESISTANCE

A great variety of physiological and biochemical changes during the initiation and development of infection have been documented. Many of these are expected from the location and nature of the infection. Root and vascular pathogens frequently disturb the flow of water (transpiration) and nutrients (translocation) in the plant and therefore cause wilting. Disturbance of the plasma membrane leads to electrolyte leakage and cell death. Obligate pathogens such as rusts and smuts that redirect nutrients without killing the cells alter the source/sink relationships of the plant. Viruses subvert the transcriptional and translational machinery of infected cells. Increases in respiration and changes in photosynthesis are frequently seen after infection by many pathogens. Stimulation of ethylene synthesis is another common response to infection as well as to other stresses.

Biochemical plant pathologists are especially interested in those plant responses that are critical to determining whether a particular disease interaction, once initiated, will lead to resistance or susceptibility. It is important to keep in mind that 'resistance' does not necessarily mean lack of tissue damage. Rapid localized cell death is a component of many interactions termed resistant. Conversely, many compatible reactions provoke tissue damage only late in the process. Susceptibility and resistance are basically economic and not biochemical terms. Regardless of what happens during infection at the cellular level, if the overall yield is not reduced, the plant is considered resistant.

13.4.1   Constitutive resistance

Constitutive resistance factors (also known as preformed factors) are morphological and chemical entities that are present prior to pathogen attack. The cuticle can be considered a constitutive first line of defense against environmental stress including pathogens. A number of antimicrobial secondary metabolites are constitutively produced by plants, such as cyanogenic glucosides and saponins. These preformed chemicals are also known as phytoanticipins. In order to parasitize oat, *Gaeumannomyces graminis* var. *avenae* requires an enzyme, avenacinase, which deglucosylates avenacin, a constitutive fungitoxic saponin (see Fig. 13.5). *G. graminis* var. *tritici*, which lacks this enzyme and is therefore sensitive to avenacin, cannot attack oat (Bowyer *et al.*, 1995). Tomato fruits contain a related saponin, tomatine, which is degraded by some fungi by an enzyme closely related to avenacinase.

13.4.2   Induced resistance

Induced resistance refers to defense processes that are triggered only by attempted infection. Many induced or 'active' responses of plants to attempted infection have been documented. The major ones are biosynthesis of low-molecular-weight antimicrobial secondary metabolites (phytoalexins), localized rapid cell death (the

hypersensitive response), and synthesis of novel proteins that inhibit pathogens (pathogenesis-related proteins). These responses are collectively referred to as 'defense' responses. However, most such defense responses have not been rigorously tested for an actual role in defense.

Induced resistance is generally considered to be more important than constitutive resistance. This is supported by many studies showing that the treatment of plants with heat or with inhibitors of protein or RNA synthesis can render normally resistant tissues reversibly susceptible, which indicates that *de novo* protein synthesis, for example, of PR-proteins or phytoalexin biosynthetic enzymes, is necessary for resistance.

The timing and extent of inducible defense responses in compatible vs. incompatible reactions also supports the importance of active defenses. Typically, active defense responses occur more quickly in an incompatible reaction and correlate in timing with the onset of a macroscopic resistance response and cessation of pathogen growth (Fig. 13.4). Active defense responses are also induced in compatible reactions (Fig. 13.4), albeit with different kinetics. This fact has focused interest on the timing of the plant response as being critical in determining the final outcome of any interaction. If a plant's inducible defenses are triggered quickly, it will be resistant, but if too slowly, the spread of the pathogen is not limited and disease results. To be effective, then, a pathogen can be seen as having two choices: either to kill the host cells so quickly that they cannot mount an active defense, or to avoid triggering the active defense responses altogether (see Walton & Panaccione, 1993).

Salicyclic acid is strongly implicated in the induction of active defense responses in several plant species against several types of pathogens. Salicylic acid levels rise dramatically following pathogen challenge. Expression of a bacterial gene that encodes salicylate hydroxylase (which degrades salicylic acid to catechol) in transgenic plants leads to increased disease susceptibility to viruses, bacteria, and fungi, suppression of active defense responses such as PR-protein accumulation (see section 13.4.2d), and attenuation of systemic-acquired resistance (see section 13.4.2f).

(a)   *Hypersensitive reaction*

A common response of plants to infection by incompatible fungi, bacteria, nematodes or viruses is the hypersensitive reaction (HR), characterized by localized and rapid cell death (necrosis). Complete non-pathogens such as *Escherichia coli*

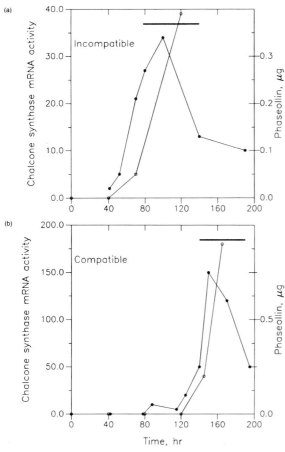

Figure 13.4 Expression of chalcone synthase mRNA (●), accumulation of the isoflavonoid phytoalexin phaseollin (○) and disease progress in an incompatible and a compatible interaction. The host in both cases is *Phaseolus vulgaris* (common bean) cv. Kievitsboon Koekoek. The incompatible pathogen (a) is race β of the filamentous fungus *Colletotrichum lindemuthianum* and the compatible pathogen (b) is race γ of the same fungus. The heavy bar in (a) indicates the time from the start of the hypersensitive response to its full development and that in (b) indicates the onset of disease symptoms through the development of extensive water soaking and spreading of lesions. Chalcone synthase induction and phaseollin accumulation occur in both reactions, albeit at different times, and are associated with tissue damage. Induction of chalcone synthase mRNA occurs just prior to tissue damage and phaseollin accumulation in both cases, and is transient. Note that although chalcone synthase mRNA induction and phaseollin accumulation occur later in the compatible interaction, the final levels reached are much higher. This is due to the fact that the incompatible reaction is restricted to the cells immediately surrounding the infection site, whereas the tissue damage occurring ultimately in the compatible reaction is spread over a large area. Chalcone synthase mRNA levels are expressed as percentage incorporation ($\times 1000$) of [^{35}S]methionine into *in vitro* translation products immunoprecipitable with anti-chalcone synthase antibody. Only the data for tissue adjacent to the infection site are shown. Phaseollin accumulation is expressed per inoculated site. (Redrawn with permission from J.N. Bell *et al.*, 1984.)

do not elicit the HR. The HR is restricted to tissue immediately surrounding the site of inoculation. Typically, the HR develops within 24 h of attempted infection, whereas cell death due to the development of the symptoms of disease (compatibility) takes several days. Cell death alone can be rationalized as an effective defense against obligate pathogens such as viruses and rust and smut fungi, but it is not clear how it prevents the growth of facultative saprophytes. However, the HR is usually accompanied by other defense responses such as phytoalexin and PR-protein accumulation (see sections 13.4.2b and 13.4.2d), which may be the actual cause of resistance. It has been proposed that HR-associated cell death (or, more precisely, cell morbidity) is the trigger of phytoalexin biosynthesis. However, it has also been argued that phytoalexins, which are known to be phytotoxic, actually cause the HR-associated cell death.

There have been many attempts to define more precisely the biochemical attributes characteristic of the HR. Within one hour of treatment of tobacco cells with the HR-inducing incompatible

bacterium *Ps. syringae* pv. *pisi*, a passive K$^+$/H$^+$ exchange across the plasmalemma is provoked (Atkinson *et al.*, 1985). The resulting perturbation of the ionic balance of cells might be the cause of or contribute to their death. Rapid generation of activated oxygen species is associated with very early stages of the HR. Generation of superoxide and hydrogen peroxide, as well as increases in superoxide dismutase and peroxidase, are seen in plants and cell cultures treated both with intact incompatible bacteria and with elicitors (see section 13.4.2c). The critical consequences of the production of activated oxygen are postulated to include oxidation of membrane lipids leading to host cell death and/or the generation of secondary signals, the initiation of cell wall lignification and cell wall protein cross-linking (see section 13.4.2e) and direct injury to the pathogen (Tenhaken *et al.*, 1995).

The HR is widely used as an assay in molecular genetic studies of bacterial pathogens. When a leaf is inoculated with high concentrations of an avirulent (incompatible) bacterium, confluent necrosis, an easily observable macroscopic reaction, develops within 24 h. Successful pathogens typically cause watersoaking lesions, accompanied by bacterial multiplication, in 48–72 h. Bacterial *avr* and *hrp* genes have been cloned on the basis of the HR assay (see sections 13.3.4a, b).

(b)   *Phytoalexins*

Phytoalexins are antimicrobial low-molecular-weight compounds (secondary metabolites) that are synthesized *de novo* following pathogenic attack (Bailey & Mansfield, 1982). Hundreds of phytoalexins have been characterized. They are especially widespread in the Fabaceae (Leguminosae) and Solanaceae but are also found in about 15 other families. Phytoalexins are not cosmopolitan; they have not been reported from the Cucurbitaceae and in the Poaceae (Gramineae) they occur in oats, sorghum and rice, but apparently not in maize.

Phytoalexins are chemically diverse, but a large number of them are products of the shikimic acid (phenylpropanoid) pathway from which many other plant secondary metabolites (particularly the flavonoids), lignin and anthocyanins are also derived. Flavonoid phytoalexins include kievitone and phaseollin from bean, pisatin from pea and glyceollin from soybean (Fig. 13.5). Key early enzymes in this pathway include phenylalanine ammonia lyase (PAL), cinnamate 4-hydroxylase (C4H), 4-coumarate-CoA ligase (4CL), and chalcone synthase (CHS). The genes for all of these enzymes have been cloned. Chalcone synthase is the first committed step separating the biosynthesis

of flavonoid phytoalexins from that of lignin precursors. Gossypol from cotton, casbene from castor bean, and rishitin from plants in the Solanaceae are examples of phytoalexins biosynthesized by the mevalonic acid pathway (Fig. 13.5).

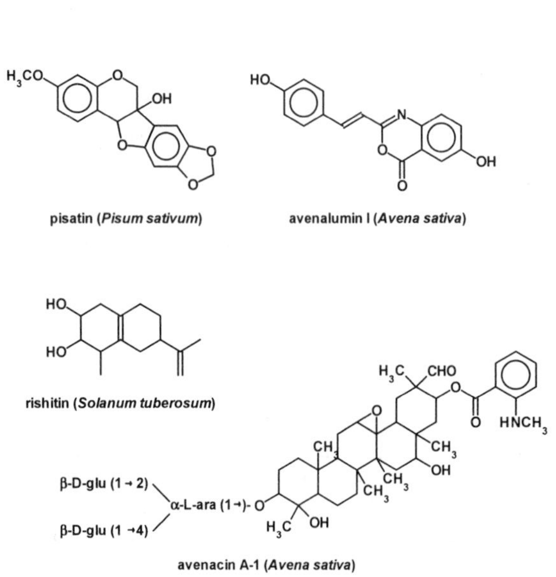

There has been extensive study of the accumulation, distribution, enzymology, and molecular biology of phytoalexin synthesis following pathogen attack. Phytoalexins are typically synthesized in moribund cells surrounding the site of infection and accumulate to high levels in dead and dying cells. Careful quantitation of the concentrations of phytoalexins in thin sections surrounding the site of infection has found levels sufficiently high to inhibit growth of fungi and bacteria, as well as to kill plant cells. In sorghum cells challenged with *Colletotrichum graminicola*, pigmented phytoalexins related to anthocyanins accumulate in intracellular vesicles. When these vesicles fuse and burst, high concentrations of phytoalexins are released, killing both the plant cells and the invading fungus (Snyder & Nicholson, 1990).

Phytoalexin accumulation is regulated by levels of the relevant biosynthetic enzymes. Increases in the levels of phytoalexin biosynthetic enzymes themselves are due to increased expression of the genes encoding the enzymes. When phytoalexin synthesis is stimulated with elicitors (see section 13.4.2c), *de novo* transcription of the genes encoding phytoalexin biosynthetic enzymes starts within minutes. Consistent with the localized cellular distribution of phytoalexins themselves during pathogen penetration, *de novo* induction of the phytoalexin biosynthetic enzyme genes is also localized to the cells surrounding the site of infection. Induction of phytoalexin genes and enzymes is frequently transient, reaching a peak some hours after attempted infection and then declining (Fig. 13.4).

The experimental evidence that directly addresses the importance of phytoalexins in disease resistance is still incomplete. Although phytoalexins and the mRNAs for their biosynthetic enzymes (such as PAL, C4H, and CHS) frequently accumulate more quickly and to higher localized levels during an incompatible reaction (Fig. 13.4), it is not infrequently observed that phytoalexins and/or the appropriate mRNAs accumulate at a similar rate and to similar levels during both compatible and incompatible reactions. The strongest evidence to date for the importance of phytoalexins has come from study of the ability of some pathogenic fungi to detoxify phytoalexins. Virulence of the pea pathogen *Nectria haematococca* is genetically correlated

Figure 13.5   Examples of phytoalexins (pisatin, rishitin and avenalumin) and a phytoanticipin (avenacin A-1).

with ability to detoxify the pea phytoalexin pisatin (Fig. 13.5) by cytochrome *P*450-catalyzed demethylation.

Alteration of the expression of phytoalexin biosynthetic genes is now feasible, through overexpression by introduction of multiple copies, constitutive expression using strong promoters, or antisense technology. Such experiments are technically demanding because many of the plant species whose pathology and phytoalexin anabolism are best known are difficult to transform. Also, since the phytoalexin biosynthetic genes exist as gene families, the members of which might have different functions and be induced by different treatments, it is necessary first to determine which member of a family is important, and then to differentially block its expression. Therefore, the role that phytoalexins play in resistance is still open.

(c) *Elicitors*

What components of pathogens are responsible for triggering active defense responses in plants? Elicitors were originally defined as compounds of microbial origin that induce phytoalexin accumulation. The term is now widely used for compounds that induce some or all of a series of putative plant defense responses, including the HR, electrolyte leakage, necrosis, ethylene biosynthesis, and pathogenesis-related protein synthesis. Many compounds can act as elicitors, including UV irradiation, glutathione, heavy metal ions, inhibitors of respiration such as sodium fluoride, cyanide, and 2,4-dinitrophenol, detergents, and inhibitors of protein and RNA synthesis. A particular elicitor does not necessarily elicit all defense responses. Ethylene, for example, triggers PR-protein synthesis but not phytoalexin accumulation.

Of relevance to plant pathology are elicitors that might actually function during pathogenesis. Fungi, including dead spores and extracts, and to a lesser extent bacteria and viruses, can act as elicitors of phytoalexin accumulation. Attempts to fractionate micro-organisms and obtain the compounds that induce phytoalexins have resulted in the identification of a number of active macromolecules, including proteins (including both structural proteins and enzymes), glycoproteins, polysaccharides, and lipids. The best-characterized pure polysaccharide elicitor is a hepta-β-glucoside from the cell walls of the oomycetous fungus *Phytophthora megasperma* that is active at 10^{-9} M (Darvill & Albersheim, 1984). Deacetylated chitin (chitosan) from fungal cell walls is an elicitor.

Plant enzymes such as endo-β1,3-glucanase release elicitor-active polysaccharides from fungal cell walls. Arachidonic acid from the walls of *Phytophthora infestans* is an example of a lipid elicitor.

Fragments of the plant cell wall itself also act as elicitors. Degradative enzymes from pathogens, such as polygalacturonase, release heat-stable polysaccharide elicitors from walls. It has been suggested that these 'constitutive' or 'endogenous' elicitors might be important in amplification of pathogen signals or in acquired resistance (see section 13.4.2f).

The vast majority of elicitors are non-selective, i.e. they are present in both virulent and avirulent races of a pathogen, or act equally well on susceptible and resistant plant tissues. The existence of host-specific elicitors has been widely postulated since this would give a simple biochemical explanation for the genetic dominance of avirulence that is seen in many gene-for-gene interactions. The corresponding resistance genes have been postulated to be receptors for host-selective elicitors. Recently, two elicitors have been demonstrated to be host-selective. One is the gene product of the *avr*9 gene of *Cladosporium fulvum*. This 28-amino acid peptide is produced just by races of *C. fulvum* that are avirulent on tomato varieties containing the *Cf* 9 resistance gene. The *avr*9 gene product causes necrosis only on tomato varieties containing *Cf* 9 (De Wit, 1992). The second known host-selective elicitor is controlled by the *avrD* gene of the bacterium *Ps. syringae* pv. *tomato*. When expressed in *E. coli*, *avrD* directs the synthesis of two compounds, called syringolides A and B, which are small secondary metabolites. The differential necrosis caused by syringolide on different soybean cultivars matches the differential induction of HR by bacteria harboring *avrD* on those same cultivars (Keen *et al.*, 1990).

Molecular studies have characterized in some detail the process of *de novo* gene expression that occurs after treatment of plant tissues with elicitors. Elicitor receptors, such as the membrane-localized receptor for the hepta-β-glucoside derived from cell walls of *P. megasperma*, are being characterized. Evidence based mainly on the use of specific metabolic inhibitors implicates protein kinase and phosphatases, perhaps mediated by activated oxygen (see section 13.4.2a) in the pathway from initial binding of an elicitor to defense gene activation (see section 13.4.2d).

In many cases, the distinction between an elicitor and a toxin is not clear-cut. This situation arises because cell death is a characteristic of both resistance (i.e. the HR) and susceptibility (i.e. most diseases) eventually. Likewise, symptoms, such as electrolyte leakage, ethylene production and necrosis, are used in different disease inter-actions as indicators of both resistance and sus-ceptibility. In cases in which a microbial com-pound inducing one of these symptoms (regardless of whether it be a secondary metabolite, a lipid, a protein, or a polysaccharide) cannot be correlated with either compatibility or incompatibility, the labels 'elicitor' and 'toxin' are arbitrary. Some compounds seem equally at home with both labels. For example, the host-selective toxin victorin from *Cochliobolus victoriae* is an effective host-selective elicitor of the avenalumin phytoalexins (Fig. 13.5) in oats.

(d) *Pathogenesis-related proteins*

The term pathogenesis-related (PR-) protein is now used to include all antimicrobial plant proteins that are induced following pathogen attack or other stress. The term was originated to describe a group of proteins that are induced in tobacco plants by infection with TMV and that are differentially extractable with acidic buffers. It has been extended to include proteins induced by bacteria and fungi on both dicotyledons and monocotyledons. PR-proteins can also be induced by ethylene and salicylic acid and by various elicitors. Proteins with no known antimicrobial activity but whose expression is enhanced by pathogen infection are also sometimes called PR-proteins. Acidic PR-proteins typically accumu-late in the cell wall and basic ones in the vacuole. In recent years a number of these proteins and the genes that encode them have been isolated. Insofar as PR-proteins are general resistance factors, manipulation of their genes might have practical applications for the development of new disease-resistant crops.

Four of the classic tobacco PR-proteins are β1,3-glucanases and four others have chitinase activity. Although viruses and bacteria do not contain the substrates for these enzymes, cell walls of higher fungi are composed predominantly of β1,3-glucan and chitin. The former occurs only sparsely in plants and the latter not at all, suggesting that these enzymes serve a defensive and not endogenous function. Other types of PR-proteins include ones related to thaumatin, a sweet-tasting, basic protein; the thionins, cysteine-rich proteins localized in the cell walls of many cereals; zeamatin and related proteins; and proteins with homology to lectins. Many PR-proteins have no known enzymatic function. Expression of some PR-proteins is also under development- and tissue-dependent control, and others are also induced by wounding, hormones, and abiotic stresses.

Expression of particular PR-proteins has been experimentally altered in several systems to test their involvement in disease resistance. In some cases transgenic plants have increased resistance, e.g. tobacco plants expressing PR-1a against two oomycetous fungi (Alexander *et al.*, 1993), and tobacco expressing a chitinase, a β1,3-glucanase, and a ribosome-inactivating protein against *Rhizoctonia solani* (Jach *et al.*, 1995), whereas in other cases no change in resistance is seen (Alexander *et al.*, 1993).

(e) *Structural alterations of the plant cell wall*

Several rapid and striking alterations of the plant cell wall occur during actual or attempted penetration by fungal pathogens. One response is the formation of papillae, which are knob-like appositions that form on the inner surface of the plant cell wall underneath penetrating fungal spores and in response to some bacterial patho-gens. They are made in both roots and leaves of monocotyledons and dicotyledons and typi-cally contain free phenolics, lignin, and callose (β1,3-glucan). Studies using inhibitors of papilla formation support the idea that, at least in some cases, the rate and extent of papilla formation can contribute to active resistance (Sherwood & Vance, 1980).

Increased lignification of the cell wall is another response frequently associated with unsuccessful penetration and the hypersensitive response. In some cases, elicitors of phytoalexin accumula-tion can also induce lignification; the early steps of the phenylpropanoid biosynthetic pathway are common to isoflavanoid phytoalexins and to lig-nin. Since lignified cell walls are mechanically stronger and more resistant to digestion by cell wall degrading enzymes, lignification could physi-cally restrict the pathogen. The biochemical events that lead to increased lignification are unclear. One possibility currently being studied is that hydrogen peroxide reduction catalyzed by cell wall peroxidases generates free radicals of coniferyl alcohol, which then spontaneously crosslink to form lignin.

Plants contain a number of wall glycoproteins that are rich in proline, glycine, and/or hydroxy-proline (these last are also called extensins). Like

Table 13.1 Cloned disease resistance genes

Gene	Plant	Pathogen and *avr* gene specificity	Deduced structure[a]	Cellular location[b]
Hm1	maize	*Cochliobolus carbonum* race 1	carbonyl reductase	cytoplasmic
Pto	tomato	*Pseudomonas syringae* pv. *tomato* (*avrPto*)	PK	cytoplasmic
RPS2	*Arabidopsis*	*Ps. syringae* (*avrRpt2*)	LRR, NBS, LZ	cytoplasmic
N	tobacco	tobacco mosaic virus	LRR, NBS	cytoplasmic
Cf-9	tomato	*Cladosporium fulvum* race 9	LRR	plasma membrane/ extracellular
L⁶	flax	*Melampsora lini* (flax rust)	LRR, NBS	plasma membrane
RPM1	*Arabidopsis*	*Ps. syringae* (*avrRpm1* and *avrB*)	LRR, NBS, LZ	cytoplasmic
Xa21	rice	*Xanthomonas oryzae* pv. *oryzae* race 6	LRR, PK	plasma membrane/ extracellular

[a] PK, protein kinase; LRR, leucine-rich repeats; NBS, nucleotide binding site (P-loop); LZ, leucine zipper.
[b] Tentative, based on deduced amino acid sequence.

lignin, these structural glycoproteins strengthen the cell wall and may, therefore, restrict pathogen invasion. Wall glycoproteins are found in many plant tissues and cell types and their genes are under various kinds of developmental and environmental control. However, the genes of some are also induced in response to infection and elicitors. One hydroxyproline-rich glycoprotein becomes quickly insolubilized, presumably by crosslinking, within 30 min of wounding or elicitor treatment in a manner independent of RNA and protein synthesis (Tenhaken *et al.*, 1995).

(f) *Acquired resistance*

In some cases, inoculation of a plant with either a virulent or an avirulent pathogen can give resistance to a subsequent 'challenge' inoculation with a virulent pathogen. Acquired resistance can be restricted to the area near the site of the original inoculation, or it can be systemic (SAR). Acquired resistance can also be induced with biotic and abiotic compounds. Induced resistance is nonspecific: inoculation of one leaf of cucumber with the ascomycetous fungus *Colletotrichum lagenarium*, tobacco necrosis virus, or *Pseudomonas lachrymans* induces resistance to *C. lagenarium* in other leaves several days later. Localized acquired resistance might be due to some or all of the known defense responses of plants, including PR-protein, phytoalexin synthesis, increased papilla formation and lignification of the cell wall. Necrosis, caused either by the HR or by disease, appears to be required for induction of acquired resistance, and, like elicitors in general, the chemical compounds that induce acquired resistance tend to be phytotoxic.

Significant evidence indicates that salicylic acid is required for SAR, as it is for induction of local resistance responses, although it is probably not the actual signal that moves from the challenged leaf to the responding leaf (Ryals *et al.*, 1995). Salicylic acid binds to and inhibits catalase, on the basis of which it has been proposed that peroxides and other activated oxygen species are critical to acquired resistance and to the action of salicylic acid (Chen *et al.*, 1995).

(g) *Disease resistance genes*

Biochemical approaches have rarely identified the function of plant genes controlling specific resistance to pathogens (R genes). In the past few years several resistance genes have been isolated using techniques that do not rely on any assumptions about the nature of the gene products, such as transposon tagging and chromosome walking ('positional cloning').

Specific resistance genes against viral, fungal and bacterial pathogens have been cloned (Table 13.1). *Hm1* encodes a carbonyl reductase that detoxifies the host-selective toxin HC-toxin (see section 13.3.2). Many other cloned R genes, all of which interact with specific *avr* genes, are remarkable for sharing a common feature of leucine-rich repeats (Grant *et al.*, 1995; Song *et al.*, 1995; Staskawicz *et al.*, 1995). The polygalacturonase-inhibiting protein of bean (see section 13.3.1) also contains leucine-rich repeats (Toubart *et al.*, 1992). Proteins containing leucine-rich repeats, as well as protein kinases, are known to interact with other proteins, and this leads to models based on signal transduction pathways in other organisms such as protein kinase cascades (Zhou *et al.*, 1995). Elucidation of these pathways will be a major area of research in biochemical plant pathology in the future.

REFERENCES

Alexander, D., Goodman, R.M., Gutrella, M., Glascock, C., Weymann, K., Friedrich, L., Maddox, D., Ahlgoy, P., Luntz, T., Ward, E. & Ryals, J. (1993) *Proc. Natl. Acad. Sci. U.S.A.* **90**, 7327–7331.

Atkinson, M.M., Huang, J-S. & Knopp, J.A. (1985) *Plant Physiol.* **79**, 843–847.

Arlat, M., Van Gijsegem, F., Huet, J.C., Pernollet, J.C. & Boucher, C.A. (1994) *EMBO J.* **13**, 543–553.

Bailey, J.A. & Mansfield, J.W. (eds) (1982) *Phytoalexins*. John Wiley, New York, 334pp.

Ballio, A. & Graniti, A. (eds) (1991) *Experientia* **47**, 751–826.

Bell, J.N., Dixon, R.A., Bailey, J.A., Rowell, P.M. & Lamb, C.J. (1984) *Proc. Natl. Acad. Sci. USA* **81**, 3384–3388.

Bonas, U. (1994) *Curr. Topics Microbiol. Immunol.* **192**, 79–98.

Bowyer, P., Clarke, B.R., Lunness, P., Daniels, M.J. & Osbourne, A.E. (1995) *Science* **267**, 371–374.

Chen, Z., Malamy, J., Henning, J., Conrath, U., Sánchez-Casas, P., Silva, H., Ricigliano, J. & Klessig, D.F. (1995) *Proc. Natl. Acad. Sci. U.S.A.* **92**, 4134–4137.

Dangl, J.L. (1994) *Curr. Topics Microbiol. Immunol.* **192**, 99–118.

Darvill, A.G. & Albersheim, A. (1984) *Annu. Rev. Plant Physiol.* **35**, 243–275.

De Wit, P.J.G.M. (1992) *Annu. Rev. Phytopathol.* **30**, 391–418.

Ding, B., Haudenshield, J.S., Hull, R.J., Wolf, S., Beachy, R.N. & Lucas, W.J. (1992) *Plant Cell* **4**, 915–928.

Ellingboe, A.H. (1976) *Encycl. Plant Physiol., New Series* **4**, 761–778.

Fenselau, S. & Bonas, U. (1995) *Mol. Plant–Microbe Interact.* **8**, 845–854.

Grant, M.R., Godiard, L., Straube, E., Ashfield, T., Lewald, J., Sattler, A., Innes, R.W. & Dangl, J.L. (1995) *Science* **269**, 843–846.

Gross, D.C. (1991) *Annu. Rev. Phytopathol.* **29**, 247–278.

Hamilton, W.D., Axelrod, R. & Tanese, R. (1990) *Proc. Natl. Acad. Sci. U.S.A.* **87**, 3566–3573.

He, S.Y., Huang, H-C. & Collmer, A. (1993) *Cell* **73**, 255–1266.

Heath, M.C. (1991) *Phytopathology* **81**, 127–130.

Hoch, H.C., Staples, R.C., Whitehead, B., Comeau, J. & Wolf, E.D. (1987) *Science* **235**, 1659–1662.

Howard, R.J., Ferrari, M.A., Roach, D.H. & Money, N.P. (1991) *Proc. Natl. Acad. Sci. U.S.A.* **88**, 11281–11284.

Jach, G., Görnhardt, B., Mundy, J., Logemann, J., Pinsdorf, E., Leah, R., Schell, J. & Maas, C. (1995) *Plant J.* **8**, 97–109.

Johal, G.S. & Briggs, S.P. (1992) *Science* **258**, 985–987.

Keen, N.T. (1982) *Adv. Plant Pathol.* **1**, 35–82.

Keen, N.T., Tamaki, S., Kobayashi, D., Gerhold, D., Stayton, M., Shen, H., Gold, S., Lorang, J., Thordal-Christensen, H., Dahlbeck, D. & Staskawicz, B. (1990) *Mol. Plant–Microbe Interact.* **3**, 112–121.

Levings, C.S. III & Siedow, J.N. (1992) *Plant Mol. Biol.* **19**, 135–147.

Meeley, R.B., Johal, G.S., Briggs, S.P. & Walton, J.D. (1992) *Plant Cell* **4**, 71–77.

Ryals, J., Lawton, K., Delaney, T.P., Friedrich, L., Kessmann, H., Neuenschwander, U., Uknes, S., Vernooij, B. & Weymann, K. (1995) *Proc. Natl. Acad. Sci. U.S.A.* **92**, 4202–4205.

Sherwood, R.T. & Vance, C.P. (1980) *Phytopathology* **70**, 273–279.

Snyder, B.A. & Nicholson, R.L. (1990) *Science* **248**, 1637–1639.

Song, W.-Y., Wang, G.-L., Chen, L.-L., Kim, H.-S., Pi, L.-Y., Holsten, T., Gardner, J., Wang, B., Zhai, W.-X., Zhu, L.-H., Fauquet, C. & Ronald, P. (1995) *Science* **270**, 1804–1806.

Staskawicz, B.J., Ausubel, F.M., Baker, B.J., Ellis, J.G. & Jones, J.D.G. (1995) *Science* **268**, 661–667.

Tenhaken, R., Levine, A., Brisson, L.F., Dixon, R.A. & Lamb, C. (1995) *Proc. Natl. Acad. Sci. U.S.A.* **92**, 4158–4163.

Toubart, P., Desiderio, A., Salvi, G., Cervone, F., Daroda, L., De Lorenzo, G., Bergmann, C,., Darvill, A.G. & Albersheim, P. (1992) *Plant J.* **2**, 367–373.

Walton, J.D. (1994) *Plant Physiol.* **104**, 1113–1118.

Walton, J.D. & Panaccione, D.G. (1993) *Annu. Rev. Phytopathol.* **31**, 275–303.

Wei, Z-M., Laby, R.J., Zumoff, C.H., Bauer, D.W., He, S.Y., Collmer, A. & Beer, S.V. (1992) *Science* **257**, 85–88.

Zhou, J., Loh, Y-T., Bressan, R.A. & Martin, G.B. (1995) *Cell* **83**, 925–935.

FURTHER READING

Agrios, G.N. (1988) *Plant Pathology*, 3rd edn. Academic Press, New York, 803pp.

Bowles, D.J. (1990) Defense-related proteins in higher plants. *Annu. Rev. Biochem.* **59**, 873–907.

Callow, J.A. (ed.) (1983) *Biochemical Plant Pathology*. John Wiley, Chichester, 484pp.

Daniels, M.J., Downie, J.A. & Osbourn, A.E. (eds) (1994) *Advances in Molecular Genetics of Plant–Microbe Interactions*, Vol. 3. Kluwer Academic, Dordrecht, 414pp.

Dixon, R.A. & Harrison, M.J. (1990) Activation, structure, and organization of genes involved in microbial defense in plants. *Adv. Genet.* **28**, 165–234.

Durbin, R.D. (ed.) (1981) *Toxins in Plant Disease*. Academic Press, New York, 515pp.

Larkins, B., Jackson, A. & Taylor, C. (eds) (1996) *Plant Cell*, October 1996. (A special issue devoted to plant–microbe interactions.)

Ryan, C.A. & Jagendorf, A. (eds) (1995) Self-defense in plants. *Proc. Natl. Acad. Sci. U.S.A.* **92**, 4075–4205. (A series of papers from a colloquium, including articles on diverse biochemical aspects of plant/microbe interactions.)

Smith, C.J. (ed.) (1991) *Biochemistry and Molecular Biology of Plant–Pathogen Interactions*. Clarendon Press, Oxford, 291pp.

14 Biochemical Plant Ecology

Jeffrey B. Harborne

14.1 INTRODUCTION

Much plant biochemistry is carried out on plants that are growing under optimal conditions in a glasshouse. Even where field-grown plants are studied, such organisms are unusually well protected from stress. They are protected from nutritional inadequacies (by fertilization), from drought (by irrigation), from disease and pests (by spraying), from other plants (by herbicide treatment) and from mammalian herbivores (by fences). Few wild plants are so fortunate. Indeed, the natural vegetation that proliferates in uncultivated parts of the land surface has had to struggle to survive adversity. The evolutionary history of the plant kingdom is a story of constant adaptation to environmental stresses and to the threat of uncontrolled herbivory by grazing animals. Yet in spite of such environmental pressures, there is scarcely any habitat in the world where plants are not to be found.

That plants have survived so well is due to their enormous flexibility to adapt to the diversity of growth conditions and environments that are present on this planet. Plants have adapted by modifying morphological and anatomical features (e.g. spines instead of leaves in desert habitats), through physiological adaptation (e.g. by reducing transpiration under a hot midday sun) or by biochemical means. In recent years, increasing attention has been paid to the ways that plants are biochemically adapted to their differing environments. Biochemical adaptation may involve both primary and secondary metabolism. It may be focused on environmental variables (drought, frost, salinity, heavy metal toxicity) or on the response of plants to other forms of life, whether they be bacteria, fungi, insects, molluscs or grazing mammals.

The study of biochemical adaptation of plants to ecological factors is the subject matter of ecological biochemistry, or chemical ecology as it is often termed (Harborne, 1993). The purpose of this chapter is to provide a general introduction to such biochemistry. The emphasis will naturally be on the plant and its responses, although much of the literature of plant–animal interactions has been concerned with animals (especially insects) rather than with plants. Plant–microbial interactions will not be considered here since they are the subject matter of Chapter 13.

The first section of this chapter will deal with plant adaptation to climatic and edaphic factors in the natural environment. The second section will consider the various toxins produced constitutively by plants in response to pressures of herbivory. The third section will discuss how far the distribution patterns of these toxins which accumulate in plants are correlated with a defensive role. The fourth and final section will be devoted to the ways that plants can respond dynamically to the damaging effects of grazing animals by producing new chemicals, becoming unpalatable in the process.

14.2 PLANT RESPONSES TO THE ENVIRONMENT

The biochemical responses of plants to various forms of environmental stress are listed in Table 14.1. Each response involves one or more changes

Table 14.1 Biochemical responses of plants to differing environmental stresses

Stresses	Biochemical responses
High temperatures (moist subtropics)	Long term: C_4 (Hatch–Slack) photosynthesis Short term: Heat-shock proteins
Drought (deserts)	Crassulacean Acid Metabolisms Increase in abscisic acid production Increase in proline or pinitol levels
Low temperatures (arctic or alpine areas)	Accumulation of sugars and/or polyols Synthesis of anti-freeze proteins Increases in the proportion of unsaturated fatty acids in the chloroplast membrane lipids
Salt (sea coast, salt marsh)	Accumulation of cytoplasmic osmotica (proline, glycinebetaine)
Heavy metals (lead, zinc or nickel containing soils)	Modification of the enzymes at the root surface Binding of metal to sites in the root cell wall Accumulation/detoxification as peptide chelates Accumulation/detoxification as organic acid chelates (e.g. nickel citrate)

in the biochemistry of the plant cell. Thus it may require the development of a new metabolic pathway. This happens in the case of photosynthetic adaptation to subtropical and tropical climates, when the Hatch–Slack C_4 cycle comes into operation (Chapter 2). It may result in the accumulation of a low-molecular-weight metabolite, from the carbohydrate or amino acid pool, which is otherwise only a trace component. An example is the imino acid proline, which may accumulate in both drought- and salt-stressed plants. Such a compound may have a special protective, e.g. osmotic, effect within the plant cell. This is true of proline and glycinebetaine, zwitterionic substances which are both formed under saline stress.

Another response to stress may be a change in plant hormone levels. For example, plants subject to drought commonly produce higher levels of abscisic acid in the guard cells of the leaf, in order to close the stomata and reduce the loss of water by transpiration. Related analogs may increase, instead of abscisic acid, in a few plants; thus phaseic acid increases in concentration and produces stomatal closure in drought-stressed vine plants, *Vitis vinifera*.

A further change in biochemistry may be the synthesis of special proteins (e.g. heat-shock or anti-freeze proteins) which are not otherwise detectable in the unstressed plant. Such proteins may have a special role in limiting the damage within the cell or within the cell membranes, caused by that particular stress. In the case of

adaptation to heavy metal (zinc, lead or copper) contamination in the soil, plants may respond by the synthesis of proteins or peptides, which have the ability through the presence of cysteine residues, to chelate those metals. Chelation of heavy metals with organic acids such as citric or oxalic is also possible in such plants.

The biochemical responses to stresses in plants (Table 14.1) are probably more complex than is suggested here. There may be more subtle changes in cellular biochemistry that are not so readily detected as those listed in Table 14.1. Biochemical responses are often correlated with physiological or anatomical adaptations, which have often been studied in much greater detail by botanists (e.g. Fitter & Hay, 1987; Crawford, 1989).

The biochemical responses to stress in plants are often paralleled by similar responses in animals. The ability to respond to mild temperature shock by the synthesis of heat-shock (HS) proteins within two hours of the shock is a general one and has been observed in microbes and animals as well as in plants. Typically in a plant such as the soybean, a two-hour preincubation at 40°C will protect against a two-hour treatment at 45°C, which otherwise would be lethal (Schoeffl *et al.*, 1984). Two groups of HS protein are produced of low and high molecular ranges and they persist as long as the increases in external temperature are maintained. These HS proteins are channeled within the cell to particular sites (e.g. the nucleus) to provide the necessary protection from the temperature increase. The ecological significance of the HS protein response in plants is still not entirely clear, although it would appear to be part of the armoury of desert plants to enable them to withstand high daytime temperatures.

A divergence in biochemical response between plants and animals is apparent in the adaptation to heavy metal poisoning. Thus plants detoxify heavy metals by chelating them with small peptides, called phytochelatins, which are produced in response to this stress (Grill *et al.*, 1987). These phytochelatins have the general formula: (glutamyl cysteinyl)$_n$ glycine, where $n = 2$–8, and are synthesized from glutathione. By contrast, animals sequester potentially poisonous levels of metals such as zinc, copper or lead by chelation with proteins, called metallothioneins, which have other amino acid residues in addition to cysteine, glutamic acid and glycine. It should be remembered that adaptation to heavy metal poisoning in plants may also involve excluding the metal from entry into the plant or by binding it to the cell wall of the root surface (Table 14.1).

Plants thus vary in their adaptive response to soils with toxic levels of heavy metals and the response may vary depending on which metal is present. The remarkable facility of plants to adapt by natural selection to high concentrations of toxic metals is well illustrated by the grass, sheep's fescue, *Festuca ovina*, which colonizes waste tips from lead and zinc mining within a few years of the mining operation. Plants are also found growing on soils which naturally have unusually high levels of other potentially toxic elements, including copper, cadmium and nickel. Such plants adapt biochemically by exclusion, binding at the cell wall or more commonly by chelating the cation and transporting it within the plant to a suitable compartment (e.g. the vacuole). Plants which accumulate metals in this way are useful indicator plants for showing where valuable mineral deposits may be found under the soil. For example, the shrub *Hybanthus floribundus* is an indicator for nickel deposits in Australia and it can accumulate up to 22% of its ash as this metal.

Most plants are killed if they are grown in the presence of small amounts of sodium chloride and yet adapted plants, called halophytes, grow and flourish in saline habitats where there may be as much as 20% NaCl in the soil. How is this achieved? The answer lies in controlling the intake and movement of NaCl within the plant, so that cell metabolism is not subject to salt stress. One major adaptation is the accumulation in the cytoplasm of non-toxic solutes, such as proline or glycinebetaine. These molecules exert a protective effect on osmotic regulation by balancing the decreased water potential caused by the accumulation in the vacuole of sodium and chloride ions, which have moved up into the plant from the soil (Wyn Jones, 1985).

From NMR measurements, it has been possible to show that in salt-stressed spinach leaves, the concentration of glycinebetaine in the chloroplast is ten times that of the whole leaf. In fact, only half the protective glycinebetaine is concentrated in the chloroplast, the other half being in the cytoplasm. Localization in these organelles within the cell allows the zwitterionic molecule of glycinebetaine (Fig. 14.1) to produce an osmotic potential sufficient to balance that due to the storage of sodium cations in the vacuole.

The concentrations of these cytoplasmic osmotica in halophytes may be relatively high. Thus in the halophyte *Triglochin maritima*, as much as 10–20% of the shoot dry weight is made up of proline, whereas in *Suaeda monoica* glycinebetaine levels can reach 0.5 g per 10 g dry weight. The cost in terms of increased demand for amino acid nitrogen can be considerable and this may explain why some recycling of these osmotica may occur when salt stress is reduced seasonally within a salt marsh. It may also explain why other halophytes occasionally produce non-nitrogen-containing osmotica, such as the polyol sorbitol or the sulphonium salt, *S*-dimethylsulphoniumpropanoic acid (Fig. 14.1).

A correlation between lipid fatty acid levels and freezing tolerance has long been proposed but it is only recently that compelling experimental evidence has been produced supporting such an idea. An increase in the ratio of unsaturated fatty acid to saturated fatty acid in membrane lipids has been correlated with freezing tolerance in several dicotyledonous plants. Such an increase is beneficial to the plant because it maintains the fluidity of the membrane at the lower temperatures. These changes are restricted to the phosphatidylglycerol

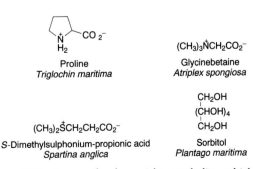

Figure 14.1 Low-molecular-weight metabolites which accumulate in salt-tolerant plants.

fraction (see Chapter 6) of chloroplast membranes. Plants with a high proportion of *cis*-unsaturated fatty acids (e.g. linoleic) such as spinach are resistant to chilling, whereas species like squash *Cucurbita pepo* with only a small proportion of *cis*-unsaturated acids are susceptible to chilling (Murata *et al.*, 1992). Monocot plants may operate a different system of protection. Thus, a decrease in the proportion of *trans*-Δ^3-hexadecenoic acid has been observed in the chloroplast membranes of wheat and rye plants tolerant to freezing (Huner *et al.*, 1989). Such protection of the chloroplast membrane allows photosynthesis to resume once the external temperature moves above freezing.

14.3 PLANT RESPONSE TO HERBIVORY: TOXIN PRODUCTION

14.3.1 Chemical defenses

By and large, plants are resistant to herbivory. If they were not, there would be few plant species remaining, considering the potential for exponential population growth among animals. As Feeny (1975) has put it, the most conspicuous non-event in the history of the angiosperms is the failure of animals to attack plants on a wide scale. Green plants still dominate our landscape, in spite of the proven ability of animals to overeat and destroy them. It follows that all plants must be broadly repellent to animals as food, in spite of their nutritional qualities, and hence toxic in the widest sense. It is the selective ability of phytophagous insects and grazing animals to overcome these defense mechanisms that allows the limited feeding that we witness today.

Plant defense is based on both physical and chemical factors. Physical defense mechanisms against herbivores are readily appreciated: tough epidermises, cuticular deposits, spines, thorns, prickles and stinging hairs. Defense may be purely strategic, as in the case of grasses which adapt to grazing by clinging close to the soil and by vegetative reproduction under the soil. Trees in temperate climates avoid herbivory by deer and hare by growing most of their leaves out of reach of these grazers. Nevertheless, chemical defenses are critically important and are provided by toxins and repellent substances of one type or another within the plant itself. These toxins have had and continue to have a key role in protecting plants from overgrazing.

It is a popular myth that humans can consume wild plant species with impunity. So-called 'herbal' plants may contain dangerous levels of poisonous agents; the leaf of the comfrey, *Symphytum officinale*, for example, contains pyrrolizidine alkaloids which are cumulative liver poisons. Furthermore, comparative analyses of the secondary constituents of cultivated plants and their wild relatives show in many cases that the toxic agents have either been eliminated or reduced in concentration as part of the domestication process. The potato, *Solanum tuberosum*, is a good example of a staple modern food plant, where plant breeding has successfully eliminated poisonous steroidal alkaloids from the tuber tissues. The tubers of wild potato species still retain toxic levels of alkaloid, as do the leaves, sprouts, flowers and fruits of the domestic potato.

It is suggested here that all green plants have the potential to be toxic to herbivores, due to the secondary metabolites that accumulate in them. The toxicity of a plant chemical is relative, dependent on the dose taken over a given time period, the age and state of health of the animal and the mechanism of absorption and mode of excretion. Plant toxicity can be overcome by the deliberate induction of detoxification systems within the grazing animal, so that what appears to be a toxic plant may be consumed safely by an adapted herbivore.

Not all plant parts are necessarily toxic (see also section 14.4) and indeed fruits are usually free of toxins, since they are provided as food to a herbivore for seed dispersal. Even ripe fruits may contain substances harmful to some herbivores. The presence of compounds such as hydrolyzable tannins in *Caesalpinia coriaria* fruits will deter the wrong seed disperser from eating the fruit (Janzen, 1978). By contrast with the fruit tissue itself, seeds within the fruit are usually protected by toxins, so that they are not eaten in the same way. 'Bitter almond' seeds, for example, contain lethal quantities of the cyanogenic glycoside, amygdalin. It is only the selection of the 'sweet almond', a mutant lacking this defense, that provides mankind with a useful food source.

14.3.2 Plant toxins

The plant toxins that have been characterized so far are many and various (Table 14.2) but the great majority are secondary metabolites. Some are general toxins, whereas others are targeted towards particular groups of herbivore. The major classes of secondary metabolite that are toxic to animals are listed in Table 14.2, with examples and with indications of toxic effects.

Table 14.2 The major classes of plant toxins

Class of compund[a]	Example	Toxicity
Alkaloids (10 000)	Senecionine in ragwort, *Senecio jacobaea*	Cumulative poison in cattle (hepatotoxin)
Cardenolides (200)	Ouabain in *Acokanthera ouabaio*	Heart poison; LD_{50} in rats $14\,mg\,kg^{-1}$
Cyanogenic glycosides (60)	Amygdalin in *Prunus amygdalus*	Universal toxin; fatal dose in humans ca. 50 mg
Furanocoumarins (400)	Xanthotoxin in *Pastinaca sativa*	Molluscicide; toxic to toads; allergenic in humans
Glucosinolates (150)	Sinigrin in *Brassica oleracea*	Damages thyroid, liver and kidney in cattle
Iridoids (250)	Aucubin in *Aucuba japonica*	Toxic to mammals, birds and insects
Isoflavonoids (1000)	Rotenone in *Derris elliptica*	Insecticide and fish poison; LD_{50} in mice $2.8\,mg\,kg^{-1}$
Non-protein amino acids (400)	β-Cyanoalanine in *Vicia sativa* seed	Neurotoxin; LD_{50} in rats $13.4\,mg\,kg^{-1}$
Peptides (50)	Viscotoxin in *Viscum album* berry	Toxic to heart muscle in mammals
Polyacetylenes (650)	Oenanthetoxin in *Oenanthe crocata* roots	Toxic to sheep and cattle
Proteins (100)	Abrin in *Abrus precatorius* seed	Fatal dose 0.5 mg in humans
Quinones (800)	Hypericin in *Hypericum perforatum* leaf	Causes facial eczema in sheep
Saponins (600)	Lemmatoxin in *Phytolacca dodecandra*	Toxic to snails, LD_{50} $1.5\,mg\,l^{-1}$
Sesquiterpene lactones (3000)	Hymenoxin in *Hymenoxys odorata*	Livestock poison

[a]Approximate numbers of known structures are given in parentheses.

These toxins can be broadly divided on biosynthetic grounds into nitrogen-based and non-nitrogenous toxins. The best-known nitrogen-based compounds are the alkaloids, which are biosynthetically derived from amino acid precursors. These substances have been used since time immemorial for poisoning purposes. Although the general toxicity of plant alkaloids in mammals is widely recognized, their teratogenic effects have only been recorded in more recent times. Adult female cattle and sheep may imbibe alkaloids in their diet in insufficient amounts to cause their death, but as a result of feeding on alkaloid-containing plants such as species of *Lupinus*, congenital defects may occur in the offspring. The malformed offspring usually suffer skeletal damage and defects of the digits or of the palate and have a low survival rate. Other nitrogen-based toxins are the non-protein amino acids, cyanogenic glycosides, glucosinolates, certain peptides and proteins. The biosynthetic origins and biological properties of these various metabolites are discussed in more detail in Chapter 12.

The non-nitrogenous toxins in plants are mainly of terpenoid origin. Thus iridoids, sesquiterpene lactones, cardenolides and saponins are biosynthesized from mevalonolactone and isopentenyl pyrophosphate (Chapter 11). Other classes of terpenoid not listed here also have toxic members (e.g. the monoterpene, thujone and the diterpene, acetylandromedol of *Rhododendron* leaves). Isoflavonoids, quinones and furanocoumarins listed in Table 14.2 are of shikimic acid origin and their biosynthesis and properties are discussed in more detail in Chapter 10.

14.3.3 The cost of chemical defense

The synthesis of secondary metabolites is costly to the plant, requiring as it does a steady flow of precursors from primary metabolism, together with enzymes and energy-rich cofactors (ATP, NADPH, etc.) to drive the biosynthetic pathways. Photosynthesis normally ensures a more than adequate supply of precursors for carbon compounds (e.g. terpenoids). By contrast, nitrogen uptake by the plant is limited and the synthesis of nitrogen compounds (e.g. alkaloids) can compete for precursor with protein synthesis. Indeed, the cost of synthesis of alkaloids has been estimated at 5 g of photosynthetic CO_2 per gram of toxin compared to a value of 2.6 g for a phenolic (Gulmon & Mooney, 1986). This cost has to be balanced against the cost of new plant growth. Thus, all plants face a dilemma when attacked by herbivores expressed in the words of Herms & Mattson (1992): to grow or to defend.

The competition for resources may be met in different ways and various theories have been put forward in recent times to explain the phenotypic variation that occurs in secondary metabolism and hence in chemical defense. The growth–differentiation balance (GDB) hypothesis

(Tuomi *et al.*, 1990) has as its fundamental premise the existence of a physiological trade-off between growth and differentiation, the latter term including secondary metabolism synthesis. This hypothesis suggests that perennial plants might be divided into two groups: first, growth-dominated plants, with rapid growth, poor chemical defense but with a highly inducible resistance system (see section 14.5) and second differentiation-dominated plants with a slow growth rate, well defended with high levels of toxin but with poorly developed inducible resistance. There is evidence supporting such a dichotomy in growth characteristics (Herms & Mattson, 1992).

These hypotheses help to explain the many differences that are apparent in the chemical armory of flowering plants. They also indicate why the concentrations of secondary metabolites can increase in plants as a result of environmental stress. Growth on low-nitrogen soils or under drought stress can cause plants to stop producing new leaves, with the precursors and energy thus released flowing into secondary synthesis. Experiments with the composite plant *Heterotheca subaxillaris*, which is defended by monoterpenoids, illustrate the changes that can occur in growth and metabolism. The younger, softer leaves have higher terpene levels than the older, tougher leaves. Feeding leaves taken from plants growing in nutrient-rich soils to the generalist insect *Pseudoplusia includens* results in a larval survival rate of 78 and 98% according to leaf age (Mihalaik & Lincoln, 1989). Plants growing in soils with low nitrate levels immediately become more resistant to insect feeding and larval survival rates are reduced to 14% on young leaves and 38% on old leaves.

In this case, when nutrient levels in the soil are adequate for growth, the plant accepts a degree of insect feeding on the leaves and maintains monoterpene production at a moderate level, principally to protect the very young leaves. However, nitrate stress induces increases in terpene production so that insects can no longer feed successfully. This change can be seen as a dynamic response to herbivory when the plant cannot grow new leaves and is then more vulnerable to leaf loss.

14.4 CONSTITUTIVE DEFENSE MECHANISMS

14.4.1 Localization of toxins at the plant surface

If secondary compounds do have a protective function against herbivory, then they are most likely to be located where they are most readily perceived by animals, namely at the plant surface. There is increasing evidence that this is so for many different kinds of plant; toxins are indeed often concentrated at or near the plant surface. Secondary compounds have been detected variously in glandular hairs (or trichomes), in leaf waxes, in leaf resins and in bud exudates. Even when they are present in the leaf vacuole, they are often found specifically in the epidermal cells of the upper surfaces. This is true, for example, of the localization of quinolizidine alkaloids in the leaves of lupin plants (Wink, 1983). Again, in the case of lactiferous plants, toxins are found in the latex and are detected by animals as the latex exudes from the leaf following damage.

Localization at the surface particularly affects insect feeders and many of the adverse effects of surface toxins are on insect pests. One of the most bizarre types of trichome defense has been observed in the wild potato species *Solanum berthaultii* (Table 14.3). The defense is two-pronged, with different agents secreted in morphologically distinct trichomes A and B, and is aimed primarily at aphids. Type B trichomes contain a volatile signal, E-β-farnesene, which occurs with other sesquiterpenes. Farnesene is actually a well-known aphid alarm pheromone and its release by the plant causes a similar

Table 14.3 Trichome toxins of *Solanum* leaves[a]

Type of trichome	Toxins present	Effect on insect
A	Polyphenol/phenolase system → sticky exudate on disruption	Aphid arrested, stops feeding and starves
B	Sucrose acid esters (e.g. 3,4-di-isobutyl-6-caproyl sucrose)	Aphid arrested, stops feeding and starves
B	β-Caryophyllene and E-β-farnesene	Aphid alarm pheromone disrupts feeding

[a]Detected mainly in *Solanum berthaultii* and/or *S. tuberosum* leaf trichomes; sucrose esters also occur widely in many solanaceous plants in their trichome secretions.

disturbance to aphid feeding. The second defense in the A-type trichome consists of a phenolic-containing exudate, linked to a phenolase/peroxidase enzyme system. When an aphid lands on the leaf, it disrupts the droplet, the enzyme then reacts with the phenolic substrate and a brown, sticky residue is formed. This immobilizes the aphid, prevents it from feeding and causes its death from starvation (Gregory *et al.*, 1986). The sticky exudate is reinforced by a mixture of non-volatile sucrose esters, which are secreted along with the alarm pheromone in type B trichomes. This defense must be effective since *S. berthaultii* is virtually free of insect attack; indeed, potato breeders have considered hybridizing this species with *S. tuberosum* in order to improve the aphid resistance of the domestic potato.

The range of substances found in trichomes is considerable and includes some water-soluble compounds, as in the case of the tomato. Here, two classes of toxin are present: the simple hydrocarbon 2-tridecanone and the phenolic compounds, rutin and chlorogenic acid. These substances have adverse effects on tomato pest insects, such as *Heliothis zea*. The concentrations of these trichome constituents are much higher in the wild tomato species from which the cultivated tomato is derived. Such toxins secreted in the leaf glands do not necessarily have to kill insect herbivores. All they need to do is to generally weaken them, retard their growth or delay their pupation. As a consequence, the insects will be more vulnerable to disease, to predation and to the environment (Stipanovic, 1983) and the plant will benefit from the reduced herbivory.

One final point needs to be mentioned about trichome constituents: they are almost inevitably multifunctional in their effects. Many of the same substances which are toxic to insects are anti-microbial (see Chapter 13) and also have deleterious effects on mammalian herbivores. The allergenic effects in man of the trichome constituents of *Parthenium hysterophorus* and *Primula obconica* are well known. Many other allergenic plants have similarly yielded, on chemical examination, related toxins in the trichomes (Rodriguez *et al.*, 1984).

Leaf waxes provide a second line of defense in some plants, since they may be a barrier to certain insect feeding. More importantly, at least 50% of angiosperm species contain 'extra' secondary constituents (e.g. sterols, methylated flavonoids, etc.) at the leaf surface mixed in with the wax. These substances almost certainly have a protective role, although less work has been done with them than with the trichome compounds. There is

also evidence that the actual constituents of the leaf wax may, on occasion, be repellent to insects. Certain varieties of *Sorghum* are distasteful to *Locusta migratoria*, because of the presence of leaf alkanes of chain length C_{19}, C_{21} and C_{23} and of fatty acid esters of chain length C_{12}–C_{18}. These and other effects of surface wax constituents on insects are reviewed by Woodhead & Chapman (1986).

A third line of defense in plants from insect predation is latex production. Latex has been reported in over 12 000 plant species and one of its main functions would appear to be to protect plants from herbivory. The effectiveness of latex, a white viscous liquid consisting of a suspension of rubber particles, as a feeding deterrent is often reinforced by the presence of terpenoid toxins. One particularly unpleasant series of latex constituents are the phorbol esters of Euphorbiaceae. These were first isolated from croton oil, *Croton tiglium*, but they occur widely in the family. Although most notorious as skin irritants and co-carcinogens in mammals, these diterpene esters have recently been shown to be toxic to insects as well.

The effectiveness of sticky latex as an insect feeding deterrent is nicely illustrated by the case of the larvae of the Monarch butterfly, *Danaus plexippus*, when feeding on its food plant, the milkweed *Asclepias syriaca*. Here the larvae carefully cut the leaf veins before feeding. This releases the latex as a series of white blobs along the leaf. The larvae then feed safely on the outer parts of the leaf, avoiding the latex in the process.

The value of a sticky latex as an insect feeding deterrent may often be enhanced by the presence of toxic secondary substances within the flow of the latex. For example, the latex of chicory is rich in bitter-tasting sesquiterpene lactones (lactupicrin and 8-deoxylactucin, Fig. 14.2) and feeding experiments against the locust *Schistocerca gregaria* show that these are antifeedants at a concentration of 0.2% dry wt. Analyses of the concentrations of these lactones in roots, stems

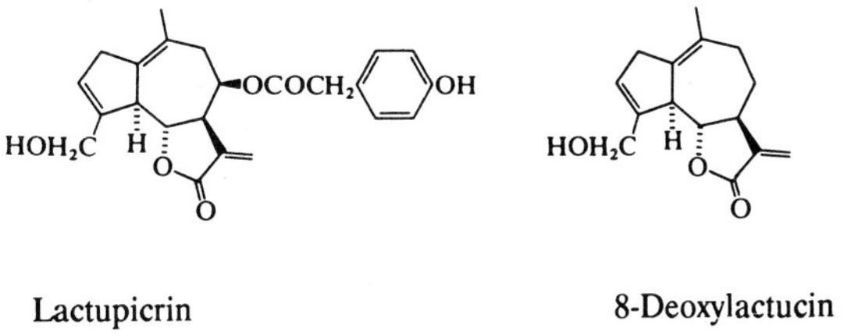

Lactupicrin   8-Deoxylactucin

Figure 14.2 Feeding deterrents of chicory root, stem and leaf.

Table 14.4 Accumulation of sesquiterpene lactones in chicory plants during the growing season

Month of analysis	% dry weight of total sesquiterpene lactones in			
	Roots	Base leaves	Mid-stem leaves	Top-stem leaves
March	0.22[a]	0.06	—	—
April	0.38[a]	0.10	—	—
May	0.52[a]	0.21	0.27[a]	—
June	0.81[a]	0.40[a]	0.32[a]	0.24[a]
July	0.11	0.17	0.22[a]	0.45[a]
August	0.25[a]	0.25[a]	0.15	0.38[a]
September	0.68[a]	0.22[a]	0.32[a]	0.28[a]

The lactones measured were 8-deoxylactucin, lactupicrin and lactucin (a minor component).
[a]Concentrations which significantly deter feeding by locusts. Data from Rees & Harborne (1985).

and leaves of chicory show that the levels are highest in the actively growing regions of the plant but they rarely drop below the threshold value required for deterrency (Table 14.4). The chicory plant is remarkably free from insect pests and there is evidence here that the latex constituents comprise a major source of defense. The extreme bitterness of lactupicrin and 8-deoxylactucin to mammals undoubtedly provides some protection from other herbivores too (Rees & Harborne, 1985).

Very convincing evidence for the importance of the localization of secondary constituents at or near the plant surface comes from detailed investigations of the browsing behavior during the winter months of the snowshoe hare, *Lepus americanus* on the small number of trees available to it in its natural habitat in Alaska (Table 14.5). Deterrent chemicals, either terpenoids or lipophilic phenolics, are specifically present in concentration in a resinous coating on the plant surface. Furthermore, they are only produced in quantity when they are needed, i.e. at particular stages in the life cycle, at certain seasons, and in those tissues (e.g. reproductive organs) which require most protection. Otherwise, selective feeding takes place and the snowshoe hare exists in a state of equilibrium with its host trees.

The most dramatic example of chemical protection occurs in the paper birch where juvenile growth-phase internodes are made unpalatable to the hare by the enormous concentration (up to 30% of the dry weight) of the triterpenoid papyriferic acid. This is 25 times the level that is found in mature internodes. The unpalatibility of the juvenile tissues could not be explained by differences in the levels of any other chemicals (e.g. inorganic nutrients or phenolics). Furthermore, the triterpene was clearly identified as a feeding deterrent by feeding it to hares in oatmeal at 2% of the dry weight. The triterpene is deposited as a resinous solid on the surface of the juvenile twigs and combined with volatiles in the resin, renders the juvenile tree highly deterrent to grazing by the hare (Reichardt *et al.*, 1984). There is provisional evidence that papyriferic acid also protects the tree from browsing by moose and rodents. Rejection of plant tissue containing papyriferic acid by these herbivores is probably based on its potential toxicity and this is underlined by the fact that in laboratory tests, papyriferic acid at a concentration of $50\,mg\,kg^{-1}$ will kill mice.

The snowshoe hare also feeds selectively on *Alnus crispa*, and here chemical protection is allocated to those organs which are dormant during the winter, i.e. the buds and the staminate catkins, in preference to the more expendable internodal tissues. The chemical agent in *Alnus crispa* is pinosylvin methyl ether, at concentrations of $2.6 \pm 0.2\%$ in buds, $1.7 \pm 0.1\%$ in staminate catkins and $0.05 \pm 0.0\%$ of the dry weight in internodes. Again, the stilbene is present in the resin, where it immediately deters feeding. The mountain hare (*Lepus timidus*) of northern Europe also grazes selectively on trees in the same way as the snowshoe hare. Species of *Populus* and *Salix* are the main winter food and juvenile trees are protected by enhanced levels of phenolic glycoside (mainly salicin). These concentrations decline in

Table 14.5 Specifically-located toxins in trees grazed by the snowshoe hare

Tree	Antifeedant/toxin	Localization and concentration (% dry wt)
Alnus crispa, green alder	Pinosylvin methyl ether	Buds (2.6%) and staminate catkins (1.7%)
Betula resinifolia, paper birch	Papyriferic acid	Juvenile internodes (30%)
Ledum groenlandicum, Labrador tea plant	Germacrone	Leaves and internodes (0.3%)
Picea glauca, white spruce	Camphor	Juvenile tissue (× 4 concn. in mature tissue)
Populus balsamifera, balsam poplar	Cineole and α-bisabolol	Buds

mature tall willow species but remain high in low-growing willows, as an adaptation against mammals browsing from ground level.

14.4.2 Timing of toxin accumulation

One of the problems in establishing a defense function for secondary constituents is the bewildering fluctuations that can occur in the amounts that are present in the plant at any given time. It is true that some metabolites are present at an appreciable level throughout the life cycle of the plant (e.g. the sesquiterpene lactones in chicory, see Table 14.4), but others can vary considerably in their concentrations. This is particularly true of the plant alkaloids. And yet, these variations would be explicable if the pattern of accumulation correlated with a proposed defense strategy, i.e. the compounds were present in highest concentration when the plant is most vulnerable to predation. There is now rather good evidence in the case of the coffee plant, *Coffea arabica*, that the high levels of purine alkaloid present coincide precisely with those periods when the plant is most open to herbivory.

The alkaloid caffeine, which is responsible for the stimulating effects of the coffee we drink, is not restricted to the coffee bean but occurs throughout the plant. Its levels have been carefully monitored at critical points in the life cycle. Thus, during germination, the coffee seeds are covered by a solid protective endocarp so that at first there is no large concentration of alkaloid in the developing shoot. However, as the endocarp decays, the concentration of caffeine in the young seedling more than doubles. During leaf development, the concentration of alkaloid increases to a high level (4% of the dry wt) just at the time the young soft leaf is most vulnerable to grazing. However, as the leaf matures, the rate of biosynthesis decreases exponentially, from 17 to 0.016 mg day^{-1} g^{-1} of leaf tissue, and this low level is then maintained when the leaf is fully expanded (Frischknecht *et al.*, 1986). Eventually, the leaves at senescence are essentially alkaloid-free and it is likely that the caffeine is recycled within the plant so that the nitrogen can be incorporated into protein synthesis. Finally, during fruit development, when the pericarp is soft, there is a resurgence of caffeine biosynthesis, with accumulation of 2% dry wt. As the endocarp differentiates and hardens, the levels decrease and there is about 0.24% caffeine present at the end of the ripening process.

That caffeine is defensive in the coffee plant is apparent from feeding experiments with insects. It may either have a lethal effect, killing larvae of the tobacco hornworm, *Manduca sexta*, at a dietary concentration of 0.3% or it may cause sterility, as happens to the beetle *Caliosobruchus chinensis* at a concentration of 1.5% (Nathanson, 1984).

A related pattern of pyrrolizidine alkaloid accumulation has been observed in the groundsel, *Senecio vulgaris*, where, for example, the epidermal cells of the stem contain ten times the alkaloid levels of the remaining cells. During flowering, more than 85% of the total alkaloid is concentrated in the inflorescence to protect it from herbivory (Hartmann *et al.*, 1989). A similar pattern has also been discovered in *Ipomoea parasitica* which contains ergot alkaloids. These are concentrated in the seed (0.39%), in the seedling (0.45%) and in the flowerhead (0.08%), with much less occurring in the vegetative tissues (0.02%) (Amor-Prats & Harborne, 1993). Due to their nitrogen content these alkaloids could be considered as both primary and secondary metabolites. There is a clear advantage to the plant to reclaim the nitrogen locked up in the alkaloid molecule as soon as the alkaloid is no longer needed as a defensive agent.

Timing of accumulation of toxin is also important in the case of the protective role of plant tannins. These phenolic substances are generally feeding repellents because of their distinctive astringent taste and ability to bind to protein. They tend to accumulate over time in the woody plants and trees that produce them. In the oak leaf, for example, they are produced in moderate amounts in the young leaf (ca. 1% dry wt) in the spring but increase fourfold in concentration in the summer. As a result, the larvae of the winter oak moth, *Operophthera brumata*, and other lepidopteran feeders, are forced to stop feeding and move off to other plants (Feeny, 1975). This allows the oak, even if heavily defoliated, to produce new mid-season leaves, with high tannin levels, that photosynthesize and build up energy to store for the winter months ahead.

14.4.3 Variability in palatability within the plant

Besides variation with time of growth (see above), variation between tissues is another common feature of secondary metabolite accumulation in plants. Variation between leaf and flower or root and stem is often manifest and may reflect the movement of secondary compounds within the plant (alkaloids may be synthesized in the root and transported to the leaf) or the fact that some parts of the plant are more expendable than others. Floral tissues, for example, may not be obviously protected from predation because they are generally so short-lived compared to leaves. Finally, there is the possibility of variation within the same tissues. This may be apparent especially in leaves of long-lived plants, i.e. trees, which are the most vulnerable of all plant tissues to herbivore attack. Deciduous trees produce leaves year after year, sometimes for centuries so that insect populations can 'home in' on a readily available and predictable food resource. Since insect predation has the potential to remove such a tree, there must be some other factor which provides resistance in the plant. The most recent suggestions of ecologists are that one protective factor is variable defense.

The experimental evidence in favor of chemical variation within leaf tissues is of two sorts: the patterns of insect grazing on trees and chemical analyses within tree canopies. A study of insect grazing on some trees, such as birch and hazel, shows that there is an overdispersion of grazing initiations and that a proportion of leaves receive a low level of grazing. Such a pattern would be consistent with between-leaf variations in palatability. This idea has been supported by chemical analyses of leaves within the canopy in *Betula lutea* and *Acer saccharum*, which indicate significant variations in chemical parameters from leaf to leaf on the same branch (Schultz, 1983). Palatability is principally defined by tannin levels, total phenolic content, toughness (or degree of lignification), water content and nutrient content and all these measurements can change from leaf to leaf. Variation of this type may be due in part to genetic factors, the light regime, nutrient status of the tree, position of leaf on the tree and so on.

There may also be considerable variations in leaf chemistry in different individuals within a tree population. This occurs, for example, in the neotropical tree *Cecropia peltata* (Moraceae), where leaves of some individuals are rich in tannin and are low in herbivore damage and the leaves of others are low in tannin and suffer considerable insect feeding. The tannin levels in the leaves vary from 13 to 58 mg g^{-1} dry wt (Coley, 1986). Tests with the armyworm, *Spodoptera latifascia*, showed that there was reduced feeding on leaves of the high-tannin individual. There was, however, a significant cost to the tree in terms of tannin production and the high-tannin trees produced fewer leaves than the low-tannin trees, growing under the same conditions.

For the insect seeking food on trees, there are considerable risks, since having to spend more time moving around within the canopy must increase the chance of it being the victim of one of its predators. Certainly, within-leaf variation, where it occurs, must place a constraint on the insect grazer. As Schultz (1983) has put it:

> The situation can be said to resemble a 'shell game', in which a valuable resource (suitable leaves) is hidden among many other similar-appearing but unsuitable resources. The insect must sample many tissues to identify a good one. The location of good tissues may be spatially unpredictable, and may even change with time. For a choosy or discriminating insect, finding suitable food in an apparently uniform canopy could be highly complex.

14.5   INDUCED PHYTOCHEMICAL RESPONSE TO HERBIVORY

14.5.1   Phytochemical induction in plants

One way that a plant might reduce the metabolic costs of synthesizing and storing toxins for chemical defense is only to produce the defensive chemicals when they are actually needed, i.e. in direct response to herbivore attack. Such a mechanism was first described from experiments in plant pathology, when it was discovered that plants can respond to fungal infection by the *de novo* synthesis of antifungal agents at the site of attack. These so-called 'phytoalexins' were produced within a few hours of infection and reached a maximum within 48 h of the invasion (see Chapter 13). It is now apparent that insect feeding on plant tissues can produce a related dynamic response to damage. With insect feeding, however, the effect is systemic and the whole plant is alerted to the fact of invasion. Insect feeding can be mimicked by mechanical damage, i.e. punching a hole in a leaf, and the response is quite separate from the 'wound response', which is a localized repair mechanism of plants.

Three distinctive types of herbivore-induced defense have been recognized so far. The first, the so-called 'PIIF induction', produces completely new defense agents, the synthesis of proteinase inhibitors, within 48 h. These cause the leaf to become unpalatable and insect feeding is arrested. The second type of induced defense is the increase in the synthesis of a single class of toxin, which is already being produced in lower amounts before attack. Again, the effect is the production of unpalatable leaves, the effect increasing over several days after the induction. The third and most remarkable type of induced defense is the production of volatile chemicals at the feeding site on the leaf, which has the apparent effect of attracting insect parasitoids. These parasitoids then visit the plant and destroy the herbivore. Each of these three types of herbivore-inducing defense will now be considered.

14.5.2 *De novo* synthesis of proteinase inhibitors

It has been shown that a Colorado beetle feeding on potato or tomato leaves can cause the rapid accumulation of proteinase inhibitors, even in parts of the plant distant from the site of attack. The process is mediated by a proteinase inhibitor inducing factor, known as PIIF, which is released into the vascular system. Within 48 h of leaf damage, the leaves may contain up to 2% of the soluble protein as a mixture of two proteinase inhibitors. Subsequently, the presence of the proteinase inhibitors in the leaf is detected by the beetle, which avoids further feeding and moves onto another plant (Fig. 14.3).

In theory, the inhibitors, if taken in the diet, will have an adverse effect on the insects' ability to digest and utilize the plant protein, since they inhibit the protein-hydrolyzing enzymes, trypsin and chymotrypsin. In fact, proteinase inhibitors are well known as constitutive constituents of many plant seeds, where they have a similar protective role in deterring insect feeding.

PIIF induction is also brought about by mechanical wounding of plant tissue so that it is not yet entirely clear how specific this effect is to herbivore grazing. The nature of the chemical signal PIIF has been explored in tomato plants and it is a small protein, called systemin. This protein is 10 000 times more active than oligosaccharides, which also have the ability to trigger this defense system (Ryan, 1992). A volatile chemical, the fatty acid metabolite methyl jasmonate, may also be involved in the signaling process (Farmer & Ryan,

1990). As PIIF-like activities have been detected in extracts of 37 plant species representing 20 families, this mechanism may well be a general one. Ecological experiments have also shown that PIIF induction in the tomato reduces the grazing by larvae of the armyworm *Spodoptera littoralis* within 48 h, with avoidance being most pronounced on the young leaves (Edwards *et al.*, 1992).

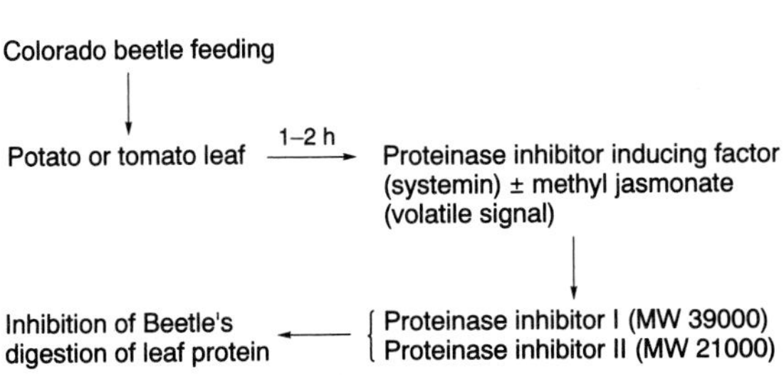

14.5.3 Increased synthesis of toxins

A related form of induced defense, apparently quite distinct from the PIIF system, has been observed in a variety of plants. The effect is relatively rapid and the leaves become unpalatable to animals within a matter of hours or a few days. It may be short term, disappearing after the insect has stopped feeding, or long term, extending over to the following season in trees. The chemical changes involve an increase in the concentration of existing toxins (Fig. 14.4), sufficient to lead to herbivore avoidance.

Such increases in toxin synthesis have been observed in two alkaloid-containing plants. One is the wild tobacco species *Nicotiana sylvestris*, which contains nicotine and nornicotine as the major alkaloids. Larval feeding induced a 220% increase in alkaloid content throughout the plant

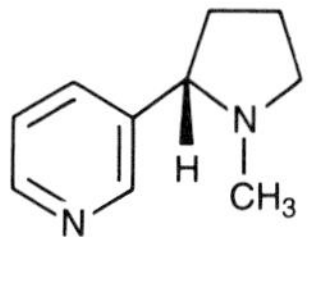

Nicotine
(*Nicotiana sylvestris*)

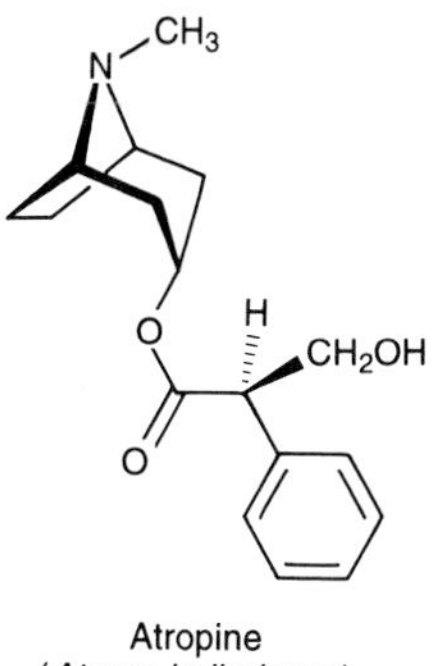

Atropine
(*Atropa belladonna*)

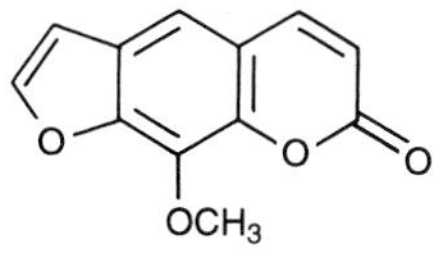

Xanthotoxin
(*Pastinaca sativa*)

Figure 14.4 Chemicals which increase in concentration in plant leaves in response to insect herbivory.

Colorado beetle feeding

Potato or tomato leaf —1–2 h→ Proteinase inhibitor inducing factor (systemin) ± methyl jasmonate (volatile signal)

Inhibition of Beetle's digestion of leaf protein ← { Proteinase inhibitor I (MW 39000) Proteinase inhibitor II (MW 21000)

Figure 14.3 *De novo* synthesis of proteinase inhibitors in potato and tomato leaves.

over a period of 5–10 days. Mechanical damage which avoids cutting the secondary veins produced a smaller response (170%). In fact, the tobacco hornworm *Manduca sexta* when feeding on the tobacco leaf avoids cutting through the secondary veins. It thus avoids triggering off the fullest response in the leaf, which can be as much as 400% of the control if the simulated damage includes damaging the vein. The nicotine alkaloids are synthesized in the roots and transported up into the leaf and this was apparent in experiments in which pot-bound plants with confined roots failed to show any significant alkaloid increase after mechanical damage (Baldwin, 1988). Similar experiments with the tropane alkaloids in leaves of *Atropa acuminata* showed a maximum increase of 153–164% over the control 8 days after mechanical damage or slug feeding. Repeated mechanical damage at 11-day intervals initially increased the response to 186% of the control but this effect fell off with time. Further experiments showed that only 9% of the leaf area needed to be removed mechanically or by animal feeding to produce the maximum response (Khan & Harborne, 1991).

Another well-investigated example of induced chemical defense is the wild parsnip, *Pastinaca sativa*, which produces five furanocoumarins in the leaves. Artificial damage increased furanocoumarin synthesis to 162% of the control, whereas feeding by the generalist insect *Trichoplusia ni* increased it to 215%. Furthermore, larvae of *T. ni* grew very slowly on induced leaves, and larvae on an artificial diet supplemented with furanocoumarins were similarly affected (Zangerl, 1990). The response of oil seed rape, *Brassica napus*, to insect infestation or leaf damage is quite distinctive and involves the massive accumulation of indole glucosinolates, which are barely detectable in the control. There is a corresponding reduction in the amounts of the aliphatic glucosinolates of the plant, but the total titer of glucosinolate does appear to increase under these treatments (Koritsas *et al.*, 1989).

Other examples where induced changes in protective chemistry have been recorded are given in Tallamy & Raupp (1991). For every plant that shows a positive response, there is another plant where no detectable change in palatability occurs. Rapid inducible resistance appears to be weak in or absent from the leaves of slow-growing plants (section 14.3.3). Environmental factors also determine the magnitude of the response. Finally, the response may disappear as the plant grows older and concomitantly more resistant to grazing. For example, two-year-old trees of *Pinus contorta* respond to defoliation by

increasing the concentration of both terpenes and tannins in the needles, whereas ten-year-old trees fail to show any increases.

14.5.4 Release of predator-attracting volatiles

An even more interesting and remarkable plant–animal interaction involving induced chemical changes has been observed by Dicke *et al.* (1990). In response to herbivory, some plants have developed the means to release volatile chemicals (Fig. 14.5), which are particularly attractive to parasitoids of their herbivores, which then visit the plant and destroy the herbivores. As Dicke puts it, plants may 'cry for help' when attacked by spider mites and predatory mites come to the rescue. Much research has been conducted on the spider mite *Tetranychus urticae*, the predatory mite *Phytoseiulus persimilis*, and the host plants. The chemicals released seem to be plant species-specific. Cucumber plants infested by the spider mite release β-ocimene and 4,8-dimethyl-1,3,7-nonatriene (see Fig. 14.5) and are only moderately attractive to the predatory mites, whereas lima beans release a cocktail of linalool, β-ocimene, the nonatriene and methyl salicylate which is highly attractive. A further advantage to the plant world is that the volatile released may alert uninfested neighboring plants so that they become better protected from spider mite attack. Thus cotton seedlings, when infected by these mites, release volatile cues which both attract predatory mites and also alert neighboring plants to withstand herbivore attack.

The systematic release of volatile chemicals which mediate in plant–herbivore–predator interactions has been observed in other plant systems. Thus, it has been recorded that corn (*Zea mays*) seedlings respond to beet armyworm (*Spodoptera exigua*) attack by releasing volatiles, which attract parasitic wasps, *Cotesia marginiventris*, to attack

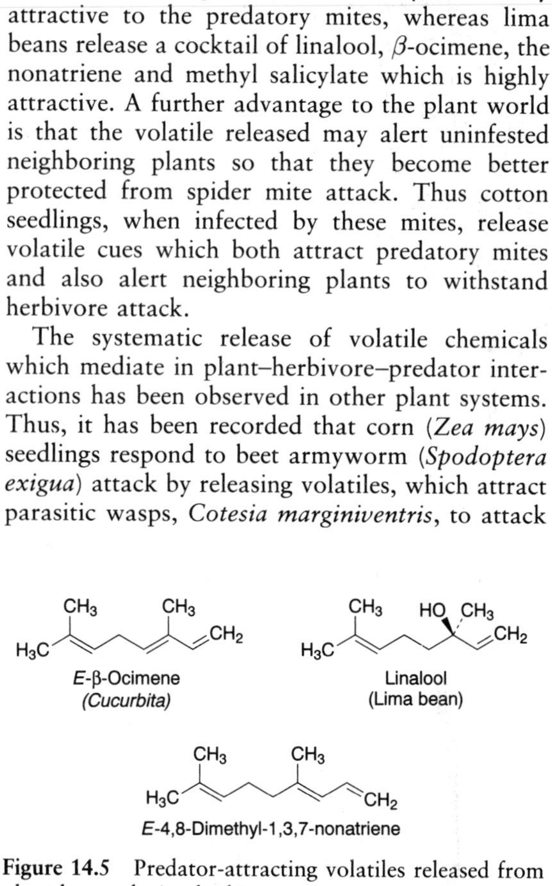

Figure 14.5 Predator-attracting volatiles released from plant leaves during herbivory.

the herbivore. The response occurs throughout the plant and not only at the site of damage. The chemicals involved include linalool, which is released at the rate of $1\,ng\,h^{-1}$ before damage and at the rate of $110\,ng\,h^{-1}$ 6 h after armyworm attack (Vet & Dicke, 1992).

A similar tritrophic system exists in the case of the soybean plant, the soybean looper *Pseudoplusia ineludens* and its parasitoid *Microplitis demolitor*. Here the volatiles again include linalool, but the more important attractants are guaiacol and 3-octanone. These latter two compounds do not appear to be released in appreciable amounts from the plant, but are released from the insect frass and are formed within the larvae from dietary sources. In other plants such as cotton and cowpea, the release of green leaf volatiles (e.g. *E*-2-hexenal and *E*-2-hexen-1-ol) appears to be sufficient to attract parasitic wasps to attack leaf-feeding caterpillars.

14.6 CONCLUSION

Other examples of plant–animal interactions that are mediated by plant secondary metabolites can be found in Feeny (1975), Harborne (1993) and Herms & Mattson (1992). Plants may concentrate their chemical defense by accumulating one class of toxin (e.g. a mixture of pyrrolizidine alkaloids) and others may synthesize several types (e.g. flavonoids, glucosinolates and cardenolides). They may produce one major toxin (e.g. morphine, as in *Papaver somniferum*) or equal amounts of several structures (e.g. five furanocoumarins in *Pastinaca sativa*), which synergize to provide an effective defense against insect attack.

In spite of the wealth of secondary metabolites that may be present in a plant, no physical or chemical defense against herbivory is absolute. There is always some phytophagous insect or mammalian herbivore that can overcome that defense, usually by detoxification and further metabolism of the plant toxins. Such detoxification is expensive to the animal in terms of energy sources and the synthesis of new enzymes, so that there may be a limit to this. In the case of overcoming dietary plant tannins, mammals produce special salivary proline-rich proteins to complex with the tannins and make them safe.

Some animals (e.g. goats) respond to the presence of toxins in plants by eating small amounts of many plant species to overcome this problem. Other animals, such as geese, practise 'geophagy' and mix their plant material with clay in order that the clay binds the toxins present. Occasionally, animal detoxification produces a new metabolite which is more harmful than the original plant toxin. This happens with the pyrrolizidine alkaloids, where the new metabolite, a pyrrole, is hepatotoxic. It is not surprising therefore to find that these alkaloids are responsible for at least 50% of all cattle deaths attributable to eating poisonous plants.

The elaboration of animal-targetted toxins by plants has been likened to a coevolutionary 'arms race' between plants and animals for mutual survival. As a result of co-evolution, a balance is achieved in natural populations where animals are able to exert restricted and selective herbivory on their food plants. Likewise, plants may regularly lose leaves to herbivores and yet be sufficiently defended by their second chemistry to flourish, flower and reproduce. Much remains to be done in the future to understand precisely the biochemistry of plant defense and of animal response.

REFERENCES

Amor-Prats, D. & Harborne, J.B. (1993) *Chemoecology* **4**, 55–61.

Baldwin, I.T. (1988) *Oecologia* **77**, 378–381.

Coley, P.D. (1986) *Oecologia* **70**, 238–241.

Crawford, R.M.M. (1989) *Studies in Plant Survival.* Blackwells, Oxford.

Dicke, M., Sabelis, M.W. & Takabayashi, J. (1990) *Symp. Biol. Hung.* **39**, 127–134.

Edwards, P.J., Wratten, S.D. & Parker, E.A. (1992) *Oecologia* **91**, 266–272.

Farmer, E.E. & Ryan, C.A. (1990) *Proc. Natl. Acad. Sci. U.S.A.* **87**, 7713–7716.

Feeny, P. (1975) In *Coevolution of Animals and Plants* (L.E. Gilbert & P.H. Raven, eds), pp. 3–19. University of Texas Press, Austin, Texas.

Fitter, A.H. & Hay, R.R.M. (1987) *Environmental Physiology of Plants*, 2nd edn. Academic Press, London.

Frischknecht, P.M., Ulmer-Dufek, J. & Baumann, T.W. (1986) *Phytochemistry* **25**, 613–616.

Gregory, P., Ave, D.A., Bouthyette, R.J. & Tingey, W.M. (1986) In *Insects and the Plant Surface* (B. Juniper & T.R.E. Southwood, eds), pp. 173–184. Edward Arnold, London.

Grill, E., Winnacker, E.L. & Zenk, M.H. (1987) *Proc. Natl. Acad. Sci. U.S.A.* **84**, 439–443.

Gulmon, S.L. & Mooney, H.A. (1986) In *On the Economy of Plant Form and Function* (T.J. Girmish, ed.), pp. 681–698. Cambridge University Press, Cambridge.

Harborne, J.B. (1993) *Introduction to Ecological Biochemistry*, 4th edn. Academic Press, London.

Hartmann, T., Ehmke, A., Sonder, H., Borstel, K.V., Adolph, R. & Toppel, G. (1989) *Planta Med.* **55**, 218–219.

Herms, D.A. & Mattson, W.J. (1992) *Q. Rev. Biol.* **67**, 283–335.

Huner, N.P.A., Williams, J.P., Maissan, E.E., Myscich, E.G., Krol, M., Laroche, A. & Singh, J. (1989) *Plant Physiol.* **89**, 144–150.

Janzen, D.H. (1978) In *Biochemical Aspects of Plant and Animal Coevolution* (J.B. Harborne, ed.). Academic Press, London.

Khan, M.B. & Harborne, J.B. (1991) *Biochem. System. Ecol.* **19**, 529–534.

Koritsas, V.M., Lewis, J.A. & Fenwick, G.R. (1989) *Experientia* **49**, 493–495.

Mihalaik, C.A. & Lincoln, D.E. (1989) *J. Chem. Ecol.* **15**, 1579–1588.

Murata, N., Nishizawa, O.I., Higaski, S., Hayashi, H., Tasaka, Y. & Nishida, I. (1992) *Nature* **356**, 710–713.

Nathanson, J.A. (1984) *Science* **226**, 184–186.

Rees, S.B. & Harborne, J.B. (1985) *Phytochemistry* **24**, 2225–2231.

Reichardt, P.B., Bryant, J.P., Clausen, T.P. & Wieland, G.D. (1984) *Oecologia* **65**, 58–69.

Rodriguez, E., Healey, P.L. & Mehta, I. (eds) (1984) *Biology and Chemistry of Plant Trichomes.* Plenum Press, New York.

Ryan, C.A. (1992) *Plant Mol. Biol.* **19**, 123–133.

Schoeffl, F., Lin, C.Y. & Key, J.L. (1984) *Annu. Proc. Phytochem. Soc. Eur.* **23**, 129–140.

Schultz, J.C. (1983) In *Plant Resistance to Insects* (P.A. Hedin, ed.), pp. 37–54. American Chemical Society, Washington DC.

Stipanovic, R.D. (1983) In *Plant Resistance to Insects* (P.A. Hedin, ed.), pp. 69–103. American Chemical Society, Washington DC.

Tallamy, D.W. & Raupp, M.J. (eds) (1991) *Phytochemical Induction by Herbivores.* Wiley, New York.

Tuomi, J., Niemela, P. & Siren, S. (1990) *Oikos* **59**, 399–410.

Vet, L.E.M. & Dicke, M. (1992) *Annu. Rev. Entomol.* **37**, 141–172.

Wink, M. (1983) *Z. Naturforsch.* **38c**, 905–909.

Woodhead, S. & Chapman, R.F. (1986) In *Insects and the Plant Surface* (B. Juniper & T.R.E. Southwood, eds), pp. 123–136. Edward Arnold, London.

Wyn Jones, R.G. (1985) *Chem. Britain* 454–459.

Zangerl, A.R. (1990) *Ecology* **71**, 1926–1932.

15 Plant Cell Biotechnology

M.D. Brownleader and P.M. Dey

15.1 INTRODUCTION

Plant-derived biomass provides as much as 15% of the world's energy needs. Industrial conversion of lignocellulose to alcohols is a typical use of plant biomass. In Brazil and Kenya, ethanol is used as a fuel in combination with petroleum. In Kenya, switching of 400 000 hectares from food to sugar cane production has led to an undesirable increase in food import, although they do not have such a large oil import burden.

A third of the total world crop losses are thought to be due to plant disease. Well-being of humankind is closely associated with an ability to cultivate plants productively. In Ireland during the potato famine of the 1850s, 1 million people starved and 2 million people were forced to emigrate to North America as a direct consequence of the loss of the potato crop by the fungal pathogen, late blight. Breeders and biotechnologists have employed either conventional plant breeding programmes or recombinant DNA technology to enhance plant characteristics such as biomass yield, herbicide resistance, drought tolerance, plant disease resistance, prolonging the shelf-life of fruits and vegetables and increasing their nutritional value. However, conventional plant breeding may take at least 10 years to cultivate resistant crops and the desired genetic trait may not be transferred to the new variety. Genetic transformation has further advantages over conventional plant breeding because the desired gene(s) can be transferred to another plant without the concomitant transfer of undesirable characteristics, the genes and their products are better characterized and gene transfer can transcend sexually incompatible plant species. However, it may not be possible to genetically transform and regenerate specific crop cultivars and there will always be environmental, ethical and ecological considerations.

Genetic manipulation and modern plant breeding techniques could, in future, remove toxic components of plants. Castor beans (*Ricinus communis*) contain the useful industrial seed oil, triricinolein. Due to the presence of a very toxic lectin and the alkaloid, ricinine, which obviate its use as a food product, castor bean plants are not grown as a crop in North America. Many conventional breeding programs are designed to increase crop biomass by enhancing photosynthesis and many breeding programs select crop species with enhanced rates of photosynthetic CO_2 fixation. Many factors regulating CO_2 fixation could be enhanced by genetic modification. A reduced ribulose bisphosphate (RuBP) oxygenase to RuBP carboxylase activity ratio would enhance net photosynthesis by reducing non-essential photorespiration without a large reduction in the rate of RuBP carboxylation. In addition, scientists have isolated a gene from thermophilic bacteria which codes for an enzyme that fixes CO_2. They transformed *Escherichia coli* with the modified gene and found that it fixed CO_2 two times faster than plant RuBisCo. If plants could capture CO_2 faster, without risking dehydration, plants could then grow in arid regions of the world. World reserves of grain dropped to an unprecedented low level of 230 million tonnes in 1995 which is enough to feed the world for only 47 days at current consumption rates and well below the minimum required for world food security. The global population is increasing by 90 million each year and is expected to rise to 7.2 billion by the year

PLANT BIOCHEMISTRY
ISBN 0-12-214674-3

2010. No one is certain whether this low harvest is a cyclical fluctuation or whether the earth is unable to feed such a fast-growing population. The reasons for the low level of cereal production are not disputed and are the culmination of arable land being taken out of production, the weather with droughts or heavy rainfall occurring at the wrong time in important cereal growing regions, and the affluent population of China changing their diets and switching to alternative Western-style food products such as chicken, pork and beer. Perhaps some good harvests in the next few years will reverse this declining trend. Cereal production in some African countries such as Zimbabwe, which have suffered from drought in the past years, have increased to such an extent they may soon be able to export maize in the future. There is a widespread belief that very little increase in yield will be achieved by adding fertilizers to land already saturated with them. Plant biotechnology and genetic engineering may in future, be able to improve important plant characteristics such as plant resistance to diseases and herbicides or drought tolerance and begin to restore the level of grain reserves.

Plants such as tomato, carrot and potatoes are cultured for, (1) year-round production, and (2) rapid production of new disease-free varieties by conventional genetics. Mass production of genetically modified cells derived from protoplast fusion or recombinant DNA technology is also extremely important. Long-term storage of stable genetic material, reducing storage requirements and decreasing maintenance costs are important considerations for plant micropropagation. Plant cells can also be used to produce an array of low volume/high value chemicals such as pharmaceuticals, dyes and flavors that are normally derived from the whole plant and are estimated to produce 25% of prescribed medicines and many fine chemicals.

Even the most hard-nosed physicists admit now that a flap of a butterfly's wings can affect the weather system hundreds of kilometres away. Biological systems are highly interactive and dynamic. Understanding these complex mechanisms *in vivo* is proving difficult because physiological responses are the manifestation of several intricate biochemical reactions. Plant cell cultures therefore provide a precisely controlled and less complex model system for experimental analysis than the intact plant.

This chapter focuses on particular aspects of an increasingly expanding area of plant biotechnology. The impact of plant biotechnology on society in the future will be incalculable when environmental limitations of a growing global population blend with changing dietary life styles.

15.2 PLANT CELL CULTURE

Plant tissue culture refers to the culture (incubation in a nutrient medium) of cells and plant organs such as roots, shoot tips and leaves. However, we will primarily refer to plant cell culture because significantly more information is available and cell cultures have produced commercial quantities of secondary metabolites. Root cultures have allowed investigations into root systems that are otherwise difficult to study buried and intact in the soil. Shoot tips, leaves and flower parts are, however, unlike excised root cultures because they are only capable of limited development in culture.

In plants, most co-ordinated cell division takes place in the meristems. Non-dividing parenchyma and the cells in meristems, vascular cambial tissues and embryonic tissues are at early stages of development and exist in an 'undetermined' state which can rapidly proliferate (dedifferentiate) to produce calli. A callus is a disorganized mass of mainly undifferentiated cells. Calli are grown in semi-solid nutrient media whereas suspension cultures are maintained in liquid media. Callus culture can be maintained on an appropriate plant tissue culture medium containing nutrients, vitamins and hormones. Under suitable cultural conditions, cell masses known as calli can form shoots and roots and eventually regenerate into whole plants. Plant cells, in contrast to animal cells, are not attachment-dependent. A callus can be disrupted mechanically, chemically or enzymatically into a uniform single cell suspension and small clumps that can be grown in a liquid suspension culture.

Small scale cell culture requires basic laboratory apparatus. A laminar airflow should be used to provide an aseptic environment for all culture manipulations because all plant tissue culture media can support the growth of many fast-growing bacteria and fungi that will destroy the slow-growing plant tissues. Even slow-growing contaminants can produce toxins which may affect plant cell growth. Components of typical plant tissue culture media include (1) inorganic macronutrients (e.g. Fe, Mg, Ca, K, P, N), (2) inorganic micronutrients (e.g. Mn, Cu, Zn, B, Na, Cl, I, S, Mo, Co, Al and Ni), (3) organic nitrogen sources (e.g. glycine, inositol), (4) vitamins (e.g. nicotinic acid, pyridoxine and thiamine), (5) carbon source (e.g. glucose, sucrose), (6) plant growth regulators (e.g. auxin, cytokinin), (7) optional organic compound (e.g. casein hydrolysate, yeast extract), (8) antibiotics such as kanamycin and (9) a gelling agent (0.5–1.0% w/v good quality, bacteriological grade agar) if the medium is to be semi-solid. Plant cell culture media are less expensive and less complex than mammalian cell culture media. Premade culture media are commercially available. However, plant growth regulators such as auxins and cytokinins tend to be omitted from the basic medium formulation because culture growth and differentiation is affected by these growth regulators which help to maintain dedifferentiated cell growth and promote cell division respectively. Synthetic auxins such as 2,4-D (2,4-dichlorophenoxyacetic acid) and NAA (1-naphthaleneacetic acid) can replace the naturally occurring auxin IAA (indole-3-acetic acid) because IAA is readily oxidized by plant cells. BAP (6-benzylaminopurine) and kinetin (6-furfurylaminopurine) are the most commonly used cytokinins.

15.2.1 Preparation of explants

A wide range of plant organs and tissues can be used as a source of explants for the initiation of callus cultures. Meristems are usually free of virus infection because they are devoid of vascular tissue that transport viruses; they have active mitotic division. Heat treatment is commonly used to inactivate the virus prior to culturing meristems. In many cases, 34–36°C will inactivate the virus without any untoward effects upon the plant. The smallest possible meristem explant should be cultured in order to reduce the possibility of infection. However, small explants do not regenerate roots as well as large meristems. There are no hard and fast rules for the optimal explant size. Some explant sources are preferable if regeneration of the whole plant is required from established embryonic cultures. Explant source may also be critical for secondary product biosynthesis.

15.2.2 Formation of callus and suspension cultures

The explant must first be sterilized. The most common sterilizing agents are 1–2% sodium hypochlorite, 9–10% calcium hypochlorite (10–40 min, young stems, petioles, roots, fruits) and 70% ethanol (2–5 min, leaves, seeds). It is always preferable to sterilize the seed prior to germination in aseptic conditions on 0.8–1.0% (w/v) agar in conical flasks or test tubes. After germination, the appropriate organ (cotyledon, hypocotyl, stem or root) can be excised from the plantlet with a sharp, sterile scalpel and transferred to an agar-based plant tissue culture medium. If the explant is to be taken from a plant root, hypocotyl or a woody

stem segment, the whole organ or tissue is surface-sterilized before excising a piece of tissue from the inner undamaged and uncontaminated section.

The callus may take many months to form. The established callus should be subcultured at regular 4–6 week intervals. A suitable temperature for incubating cultures is between 20 and 25°C and they should be left either in the dark or under very low level illumination. Transfer of callus to a liquid medium tends to form suspension cultures readily. The flasks with suspension culture are normally agitated on an orbital shaker between 60 and 150 rpm. When the newly established suspension culture has reached a suitable cell density for subculture (usually every 7 days) the cells are simply poured into a fresh medium. A large inoculum is normally necessary to retain active growth following transfer and, in some cultures, most of the growth occurs on the surface of small clumps. Growth of callus or suspension culture can be monitored by an increase in weight, cell number and/or cell viability. Fresh (wet) weight is measured after removing as much water as possible by either filtration or centrifugation (200 g, 5–10 min). Dry weight is measured after drying samples in a 60–80°C oven until there is no further weight change. Callus cultures and suspension cultures consisting of cell aggregates must be disrupted and cell numbers can then be determined with a hemocytometer. Packed cell volume (PCV) is also related to cell number. The entire content of a flask is transferred to a graduated centrifuge tube and spun at 200 g for 10 min. The PCV is the percentage of the volume of the pellet after centrifugation to the total culture volume. Vital stains such as Neutral red or Evan's blue coupled to light microscopy or fluorescein diacetate coupled to fluorescence microscopy can be used to assess cell viability. With fluorescein diacetate, viable cells emit a green/yellow fluorescence, whereas non-viable cells remain unstained.

15.2.3 Plant regeneration from callus

In many callus tissues, the relative levels of auxin and cytokinin control shoot and root formation. An undifferentiated callus can be regenerated into a whole plant by simply altering the concentration of growth regulators. Only roots are formed when the auxin/cytokinin ratio is high and shoots are initiated when the cytokinin/auxin ratio is high (Fig. 15.1). Intermediate levels of both hormones produce completely disorganized callus growth. Cereals and small grains as well as important dicotyledonous species, such as soybean and

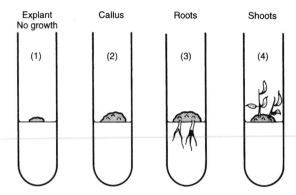

Figure 15.1 Effect of plant growth regulators on development of tobacco explant on agar medium. Regulator composition (A, auxin; C, cytokinin); (1) No A + average C, (2) high A + average C, (3) high A + low C, (4) low A + high C.

cotton, appear to be unresponsive to tissue culture manipulations. Undifferentiated tissue or callus can be produced from 'recalcitrant' species but regeneration of the whole plant was virtually impossible. Cell divisions usually occur in a random fashion in the callus generating disorganized tissue with no obvious form or polarity. Under certain conditions, shoots and root meristems or even embryo-like structures called embryoids can develop. These organized structures often form plantlets and then whole plants. Embryos are a type of callus with a smooth, rounded surface and containing cells with a dense cytoplasm. These meristem-like cells proliferate slowly, are not completely undifferentiated and regenerate into whole plants. Some cells within callus masses are capable of forming embryos or organized meristems. As these cells retain all the genetic information required for normal plant development, they are referred to as 'totipotent'.

15.2.4   Embryogenesis

There are four developmental stages of embryogenesis. During the *initiation* phase, cells are induced to form somatic embryos. This is followed by a *proliferative* stage where these cells undergo multiplication with very little or no maturation of the embryogenic tissue. During *maturation*, cells are transferred to a medium devoid of auxin which may contain abscisic acid, proline or sucrose. Immature embryos may not germinate. In fact, *germination* is quite sporadic and some embryos never germinate to form roots and shoots. The plantlets are transferred to a soil environment, acclimatized and moved to a glasshouse.

15.2.5   Protoplast isolation, culture and plant regeneration

Protoplasts (wall-less cells) are a starting point for genetic modification of plant cells because they are ideal recipients of foreign DNA or larger particles (organelles, bacteria, etc) and can fuse with different plant protoplasts. Genes for disease resistance can be transferred from one species to another in order to widen the genetic base for plant breeding. All plant cells theoretically contain the full complement of genetic information required for whole plant development. Although, most explants can be used as a source of protoplasts, isolating viable protoplasts capable of sustained division and plant regeneration is still restricted to a small number of species/explant combinations.

Cellulase and pectinase are normally added to plant cells in a solution of high osmotic pressure to prevent the released protoplasts from bursting (up to 15% mannitol or sorbitol). After overnight incubation to increase the protoplast yield, cell debris are removed by filtration and/or centrifugation at very low speed ($\sim$100 g) and protoplasts are collected from the surface. Protoplasts are normally washed, counted by a hemocytometer to find an effective inoculum density for growth and viability determined by using vital dyes or fluorescein diacetate coupled with light or fluorescence microscopy, respectively.

Protoplasts prefer to be embedded in a semi-solid medium rather than cultured in a liquid medium. Agar or agarose is believed to provide the structural support facilitating cell wall development. Protoplasts will not start to divide until new walls have formed. Protoplast cultures are usually static and require a temperature of about 25–30°C under low intensity continuous illumination. Protoplasts are cultured by three main ways: (1) embedded in agar; (2) plated on a liquid medium over an agar/agarose-based medium; and (3) by the hanging drop culture method. Small droplets containing the protoplasts are placed on to the lids of Petri dishes. The lid with the droplets of protoplasts is inverted and placed over the Petri dish containing culture medium.

(a)   *Somatic hybridization*

Once the protoplasts have been isolated, they are fused, hybrids selected, plantlets regenerated by embryogenesis and organogenesis and the regenerated plants subsequently analyzed. Protoplast fusion can be induced either chemically or electrically. The plasma membrane is temporarily destabilized resulting in pore formation and cytoplasmic linkages between adjacent protoplasts.

In chemical fusion, high concentrations of polyethylene glycol, dextran or polyvinyl alcohol, sometimes with a high pH/Ca^{2+} buffer or a high pH/hypotonic buffer are required to induce pore formation and rapid fusion (less than 30 min) between membranes of adjacent cells. Electro-fusion is often preferred to chemical fusion because it is more reproducible and is not so damaging to the protoplasts but does demand sophisticated and expensive equipment. It uses an alternating current to align protoplasts and initiate close membrane contact and a short direct current to induce membrane breakdown.

Fusion is random and generates a mixture of unfused homokaryon protoplasts, fused homokar-yons and fused heterokaryons. The desired hybrid colonies must then be selected which can be difficult if the desired fusion occurs at a very low frequency. Selection parameters include morpho-logical characteristics such as pigmentation, and cytological criteria such as chromosome number. Hybrid cells can outgrow cell colonies derived from parent protoplasts which can therefore be isolated and transferred to a plant regeneration medium. The cells will regenerate a new wall and then form hybrid whole plants. The somatic hybrid plants are then characterized morphologi-cally, cytologically and biochemically by compara-tive analysis between the parent species and the regenerated hybrid plants. Routine biochemical analysis includes isoenzyme analysis, resistance of plants to microbial infection, herbicide and fungal toxin sensitivity.

No new crops have been successfully produced by protoplast fusion techniques although some crop species, such as rape, cabbage, millet, are amenable to protoplast isolation and regeneration. Protoplast isolation and subsequent regeneration has not proved to be very successful with legumes and cereals. However, most early fusion experi-ments did focus on the production of distant somatic hybrids that could not be generated by conventional sexual hybridization.

15.2.6 Somaclonal variation ('sports')

In contrast to shoot-tip propagation the calli which were originally genetically homogenous yield plantlets that can show considerable varia-tion in characteristics such as aneuploidy, sterility and morphology. Progeny that differ significantly from the parent are called 'somatic variants'. Introduction of variability in regenerated plants

can be agriculturally useful. These somatic vari-ants have established new and important varieties such as the naval orange or nectarines. Potato and sugar cane 'sports' were found to be resistant to disease in contrast to the susceptible parent plants. Herbicide resistance, cold- and salt-tolerance are additional favorable traits that can be found in 'sports'.

Systematic comparisons were made of plants regenerated from callus of five cultivars of *Pelargonium* (geranium) species. Plants from geranium stem cuttings *in vivo* were uniform, and plants from *in vivo* root and petiole cuttings and plants regenerated from callus were quite variable when compared to the parent plants with regard to plant and organ size, leaf and flower morphology, essential oil characteristics and anthocyanin pigmentation.

Callus cultured for several weeks, will show signs of ageing, monitored as deceleration of growth, necrosis (cell death) or browning, and finally desiccation. Ageing usually results from exhausted nutrients, inhibition of nutrient diffu-sion, evaporation accompanied by an increased concentration of some medium constituents and accumulation of toxic metabolites. When callus or cell suspensions are maintained for long periods, their morphogenetic ability is either lost or significantly reduced under the same culture conditions. Although factors affecting morpho-genetic expression have not been precisely estab-lished, disorganized growth results in aneuploidy and chromosome mutation which is thought to lead to a change in or loss of morphogenetic capacity. Changes in nuclear chromosomes (both number and rearrangements), recombination, mutation of single genes or modification of extra-nuclear genes in mitochondria and chloroplasts can account for some variation detected in regenerated plants. The phenotypic variability in regenerated plants probably reflects both genetic differences and tissue culture-induced changes.

Callus is produced in the intact plant as a natural wound response and only in a few instances are root and shoot initials produced from the callus itself in the intact plant. The callus proliferates with the use of growth regulators such as auxin and cytokinin. It is therefore not surprising that callus and cell suspension culture will have abnormal developmental pathways. Transient variation in the regenerated plant may have an epigenetic cause where there is no change in genetic composition of the cell. These epigenetic changes are usually induced by culture conditions and are not useful for crop improvement because they are not expressed in progeny of regenerated plants.

15.2.7 Cryopreservation

Preservation of callus at low temperature (5–10°C) for long periods of time (6–12 months) without transfer is not recommended to prevent somatic variation. Cells of flax frozen at −50°C remained viable. However, −50°C is not a sufficiently low temperature for stable, long-term storage. Cryopreservation therefore involves freezing the cells in the presence of a cryopreservant (glycerol, dimethylsulfoxide, or methanol) to a very low temperature of −196°C (temperature of liquid nitrogen). After storage, the samples are thawed to ambient temperature. The freezing and thawing protocol must always try and ensure that the biological material is exposed to as little physical stress as possible. Plant cells are cryopreserved because of genetic instability and reproducible long-term storage of germ plasm, risk of losing cells through microbial contamination or large expense of maintaining stock cultures.

(a) *Pregrowth period*

Culture conditions should be optimized before cryopreservation. Uniform cell suspensions comprising cytoplasmic meristematic cells are more suitable than vacuolated cells with a high water content. Before freezing, there is a pregrowth period that ensures optimal recovery after freezing. The pregrowth period does tend to be combined with cold hardening treatment or the use of additives, e.g. abscisic acid, that enhance plant stress tolerance. Relatively low concentrations of dimethylsulfoxide, for example, during the pregrowth period often enhances shoot-tip recovery after cryopreservation.

(b) *Cryoprotective treatments*

The cryoprotective treatments include first, vitrification where a high concentration of osmotically active compounds (sugars/polyols) is used to prevent ice crystal formation when biological material is exposed to low temperatures. The concentrated solution does not allow ice crystal nucleation, and water exists in an amorphous 'glassy' state. Second, penetrating chemical cryoprotectants reduce the toxic effects of a cold-induced concentration of solutes, and non-penetrating cryoprotectants protect the plant tissue by osmotic dehydration. These cryoprotectants also alter the biochemical properties of membranes and therefore enhance freeze tolerance of the plant tissue. Third, cells which are

dehydrated have a much lower water content available for ice formation and may be immersed in liquid nitrogen without the use of chemical cryoprotectants. The plant tissue is desiccated over silica gel. Fourth, the plant tissue is encapsulated in calcium alginate beads, pregrown in liquid culture media containing high concentrations of sucrose and then transferred to a sterile air flow cabinet and desiccated further. Tissues can then tolerate liquid nitrogen without using chemical cryoprotectants.

(c) *Freezing, storing and thawing*

Samples can be immersed in liquid nitrogen (rapid freezing) or chilled gradually at a precisely controlled rate (usually −1°C min^{-1}) to an intermediate temperature (e.g. −35°C) in a cryostat and then transferred directly to liquid nitrogen and maintained in storage dewars containing liquid nitrogen. The temperature and rate of thawing of plant tissue is dependent upon the freezing method and this is very important in vitrified tissues because ice crystals can form during thawing.

Thawing is usually achieved by plunging cryovials into sterile water at 40–45°C. The samples are removed when the ice has melted. Tissues frozen by encapsulation and/or dehydration are usually thawed at ambient temperatures and vitrified and excessively dehydrated tissues demand that the osmoticum concentration is lowered sequentially throughout the thawing and early recovery processes.

The plant tissue is assessed by viability studies, morphological changes (leaf expansion, callusing, embryo development), regeneration of shoots and plants and re-establishment of normal metabolism, Recovery of plant tissue is frequently induced with hormones.

15.3 MICROPROPAGATION

Plant tissue culture is an experimental model system confined to discrete liquid or semi-solid nutrient media whereas micropropagation involves *in vitro* propagation of the selected genotype and ultimate establishment of the plant in the field or a glasshouse. Horticultural practice has traditionally used vegetative cloning of plant parts (shoots, leaves) for selective and rapid propagation of many plant species. After establishment of the plant in culture, the horticulturist will want to produce large numbers of the plantlets and establish them in soil.

Micropropagation is concerned with mass multiplication of plants in a glasshouse and is often associated with stem, root cuttings etc, and a soil environment. In 1991, there were approximately 600 million micropropagated plants produced worldwide. These were mainly pot plants, cut flowers and fruit trees. There are five stages in micropropagation:

1. *Preparation of explant*: the plant material for *in vitro* culture is prepared and explants are obtained which require less aggressive sterilization.
2. *Formation of callus*: for most micropropagation work, the explant of choice is an apical or axillary bud. If one wants to produce a virus-free plant from an infected individual, it is obligatory to start with the submillimetre shoot tip. There are several rules that should be remembered. Aerial plant parts are less contaminated than underground parts, interior parts are less contaminated than exterior parts, the smaller the explant, the smaller the risk of contamination; and regeneration capacity is usually inversely proportional to age and size of the explant and to the age of the explant source.
3. *Shoot proliferation*: a high cytokinin/auxin ratio induces shoot formation and a high auxin/cytokinin ratio induces root formation. In many cases, a cytokinin alone is enough for optimal shoot multiplication. Meristematic centers are induced and developed into buds and/or shoots. After 4–8 weeks, the original explant is transformed into a mass of branched shoots or a cluster of basal shoots. The small shoots or clusters are then re-planted on to a fresh medium (in which the cytokinin level can be increased) to increase shoot multiplication.
4. *Shoot elongation and root formation*: adventitious and axillary shoots lack roots in the presence of cytokinin. Instead of transplanting the shoots to fresh medium, liquid media can be added to established cultures (double layer technique). Auxins such as NAA are usually required to induce rooting and activated charcoal can be added to the liquid medium in order to absorb any residual cytokinins.
5. *Transfer to a glasshouse*: the micropropagated plantlet must acclimatize to the environment of the glasshouse with appropriate humidity and temperature control. Acclimatization can proceed *in vitro* with bottom cooling, reducing the relative humidity in the head space of the container. The culture vessels are uncapped and placed in the glasshouse several days prior to

the removal of the plants from the culture medium. The plants are washed to remove agar because the agar will serve as a substrate for growth of disease-causing organisms.

The plants require high humidity in the first few days which can be achieved using polyethylene tents or humidifiers. The transition from culture to glasshouse is facilitated with the use of a screen to prevent temperature peaks, and a lower light intensity.

Advantages of cloning plants are the speed of plant multiplication and the quantity and uniformity of generated plants. In some species such as ferns, orchids and ornamental trees propagation *in vitro* is considerably faster than conventional propagation.

15.3.1 Case study: oil palm plants

The family of commercially important *Palmae* include date palm (*Phoenix dactylifera* L.), oil palm (*Elaeis guineensis* Jacq.) and coconut palm (*Cocos nucifera* L.). Oil palm provides about 14% of the world's consumption of vegetable oils and together with palm kernel oil is only superseded by soybean oil. Improvement of this species by conventional breeding techniques is limited by the fact that it is long-living. It also has a juvenile phase of many years before the productivity of the tree can be evaluated. Palm breeding is slow because the oil palm tree does not bear fruit until about 3 years after germination. In addition, oil palm is a naturally out-breeding species and there is significant heterogeneity of the plantations.

As a perennial crop and all-year production, an oil palm plantation can produce up to 6 tonnes of oil per hectare under good management practise in favorable climatic conditions. However, there can still be considerable yield improvement of individual palms. *In vitro* culture of oil palm can generate homogeneous, agronomically useful plants in sufficient quantity.

The fruit of the oil palm is known as 'dura' and this fruit has a hard shell enclosing the kernel (Fig. 15.2). There is no horticultural method available for clonal propagation of oil palms as they do not produce offsets and there is no method for establishing cuttings. Unilever began to culture oil palms in 1968. The first palms were grown in tissue culture in 1975 and were sent to Malaysia in early 1976 for planting in the field. Although micropropagation of oil palm began over 20 years ago, we are only recently gathering information about this program.

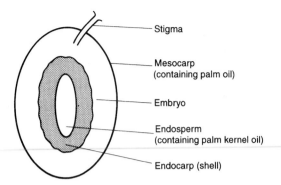

- Stigma
- Mesocarp (containing palm oil)
- Embryo
- Endosperm (containing palm kernel oil)
- Endocarp (shell)

Figure 15.2 Schematic diagram of a longitudinal section of oil palm fruit.

Disorganized callus (from root or leaf) was induced with growth regulators and maintained indefinitely. Leaf explants are present in large amounts and can be removed from the plant with relatively little deleterious effect. A secondary callus was formed which developed into adventitious embryos (Fig. 15.3). Development of secondary callus from primary callus and formation of embryoids occurred randomly and at low frequency. The embryoids were transferred to a proliferation medium for the large-scale multiplication of plants. Shoot formation accompanied somatic embryoid development. This was followed by auxin-induced root formation, hardening of the plant and subsequent planting in soil. New research work has been initiated to produce isolated embryos in liquid medium.

Initially, no significant variability in the mature oil palm plants was observed. However, abnormal flowering has since been found and its occurrence appears to be related to the period of embryoid proliferation. Almost all embryogenic material propagated *in vitro* for 3 years results in abnormal flowering. A large number of palms planted since 1983 have produced abnormal flowers, particularly in plants subjected to large-scale propagation. This led researchers to wonder whether abnormal flowering was a reaction to large-scale propagation. A French agronomic organization (Institut de Recherche pour les Huiles et Oléagineux; IRHO) have also found this abnormal flowering in their cloned oil palm plants and the low proportion of affected flowers coupled to reversibility suggests that the trigger is epigenetic and associated with the plant growth regulators.

Oil has been analyzed from several bunches from several different palms. Fatty acids, triacylglycerols, carotene content and iodine levels were found to be different between each clone. However, after a report on the incidence of mutation in micropropagated clones, production of oil palm in Asia has declined dramatically. The abnormal flowering must be monitored closely before commercial opportunities are exploited. Perhaps different types of oil can be developed from the present palm oils. However, efforts put into altering the quality of the palm oil and increasing the yield of oil palm fruit will almost certainly increase the price of palm oil in the short term which will have an impact on the economies of underdeveloped countries.

15.4 COMMERCIAL EXPLOITATION OF PLANT BIOTECHNOLOGY

Particular genetic traits can be introduced into modern cultivars to improve nutritional quality, herbicide resistance, disease resistance and plant yield etc. $600 million of damage to rice crops occurs each year through flooding. This is a major problem in countries such as Bangladesh and Thailand which are prone to floods. Traditionally, taller varieties of rice have been grown to counter flooding. However, the yield of rice is up to 40% lower than in the dwarf varieties because nutrients

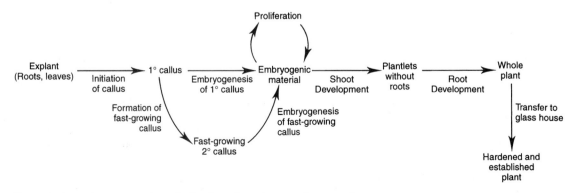

Figure 15.3 Micropropagation of oil palm. 1° = primary; 2° = secondary.

are channeled into growth rather than grain production. Geneticists have recently identified a genetic marker of the gene that is believed to confer tolerance to flooding in rice. This marker will now make it easier to breed flood-tolerant rice because a simple genetic marker test will reveal the presence of the flood-tolerant gene. New rice varieties grown after crossing flood-tolerant varieties with susceptible varieties can be monitored easily for flood tolerance without a laborious field trial that can take up to 2 months. In addition, farmers can also kill undesirable weeds by simply flooding the rice fields.

Today growers harvest their tomato crop firm and green so that it can be washed, sorted, packed and shipped without suffering extensive bruising. Consumers buy nearly 3 billion pounds of tomatoes every year in USA and they would not buy a tomato that is bruised and oozing ketchup. Before shipping, commercial packers bathe the tomatoes in ethylene gas for several days to induce ripening. Tomatoes that ripen on the vine are too soft to ship over long distances. Prematurely plucked and artificially-ripened supermarket tomatoes do not have particularly fresh flavors and lack the colour of a ripe freshly-picked tomato. The ideal mass-produced tomato should be left on the vine to build up the sugars and acids critical to fresh taste but remain firm enough to handle without damage. Using antisense technology, Calgene Fresh Inc. (USA) has permanently introduced a backward (antisense) copy of the gene for polygalacturonase (a fruit-softening enzyme together with pectin methylesterase that breaks down cell wall pectin) with the aid of *Agrobacterium tumefasciens*. Both copies of the gene produce mRNA. However, antisense RNA molecules bind to the normal-sense RNA preventing the tomato from making the usual amount of polygalacturonase. The new 'Flavr Savr' tomato will therefore resist softening and have a fresh flavor and the extended shelf-life demanded by the consumer. The genetically-modified delayed ripening tomato is the first genetically-engineered whole food product to be marketed in USA. The US Food and Drug Administration (FDA) has recently concluded 'that "Flavr Savr" tomatoes have not been significantly altered and are as safe to eat as other tomatoes'. Natural delayed-ripening tomato mutants are also being evaluated and are expected to compete with the 'Flavr Savr' tomatoes.

A fungal disease called witches'-broom could have decimated Brazil's cocoa industry. Witches'-broom causes the abnormal bushy growth of shoots. However, researchers at Ceplac in the North Eastern state of Bahia, produced a new resistant cocoa tree called Theobahia by crossing a local cocoa variety with a wild Amazonian variety. In the early controllable stages of infection, the fungus attacks branches but in its advanced stage, cocoa pods are destroyed. Brazil is the world's second largest producer of cocoa and annual production has dropped from about 360 000 tonnes to 240 000 tonnes in 1994. Rare animal species have also benefited from selective plant breeding programs because cocoa trees survive under the shade of taller trees.

Genetic manipulation and chemical technology have undoubtedly enhanced management practices in weed control. Herbicides destroy weeds which reduce the target plant yield by increasing competition for water, light and nutrients. Weed seed contamination can also result in poor flavors following processing. Important classes of herbicides such as the triazines and the sulfonylureas are more toxic to the weeds than to the specific crop due to the crop plant metabolizing the herbicide. Other herbicides are physically prevented from passing through the waxy cuticle or the crop plant can simply sequester the herbicide into an internal compartment. If plants could be genetically modified to be resistant to a herbicide, farmers could choose more effective, selective and low environmental-impact herbicides with broader weed activity profiles. In fact, the first sulfonylurea-resistant mutant plants were selected from tobacco cells in tissue culture and subsequently regenerated into whole plants.

Guar gum and locust bean gum are used in many industries such as the paper, food, cosmetic and pharmaceutical industries. They are good promoters of gelling when mixed with gelling polysaccharides such as caragenan, agar and xanthan. Guar gum is however, a less effective gel promoter than locust bean gum because locust bean gum has a lower galactose/mannose ratio than guar gum (G/M 1:2 for guar gum compared to G/M 1:4 for locust bean gum) and it is the extended galactose-deficient regions that are believed to interact with the gelling polysaccharides. Guar is an annual crop and can be grown in semi-arid conditions, whereas carob (locust bean) trees may take up to 20 years to reach full maturity and are grown in highly intensively farmed Mediterranean regions. Guar gum is therefore cheaper than locust bean gum because of competition from other crops. To date, fungal α-galactosidases have been successfully purified which specifically remove galactose moieties from guar gum to yield a polysaccharide with functional properties similar to those of locust bean gum (see Dey *et al.*, 1993).

Plants are a source of high added value, low volume compounds for the pharmaceutical, perfume, and fine chemical industries. About 25% of all drugs prescribed are derived from plants, e.g. quinine (*Chinchona ledgeriana*), codeine (*Papaver somniferum*) and digitoxin (*Digitalis lanata*). Almost 1500 new plant compounds are reported in the literature each year. The majority of compounds of commercial importance are secondary metabolites. The fragrance, Jasmine, derived from *Jasminum* costs in excess of $5000 kg^{-1}. Castanospermine is a glycosidase inhibitor with anti-HIV activity and is purified from the plant, *Castanospermum australe*. The anti-malarial agent, quinine is purified from *Chinchona ledgeriana* and costs more than $100 kg^{-1}.

Large amounts of plant tissue need to be processed in order to isolate a small amount of drug from the wild plant. Potential consequences of this action, without clonal propagation and plant breeding programs will be loss of some rare plant species. However, in some cases, when a medicinal drug has been isolated, e.g. steroid-producing species of *Dioscorea* yams, micropropagation has increased the occurrence of the plant species. There are two key questions that need to be addressed before a plant product can enter a biotechnology-based drug developmental program. First, will the projected investment generate a high commercial return and second, are technical difficulties inherent in conventional production overcome by the biotechnological approach?

15.4.1   Mass plant cell culture

Callus is too slow-growing and heterogeneous to be useful for commercial production purposes. Suspension cultures are therefore used for mass culture. The ability to grow substantial volumes of plant cells under conditions similar to those used for microbial fermentation has enabled plant cell culture to be an alternative to field-based agricultural production. Field-grown product formation is fraught with many difficulties such as, (1) erratic supply due to the weather and the local political environment, (2) effects of pests and diseases, and (3) seasonal supply which can vary in both quality and quantity.

In the 1970s Japanese scientists grew tobacco cells in 6500 l modified stirred-tank bioreactors. Air-lift bioreactors have been used to grow up to 1500 l plant cell suspensions and are preferred to stirred-tank bioreactors because distorted plant cells have much reduced cell division. The two largest plant cultures to date (75 000 l and 5000 l)

have been grown in stirred-tank bioreactors and reflects the availability of stirred-tank bioreactors over air-lift bioreactors. The main problem with air-lift reactors is associated with the gas regime. Air-lift reactors require a gas stream for mixing and to meet physiological cellular requirements for oxygen and carbon dioxide. 'Overgassing' has been found to cause reduced plant cell growth. Plant cells are large (40–200 μm in length and 20–40 μm in diameter). Generally, individual cells are not found and the cells form aggregates with diameters up to 1000 μm. Difficulties with plant cell mass culture include high settling rate, difficulties of taking samples and diffusion limitations inside larger aggregates. Plant cells grow more slowly with doubling times of at least 2 days resulting in long fermentation times and therefore increasing the possibilities of infection. Due to their slow growth rate, plant cells have a low oxygen demand compared to micro-organisms and more controlled aeration is required. Cultured plant cells and microbial cells probably have markedly different growth rates because plant cells can be anything up to 100 times larger than microbial cells and they also have a large number of chromosomes to duplicate during cell division (Table 15.1). Microbial cells usually prefer a constant pH during their growth and production phases. Plant cells do not tolerate a fixed pH value. Usually plant cell culture starts at a pH value around 5 and ends at pH 6–6.5. Due to the low metabolic rate of plant cells, foaming and adhesion of plant cells to fermentation vessels are not major difficulties. The plant cell culture medium is more expensive than inexpensive beet molasses which are often used for micro-organisms. Because of their size, rigid cellulose-based cell wall and large vacuole, plant cells are shear-sensitive. In stirred-tank bioreactors the turbine impeller required for good mixing and to break up air bubbles in order to achieve high aeration does produce a high shear (Fig. 15.4). Although plant

Table 15.1   Differences between plant and microbial cells

Characteristics	Plant cells	Microbial cells
Aggregation	Normal	Mostly single cells
Size	Large	Small
Doubling time	Days	<1 hour
Inoculum	At least 10–20% of total volume	Low
Shear	Sensitive	Insensitive
O$_2$ requirement	Low	High

Airlift (Draught tube) Bioreactor

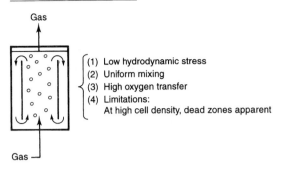

(1) Low hydrodynamic stress
(2) Uniform mixing
(3) High oxygen transfer
(4) Limitations:
    At high cell density, dead zones apparent

Stirred Tank Bioreactor

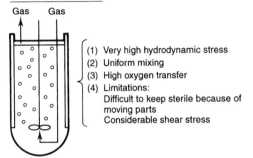

(1) Very high hydrodynamic stress
(2) Uniform mixing
(3) High oxygen transfer
(4) Limitations:
    Difficult to keep sterile because of moving parts
    Considerable shear stress

Figure 15.4 Comparisons of airlift and stirred tank bioreactors.

cell cultures have a low oxygen demand, the gaseous regime, e.g. CO_2 levels, can influence both growth and product formation. Consequently, fermentation run times increase and often secondary metabolite product induction is achieved by exposing cells to a second medium. An ideal plant bioreactor must have a moderate aeration rate, low shear, fine control of the gaseous regime and provide a good mixing environment.

Slow-growing plant cells can also be cultured in an immobilized plant cell bioreactor. The plant biomass can be separated from the medium which extends the usable life of the plant cells, protects them from shear and sometimes stimulates secondary product formation. The disadvantages of immobilizing plant cells are that, (1) the method of immobilization may be too severe for plant cells, and (2) plant secondary products are often stored in the vacuole and not released into the medium. Severe damage to the plant cells is necessary to release intracellular secondary products. The accumulation and excretion of secondary products may be controlled by the external pH. Chemicals altering the selective permeability of the plant membranes would be highly desirable so that living plant cells can release their secondary products into the medium. However, this will increase the production costs and obviate the use of the same plant cells for further product formation. Cultured plant cells contain unique compounds and this often offsets higher production costs associated with bacterial or fungal batch cell cultures.

High-yielding undifferentiated cell lines also overcome both seasonal and tissue-specific production and can be selected to produce a significantly higher concentration of a particular secondary product than is found in the intact plant. In addition, specialized plant cells such as oil glands are required to synthesize some secondary products. Very high levels of secondary metabolites tend to be produced by rapidly growing non-aggregating cell suspensions during the logarithmic growth phase. However, product formation may be higher in slow-growing aggregated cells (because of nutrient stress) in static cultures because of cellular differentiation. Although organ culture in small flasks has been achieved, no fermentor has been developed for the realistic culture of roots, embryos, leaves or shoots in large volumes.

(a) Case study: shikonin production

The plant cell culture production of the valuable chemical shikonin is the most successful and widely quoted example. Shikonin is a red naphthoquinone pigment used as a dye, a medicine and in the cosmetic industry and is derived from *Lithospermum erythrorhizon* cell cultures. The purple root of *L. erythrorhizon* is a perennial herb found in Japan, Korea and China and has been used as a specific remedy for burns, wounds, frostbite and skin ulcers and anal hemorrhage etc, as well as a precious natural dye for silk and cosmetics. The shikonin derivatives are antibacterial, anti-inflammatory and also promote wound healing. Depletion of the wild plant species, difficulty of cultivation and high demand for the pigment convinced a Japanese petrochemical company, Mitsui, to try and produce shikonin commercially from *Lithospermum* cell cultures. Basic studies with the *Lithospermum* callus cultures were undertaken in order to improve the productivity and stability of the cultured cells so that requirements for large-scale production of shikonin are met. Cell lines with a higher pigment content were isolated by the cell aggregate cloning method.

Factors such as temperature, light intensity, supply of growth regulators, carbon and nitrogen source were precisely controlled in order to select high-producing cultured plant cells. Nitrate and ammonia are the common nitrogen sources used in tissue culture media. Ammonia has been found to inhibit shikonin production at levels of only 3% of total nitrogen. White's medium which does not contain ammonia but uses nitrate as the sole nitrogen source enabled the cells to synthesize shikonin despite inferior cell growth. All components of White's medium were thoroughly examined to determine the optimal concentration. A new liquid 'production' medium based on White's medium and called M-9 medium was formulated. Although the M-9 medium is superior to Linsmaier–Skoog (LS) medium for shikonin yield, it is inferior to LS for promoting cell growth because it does not contain nutrients such as ammonium and phosphate. LS medium was examined for improved growth and a new medium called MG-5 medium was formulated. Cells were then transferred to M-9 medium for shikonin production. A two-stage MG-5/M-9 dual culture protocol was used. Cells were first grown in a 'growth' medium without producing shikonin and then transferred to a 'production' medium for optimal shikonin yield. Increasing the inoculum size when the medium concentration and oxygen supply were high and choosing an improved high-yielding strain also increased the shikonin yield.

(b) *Industrial production of shikonin*

Large-scale culture set-up is observed in Fig. 15.5. The cell culture was first grown in MG-5 medium for cell proliferation for 9 days in a first-stage tank (200 l). The used 'growth' medium was filtered off and the 'production' medium M-9 added to the cells in a second large tank for shikonin production. After 14 days in the second tank, the medium was filtered off and the pigmented cells harvested.

Shikonin production by plant cell culture has been compared to plant cultivation and to chemical synthesis. Plant cell culture has proved to be more economical than chemical synthesis and more productive than plant cultivation. The chemical composition of shikonin isolated from cell culture is similar to that of crude drug derived from intact roots and with less fluctuation in quality currently observed in the market. Shikonin derivatives extracted from cultured cells can be readily hydrolyzed with potassium hydroxide and crystallized to give pure shikonin. Cell culture is about 800 times more productive than plant cultivation. A 750 l culture tank with 600 l working volume capable of producing $2 g l^{-1}$ shikonin in 2 weeks is equivalent to 17.6 hectares of *Lithospermum* plants grown in a field. Shikonin produced by large-scale culture of *Lithospermum* cells has been commercially exploited for cosmetics such as lipsticks, lotions and soaps in Japan since 1984.

Agricultural biotechnology has spurred considerable practical achievements with herbicide-resistant crops, plant disease control, enhanced nutritional quality and crop yield. Genetic investigations have been possible because specific target reactions are being identified through biochemical and physiological investigations. However, there are currently serious concerns about the release of genetically modified organisms into the environment because 'foreign' genes may pass to other organisms. There would need to be considerable risk assessment, long-term monitoring and tight control of these genetically modified organisms because they cannot be recalled once they have been released. The predicted increase in the world population in the next 20–30 years will require increased 'sustainable' agricultural productivity without unacceptable destruction of the environment.

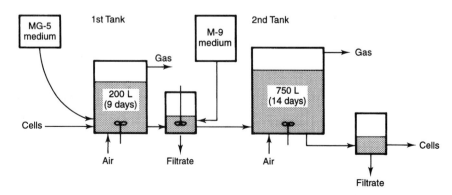

Figure 15.5 Schematic diagram for large-scale growth of cells for shikonin production.

15.5 DEFINITIONS

Adventitious: Organ development from abnormal positions on the plant, e.g. shoots from roots, leaves or callus and embryos from any cell other than a zygote.

Aneuploidy: Loss or gain of chromosomes resulting in abnormal chromosome complements at metaphase usually leading to a less-fertile regenerated plant.

Auxins: Plant growth regulators which induce cell elongation, apical dominance and root initiation, e.g. IAA and NAA.

Callus: Cell mass consisting of dedifferentiated cells produced mostly as a consequence of wounding or culture.

Cryopreservation: Ultra-low temperature storage of biological tissues.

Cytokinins: Plant growth regulators which induce cell division, cell differentiation and shoot differentiation, e.g. kinetin, benzylaminopurine (BAP) and zeatin.

Immobilized plants cells: Cells entrapped in a matrix comprising alginate agarose, carrageenan, polyurethane or polyester.

Juvenility: Immature condition of a seedling that prevents persistent flowering or sexual gametogenesis.

Meristem: Localized region of active cell division in plants. Apical meristems in the flowering plants occur at the tips of the stems and roots.

Protoplast: Plant cell with cell wall removed either mechanically or enzymatically.

Recalcitrant: In this respect, a plant species which is not amenable to plant tissue culture manipulations.

Secondary metabolites: Compounds not involved in primary metabolism, e.g. respiratory intermediates, and are involved in, for example, plant defense, such as phytoalexins and phenolics, flower pigmentation or are simply waste products of primary metabolism.

Totipotency: The ability of individual plant cells to regenerate whole plants.

FURTHER READING

Alberts, B., Bray, D., Lewis, J., Raff, M., Roberts, K. & Watson, J.D. (1993) *Molecular Biology of the Cell.* Garland, New York, London.

Boulter, D. (1995) Plant biotechnology: facts and public perception. *Phytochemistry* 40, 1–9.

Butcher, D.N. & Ingram, D.S. (1976) *Plant Tissue Culture: Studies in Biology.* Edward Arnold, London.

Chet, I. (1994) *Biotechnology in Plant Disease Control.* Wiley-Liss, New York.

Debergh, P.C. & Zimmerman, R.H. (1993) *Micropropagation: Technology and Application.* Kluwer Academic Publishers, Dordrecht.

Dey, P.M., Patel, S. & Brownleader, M.D. (1993) Induction of α-galactosidase in *Penicillium ochrochloron* by guar (*Cyamopsis tetragonoloba*) gum. *Biotechnol. Appl. Biochem.* 17, 361–371.

Dixon, R.A. & Gonzales, R.A. (1994) *Plant Cell Culture: a Practical Approach.* IRL Press at Oxford University Press. Oxford.

Dodge, A.D. (1989) *Herbicides and Plant Metabolism.* Cambridge University Press, Cambridge.

Fowler, M.W. (1986) Process strategies for plant cell cultures. *Trends Biotechnol.* 8, 214–219.

Giles, K.L. & Morgan, W.M. (1987) Industrial-scale plant micropropagation. *Trends Biotechnol.* 5, 35–40.

Kung, S-D. & Arntzen, C.J. (1989) *Plant Biotechnology.* Butterworth, Boston.

Mantell, S.H., Mathews, J.A. & Mckee, R.A. (1985) *Principles of Plant Biotechnology: an Introduction to Genetic Engineering in Plants.* Blackwell Scientific, Oxford.

Marsh, J. (1990) Workshop: getting the best from plants. *Trends Biotechnol.* 8, 139–140.

Mazur, B.J. & Falco, S.C. (1989) The development of herbicide resistant crops. *Annu. Rev. Plant. Physiol. Mol. Biol.* 40, 441–467.

Panda, A.K., Mishra, S., Bisaria, V.S. & Bhojwani, S.S. (1989) Plant cell reactors – a perspective. *Enzyme Microb. Technol.* 11, 386–397.

Stafford, A., Morris, P. & Fowler, M.W. (1986) Plant cell biotechnology: A perspective. *Enzyme Microb. Technol.* 8, 578–587.

Vasil, I.K. (1984–1988) *Cell Culture and Somatic Cell Genetics of Plants*, Vols 1–5. Academic Press, Orlando.

Wang, K., Estrella, A.H. & Montagu, M.V. (1995) *Plant and Microbial Biotechnology Research* series-3. Press Syndicate of University of Cambridge, Cambridge.

Zaitlin, M., Day, P. & Hollaender, A. (1985) *Biotechnology in Plant Science.* Academic Press, New York, London.

Index

Note: All numbers indicate page numbers. Figures are indicated by *italic numbers*, and Boxes and Tables by **bold numbers**

abrin, **507**
abscisic acid
  biological role, 422
  biosynthesis, 422, *423*
  metabolism, 422
  occurrence, 422
  structure, *422*
acceptor-side photoinactivation
  (of PSII), 76–7
Acer saccharum, 512
acetate/malonate pathway, 387,
  413
  see also polyketide pathway
acetate/mevalonate pathway, 387,
  418
Acetobacter xylinum, cellulose
  biosynthesis, 223–4
acetohydroxy acid
  isomeroreductase, 302
acetohydroxy acid synthase, *300*,
  301
acetophenones, **388**
acetyl-ACP, formation, *245*
acetylandromedol, 507
acetylcholine, 449
acetyl CoA
  in alkaloid synthesis, *463*
  in citric acid cycle, 121, *122*
  conversion to carbohydrate, *21*,
    125, *126*, 258, *261*
  in fatty acid synthesis, *245*
  formation, *21*, 114, *122*, 244,
    245, 258, 479
  in isoprenoid biosynthesis, *418*
  in polyketide cyclization, 414
acetyl-CoA carboxylase, 244, *245*
  factors affecting activity, **267**
  inhibition by herbicides, 271
acetylene reduction assay (for
  nitrogen fixation), 276
N-acetylglutamate kinase, 304
N-acetylglutamate synthase, 304
acid phosphatases, 31
acid proteases, 25
aconitase, 21, *123*, 124, *261*
aconitine, 465, **468**, **470**
Aconitum napellus, 161
acquired resistance, 501
acridone alkaloids, *461*
acronycine, *461*, **470**
ACR-toxin, *493*
actin proteins, 22–3, *23*
acylation, proteins, 347

acyl-CoA, 479
acyl-CoA oxidase, 257
acyl-CoA substrates, 246, 258
acyl-CoA sucrose derivatives, 158
acylglycosides, flavonoid, 406
acyl hydrolases, 254
adenine, in DNA/RNA, 315, 482
adenosine-5-monophosphate, 472,
  473, *481*, 483
adenosine triphosphate, 484
S-adenosylhomocysteine, 475
S-adenosylmethionine, 474–5
  biological functions, 296, 474–5
  biosynthesis, *293*, 296, 476
S-adenosylmethionine
  N-methyltransferases, 452
S-adenosylmethionine synthetase,
  296–7
ADP-glucose, in synthesis of
  starch, 179, *180*
ADP-glucose pyrophosphorylase,
  96, 101–2, 179, 180–2
adrenaline, 452
agmatine, *446*
Agrobacterium spp.
  A. rhizogenes, 472
  A. tumefaciens, 45–6, 377, *379*,
    472, 488
agrocinopine-A, 158
AIP (2-amino-indan-2-phosphonic
  acid), 393
air-lift bioreactors, 526
ajmalicine, *461*, 466, **467**, 471
ajmaline, 466
ajugose, 168
alanine-aldehyde
  aminotransferase, 444
alanine aminotransferase, 292–3
Albersheim cell wall model, 221
albicanol, *421*
albizziine, *441*, 442
alcohol dehydrogenase, 114
aldehydes, characteristic smell, 259
alditol galactosides, 161
aldolase, 85, *112*, 113
aldol condensation, *414*
alfalfa *see Medicago sativa*
algae, 1
alizarin, 413, *414*
  biosynthesis, 413, *415*
alkaloids
  allelochemical effects, 469, **470**,
    507, **507**

biological functions, 469–71
biosynthesis, 459–65, *466*, *467*
  regulation, 465
compartmentation of
  biosynthesis and
  storage, 465, **467**
definitions, 458–9
detection, 459
diurnal variations, 468
evolution, 459
factors influencing distribution
  and concentration, 466, 468
occurrence, 459, 460–3, **467**, 507
recycling/turnover, 469
storage sites, 11, 465–6, **467**
transport, 468–9
allantoinase, 22
allelochemical effects
  alkaloids, 469, **470**
  amines, 451
  factors affecting effectiveness,
    440
  non-protein amino acids, 442–3
allelopathy, 389, 443, 469
allicin, biosynthesis, *443*
Allium spp., non-protein amino
  acids, 443
allyl isothiocyanate, *453*
almonds, 506
Alnus crispa (green alder), 510,
  510
alpha-cellulose, meaning of term,
  209
alpigenine, *470*
Alternaria spp., toxins, 493, *493*
alternative oxidase, 15, *16*, 127,
  131–2
α-amanitin, sensitivity of RNA
  polymerases, 329–30
amanitine, **470**
Amaryllidaceae alkaloids, *462*
amentoflavone, 388, *389*, 411
amine oxidases, 447, 448
amines, 443–53
  aliphatic monoamines
    biological functions, 445–7
    biosynthesis, 444–5
    occurrence, *444*
    structures, *444*
  aromatic amines, 448–53
  diamines
    biological function, 448
    biosynthesis, 447–8